FLOOD GEOMORPHOLOGY

FLOOD GEOMORPHOLOGY

Edited by

VICTOR R. BAKER
Department of Geosciences, The University of Arizona

R. CRAIG KOCHEL
Department of Geology, Southern Illinois University at Carbondale

PETER C. PATTON
Department of Earth and Environmental Sciences, Wesleyan University

WILEY

A WILEY-INTERSCIENCE PUBLICATION

JOHN WILEY & SONS

New York / Chichester / Brisbane / Toronto / Singapore

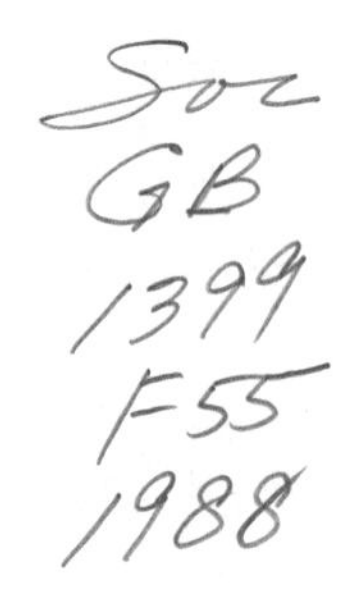

Library of Congress Cataloging in Publication Data:

Flood geomorphology/edited by Victor R. Baker, R. Craig Kochel, Peter C. Patton.
p. cm.

"A Wiley-Interscience publication."
Bibliography: p.
ISBN 0-471-62558-2
1. Floods. 2. Geomorphology. I. Baker, Victor R. II. Kochel, R. Craig. III. Patton, Peter C.

GB 1399.F55 1987
551.48′9—dc19
87-27920 CIP

Printed in the United States of America

10 9 8 7 6 5 4 3 2 1

CONTRIBUTORS

VICTOR R. BAKER, Department of Geosciences, University of Arizona, Tucson, Arizona

G. ROBERT BRAKENRIDGE, Department of Geological Sciences, Wright State University, Dayton, Ohio

WILLIAM B. BULL, Department of Geosciences, University of Arizona, Tucson, Arizona

MICHAEL CHURCH, Department of Geography, The University of British Columbia, Vancouver, British Columbia, Canada

JOHN E. COSTA, U.S. Geological Survey, David A. Johnston Cascades Volcano Observatory, Vancouver, Washington

THOMAS DUNNE, Department of Geological Sciences, University of Washington, Seattle, Washington

WILLIAM L. GRAF, Department of Geography, Arizona State University, Tempe, Arizona

AVIJIT GUPTA, Department of Geography, National University of Singapore, Singapore

BRUCE P. HAYDEN, Climate-Ecosystems Dynamics Group, Department of Environmental Sciences, University of Virginia, Charlottesville, Virginia

KATHERINE K. HIRSCHBOECK, Department of Geography and Anthropology, Louisiana State University, Baton Rouge, Louisiana

CLIFF R. HUPP, U.S. Geological Survey, Nashville, Tennessee

JAMES C. KNOX, Department of Geography, University of Wisconsin, Madison, Wisconsin

R. CRAIG KOCHEL, Department of Geology, Southern Illinois University at Carbondale, Illinois

PAUL D. KOMAR, College of Oceanography, Oregon State University, Corvallis, Oregon

PETER L. KRESAN, Department of Geosciences, University of Arizona, Tucson, Arizona

JIM E. O'CONNOR, Department of Geosciences, University of Arizona, Tucson, Arizona

PETER C. PATTON, Department of Earth and Environmental Sciences, Wesleyan University, Middletown, Connecticut

DALE F. RITTER, Department of Geology, Southern Illinois University at Carbondale, Illinois

ASHER P. SCHICK, Department of Physical Geography, Institute of Earth Sciences, Hebrew University, Jerusalem, Israel

ROBERT H. WEBB, U.S. Geological Survey, Tucson, Arizona

GARNETT P. WILLIAMS, U.S. Geological Survey, Denver, Colorado

To Bren, Elsie, and Pauline

PREFACE

Why a book on flood geomorphology? Is not the scientific study of floods the proper concern of hydrologists rather than geomorphologists? Are there not numerous extensive reviews of flood hydrology?

A prominent flood hydrologist recently provided the following definition for "scientific." "Scientific analysis," he stated, "implies that all the necessary equations can be written down and programmed for a digital computer." Readers sharing this view may save considerable time by now placing this book aside.

We believe that science cannot be so arbitrarily defined in its methodology as by our hydrologist friend. Moreover, true science is concerned with understanding nature no matter what the methodology. In our view, if the wrong equations are programmed because of inadequate understanding of the system, then what the computer will produce, if believed by the analyst, will constitute the opposite of science. For that reason this book will concentrate less on computers and equations, and more on flooding, a phenomenon that we would like to understand.

For those readers who remain concerned with definitions, flood geomorphology can be considered to be the study of the role of floods in shaping the landscape, including the analysis of flood causes, flood processes, resistance factors to flood-induced landscape change, and changes in flood-related processes and forms through time. Geomorphologists are also interested in how the landscape affects flood processes. Indeed, over varying time scales, in varying climatic and physiographic settings, floods and riverine landscapes constitute complex mutually interactive systems. The study of those systems, with the goal of understanding thresholds for response, feedback elements, and other intricacies, constitutes the core of flood geomorphic research.

The answers, then, to the questions posed in the first paragraph center on the need for a scientific treatment of the subject, as opposed to an engineering or statistical one. Nearly all the interest by hydrologists in the phenomenon of flooding has centered on the problem of estimating probabilities. Hundreds of papers on this subject have been published in the last decade in journals such as *Water Resources Research*, *Journal of Hydrology*, and *Water Resources Bulletin*. Despite the accelerating publication of papers on flood frequency analysis, no doubt aided by the "computer revolution" in hydrology, there is a nagging concern about the scientific value of such studies. This is born of abundant evidence, as stated by D. W. Braben in the journal *Nature* (Vol. 316, pp. 401–402), "Research driven by technique . . . seems to be a poor bet, since almost invariably the technician's skill is a solution looking for a problem."

The general thesis of this volume is that the flooding process is of intrinsic scientific interest, regardless of concerns for designing structures or estimating risk. Moreover, if there is to be a science of floods, then it is impossible to divorce study of the flooding process from study of the landscape that affects and is in turn affected by that process. As with all science, such a view may lead to discoveries with unforeseen practical application. Such a discovery emerged in the past decade for flood geomorphology. It was found that some landscapes, through erosional scars or sedimentary deposits, preserve a record of past floods that can yield important information for engineering design and risk estimation. While similar evidence has been known to geomorphologists since the nineteenth century, the recent discovery is that great accuracy and completeness can be achieved in certain geomorphic settings for the reconstruction of ancient floods. This book describes such studies under the rubric, "paleoflood hydrology."

The accelerating scientific interest in flood geomorphic research is reflected by several recent events. The term *paleoflood hydrology*, introduced in papers by the editors in 1982, served as the subject of a symposium at the

December 1984 American Geophysical Union annual meeting. The same topic reappeared at the U.S.–China Bilateral Symposium on the Analysis of Extraordinary Flood Events held in Nanjing, China, in October 1985. Chinese researchers, unlike most American hydrologists, placed an emphasis on the discovery of evidence for ancient floods of great magnitude. Interest continued at the International Symposium on Flood Frequency and Risk Analysis, held in May 1986, in Baton Rouge, Louisiana. The paleohydrologic flood analysis sessions were among the most popular symposium events. In the closing session on research needs, an imperative statement was made that flood analysis can be broadened beyond the limits of civil engineering and statistics to include aspects of geomorphology and climatology.

In May 1987, the British Geomorphological Research Group held a workshop entitled "The Hydrology, Sedimentology and Geomorphological Implications of Floods." Professor John Lewin (University College of Wales), in an invited address entitled "Floods in Fluvial Geomorphology," observed that the increased interest in flood geomorphology reflects a major paradigm change for fluvial geomorphology. During the period from 1950 to 1970, the prevailing emphasis in fluvial geomorphology was on the role of bankfull discharges, which were presumed to have recurrence intervals of 1 to 2 years. Starting in the 1970s interest accelerated in the varying morphogenetic impacts of rare floods. This in turn has led to new views of floodplain genesis and on the role of floods in the total operation of fluvial geomorphic systems.

The role of rare, great floods was even more boldly asserted in September 1987 at the Eighteenth Annual Geomorphology Symposium entitled "Catastrophic Flooding." In October the 1987 Kirk Bryan Award of The Geological Society of America was given to Dr. Richard B. Waitt, Jr., for his paper "Case for Periodic, Colossal Jökulhlaups from Pleistocene Glacial Lake Missoula" published in 1985 in *The Geological Society of America Bulletin.* In 1980 the highest award of that society, the Penrose Medal, was given to Professor J Harlen Bretz for his discovery of the same great glacial floods. Bretz argued for the geomorphic significance of floods half a century before general acceptance by the earth science community. For "Doc" Bretz the emerging paradigm of flood geomorphology is yet another indication of his scientific vision.

This book is the evolutionary product of years of research by its authors and editors. By chance we had been attracted to the scientific study of the unusual extreme of high-magnitude events in fluvial geomorphology. We found that traditional approaches from geomorphology, hydrology, and engineering were not wholly satisfactory in producing a complete understanding of flood phenomena. We were forced to develop tools and achieve personal experience for creating a broad perspective on floods and their effects in time and space. The editors, initially including Dr. John E. Costa, considered writing a textbook. However, despite their heurisitc and organizational advantages, textbooks tend to create an overly systematized view of an actively evolving research area. Because of its accelerating scientific growth, flood geomorphology is an obvious candidate for research symposia and published proceedings. However, symposia proceedings vary considerably in the breadth and quality of subject coverage. We attempted a compromise to avoid these problems. To retain the flavor of actively evolving research, we contacted various active researchers as chapter authors. For complete coverage not achieved in most symposia volumes, we solicited contributions on most of the major topics for the newly defined subdiscipline: flood geomorphology.

The book is organized into five parts. Part I considers the external controls and geomorphic measurements that relate to flood phenomena in general. Part II analyzes flood erosional and depositional processes. In Part III the detailed interactions of climate, landscapes, and floods are explored. Various chapters describe flood phenomena across the broad range of geomorphic settings. Finally, Part V gives examples of the interactions of floods with human activity on the landscape.

Because most flood research concerns rainfall–runoff flooding, this topic receives major emphasis in this volume. Nevertheless, whole chapters treat floods from dam failures and floods related to snow and ice. Coastal flooding by storm surges is not considered. The book also emphasizes modern and Holocene floods, that is, flood phenomena of the past 10,000 years. Such floods are most easily documented by their impact on the landscape; so their study is appropriate as the starting point of a new subdiscipline. Nevertheless, what is learned of Holocene flood processes can be of great use in the study of ancient floods in the geological record.

In attempting to present the core of flood geomorphic research in one volume, certain exciting though peripheral topics had to be excluded. Restriction to the Holocene time period means that the fascinating late glacial floods and jökulhlaups associated with the end of the Pleistocene, or Ice Age, are not analyzed. Also missing is a treatment of flood-related features on the planet Mars. There is no question that these phenomena are important concerns in flood geomorphology. However, they are analyzed in other recent books, whereas similar treatment has not been afforded the core subdiscipline.

The authors who contributed the 27 chapters in this volume have a collective experience of approximately 400 years of work in flood geomorphology. Only 3 of 21 authors have affiliations outside the United States. Despite this bias, the authors have broad experience in many flood geomorphic environments. The immense variety of the American landscape alone nearly suffices in this regard. Nevertheless, the reader will find various chapters relating extensive flood geomorphic experience from Australia, Canada, China, India, Israel, Germany, Jamaica, and New Zealand. The result of the temporal and spatial scale of the authorship is a book that conveys immense direct experience with a phenomenon whose study is difficult because of its rarity. However, in order to understand a phenomenon, we must observe and measure its entire range of activity. The lack of data on modern floods is therefore extensively supplemented by reconstructions of ancient floods.

Victor R. Baker
R. Craig Kochel
Peter C. Patton

Tucson, Arizona
Carbondale, Illinois
Middletown, Connecticut

January 1988

ACKNOWLEDGMENTS

Although this book was a personal project, it would not have been possible without the research perspective gained through sponsored projects over many years. Especially important have been a series of National Science Foundation grants, beginning with grant GA-21478 in 1970 to study "Paleohydrology of the Lake Missoula Floods." In 1977–1983 the National Science Foundation supported an extensive study, "Holocene Paleohydrology of the Southwestern United States," with grants EAR-7723025, EAR-8100391, and EAR-81-8119981. In 1983–1984 work shifted to Australia, supported by National Science Foundation grant EAR-8300183, "Paleoflood Hydrology of Arid and Savanna Regions."

Our shared research on floods in Texas began with a project in 1973 sponsored by the Bureau of Economic Geology, The University of Texas at Austin. It continued with support by the National Aeronautics and Space Administration (Contract No. 9-13314) and the National Weather Service (Contract No. A-35460). Work on floods in Arizona began in 1982, and was supported for three years by the U.S. Department of Interior Water Resources Institute Program. More recently, studies of Arizona paleofloods were funded by the Salt River Valley Water Users' Association.

Over the years our flood studies have been inspired by many outstanding scientists. In addition to the chapter authors, we have been stimulated by our associations with L. R. Beard, William C. Bradley, J Harlen Bretz, John T. Hack, Luna B. Leopold, Thomas Maddock, Jr., Walter H. Moore, John H. Moss, Geoff Pickup, Brian R. Reich, Stanley A. Schumm, Russell G. Shepherd, David T. Snow, Jery R. Stedinger, and M. Gordon Wolman.

Individual chapters in this book were reviewed by many specialists, notably by the authors of other chapters in the volume. In addition to those reviewers we thank the following: Gerald G. Nanson, Peter R. Waylen, Julio L. Betancourt, V. Gardiner, Thomas W. Gardner, Patrick J. Bartlein, Alan D. Howard, V. Klemeš, Robert A. Mueller, Russell G. Shepherd, and Jery R. Stedinger.

V. R. B.
R. C. K.
P. C. P.

CONTENTS

PART IV: PALEOFLOODS

PART V: ENVIRONMENTAL MANAGEMENT

FLOOD GEOMORPHOLOGY

OVERVIEW

VICTOR R. BAKER
Department of Geosciences, University of Arizona, Tucson, Arizona

The first known floods on Earth are recorded in the oldest known rocks on the planet. Conglomerate of the 3.8-billion-year-old Isua Formation in Greenland contains flood-rounded cobbles. The Viking spacecraft images of Mars (Fig. 1) provide evidence of even more ancient flooding on that planet (Baker, 1982). The outflow channels of Mars show that immense cataclysms occurred over the billions of years of planetary history recorded in the extant Martian landscape. On Earth, one can envision the first floods eroding a barren landscape, transporting immense quantities of sediment (Schumm, 1968). Because of the lack of preserved drainage basins and stream channels, the nature of those most ancient floods is poorly known. Perhaps the poetic vision of John Milton in *Paradise Lost* can provide the inspiration to visualize them:

Immediately the mountains huge appear
Emergent, and their broad bare backs upheave
Into the clouds, their tops ascend the sky.
So high as heavied the tumid hills, so low
Down sank a hollow bottom broad and deep,
Capacious bed of waters; thither they
Hasted with glad precipitance, uprolled
As drops on dust conglobing from the dry;
Part rise in crystal wall, or ridge direct,
For haste; such flight the great command impressed
On the swift floods; as armies at the call
Of trumpet, for of armies thou hast heard,
Troop to their standard, so the watery throng,
Wave rolling after wave, where way they found;
If steep, with torrent rapture, if through plain,
Soft-ebbing; nor withstood them rock or hill,
But they, or under ground, or circuit wide
With serpent error wandering, found their way,
And on the washy oose deep channels wore,
Easy, ere GOD had bid the ground be dry,
All but within those banks, where rivers now
Stream, and perpetual draw their humid train.

The greatest freshwater flooding known on Earth occurred between approximately 17,000 and 12,000 years ago in the northwestern United States. These floods emanated from late Pleistocene Glacial Lake Missoula and carved the bizarre landscape of the Channeled Scabland (Bretz, 1923, 1928, 1969; Baker, 1973, 1981; Baker and Nummedal, 1978). Space imagery reveals the plexus of scabland channelways carved into the basalt bedrock (Fig. 2). Great cataracts, potholes, and grooves were eroded into the rock (Fig. 3). Immense boulders were transported by the flows (Fig. 4) and deposited in great flood bars. Reconstructions of the largest floods (Baker, 1973; Baker and Komar, 1987) indicate that Lake Missoula released as much as 2000 km^3 of water, which achieved peak discharges of 15–17 million m^3/s. The values of power per unit area of streambed generated by these flows were as much as 30,000 times greater than that produced in the Amazon, the world's biggest river (Baker and Costa, 1987).

FLOOD STUDIES AS A SCIENTIFIC OUTRAGE

The study of extraordinary floods follows in a long tradition of somewhat disreputable scientific activity. This lack of respect from other scientists derives from centuries of mistaken views as to what constituted proper scientific pursuits. Through the seventeenth and eighteenth centuries it was common scientific practice to try to reconcile the surface features of the Earth with cataclysmic events, such as the Noachian flood. By the nineteenth century, the efforts of James Hutton, John Playfair, and Charles Lyell had replaced the biblical-catastrophist view of earth history with a concept

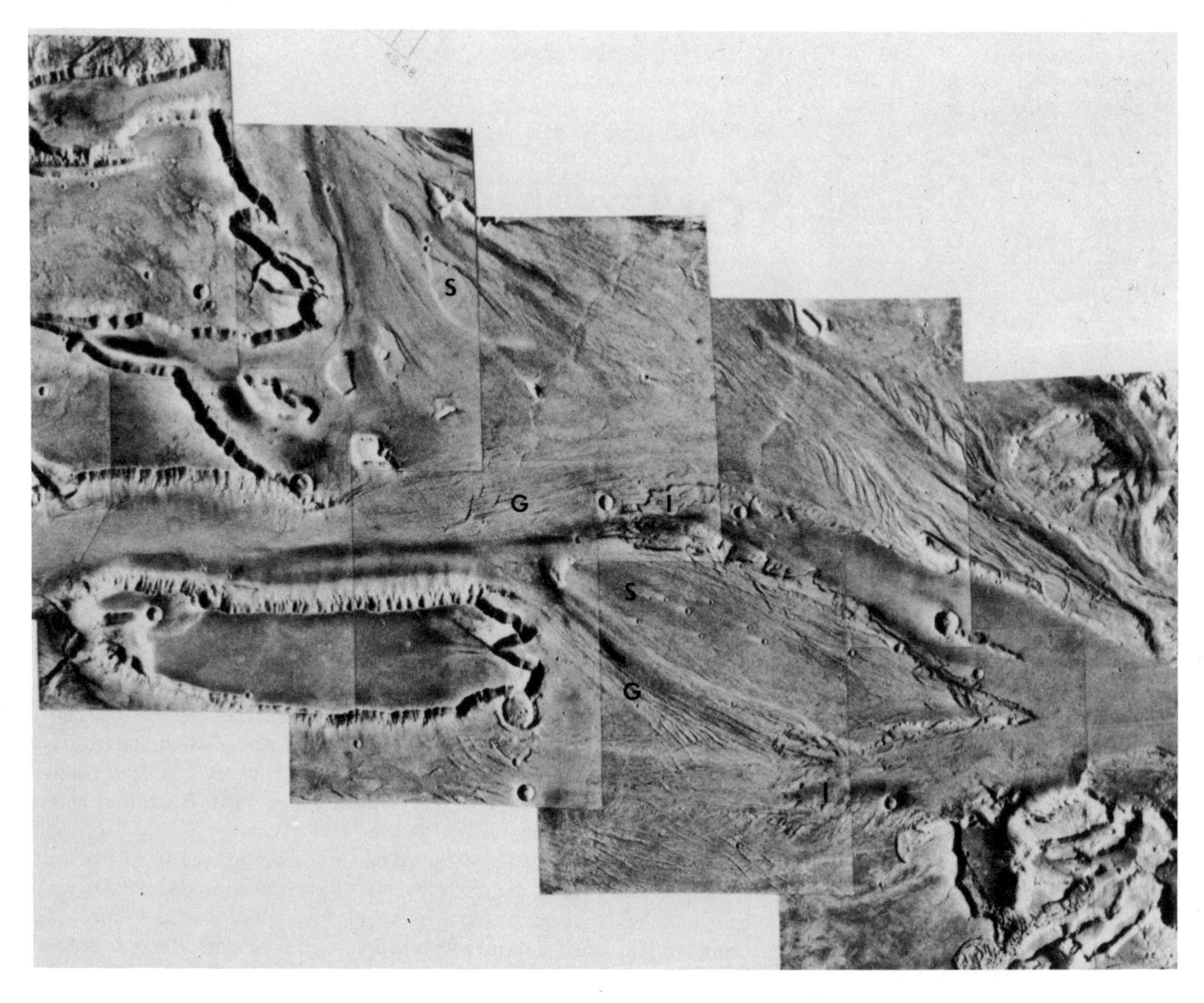

FIGURE 1. A portion of the Kasei Vallis region of the planet Mars near latitude 62°W, longitude 25°N. Features in this area indicative of cataclysmic flood processes were described by Baker and Milton (1974), Baker (1978, 1982), and the Mars Channel Working Group (1983). Erosional forms include inner channels with recessional headcuts (I), longitudinal grooves (G), and streamlined hills (S). These features display a regional conformity to a fluid-flow direction from left to right. The scene is approximatley 300 km across and was imaged by Viking orbiter spacecraft cameras (JPL Mosaic 211-5371).

FIGURE 2. Landsat image (E-1039-1814-5) taken on August 31, 1972, showing an area 185 km × 185 km centered approximately on the Cheney–Palouse scabland tract of the Channeled Scabland. The flood-scoured channelways appear as an anastomosing pattern of dark patterns within the overall bright pattern of areas underlain by loess. Cataclysmic floodwater moved from the top of the scene toward the bottom and the lower left.

FIGURE 3. Longitudinal grooves, potholes, and inner-channel cataracts produced by Missoula floodwater in the Dry Falls area of the Channeled Scabland. The cataract complex at the center of the picture measures approximately 10 km across. Compare it to Martian channel topography in Figure 1.

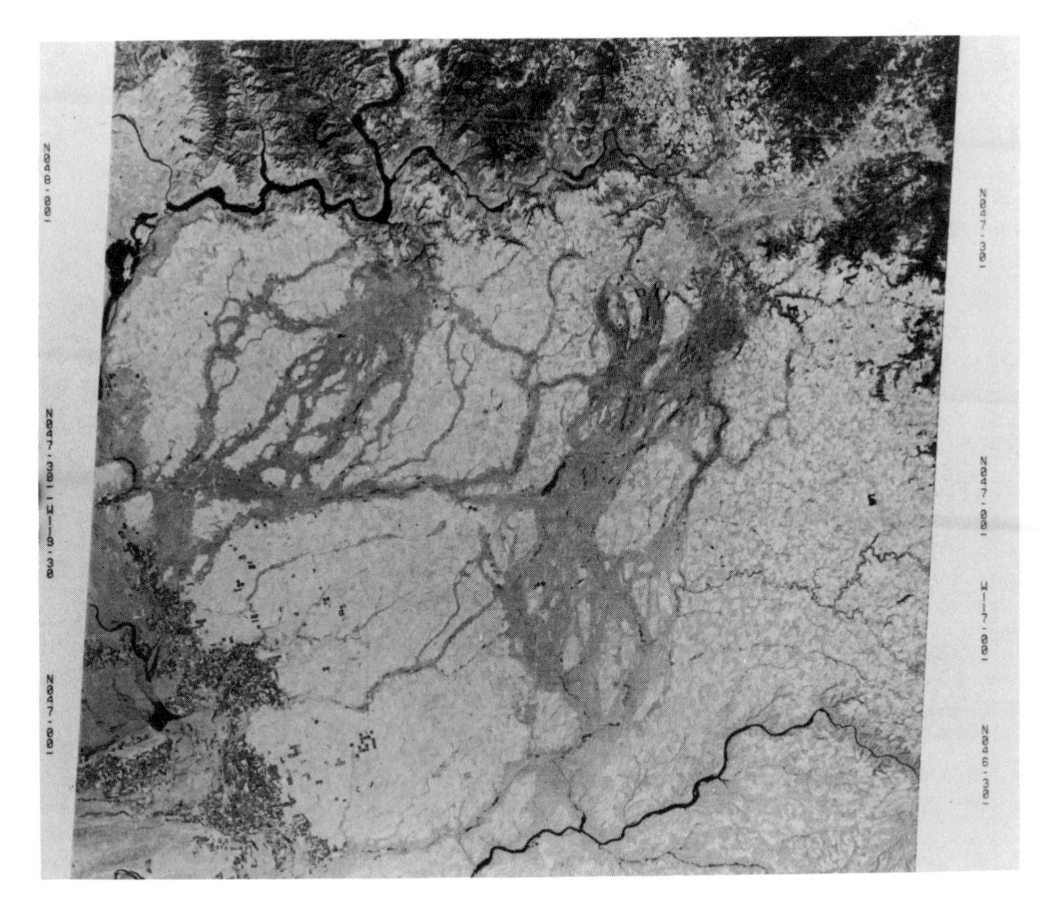

FIGURE 2.

FIGURE 3.

FIGURE 4. Boulder of basalt bedrock eroded from the Grand Coulee region by cataclysmic Missoula Flood flows. The boulder measures 18 m × 11 m × 8 m (Baker, 1973).

gradualism. The new dogma held that fluvial landform development, like science itself, proceeded slowly and with order. Catastrophist views came into disrepute.

A great misconception among many earth scientists is that this transition had something to do with uniformitarianism. As reviewed by Albritton (1967), Baker (1981), Gould (1965), Hubbert (1967), and Shea (1982), uniformitarianism in modern earth science holds merely that among competing hypotheses, the simpler hypothesis often tends to prevail. This principle, also known as Occam's razor, leads to hypotheses such as: the basic laws of nature remain invariant with time (or at least over the time period of interest). Uniformitarianism in its twentieth century form has absolutely nothing to do with whether or not a process is catastrophic. Rather, it has everything to do with whether a hypothesized cataclysm obeys the laws of physics and is consistent with the field evidence.

One of the greatest of geological controversies arose in the 1920s and 1930s because of the absolutely erroneous belief that hypotheses involving cataclysmic origins of features could be rejected merely because catastrophic processes were inconsistent with uniformitarianism. J Harlen Bretz, in a series of a dozen papers, documented the cataclysmic flood origin of the Channeled Scabland in the finest tradition of uniformitarianism (Baker, 1978, 1981; Baker and Bunker, 1985). In their righteous defense of an anachronistic, Victorian concept of scientific dignity, it was Bretz's critics who were the nonuniformitarians (Baker, 1987).

Despite this legacy, there remain vestiges of concern that the immense power and energy of extraordinary floods are something to be downplayed. How are such processes to be reconciled with the orderly, slow progression of landscape change? In science such paradoxes are known as anomalies, and it is in the study of anomalies that major advances in knowledge can occur (Kuhn, 1962).

HYDROLOGY AND GEOMORPHOLOGY

The topic of flooding has been extensively reviewed from the broad perspectives of engineering hydrology, effects on human activity, and land-use planning (Hoyt and Langbein, 1955; Waananen et al., 1977; Ward, 1978). However, geomorphic aspects of flooding have generally received limited treatment in works concentrating on flood phenomena in relation to socioeconomic concerns.

In a perceptive editorial, "Hydrology, The Forgotten Earth Science," Bras and Eagleson (1987) note that the hydrological sciences have been largely problem driven. Applications, such as hazard reduction, have been developed mainly by civil engineers. They state (p. 227), "The cultivation of hydrology as a science per se has not occurred, and there has been no established platform within the hierarchy of science on which to build a coherent understanding of the global water cycle." Among the scientific questions they raise is that of studying the soil, sediment, vegetation, and stream network geometry to elucidate basin history and to predict hydrologic response to future climate change. Thus, fluvial geomorphic studies are intricately linked with this future vision of hydrology as a science.

Ironically, the classical hydrological literature attached great importance to understanding the causes of extraordinary floods (Hazen, 1930). Why has modern hydrology moved away from such valid scientific concerns?

During the last 20 years there has been a revolution in the use of computers to predict hydrologic phenomena (Wallis, 1987). This trend has been paralleled by a decline in field research on the same phenomena. Pilgrim (1986, pp. 169S–170S) describes the result of these trends.

> In comparison with analytical studies using the computer, the uncertainties of field research such as the occurrence of suitable weather or field conditions, or of runoff events and floods, provide a considerable disincentive. In addition, there is a tendency among researchers to regard research involving complex mathematical procedures as having greater prestige than field-based research.
>
> Our analytical ability has now far outstripped our knowledge of physical processes during floods. While this knowledge has advanced, there is still a great deal about many processes that is not known. In many ways the advances in understanding that have been made have revealed and brought into focus the extent of our lack of knowledge of physical processes, and in this the computer has made a positive contribution. However, the discouragement of field-based research resulting from the attractiveness of the computer has to some extent been counterproductive. There has been a tendency for researchers to develop complex models of what they assume or imagine happens on real watersheds based on limited data. The enshrinement of procedures in sophisticated models may then lead to general acceptance that nature does actually behave in the assumed manner.

Klemeš (1987) is even more critical of overemphasis on mathematical approaches to hydrological problems. He states

> Paradoxically, it has become almost axiomatic that to be a good hydrologist means to learn how to be a mediocre mathematician or statistician. This attitude is the result of a common misconception about the role of mathematics in science, in particular that, rather than being a method for a logically consistent and parsimonious formulation of scientific insights and exploration of consequences of scientific hypotheses, it is a generator of new scientific knowledge about the outside world; that correct mathematical manipulations and proofs lead ipso facto to correct statements about nature.

Klemeš (1986, p. 187S) provides this alternative view of hydrology as science, rather than as an activity of pseudo-mathematical dilettants

> What then remains for the hydrologist to do if we take away from him the curve fitting, model calibration, the chasing of systems responses, correlations, finite elements, kriging, etc.? Perhaps, his efforts expended on the fitting of flood and drought frequency curves could be better spent in acquiring deeper knowledge of climatology, meteorology, geology, and ecology, since many hydrologic problems transcend the framework of hydrology as we know it today.

MISCONCEPTIONS OF EXTREME (UNUSUAL) EVENTS

In the past several decades geomorphologists have elucidated the marvelous complexity of fluvial systems. Recent review volumes (Knighton, 1984; Petts and Foster, 1985; Richards, 1982) highlight advances in the subject. The complexity of fluvial systems may be analogous to the complexity of ecological communities. Field biologists have long studied the latter by continuous observation over periods of several years. The problem is that considerable change in ecological communities is associated with unusual events, such as disease and natural disasters. Biologists have found that it is misleading to treat such events as mere statistical phenomena (Weatherhead, 1986). There must be a perspective on the long-term pattern of change in the ecological community. Such a perspective can rarely be achieved in a study lasting only several years.

In contrast to ecologists, many hydrologists and some geomorphologists have had an adversion to the study of the unusual. Floods are treated as statistical phenomena, often reduced to single data points (discharges) that are arrayed to produce the all-important sample sizes. The problems with this approach become most apparent when one considers the largest and rarest floods. Such extraordinary floods do not provide adequate samples of discharge data points for analysts enamored with statistics. This leads to the use of statistical models to extrapolate from the known properties (discharge data points) of common, small "floods" to the unknown discharges of extraordinary floods. An immense literature surrounds the formulation, testing, and validation of such models. In part, such work is justified by the argument that flood-river systems are so complex that one must use a statistical-systems approach in order to make useful predictions about system behavior. Often such studies are not justified at all, yet their practitioners may self-righteously consider themselves at the frontiers of flood "science."

In science, the extrapolation made possible by a model serves as a tool for probing the limits of existing knowledge, that is, as a hypothesis to be tested. As described by Klemeš (1986, p. 177S) this goal is perverted by models of the following type: " . . . whose cheaply arranged ability to fit data is presented as proof of their soundness and as a justification for using them for user-attractive but hydrologically indefensible extrapolations." Klemeš (1987) notes, ". . . if we have to extrapolate, then we should pay much more attention to the largest recorded floods than to the small ones since they better reflect the type of conditions influencing the tail behavior which we want to estimate."

The scientific study of an unusual phenomenon begins with the complete description of the processes associated with it. The phenomenon of a flood is not adequately characterized in science by a single data point, such as a peak discharge, or even by a set of data, such as a hydrograph. Moreover, we cannot reasonably expect that the properties of usual events necessarily will tell us all we need to know about unusual events. To understand unusual events, such as extraordinary floods, we simply must learn all we can about those events. We must move aggressively into the disreputable, sometimes outrageous scientific study of extraordinary floods.

AN EXAMPLE

On the morning of August 2, 1978, a stationary thunderstorm complex drenched a local area of central Texas with some of the most intense rain ever recorded in the United States. Locally at least 79 cm (31 inches) was recorded in 24 hours. The storm system was focused and anchored by especially complex meteorological phenomena (Caracena and Fritsch, 1983), including the influence of Tropical Storm Amelia.

On the upper Medina River a peak discharge of 3480 m^3/s was recorded for a drainage area of 175 km^2 (Schroeder et al., 1979). At the Pipe Creek gauging station on the Medina River (drainage area 1228 km^2) the peak flow was 15 m deep and reached 7920 m^3/s. As is often the case, the return period for this rare large event could not be adequately assessed from the perspective of the 38 years of continuous gauge records (Schroeder et al., 1979). The return period may lie somewhere between 100 and 500 years (Sullivan, 1983).

Geomorphological effects of this intense flooding included the massive destruction of stands of bald cypress (*Taxodium distichum*) that lined the low-flow banks of the Medina River (Fig. 5). Trees up to 2 m in diameter were scoured from the channel, and locally as much as 90% of tree crown cover was removed from riparian zones (Sullivan, 1983). The scour of trees exposed the underlying floodplain to macroturbulent scour (Baker, 1977), leading to extensive removal of floodplain sediments that had accumulated over centuries (cf. Nanson, 1986).

At one chute cutoff it was possible to document extensive bedrock erosion (Sullivan, 1983). Boulders up to 2.3 m in diameter were scoured from jointed bedrock ledges and transported several tens of meters downstream. In other

FIGURE 5. Flood scour of riparian vegetation and emplacement of giant current ripples (R) at Highland Waters on the Medina River in central Texas. The preflood low-flow channel, lined by bald cypress trees (top center) averaged about 30 m wide. The August, 1978, flash-flood water spread about 500 m wide across the valley bottom, flowing up to 10 m deep and at velocities of 3 to 4 m/s (Sullivan, 1983). The force of this water was sufficient to produce the extensive destruction of trees seen at the bottom of the photograph. Floodwater moved from the bottom to the top of the photographed area.

reaches, giant current ripples (Baker, 1984) were emplaced (Fig. 6). The largest ripples were composed of gravel and boulders up to 1 m in diameter and had spacings of 80 m. These bedforms were generated by flows of 3 to 4 m/s with depths of 10 m (Sullivan, 1983).

In terms of the flood-generated bed shear stress and power per unit area of bed (Baker and Costa, 1987), the central Texas extraordinary floods are more similar to the cataclysmic Missoula floods than they are to floods on large, low-gradient alluvial rivers, such as the Mississippi. This immense range of flood-geomorphic phenomena illustrates the need for a broad survey of the subject. Under what conditions do floods dominate as agents of fluvial

FIGURE 6. Giant current ripples, or transverse bars of gravel, formed on the Pedernales River in central Texas by August 1978, flash floods. Flood flow was from left to right.

landscape change? What are the detailed processes of flood erosion, transport, and deposition? Can the long-lasting effects of extraordinary floods be used to calculate their past magnitudes? What contribution can geomorphological flood studies make to flood-control management? These are some of the questions addressed in the scientific discipline of flood geomorphology.

REFERENCES

Albritton, C. C., (1967). Uniformity, the ambiguous principle. *In* "Uniformity and Simplicity" C. C. Albritton, ed. pp. 1–2 *Geol. Soc. Am. Spec. Pap.* **89**.

Baker, V. R. (1973). Paleohydrology and sedimentology of Lake Missoula flooding in eastern Washington. *Geol. Soc. Am. Spec. Pap.* **144**, 1–79

Baker, V. R. (1977). Stream channel response to floods, with examples from central Texas. *Geol. Soc. Am. Bull.* **88**, 1057–1071.

Baker, V. R. (1978). The Spokane Flood controversy and the Martian outflow channels. *Science* **202**, 1249–1256.

Baker, V. R., ed. (1981). "Castastrophic Flooding: The Origin of the Channeled Scabland." Dowden, Hutchinson and Ross, Stroudsburg, Pennsylvania, 1–360.

Baker, V. R. (1982) "The Channels of Mars." Univ. of Texas Press, Austin, Texas, 1–198.

Baker, V. R. (1984). Flood sedimentation in bedrock fluvial systems. *In* "Sedimentology of Gravels and Conglomerates," E. H. Koster and R. J. Steel, eds., pp. 87–98, Can. Soc. Petro. Geol. Mem. **10**.

Baker, V. R. (1987). The Spokane Flood debate and its legacy, *In* "Geomorphic Systems of North America" W. H. Graf, ed. The Geology of North America, Centennial Special Vol. 2, pp. 416–423, *Geol. Soc. Am.*

Baker, V. R., and Bunker, R. C., (1985). Cataclysmic late Pleistocene flooding from glacial Lake Missoula: A review. *Quat. Sci. Rev.*, **4**. 1–41.

Baker, V. R., and Costa, J. C. (1987). Flood power. *In* "Catastrophic Flooding" (L. Mayer and D. Nash, eds.) pp. 1–24, Allen and Unwin, London.

Baker, V. R., and Komar, P. D. (1987). Cataclysmic flood processes and landforms. *In* "Geomorphic Systems of North America." (W. H. Graf, ed.) pp. 423-443, The Geology of North America, Centennial Special Vol. 2.

Baker, V. R., and Milton, D. J. (1974). Erosion by catastrophic floods on Mars and Earth. *Icarus* **23**, 27–41.

Baker, V. R., and Nummedal, D., (eds.) (1978). "The Channeled Scabland: A guide to the geomorphology of the Columbia Basin." NASA, Washington, D.C., 1–186.

Bras, R., and Eagleson, P. S. (1987). Hydrology, the forgotten earth science. *EOS* **68** (16) 227.

Bretz, J H. (1923). The Channeled Scabland of the Columbia Plateau. *J. Geol.* **31**, 617–649.

Bretz, J H. (1928). The Channeled Scabland of eastern Washington. *Geo. Rev.* **18,** 446–477.

Bretz, J H. (1969). The Lake Missoula floods and the Channeled Scabland. *J Geol.* **77**, 505–543.

Caracena, F., and Fritsch, J. M. (1983). Focusing mechanisms in the Texas Hill Country flash floods of 1978. *Mon. Weather Rev.* **111**, 2319–2332.

Gould, S. J. (1965). Is uniformitarianism necessary? *Am. J. Sci.* **263**, 223–228.

Hazen, A. (1930). "Flood Flows." Wiley, New York.

Hoyt, W. G., and Langbein, W. B. (1955). "Floods." Princeton Univ. Press, Princeton, New Jersey.

Hubbert, M. K. (1967). Critique of the principle of uniformity. *In* Uniformity and Simplicity. C. C. Albritton, ed. pp. 3–33. *Geol. Soc. Am. Spec.* Pap. **89**.

Klemeš, V. (1986). Dilettantism in hydrology: Transition or destiny? *Water Resour. Res.* **22**, 177S–188S.

Klemeš, V. (1987). Hydrological and engineering relevance of flood frequency analysis. *In* (V. P. Singh, ed.) "Hydrologic Frequency Modeling," pp. 1–18, D. Reidel, Dordrecht, Holland.

Knighton, David, (1984). "Fluvial Forms and Processes." Edward Arnold, London.

Kuhn, T. S. (1962). The Structure of Scientific Revolutions." Univ. of Chicago Press, Chicago.

Mars Channel Working Group (1983). Channels and valleys on Mars. *Geol. Soc. Am. Bull.* **94**, 1035–1054.

Nanson, G. C. (1986). British fluvial geomorphology in the 1980s: A review of recent reviews. *Australian Geol.* **17** (1), 87–91.

Petts, Geoff, and Foster, Ian (1985). "Rivers and Landscape." Edward Arnold, London.

Pilgrim, D. H. (1986). Bridging the gap between flood research and design practice. *Water Resour. Res.* **22**, 165S–176S.

Richards, Keith (1982). "Rivers: Form and Process in Alluvial Channels." Methuen, London.

Schroeder, E. E., Massey, B. C., and Waddell, K. M. (1979). Floods in central Texas, August 1978. *Geol. Surv. Open-File Rep.* (U.S) **79-682**, 1-121.

Schumm, S. A. (1968). Speculations concerning paleohydrologic controls of terrestrial sedimentation. *Geol. Soc. Am. Bull.* **79**, 1573–1588.

Shea, J. H. (1982). Twelve fallacies of uniformitarianism. *Geology* **10**, 455–460.

Sullivan, J. E. (1983). Geomorphic effectiveness of a high-magnitude rare flood in central Texas. M.A. Thesis, University of Texas, Austin (unpublished).

Waananen, A. O., Limerinos, J. T., Kockleman, W. J., Spangle, W. E., and Blair, M. L. (1977). Flood-prone areas and land-use planning—selected examples from the San Francisco Bay region, California. Geol. Surv. Prof. Pap. (U.S.) **942**, 1–75.

Wallis, J. R. (1987). Hydrology—the computer revolution continues. *Rev. Geoph.* **25**, 101–106.

Ward, R. (1978). "Floods—A geographical Perspective." Wiley, New York.

Weatherhead, P. J. (1986). How unusual are unusual events? *Am. Naturalist*, **128**, 150.

PART I

EXTERNAL CONTROLS AND GEOMORPHIC MEASUREMENTS

The 1978 Amelia floods illustrate an interplay of meteorology and drainage basin characteristics that provides central Texas with what may be the most intense flood phenomena in the United States (Baker, 1975, 1977; Caran and Baker, 1986). Just how these characteristics combine to produce spectacular geomorphic responses is explored more fully in Chapter 11. In Part I, the chapters review the separate components of the atmospheric and landscape features that contribute to flood characteristics.

Chapter 1 by Bruce P. Hayden analyzes the global climatology of floods. The chapter contains a global classification of flood-producing climates based on the mean seasonal state of the atmosphere. Where the atmosphere is primarily characterized by barotropy, typical of tropical low latitudes, precipitation is released by synoptic convective activity (e.g., tropical storms), local convective activity (thunderstorms), and rising circumequatorial air at the Intertropical Convergence Zone. Where the atmosphere is baroclinic, typical of higher latitudes, precipitation is released by cyclones and fronts. At even higher latitudes, rainfall intensity and totals become modest, and snow accumulation and melting become the major climatic controls on flooding.

Hayden's global mapping of floods and climatic types raises some fascinating questions concerning floods and climatic change. The El Niño phenomenon of the Peruvian coast is a case-in-point. El Niño is an anomalous warming of coastal waters off central-western South America, generally appearing in late December. The phenomenon is driven by a climatic shift known as the Southern Oscillation (SO). Essentially the combined El-Niño–Southern Oscillation (ENSO) is a large-scale seesaw of atmospheric mass between the Pacific and Indian Oceans in the tropics and subtropics. It is associated with a whole family of related phenomena on a global scale. These include shifts in the Intertropical Convergence Zone, variations in the Indian monsoons, and "teleconnections" that link atmospheric and oceanic conditions in the tropical Pacific to weather over North America and elsewhere. The connection with slow-moving ocean currents slows the period of ENSO variation to several years. Moreover, there are major, longer term variations in the intensity of ENSO development.

During the 1972–1973 ENSO, regional flooding was experienced in Peru, California, and the Philippines. Coincident droughts occurred in Hawaii, New Guinea, and the USSR. The more spectacular 1982–1983 ENSO was marked by floods on the Pacific coast of United States and on islands of the eastern Pacific. Coincident droughts occurred in Australia and the Sahel. Historical studies indicate that the 1877–1878 ENSO resembled that of 1982–1983.

These global controls show that flooding need not be perceived as a purely random phenomenon. In regions affected by ENSO shifts, some time periods are more prone to high-magnitude flood occurrence than other time periods. Scientific flood studies seek to understand the clumping of floods in time and the genetic mechanisms therefor.

Within the framework of their global distribution, floods manifest themselves locally as various magnitudes of flow with a frequency distribution in time. The study of flood magnitude–frequency relationships has been the traditional concern of engineering hydrologists. Potter (1987), in a review of recent research, describes the focus of most flood frequency studies as follows: "statistical estimation of flood quantiles (discharges with specified probabilities of exceedence) for use in water-resource decision-making (e.g., floodplain zoning or design of engineering works)." However, in recent years there has been renewed interest in studies of physical characteristics of flooding spurred on by the past neglect of physical process studies of hydrologists (Pilgrim, 1986). As stated by Potter (1987), "In spite of the increasing sophistication of the statistical tools which we use to analyze flood frequency, we have not made comparable progress in understanding the physical processes affecting flood frequency." Dooge (1986), Klemeš (1986), and Pilgrim (1986) discuss reasons for this state of affairs.

Potter (1987) describes two main areas of accelerating research on the physical processes affecting flood frequency: (1) discriminating the distinct mechanisms that produce floods, and (2) understanding the role of the drainage network in flood generation. The first of these major areas is reviewed here in Katherine K. Hirschboeck's Chapter 2, "Flood Hydroclimatology." The second major area is reviewed in Peter C. Patton's Chapter 3, "Drainage Basin Morphometry and Floods."

Chapter 2 represents a reaction to what has become the norm for flood frequency analysis. The common sense of science dictates that floods be treated as real-world physical events occurring within a context of climatic variations in magnitude and frequency. Unlike hydrometeorology, which focuses on short-term interactions of the atmosphere and hydrosphere, hydroclimatology analyzes such interactions on broader spatial and temporal scales. Hirschboeck utilizes the context of hydroclimatology to re-evaluate various assumptions and models that are presently used to describe how floods vary over time in relation to climate. In moving toward defining the discipline of flood hydroclimatology, Chapter 2 provides a philosophical review of much of contemporary flood hydrology.

The magnitude and frequency of rainfall–runoff floods are most influenced by the climatological factors that dictate precipitation (Chapters 1 and 2). However, once the precipitation reaches the ground, its conversion to flood flow depends on the physical characteristics of drainage basins and stream channels. Geomorphologists and hydrologists therefore developed a common concern with drainage basins, which can be considered to be the fundamental geomorphic units of the Earth's surface (Leopold et al., 1964).

Chapter 3 reviews various quantitative geomorphic measures of drainage basin form. These morphometric parameters can be related to stream runoff through various models. These models have predictive capability, which is required in applied hydrology and desirable as a goal in geomorphology (Shreve, 1979). However, an adequate theory relating quantitative geomorphology to flood response is still lacking. Numerous internal complexities characterize drainage basin systems both temporally and spatially. Progress is possible in several areas. Small, well-instrumented basins may be used to develop an empirical data base with which to calibrate models (Woolhiser, 1973, 1982). Remote sensing technology has an immense potential to expand the spatial scale for determining hydrologic parameters and for monitoring their temporal variation. Remotely sensed spectral data and digital topographic data may be combined in computer models that will induce a renaissance in quantitative drainage basin geomorphology.

Another approach to learning more about flood-causing factors is to study the floods themselves. Although this is an obvious goal, it has not proven as easy to achieve as for other fluvial phenomena. The largest, most energetic floods, for which there is the greatest scientific and engineering concern, are also the rarest and most difficult to investigate. Such events have exceedingly low probabilities of occurring in the small number of well-instrumented watersheds. Moreover, even when the improbable occurs, the most energetic floods often destroy the instrumentation.

Nevertheless, every year great, rare floods occur on the planet's land surface in various climatic, geomorphic, and socioeconomic settings. The effects of these floods are manifest immediately after their passage. Chapter 4 by Garnett P. Williams and John E. Costa provides a standard list of features for geomorphologists to observe, measure, or estimate after a flood. Such a flood "audit" must be standardized for consistency in flood investigations. Among other things, the field investigator needs to consider hydraulic studies, evidence of debris flows, and the monitoring of channel and floodplain recovery.

Williams and Costa draw the analogy between postflood studies and the Vigil Network. The Vigil Network inspired by Luna B. Leopold, has served to make long-term, continuing measurements of landscape change at specific study sites. The concept of flood audits involves a specific phenomenon, which is known to be of major geomorphic importance, but for which we have relatively few direct measurements. The challenge to geomorphologists is to make the appropriate measurements when the opportunity occurs of a cataclysmic flood somewhere in a broad region. Given enough such studies, in enough geomorphic settings, there will eventually be an adequate data base with which to assess the variable role of floods in landscape change.

REFERENCES

Baker, V. R. (1975), Flood hazards along the Balcones Escarpment in central Texas—Alternative approaches to their recognition, mapping, and management. *Univ. Texas, Bur. Econ. Geol. Circ.* **75-5**, 1–22.

Baker, V. R. (1977). Stream-channel response to floods with examples from central Texas. *Geol. Soc. Am. Bull* **88**, 1057–1071.

Caran, S. C., and Baker, V. R. (1986). Flooding along the Balcones Escarpment, central Texas. *In* "The Balcones Escarpment: Geology, Hydrology, Ecology and Social Development in central Texas. P. L. Abbott and C. M. Woodruff Jr., eds. Earth Enterprises, Inc., Austin, Texas, 1–14.

Dooge, J. C. I. (1986). Looking for hydrologic laws. *Water Resour. Res.* **22**, 46S–58S.

Klemeš, V. (1986). Dilettantism in hydrology: Transition or destiny? *Water Resour. Res.* **22**, 177S–188S.

Leopold, L. B., Wolman, M. G., and Miller, J. P. (1964). "Fluvial Processes in Geomorphology." W. H. Freeman, San Francisco.

Pilgrim, D. H. (1986). Bridging the gap between flood research and design practice. *Water Resour. Res.* **22**, 165S–176S.

Potter, K. W. (1987). Research on flood frequency analysis: 1983–1986. *Rev. Geophy.* **25**, 113–118.

Shreve, R. L. (1979). Models for prediction in fluvial geomorphology. *Math. Geol.* **11**, 165–174.

Woolhiser, D. A. (1973). Hydrologic and watershed modeling—state of the art. *Trans. Am. Soc. Agri. Eng.* **16**, (3), 553–559.

Woolhiser, D. A. (1982). Physically based models of watershed runoff. *In* "Rainfall–runoff relationship." pp. 189–202 Water Resources Publications (V. P. Singh, ed.) Fort Collins, Colorado.

1

FLOOD CLIMATES

BRUCE P. HAYDEN
Climate-Ecosystems Dynamics Group, Department of Environmental Sciences, University of Virginia, Charlottesville, Virginia

INTRODUCTION

An examination of the literature reveals neither a global classification of flooding nor a regionalization or map of flood climate types on a global basis. Our goal here is to generate such a classification and a map of the resulting flood climate types and to detail the basis for its construction.

A coherent global climatology of flooding is not easily constructed from stream discharge frequency and magnitude statistics. The difficulty arises, in part, because the frequency and magnitude of floods vary between and within drainage networks because of variability in basin characteristics. This variability is further complicated by the diversity of weather systems that may give rise to flooding, each with its own characteristic return interval spectrum. Although the variability in basin characteristics prohibits regionalization on a global basis, the meteorological causes and potential for flooding can be classified and regionalized. In this chapter flood climate regions are delineated on the basis of meteorological causation. Subregionalizations based on basin characteristics might then prove possible.

PRIMARY RESERVOIR OF FLOODWATERS

The primary reservoir of floodwaters is the atmosphere. The water stored on the land as snow and ice is the secondary reservoir. The potential for flooding depends on the content of these reservoirs and on the rates and duration of discharges. The regionalization scheme proposed here focuses first on the content of the reservoirs and then on the meteorological mechanisms responsible for discharge from the reservoirs.

The atmosphere contains about 1.4×10^4 km^3 of water. This is only a small fraction (10^{-5}) of the total water inventory of the earth (Lamb, 1972). On a global average basis there is only about 23 mm of water in the air. In contrast, global precipitation averages about 1000 mm/yr. Evaporation effectively replenishes the total atmospheric water about 40 times per year. Total annual evaporation is approximately equal to the water that resides in lakes and streams at any time (0.3×10^6 km^3) (Lamb, 1972).

Most of the water in the atmosphere is contained within the troposphere. Within this layer of the atmosphere the processes that give rise to precipitation take place. The troposphere in tropical latitudes extends from the surface to a height of 17 km, and in polar latitudes to about 11 km. The actual volume of the atmospheric reservoir thus declines with latitude. Average tropospheric temperatures also decline with latitude. Therefore the atmosphere's capacity to hold water vapor also declines, because the saturation vapor pressure of the air increases as air temperature increases. Warm air can hold more moisture than can cold air. In addition, evaporation is a temperature-dependent process, so the primary areas of atmospheric recharge occur in the low latitudes where temperatures are high. More then 60% of the water evaporated into the atmosphere occurs between 30°N and 30°S (Lamb, 1972). In contrast, only about 5% of the total evaporation takes place poleward of the 50th parallel. This latitudinal inequality is complicated by the distribution of land and sea. More than 60% of the evaporation occurs over the oceans. Vigorous circulation of the atmosphere reduces the equator-to-pole and ocean-to-land contrasts in the total water content of the atmosphere but does not eliminate them.

Geographic variations in available water within the atmospheric reservoir can be assessed using the parameter total water content. The total water content of the atmosphere is that depth of water that would result if the water content of a column of the atmosphere extending to the top of the troposphere were condensed to liquid. In polar continental regions in winter this depth may be as little as 5 mm, whereas in tropical maritime areas values around 50 mm are common all year. Clearly if air were not advected, converged, and uplifted, rain in amounts sufficient to directly cause flooding would never occur.

Barotropic and Baroclinic Atmospheres

Earth's troposphere may be divided into two broad classes: baroclinic and barotropic. The distinction is somewhat complex but involves the root cause of rotational circulation. It is thus important to detail the difference. The atmosphere is said to be baroclinic when solenoids are present. Solenoids arise when gradients of pressure and temperature intersect. A simple example of a solenoid would be a saucepan filled with water but heated only on the bottom outer rim. Temperature declines both toward the center of the pan and upward toward the surface of the water in the pan. On the other hand, pressure just declines upward as dictated by the hydrostatic relationship. Temperature gradients and pressure gradients intersect. When such intersections are present, a circulation is initiated. In the pan water rises around the side walls of the pan; it flows toward the center, sinks to the bottom, and then flows toward the heated rim. The baroclinic condition of solenoidal circulation is largely absent in the barotropic atmosphere. Temperature and pressure gradients are weak and nearly parallel. Solenoidal circulations are poorly developed. Alternative means of generating circulations are needed.

A barotropic atmosphere is more typical of tropical low-latitude regions, and a baroclinic atmosphere is typical of higher latitudes with contrasting air masses. In a barotropic atmosphere pressure is constant on constant-density surfaces. This condition prevails where horizontal thermal gradients are small. A consequence of small horizontal thermal gradients is the absence of large changes in wind speed with height. So, the barotropic atmosphere is characterized both by little horizontal temperature contrast and also by little vertical wind shear. Once upward motions are started due to surface heating or airstream convergence, or by orographic effects, it is the release of latent heat that drives convection, vertical motions, and precipitation in the barotropic atmosphere.

Vertical motions give rise to condensation and upward-growing clouds in both the barotropic and baroclinic atmospheres. However, with the absence of significant wind shear with height in the barotropic atmosphere, the upward motions are nearly vertical. These convective clouds grow as great vertical chimneys as the clouds are not sheared off by strong winds aloft. Convection through a deep layer of the atmosphere is possible, and high rainfall rates can be generated.

The warm moist air typical of the low-latitude barotropic atmosphere tends to be unstable or nearly unstable. Upward motions give rise to a temperature decline, condensation, and the release of latent heat. The rising parcel of air becomes less dense, lighter than the surrounding air, and thus more buoyant than its surroundings. Upward motion therefore induces further upward motion. It is this autoconvective process that sustains the vertical motions and upward cloud growth that typifies the barotropic atmosphere of the low latitudes. Small vertical motions in the surface layer can give rise to the towering cumulonimbus clouds and the deep convection that characterizes the tropical atmosphere. This cumulus cloud scale convective chimney is the fundamental rainmaking element of the barotropic atmosphere. Even great tropical hurricanes can be viewed as an organized structure, a vortex, of convective cloud elements.

In a baroclinic atmosphere, typical of the higher latitudes, pressure is not constant on surfaces of constant density. Accordingly, horizontal thermal contrasts are often sharp, and wind speeds increase with height in proportion to the strength of the horizontal thermal contrast. The intersections of pressure and temperature surfaces give rise to solenoidal circulations in the horizontal plane. Convergences in the horizontal circulation in turn give rise to vertical motions, cooling, condensation, and precipitation. The middle- and high-latitude baroclinic atmosphere is driven by the hydrodynamics of horizontal flows, which in turn sustain the vertical motions that give rise to precipitation. However, rainfall rates are modest compared to those generated by the deep convective processes of the barotropical atmosphere. This difference is in part counteracted by the longevity of baroclinic weather systems and thus the longer durations of baroclinic rainfall.

In a general sense the earth's atmosphere may be divided into low-latitude barotropic and high-latitude baroclinic portions. The boundary between the two varies on both seasonal and synoptic time scales but is roughly the northern limit of tropical air masses. For our purposes the 25-mm contour of total atmospheric water content marks this transition and separates the atmosphere into two parts with fundamentally different driving forces in the precipitation process. It is of importance to note that this division also delimits those regions where high rainfall rates of short duration are the norm from those areas with modest rainfall rates but longer durations. Discharges from the atmospheric reservoir are fundamentally different in the barotropic and baroclinic atmospheres.

Figure 1 shows the January and July 25-mm contours of atmospheric water content. The position of the line follows that of Lamb (1972) except in the area of North Africa where barotropic conditions are found but water

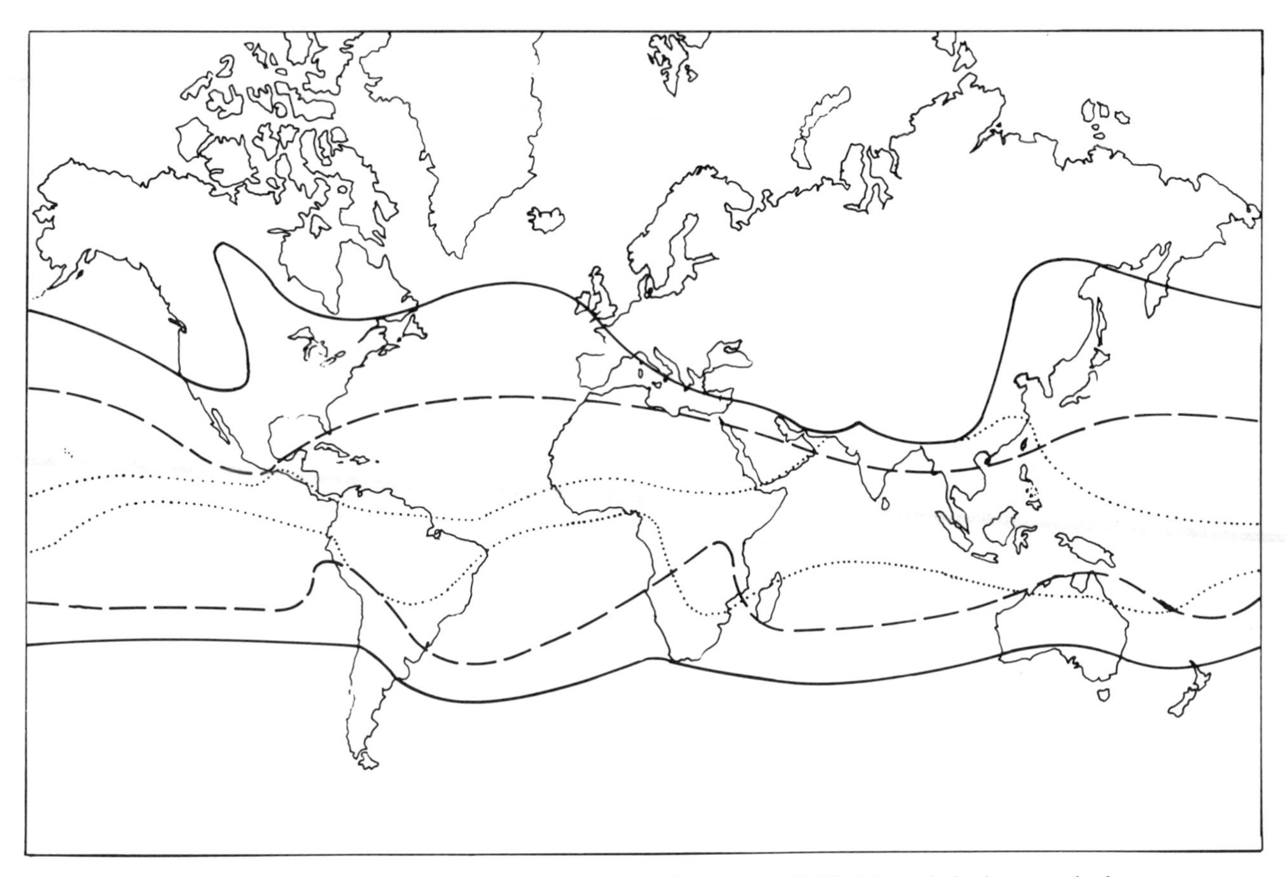

FIGURE 1. The mean winter (dashed line) and summer (solid line) boundaries between the low-latitude barotropic and high-latitude baroclinic atmospheres. The four lines are after Lamb (1972) and represent the 25-mm atmospheric water content contours in January and July. The dotted lines are the July (upper dotted line) and January (lower dotted line) positions of the ITCZ (after Atkinson and Sadler, 1970).

contents fall below 25 mm. The contour is adjusted in this region to reflect the distribution of barotropic conditions. Also shown in Figure 1 are the January and July positions of the convergences of the trade winds from the two hemispheres. This circumequatorial band of convergence and cloudiness is known as the Intertropical Convergence Zone (ITCZ), discussed in a subsequent section. The winter and summer 25-mm contours of atmospheric water taken separately define three broad zones. In the high latitudes, where mean water content is generally less than 25 mm all year, we have the perennially baroclinic zone. In the low latitudes water contents exceed 25 mm all year, and we have the perennial barotropic zone. In the middle latitudes barotropic and baroclinic conditions alternate on seasonal and synoptic time scales. The 25-mm contour in July also clearly indicates the monsoon invasion of moist air in eastern Asia and in the Northern Hemisphere summer as well as the midsummer penetration of maritime tropical air and barotropic conditions into the high plains of North America.

In the north of Africa the poleward limit in January of barotropic conditions is not well delineated by the 25-mm contour. The air over North Africa, while usually barotropic, has less than 25 mm of water in it. The dashed line in Figure 1 along the coast of North Africa represents the mean winter poleward limit of barotropic conditions. This then is a departure from the 25-mm contour of Lamb (1972). The air over arid North Africa as well as that over the Outback of Australia and the Rajastan desert of northwest India is then sufficiently moist to support considerable rainfall and flooding. However, the rainfall is not often realized because of the absence of the meteorological mechanisms to initiate, support, and sustain the vertical motions needed for condensation and precipitation.

Weather Systems of the Barotropic Atmosphere

The horizontal convergence and heating in the surface layer of the barotropic atmosphere gives rise to the vertical motions needed to support condensation and precipitation.

The convective cloud is the basic element of rain production in the barotropic atmosphere. Figure 2 shows the global distribution of thunderstorms, fully matured convective systems, by broad frequency classes: more than 100, between 50 and 100, between 20 and 50, and less than 20 thunderstorms per year. Frequent thunderstorms on an average annual basis are typical of both the perennial and the seasonal barotropic conditions. Populations of thunderstorms are, however, often organized into larger weather systems, and it is these larger scale assemblages of convective elements that are of greatest importance where other than highly localized flooding is concerned. The dominant rain-producing organized weather systems of the barotropic low latitudes are: (1) the convergence of the trade wind streams between the two hemispheres (the ITCZ) (2) cyclonic curvatures of the pressure field (tropical storms and easterly waves), and (3) orographic uplift as airstreams rise over the relief of the land. Barotropic regions where these weather systems are infrequent or weak, such as North Africa, tend to be arid in spite of the barotropic conditions of the atmosphere.

The Intertropical Convergence Zone. The Intertropical Convergence Zone (ITCZ), sometimes referred to as the meteorological equator and the doldrums, marks the convergence of wind streams from the two hemispheres. The convergence and rising air results in a circumequatorial band of cloudiness. This band is clearly evident in satellite photographs (Fig. 3). The zone of intertropical convergence and the belt of raininess that results goes through an orderly annual north–south excursion. The day-to-day position and movement of the ITCZ are difficult to forecast with precision. It is in its most northward position during the Northern Hemisphere summer and is farthest southward during the Southern Hemisphere summer. Because of this latitudinal variation in the ITCZ, a pronounced seasonality in rainfall results over much of the intertropics. The generic term *monsoon* is often associated with this summer rainy season. The specific term *Monsoon* is generally reserved for the pronounced Asian monsoon. Figure 1 shows the January and July limits of the ITCZ. The intertropics defined by the seasonal swings of the ITCZ encompass most of what is commonly known as the moist tropics. However, in some regions such as eastern Africa and the southern coast of the Arabian peninsula, the strength of the convergence is often weak, upward motions small, and rainfall slight. Where there are highlands present within this intertropical zone, uplift may be enhanced by orographic pro-

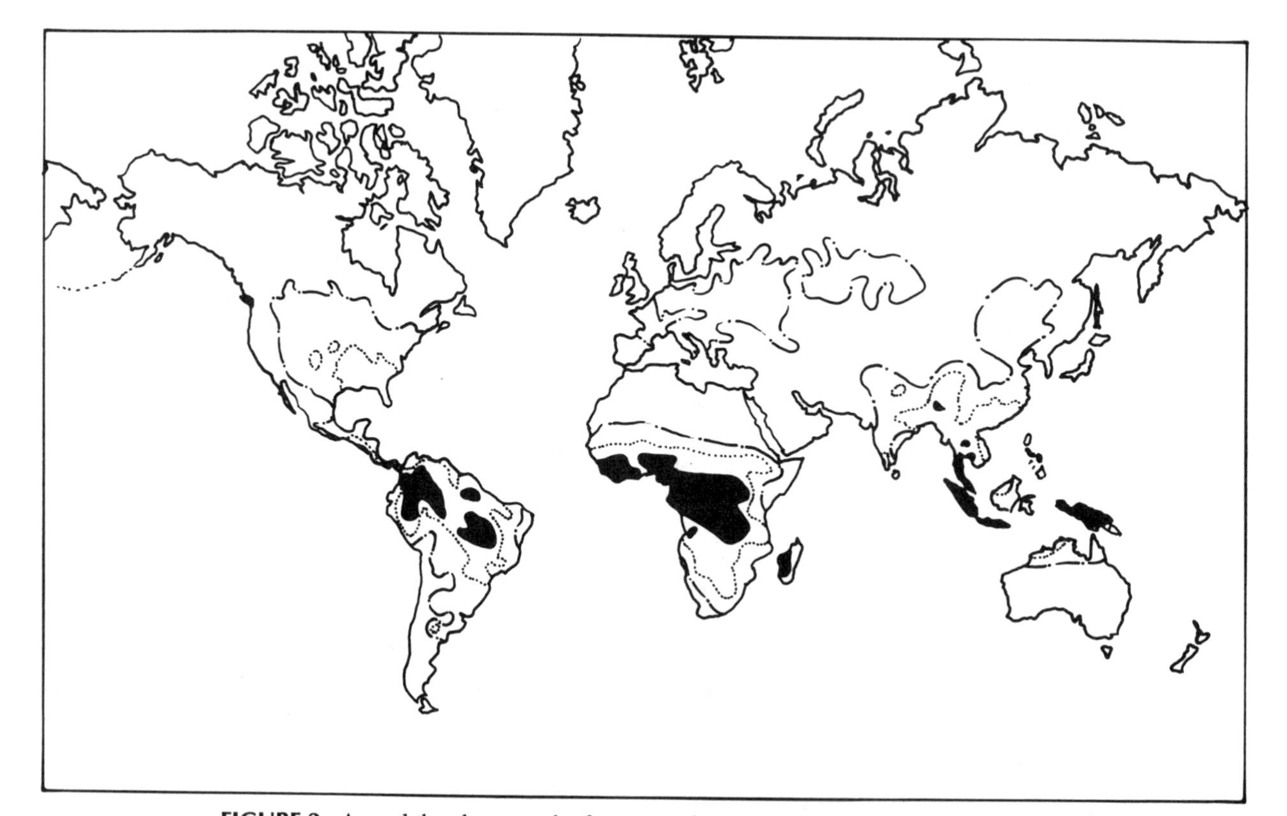

FIGURE 2. Annual thunderstorms by frequency classes. Regions with 100 or more thunderstorms per year are black. The dotted line encompasses regions with more than 50 thunderstorms per year and the dashed-dotted line encompasses regions with at least 20 thunderstorms per year (Lamb, 1972).

FIGURE 3. The ITCZ on 21 July 1975. The dashed line follows the axis of brightest cloud masses. The photograph is an infrared image and the tallest and coldest cloud tops are the brightest areas. Regions of greatest convection are clearly seen.

cesses, and great quantities of rain may result. At Cherrapunji, India (in June 1876) 2500 mm (100 in.) of rain fell in about 100 hr. Most of the rains that give rise to Nile flooding fall on the Ethiopian highlands as the ITCZ passes northward during the Northern Hemisphere summer. In years when the convergence is weak or fails to penetrate that far north, drought occurs. The great drought and famine of 1981–1985 marks an episode in the erratic, unreliable history of the ITCZ in this region.

Off the west coast of South America the normal seasonal excursions of the ITCZ encompass only a few degrees of latitude and are largely confined to Colombia and Equador. However, in years of the El Niño, warm tropical waters extend southward along the coast as does the ITCZ. Heavy convective rains may then come to the extremely arid Peruvian coast. In years of the El Niño, floods are common in Peru.

Cyclonic Motions. On occasion, large-scale cyclonic curvatures of the wind streams occur within the perennially barotropic atmosphere of the lower latitudes. The most common weather system of this type is the easterly wave (Fig. 4). The easterly wave is a wavelike or sinusoid deformation in the trade wind flow pattern. Convective clouds (thunderstorms) become organized at the scale of the easterly wave and may give rise to intense prolonged periods of rainfall. These easterly waves tend to move from east to west within the trade wind zone, and the greatest rainfalls occur when the easterly wave passes over islands or continental margins. Here the heated land surfaces and the roughness of the landscape enhance upward motions, and the convective elements intensify.

The most important type of cyclonic motion in the barotropic atmosphere is the tropical cyclone. Tropical cyclones frequently arise from the organized system of convective elements of an easterly wave. It is not known why a few of the easterly waves intensify into tropical storms nor why most do not. Those tropical cyclones that intensify to the level where winds of 116 km/hr (72 mph) or greater are generated are known as hurricanes. This is the name generally applied to those tropical storms of great intensity in the North American sector. Over most of the Pacific Ocean basin they are referred to as typhoons. In the Indian Ocean area they are called cyclones. In general these storms do not form adjacent to the equator where the Coriolis parameter is small and closed circulation systems uncommon. The genesis of tropical storms requires an ample supply of latent energy, and thus the region of genesis is restricted to the subtropical oceans where sea surface temperatures exceed 27°C. An additional constraint is that the atmosphere be largely free of vertical wind shear. Although the barotropic atmosphere in general tends to have little shear, this is not true everywhere. The south Atlantic Ocean region in the subtropics tends to have a wind shear in the vertical, and tropical storms are essentially unknown there.

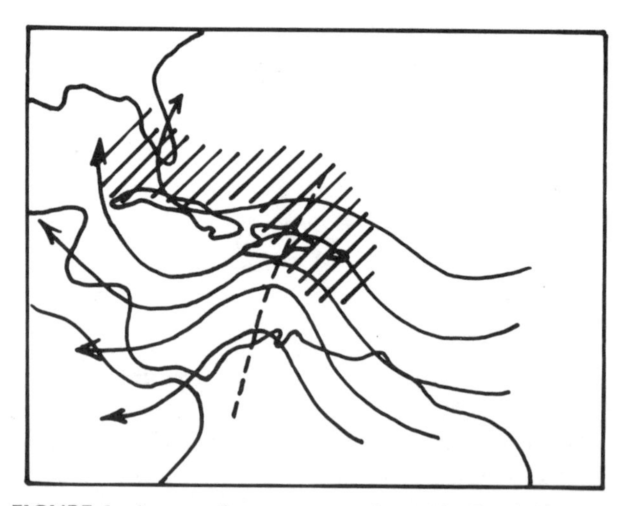

FIGURE 4. An easterly wave centered over the Dominican Republic on August 13, 1969. The dashed line indicates the axis of the easterly wave trough. The arrows indicate the prevailing airstreams and the cross-hatched area is the region of cloud masses.

The absence of shear permits deep convection. The hurricane is a spiral-form organization of hundreds of convective elements (Fig. 5). Strong upward motions and the release of latent heat support the vigorous horizontal cyclonic motions of the tropical storm. Should the top of the convective chimneys of the thunderstorms be "blown off" because of increasing velocities aloft, the hurricane would tend to dissipate quickly.

Once formed, tropical cyclones tend to move westward within the trade wind belt, then northward and eastward into baroclinic regions. Here tropical cyclones may interact with midlatitude frontal systems and benefit from the hydrodynamics of the baroclinic atmosphere. While some of its characteristics as a tropical system may then disappear, great quantities of rain may result. Hurricane Camille in 1969 and Agnes in 1972 did just this and resulted in 250–500 mm of rain on the Atlantic Coastal Plain. The highest rainfall totals from both storms occurred when the most baroclinic portion of the atmosphere intersected the remnants of the tropical storm. Similar conditions are also common in southeastern China. Figure 6 shows the distribution of the trade winds where easterly waves occur and the major tropical cyclone tracks.

Some otherwise arid areas, such as Arizona, only rarely experience tropical storms. Thus these landscapes are not "equilibrated" to the copious rainfalls that result from even modest tropical storms. Modest tropical storms may result in record floods in such areas. Such circumstances occur in Arizona where tropical storms may give rise to extensive flooding. Chapter 2 (on flood hydroclimatology) is particularly instructive. The distribution of tropical storms shown in Figure 6 must thus be viewed in two ways. First, there are those regions that because of the high frequency of tropical cyclones may well experience the most intense tropical storms and are prone to frequent hurricane flooding. Second, there are those regions that only rarely experience the effects of tropical storms, but when a storm track on occasion enters the region, major flooding occurs.

Orographic Uplift. Upward motions and strong convective activity are often directly coupled to the terrain over which airstreams must flow. In the barotropic atmosphere these upward motions are accompanied by condensation and the release of latent heat, and the air becomes more unstable. Rising air becomes less dense than the surrounding air, and further upward motions follow. This autoconvectivity is enhanced over highlands as solar heating of the land

FIGURE 5. Tropical storm Blanch off the Carolina coast on July 27, 1975. Although storms at this stage are not classed as a full hurricane, the spiral bands of convective clouds are clearly evident.

and, in turn, the air further increases the upward motions and the chances of rain. Orographic rainfall often has a strong diurnal component, with rainfall maxima following the period of strongest solar heating. Orographic rainfalls are regionally organized where north–south land masses, mountain chains, and archipelagos intersect the persistent east-to-west trade winds of the subtropical latitudes.

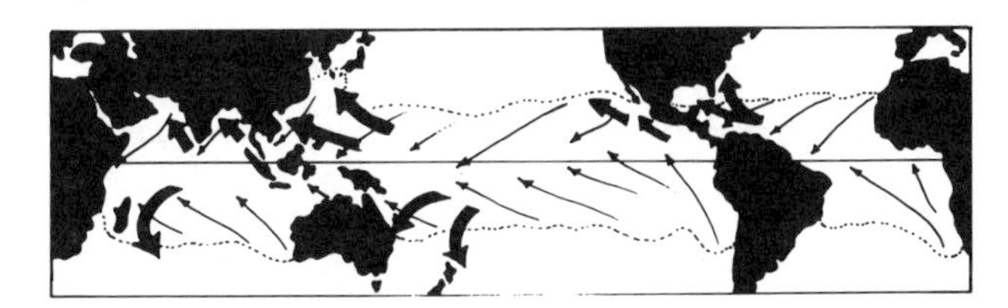

FIGURE 6. The zone of easterly waves and trade wind disturbances. The thin arrows indicate the general direction of the trade winds (after Crowe, 1949) and the large black arrows the major hurricane tracks (after Simpson and Riehl, 1981).

In the middle latitudes the atmosphere tends to be barotropic in the summer half of the year. Convective precipitation is often associated with thunderstorms. Usually, serious flooding from individual thunderstorms is restricted to arid regions. However, in regions where orographic effects may be present, thunderstorms may merge to form what is now called mesoscale convective clusters. Rainfall rates from these merged thunderstorms are like those of the deep barotropic atmosphere of the tropics and occur because of the dynamic interaction of the storm and the terrain. The mesoscale convective cluster may be stationary for a long time. The Rapid City, South Dakota, flood of

1972 and the Johnstown, Pennsylvania, flood of 1979 were devastating floods of this type. The mesoscale convective cluster is typical of areas where barotropic and baroclinic conditions alternate on seasonal and synoptic time scales. While little is known about these newly recognized atmospheric phenomena, they are easily detected from satellite images, and we are sure to learn more about their climatology in the years ahead.

Although the inclusion of every element of relief on a map of flood climate regions would seem advantageous, the map quickly becomes filled with details inappropriate to a global map. Accordingly only the major terrain elements will be shown, and then only in a stylized fashion. The detail of terrain relative to the flood climate classification is left to the reader to deal with on a region by region basis. However, it is essential that a map of the climatology of flooding include the major elements of relief of the landscape.

Weather Systems of the Baroclinic Atmosphere

In a baroclinic atmosphere wind shears in the horizontal and vertical planes arise from strong thermal contrasts and the solenoidal fields associated with intersecting temperature and pressure surfaces. Convergences and divergences at various levels in the atmosphere give rise to vertical motions, condensation, and precipitation. This is not to say that the release of latent heat is an unimportant process in the baroclinic atmosphere, but rather that the initiating forces are more hydrodynamic than thermodynamic in nature.

Strong horizontal thermal contrasts in the baroclinic part of the earth's atmosphere tend to be associated with fronts. At these fronts masses of air from different source regions converge. The polar front exhibits the strongest temperature contrast. The polar front usually separates the barotropic tropical atmosphere from the high-latitude baroclinic atmosphere. In the region of the polar front, north–south thermal contrasts are greatest, and the westerly winds increase rapidly with height. Aloft there is usually a jet stream associated with the polar front. The polar front marks a region of great hydrodynamic instability. It is in this region the frontal cyclones commonly develop.

These polar front storms have access to tropical barotropic air to the south and the latent energy available from that moist air. Storms along the polar front may reach great intensity and size. The strong upward motions, sustained by convergence toward the storm center at the surface and divergence at the top of the storm supported by the jet stream aloft, may result in heavy precipitation.

These frontal storms vary in size from several hundred kilometers to more than a thousand kilometers in diameter. They tend to move from west to east and/or southwest to northeast. The "footprint" of precipitation that results is of the same size class as the large drainage basins of the middle latitudes. Precipitation from these storms and their associated frontal zones generally has durations of 12–36 hr. It is significant that frontal storms tend not to be solitary, but rather several storms form along the frontal zone at separation distances of several hundreds of kilometers. Figure 7 illustrates such a family of frontal cyclones. This is significant for flooding in the middle latitudes because sequences of storms and rainfall are common. The first may saturate the ground and subsequent storms result in flooding. In the middle latitudes storms within a family tend to pass a given region about every 3.5 days. It is common, however, for every other storm to be more vigorous in rain production. As a result a weekly cycle of precipitation is observed. Namias (1966) referred to this weekly cycle as a quasi-periodic phenomenon. A more general term of the regular recurrence is the *synoptic cycle*. It is not uncommon for rain to occur for 10–15 consecutive Tuesdays. It seems, however, that 10–15 consecutive weekends of rain are more common, but this is the result of our memory of dashed plans for our leisure time. Over the long term there is no day of preference within the week, that is, the phase angle of this quasi-periodic phenomenon varies from year to year.

Polar front cyclones are largely restricted to the zone of the seasonally baroclinic atmosphere (see Fig. 1). However, even a casual examination of a weather map for the Northern Hemisphere will reveal other fronts north of the polar front. Like the polar front, these other fronts separate masses of air from different source regions and thus air masses with differing thermodynamic properties. Unlike the storms of the polar front, when over continental areas, the frontal cyclones of these other fronts rarely have access to masses of air with sufficient moisture to result in heavy rainfall and flooding. However, if there is orographic enhancement of condensation and precipitation, heavier rains may result. For flooding to occur, the relatively modest precipitation from these frontal cyclones must be stored on the land as snow and ice, accumulated from storm to storm during the winter season. In short, to the north of the zone of seasonally baroclinic conditions, that is, the zone of polar front storms, flooding is usually a spring and early summer phenomenon. In polar regions the atmosphere is sometimes barotropic, that is, temperature and density contours intersect; however, the Coriolis parameter is extreme and horizontal circulations of a dynamic rather than a thermodynamic nature dominate. In addition the air in polar regions has low water content and cannot support heavy precipitation.

THE SECONDARY RESERVOIR OF FLOODWATERS

The secondary reservoir of floodwaters is the accumulated snow and ice on the land surface. Flooding from this

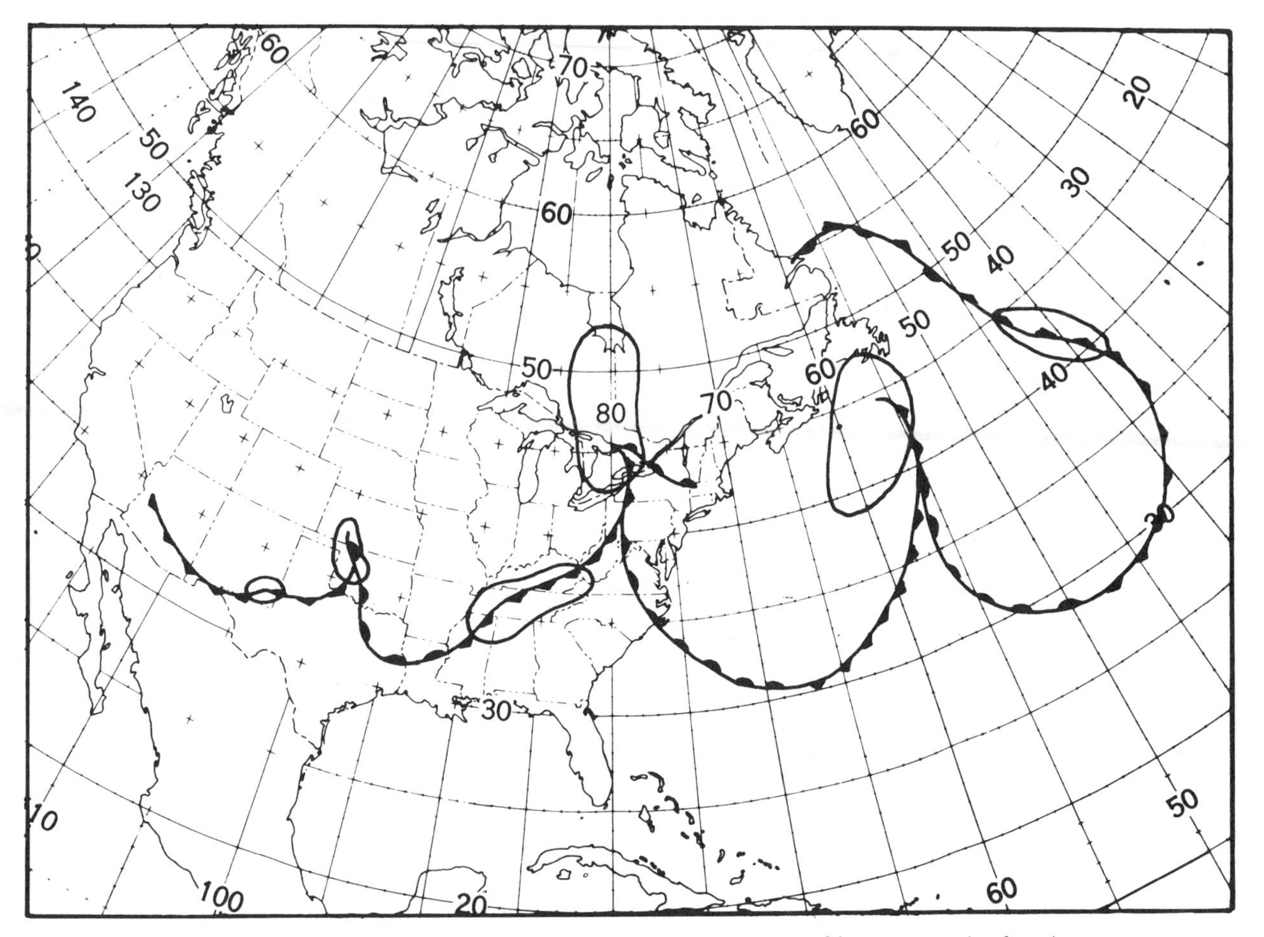

FIGURE 7. A family of cyclones on the polar front. The centers of low pressure (cyclones) are indicated by circles (closed isobars). In regions where the air of tropical origin is moving poleward, a warm front (half circles on the front) is indicated. where air of higher latitude origin is moving southward, a cold front (triangles on the front) is indicated. The family of cyclones progresses from west to east.

reservoir is dependent on the amount of snow cover and the rate and frequency of meltwater discharge. In some regions, such as Greenland and the Antarctic, the snow and ice cover is perennial, and discharges during the summer ablation season are inadequate to cause flooding. In addition there are regions with a winter snow cover and a spring melt that do not flood because there is inadequate snow cover. Extensive areas of the Canadian and Asian Arctic fall in this category. However, many other high-latitude land areas have sufficient snow cover and a spring melt that result in an annual flooding. In the zone we called the seasonally baroclinic, snow cover, while often substantial, is ephemeral. Furthermore, melting is often associated with rainstorms following snowstorms, resulting in rapid and large discharges. Within winter and spring flooding is common in this region.

Flooding for those regions that are perennially baroclinic or seasonally barotropic is often associated with discharges from the secondary reservoir of floodwaters. The several regions discussed above are classified here with the following criteria: (1) perennial ice cover, (2) winter snow cover less than 50 cm with durations that exceed 50 days, (3) winter snow cover of more than 50 cm with durations that exceed 50 days, and (4) those regions with any snow cover duration between 10 and 50 days. Figure 8 shows the geographic extent of each of the four classes. Details of ice and snow cover associated with mountains and islands are omitted.

FLOOD CLIMATE REGIONS

The flood climate regions discussed here are derived from the union of Figures 1, 2, 6, and 8. The composite and the resulting flood climate regions are shown in Figure 9. Sixteen types are delineated. The figure caption gives the definitions of the boundaries between regions, and the legends on the map are keys to the symbolic nomenclature.

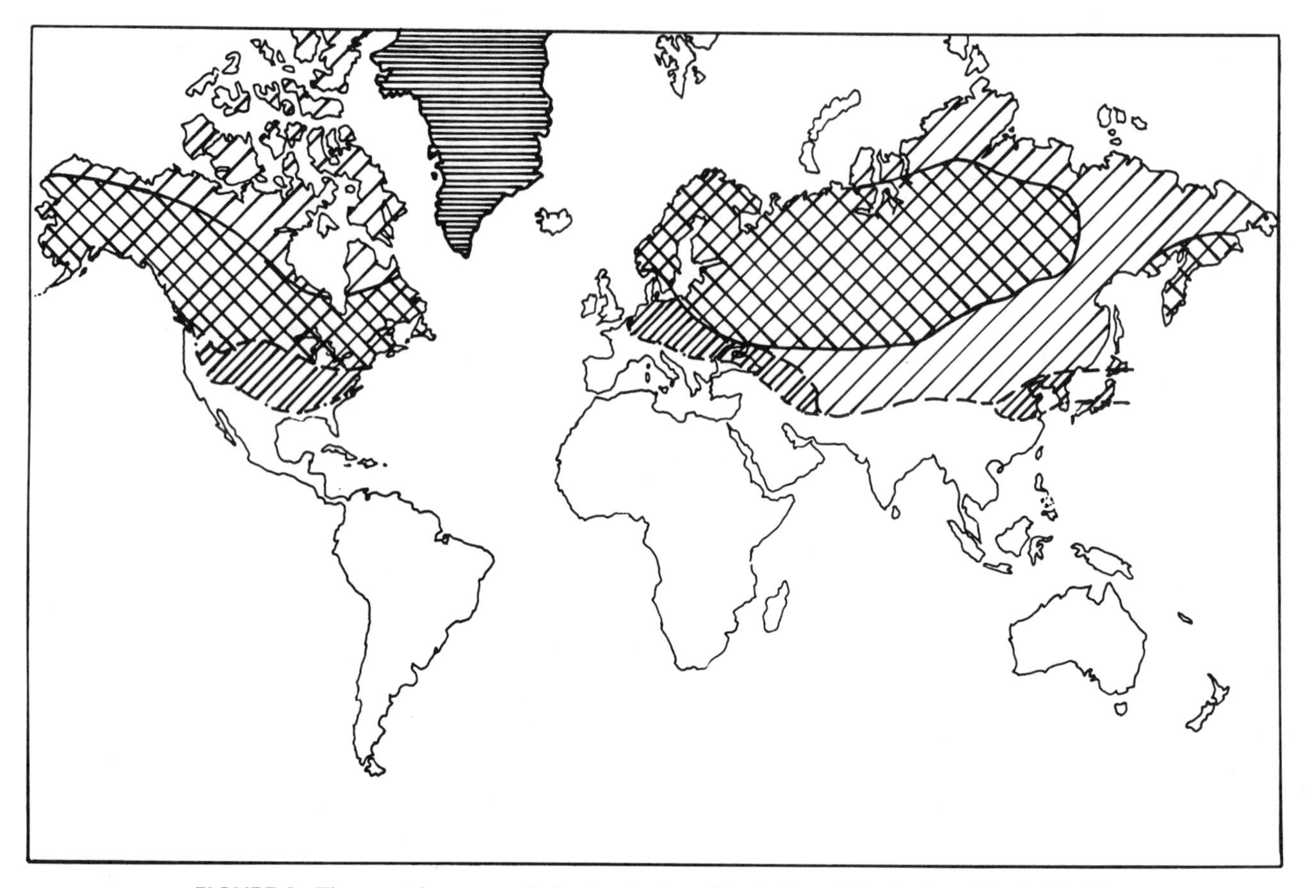

FIGURE 8. The secondary reservoir for floodwaters. The horizontal hatched area is glacial. The cross-hatched area has at least 50 days of seasonal snow cover and at least 50 cm total snowfall. The wide-spaced, diagonal-shaded area has at least 50 days of seasonal snow cover but less than 50 cm total snowfall. The narrow diagonal-shaded areas have less than 50 days of seasonal snow cover and more than 10 cm of snowfall. Data from Lamb (1972). Snowfall and snow cover in mountainous areas are ignored.

The primary reservoir may be characterized either by atmospheric barotropy (T) or baroclinicity (C). These states may be present in all months, that is, perennial (p) or only present on a seasonal basis (s). When barotropic conditions are present, discharge from the reservoir may be due to the intertropical convergence zone (z), organized convective activity at the synoptic scale (o) or unorganized convective activity (u) as in the case of individual thunderstorms. In the baroclinic atmosphere discharge from the atmospheric reservoir of water results from cyclones and fronts. These systems may be present in all seasons (p) or for only part of the year (s). In the high latitudes where rainfall intensity and totals are modest, accumulation of snow on the surface (S) and subsequent discharge is the main mechanism of flooding. When snow accumulates over the winter season and then melts in the spring, the subscript (s) is used and indicates seasonal snowmelt and flooding. When the accumulated snow is discharged within the winter season on a periodic basis, the subscript (e) is used and indicates the ephemeral nature of the snow cover in this reservoir. The potential for flooding at the end of the winter snow accumulation season is dependent on the depth of snow cover. Areas with sufficient snow cover (50 cm or more) are indicated by double asterisks (**) and areas with less than 50 cm snow cover by single asterisk (*).

DESCRIPTION OF FLOOD CLIMATE REGIONS

The letter notations used in this section to designate different regions are shown in Figure 9.

Tpz. —In this region perennially barotropic conditions prevail, and the intertropical convergence zone provides the upward motions needed to initiate the precipitation. The ITCZ undergoes a north–south movement, and so there is usually a rainy season and a drier season. The exception to this seasonality occurs along the west coast of South America in Colombia. Here the north–south excursions of the ITCZ are small and it is rainy in all months of the year. While the zone is circumglobal, there are strong longitudinal variations in the amount of precipitation realized as the ITCZ is not equally vigorous at all longitudes.

Tsz. —In Tsz the seasonality of precipitation is twofold. The ITCZ is present for part of the year as it reaches its extreme southern position in January and with it the rainy season. Six months later the ITCZ is far to the north, and the condition of barotropy, as defined by atmospheric water contents in excess of 25 mm, disappears. During this dry season even individual convective thunderstorms are uncommon. Tsz occurs only in southern Africa, and unlike similar areas elsewhere on earth, it is distant from the trade wind zone and organized disturbances like easterly waves and tropical storms.

Tszo. —Like Tsz this region has a pronounced seasonality in discharge from the atmospheric reservoir of water. Conditions are seasonally barotropic as tropospheric water content falls below 25 mm in winter, and the Tszo regions are poleward of the winter positions of the excursions of the ITCZ. However, unlike the Tsz region of southern or south central Africa, the Tszo region also gets heavy rainfalls from organized convective systems, that is, easterly waves in the trade wind streams and tropical storms. Tszo regions occur in Southeast Asia, the subcontinent of India, and northern Australia.

Tpo. —Tpo regions are barotropic all year and are poleward of the seasonal limits of the ITCZ. Flood-producing rainfalls arise from organized systems of convective elements, that is, easterly waves in the trade wind streams and tropical storms. Tpo regions are within the trade wind

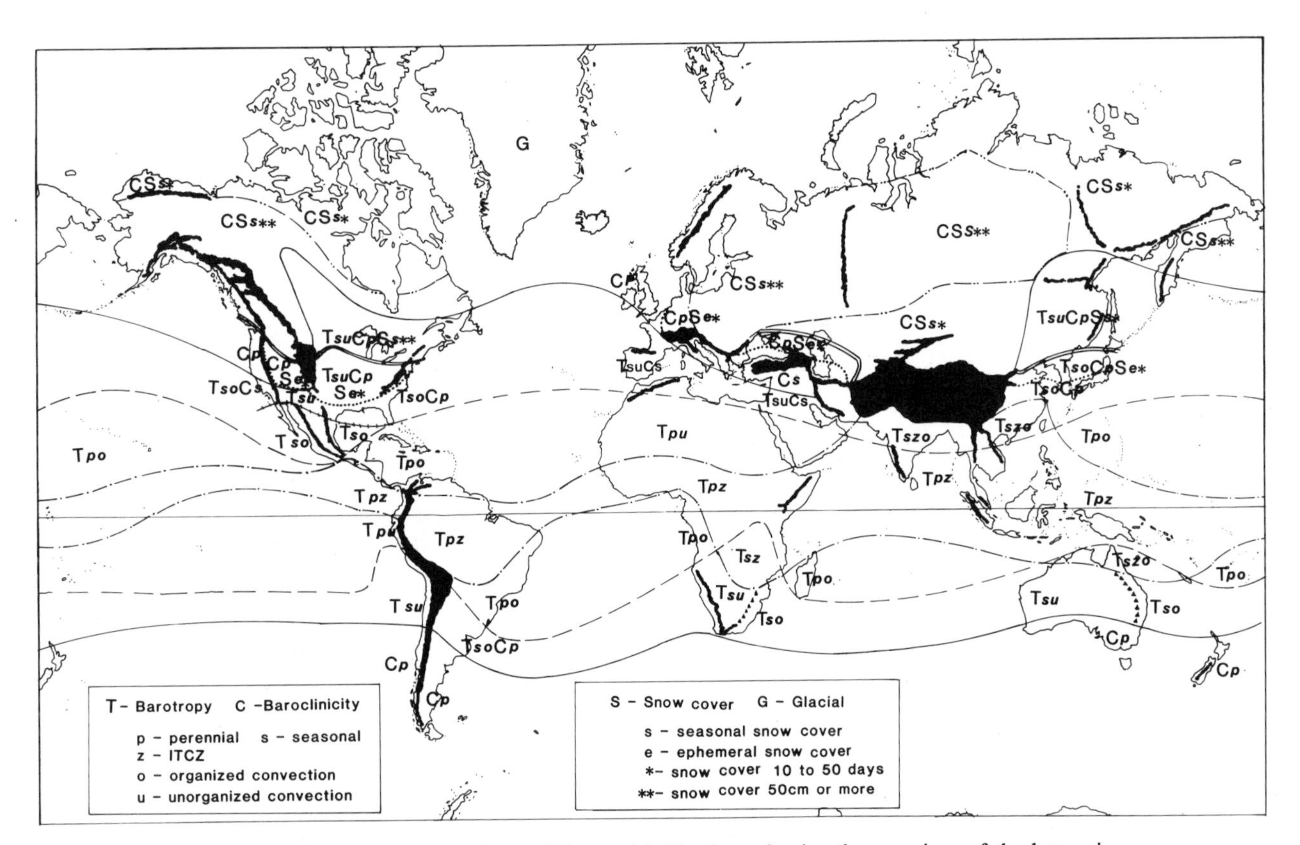

FIGURE 9. Flood climate regions of the world. The legends give the meanings of the letters in the symbolic notation. The solid lines indicate the poleward limits of barotropic conditions in summer and the dashed lines the same limit in winter. The dash-dot lines are the January and July positions of the ITCZ. The dotted line marks the equatorward limit of winter snow cover durations of 10 days or more. The double solid line indicates the equatorward limit of snow cover durations of 50 days or more. The dash-dot-dot line indicates regions with more than 50 days of seasonal snow cover and more then 50 cm of snow. The cross-hatched solid line marks the equatorward limit of frontal cyclones in the North American sector. Solid areas indicate major mountainous regions.

zone, and precipitation enhancement of upward motions by orography and heated land surfaces are important in the initiation of precipitation and in flooding. Many islands in oceanic areas fall within Tpo regions. These regions tend to have a strong diurnal cycle in rainfall as daily solar heating is an important forcing process. Rainfalls are most common in the late afternoon and just after sunset.

Tpu. —Tpu regions are also barotropic in all months, but unlike the Tpo areas easterly waves and tropical storms are uncommon or absent. Convective elements are usually discrete or individual thunderstorms. These regions are generally semi-arid or arid and streams frequently ephemeral. Orography is, however, an important aspect where mountains occur in Tpu regions. Unusual conditions may produce flooding in these regions. The Tpu region along coastal Peru may go for decades without sufficient rain to cause flooding and then experience extensive flooding as the ITCZ moves unusually far to the south. This is most likely to occur in El Niño years when warm waters and the ITCZ extend southward along the Peruvian coast. The Tpu region that covers much of North Africa and the southern portion of Saudi Arabia may on infrequent occasions get rains from an unusual northward excursion of the ITCZ in summer or a frontal cyclone in winter. When these rare events occur, significant flooding may occur as the landscape is not adjusted to such precipitation.

Tso. —Tso regions are seasonally barotropic and seasonally baroclinic. During the barotropic part of the year, easterly waves and tropical storms are the most likely discharge mechanism that result in flooding. During the baroclinic part of the year frontal cyclones are the usual source of precipitation for flooding. Tso areas are generally arid or semi-arid and not well adjusted to the copious rainfalls that can arise from tropical storms. The Gulf of California and the central east coast of Australia are two examples of the Tso flood climate type.

Tsu. —Tsu regions like Tso areas are seasonally barotropic (summer) and seasonally baroclinic (winter). During the barotropic part of the year, rain from individual thunderstorms, sometimes orographically enhanced, is the main source of precipitation. During the winter baroclinic period, only modest rainfalls result from fronts and cyclones. The regions are arid and thunderstorms are the main agent of flooding. Thunderstorms in these regions are initiated over the heated surface and along fronts that pass through the region. Central Australia and portions of coastal Chile and Peru are flood climate regions of this type.

TsoCp. —This type of region is predominantly barotropic in the summer season, but fronts and frontal cyclones are present in all months of the year. Tropical storms are common in these regions. The piedmont and coastal plain of the eastern United States is an example. Summer barotropic thunderstorms are often associated with the frontal systems. Mesoscale convective clusters, merged thunderstorms, and severe hail and tornado-producing storms are frequent in this zone. Frontal cyclones in this region may result in heavy rains as they usually involve moist air of tropical origin. Rainfall rates may be high and storms often occur in sequences. Floods occur in all seasons of the year.

TsoCs. —An example of this region is restricted to southern California and northern Baja, Mexico. Barotropic conditions prevail in the summer, and this coastal region is affected by tropical storms. Summer floods of great magnitude, while uncommon, may result from these tropical storms. Otherwise flooding is restricted to the winter when cyclones from the Pacific are common along the California coast. Were it not for the presence of tropical storms this area would classify as TsuCs like the Mediterranean region and southwestern coastal Australia.

TsoCpSe.* —This region has all the properties of the adjacent TsoCp region but also has winter snow accumulations that are generally short-lived and can result in winter flooding. Tropical storms are common and give rise to heavy flooding in this region. The upper piedmont east of the Blue Ridge Mountains would fall into this category, but the resulting region is small relative to the other regions mapped. Careful inspection of Figure 9 reveals an area between the dotted line and the mountains of New Jersey, Pennsylvania, Maryland, Virginia, and North Carolina. This thin strip would classify as TsoCpSe*.

TsuCpSe.* —This region is similar to TsoCpSe* except that tropical storms are rare or absent. An example of this type region is the central United States east of the Rocky Mountains and includes the grasslands and much of the deciduous forest of the eastern United States. Like TsoCpSe* there is a diversity of weather systems that may give rise to flooding. It is possible for remnants of dissipated hurricanes to enter these regions and become associated with fronts, but they will differ little from frontal cyclones in these regions that have access to moist air from the south. Flooding occurs in all seasons. Flooding may occur during winter as the snows from earlier storms melt as new rains fall.

TsuCpSs.* —This region is like TsoCpSe* except that the winter snows usually do not melt and cause midwinter flooding, but there is the potential for an annual flooding at the end of the winter season. The regional example is found in eastern China. The spring flooding is modest because the snow cover is not great. Moist barotropic air masses from the low latitudes provide the moisture for summer thunderstorms that can give rise to local flooding and also provide the moisture for greater precipitation from summertime frontal cyclones.

*TsuCpSs***. —Here winter snow cover is great enough to result in significant spring flooding. Fronts and cyclones occur all year. Barotropic conditions and severe convective thunderstorms occur from time to time during the summer. This type is restricted to North America where warm moist maritime tropical air penetrates far into Canada.

Cp. —In these regions baroclinic conditions prevail all year and flooding arises from frontal cyclone precipitation. Although there may be a seasonality in the frequency of these storms, which are most frequent in winter, they occur in all months of the year. These regions are found in the upper middle latitudes on the western margins of the continents and have access to relatively moist air from the cool oceanic areas. Most of these regions also have mountain ranges, and orographic enhancement of precipitation is important in generating sufficient rainwater for flooding or snowfall accumulation in the mountains. Because these areas are in close proximity to the temperate ocean, the climate is moderate even in winter and snowfalls, except in the mountains, are rare and quickly melt.

*CpSe**. —These areas, like Cp regions, are baroclinic all year, and precipitation arises from frontal cyclones. Unlike Cp regions, CpSe* areas are considerably colder in winter and have snowfalls that may give rise to within-winter flooding as these snows melt. Where elevations are high in CpSe* regions, snows may accumulate throughout the winter season and, springtime flooding from the highlands is common.

*CSs***. —This is a perennial baroclinic region, but rainfall rates and durations from frontal cyclones are usually modest. This low productivity of precipitation is due largely to the inadequate supply of atmospheric moisture. Moist air from barotropic regions is rare. Rainfall from cyclones is generally modest and rarely results in flooding. In these regions the effect of many cyclones must be aggregated by snowfall accumulations. Snowfall is substantial and the snow cover season is long. Flooding is associated with the spring snow melting.

*CSs**. —Like CSs**, conditions are perennially baroclinic but cyclones do not produce enough rainfall to give rise to flooding without accumulations from one snowfall to the next. However, in CSs* regions total snowfalls are modest. In regions of this type in the high latitudes where evaporation is small the spring snowmelt may result in standing water on the tundra but little stream flooding. The CSs* region in central Asia has but a modest spring snowmelt. This region is arid, and, although thunderstorms are infrequent as are well-developed cyclones, these systems are the primary agents of flooding. Ephemeral streams are common in this area of central Asia.

RELATIONSHIPS TO OTHER CLASSIFICATIONS

The flood climate regions delineated in this chapter bear considerable similarity to charts of the vegetation cover classes based on climatic relationships (see Emanuel et al., 1985) as well as to the more standard classifications of vegetation cover such as that of Udvardy (1975). It is not surprising that this correspondence occurs. Vegetation cover is in part a function of rainfall, temperature, snow cover, and the seasonality of these parameters. The correspondence, however, is not exact. Ideally the vegetation cover reflects the mean of conditions of recent centuries and millennia.

Similarities with the more traditional climate classifications of Koppen and Trewartha are apparent in a comparison of Figure 9 with Köppen (1936) and Trewartha's (1968) maps. In their classifications temperature plays a major role. In the flood climate classifiation presented here barotropic and baroclinic conditions of the atmosphere are an "unintended" proxy of temperature.

Similarities of the present classification to classifications of climates based on the circulation are especially encouraging. The reader is encouraged to compare the present flood climate classifiction with Dietrich's (1963) classification of the climate of the ocean; the classification of coastal and marine regions of the world by Hayden et al (1984); and, Bryson's (1966) classification of North American climates based on air masses and streamlines. These classifications and the one generated for this chapter all had the goal of deriving a classification based on atmospheric dynamics.

MEAN VERSUS EXTREME CONDITIONS

Like most climatologies and climate classifications the one proposed here is based on mean conditions. Flooding, by its very nature, is a significant departure from mean conditions. The classification of flood climate regions might well have the term *potential* appended in front. The flood climate regions delineated should have similar potential for flooding. Clearly there will be variations within the delineated regions in the realization of flooding. Basin morphologies, soils, and vegetation cover all vary within regions.

The boundaries between the flood climate regions proposed here reflect the presence or absence of weather systems and their movement as well as the circulation of the atmosphere. The mapped boundaries are mean boundary positions. The actual positions vary from day to day. We have accounted for the seasonal variations in mean positions of the boundaries but have not included the year-to-year differences in the mean boundary positions. The boundaries are not fixed entities. When the ITCZ, for example, moves further northward across North Africa, rain is delivered to the Sahel and the southern margins of the Sahara Desert. Flooding at the times of these extreme movements of the boundaries is especially likely. Tropical storms only rarely penetrate deep into Arizona from the Gulf of California, but, if they do, extreme flooding can result. These landscapes are just not adjusted to such high rainfall rates or totals.

IMPLICATIONS TO PALEOHYDROLOGY AND PALEOGEOMORPHOLOGY

On longer time scales we should also expect to see modifications of the boundaries between flood climate regions. Records of the Holocene and the Pleistocene clearly indicate different hydrologic regimes than those we experience today. During glacial times baroclinic conditions would have extended further equatorward than at present. Tropical cyclone frequencies were probably lower than today (Wendland, 1977) and the extent and duration of winter snow cover were markedly different.

Based on charts published by Joussaume et al (1984), rainwater from the ITCZ has a O–18 isotopic content of about −2 parts per thousand; the rainwater from Tpo regions between −2 and −3 or −4 parts per thousand. In general rainwater taken from north of the polar front zone in winter has O–18 contents that exceed −8 parts per thousand and may reach −28 parts per thousand. Fossil water O–18 may well provide a marker diagnostic of flood climate conditions and as such may help in deciphering the record of Holocene hydrology and geomorphology. Preliminary comparisons of atmospheric vapor O–18 contents using atmospheric general circulation models compare most favorably with observed O–18 contents of rainwater (Joussaume et al., 1984). It may thus be possible to link general circulation model outputs of Holocene climates to the flood climate regions proposed here. Proxies of paleoflood frequencies may in turn help in verifying reconstructions of past climates.

CONCLUSION

The flood climate regions are based on the potential available floodwater in the atmosphere and on the land and on the weather systems that result in the discharge from the atmospheric reservoir and the recharging of the reservoir of snow and ice on the land. The presence of mountains that orographically enhance precipitation and tend to have winter storage of snows complicate the latitudinal and longitudinal symmetry of the distribution of the flood climate regions. Nonetheless, the resulting regions should be internally homogeneous in terms of the kinds of flood-generating events, moisture availability, and other aspects of the water resources of the regions defined.

The goal of this work was to develop a globally consistent overview of the flooding. In this regard it should stand as a valuable addition to the often used charts of global climate, vegetation, soils, and fluvial geomorphologic features.

REFERENCES

Atkinson, G. D., and Sadler, J. C. (1970). Mean-cloudiness and gradient-level-wind charts over the tropics. *Air Weather Serv., USAF, Tech. Re.* **215** 1–149.

Bryson, R. A. (1966). Airmasses, streamlines and the boreal forest: *Geogr. Bull.* **8**, 228–269.

Crowe, P. R. (1949). The trade wind circulation of the world: *Trans. Inst. Br. Geogr.* **15**, 38-56.

Dietrich, D. (1963). "General Oceanography: An Introduction:" Wiley-Interscience, New York.

Emanuel, W. R., Shugart, H. H., and Stevenson, M. P. (1985). Climatic change and the broad-scale distribution of terrestrial ecosystem complexes: *Clim. Change* **7**, 29-44.

Hayden, B. P., Ray, G. C., and Dolan, R. (1984). Classification of coastal and marine environments: *Environ. Conserv.* **11**, 199-207.

Joussaume, S., Sadourny, R., and Jouzel, J. (1984). A general circulation model of water isotope cycles in the atmosphere: *Nature (London)*, **311**, 24-29.

Köppen, W. (1936). Das Geographishe System Der Klimate, Handbuch Der Klimatologie (W. Köppen and R. Geiger, eds.) Berlin, Gebrüder Von Borntraeger.

Lamb, H. H. (1972). "Climate: Past, Present and Future," Vol. 1: Methuen, London.

Namias, J. (1966). A weekly periodicity in eastern U.S. precipitation and its relation to hemispheric circulation: *Tellus* **18**, 731-744.

Simpson, R. H., and Riehl, H. (1981). "The Hurricane and Its Impact:" Louisiana State Univ. Press, Baton Rouge.

Trewartha, G. T. (1968). Introduction to Climate, New York, McGraw-Hill.

Udvardy, M. D. F. (1975). A classification of the biogeographical provinces of the world: Morges, Switzerland. *IUCN Occas. Pap.* **18**, 1–49.

Wendland, W. M. (1977). Tropical storm frequencies related to sea surface temperatures: *J. Appl. Meteorol.* **16**, 477-481.

2

FLOOD HYDROCLIMATOLOGY

KATHERINE K. HIRSCHBOECK
Department of Geography and Anthropology, Louisiana State University, Baton Rouge, Louisiana

INTRODUCTION

Until very recently, the status quo in many forms of flood analysis has been to treat events in a hydrologic time series as a set of varying time-ordered numerical values. The methodologies that have been developed and refined through the years to manipulate, model, and predict flood values have become increasingly sophisticated. In some circles, however, the obvious fact that these values represent a response to varying processes in the physical world has tended to become less important than the urge to statistically model flood values in search of the best fit of the observed data and therefore (ideally) the best predictive capability for future flows:

> The main emphasis in stochastic analysis of hydrological processes, which basically is the domain of pure hydrology, has been on the fitting of various preconceived mathematical models to empirical data rather than on arriving at a proper model from the physical nature of the process itself. The empirical data representing a hydrologic event are treated as a collection of abstract numbers that could pertain to anything or to nothing at all. Their hydrologic flavor, the physical substance that makes, for instance, a precipitation record an entity entirely distinct from, say, a record of stock market fluctuations, is not reflected in the analysis. Thus what we usually find is not, in fact, statistical or stochastic hydrology but merely an illustration of statistical and probabilistic concepts by means of hydrologic data. Such an approach can hardly contribute to the hydrological knowledge.
>
> In trying to improve this situation, the main problem is to find the ways in which the physical features of a phenomenon can be introduced into the analysis.
>
> —(Klemeš, 1974, p. 2)

The cross-discipline of hydroclimatology is an approach to studying hydrologic events within their climatological context. By focusing on atmospheric inputs to flooding, hydroclimatology provides one way to integrate the physical sources of variability in a hydrologic time series with the statistical properties of the varying series itself, thus both enhancing our understanding of the flooding process and improving the quantitative assessment of its variability. In a hydroclimatic approach to flood analysis the events recorded in a flood series are viewed not only as numerical values, or as isolated hydrologic occurrences, but as real-world physical events occurring within the context of a history of climatic variations in magnitude and frequency. The physical basis of the approach emerges when these events are analyzed within the spatial framework of regional and global networks of changing meteorologic features and circulation patterns.

SPATIAL AND TEMPORAL SCALES OF HYDROCLIMATIC ACTIVITY

Flood-producing atmospheric circulation patterns operate within a space–time domain that at times is quite different from the domain of hydrologic activity within a drainage basin. Figure 1 depicts the characteristic spatial and temporal scales at which selected meteorologic, climatologic, and hydrologic phenomena vary. The figure displays the variety of scales over which climatic activity can generate flooding, ranging from small downpours that quickly fill culverts and drainage ditches, to global-scale circulation anomalies that have the ability to steer one major storm after another into an area along the same persistent track. This wide range of interactions between the atmosphere and the hydrosphere illustrates the concept of *proximate* versus *ultimate*

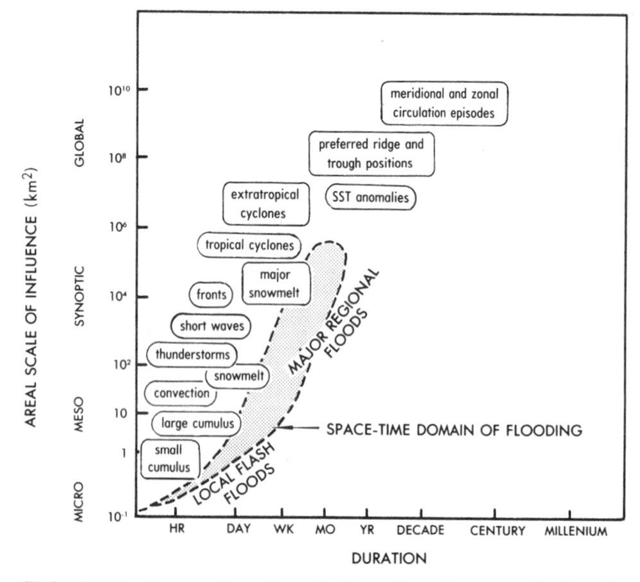

FIGURE 1. Space–time domain for selected meteorologic, climatologic, and hydrologic phenomena. The spatial domain is represented by the typical areal scale of influence of each phenomenon, in kilometers squared. The temporal domain is represented by the typical duration of each phenomenon.

causes for flooding. The proximate, or most immediate climatic causes of flooding are readily apparent to any observer of the small-scale, short-duration relationship between rainfall and subsequent runoff. Climatic activity operating at much larger spatial scales and longer temporal scales is less often perceived as a source of flooding, yet the configuration of the atmosphere at these scales provides the ultimate framework from which the more immediate causes of flooding are generated.

In Figure 1 the spatial dimension is approximated by the typical areal extent of influence exhibited by different types of features, while the time dimension is represented by the characteristic duration of each type of event. For example, a typical summer thunderstorm with a diameter of 10 km covers an area of about 100 km^2 at any one time and has a life span of several minutes to over an hour, while a mature extratropical cyclone with a diameter of 1000 km affects about 10^6 km^2 at any one time, and has an influence over an even larger area during its characteristic life span of 3–6 days. At larger spatial scales, in the upper atmosphere, long wave ridges and troughs with dimensions of 10^6 km^2 or greater migrate around the globe from day to day, at times reforming again and again in a preferred location to produce a pattern that may persist throughout a season or longer. This persistence of certain upper-air patterns has been linked to anomalous pools of warm and cool sea surface temperatures that may extend over areas as large as 10^6 km^2 and persist for several months to over a year (Namias, 1974; Namias and Cayan, 1981). At the global scale the meandering pattern of the circumpolar vortex of upper-air winds (10^7 km^2) is reflected in hemisphericwide features. These include the position of the polar front, the jet stream, and the tendency toward zonality or meridionality in the long wave pattern (Fig. 2).

The spatial domain of flooding, unlike that of climate, is constrained by the areal dimensions of a given drainage basin or by a group of basins responding similarly within a hydroclimatically homogeneous region. The upper spatial limit of flooding in Figure 1 has been chosen as 10^6 km^2 to reflect the areal dimensions of the Mississippi River basin, a reasonable upper limit for widespread regional flooding resulting from large-scale climatic inputs. The temporal domain of flooding is related to either the length of time specific flood-generating atmospheric phenomena are positioned over a basin or the interval of time during which a series of flood-producing events affects a basin. The lag time between an atmospheric input and the corresponding hydrologic output is heavily dependent on factors internal to a drainage basin system such as basin area and shape, channel form and roughness, drainage density, vegetative cover, permeability of the substrate, and land use (i.e., type of agricultural practice, degree of urbanization, etc.). In general, however, the duration of flooding will nearly always exceed the duration of the atmospheric input that generated the flood. This is depicted in Figure 1 by the shift in the domain of flooding toward longer durations than those of atmospheric phenomena at the same spatial scale.

Hydrometeorology and Hydroclimatology

The cross-discipline of *hydrometeorology*, which Bruce and Clark (1980) define as "an approach through meteorology to the solution of hydrologic problems" (p. 2), has evolved to analyze the relatively short-term interactions between the atmosphere and hydrosphere at micro-, meso-, and synoptic spatial scales of influence (Fig. 1). These scales of inquiry are extremely effective for a variety of flood climate studies, including predicting and analyzing flash-flood events (Maddox et al., 1979, 1980), developing real-time river forecast models for specific drainage basins (Georgakakos and Hudlow, 1984), and identifying and compiling the characteristic synoptic features that generate flooding in selected areas (Hansen and Schwarz, 1981).

The cross-discipline of *hydroclimatology* encompasses larger scale interactions between the atmosphere and the hydrosphere and has been defined by Kilmartin (1980) as the "modeling of long-term climatic fluctuations in water resources systems analysis" (p. 166). In his call for the active development and growth of hydroclimatology, Kilmartin cited the need to close some gaps between the closely related disciplines of hydrology, meteorology, cli-

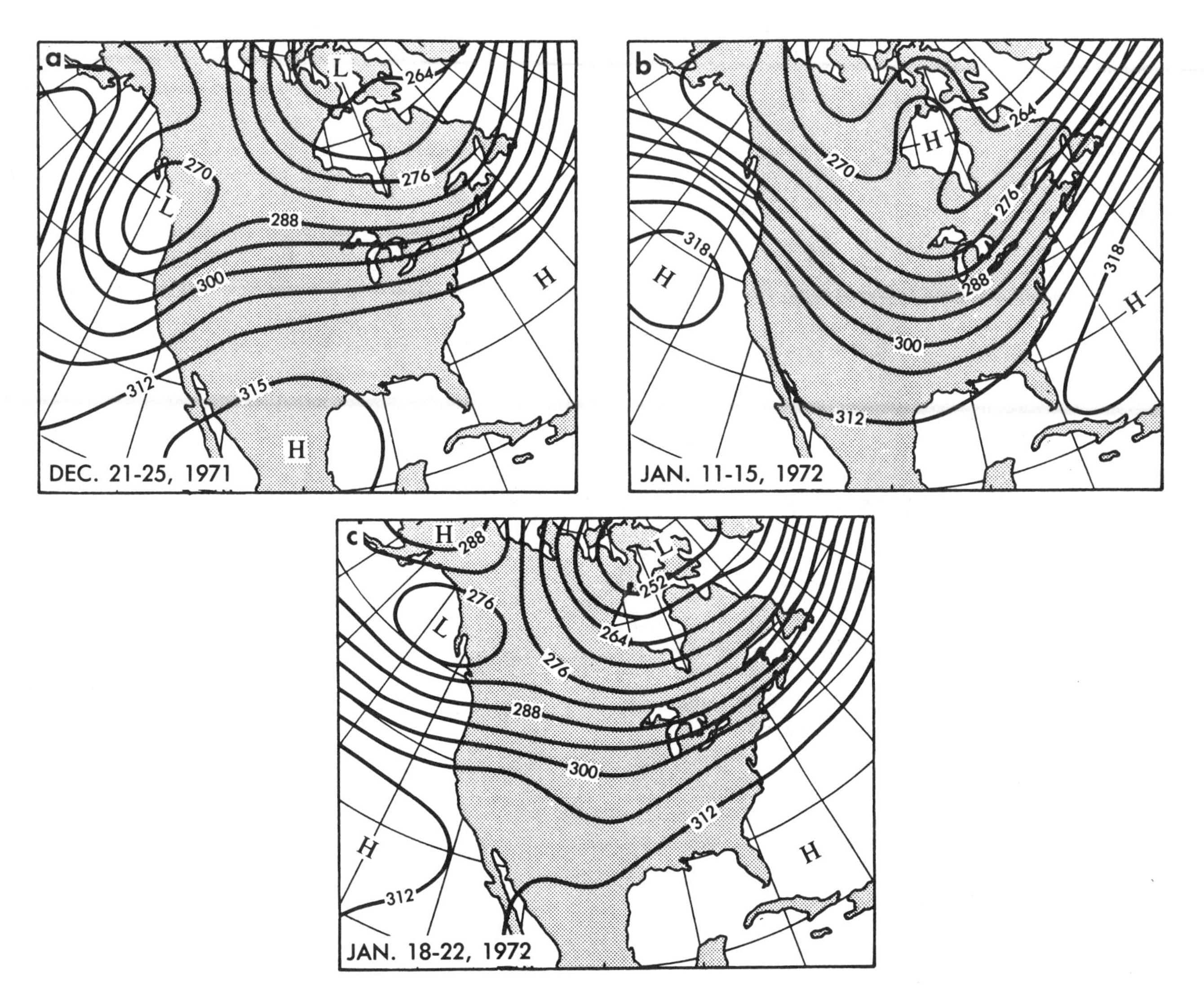

FIGURE 2. Three types of large-scale upper-air patterns that occurred during the winter of 1971–1972. The contoured values are heights (in dekameters) of the 700-mb pressure surface averaged for weekly periods. The wave pattern, defined by the steepest gradient in the contours, tends to separate cold polar air masses from warmer tropical air masses. Upper-air winds and steering currents, such as the jet stream, flow parallel to the contours. (*a*) High-amplitude meridional wave pattern over western United States. A deep trough of upper-air low pressure is situated over the West Coast. Synoptic studies have demonstrated that associated surface low-pressure systems are most likely to develop under the eastern sides of upper-air troughs. (*b*) A slightly lower-amplitude meridional wave pattern. The broad ridge of upper-air high pressure off the West Coast and broad trough over eastern United States is a preferred pattern in winter. The ridge tends to develop frequently over western North America in winter due to a combination of factors including sea surface temperatures, the presence of a blocking high-pressure system in the North Pacific, and the anchoring effect of the western Cordillera mountain system. (*c*) A zonal (west–east) wave pattern interrupted slightly by a shallow trough over Texas and Oklahoma.

matology, oceanography, and statistics, and to expand and develop the approach to hydrologic problems that is already found in the cross-discipline of hydrometeorology.

Kilmartin's emphasis was not on floods but on long-period events in hydrology, for he was specifically concerned with past and future climatic changes and their effects on long-term reservoir storage and water supply. The term *hydroclimatology*, however, can be equally applied to the analysis of short-period events, such as those in an annual or partial duration flood series, by examining how these events vary temporally and spatially in response to longer term climatic variations.

The importance of climatic variability as a source of nonstationarity and nonhomogeneity in streamflow regimes has been both recognized and hotly debated through the years (Yevjevich, 1968; National Research Council, 1977; Willeke, 1980). Nevertheless, assumptions of stationarity and homogeneity are inherent in most of the current methods of flood series analysis. Traditionally, much effort has been placed on identifying nonclimatic sources of streamflow variability that originate within the drainage basin due to such factors as land-use changes, channel modifications, or complex responses. This emphasis has prevailed despite the fact that the initial hydrologic variability imparted to a catchment by climate always precedes any subsequent variations that may arise in the basin itself.

Violations of the stationarity and homogeneity assumptions may occur from a variety of climatic factors. Decadal-scale climatic persistence or infrequent anomalous extreme events can strongly bias a 30- or 40-yr flood record, even though no actual "climatic change" is perceived to have transpired. Large-scale, long-period climatic processes operating at regional and global spatial scales also have a profound impact on flooding variability over time. Some streamflow regimes, such as those in arid regions or climatic transition zones, are especially sensitive to variations in climate, and hence by their very nature are most susceptible to violations of the stationarity and homogeneity assumptions. It is important, therefore, that climate fluctuations not be "filtered out" or removed, but thoroughly examined, when modeling flood variability in such streams.

FLOOD HYDROCLIMATOLOGY

The cross-disciplines of both hydrometeorology and hydroclimatology are essential for understanding the interactions between the atmosphere and the hydrosphere. Hydrometeorology has more traditionally been applied to the analysis of floods, but the broader perspective of hydroclimatology provides new insights into flood variability by synthesizing and integrating information emerging from hydrometeorologic studies. The subtle differences between a hydroclimatic and a hydrometeorolgic approach to the analysis of floods can be explained in part by a comparison of *climate* and *weather*. Fairbridge (1967) presents the following definitions:

> Climatology is that branch of atmospheric science which deals with the climate, i.e., the statistical synthesis of all weather events taking place in a given area in a long interval of time. It is customary to describe the climate by the seasonal variation of various meteorological elements and their characteristic combinations.
>
> —(pp. 217–218)

> Weather is defined as a state or condition of the atmosphere at any particular place and time. . . . Weather is specifically distinguished from climate, which represents a regional or global synthesis of weather extended through time on the scale of years, rather than minutes or hours.
>
> —(p. 1114)

The key phrases here are "synthesis of events," "seasonal variation," "long interval of time," "characteristic combinations, " "frequency of events," and "regional or global." Hydroclimatology places a hydrologic event in the context of its history of variation—in magnitude, frequency, and seasonality—over a long period of time and in the spatial framework of the regional and global network of changing combinations of meteorologic elements such as precipitation, storm tracks, air masses, and other components of the broad-scale atmospheric circulation.

Flood hydroclimatology, therefore, has as its foundation the detailed focus of hydrometeorologic-scale atmospheric activity, while at the same time seeking to place this activity within a broader spatial and temporal, "climatic" perspective. Large-scale anomaly patterns, global-scale controls, long-term trends, and regional relationships in flooding might be overlooked if analysis is limited to the hydrometeorologic space–time domain alone, whereas these same patterns, controls, and relationships are readily detected at the broader hydroclimatic domains of analysis. In effect, the complete spectrum of atmospheric activity depicted in Figure 1 has the potential for generating flooding, either directly or indirectly.

Hydrometeorologic-scale Activity

Microscale and small mesoscale atmospheric activity such as convectional showers, isolated or small thunderstorms, and squall line disturbances tend to have a localized or limited regional areal extent of influence of less than 1–1000 km^2 and a storm life of a few minutes to one or two hours. These events are most likely to produce local flash flooding of small areal extent (Fig. 3).

Larger mesoscale features such as severe thunderstorms, multiple squall lines, extensive moist and unstable layers in the atmosphere, and shortwave troughs have the capabilities of producing major precipitation events of great intensity over relatively large areas. Atmospheric activity at this scale has been responsible for many catastrophic flash floods, such as the Johnstown, Pennsylvania, flood of July 1977 (Fig. 4).

Macroscale (synoptic) features such as major fronts, tropical storms, and extratropical cyclones affect much larger areas of 1000 to 1,000,000 km^2 during their longer life spans of several hours to several days. These features at times are associated with flash flooding when they provide the necessary synoptic situation for locally intense meso-

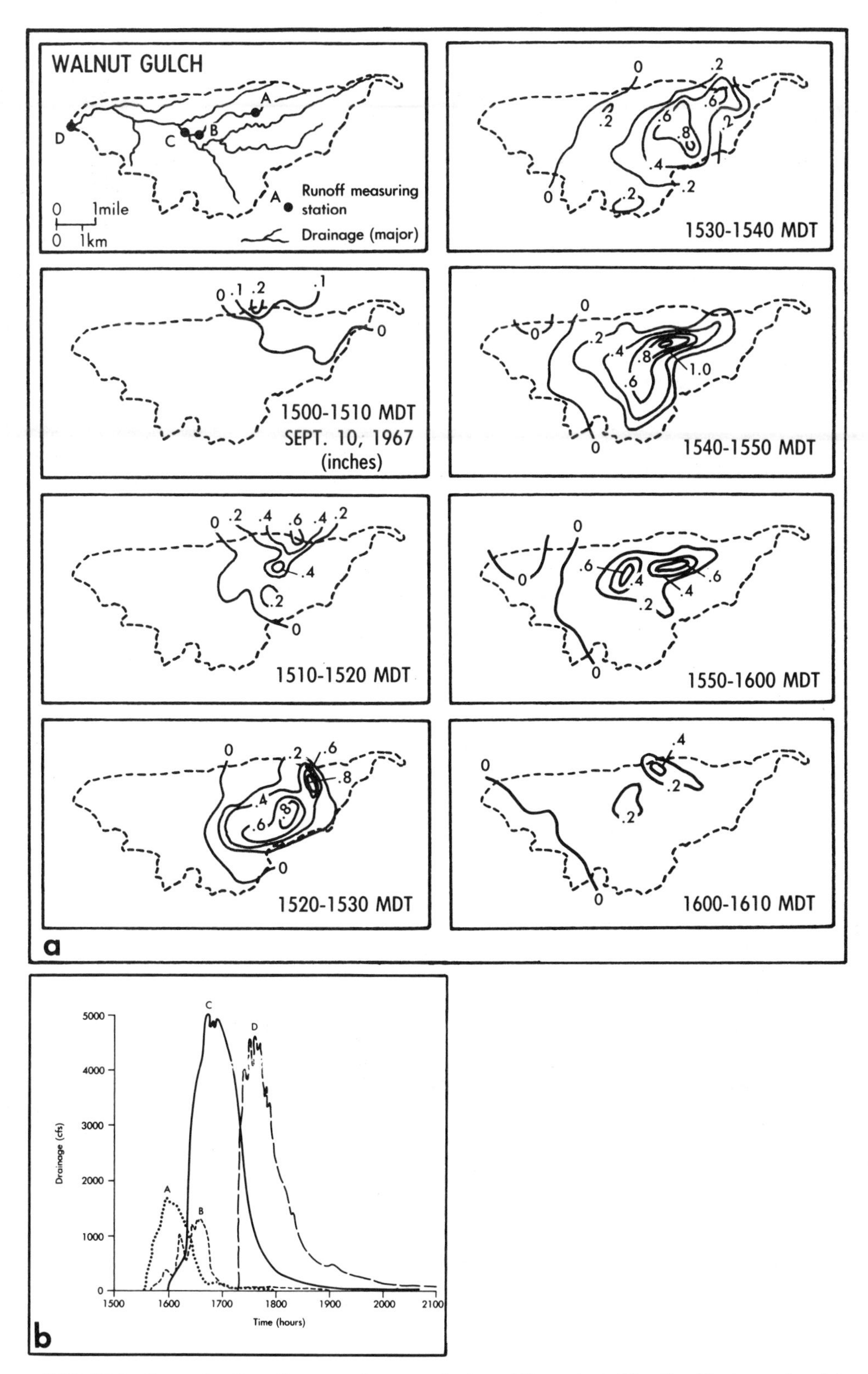

FIGURE 3. Example of small mesoscale atmospheric activity and resulting flood hydrographs. (*a*) Isohyetal maps (in inches) for 10-min intervals showing the movement of a thunderstorm across the Walnut Gulch watershed (93 km^2) near Tombstone, Arizona, from 15:00 to 16:10 MDT on September 10, 1967. Several localized cells of high-intensity rainfall developed and dissipated during the course of the storm, which lasted a little over an hour. (*b*) Discharge hydrographs resulting from the September 10th storm for four subwatersheds within the Walnut Gulch catchment. Over 80% of the total annual 1967 runoff was generated by this one storm (Source: Osborn and Renard, 1969).

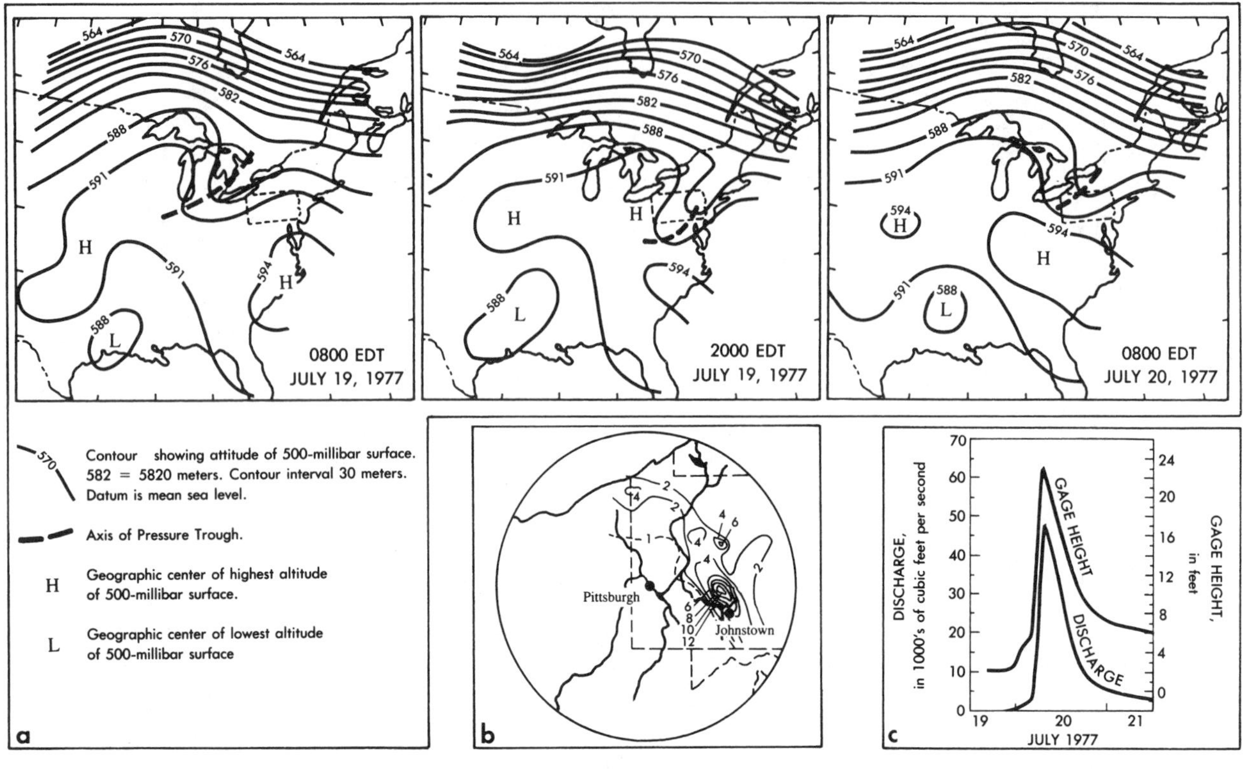

FIGURE 4.

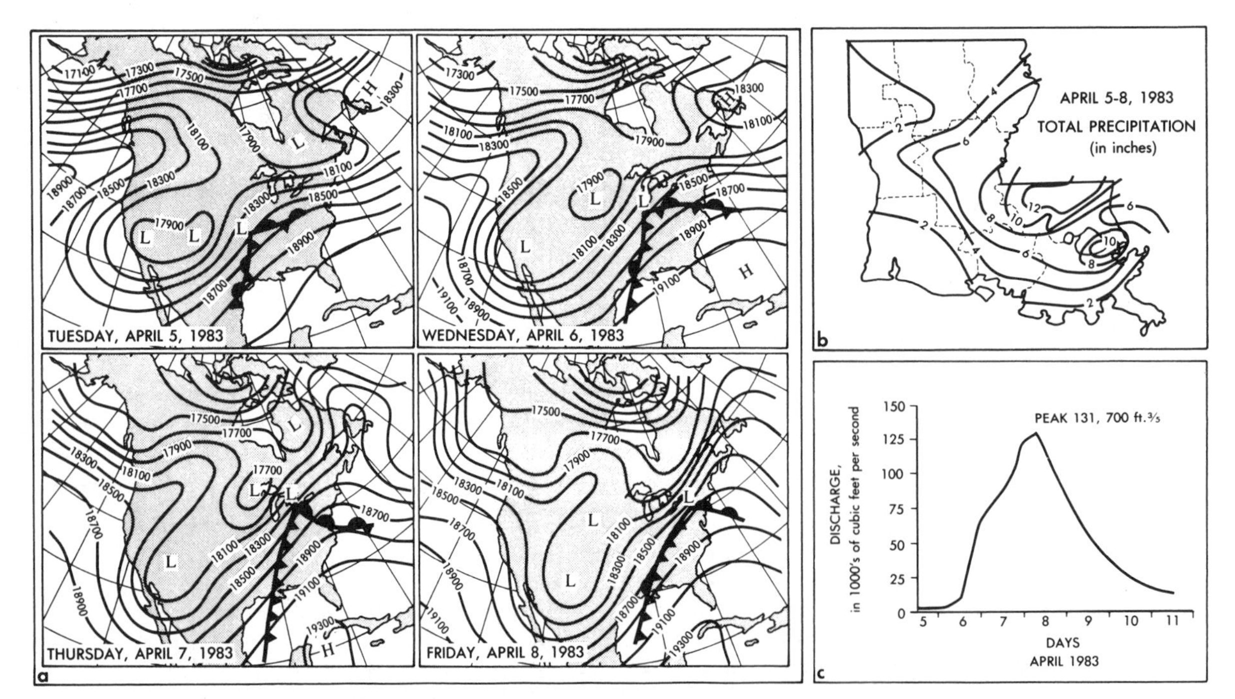

FIGURE 5.

scale activity to develop (Maddox et al., 1979, 1980; Huff, 1978). However, in addition to flash flooding, synoptic-scale events also have the ability to produce long-duration, widespread flooding throughout a large drainage basin or in several basins across a region. Precipitation generated by macroscale systems is characterized by alternating periods of high and low rainfall intensities, persisting either continuously or intermittently for several hours or days. The widespread nature of these storms, coupled with their complex intensity–duration properties leads to an overall tendency for flood hydrographs with slower rise times and longer periods with streamflow at flood stage, than are found in the smaller-scale, flashier events (Fig. 5).

Larger Scale Hydroclimatic Activity

Although less obvious, larger scale and longer duration climatic events that take place beyond the space–time domain of most hydrometeorologic activity also have a significant impact on flooding. The most widely recognized of these is the seasonal accumulation and melting of winter snow, a process that may contribute to runoff volumes for several days or weeks and be generated from an area as large as the entire upper Mississippi River basin. Spatially, snowmelt flooding can occur in any sized drainage basin, but in large basins the area affected by snowmelt may be extensive because the snowmelt process has the capability to induce flooding at downstream reaches of a basin that are far removed from the area that directly received the original precipitation input. Temporally, the runoff resulting from snowmelt reflects the cumulative climatic events of several weeks or months prior to the runoff event itself, even though the actual snowmelt may occur rapidly over a 1- or 2-day period. Furthermore, the rate of snowmelt is a function of climatic factors other than the amount of snow itself, such as air temperature, solar radiation, cloud cover, and rainwater falling on the snow surface. Because of the space–time domain of the processes that contribute to snowmelt floods, they are most appropriately analyzed from a larger scale, longer duration, climatic perspective (Fig. 6).

Other large-scale climatic effects on flooding include anomalous configurations in the upper-level circulation, sea surface temperature anomalies, and decadal-scale circulation episodes. Although the direct link to flooding from climatic activity at these scales is less well understood, they provide the key for identifying the climatic scenarios within which widespread flood-generating hydrometeorologic activity is likely to develop.

Anomalous Circulation Patterns. In many cases floods result simply from excessive amounts of precipitation or snowmelt, or from an unusual intensity in an otherwise typical hydrometeorologic circulation mechanism, such as a front, squall line, mesoscale convective complex, or synoptic-scale cyclone. Occasionally, however, certain floods are associated with very atypical patterns in the atmospheric circulation. These anomalies can be in the form of (1) an unusual combination of several common mechanisms occurring together, (2) an unusual location or unseasonal occurrence of an otherwise typical circulation mechanism, (3) the unusual persistence of a specific circulation pattern, or (4) a rare configuration in the upper-air pattern itself.

Figure 7 depicts an example of the first type of anomaly—an unusual combination of circulation features—that resulted in widespread flooding throughout central and southern Arizona. During September 4–6, 1970, several

FIGURE 4. Example of large mesoscale atmospheric activity and resulting flood hydrograph. (*a*) 500-mb charts for July 19–20, 1977, showing the movement of a short-wave trough over western Pennsylvania. The mesoscale trough triggered widespread thunderstorms across Pennsylvania and was associated with two major squall lines that moved across the state. (*b*) Total observed rainfall (in inches) over western Pennsylvania from 0800 EDT July 19–0800 EDT July 20, 1977. (*c*) Discharge hydrograph for Stony Creek at Ferndale, Pennsylvania (726 km^2), about 2 km upstream from the Johnstown city limits. Many deaths and extensive property damage resulted from this event, estimated at a recurrence interval of 100 yr (from Hoxit et al., 1982).

FIGURE 5. Example of macroscale atmospheric activity and resulting flood hydrograph. (*a*) 500-mb charts for April 5–8, 1983, showing the position of major surface fronts. Widespread flooding across southeastern Louisiana and southern Mississippi was associated with the stationary front. High pressure over southeastern United States prevented the upper-air trough to the west and its associated surface front from moving eastward, causing the system to remain in the Gulf Coast area for several days. (*b*) Total precipitation (in inches) over Louisiana for the April 5–8 storm event (Source: Muller and Faiers, 1984). (*c*) Discharge hydrograph from Bogue Chitto near Bush, in southeastern Louisiana (1952 km^2). The peak discharge of this event was more than twice the previous maximum and was estimated at a recurrence interval of greater than 100 yr (Source: Carlson and Firda, 1983).

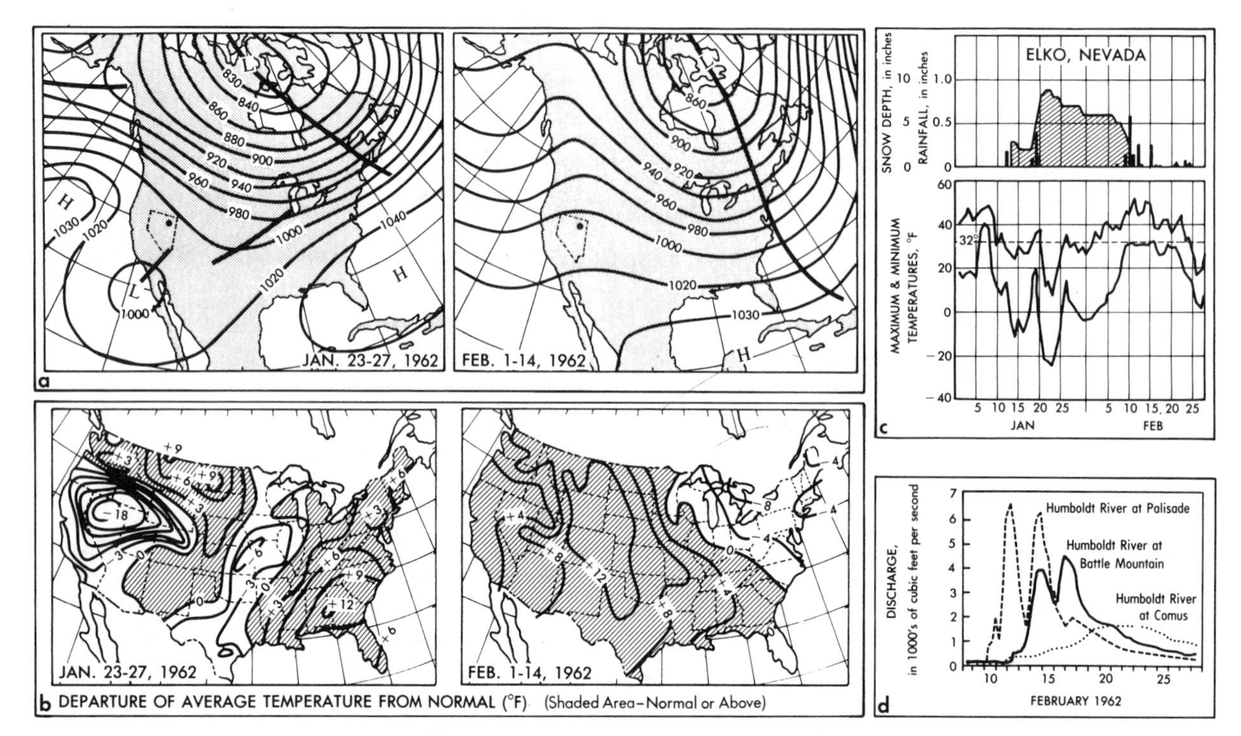

FIGURE 6.

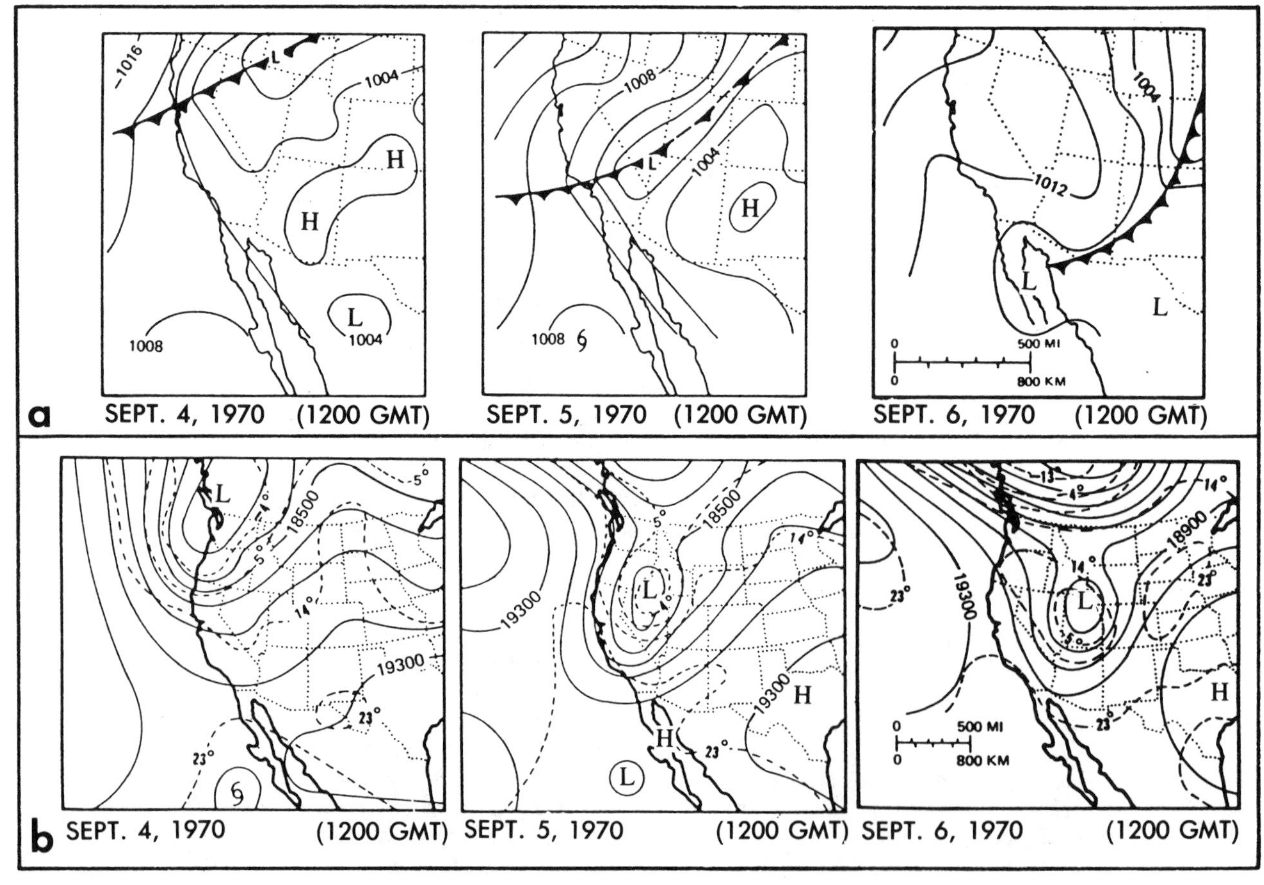

FIGURE 7.

major synoptic features occurred simultaneously and introduced excessive amounts of precipitable water vapor into the Southwest. At the same time they provided the necessary uplift triggering mechanisms to release this moisture. The features included a deep upper-air trough, an incipient cutoff low circulation, a surface cold front, and a tropical storm off the Baja California coast (Fig. 7). This unusual coincidence of synoptic-scale events generated severe flash flooding in small mountain watersheds throughout southern Arizona as well as widespread flooding in many larger drainage basins, with some gauges recording the highest annual flood of record. Each of the synoptic-scale operating mechanisms alone had the potential for generating a flood. However, none of them was particularly anomalous in itself. It was the combination of several mechanisms operating simultaneously that produced an anomalous circulation pattern that resulted in a major flooding episode.

An example of the second type of circulation anomaly—unseasonal atmospheric conditions—occurred in northern California during February 11–19, 1986. A movement of the ridge off the western North American coast (see Fig. 2*b*) produced a high-latitude blocking effect in the eastern North Pacific ocean and shifted the main branch of the winter jet stream to a more southerly location for over a week, a situation that is unusual for February. These events allowed massive low-pressure systems to develop and redevelop over the ocean and fed a succession of devastating storms into California over a 9-day period. The result was extensive flooding, loss of lives, forced evacuation of thousands of residents, and millions of dollars of damage.

Severe flooding in the Mississippi River basin during the spring of 1973 had its origin in the third type of circulation anomaly—the unusual persistence of a specific upper-air pattern over an extended period of time (Fig. 8). Throughout March and April the repeated development of a trough over the southern United States produced frequent and persistent episodes of southerly wind flow. As this southerly flow moved through the eastern side of the trough, it introduced moist maritime gulf air masses into the lower Mississippi valley. In addition, the strong surface convergence and divergence aloft, typically associated with the eastern sides of troughs, provided the necessary mechanisms for frontal formation, storm development, uplift, and release of the excess moisture. This extended hydroclimatic episode resulted in new records for consecutive days above flood stage for many of the main-stem Mississippi River gauging stations from southern Iowa to Louisiana (Chin et al., 1975).

Finally, some of the most unusual flood-producing conditions are those that result from the fourth type of circulation anomaly, a rare configuration in the upper-air pattern itself. In June 1972, Hurricane Agnes produced flooding that devastated the East Coast of the United States in what was called at the time, "the greatest natural disaster ever to befall the Nation" (U.S. Department of Commerce, 1973, p. 1). Although not an unusual storm in the beginning, the area covered by Agnes was exceptionally large, and its slow development and movement permitted large amounts of moisture to be entrained into the system from the deep Tropics. However, it was the influence of a highly abnormal configuration in the large-scale circulation pattern over the North Atlantic ocean that affected Agnes' unusual path

FIGURE 6. Example of large-scale climatic activity and its affect on snowmelt flooding. (*a*) Contrast between mean 700-mb charts for the weeks of January 23–27 and February 1–14, 1962. During the last week of January a deep trough of very cold air was situated over most of the far west as a surface arctic high-pressure system settled over the Great Basin, however, by the end of the second week of February, a strong ridge of warmer air had replaced the trough in the west. (*b*) Corresponding temperature departure maps for each week showing the departure of average surface temperature from the 1931–1960 normal (in °F). In late January subfreezing temperatures in the Idaho–Nevada area froze the ground to depths of as much as 3 ft under a cover of light snow. (*c*) Antecedent climatic conditions and rainfall at Elko in northeastern Nevada. (*d*) Discharge hydrographs for three stations in the Humboldt River basin in northeastern Nevada. The February 10th–15th flooding resulted from the combination of several days of low-intensity rain falling on moderate amounts of snow that had accumulated during January. The snow melted rapidly in response to the warmer temperatures and light rain, but due to the severity of the previous cold spell, the ground beneath remained frozen and exacerbated the flooding. The resulting complex hydrographs show the contributions of individual upstream tributaries and the downstream progression of the flood wave. Floods estimated at recurrence intervals of greater than 50 to over 100 yr occurred throughout northeastern Nevada and southern Idaho during this unusual hydroclimatic episode [Source for (*a*) and (*b*): Stark, 1962; Andrews, 1962; source for (*c*) and (*d*): Thomas and Lamke, 1962].

FIGURE 7. A flood-producing circulation anomaly that resulted from the simultaneous occurrence of several synoptic-scale events. (*a*) Surface charts for September 4–6, 1970. (*b*) Corresponding 500-mb charts [Source for (*a*) and (*b*): Hansen and Schwarz, 1981].

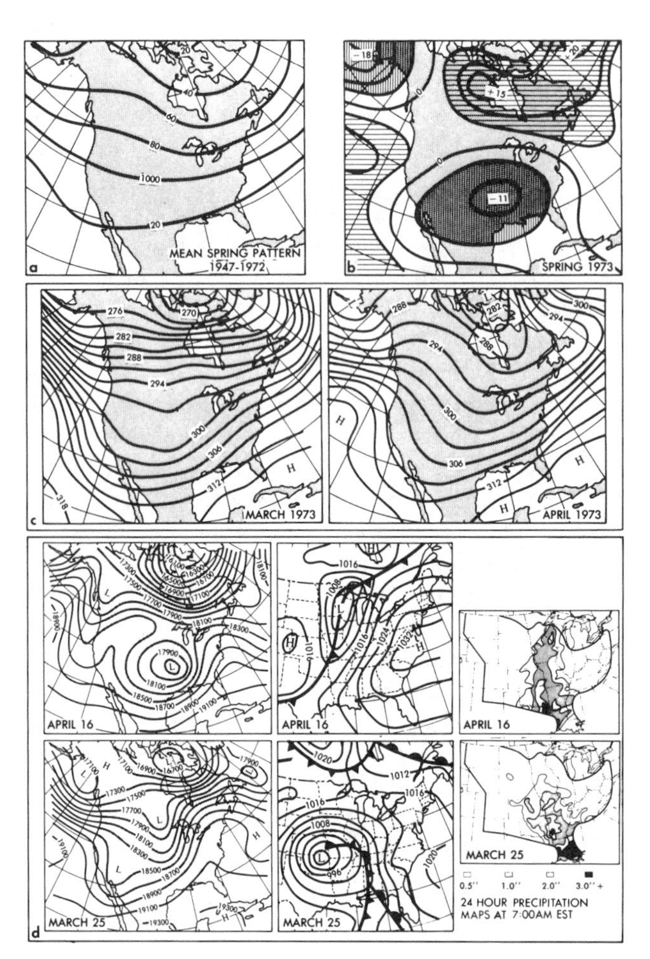

FIGURE 8. A flood-producing circulation anomaly that resulted from the repeated development of a trough-to-ridge configuration over the Mississippi River basin and the unusual persistence of this pattern for 2 months. (*a*) Mean spring 700-mb height pattern in tens of feet (based on March–May over the period 1947–1972). (*b*) Spring 1973 departure pattern from the 26-yr mean in tens of feet. Contour interval 50 ft [Source for (*a*) and (*b*): Namias, 1979]. (*c*) Mean monthly 700-mb height patterns for March and April, 1973 in dekameters. (*d*) Selected daily 500-mb and surface charts showing the position of the trough in relation to the Mississippi basin and the resultant pattern of precipitation in the basin (24-hr totals over 0.5 in.). Similarly positioned troughs were present on about 60% of the days in March and April. The associated surface low-pressure centers and fronts brought persistent heavy rainfall over the basin in March and April and set new monthly records for many stations [Source for (*c*) and (*d*): Chin et al., 1975).

and fed large amounts of moisture into the storm during its latter stages, resulting in phenomenal rains and record-breaking floods in Virginia and Pennsylvania (Fig. 9).

Sea Surface Temperature Anomalies. The June 1972 circulation anomaly associated with Agnes and its attendant anomalously warm Atlantic sea surface temperatures were the culmination of global-scale events that had actually been developing over a much longer span of time. According to Namias (1973), the June anomaly pattern that influenced the track of Agnes began to establish itself as early as February or March, due in part to the external forcing of abnormally warm sea surface temperatures (SSTs) and positive feedback effects between the ocean and the atmosphere. Such air–sea interactions—along with the tendency for major pressure centers in the global circulation to exhibit intercorrelations or "teleconnections"—form the basis for many long-range forecasting techniques. They also present an argument for possible long-term, large-scale hydroclimatic controls on major flooding episodes around the world.

El Niño, the anomalous warming of sea surface temperatures from coastal Peru westward along the equator, and the Southern Oscillation, a related atmospheric pressure shift in the western South Pacific ocean, are the most frequently cited large-scale SST factors to be linked to flood events at diverse locations across the globe such as Peru, Bolivia, and Ecuador in South America, the Pacific coast of North America from Oregon to Baja California, the Colorado River basin, and coastal areas of the Gulf of Mexico (Quiroz, 1983; Rasmusson, 1985). In other parts of the world, however, the El Niño/Southern Oscillation phenomenon has been associated with the occurrence of droughts. Continued research is needed to better understand the global-scale atmospheric teleconnections and air–sea interactions that indirectly affect flooding over such vast spatial scales, especially in terms of how they interface with local mesoscale and macroscale processes that have a more proximate flood-generating impact.

Long-Period Circulation Episodes. Episodic tendencies in the overall pattern of the circumpolar upper-air waves constitute the largest spatial and longest temporal scales to have a potential hydroclimatic impact on flooding. Although the ridges and troughs that form the upper-air wave pattern may adjust into high- and low-amplitude patterns on a daily, weekly, monthly, or seasonal basis, over the last 100 years extended intervals of time characterized by more zonal circulation patterns have alternated with periods characterized by more meridional patterns. These circulation episodes, often several decades in length, have been documented in a variety of ways by researchers who have used both subjective and objective means to classify large-scale patterns and adjustments in the atmosphere over time (Dzerdzeevskii, 1963, 1969; Kutzbach, 1970; Kalnicky, 1974; Knox et al., 1975; Lamb, 1977; see also Barry and Perry, 1973, pp. 365–377). Circulation adjustments at these decadal scales have their greatest hydroclimatic impact in generating trends and variations in flood series over time (Fig. 10).

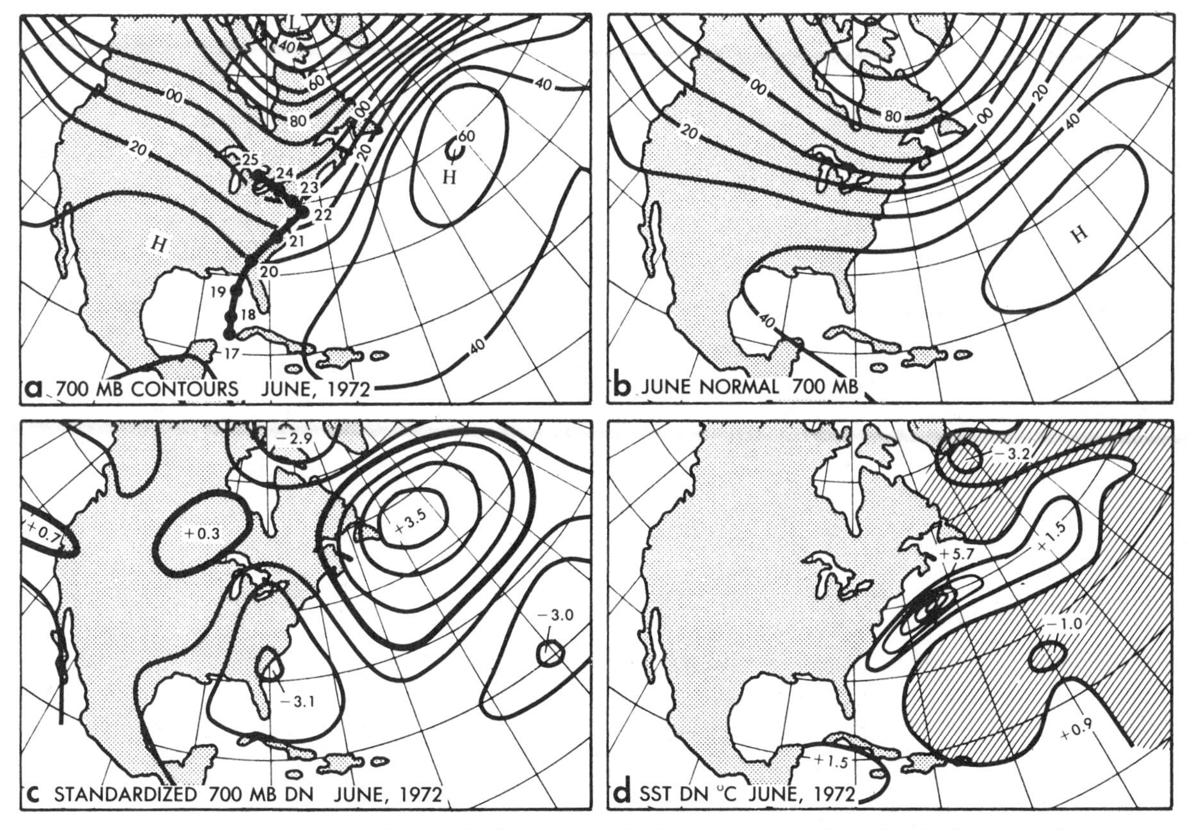

FIGURE 9. A flood-producing circulation anomaly due to a rare configuration in the upper-air pattern. (*a*) June 1972 700-mb height pattern showing path of Agnes and a deeply developed trough and ridge pattern in the western North Atlantic Ocean. (Contours are labeled in tens of feet with hundreds omitted. Path of Agnes from June 17–25 is shown with a solid line.) (*b*) Normal June pattern based on the mean of Junes from 1947–1963. (*c*) Deviations of the June, 1972, 700-mb height pattern from the normal pattern. (Isopleths are in standard deviations with a contour interval of 1.) The unusual ridge (blocking high) in the far North Atlantic helped to steer the path of Agnes due north. (*d*) Sea surface temperature departures from normal for June 1–26. Air masses feeding into Agnes moved over these anomalously warm waters and picked up excessive amounts of moisture (Source: Namias, 1973).

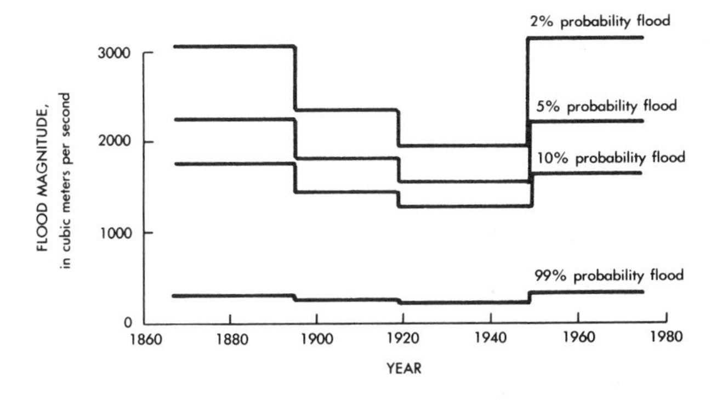

FIGURE 10. Affect of decadal-scale circulation episodes on the partial duration flood series of the Mississippi River at St. Paul. Separate estimates of flood probability were computed for different sections of the record using the standard log-Pearson Type III method. The boundaries defining the subperiods are not arbitrary, but have a distinct physical basis because they represent significant dates of change in prevailing global circulation patterns. Meridional circulation patterns were more frequent before about 1895 and since 1950, but zonal circulation patterns were more common between 1895 and 1950 (Source: Knox, 1983).

FLOOD ANALYSIS WITH A HYDROCLIMATIC PERSPECTIVE: AN OVERVIEW

From the preceding discussion it is clear that climate affects flooding across a wide spectrum of spatial and temporal scales. These interactions between climatic and hydrologic processes have been recognized since the hydrologic cycle was first conceptualized; yet an examination of the hydrologic literature shows that the climatic factor has been incorporated into standard models and techniques of flood analysis only in limited degrees, most of which have been concentrated in the hydrometeorologic time–space domain.

A convenient way of describing different approaches to hydrologic modeling is to label the model as either deterministic or stochastic. Deterministic approaches to flood analysis, many of which are based on rainfall–runoff relationships, frequently employ climatic information in their models. Stochastic approaches, on the other hand, are more likely to be runoff based, using hydrologic data alone and only indirectly considering climate. Criticisms often raised against deterministic methods of analyzing floods are that modeling of the climatic-hydrologic relationship generally involves only the means of variables and that the models are based on constraining underlying assumptions about the environment that oversimplify the processes operating in the real world and neglect the large number of random factors that can affect the responses of natural hydrologic systems (Yevjevich, 1974). Conversely, stochastic approaches to flood analysis tend to be constrained at the opposite end of the spectrum by having as their basis the underlying assumption that all hydrologic processes should be viewed as "random numbers of random variables." Although the randomness assumption allows hydrologic processes to be modeled and described on the basis of probability density functions, in practice the assumption has tended to somewhat limit the scope of certain schools of stochastic hydrology by focusing analyses on the hydrologic time series itself and appearing to eliminate the need to examine hydrologic events from any sort of physically based viewpoint, especially in terms of the climatic origins of the events. In recent years the distinction between deterministic and stochastic approaches has become blurred since many models contain elements of both. Physically based stochastic models of rainfall and runoff, in particular, have ushered in a new direction in hydrology by effectively combining stochastic and deterministic methodologies (e.g., Eagleson, 1972, 1978; Chan and Bras, 1979; Waymire et al., 1984).

The following sections present an overview of the various ways in which climate has been incorporated into deterministic, stochastic, and physically based approaches to flood analysis.

Deterministic Approaches and Climate

The basis of a deterministic approach to flood analysis is that "floods are physical phenomena which result from an input of precipitation into a drainage basin, the flood magnitude varying with the nature of both the precipitation and the drainage basin" (Ward, 1978, p. 71). Although determinism is not synonymous with causality (Klemeš, 1978), deterministic models are often developed with causality in mind or used to explore the possibility of causal relationships. It is therefore within the deterministic approach that much effort has been placed on examining the relationship between climatic inputs and runoff responses. Climate-based deterministic methods of flood analysis include rainfall–runoff models, probable maximum precipitation, and water budget analysis.

Rainfall–Runoff Models. Most deterministic methods are designed to calculate or predict a hydrologic output, such as a flood hydrograph, from a given or predetermined climatic or meteorologic input, such as rainfall duration and intensity. The fact that the cross-discipline of hydrometeorology evolved contemporaneously with the expansion of rainfall–runoff models in the twentieth century reflects the prominence of the role of climate and meteorology in these models.

Comparative discussions of the many deterministic models that have been developed for runoff analysis have been presented in several recent review papers. An overview of rainfall–runoff models is given by Linsley (1982), deterministic surface water routing models are described by Dawdy (1982), physically based and process-oriented models are discussed by Woolhiser (1982) and Dunne (1982), and some deterministic models that specifically focus on the analysis of peak flows are reviewed by Feldman (1980). The common thread among most of these approaches—from the earliest simple mathematical calculations of peak discharge using the "rational formula" to the sophisticated modeling of runoff by solving partial differential equations for three-dimensional, time-varying flow—is the inclusion of climatic or meteorologic variables as important inputs in the analysis.

Probable Maximum Precipitation and Hydrometeorologic Studies. A method that exemplifies the deterministic hydrometeorologic approach to flood analysis is the *probable maximum precipitation* technique, used to determine the design flood for a river basin (see Myers, 1969; Miller, 1973). Probable maximum precipitation (PMP) is defined as "theoretically the greatest depth of precipitation for a given duration that is physically possible over a given size storm area at a particular geographical location at a certain time of the year" (Interagency Advisory Committee on Water Data, 1985, p. 124). The methodology was developed in the United States in the mid-1930s and remains in use today, particularly as a guide for the design specifications of large dams. In the PMP approach information obtained from extensive meteorological analyses of major storms and atmospheric conditions known to have produced flooding is used to develop regional maps depicting precipitation depth estimates for storms of different duration. The PMP estimates are then converted to estimates of the probable maximum flood for a particular drainage basin through a deterministic model such as the unit hydrograph.

The PMP method has aroused a great deal of controversy between the deterministic and stochastic schools of thought. The crux of the argument against PMP by some stochastic hydrologists is that the method is not based on any concept of probability or statistics (despite its name) and that it implies that there are definable upper limits to the meteorologic processes operating in an area (Yevjevich, 1968; Benson, 1973; discussion in Schulz et al., 1973, pp. 95–113). Proponents of PMP have responded that the method provides a useful bench mark for determination of the design flood for a dam, especially when existing records are inadequate for using statistical flood frequency methods to estimate extreme floods (discussion in Schulz et al., 1973, pp. 95–113). E. M. Laurenson, commenting during a discussion session on PMP, summed up the controversy as follows:

> The argument that has gone on over the past twenty years between determinists and probabilists on the question of PMP has been most destructive, because it has forced people into opposing camps and into positions they feel they must maintain. . . .
>
> The only hope for advancement in the area of estimation of extreme floods is in a combination of the deterministic approach to those aspects of the precipitation and runoff processes where we have physical knowledge and the probabilistic approach to those aspects which cannot be described in terms of cause and effect.
>
> —(E. M. Laurenson in discussion in Schulz et al., 1973, p. 105)

As Laurenson suggests, deterministic hydrometeorologic studies of runoff phenomena have greatly increased our understanding of the physical processes that produce floods. In the arid, semi-arid, and mountainous West, the complexities of the storm–flood relationship are especially difficult to analyze. Several studies have contributed in this area, notably National Weather Service Hydrometeorological Report No. 50 by Hansen and Schwarz (1981) on the meteorology of important rainstorms in the Colorado River and Great Basin drainages, and the National Oceanic and Atmospheric Administration (NOAA) studies of R. A. Maddox and his colleagues (Maddox and Chappell, 1978; Maddox et al., 1979, 1980) on meteorological characteristics of flash floods in the western United States.

Water Budget Analysis. Studies of flooding from a meteorological perspective have advanced our knowledge of the physical causes of individual flood events. However, the deterministic hydrometeorologic approach is usually not directed toward synthesizing this information over sufficiently large spatial or sufficiently long temporal scales to present a picture over time of the *variability* of occurrence of the hydrometeorologic phenomena. Here lies the strength of hydroclimatic methods for analyzing floods. One such technique is water budget analysis, an environmental systems approach to the hydrologic cycle that studies the income, outgo, and storage of water at the surface of the earth (Muller, 1976; Mather, 1978). The water budget can be computed at various spatial scales and over daily, monthly, seasonal, or yearly time scales, depending on the needs of the analysis. The approach is especially useful for distinguishing climatic from nonclimatic sources of runoff variability because it can compute expected runoff solely on the basis of surpluses in the climatically derived water budget and thereby assess the amount of streamflow variability that is related to climatic variability (Fig. 11).

Stochastic Approaches and Climate

The stochastic approach in hydrology has been defined as "the manipulation of statistical characteristics of hydrologic variables to solve hydrologic problems, on the basis of the stochastic properties of the variables. A stochastic variable is defined as a *chance* variable or one whose value is determined by a probability function" (Committee on Surface-Water Hydrology, 1965, p. 77). Some hydrologists make a distinction between the terms *stochastic* and *probabilistic*, the former referring to the treatment of variates as time dependent and the latter as time independent (Chow, 1964). Most, however, avoid this distinction and refer to a stochastic process as one that "evolves, entirely or in part, according to a random mechanism" (Kisiel, 1969a, p. 15), and use the terms *stochastic*, *probabilistic*, and *random* interchangeably as synonyms for any process that is governed by the laws of chance (Yevjevich, 1974). According to Klemeš (1983), although the usage of the term *stochastic* is not uniform among hydrologists, "the prevailing view is that whenever some variables or parameters in mathematical formulations of hydrologic processes or relationships are defined as variates (random variables), the formulations belong under the label of stochastic hydrology" (p. 695).

Since the basis of a stochastic approach is the modeling of a process according to the laws of chance, climate—

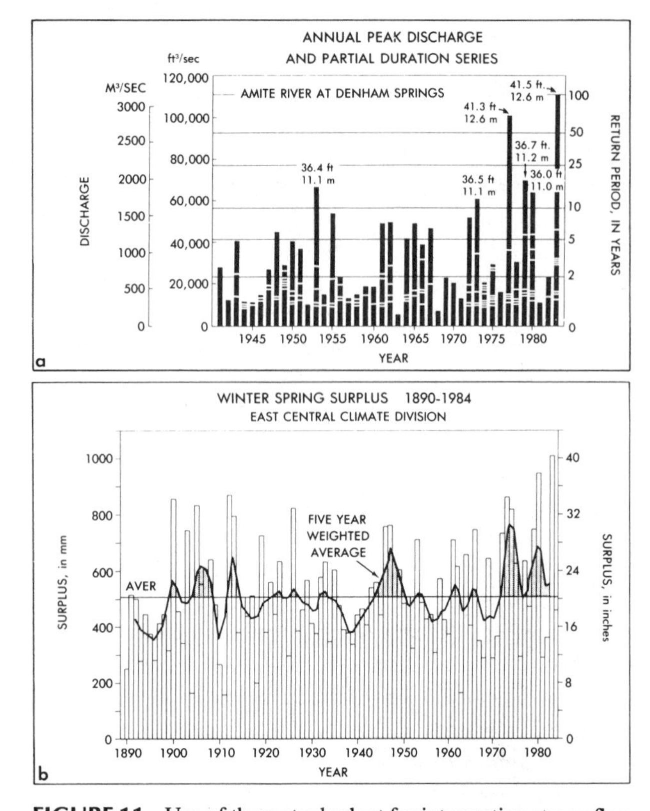

FIGURE 11. Use of the water budget for interpreting streamflow variability. (*a*) Annual and partial flood series for the Amite River in eastern Louisiana. Partial series peaks are shown with white horizontal bars. Most of the annual floods occur in spring. A trend toward larger floods is evident in recent years. Nonclimatic factors, such as land-use changes or measurement error, are often proposed as explanations for trends like this one in a flood series. (*b*) Time series of the winter–spring surplus available for runoff in east-central Louisiana, computed from climatic data. The plot shows that the 11-yr period between 1973 and 1983 was extremely wet in winter and spring and generated some of the largest seasonal moisture surpluses in the entire 96-yr record. This additional hydroclimatic information lends support to a climatic explanation for the recent trend toward increasing flood peaks (Source: Muller and McLaughlin, 1987).

when viewed as a deterministic component—is generally not involved in most stochastic analyses of hydrologic processes. Furthermore, in some cases, when the effects of a climatic control may be influencing a hydrologic series by producing a trend or a jump in the record, this deterministic component of the series is often removed (see Yevjevich, 1972). In general, climatic information enters into most stochastic analyses of hydrologic processes mainly by offering either support or reservations concerning the underlying assumptions of stationarity and homogeneity in the hydrologic data. The climatic basis for these assumptions is important in both probabilistic flood frequency analyses and in stochastic time series analyses of long-term hydrologic variability.

Flood Frequency Analysis. Stochastic approaches to flood analysis were formally introduced in 1914 by Fuller when he presented a discussion of flood frequencies and used the return period as a measure of the probability of recurrence of floods of different magnitudes (Fuller, 1914). This concept soon expanded into the use of theoretical probability distribution functions to describe the actual frequency distribution of the floods (Foster, 1924; Hazen, 1930; Gumbel, 1941) and is still in active use today in what might be referred to as standard probabilistic flood frequency analysis (U.S. Water Resources Council, 1981).

Numerous methods for calculating the best estimates of flood frequency have evolved over the past 70 years and are reviewed by Jarvis and others (1936), Benson (1962), and Reich (1976). In 1968 Benson reported on a study by a federal interagency group that compared the most commonly used methods of flood frequency analysis (Benson, 1968). This group recommended that all U.S. government agencies adopt the log-Pearson Type III distribution as their base method in order to achieve a uniform procedure for computing flood estimates. Although the recommendation set off a major controversy (Kisiel, 1969b; Benson, 1969; Reich, 1977), it did result in a standard procedure for flood analysis in the United States, first outlined by the U.S. Water Resources Council in 1967, revised in 1976, and revised again as Bulletin 17B in 1981. The recommendation also set off a flurry of reactivated interest in methods of flood frequency analysis that persists to the present day. [For overviews of some of these more recent approaches, see Greis (1983), Cunnane (1987), Stedinger and Cohn (1986), and Tasker (1987).] Despite the continued interest in improving flood frequency estimates, climatic information has not often been utilized in this area except as a factor in the underlying assumptions of the analysis. Bulletin 17B addresses the role of climate and climatic variability in terms of the conditions necessary for a valid statistical procedure. These assumptions are "that the array of flood information is a reliable and representative time sample of random homogeneous events" (U.S. Water Resources Council, 1981, p. 6). Possible violations of these conditions that are related to climate are subsequently elaborated in the bulletin:

> ***Climatic Trends***
>
> There is much speculation about climatic changes. Available evidence indicates that major changes occur in time scales involving thousands of years. In hydrologic analysis it is conventional to assume flood flows are not affected by climatic trends or cycles. Climatic time invariance was assumed when developing this guide.
>
> —(p. 6)
>
> ***Mixed Populations***
>
> At some locations flooding is created by different types of events. For example, flooding in some watersheds is

> created by snowmelt, rainstorms, or by combinations of both snowmelt and rainstorms. Such a record may not be homogeneous and may require special treatment.
>
> —(p. 7)

The integration of flood hydroclimatology with traditional flood frequency analysis can effectively address these possible violations by examining the physical causes of flood events and the nature of climatic anomalies that produce unusual "outliers," mixed populations, or trends in flood series. For example, the space–time domains of various kinds of hydroclimatic activity (Fig. 1) can be used to explore Bulletin 17B's assumption of "climatic time invariance" by examining the sensitivity of hydrologic systems to climatic variations at different time scales.

Time-dependent Stochastic Approaches. While the probabilistic methods of flood frequency analysis were being refined, compared, and promulgated in the 1960s and 1970s, the techniques of time-dependent stochastic analysis of hydrologic processes were also receiving attention (e.g., Yevjevich, 1963, 1972; Fiering, 1967; Mandlebrot and Wallis, 1968; Kisiel, 1969a). These time series methods were developed to analyze or synthesize hydrologic processes on long time scales. However, the impact of climatic variability on hydrologic time series was often not appreciated:

> In the analysis of the longest series of annual values of precipitation and runoff, no statistically significant climatic changes could be detected. There is, however, a possibility that some slow changes of yet unidentified characteristics may be taking place, mainly because of man-made factors. . . . These new factors are bound to change the climate and also to affect some hydrologic phenomena along the water cycle. However, an overemphasis on these changes has given them a distorted importance. The consequence has been to retard investigation of the basic structure of hydrologic time series by proper techniques. The "warming up" and "cooling off" periods over some areas, or the advance or retreat of glaciers, may be only part of a random fluctuation in the time series of temperature and volume of ice in glaciers. The "climatic change complex" . . . has diverted the efforts of hydrologic investigations to less productive scientific areas.
>
> —(Yevjevich, 1968, pp. 228–229)

These comments are symbolic of a bias among some stochastic hydrologists against any approach to the analysis of a sequence of hydrologic data that is dominated by a "ruling hypothesis" that seeks to deterministically explain all hydrologic variability as a function of climatic changes and therefore essentially denies the basis of the stochastic approach: that a hydrologic process is a random or stochastic process. Given the history of a rash of hydrologic determinists in search of hidden periodicities and cycles in the early 1900s, which culminated in what has been described as "the largest historic failure in the analysis of hydrologic processes" (Yevjevich, 1974, p. 229), the wariness of some stochastic hydrologists toward climatic change explanations for hydrologic variability is understandable.

In recent years the relationship between climatic variability and hydrologic series has been more responsibly explored by using various time series analysis techniques (e.g., Schaake and Kaczmarek, 1979; Lettenmaier and Burges, 1978; Meko and Stockton, 1984). Much of this work has been done in the realm of water resources systems analysis for the purpose of providing better estimates of long-term storage and water availability, given the possibility or reality of climatic changes or long-term climatic fluctuations. Of great interest in this area is the potential for modeling two types of phenomena that have been observed in hydrologic time series and have been linked to climate: the *Joseph* and *Noah* effects, so named by Mandelbrot and Wallis (1968).

Joseph Effects. The phenomenon described as a *Joseph effect* is the occurrence, on occasion, of very long periods of low flows (or precipitation), or very long periods of high flows (or precipitation). The term was inspired by the biblical story of Joseph, whose Egyptian reign included seven years of plenty followed by seven years of famine. Models that can account for or describe the extended wet and dry episodes often seen in hydrologic series are considered to be in the Joseph realm and have been used to explore the relationship of droughts and long-term persistence to the statistical properties of time series.

Kilmartin (1980) saw a great need for the application of a hydroclimatologic approach to the analysis of Joseph effects because "the 'Joseph event' is not merely a within-basin process, it is a basin response to a major anomaly in the atmospheric circulation, an anomaly that is often near hemispherical in areal magnitude" (p.161). Indeed, the most active current research on the relationship between climatic variability and hydrologic variability is largely concentrated in the Joseph realm of time series analysis of long-term climatic and hydrologic fluctuations and the consequences of these fluctuations for future water supplies.

Noah Effects. In contrast with the long-period Joseph effects, the term *Noah effect* refers to the short-term rare occurrences in nature of extremely high flows (or precipitation) and, of course, was inspired by the biblical story of Noah and the great flood. Time series modeling of the Noah effect has posed problems of a different nature than the Joseph effect.

Statistically, the phenomenon has usually been analyzed by using extreme value theory or other probabilistic methods for modeling outlier behavior in the tails of highly skewed distributions (see, e.g., Kottegoda, 1984). The huge recurrence interval that should, in theory, be attached to a true Noah event renders most time series techniques that use available hydrologic records useless for examining the phenomenon. Dendrochronologic reconstructions of annual or seasonal runoff—particularly effective for studying Joseph events (Stockton, 1975; Stockton and Boggess, 1980)—are generally not sufficiently sensitive to the extremes in a series to be of major use in the study of Noah events. However, some success has been achieved in evaluating the recurrence intervals of extremely large floods by using both historical and paleoflood data (e.g., Costa, 1978; Baker, 1982; Stedinger and Cohn, 1986).

Kilmartin (1980) indicates that the study of extremely large flows has traditionally fallen within the (more deterministic) realm of hydrometeorology, especially in the probable maximum flood approach. However, a hydroclimatic approach to the analysis of Noah events, outliers, or the extremes of a flood probability distribution can contribute important information about the synoptic circulation anomalies and characteristic climatic patterns that affect the recurrence intervals and probabilities of such unusual events.

Physically Based Approaches and Climate

Apart from some of the studies mentioned above, most stochastic hydrologists who analyze floods have not focused on incorporating climate or climatic variability into their models. Recently, however, a new outlook has emerged among hydrologists who are calling for a more physically based stochastic hydrologic analysis (Klemeš, 1978, 1982). The physically based approach has proceeded along two closely related avenues of inquiry: detailed conceptual rainfall–runoff modeling of the dynamic hydrologic physical system (hydrodynamical models) and attempts at arriving at a conceptual basis for the probability laws and distribution functions that emerge from the physical properties and interactions of hydrologic variables.

Hydrodynamic Models. One type of physically based approach to flood analysis can be found in the so-called hydrodynamic models (Eagleson, 1972, 1978). The underlying purpose of such models is to derive the probability distribution of peak streamflows, not on the basis of extreme value theory (Gumbel, 1941), but by conceptualizing the streamflow process as a sequence of kinematic waves that originate from a given joint probability distribution of rainfall intensity and duration, and are transformed into streamflow waves through a modeled catchment process. The model effectively integrates several of the deterministic and stochastic approaches previously discussed in this overview by aiming to describe the continuum from rainfall to flood with a set of differential equations, culminating in a derivation of the probability distribution of the resulting flood peaks. Another model with a similar physically based, combined deterministic-stochastic approach can be found in the "geomorphoclimatic" model of Rodriguez-Iturbe (1982) that seeks to "couple" the geomorphic parameters of a drainage basin with rainfall intensity and duration to arrive at an instantaneous unit hydrograph that is conceived as a stochastic response function dependent on both climate and geomorphology.

This kind of comprehensive modeling represents an exciting turn of events in hydrology and holds out the possibility of an ever-deepening understanding of hydrologic processes as they occur in nature:

> The hydrodynamical model, which aims at describing the prototype by a set of differential equations, is often viewed as an ideal of perfection and rigor, the final goal of conceptual hydrologic modeling. It is argued that such a model would have a general applicability since, by being able to describe the streamflow process in terms of the basic equations of mechanics, it would readily facilitate the derivation of all the commonly used coarser representations such as the series of mean daily, monthly, and annual flows, as well as any specific properties like those of maximum and minimum flows, drought periods, and flood volumes.
>
> —(Klemeš, 1978, p. 302)

The difficulties in developing such models, however, often lead to simplification or lumping of input parameters, estimation of critical parameters for which no actual measurements are available, and assumptions for ease of analysis that may or may not hold up in the real world. As Klemeš (1983) states, the hydrodynamic approach

> . . . faces formidable difficulties of at least two kinds. The first is the constraint of mathematical tractability which may enforce simplifications and approximations whose physical plausibility is in doubt. The second is the incompleteness of our knowledge of phenomena at the starting level which brings about the necessity of filling gaps with unverified assumptions whose effect can distort the plausibility of the final product.
>
> —(Klemeš, 1983, p. 7)

Given these factors, and the complexity and expense of developing such rigorous models, their greatest value is their immense contribution to the conceptualization of fundamental hydrologic processes, rather than their practical usefulness in providing computed end products.

Physically Based Distribution Functions. Another way in which stochastic hydrologists have attempted to incorporate more deterministic aspects of the physical world, such as climate, into their models is by probing the possibility of a physical basis for the particular shape of the various probability distribution functions (PDFs) that describe climatic approach, however, it is possible to carefully lyze the events in a flood series in terms of their climatic gins so as to evaluate $X(t)$ by investigating the following ernative assumptions:

1. Floods occurring as a result of different atmospheric mechanisms belong to different populations. The dominance of these different populations at different times, t, will determine the shape of the overall frequency distribution of the flood series.
2. Differences among the populations described in assumption 1 are due to either varying means as in Figure 14*a*, changing variances as in Figure 14*b*, or both as in Figure 14*c*. The mean and variance associated with each theoretical distribution has a physical basis that is linked to the nature of the atmospheric mechanisms operating at t.
3. A shift in general atmospheric circulation patterns or the anomalous persistence of certain patterns will be reflected in a flood series by a shift to a different theoretical distribution for the random variable $X(t)$ in the series.

Although research along these lines is at its early stages, two previously mentioned studies have demonstrated the usefulness of viewing the flood series model from a hydroclimatic perspective to explore the assumptions outlined above.

Assumptions 1 and 2 were examined by grouping flood events generated by similar hydroclimatic mechanisms into climatically homogeneous subsets (Hirschboeck, 1985, 1987). It was found that mixed distributions could be hydroclimatically defined in a flood series and that the shapes of the frequency distributions of different hydroclimatic subgroups had a physical basis that could be linked to the nature of the flood-generating mechanism.

Knox (1983) used dates marking episodic adjustments in large-scale atmospheric circulation patterns to subdivide a flood series into meaningful hydroclimatic episodes. His results, depicted in Figure 10, bear a resemblance to the nonstationary models in Figure 14 and lend support to assumption 3.

Other Applications of Flood Hydroclimatology

Numerous other applications of flood hydroclimatology can be envisioned to analyze flood time series and evaluate basic assumptions on how floods vary over time and space. In standard flood frequency analysis the hydroclimatic approach has potential for examining outliers and determining the usefulness of short flood records for representing long-term flood variability. In regional analysis (Tasker, 1987) hydroclimatology provides an effective means for grouping streams and gauging stations that covary spatially in their response to various climatic inputs.

In the development of physically based stochastic models, hydroclimatology can contribute to an understanding of how the atmosphere and hydrosphere interact at various spatial and temporal scales. Some research along these lines is already in progress; (e.g., see Waymire et al., 1984, who incorporated information on the dynamics of extratropical cyclones into a physically realistic stochastic model of mesoscale rainfall intensity).

Finally, because flood hydroclimatology encompasses long-period climatic variations as well as short-period hydrometeorological events, it provides a climatic framework for meshing paleoflood studies and other aspects of flood geomorphology with the relatively short time scales of gauged flood records. For example, knowledge of the types of circulation features that currently generate floods at a given station (Fig. 12) can be applied to paleoflood data collected for the same station. With the aid of circulation models and climatic reconstructions, the paleoflood event can be linked to the most probable flood-generating mechanism and evaluated on the basis of the modern sample frequency distribution of all floods produced by that mechanism.

CONCLUSION

Flood hydroclimatology analyzes floods as real-world physical events occurring within the context of time-varying climatic conditions and within a spatial framework of local, regional, and global networks of changing atmospheric circulation patterns. Although climate has been incorporated into flood analyses in a variety of ways, given the broad space–time domain of hydroclimatic activity, there is a great need to re-evaluate certain assumptions about how floods vary over time in relation to climate and to reexamine other current issues in flood hydrology from a new hydroclimatic perspective.

ACKNOWLEDGMENTS

I wish to thank V. Baker, R. Jarrett, and R. Muller for their valuable comments, suggestions, and encouragement. Special thanks go to P. Bartlein, V. Klemeš, and J. Stedinger, whose very careful reviews greatly improved the manuscript. I am also indebted to Mary Lee Eggart of the LSU Cartographic Section, who prepared all the figures.

REFERENCES

Andrews, J. F. (1962). The weather and circulation of February, 1962—a month with two contrasting regimes associated with a sharp drop in zonal index. *Mon. Weather Rev.* **90**, 203–210.

streamflow behavior. These studies have directed most of their attention to the timing of floods in relation to each other and the resulting PDFs that represent this process (Todorovic and Zelenhasic, 1970; Denny et al., 1974; Gupta et al., 1976; Todorovic, 1978). Interest in the timing of large floods has nurtured a further interest in the climatic factors that influence hydrologic variability, especially in terms of the seasonality of flows. A good example of this sensitivity toward climate in a stochastic analysis of flows can be found in Denny et al. (1974) where, by incorporating observations on the nature and timing of climatic inputs to streamflow, the authors developed the underlying assumptions for a Markov analysis of stream behavior and imparted a physical basis to their model.

Mixed Distributions. One of the most frequently cited areas of potential hydroclimatic flood research using physically based distribution functions is the problem of mixed distributions or multiple populations in hydrologic time series. Although homogeneity in the flood series is a basic underlying assumption for the probabilistic determination of flood magnitudes and frequencies, wherever this assumption is stated in the literature, it is often followed by a disclaimer. A typical example, found in Bulletin 17B, states that due to differences in the climatic processes involved in the generation of floods—rainfall, snowmelt, tropical storms, and so on—multiple populations or mixed distributions may be present in the data.

Despite the almost universal recognition that some observed flood samples may not be drawn from a single, climatically homogeneous population, only a handful of researchers have seriously devoted their efforts to the analysis of this problem. Potter (1958) was one of the first to discuss the evidence for two or more distinct populations of peak runoff (as seen in dogleg flood frequency curves), and he proposed possible climatic causes for the multiple populations. Singh (1968, 1974) presented a methodology for mathematically simulating mixed distributions in hydrologic samples, but although he referred to climate as a probable cause of multiple populations, his approach was to objectively search the streamflow data alone to define a mixture of distributions, rather than to decompose the data on the basis of additional climatic information. Other studies have attempted to identify mixed distributions in streamflow series by separating the flood record into seasonal subpopulations (Guillot 1973; Browzin et al., 1973).

Klemeš (1974) discussed mixed distributions and emphasized that the concept is meaningful "only if physically justified and if the component subsamples can be separated on physical grounds" (p. 6). Accordingly, Jarrett and Costa (1982), Elliott et al. (1982), and Waylen and Woo (1982) moved beyond the simple seasonal division of a flood series and looked at the differences between rainfall- and snowmelt-generated floods to examine the problem of mixed distributions in hydrologic data. By detailed examination of both streamflow and weather records, these researchers were able to subdivide flood series into rainfall and snowmelt "populations" so that separate flood frequency curves could be developed from each subset of data.

The analysis of mixed distributions can be taken a step further by using flood hydroclimatology to identify the various synoptic atmospheric circulation mechanisms and patterns that generate each flood event in a series (Fig. 12). When events in a flood series are separated into climatically homogeneous subgroups, the shape of the sample frequency distribution can be interpreted in terms of the physical processes that generated the sample. This is especially appropriate in climatically sensitive regions or in climatic transition zones where floods evolve from a variety of different processes that may be exhibited in complex frequency functions. For example, Figure 12 shows that the largest annual floods on the Salt River in central Arizona are associated with winter frontal passages and tropical storm/cutoff low circulations, while snowmelt floods and summer monsoon floods are of less importance in shaping the upper tail of the sample distribution. In the Santa Cruz River to the south, floods generated by summer monsoon circulation patterns control the basic shape of the sample distribution; however, infrequent but extreme floods generated by winter frontal passages and tropical storm/cutoff low circulations maintain an important influence on the upper tail. This new hydroclimatic information, applied to a mixed distribution, holds the promise of both enhancing our understanding of the flooding process and potentially improving flood frequency estimates by determining the physical basis for events in the upper tails.

HYDROCLIMATIC INTERPRETATION OF THE FLOOD TIME SERIES MODEL

The preceding sections have described the spatial and temporal scales at which flooding and climate interact and the various ways in which climatic information has been integrated into the analysis of floods. Climate can also be applied to the theoretical interpretation of flood series models. One way to interpret an annual flood series is to consider each peak to be an independent observation, without considering the time sequence of the flood events. Many flood frequency analysis techniques proceed under this assumption, and the role of climate is presumed to be time invariant.

However, there is another type of flood series model that seeks to describe, either conceptually or mathematically, the underlying process that determines how floods vary *over time*. This standard time series model is based on the concept of a time-dependent stochastic process and is usually assumed to be a stationary model. The dynamic nature of hydroclimatic activity, operating at long- and short-term

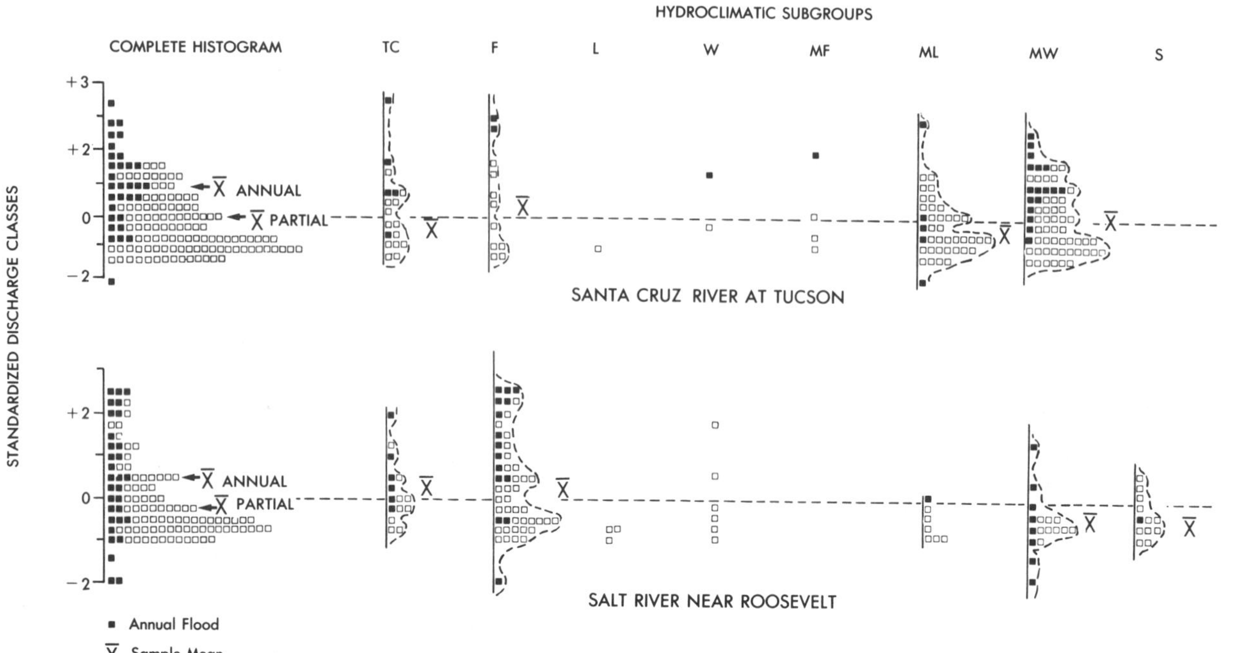

FIGURE 12. Decomposition of two partial-duration flood series into hydroclimatic subgroups to define mixed distributions in the record. The floods were separated into groups on the basis of the atmospheric mechanism that generated each flood: (TC) tropical storm/cutoff low event, (F) frontal passage, (L) local cool season precipitation, (W) widespread nonfrontal cool season precipitation, (MF), (ML), and (MW) "monsoon season" frontal, local, and widespread precipitation, respectively, (S) snowmelt. The flood discharges are in the form of standardized dimensionless z scores. (Source: Hirschboeck, 1987).

time scales, implies that the stationarity assumption may not always be valid. Hence flood time series should also be viewed from the perspective of alternative, nonstationary models that have a hydroclimatic basis.

Standard Flood Time Series Model

When modeling flood time series, the observed record of peak discharges measured at a gauging station over a finite period of time is generally viewed as a sample drawn from the population of all possible floods occurring during an undefinable length of total time (T), which theoretically goes to infinity. The standard model for a time-dependent process is depicted in Figure 13 and is described stochastically as follows:

> Mathematically speaking, a *stochastic process* is a family of random variables $X(t)$ which is a function of time (or other parameters) and whose variate x_t is running along in time t within a range T. Quantitatively, the stochastic process, which may be discrete or continuous, can be sampled continuously or at discrete or uniform intervals of $t = 1, 2, \ldots,$ and the values of the sample form a sequence of $x_1, x_2, \ldots,$ starting from a certain time and extending for a period of T. This sequence of sampled values is known as a time series. . . .
>
> The random variable $X(t)$ has a certain probability distribution. If this distribution remains constant throughout the

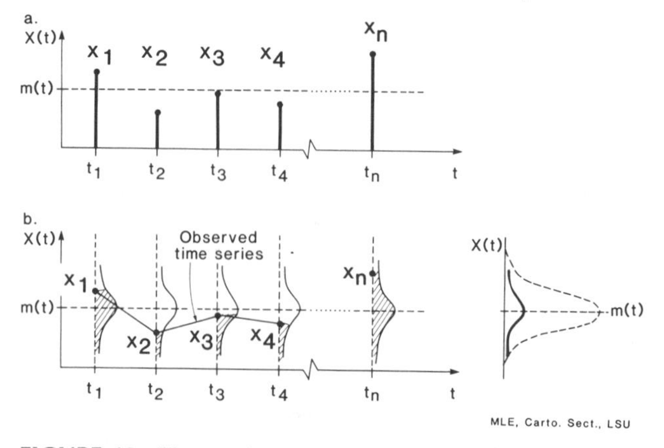

FIGURE 13. The standard stationary stochastic process model for a flood series. (*a*) Generalized version of an observed flood series. (*b*) Conceptual representation of the stationary stochastic process with time invariant mean and variance (modified from Kisiel, 1969a).

> process, the process and the time series are said to be *stationary*. Otherwise they are *nonstationary*. For example, a virgin flow with no significant change in river-basin characteristics or climatic conditions for the period of record is considered as a stationary time series. If it is affected by man's activities in the river basin or nature's large accidental or slow modifications of the rainfall and runoff conditions, the recorded or historical flow is a nonstationary time series. Since a nonstationary process is very complicated mathematically, hydrologic processes are generally treated as stationary.
>
> —(Chow, 1964, p. 8–9)

For the stationary stochastic process in Figure 13 to be valid in a hydroclimatic sense, the overall climatic environment from which floods are generated must remain essentially the same over time so that the probability distributions for the random variable $X(t)$ at each successive t will remain constant. If stationarity holds, interpretation of Figure 13b suggests that at certain times, t_n, for example, the observed flood value represents an extremely rare event because it falls well within the uppermost tail of the theoretical distribution of all possible floods at time t_n. Other floods, such as x_2, are much smaller than would be expected on the basis of the theoretical distribution and would be considered unusually low for peak events at time t_2.

Alternative Hydroclimatic Interpretations

Despite the elegance and pervasive use of the stationary stochastic model, when flood events are interpreted hydroclimatically, the validity of the stationarity assumption must be seriously questioned due to the history of variation of regional and global networks of changing meteorological features and circulation patterns. Within the framework of the space–time domain of hydroclimatic activity, three alternative flood series models are likely to be better representations of the actual flooding process over time (Fig. 14). In these nonstationary processes it is clear that the statistical properties of the random variable $X(t)$ differ from one realization of $X(t)$ in time to another. For example, in the time-varying mean model (Fig. 14a), flood event x_n, which was previously interpreted as being so rare because of its upper tail location (Fig. 13b), must now be interpreted as an event with a high likelihood of being equaled or exceeded, given the new position of the theoretical distribution at time t_n. In the time-varying variance model (Fig. 14b), the large x_n and the small x_2 events are no longer located in the extremes of the upper and lower tails of their respective populations as they were in Figure 13b. In the model depicting both time-varying mean and variance (Fig. 14c), floods x_2 and x_n lie close to the means of their respective populations. Conversely, flood x_4 now reflects an extreme in the lower tail of its theoretical distribution.

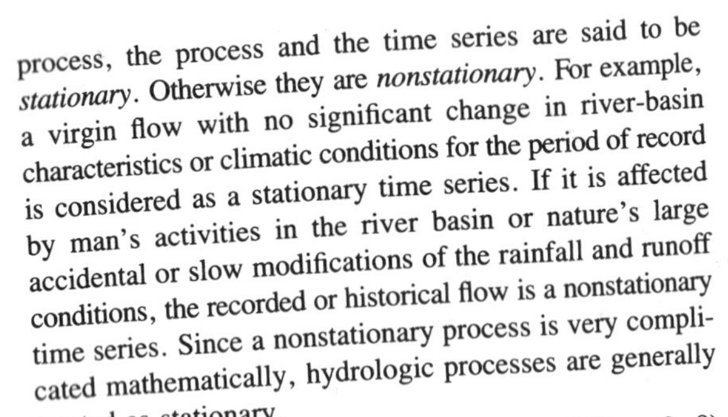

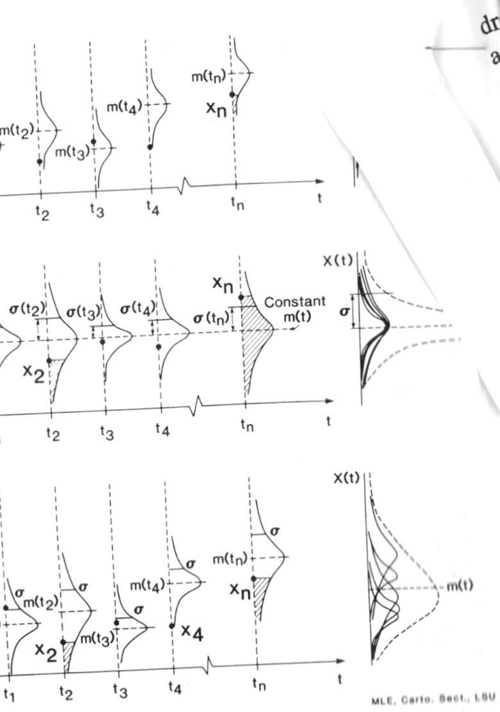

FIGURE 14. Alternative nonstationary stochastic process models for a flood series. (*a*) Time-varying mean. (*b*) Time-varying variance. (*c*) Time-varying mean and variance. The composite distribution at the right of each time series represents the theoretical distribution for the complete flood series. Although the distributions at each t are depicted as normal, skewed distributions could also be substituted into the model (modified from Kisiel, 1969a).

Although not depicted in Figure 14, another aspect of variation in a time series is the tendency for a specific feature, such as the mean, variance, or distribution shape, to persist for a certain time, and then change. In this case the intervals between changes and the times of change themselves become important variables (e.g., Fig. 10).

It is interesting to note that of the four models proposed for a flood series the models in Figure 14a and especially Figure 14c depict theoretical distributions for the overall flood series that look most like the typical positively skewed flood frequency distributions of many arid, semi-arid, and humid streams. Therefore, although the standard stationary model in Figure 13 is the most easily analyzed, it may not be the best model for studying flood processes in the real world. In fact many stochastic hydrologists concede that, in the physical world, most hydrologic time series are nonstationary with a mean and variance that tend to vary with time (Kisiel, 1969a).

In traditional flood series analyses, because the actual distributions of the random variable $X(t)$ over time are unknown, stationarity and homogeneity have been the most convenient assumptions to make and have eased the mathematics involved in analyzing time series. In a flood hy-

dr
an
o
a

Baker, V. R. (1982). Geology, determinism, and risk assessment. *In* "Scientific Basis of Water-Resource Management," Stud. Geophys. pp. 109-117. Nat. Acad. Press, Washington, D.C.

Barry, R. G., and Perry, A. H. (1973). "Synoptic Climatology." Methuen, London.

Benson, M. A. (1962). Evolution of methods for evaluating the occurrence of floods. *Geol. Surv. Water-Supply Pap. (U.S.)* **1580-A**, 1–30.

Benson, M. A. (1968). Uniform flood-frequency estimating methods for federal agencies. *Water Resour. Res.* **4**, 891–908.

Benson, M. A. (1969). Reply to Comments on 'uniform flood-frequency estimating methods for federal agencies.' *Water Resour. Res.* **5**, 911.

Benson, M. A. (1973). Thoughts on the design of design floods. *In* (E. F. Schulz, V. A. Koelzer, and K. Mahmood, eds.), "Floods and Droughts" pp. 27–33. Water Resources Publications, Fort Collins, Colorado.

Browzin, B. S., Baumbusch, C. A., and Pavlides, M. G. (1973). Significance of the genesis of floods on probability analysis. *In* (E. F. Schulz, V. A. Koelzer, and K. Mahmood, eds.), "Floods and Droughts" pp. 450-461. Water Resources Publications, Fort Collins, Colorado.

Bruce, J. P., and Clark, R. H. (1980). "Introduction to Hydrometeorology." Pergamon, Oxford.

Carlson, D. D., and Firda, G. D. (1983). Floods of April 1983 in southern Mississippi and southeastern Louisiana. *Geol. Surv. Open-File Rep. (U.S.)* **83-685**, 1–32.

Chan, S. O., and Bras, R. L. (1979). Urban stormwater management: Distribution of flood volumes. *Water Resour. Res.* **15**, 371–382.

Chin, E. H., Skelton, J., and Guy, H. P. (1975). The 1973 Mississippi River basin flood: compilation and analyses of meteorologic, streamflow, and sediment data. *Geol. Surv. Prof. Pap. (U.S.)* **937**, 1–137.

Chow, V. T. (1964). "Handbook of Applied Hydrology." McGraw-Hill, New York.

Committee on Surface-Water Hydrology (1965). Research needs in surface-water hydrology. *J. Hydraul. Div., Am. Soc. Civ. Eng.* **91**, (HY1), 75–83.

Costa, J. E. (1978), Holocene stratigraphy in flood frequency analysis. *Water Resour. Res.* **14**, 626–632.

Cunnane, C. (1987). Review of statistical models for flood frequency estimation. *In* "Hydrologic Frequency Modeling" (V. P. Singh, ed.), pp. 47–93. D. Reidel, Boston.

Dawdy, D. R. (1982). A review of deterministic surface water routing in rainfall-runoff models. *In* "Rainfall-Runoff Relationships" (V. P. Singh, ed.), pp. 23–36. Water Resources Publications, Fort Collins, Colorado.

Denny, J. L., Kisiel, C. C., and Yakowitz, S. J. (1974). Procedures for determining the order of dependence in stream flow records. *Water Resour. Res.* **10**, 947–954.

Dunne, T. (1982). Models of runoff processes and their significance. *In* "Scientific Basis of Water-Resource Management," Stud. Geophys., pp. 17–30. Nat. Acad. Press, Washington, D.C.

Dzerdzeevski, B. L. (1963). Fluctuations of general circulation of the atmosphere and climate in the twentieth century. *Arid Zone Res.* **20**, 285–295.

Dzerdzeevski, B. L. (1969). Climatic epochs in the twentieth century and some comments on the analysis of past climates. *In* "Quaternary Geology and Climate," (H. E. Wright, ed.), Publ. No. 1701, pp. 49–60. Nat. Acad. Sci., Washington, D.C.

Eagleson, P. S. (1972). Dynamics of flood frequency. *Water Resour. Res.* **8**, 878–898.

Eagleson, P. S. (1978). Climate, soil and vegetation. 5. A derived distribution of storm surface runoff. *Water Resour. Res.* **14**, 741–748.

Elliott, J. G., Jarrett, R. D., and Ebling, J. L. (1982). Annual snowmelt and rainfall peak-flow data on selected foothills region streams, South Platte River, Arkansas River, and Colorado River Basins, Colorado. *Geol. Surv. Open-File Rep. (U.S.)* **82–426**, 1–88.

Fairbridge, R. W., ed. (1967). "The Encyclopedia of Atmospheric Sciences and Astrogeology." Reinhold, New York.

Feldman, A. D. (1980). Flood hydrograph and peak flow frequency analysis. *In* "Improved Hydrologic Forecasting—Why and How," Proc. Eng. Found. Conf., 1979, pp. 1–19. Am. Soc. Civ. Eng., New York.

Fiering, M. B. (1976). "Streamflow Synthesis." Harvard Univ. Press, Cambridge, Massachusetts.

Foster, H. A. (1924). Theoretical frequency curves. *Trans. Am. Soc. Civ. Eng.* **87**, 142.

Fuller, W. E. (1914). Flood flows. *Trans. Am. Soc. Civ. Eng.* **77**, 564.

Georgakakos, K. P., and Hudlow, M. D. (1984). Quantitative precipitation forecast techniques for use on hydrologic forecasting. *Bull. Am. Meteorol. Soc.* **65**, 1186–1200.

Greis, N. P. (1983). Flood frequency analysis: A review of 1979-1982. *Rev. Geophys. Space Phys.* **21**, 699–706.

Guillot, P. (1973). Application of the method of gradex. *In* (E. F. Schulz, V. A. Koelzer, and K. Mahmood, eds.), "Floods and Droughts" pp. 44–49. Water Resources Publications, Fort Collins, Colorado.

Gumbel, E. F. (1941). The return period of flood flows. *Ann. Math. Stat.* **12**, 163–190.

Gupta, V. K., Duckstein, L., and Peebles, R. W. (1976). On the joint distribution of the largest flood and its time of occurrence. *Water Resour. Res.* **12**, 295–308.

Hansen, E. M., and Schwarz, F. K. (1981). "Meteorology of Important Rainstorms in the Colorado River and Great Basin Drainages," Hydrometeorol. Rep. No. 50. U.S. Dept. of Commerce, Natl. Oceanic Atmos. Admin., Washington, D.C.

Hazen, A. (1930). "Flood Flows: A Study of Frequencies and Magnitudes." Wiley, New York.

Hirschboeck, K. K. (1985). Hydroclimatology of flow events in the Gila River basin, central and southern Arizona: Ph.D. Dissertation, Department of Geosciences, University of Arizona, Tucson (unpublished).

Hirschboeck, K. K. (1987). Hydroclimatically-defined mixed distributions in partial duration flood series. *In* "Hydrologic

Frequency Modeling" (V. P. Singh, ed.), pp. 192–205. D. Reidel, Boston.

Hoxit, L. R., Maddox, R. A., and Chappell, C.F. (1982). Johnstown-western Pennsylvania storm and floods of July 19-20, 1977. *Geol. Surv. Prof. Pap. (U.S.)* **1211**, 1–68.

Huff, F. A. (1978). "The Synoptic Environment of Flash Flood Storms," Prepr., Conf. Flash Floods: Hydrometeorol. Aspects, pp. 10–16. Am. Meteorol. Soc., Boston, Massachusetts.

Interagency Advisory Committee on Water Data (1985). Draft report on the feasibility of assigning a probability to the probable maximum flood (p. 124). Prepared by the Working Group on PMF Risk Assessment, under the Direction of the Hydrology Subcommittee, Interagency Advisory Committee on Water Data, June, 1985.

Jarrett, R. D., and Costa, J. E. (1982). Multidisciplinary approach to the flood hydrology of foothills streams in Colorado. Int. Symp. Hydrometeorol., Am. Water Resour. Assoc., pp. 565–569.

Jarvis, C. S. et al., (1963). Floods in the United States. *Geol. Surv. Water-Supply Pap. (U.S.)* **771**, 1-497.

Kalnicky, R. A. (1974). Climatic change since 1950. *Ann. Assoc. Am. Geogra.* **64**, 100–112.

Kilmartin, R. F. (1980). Hydroclimatology—a needed cross-discipline. *In* "Improved Hydrologic Forecasting—Why and How," Proc. Eng. Found. Conf., 1979, pp. 160–198. P. Am. Soc. Civ. Eng., New York.

Kisiel, C. C. (1969a). Time series analysis of hydrologic data. *Adv. Hydrosci.* **5**, 1–119.

Kisiel, C. C. (1969b). Comments on 'uniform flood-frequency estimating methods for federal agencies' by Manual A. Benson. *Water Resour. Res.* **5**, 910.

Klemeš, V. (1974). Some problems in pure and applied stochastic hydrology. *Misc. Publ.—U.S. Dep. Agric.* **1275**, 2–15.

Klemeš, V. (1978). Physically based stochastic hydrologic analysis. *Adv. Hydrosci.* **11**, 285–356.

Klemeš, V. (1982). Empirical and causal models in hydrology. *In* "Scientific Basis of Water-Resource Management," Stud. Geophys., pp. 95–104. Natl. Acad. Press, Washington, D.C.

Klemeš, V. (1983). Conceptualization and scale in hydrology. *J. Hydrol.* **65**, 1–23.

Knox, J. C. (1983). Responses of river systems to Holocene climates. *In* "Late-Quaternary Environments of the United States" (H. E. Wright, Jr., ed.), Vol. 2, pp. 26–41. Univ. of Minnesota Press, Minneapolis.

Knox, J. C., Bartlein, P. J., Hirschboeck, K. K., and Muckenhirn, R. J. (1975). The response of floods and sediment yields to climatic variation and land use in the Upper Mississippi Valley. *Univ. Wis. Inst. Environ. Stud. Rep.* **52**, 1–76.

Kottegoda, N. T. (1984). Investigation of outliers in annual maximum flow series. *J. Hydrol.* **72**, 105–137.

Kutzbach, J. E. (1970). Large-scale features of monthly mean Northern Hemisphere anomaly maps of sea-level pressure. *Mon. Weather Rev.* **98**, 708–716.

Lamb, H. H. (1977). "Climate: Present, Past and Future: Climatic History and the Future," Vol. 2. Methuen, London.

Lettenmaier, D. P., and Burges, S. J. (1978). Climate change: Detection and its impact on hydrologic design. *Water Resour. Res.* **14**, 679–687.

Linsley, R. K. (1982). Rainfall-runoff models—An overview. *In* "Rainfall-Runoff Relationships" (V. P. Singh, ed.), pp. 23–36. Water Resources Publications, Fort Collins, Colorado.

Maddox, R. A., Canova, F., and Hoxit, L. R. (1980). Meteorological characteristics of flash flood events over the western United States. *Mon. Weather Rev.* **108**, 1866–1877.

Maddox, R. A., and Chappell, C. F. (1978). Meteorological aspects of twenty significant flash flood events. *In* "Conference on Flash Floods: Hydrometeorological Aspects and Human Aspects," Prepr., pp. 1–9. Am. Meteorol. Soc., Boston, Massachusetts.

Maddox, R. A., Chappell, C. F., and Hoxit, L. R. (1979). Synoptic and mesoscale aspects of flash flood events. *Bull. Am. Meteorol. Soc.* **60**, 115–123.

Mandlebrot, B., and Wallis, J. R. (1968). Noah, Joseph and operational hydrology. *Water Resour. Res.* **7**, 543–553.

Mather, J. R. (1978). "The Climatic Water Budget in Environmental Analysis." Lexington Books, Lexington, Massachusetts.

Meko, D. M., and Stockton, C. W. (1984). Secular variations in streamflow in the western United States. *J. Clim. Appl. Meteorol.* **23**, 889–897.

Miller, J. F. (1973). Probable maximum precipitation— the concept, current procedures and outlook. *In* "Floods and Droughts" (E. F. Schulz, V. A. Koelzer, and K. Mahmood, eds.), pp. 50–61. Water Resources Publications, Fort Collins, Colorado.

Muller, R. A. (1976). Comparative climatic analyses of lower Mississippi River floods: 1927, 1973, and 1975. *Water Resour. Bull.* **12**, 1141–1150.

Muller, R. A., and Faiers, G. E., eds. (1984). "A Climatic Perspective of Louisiana Floods During 1982-1983, Stud. Geogr. Anthropol. Geoscience Publications, Louisiana State University, Baton Rouge.

Muller, R. A., and McLaughlin, J. D. (1987). More frequent flooding in Louisiana: Climatic variability? *In* "Flood Hydrology" (V. P. Singh, ed.), pp. 39–54. D. Reidel, Boston.

Myers, V. A. (1969). The estimation of extreme precipitation as the basis for design floods, resume of practice in the United States. *In* "Floods and Their Computation," Proc. Leningr. Symp., Vol. 1, pp. 84–104. Int. Assoc. Sc. Hydrol., UNESCO, World Meteorol. Organ., Rome.

Namias, J. (1973). Hurricane Agnes—an event shaped by large-scale air-sea systems generated during antecedent months. *Q. J. R. Meteorol. Soc.* **99**, 506–519.

Namias, J. (1974). Longevity of a coupled air-sea-continent system. *Mon. Weather Rev.* **102**, 638-648.

Namias, J. (1979). "Northern Hemisphere Seasonal 700mb Height and Anomaly Charts, 1947-1978, and Associated North Pacific Sea Surface Temperature Anomalies," Calif. Coop. Ocean. Fish. Invest. Atlas No. 27. Mar. Life Res. Group, Scripps Inst. Oceanogr., La Jolla, California.

Namias, J. and Cayan, D. R. (1981). Large-scale air-sea interactions and short-period climatic fluctuations. *Science* **214**, 869–876.

National Research Council (1977). "Climate, Climatic Change, and Water Supply." Nat. Acad. Sci., Washington, D. C.

Osborn, H. B., and Renard, K. G. (1969). Analysis of two major runoff-producing Southwest thunderstorms. *J. Hydrol.* **8**, 282–302.

Potter, W. D. (1958). Upper and lower frequency curves for peak rates of runoff. *Trans. Am. Geophys. Union* **39**, 100–105.

Quiroz, R. S. (1983). The climate of the "El Niño" winter of 1982-83—a season of extraordinary climatic anomalies. *Mon. Weather Rev.* **111**, 1685–1706.

Rasmusson, E. M. (1985). El Niño and variations in climate. *Am. Sci.* **73**, 168–177.

Reich, B. M. (1976). Magnitude and frequency of floods. *CRC Crit. Rev. Environ. Control* **6**, 297–348.

Reich, B. M. (1977). "Lysenkoism" in U.S. flood determination. Paper presented at Special Session on Flood Frequency Methods, American Geophysical Union Surface Runoff Committee, San Francisco, California, December.

Rodriguez-Iturbe, I. (1982). The coupling of climate and geomorphology in rainfall-runoff analysis. *In* "Rainfall-Runoff Relationships" (V. P. Singh, ed.), pp. 431–448. Water Resources Publications, Fort Collins, Colorado.

Schaake, J. C., Jr., and Kaczmarek, Z. (1979). "Climate Variability and the Design and Operation of Water Resource Systems," World Clim. Conf. Overview Pap. No. 12. World Meteorol. Organ., Geneva.

Schulz, E. F., Koelzer, V. A., and Mahmood, K., eds. (1973) "Floods and Droughts," Proc. 2nd Int. Symp. Hydrol., 1972. Water Resources Publications, Fort Collins, Colorado.

Singh, K. P. (1968). Hydrologic distributions resulting from mixed populations and their computer simulation. *In* "The Use of Analog and Digital Computers in Hydrology," Vol. 2, Publ. No. 81, pp. 671–681. Int. Assoc. Sci. Hydrol., Washington, D.C.

Singh, K. P. (1974). A two-distribution method for fitting mixed distributions in hydrology. *Misc. Publ.—U.S.*, Dep. *Agric.* **1275**, 371–382.

Stark, L. P. (1962). The weather and circulation of January 1962. *Mon. Weather Rev.* **20**, 167–173.

Stedinger, J. R., and Cohn, T. A. (1986). Flood frequency analysis with historical and paleoflood information. *Water Resour. Res.* **22**, 785–793.

Stockton, C. W. (1975). "Long-Term Streamflow Records Reconstructed from Tree Rings, Pap. Lab. Tree-Ring Res., No. 5. Univ. of Arizona Press, Tucson.

Stockton, C. W., and Boggess, W. R. (1980). Tree rings: A proxy data source for hydrologic forecasting. *In* (R. M. North, L. B. Dworsky, and D. J. Allee, eds.), "Unified Basin Management" Am. Water Resour. Assoc., pp. 609–624.

Tasker, G. .D. (1987). Regional analysis of flood frequencies. *In* "Regional Flood Frequency Analysis" (V. P. Singh, ed.), pp. 1–9. D. Reidel, Boston.

Thomas, C. A., and Lamke, R. D. (1962). Floods of February 1962 in southern Idaho and northeastern Nevada. *Geol. Surv. Circ. (U.S.)* **467**, 1–30.

Todorovic, P. (1978). Stochastic models of floods. *Water Resour. Res.* **14**, 345–356.

Todorovic, P., and Zelenhasic, E. (1970). A stochastic model for flood analysis. *Water Resour. Res.* **6**, 1641–1648.

U. S. Department of Commerce (1973). Final report of the disaster team on the events of Agnes. *NOAA Nat. Disaster Surv. Rep.* **73–1**, 1–45.

U. S. Water Resources Council (1981). "Guidelines for Determining Flood Flow Frequency," Bull. No. 17B. U.S. Water Resour. Counc., Washington, D.C.

Ward, R. (1978). "Floods, A Geographical Perspective." Wiley, New York.

Waylen, P. and Woo, M-K. (1982). Prediction of annual floods generated by mixed processes. *Water Resour. Res.* **18**, 1283–1286.

Waymire, E., Gupta, V. K., and Rodriguez-Iturbe, I. (1984). A spectral theory of rainfall intensity at the meso-(β) scale. *Water Resour. Res.* **20**, 1453–1465.

Willeke, G. E. (1980). Myths and uses of hydrometeorology in forecasting. *In* "Improved Hydrological Forecasting—Why and How," Proc. Eng. Found. Conf., 1979, pp. 117–124. Am. Soc. Civ. Eng., New York.

Woolhiser, D. A. (1982). Physically based models of watershed runoff. *In* (V. P. Singh, ed.), "Rainfall-Runoff Relationships" pp. 189–202. Water Resources Publications, Fort Collins, Colorado.

Yevjevich, V. M. (1963). "Fluctuations of Wet and Dry Years," Part I, Hydrol. Pap. No. 1. Colorado State University, Fort Collins.

Yevjevich, V. M. (1968). Misconceptions in hydrology and their consequences. *Water Resour. Res.* **4**, 225–232.

Yevjevich, V. M. (1972). "Stochastic Processes in Hydrology." Water Resources Publications, Fort Collins, Colorado.

Yevjevich, V. M. (1974). Determinism and stochasticity in hydrology. *J. Hydrol.* **22**, 225–238.

3

DRAINAGE BASIN MORPHOMETRY AND FLOODS

PETER C. PATTON

Department of Earth and Environmental Sciences, Wesleyan University, Middletown, Connecticut

INTRODUCTION

Drainage basins are the fundamental units of the fluvial landscape, and, accordingly, a great amount of research has focused on their geometric characteristics, including the topology of the stream networks, and the quantitative description of drainage texture, pattern, shape, and relief (Abrahams, 1984). Because drainage basins are the physical entities used to measure the volume of water and sediment produced by runoff and erosion, the analysis of basin morphometry has been extended to include the interrelationships between network characteristics and the resulting water and sediment yields. An example of this approach is the correlation of water and sediment yields with the network geometry of small streams draining diverse lithologies within the Cheyenne River basin (Hadley and Schumm, 1961). One particular goal of this research has been to define the hydrophysical significance of drainage basin characteristics in order to develop predictive models of stream runoff (e.g., Maxwell, 1960; Morisawa, 1962; Patton and Baker, 1976).

Horton, who made the first modern quantitative studies of drainage basins (Horton, 1932, 1945), provided the theoretical base for the hydrogeomorphic approach by suggesting that there were certain unvarying or intransient drainage basin characteristics that correlate to the hydrologic response of a basin. His goal was to reduce complex hydrologic phenomena to a few important quantitative measures that could then be used to predict surface runoff (Horton, 1932). The important hydrologic factors necessary to consider were (1) morphometry of the drainage net; (2) soil characteristics, particularly those related to infiltration; (3) geology as it related to structure and terrain erodibility; (4) vegetation as it affected erosion, infiltration, and surface detention; and (5) meteorologic-climatic conditions that controlled the nature of the rainfall input.

Horton was primarily interested in quantifying the first two of these factors, and his pioneering work in infiltration and the generation of overland flow led to the investigation of drainage network composition and its hydrophysical significance. Horton's seminal contribution was his description of the laws of drainage network composition (Horton, 1945). Once drainage composition was quantified, drainage network evolution could be explained through a conceptual model based on the physical processes of overland flow (Horton, 1945). His quantification of drainage networks and formulation of the laws of drainage composition spawned a generation of studies concerned with the recognition and interdependence of drainage network elements as well as the first attempts to correlate these new parameters to hydrologic phenomena (Miller, 1953; Chorley, 1957; Schumm, 1956; Melton, 1957; Maxwell, 1960; Morisawa, 1962). Horton's laws of drainage composition, subsequent minor modifications (Strahler, 1952; Bowden and Wallis, 1964), and statistical treatments of network topology (Shreve, 1966, 1967; Smart, 1969) are well known and have been discussed in detail elsewhere (Strahler, 1952, 1964; Smart, 1972; Abrahams, 1984). Figure 1 illustrates Strahler's modification of Horton's drainage network ordering scheme and the definition of other important morphometric variables discussed in this chapter.

Recently, more sophisticated models of runoff, based on the geomorphic description of the drainage basin coupled with hydrologic concepts of flood storage and flood routing have been used to create synthetic unit hydrographs. These hydrographs, which combine unit hydrograph theory with

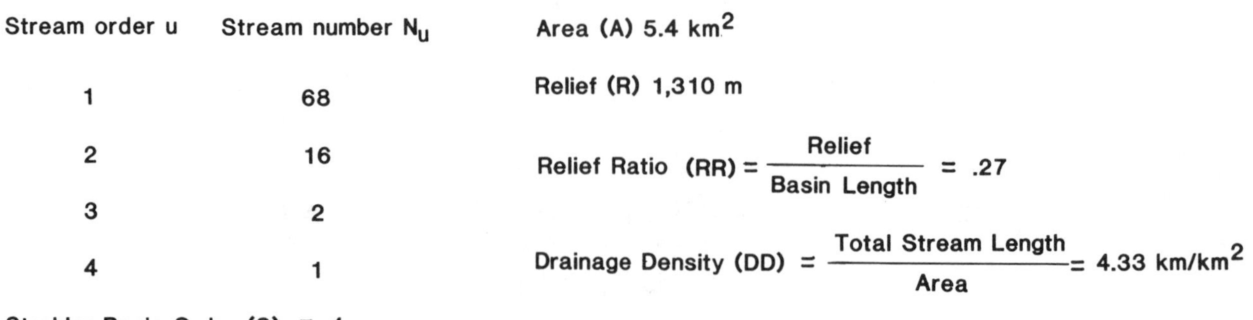

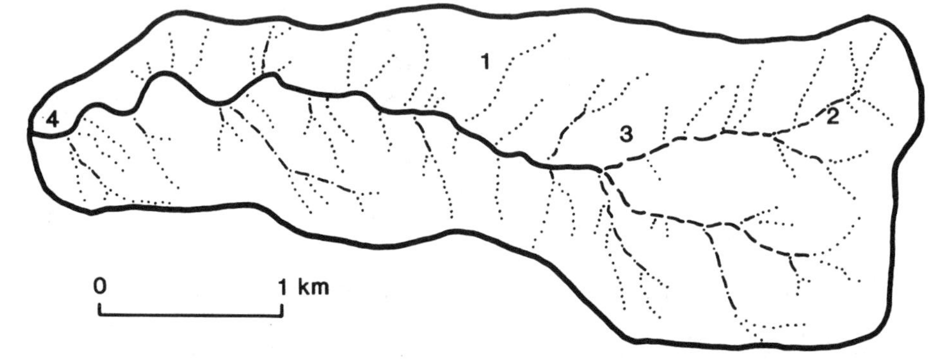

FIGURE 1. Strahler ordering system of a drainage network and definition of other morphometric parameters for Ricks Creek, Bountiful, Utah.

geomorphic parameters are termed the *geomorphic unit hydrograph*. The hydrographs are produced by combining the known probability of stream network topology with a stochastic model of storage and water transfer between stream segments to create runoff models that simulate the actual hydrologic processes.

Another approach, aimed at understanding the interrelationships between rainfall–runoff processes and drainage network morphology, has emphasized the long-term role of climate and geology in creating stream networks. The interdependence of hydrology, geology, and drainage network composition provides insight into the evolution of drainage systems and the relative geomorphic importance of hydrologic processes of differing magnitude and frequency. An understanding of the role that large infrequent flood events play in the formation of drainage networks is important to place predictive relationships on a more theoretically sound base. Unlike the predictive relationships between basin morphometry and hydrologic output, these studies attempt to analyze the various geomorphic processes that ultimately produce the drainage net.

The purpose of this chapter is to describe the relationship between basin morphometry and stream runoff, to describe models of runoff prediction based on morphometric parameters, and to demonstrate the complex interdependence between network characteristics, geomorphic processes related to flood events, and the hydrologic response of the drainage basin.

Drainage Basin Morphometry and Flood Prediction

Drainage Area–Flood Magnitude Relationships. Models of peak runoff created through regression models of drainage area and discharge date to 1865 when Dickens described the relationship between maximum flood discharge and drainage area for the Bengal region in India (Alexander, 1972). In the United states similar relationships date to the same period (Jarvis, 1936). These equations represent the first efforts to incorporate measures of the drainage system into predictive estimates of flood runoff. The equations have the form

$$Q_x = aA^b$$

indicating the power function relationship between area (A) and discharge (Q) of some recurrence interval x. Because discharge increases at a lesser rate than drainage area, the exponent is typically between 0.5 and 0.8 (Jarvis, 1936). For example, in a study of runoff and basin characteristics for the United States, Thomas and Benson (1970) found that the exponent ranged from 0.5 to 0.9 and generally decreased with increasing recurrence interval of the annual peak.

For a given region of similar physiography and climate a family of curves can be constructed relating the drainage area to the complete spectrum of flood frequency from the mean annual flood to the maximum flood of record (Thomas and Benson, 1970). However, the importance of climatic and morphometric variables is obvious when the dissimilar drainage area–flood magnitude curves from different climatic and physiographic regions are compared (Gray, 1970). Although these drainage area–flood magnitude relationships provide a rough approximation of potential flood discharge from a given size basin, they provide little insight into the physical processes of runoff generation or the evolution of drainage networks.

Drainage Density. Perhaps the most important morphometric variable invented by Horton was drainage density, the quotient of total stream length and drainage area. Drainage density is a measure of dissection that reflects the competing effectiveness of overland flow and infiltration. Horton also recognized that drainage density is an approximate measure of the length of overland flow, as one-half its inverse is the distance between a stream channel and the top of the adjacent divide. He also noted that this estimate of overland flow length is accurate only where the adjacent hillslope gradient is low, less than three times the stream gradient (Horton, 1945), and for basins with steep hillslopes a correction factor must be used for the slightly longer path of overland flow (Horton, 1945; Dingman, 1978). Horton reasoned that basins of low drainage density were the product of runoff processes dominated by infiltration and subsurface flow, whereas basins of high drainage density were the product of erosion and dissection by overland flow. Finally, for basins of comparable relief, the hydrologic response of a stream network should be directly related to drainage density because with increasing drainage density the path length of overland flow decreases while hillslope angle increases (Schumm, 1956).

Because of its apparent hydrophysical significance, drainage density has been correlated with diverse hydrologic phenomena, including stream runoff and various climatic indices. For example, drainage density has been negatively correlated with the base flow of rivers in the east-central United States (Carlston, 1963) and the Potomac River basin in Virginia (Trainer, 1963). Presumably, these correlations reflect the inverse relationship between drainage density and infiltration with greater base flow discharges associated with basins of low drainage density. Conversely, flood runoff, represented by (1) runoff intensity, the 5 yr–1 hr rainfall amount minus infiltration capacity (Melton, 1957) and by (2) the mean annual flood (Carlston, 1963), has a positive correlation that reflects the increased drainage network efficiency and more rapid hydrograph response associated with increasing drainage density.

Drainage density has also been correlated with measures of climate, including the Thornthwaites precipitation effectiveness index (Melton, 1957; Madduma Bandara, 1974) and with measures of mean annual precipitation and precipitation intensity (Abrahams and Ponczynski, 1984). The combined data on precipitation effectiveness and drainage density indicate a peak in drainage density in dry climates with little vegetative cover, a minimum in humid temperate environments because of the overriding influence of vegetation in limiting erosion, and a second region of high drainage density corresponding to the wet tropical regions where extreme rainfall totals result in extensive channel cutting (Fig. 2). General observations of mean annual precipitation with drainage density on a global scale reveal a similar trend. The trend in drainage density values with increasing mean annual precipitation is an increase in density from arid to semi-arid environments, a decrease to the lowest drainage density values in humid temperate regions, and an increase in drainage density values in the tropics (Gregory and Gardiner, 1975). These models correspond well to sediment yield curves that show high sediment production in both semi-arid regions and in the very wet, tropical, monsoonal areas (Langbein and Schumm, 1958), both regions of high drainage density.

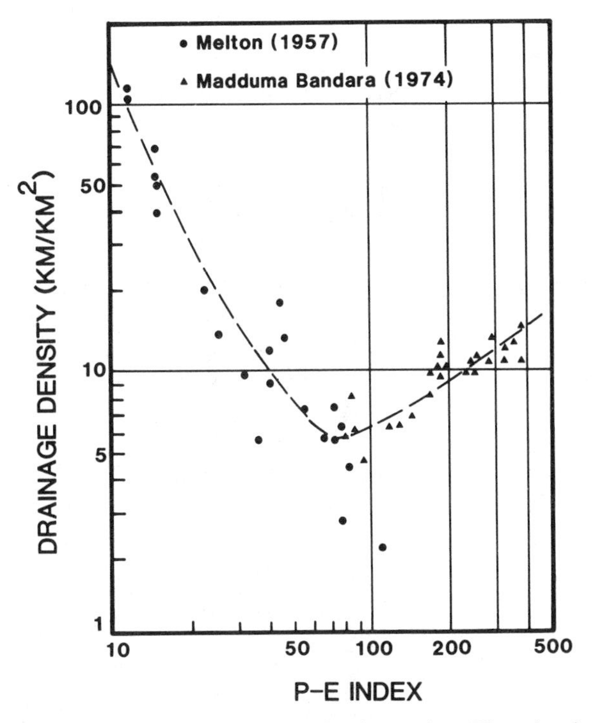

FIGURE 2. Graph of drainage density against Thornthwaites precipitation effectiveness index. Note the minimum for drainage density in humid temperate climates (modified from data in Madduma Bandera, 1974).

The correlations with actual precipitation data are less well defined. Abrahams and Ponczynski (1984) have demonstrated an inverse relationship with mean annual precipitation for drainage basins in the nonglaciated United States. Similar to the finding of Melton (1957), drainage density is highest in those environments of lowest annual precipitation. Because precipitation intensity is directly correlated with mean annual precipitation, one might expect drainage density to be negatively correlated with precipitation intensity. But, only a weak negative correlation can be found in the available data (Abrahams and Ponczynski, 1984). Nevertheless, this result is in contrast to the conclusion of Chorley and Morgan (1962) who compared the drainage density of basins in the southern Appalachian Mountains to basins in the Dartmoor region of England, both regions of similar mean annual rainfall. Their results suggested that greater precipitation intensity and greater basin slope were the reasons for the higher drainage density of the basins in the Appalachian sample. Therefore, when mean climatic conditions are similar, the magnitude and frequency distribution of individual rainfall events is the important variable that controls the relationship between drainage density and precipitation.

In spite of the above correlations and the theoretical basis for using drainage density as a measure of hydrologic response, no general model has been developed that makes use of this measure. For example, although Murphey and others (1977) demonstrated that drainage density was inversely correlated with flow duration and average hydrograph rise time, other more simplistic morphometric measures of basin size and shape were more effective in predicting the hydrologic response of the ephemeral stream basins they studied. In fact, most attempts to relate drainage density to water discharge result in low correlations that are not statistically significant (Patton and Baker, 1976). There are several problems with the use of the drainage density parameter, and they are both practical and theoretical.

One practical problem is simply that it is time consuming to interpret the drainage network from maps and photographs in order to compute drainage density. This problem has been partly solved by short-cut methods that involve estimating drainage density from the number of intersections between stream channels and a regularly spaced grid (Carlston and Langbein, 1960; McCoy, 1971; Mark, 1974). A second practical problem is associated with drainage basin interpretation, a problem that is related to the scale and resolution of the map or photograph and to the systematic method used to interpret the basin network (Morisawa, 1957; Mark, 1983). For this reason values of drainage density calculated from different sources or interpreted by different investigators for the same basin may be significantly different. This has made it difficult to develop a meaningful data base of drainage density measurements. This problem is not unique to the calculation of drainage density but exists for all the morphometric parameters that are calculated from an interpreted drainage network (Baker et al., 1975; Patton and Baker, 1976; Mark, 1983).

Perhaps one theoretical reason for the lack of success in correlating the drainage density parameter to runoff phenomena is that it is not sensitive to changes in the hydrologic response of the basin that occur during an individual storm: for example, changes in the effective drainage density, the drainage density of channels transporting runoff, changes with rainfall duration, volume, and intensity (Day, 1978). Thus a single measure of drainage density is not appropriate for all scales of runoff. For instance, Day (1978) clearly illustrated that the total length of the flowing channel network was strongly correlated with stream discharge. Similar reasoning has led to the use of drainage network volume to estimate the mean annual flood (Gregory and Ovenden, 1979), an approach that may be of considerable value in estimating flow peaks up to the bankfull stage. But as Gardiner and Gregory (1987) note, improvement of the hydrologic models using the drainage density criterion are only likely when a dynamic drainage density parameter can be coupled to a dynamic precipitation index.

Second, although Melton (1957) and Madduma Bandara (1974) demonstrated a measure of correlation between drainage density and climate, other studies have identified only a weak correlation with the prevailing climate (Abrahams and Poncynski, 1984). One reason for this may be that the development of the drainage density of a basin requires geologic time periods that are far longer than the relatively short time periods of climatic stability. Given the slow development of basin morphometry relative to climate change, it may be difficult to link a specific basin characteristic such as drainage density to a specific climatic regime. Thus, drainage basin morphometry, relict from earlier climatic regimes, may not produce deterministic hydrogeomorphic indices.

Stream Order and Basin Magnitude. Strahler stream order has been directly correlated with discharge by numerous workers (Blyth and Rodda, 1973; Stall and Fok, 1967). This high correlation is apparently related to the high correlation between drainage area and order for basins in similar climatic and geologic settings (Leopold and Miller, 1956; Patton and Baker, 1976). Also, Stall and Fok (1967) noted an increasing goodness of fit in regression analysis of discharge and stream order with decreasing frequency of the runoff event, suggesting that higher magnitude events are most important in the establishment of the drainage network. Conversely, stream order is not a sensitive measure of the topology of a basin (Shreve, 1966) in that several topologically distinct basins can exist for a given Strahler order. Because topologically distinct basins should have different hydrograph responses, a product of the different paths of water runoff, it is not surprising that other measures of basin order are more appropriate measures of flood discharge (Patton and Baker, 1976).

Shreve's (1967) method of ordering has been advocated by a number of investigators as being more descriptive of network form in relation to streamflow. The number of first-order streams (Shreve Magnitude) has been correlated with the peak discharge of streams on the Appalachian Plateau (Morisawa, 1962) and central Texas streams (Patton and Baker, 1976). In a study of morphometric parameters related to stream discharge, Shreve magnitude was the most significant variable correlated with flood discharge for widely diverse physiographic regions (Table 1) (Patton and Baker, 1976). The reason for this is the importance of first-order streams to the total number and length of streams in a basin during high flows. For example, Blyth and Rodda (1973) made a year-long study of runoff and length of flowing streams in a small catchment in southeast England. They documented that during dry periods the number of flowing first-order streams constituted less than 20% of the total flowing length of the network (Fig. 3*a*). Conversely, during periods of high flow, when the maximum extent of the network was flowing, the total length of first-order streams constituted nearly 50% of the total stream length (Fig. 3*a*). Thus, there is a dramatic shift in the stream number relationship for flowing streams within a basin as a function of discharge (Fig. 3*b*). These results suggest that measures of the maximum extent of the drainage network may be the best indices for large flood events, although the frequency of these network filling floods will vary regionally.

Drainage Basin Relief. The importance of basin relief and measures of basin slope as hydrologic parameters has long been recognized (Sherman, 1932, Horton, 1945; Strahler, 1964). With increasing relief, steeper hillslopes, and higher stream gradients, time of concentration of runoff decreases, thereby increasing flood peaks. For example, in the eastern and southern United States, Thomas and Benson (1970) found main-channel slope to be the second variable selected in stepwise multiple regression analysis of annual peak flows of all recurrence intervals. Also, in a comparison of morphometric parameters from basins of differing hydrologic response and diverse physiographic regions, two dimensionless measures of relief were significant variables in regression analyses with the maximum flood of record (Table 1) (Patton and Baker, 1976). The measures of relief were the relief ratio, the total basin relief divided by the length of the basin (Schumm, 1956), and ruggedness number, the product of drainage density and relief (Melton, 1957). Ruggedness number is particularly useful because it summarizes the interaction of relief and dissection such that highly dissected basins of low relief are as rugged as moderately dissected basins of high relief.

FLOOD HYDROGRAPHS AND BASIN MORPHOMETRY

Strahler (1964) proposed that hydrograph shape should be partly controlled by drainage basin morphometry. He speculated that basin shape and the number and internal arrangement of stream segments, as measured by changes in the bifurcation ratio, would produce noticeable changes in stream hydrographs, all other conditions being equal (Fig. 4). He suggested two extreme examples: Rotund basins with low bifurcation ratios and nearly equal path lengths of water flow would have sharp hydrograph peaks whereas elongate basins with high bifurcation ratios and greatly unequal flow path lengths would have lower hydrograph peaks but more sustained flow. The time to hydrograph peak would also vary, with the elongate basin having a shorter lag time because of the short travel time

TABLE 1 Regression Formulas for Predicting Flood Magnitudes from Drainage Basin Morphometry in Diverse Hydrogeomorphic Regions

Region	Equation	R^2	Probability
Central Texas	$Q_{max} = 17{,}369M^{0.43}(HD)^{0.54}F_1^{-0.96}$	0.85	0.001
	$Q_{max} = 36{,}650M^{0.64}(RR)^{0.54}(DD)^{-1.68}$	0.74	0.01
Southern California	$Q_{max} = 155M^{1.04}(HD)^{-0.83}F_1^{-0.73}$	0.85	0.001
	$Q_{max} = 380M^{0.89}(DD)^{-1.87}$	0.86	0.0001
North-central Utah	$Q_{max} = 23M^{0.90}(HD)^{1.19}F_1^{-1.58}$	0.72	0.005
	$Q_{max} = 38{,}618M^{2.20}(RR)^{2.51}F_1^{-3.73}$	0.83	0.005
Indiana	$Q_{max} = 424M^{0.46}(HD)^{0.73}F_1^{0.21}$	0.67	0.01
	$Q_{max} = 424M^{0.82}(RR)^{0.67}(DD)^{0.56}$	0.66	0.05
Appalachian Plateau	$Q_{max} = 100M^{0.79}(HD)^{0.19}F_1^{-0.29}$	0.92	0.0001
	$Q_{max} = 38M^{0.89}(DD)^{-0.50}$	0.91	0.0001

Symbols are as follows: M, basin magnitude; HD, ruggedness number; F_1, first-order channel frequency; DD, drainage density; RR, relief ratio; Q_{max}, maximum peak discharge (from Patton and Baker, 1976).

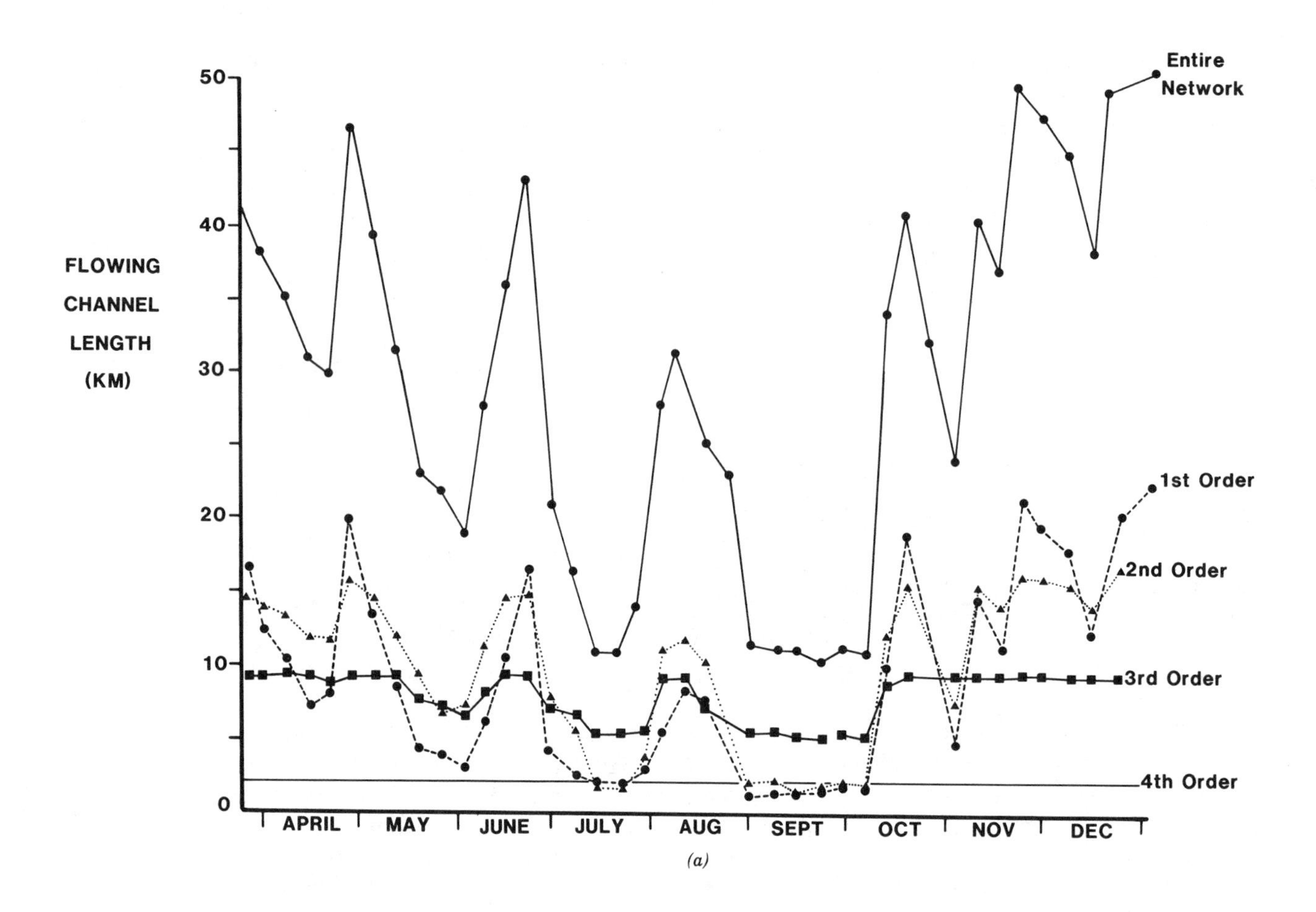

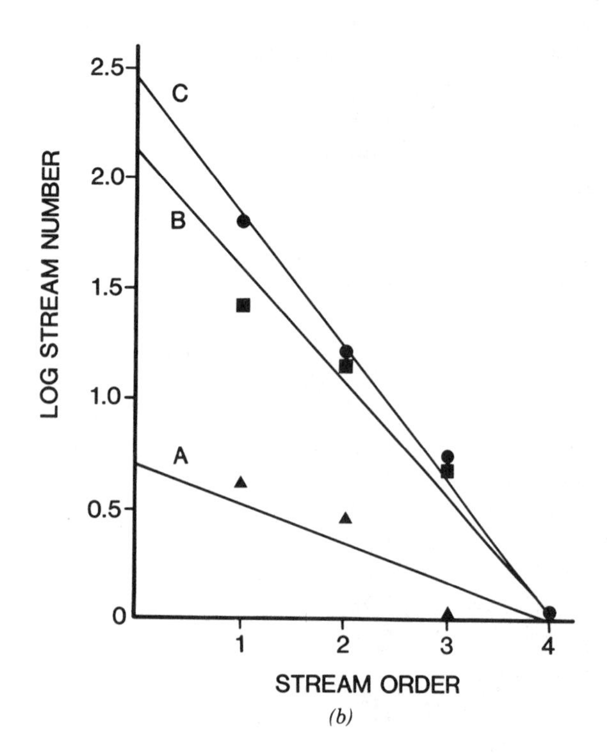

FIGURE 3. (*a*) Weekly changes in the lengths of flowing streams in the River Bay catchment at Grendon Underwood, April to December 1971. (*b*) Variations in the order number relationship as a function of stream discharge. Line A is the flowing network on June 14; line B is for May 12; and line C represents the maximum extent of the network (modified from Blyth and Rodda, 1973).

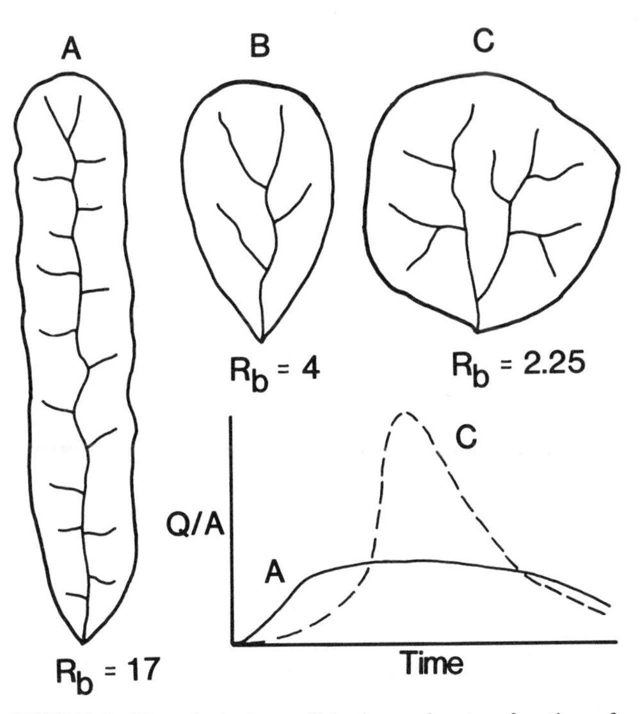

FIGURE 4. Hypothetical runoff hydrographs as a function of basin shape and bifurcation ratio (modified from Strahler, 1964).

of stream segments near the basin outlet. One shortcoming of this hypothetical model is the low variability and high stability of the bifurcation ratio even when computed for basins of greatly dissimilar pattern. Nevertheless, Strahler's (1964) general ideas about hydrograph shape and basin morphometry remain intuitively compelling, and data derived from study of unit hydrographs have begun to provide insight into the nature of these relationships.

The instantaneous unit hydrograph is the hydrograph generated by a specific depth of effective rainfall, usually one inch, uniformly distributed over a drainage basin during a specified unit of time (Sherman, 1932). Therefore unit hydrographs of similar time interval from different drainage basins have a constant hydrologic input and can be analyzed in terms of the physical geomorphic controls on hydrograph shape. Two important hydrologic measures derived from unit hydrograph theory are the lag time of the basin (the time between the centroid of the rainfall distribution and the peak runoff) and the magnitude of the hydrograph peak (Sherman, 1932). Basin characteristics, such as basin area and shape, drainage basin hypsometry, and mainstem channel length and slope, have been demonstrated to be important controls on unit hydrograph parameters and therefore on the resulting hydrograph shape (Sherman, 1932; Taylor and Schwartz, 1952; Heerdegen and Reich, 1974; Harlin, 1984). Because of the interrelationships with morphologic parameters synthetic unit hydrograph models, such as the Snyder unit graph (Snyder, 1938), have been constructed using basin characteristics to estimate the lag time and hydrograph peak for the model hydrograph. Several recent studies provide a more detailed look at the interrelationship between drainage basin morphometry and unit hydrograph shape.

In a model study of stream network development Parker (1977) was able to measure unit hydrographs for different geomorphic stages of a single basin. The study basin evolved through headward erosion because periodic base-level lowering resulted in increased drainage density and basin relief. Data on peak discharge and hydrograph lag time were compared with the changing physical condition of the basin. Because of the rapid erosional development of the basin in its early history, the hydrograph data were collected for only the more developed networks that filled the entire available basin area. The results indicated that relative peak discharge increased with increased drainage basin ruggedness (Fig. 5). The increased peak discharge was the result of the improved efficiency of the network caused by the increased relief and drainage density. At the same time basin lag time decreased as drainage density increased, reflecting the shorter path length of overland flow. Lag time also decreased as rainfall intensity increased, and the magnitude of this decrease was greatest for networks with low drainage density (Fig. 6). This last result indicates

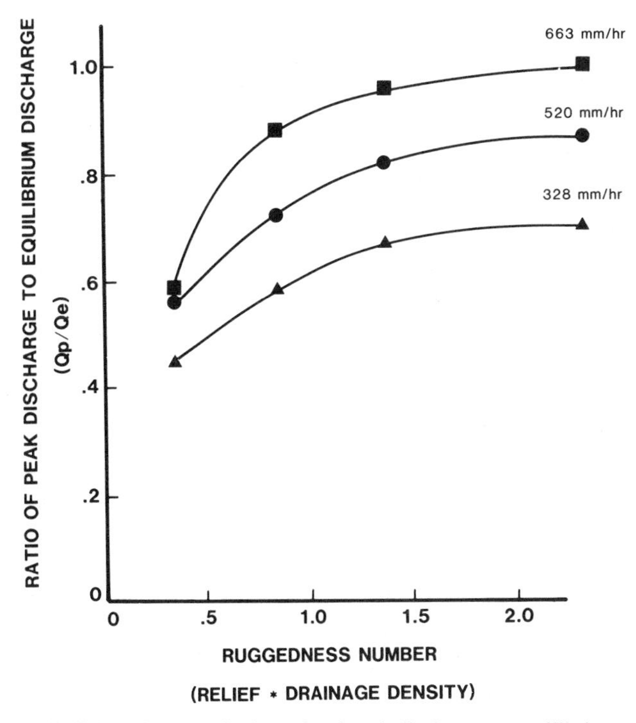

FIGURE 5. Changes in the ratio of peak discharge to equilibrium discharge rate (Q_p/Q_e) for changes in the ruggedness number. The equilibrium discharge rate is the discharge for rainfall of infinite duration while Q_p represents the peak discharge for a storm with a duration of 1 min (from data in Parker, 1977).

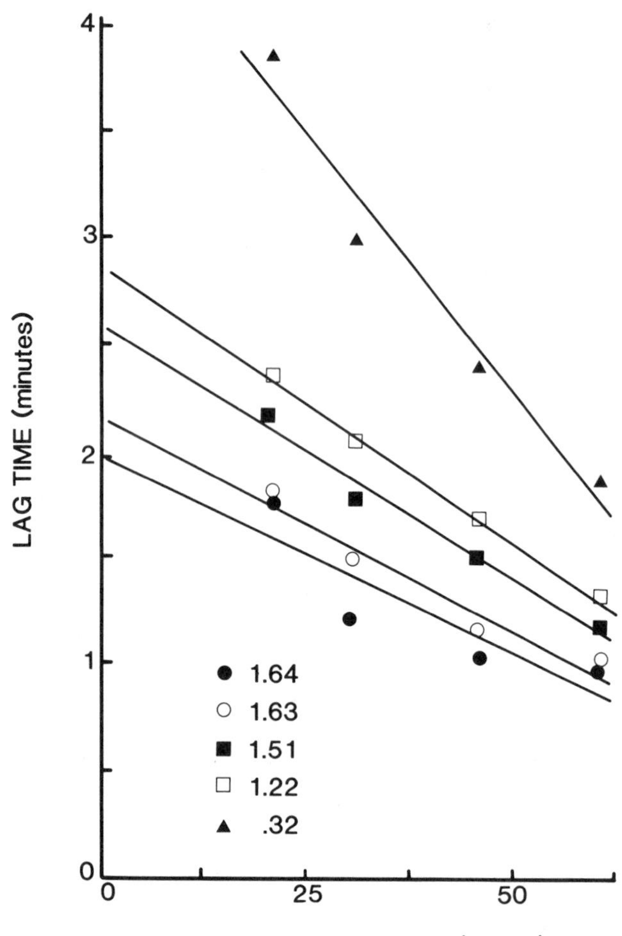

FIGURE 6. Changes in lag time of the instantaneous unit hydrograph as a function of rainfall intensity for each study basin. Values of drainage density in meters per meters squared for each network are listed on the figure. The degree of nonlinearity in the hydrograph response increases with increasing slope of the regression lines, and is greatest for the basins of lowest drainage density (from data in Parker, 1977).

that there is a nonlinear relationship between basin lag time and rainfall intensity and that this nonlinear response increases with increasing distance of overland flow. Therefore, in basins of low drainage density, the assumption of a linear unit hydrograph response for storms of varying intensity is not valid (Parker, 1977).

Using data on hydrograph lag time from a series of gauging stations nested within a larger basin, Boyd (1978) was able to demonstrate the geometric relationship between stream order and lag time (Fig. 7). The resulting equation

$$lag_u = lag_1 \cdot R_k^{u-1}$$

where the lag time of a basin of order u is equal to the product of the lag time of the first-order basin and the lag

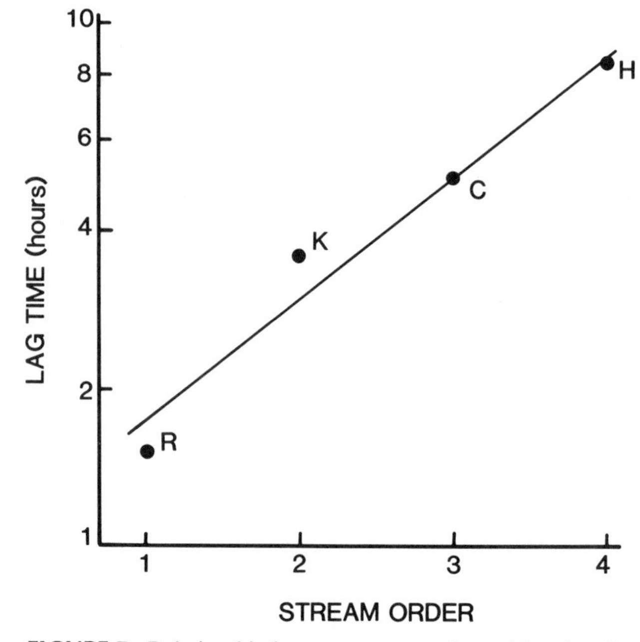

FIGURE 7. Relationship between stream order and lag time for the Hacking drainage basins (modified from Boyd, 1978). Letters refer to study basins listed on Figure 8.

ratio raised to the $u - 1$ power. The equation is similar in form to the equations for the geometric series of stream length (Horton, 1945) and stream areas (Schumm, 1956) as a function of basin order. Thus a constant average basin lag ratio can be calculated that will allow the prediction of all lag times of a basin based on the order of the basin.

Variability about the average lag ratio for measured runoff events for each gauging station approximates a lognormal distribution with positive skew (Fig. 8) similar to

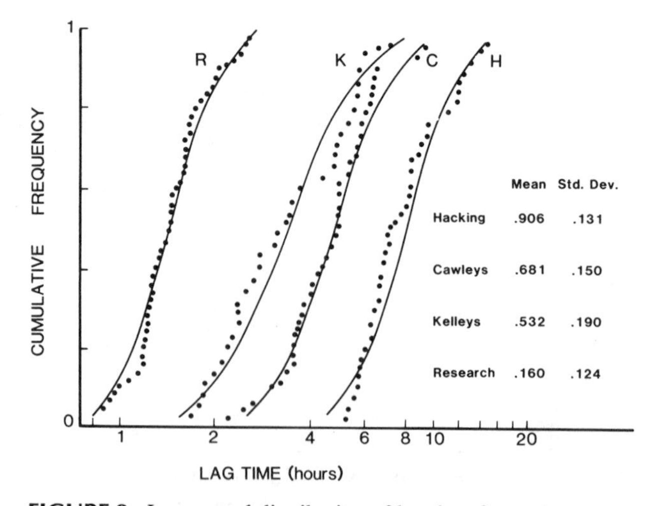

FIGURE 8. Lognormal distribution of lag time for each gauging station within the Hacking drainage basins (modified from Boyd, 1978).

the frequency distribution of both basin area and stream lengths of a given stream order (Boyd, 1978). Implicit in this relationship, but untested with actual field data, is the suggestion that the lag times of all subbasins of a given order would be distributed as the normal probability density function, a hypothesis important in assigning storage times in modeling basin runoff. Boyd developed correlations of lag time with basin magnitude and basin area to create a storage routing model.

The concept of varying lag time and storage within each element of the drainage network can be combined with a unit hydrograph model that utilizes the geometry of the real drainage network. This model, known as the geomorphic unit hydrograph (GUH), can be conceptually visualized as the cumulative time history of the advance of each randomly placed drop of effective rainfall (Rodriquez-Iturbe and Valdes, 1979; Hebson and Wood, 1982). The model parameters of the GUH are the average first-order stream length (L_1) and order ratios for area (R_A), bifurcation (R_B), and stream length (R_L) where

$$R_A = A_u / A_{u-1}$$

$$R_B = N_u / N_{u-1}$$

$$R_L = L_u / L_{u-1}$$

where A, L, and N equal the drainage area, stream length, and number of each stream segment of order u.

The GUH considers each stream order to represent a state of residence or waiting time for each drop of water, while stream junctions represent changes in state. The path an individual water drop will take is then defined by the probability of each change in state based on the probability of different stream junctions of the real drainage networks. Thus the model takes into account the topologically distinct networks within each Strahler order.

The major assumption of the GUH concerns the holding time of water within each stream order. Unfortunately, there is little actual field data on which to construct a general model of storage in each stream segment. The average holding time is assumed to be the average length of each order stream divided by the peak flow velocity where the peak flow velocity simplifies the hydrodynamics of the rainfall–runoff process to a single term (Hebson and Wood, 1982). For the purpose of the GUH the peak flow velocity is considered a constant for a stream network based on actual measurements of travel times and flood storage based on tracer studies (Pilgrim, 1976, 1977) and the low value of the exponent in the hydraulic geometry expressions for change in velocity in a downstream direction (Leopold and Maddock, 1953). The variation in holding time for a given order stream is approximated as an exponential density function, an assumption that is partially supported by the observed distribution of basin lag times for a single stream gauging station (Boyd, 1978).

The output of the GUH is calculated by determining the probability that a particle of water follows a certain path, multiplying that probability by the probability density function of the waiting time for that path, and then summing these products over all possible paths (Rodriquez-Iturbe and Valdes, 1979; Gupta et al., 1980; Hebson and Wood, 1982).

Simulations of the GUH for various network configurations of third-order drainage basins demonstrated that hydrograph shape is related to network parameters. In general, lag times decreased and hydrograph peaks increased as the average length of first-order streams decreased and as the basin area ratio and bifurcation ratio increased (Fig. 9). No attempt was made to separate these variables to determine their relative impact on the hydrograph, but it is clear that basins characterized by large numbers of short, lower order streams that flow from many small basins into a few large subbasins will have a flashy hydrograph response. Conversely, basins characterized by a relatively few, long, lower order streams and a more conservative increase in the number and size of basins will have a more sluggish hydrograph response. No clear trend was evident with stream length ratio. The simulations also indicate that the GUH model is extremely sensitive to changes in the waiting time values, which are based on estimates of the peak flow velocity.

In a test using actual rainfall–runoff hydrographs, the GUH was successful in modeling stream runoff for basins in excess of 2500 km^2 (Fig. 10*a*) but underestimated the peak discharge for a basin of 880 km^2 (Fig. 10*b*). This flaw may be the result of the assumption of a linear transfer between rainfall and runoff and the inability to precisely define the nature of the probability density function of waiting time (Gupta et al., 1980). Clearly, if the GUH is going to evolve into a successful method for estimating flood discharge, more field data on storage and travel time within a drainage network are needed.

DRAINAGE NETWORK CHANGE AND FLOODS

Although drainage basins and stream networks evolve slowly, it is possible to study their long-term evolution through scale models (Parker, 1977), probabilistic simulation models (Leopold and Langbein, 1962), and occasionally, given an adequate data base or unique geologic setting, through field studies (Ruhe, 1952; Schumm, 1956; Morisawa, 1964; Hack, 1965; Edgar and Melhorn, 1974). Observations made after rare large flood events provide additional insight into the processes by which drainage networks change.

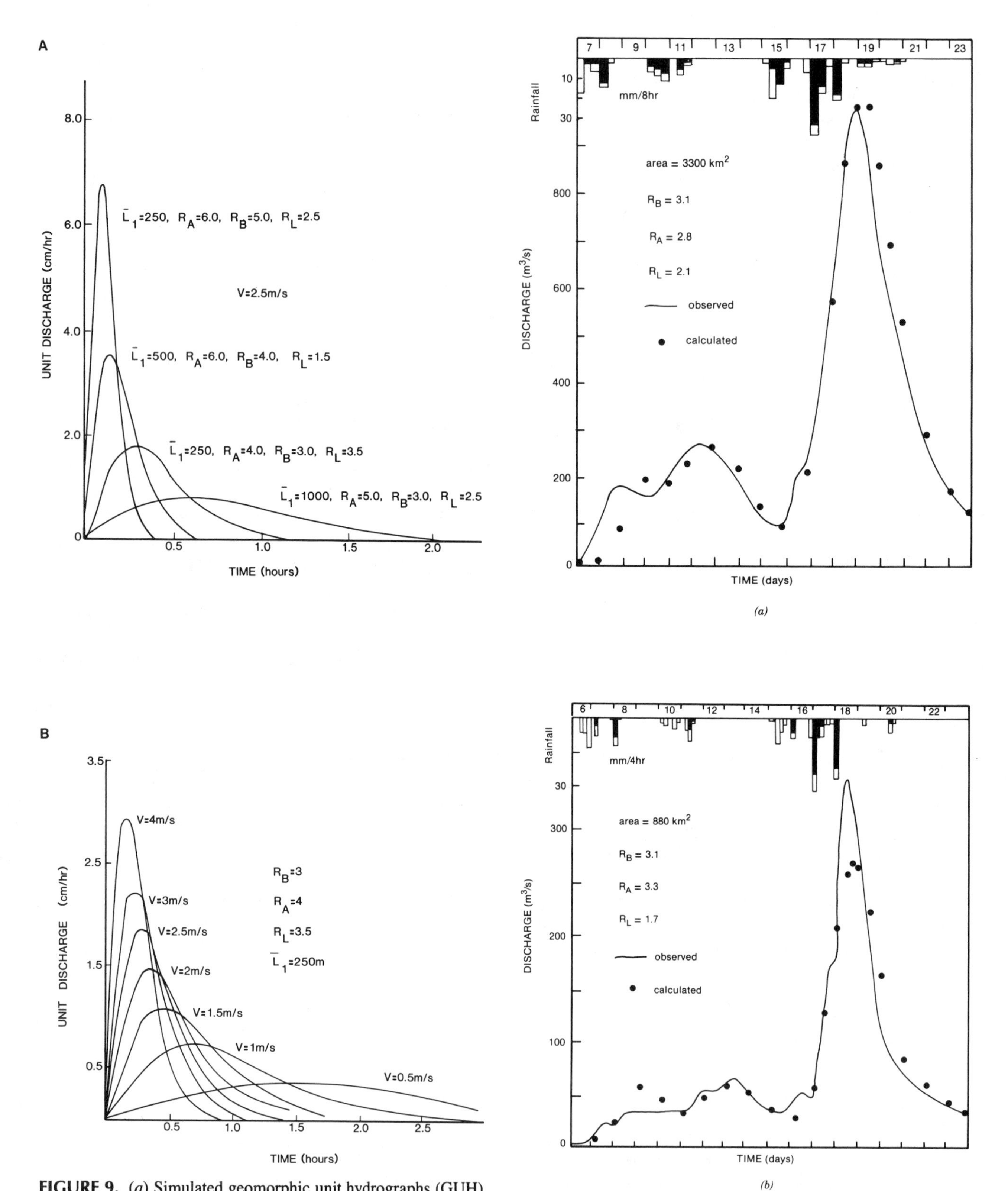

FIGURE 9. (*a*) Simulated geomorphic unit hydrographs (GUH) for a third-order basin with constant flow velocity but with varying geomorphic characteristics. L is in meters. (*b*) Simulated GUH for a third-order basin with varying flow velocity and constant geomorphic characteristics (modified from Rodriquez-Iturbe and Valdes, 1979).

FIGURE 10. Geomorphic unit hydrographs fitted to observed hydrographs for (*a*) Vermillion River basin at Danville and (*b*) the Sangamon River basin at Rewell, Illinois. Note the underestimated peak for the Sangamon River (modified from Gupta et al., 1980).

FIGURE 11. Aerial photograph of debris avalanches in the Little River basin. Photograph taken in 1955 still clearly shows the debris avalanches formed 6 yr earlier.

In areas of high relief one of the most spectacular forms of drainage network modification are debris avalanches, which often originate on hillslopes in small first- and second-order basins. Debris avalanches are recognized in all climatic regions and are not only significant geomorphic processes but are also devastating natural hazards. For example, during the heavy summer rains of January 1967, a tremendous cloudburst over the Serra das Araras escarpment in Rio de Janeiro, Brazil, triggered hundreds of debris avalanches (Jones, 1973). During this colossal mass movement event and flood, over 1000 people died. Also, Williams and Guy (1973) noted that of the 125 people who died during the Hurricane Camille floods, few died of drowning but instead were killed by injuries caused by debris avalanches.

One of the first descriptions of debris avalanches was made by Cleland (1902) who described the erosion on the slopes of Mt. Greylock, Massachusetts, following intense rainfall. The debris avalanches created shallow scars, or chutes, on the hillslopes ranging from 15 to 70 m wide and up to 500 m long. The chutes were scoured to bedrock, forming new first-order channels, and the eroded sediment was deposited as alluvial fans over the adjacent lowlands.

Hack and Goodlett (1960) studied the debris avalanches and chutes created during the June 1949 flood in the Little River basin in the Appalachian Mountains of western Virginia. Within this high-relief basin nearly 50% of the drainage area consisted of first-order basins, and many of these small basins headed in theaterlike hollows rimmed by steep slopes. The 1949 flood, which resulted from over 200 mm of rainfall, saturated the hillslopes and caused more than 100 landslides, most of which occurred on the slopes of the small first-order channels (Fig. 11). The chutes were approximately a meter deep, some headed near the slope crest, and all extended to the base of the slope and intersected the pre-existing channel. Many of the chutes developed in slight depressions on the hillslopes, which served to concentrate runoff. Six years after the flood event the chutes still appeared quite fresh and unaltered, and older chutes, created during previous floods and revegetated, could also be recognized. They concluded that the chutes represent the incipient stage of drainage extension by the creation of new hollows and new first-order streams.

Similar chutes and debris avalanches were created during floods in the James River basin in Virginia during Hurricane Camille in 1969 (Williams and Guy, 1973). The chutes developed on the steepest part of the slope profile and were preferentially developed on slopes with a north, northeast, or east aspect. Greater antecedent moisture conditions or the direction of wind-driven rain may have accounted for this preferential erosion. Like the chutes in the Little River drainage, the chutes were primarily created in the first- or second-order drainage basins and represented an extension of the drainage net.

In low-relief basins the effects of large floods on the drainage net are not well documented. Edgar and Melhorn (1974) demonstrated that over a 30-yr period between 1938 and 1968 there were measurable changes in the morphometry of several fourth- and fifth-order streams eroded into the low-relief till plains of west-central Indiana. However, the change was not uniform, with fourth-order basins losing channel segments and fifth-order basins experiencing network extension. The reasons for the change in network geometry are complex and involve long-term fluctuations in rainfall and runoff, soil characteristics, land-use practices, and the threshold and feedback processes associated with erosion and deposition in drainage basins. Unlike the rapid episodic change in network geometry in basins of high relief, drainage basin evolution in low-relief regions is apparently progressive. In order to evaluate this progressive change, more detailed long-term records are needed of network geometry and of the prevailing hydrologic and geomorphic processes.

The lasting impact of erosion by large floods on the extension and increased hydrologic efficiency of the drainage net is controlled by the recovery time of the basin (Wolman and Gerson, 1978). If basin network extension, through the creation of new first-order channels, occurs at a greater rate than processes of channel abstraction, then the response time of the basin should decrease, thereby creating a positive feedback system of enhanced flood discharge. Such a positive feedback system was suggested for small drainage basins on the Edwards Plateau in central Texas (Patton and Baker, 1976; Baker, 1977). There, small hillslope rills, apparently created by extreme rainfall events, are not abstracted from the drainage net by colluvial processes, but instead persist from one major storm to the next. In contrast, the data from small basins in low-relief regions such as Indiana (Edgar and Melhorn, 1974) indicate that abstraction of stream channels is an important process that may be tied to short-term fluctuations in rainfall and runoff. In these basins the recovery time is short because of the effectiveness of low-magnitude, high-frequency processes that fill in first-order channels with colluvium. Over historical time periods this may lead to a steady-state system with only slight variation in stream numbers about a mean, but with a varying network topology (Schumm, 1956). Finally, a negative feedback process can be envisaged where reduced relief and increased soil development cause a shift from overland flow to infiltration and subsurface storm flow with the final result being abstraction of the drainage network. Although this last model is the most hypothetical, model studies (Parker, 1977) support the trend toward abstraction with decreasing relief.

CONCLUSION

Drainage basin morphometry has been related to a wide range of hydrologic phenomena including flood discharges of varying frequency and magnitude. Recent developments have included the incorporation of unit hydrograph theory with basin topology to generate hydrographs that result from the routing of water through a unique drainage basin. The success of these models has been mixed and, in general, two problems have limited the extension of these methods. The first is a still incomplete understanding of the interrelationship between rainfall–runoff events of varying magnitude and frequency and the composition of the drainage net. That is, how does the actual contributing drainage net change with rainfall intensity and duration, and at what point does the entire drainage net contribute runoff? Answers to these questions will improve our understanding of the interrelationship between climate and morphometry as well as improve our ability to predict stream runoff from basin characteristics. The second problem is highlighted by the geomorphic unit hydrograph, a methodology that appears to have considerable potential. A lack of understanding of the distribution of lag time as a function of stream order within a basin, the nonlinear response of basins of low drainage density with varying rainfall intensity, and the sensitivity of the GUH to stream velocity limit the effectiveness of the present model. One avenue of future research should be the analysis of these phenomena for small, well-instrumented basins.

Finally, the evolution of drainage basins is governed by the interplay of flood events that work to extend the drainage net and the recovery time of the basin, which is controlled by climate and geology. Where recovery time is long, drainage basin morphometry will reflect extreme flood events, and network extension will occur. Conversely, where network recovery time is short, the drainage net will result from high-frequency flows. In these basins network extension and abstraction will result in a steady-state network of varying topology.

ACKNOWLEDGMENTS

I am grateful to V. R. Baker and V. Gardiner for their comments on an earlier version of this chapter.

REFERENCES

Abrahams, A. D. (1984). Channel networks: A geomorphological perspective. *Water Resour. Res.* **20**, 161–168.

Abrahams, A. D., and Ponczynski, J. J. (1984). Drainage density in relation to precipitation intensity in the U.S.A. *J. Hydrol.* **75**, 383–388.

Alexander, G. N. (1972). Effect of catchment area on flood magnitude. *J. Hydrol.* **16**, 225–240.

Baker, V. R. (1977). Stream-channel response to floods, with examples from central Texas. *Geol. Soc. Am. Bull.* **88**, 1057–1071

Baker, V. R., Holz, R. K., Hulke, S. D., Patton, P. C., and Penteado, M. M. (1975). "Stream Network Analysis and Geomorphic Flood Plain Mapping from Orbital and Suborbital Imagery: Application to Flood Hazard Studies in Central Texas," NASA Final Proj. Rep., Contract NAS 9-13312. N.T.I.S., Springfield, Virginia.

Blyth, K., and Rodda, J. C. (1973). A stream length study. *Water Resour. Res.* **9**, 1454–1461.

Bowden, K. L., and Wallis, J. R. (1964). Effect of stream-ordering technique on Horton's laws of drainage composition. *Geol. Soc. Am. Bull.* **75**, 767–774.

Boyd, M. J. (1978). A storage-routing model relating drainage basin hydrology and geomorphology. *Water Resour. Res.* **14**, 921–928.

Carlston, C. W. (1963). Drainage density and streamflow. *Geol. Surv. Prof. Pap. (U.S.)* **442C**, 1–8.

Carlston, C. W., and Langbein, W. B. (1960). Rapid approximation of drainage density: Line intersection method. *U. S. Geol. Surv. Water Resour. Div. Bull.* **11**.

Chorley, R. J. (1957). Climate and morphometry. *J. Geol.* **65**, 628–668.

Chorley, R. J., and Morgan, M. A. (1962). Comparison of morphometric features, Unaka Mountains, Tennessee and North Carolina, and Dartmoor, England. *Geol. Soc. Am. Bull.* **73**, 17–34.

Cleland, H. F. (1902). The landslides of Mt. Greylock and Briggsville, Mass. *J. Geol.* **10**, 513–517.

Day, D. G. (1978). Drainage density changes during rainfall. *Earth Surf. Processes* **3**, 319–326.

Dingman, S. L. (1978). Drainage density and streamflow; a closer look. *Water Resour. Res.* **14**, 1183–1187.

Edgar, D. E., and Melhorn, W. N. (1974). Drainage basin response: Documented historical change and theoretical considerations. *Purdue Univ., Water Resour. Res. Cent. Stud. Fluvial Geomorphol.* **3**, 1–196.

Gardiner, V., and Gregory, K. J. (1987). Drainage density in rainfall-runoff modelling. In "Rainfall-Runoff Relationships" (V. Singh, ed.), pp. 449–476. Water Resources Publications, Fort Collins, Colorado.

Gray, D. M., ed. (1970). "Handbook on the Principles of Hydrology." Water Information Center, Inc., Port Washington, New York.

Gregory, K. J., and Gardiner, V. (1975). Drainage density and climate. *Z. Geomorphol.* **19**, 287–298.

Gregory, K. J., and Ovenden, J. C. (1979). Drainage network volumes and precipitation in Britain. *Trans. Inst. Br. Geogr., New Ser.* **4**, 1–11.

Gupta, V., Waymire, E., and Wang, C. (1980). A representation of an instantaneous unit hydrograph from geomorphology. *Water Resour. Res.* **16**, 855–862.

Hack, J. T. (1965). Postglacial drainage evolution and stream geometry in the Ontonagon area, Michigan. *Geol. Surv. Prof. Pap. (U.S.)* **504-B**, 1–40.

Hack, J. T., and Goodlett, J. C. (1960). Geomorphology and forest ecology of a mountain region in the central Appalachians. *Geol. Surv. Prof. Pap. (U.S.)* **347**, 1–66.

Hadley, R. F., and Schumm, S. A. (1961). Hydrology of the upper Cheyenne River basin. *Geol. Surv. Water-Supply Pap. (U.S.)* **1531-B**, 186–198.

Harlin, J. M. (1984). Watershed morphometry and time to hydrograph peak. *J. Hydrol.* **67**, 141–154.

Hebson, C., and Wood, E. F. (1982). A derived flood frequency distribution using Horton order ratios. *Water Resour. Res.* **18**, 1509–1518.

Heerdegen, R. G., and Reich, B. M. (1974). Unit hydrographs for catchments of different sizes and dissimilar regions. *J. Hydrol.* **22**, 143–153.

Horton, R. E. (1932). Drainage basin characteristics. *Trans., Am. Geophys. Union* **13**, 350–361.

Horton, R. E. (1945). Erosional development of streams and their drainage basins: Hydrophysical approach to quantitative morphology. *Geol. Soc. Am. Bull.* **56**, 275–370.

Jarvis, C. S. (1936). Floods in the United States. *Geol. Surv. Water-Supply Pap. (U.S.)* **771**, 1–497.

Jones, F. O. (1973). Landslides of Rio de Janeiro and the Serra das Araras escarpment, Brazil. *Geol. Surv. Prof. Pap. (U.S.)* **697**, 1–42.

Langbein, W. B., and Schumm, S. A. (1958). Yield of sediment in relation to mean annual precipitation. *Trans., Am. Geophys. Union* **39**, 1076–1084.

Leopold, L. B., and Langbein, W. B. (1962). The concept of entropy in landscape evolution. *Geol. Surv. Prof. Pap. (U.S.)* **500-A**, 1–20.

Leopold, L. B., and Maddock, T., Jr. (1953). The hydraulic geometry of stream channels and some physiographic implications. *Geol. Surv. Prof. Pap. (U.S.)* **252**, 1–57.

Leopold, L. B., and Miller, J. P. (1956). Ephemeral streams—hydraulic factors and their relation to the drainage net. *Geol. Surv. Prof. Pap. (U.S.)* **282-A**, 1–37.

Madduma Bandara, C. M. (1974). Drainage density and effective precipitation. *J. Hydrol.* **21**, 187–190.

Mark, D. M. (1974). Line intersection method for estimating drainage density. *Geology* **2**, 235–236.

Mark, D. M. (1983). Relations between field-surveyed channel networks and map-based geomorphometric measures, Inez, Kentucky. *Ann. Assoc. Am. Geogr.* **73**, 358–372.

Maxwell, J. C. (1960). "Quantitative Geomorphology of the San Dimas Experimental Forest, California, Tech. Rep. No. 19. Dept. of Geology, Columbia University, New York.

McCoy, R. M. (1971). Rapid measurement of drainage density. *Geol. Soc. Am. Bull.* **82**, 757–762.

Melton, M. A. (1957). "An Analysis of the Relation among Elements of Climate, Surface Properties and Geomorphology," Tech. Rep. No. 11. Dept. of Geology, Columbia University, New York.

Miller, V. C. (1953). "A Quantitative Geomorphic Study of Drainage Basin Characteristics in the Clinch Mountain Area Virginia and Tennessee," Tech. Rep. No. 3. Dept. of Geology, Columbia University, New York.

Morisawa, M. E. (1957). Accuracy of determination of stream lengths from topographic maps. *Trans., Am. Geophys. Union* **38**, 86–88.

Morisawa, M. E. (1962). Quantitative geomorphology of some watersheds in the Appalachian Plateau. *Geol. Soc. Am. Bull.* **73**, 1025–1046.

Morisawa, M. E. (1964). Development of drainage systems on an upraised lake floor. *Am. J. Sci.* **262**, 340–354.

Murphey, J. B., Wallace, D. E., and Lane, L. J. (1977). Geomorphic parameters predict hydrograph characteristics in the southwest. *Water Resour. Bull.* **13**, 25–38.

Parker, R. S. (1977). "Experimental Study of Drainage Basin Evolution and Its Hydrologic Implications," Hydrol. Pap., No. 90. Colorado State University, Fort Collins.

Patton, P. C., and Baker, V. R. (1976). Morphometry and floods in small drainage basins subject to diverse hydrogeomorphic controls. *Water Resour. Res.* **12**, 941–952.

Pilgrim, D. H. (1976). Travel times and nonlinearity of flood runoff from tracer measurements on a small watershed. *Water Resour. Res.* **12**, 487–496.

Pilgrim, D. H. (1977). Isochrones of travel time and distribution of flood storage from a tracer study on a small watershed. *Water Resour. Res.* **13**, 587–595.

Rodriquez-Iturbe, I., and Valdes, J. B. (1979). The geomorphologic structure of hydrologic response. *Water Resour. Res.* **15**, 1409–1420.

Ruhe, R. V. (1952). Topographic discontinuities of the Des Moines lobe. *Am. J. Sci.* **250**, 46–56.

Schumm, S. A. (1956). Evolution of drainage systems and slopes in badlands at Perth Amboy, New Jersey. *Geol. Soc. Am. Bull.* **67**, 597–646.

Sherman, L. K. (1932). The relation of hydrographs of runoff to size and character of drainage basins. *Trans., Am. Geophys. Union* **13**, 332–339.

Shreve, R. L. (1966). Statistical law of stream numbers. *J. Geol.* **74**, 17–37.

Shreve, R. L. (1967). Infinite topologically random channel networks. *J. Geol.* **77**, 397–414.

Smart, J. S. (1969). Topological properties of channel networks. *Geol. Soc. Am. Bull.* **80**, 1757–1774.

Smart, J. S. (1972). Channel networks. *Adv. in Hydrosci.* **8**, 305–346.

Snyder, F. F. (1938). Synthetic unit graphs. *Trans., Am. Geophys. Union* **19**, 447–454.

Stall, J. B., and Fok, Y. S. (1967). Discharge as related to stream system morphology. *In* "Symposium on River Morphology," pp. 224–235. Int. Assoc. Sci. Hydrol., Bern.

Strahler, A. N. (1952). Hypsometric (area-altitude) analysis of erosional topography. *Geol. Soc. Am. Bull.* **63**, 1117–1142.

Strahler, A. N. (1964). Quantitative geomorphology of drainage basins and channel networks. *In* "Handbook of Applied Hydrology" (V. T. Chow, ed.), pp. 4-40–4-74. McGraw-Hill, New York.

Taylor, A. B., and Schwartz, H. E. (1952). Unit-hydrograph lag and peak flow related to basin characteristics. *Trans., Am. Geophys. Union* **33**, 235–246.

Thomas, D. M., and Benson, M. A. (1970). Generalization of streamflow characteristics from drainage-basin characteristics. *Geol. Surv. Water-Supply Pap. (U.S.)* **1975**, 1–55.

Trainer, F. W. (1963). Drainage density as an indicator of baseflow in part of the Potomac River. *Geol. Surv. Prof. Pap. (U.S.)* **650C**, 177–183.

Williams, G. P., and Guy, H. P. (1973). Erosional and depositional aspects of Hurricane Camille in Virginia. *Geol. Surv. Prof. Pap. (U.S.)* **804**, 1–80.

Wolman, M. G., and Gerson, R. (1978). Relative scales of time and effectiveness of climate in watershed geomorphology. *Earth Surf. Processes* **3**, 189–208.

4

GEOMORPHIC MEASUREMENTS AFTER A FLOOD

GARNETT P. WILLIAMS
U.S. Geological Survey, Lakewood, Colorado

and

JOHN E. COSTA
U.S. Geological Survey, Cascades Volcano Observatory, Vancouver, Washington

INTRODUCTION

Stream channels are among the most dynamic components of the landscape. Through them, energy and mass (streamflow containing solid and dissolved sediment loads) are transferred from one location in the landscape to another. Channels therefore are a logical and efficient place to investigate the geomorphic effects of force, resistance, erosion, transportation, and deposition.

Any high flow of water that overtops a channel's banks is a flood. Because of the widespread use of low, flat, alluvial lands for agriculture and construction, floods frequently affect people (Fig. 1); therefore, they are important and require investigation. In the United States floods account for one-fourth to one-third of the average annual dollar losses from geologic hazards and for 80% of the annual loss of life from geologic hazards (Costa and Baker, 1981).

For the geomorphologist, hydrologist, sedimentologist, or other earth scientist, floods are exciting and interesting. Channels shift course and change size, floodplains and terraces are eroded, huge boulders can be moved, and thick piles of sediments can be deposited. A great deal of geomorphic transformation and alteration takes place in a matter of minutes to weeks, depending on the size of the river in flood and on the terrain. Unlike the relatively slow data compilation associated with other types of geomorphic investigations, such as hillslope creep, weathering, or tectonic adjustments, floods produce nearly instantaneous results. Because stream channels are ubiquitous and floods inevitable, studying the geomorphic effects of floods offers some of the most wide-reaching and efficient ways to understand present, and interpret past, earth surface processes.

PRELIMINARY STEPS

News of a big flood often evokes an urge, request, or order to launch an investigation. When beginning a flood study, several important procedural aspects need to be kept in mind.

Aerial Surveillance

The geomorphic effects of a flood can occur over a large area. Initial reports may not disclose many sites of potential interest. Therefore, a personal reconnaissance throughout the affected area is essential. The most tempting procedure is to set off in an automobile. However, flood effects may be in very inaccessible locations. Even if accessible, important erosional or depositional features might not be seen near or from the road. Thus, automobile reconnaissance can be very inefficient. Finally, some roads may be closed, either due to flood damage or because they are blocked by police. In short, the only way to be assured of locating most or all of the important sites is to survey from the air and then walk the valleys. (Some aerial photography companies may have taken postflood photographs, possibly for flood insurance studies or damage assessments; how-

FIGURE 1. Flood damage due to Hurricane Camille in Virginia, 1969. Person in channel for scale. (Robert Sigafoos photograph.)

ever, there is no guarantee of the existence, timing, or areal coverage of such photographs.)

For reasons discussed later it is important to complete the reconnaissance and begin field work as soon as possible after the flood. Thus, a second reason for surveying by air is that it is fast. Finally, many features may not be readily recognizable or photographable from the ground because of local topography, large areal expanse of a deposit, or other reasons; aerial photography may be the only way to get good all-inclusive photographs of such features, prior to detailed on-site investigations. During the aerial reconnaissance, outline areas or sites of particular interest on topographic maps or on the most recent aerial photographs.

The approximate 1985 expense for the rent of a small airplane in Colorado is $70 per hour, for the airplane, fuel, and pilot. The airplane probably will be able to fly for about 5 hr before having to refuel. In comparison, a helicopter costs about $200 per hour and would need to refuel every 2 hr or so. However, a helicopter can maneuver closer to the land surface and can land in rugged areas, which can be very useful. Either type of aircraft usually can be rented on short notice, if available. Photographs taken from these aircraft usually turn out well.

Timeliness in Getting to the Field Area

Begin field work as soon as possible after the flood. Within hours, local landowners, state agencies, and others will have bulldozers at work clearing away sedimentary deposits and landforms. Investigate first any flood deposits that infringe on human activities (especially on a major highway) because those deposits probably will not be there very long.

Another reason for starting field work quickly is the natural metamorphosis of the geomorphic effects with time. Sedimentary features, such as bedforms and high-water marks, may be eroded or changed by postflood weather. Vegetation can become reestablished within weeks in some areas. Although useful studies can be undertaken years after a flood, the greater the delay in starting field work, the more information will be irretrievably lost.

Contact with Landowners

Visit the local landowners for permission to go to any sites on their property; good rapport with landowners contributes significantly to a more successful flood study. Talk with them for a while to get their impressions of the flood. They may have been eyewitnesses to some of the flood. They can describe in detail the land condition as it was immediately before and after the flood. In some cases they may have some preflood photographs of sites that you can rephotograph for a before–after comparison. Most landowners are quite familiar with the history of any streams on their property. The landowner's name, address, and telephone number are important facts to be recorded, along with notes of the discussion.

If any sites warrant detailed measurements or observations, explain to landowners the general types of data collection and associated activities that will be done. Obtain their specific permission to install any surveying reference marks or to cut brush.

DETERMINING THE TYPE OF FLOW EVENT

When investigating a large flow event, especially in a small basin (less than about 100 km^2), it is important early in the study to establish the *type* of flow event—a water-dominated flood versus a sediment-dominated debris flow (defined later). This distinction is critical because the type of flow event to some extent determines what will be measured or estimated following the flow. The distinction usually is made on the basis of the presence or absence of postevent diagnostic debris flow evidence. Hence the next few paragraphs review the key elements of debris flows.

When poorly sorted soil and rock debris are mixed with a critical amount of water, a dense, structurally coherent slurry forms that can move rapidly down slopes and along channels, causing great destruction. These slurry flows are called debris flows (Costa and Williams, 1984); they resemble wet concrete and usually form as a result of hillslope failure during rainstorms.

Debris flows are non-Newtonian fluids, and their analysis and interpretation differ from those used for water-dominated floods. Typically, debris flows contain only 10–20% water by weight. The water is inseparably bound with a poorly sorted mixture of clay-size to boulder-size particles. Debris flows have strength and apparent viscosity much higher than those of water (Costa, 1984). The presence of shear strength, which must be overcome before debris flows can deform or flow, results in distinctive deposits and landforms that are the keys to interpretation of process.

Diagnostic postevent features of debris flows are (1) unsorted and unstratified deposits of gravel, sand, and fines; (2) marginal levees of coarse clasts, the largest of which may be at the top of the levees; (3) terminal, steep-fronted lobes of debris bordering the channel or flow path; and (4) unusually large boulders, transported at the margins of flows, which may have done little or no damage to vegetation or structures (Costa, 1984). The thickness of the lobe is proportional to the shear strength of the debris flow.

Valuable information about debris flow deposits includes their width and depth, thickness of the terminal lobe(s), channel slope, the three axes of the largest boulders, and rock type (needed for density estimations). Take several samples of the sediments for particle size analysis; it usually is practical only to include clasts up to 100 mm in diameter.

Finally, any information about the debris flow source area would be valuable. What triggered the flow? What was the nature of the debris source area? How far downstream was affected? Are any precipitation intensity data available?

HYDRAULIC ESTIMATES

Indirect Methods of Estimating Streamflow

Carefully determined estimates of the hydraulics associated with flood erosion and deposition should be an integral part of every geomorphic study after a flood. The postflood evidence in nearly all cases permits an appraisal only of values of the peak flow variables. These values are much more accurately estimated if high-water marks are still definable. Because they were not measured during the flow, these variables are estimated or determined by indirect methods. Benson and Dalrymple (1967) provide a very useful summary of field and office procedures for indirect measurements.

Where high-water marks of the flood are still present, the width of the peak flow water surface can be measured reliably. The associated cross-sectional flow area presumably is measurable also; however, this depends on the critical assumption that the stream channel at the time of measurement has not changed significantly, especially in streambed average elevation, since the time of the peak flow (e.g., no postpeak scour or fill has occurred). In some cases this assumption is risky, but often there is no choice. Mean flow depth is computed as cross-sectional area divided by water surface width.

Water discharge, slope, and mean flow velocity probably are the variables of primary interest to geomorphologists. Many indirect methods are available for estimating these variables. Some methods, however, are sophisticated enough that they probably are best left to people trained in the application of such methods. Included in this group are the methods of slope conveyance, slope area, contracted opening, flow through culverts, flow over dams and embankments, critical depth, and step backwater. Barnes and Davidian (1978) discuss these and other methods; see also Dalrymple and Benson (1967) and Davidian (1984).

At some stream sites the channel may not have definable banks due to local terrain or other reasons; in these and in other situations the geomorphologist may want to study the effects of unusually high within-channel flows. Several indirect methods for estimating water discharge can be applied by investigators who have not had an extensive hydraulic-engineering background. For high within-bank discharges Riggs (1976) proposed

$$Q = 3.39A^{1.30}S^{0.32} \quad (1)$$

where Q = water discharge in cubic meters per second,
A = cross-sectional flow area in square meters (the average of three representative cross sections), and
S = water surface slope in meters per meter.

Although originally developed for channels having gravel or coarser bed material, the equation also seems to apply to other natural channels (Williams, 1978). The approximate applicable ranges are $0.7 \leq A \leq 8500$ m^2 and $0.00004 \leq S \leq 0.08$ m/m. Riggs found his equation to have about the same accuracy (a standard error of about 20%) as the slope–area method in estimating discharge.

Riggs (1976, p. 289) recommended his equation for "nearly full natural channels without substantial overbank flow." For practical purposes this is probably bankfull, or nearly bankfull. Williams (1978) gave the following empirical equation for bankfull discharge:

$$Q_b = 4.0A_b{}^{1.21}S^{0.28} \quad (2)$$

where Q_b = bankfull discharge in cubic meter per second,
A_b = bankfull flow area (the average of three representative cross sections) in meters squared, and
S = channel slope in meters per meter.

The accuracy of this equation seems to be roughly the same as that of Riggs (standard error of about 20%), and the applicable range is about the same. Bankfull stage for Equation (2) is the level of the active floodplain—the overflow surface actively growing and forming by present alluvial processes and below which the channel usually has little or no vegetation. Other investigators have defined bankfull in many different ways, including the levels of the valley flat, various benches, tops of channel bars, and breaks in plotted trends of channel geometry variables (Williams, 1978). Consistency in the definition from one study to another is very desirable, and investigators should always specify the particular definition they used.

Where flow has traveled rapidly around a bend, the water surface rides up higher on the outer part of the bend than on the inside. This elevation difference, called superelevation, can be used to estimate the mean velocity of the flow (Chow, 1959, p. 448). The equation is

$$V = (\Delta h \; g \; R_c/W)^{1/2} \quad (3)$$

where Δh = superelevation in meters,
g = acceleration due to gravity in meters per second squared,
R_c = radius of curvature of the bend in meters, and
W = water surface width in meters.

Due to possible wave action, irregularities in cross-sectional geometry, and many other factors, the error with this method easily could be 50%.

Where the investigator has not been able to begin field work while the line of the peak flow water surface is still definable, indirect methods of somewhat lesser accuracy must be used. For these conditions such flow variables as water surface width and cross-sectional flow area probably cannot be reconstructed at all. An example of the methods in this group is Costa's (1983) equation for estimating mean flow velocity on the basis of the average intermediate axis of the five largest boulders moved by a flood:

$$V = 0.18d^{0.49} \quad (4)$$

where d is the average value of the b axis in millimeters. The equation applies to particles ranging from 50 to 3200 mm in diameter. The likely error is on the order of 25–100%.

Many other indirect methods or equations for estimating streamflow, besides those discussed here, have been used. Williams (1984) summarizes most of these.

Choosing the Measuring Reach and the Cross-Section Sites

Equations (1) and (2) and most of the standard indirect methods, such as slope area, require measured cross sections and longitudinal slope. There are five highly desirable features of a measuring reach.

First, the measuring reach must be as close as possible to, and preferably in the same reach as, the features of geomorphologic or sedimentologic interest. If the two sites are separated, the hydraulic estimates might not apply because of entering tributary inflow, loss of water into porous streambeds, retardation of flow by debris jams, and other features.

Second, well-defined high-water marks are indispensable for determining the flow cross section and the longitudinal slope. The marks should be on surfaces parallel to the main flow direction, rather than on surfaces facing into or away from the flow. Ground marks, where wave action and surge are minimal (quiet-water marks), are preferable to marks in bushes and trees, defined by debris. Finer materials (seeds, mud, or silt) are better indicators than large objects. Drift lodged in bushes and trees within the channel is not as dependable as drift on banks.

The approximate elevation of the peak flow water surface is indicated by streamside vegetation in some instances. Some vegetation may have been removed by the flow, so that new vegetation in this zone would be noticeably younger. Young trees and bushes within the flow may have been bent horizontally, whereas those of the same age at higher elevations will still be vertical. Large scars on the upstream face of tree trunks, due to debris impactions, might crudely indicate the peak water level. Again, if one can assume no significant postpeak changes in streambed elevation and cross-section size, the mean depth of the peak flow might be approximately determined by measuring the cross section up to the estimated water level (Harrison and Reid, 1967).

The third requirement is that the channel cross section along the reach be as uniform as possible in size and shape. Fourth, the reach should be as straight as possible. No reach probably will ever be perfect in these respects; nevertheless, most equations for estimating streamflow were not designed for reaches that have significant contractions (common at bridges), expansions, debris dams (common causes of backwater and hence of erroneous high-water marks), free-fall sites, and bends.

Fifth, the reach ought to be long enough to permit a reliable measurement of the slope. This means it must be long enough that the vertical fall is much greater than the vertical fluctuation in the measured elevations. Some rules of thumb are that the reach length be at least 20 channel widths or 75 times the mean flow depth (Dalrymple and Benson, 1967).

Within the measuring reach give particular care to the selection of cross-section sites. Long reaches that can be typified by one cross section are rare. Cross sections must be spaced closely enough to accurately reflect the hydraulic characteristics of both the main channel and any adjacent overflow zones. Thus, locate cross sections at major changes in bed or water surface profiles, roughness characteristics, cross-sectional shape, and dimensions (especially points of minimum and maximum cross-sectional area, width, or depth). The number of cross sections along a reach therefore depends on the uniformity of the reach; more cross sections are needed where the features just mentioned change significantly. Also, orient all cross sections at right angles to the flow, in both the main channel and overbank sections; this may necessitate angled or curved sections. Cross sections must include reliable high-water marks to delineate the flow boundaries. Benson and Dalrymple (1967), Davidian (1984), and Jarrett (1985) give further discussion and are recommended reading.

Some flow formulas used in indirect measurements require a value for Manning's n. Chow (1959, pp. 106–114) and Benson and Dalrymple (1967, pp. 20–24) explain how to select an n value (see also Williams, this volume). Other methods for selecting n include photograph comparison (Barnes, 1967) and direct use of an empirical equation (Limerinos, 1970; Jarrett, 1984; Thorne and Zevenbergen, 1985; Bathurst, 1985). Where the flow cross section is not "regular" in shape (e.g., includes overbank flow in addition to main-channel flow), or where the cross section includes significant lateral changes in roughness, subdivide the flow cross section and choose separate n values for each subsection (Davidian, 1984).

Equipment

A surprising array of equipment is needed for geomorphic investigations of floods to ensure safe and profitable field time. Many items serve multiple purposes, such as a hand level for leveling and surveying, or a folding wooden ruler, which can serve as a surveying rod, photographic scale, or ruler for measuring coarse particle sizes. Table 1 lists

TABLE 1 Equipment Checklist

Topographic maps (1:24,000 or larger)
Aerial photographs (large scale)
Barnes (1967) (if n values will be estimated by photo comparison)
Notebook (waterproof paper)
Several pencils
Level or alidade, Tripod, Rod } if no extensive walking or rugged access
Hand level (if access is difficult or rugged)
Carpenter's rule (folding type; metric units)
Brunton compass
35-mm camera
Wide-angle lens
Color slide and black-and-white film
Long-handle pointed shovel
Hand auger
Bright-colored cloth tape (at least 50 m long)
Metal tape (at least 50 m long)
Rain suit
Munsell soil color chart
Treecorer
Large sample bags (25 or more)
Leather gloves
First-aid kit
Rubber boots or waders
Whistle
Snake-bite kit
Insect repellent
Nylon cord (5–10 m)
Marking pens (waterproof)
Pocket knife
Red flagging
Two screwdrivers (for holding ends of tapes)
Canteen (filled with appropriate refreshments)
Rebars (different lengths) or wooden stakes
Geologic hammer

FIGURE 2. Young geomorphologist investigating sand deposits on forested floodplain following disastrous flooding in the Connecticut Valley, Connecticut, 1955.

the most useful and important equipment for flood studies. Nearly all of the items can be worn and carried in one backpack. Use common sense in selecting the most utilitarian types of field equipment; a collapsible, fiberglass, 7-m-long stadia rod is a good example.

SEDIMENTARY MEASUREMENTS AND OBSERVATIONS

The value of accurate documentation of sedimentary textures, thickness, and structures is inestimable combined with the hydraulic and hydrologic measurements and estimations described in the previous section. Model studies of sedimentary features deposited by floods are the studies by Krumbein (1940, 1942) in the San Gabriel Mountains, California, and by McKee and others (1967) on a Great Plains sand stream in Colorado.

Flood sediments occur in numerous settings, such as fans, splays, channel fills, overbank deposits, and backwater sites (Fig. 2). The sediment setting and probable environment of deposition need to be described as carefully as possible.

Spatial Distribution of Deposits

Documentation of the spatial distribution of deposits and their volumes using large-scale postflood aerial photographs and field work is needed. Thickness of deposits can be determined by digging or augering down to the preflood surface at selected locations. If sediments are too coarse to disturb, thickness can be estimated by leveling from the preflood surface. The volume of overbank deposition is computed from area thickness data using isopach maps constructed from thickness data or by the Theissan polygon method (see Jarrett and Costa, 1986, for an example).

Long-handled pointed shovels are best for digging in overbank sediments on floodplains; flat, short-handled shovels are best for clearing channel banks. An alternative tool, when only thickness of coarse sand and finer sediments is desired, is a hand auger. Augers can be purchased in sections ranging from 0.3 to 1 m long. Data on texture and ash layers from floodplain sediments have been recovered with a hand auger from depths of over 7.3 m (Smith, 1973). A hand level and carpenter's ruler can be used if leveling is to be done in hilly terrain; these items are lightweight and have many other practical uses in the field.

Another important piece of sedimentological information from flood sediments would be the distance that particles of distinct lithology or source traveled during the flood. Scott and Gravlee (1968) reported that a single boulder moved 3.4 km in the 1964 dam-break flood on the Rubicon River in California.

Bedforms and Other Primary Sedimentary Structures

One of the most important measurements to be made following a large flood is the description of bedforms and other primary sedimentary structures deposited by the floodwaters. Flood bedforms exhibit a remarkable range of sizes and expressions. In sandy sediments division of bedforms and other primary sedimentary structures into lower and upper flow regimes is useful (Simons and Richardson, 1966). These bedforms include ripples, dunes, plane bed, and antidunes, and other sedimentary structures such as ripple cross-laminations, cross-bedding, horizontal stratification, and upstream dipping cross-bedding (Jopling, 1963, 1966; McKee et al., 1967; Hand et al., 1969). These features can be tied directly to the flow characteristics (depth, velocity, water surface slope) of the flood (e.g., Harms and Fahnestock, 1965; Jopling, 1966).

In gravel rivers bedforms and primary sedimentary structures are more poorly understood (e.g., Church and Jones, 1982; Baker, 1984). The most successful reconstructions of flow conditions have been made from antidune regime flows (Shaw and Kellerhals, 1977; Foley, 1977). The best-known small-scale bedforms are a series of regularly spaced pebble, cobble, or boulder ridges oriented transverse to flow, called transverse ribs (McDonald and Banerjee, 1971; Koster, 1978).

Larger scale flood bedforms (bars) form in streams of varying bed material size and have been described by Baker (1984). These bedforms consist of expansion bars, eddy bars, pendant bars, and large-scale gravel dunes. Such bedforms seem to be most common in bedrock canyons where flows are confined and rapid expansions and contractions of flow boundaries occur.

Expansion bars form at abrupt channel expansions where large amounts of energy are dissipated. Expansion bars show a rapid decrease in particle size from the bar head to bar toe.

Eddy bars form at the mouths of tributaries to bedrock canyons inundated by large flood flows. They contain a variety of grain sizes and structures, including interfingering boulder gravel, laminated silts, cross-bedded granule gravel, and graded sand-silt layers. Foreset bedding indicates varying directions of bedload transport resulting from swirling eddies that deposited the bars.

Pendant bars are streamlined mounds of flood gravel deposited in flow separations that can develop downstream from large flow obstructions such as bedrock outcrops, large boulders, trees, or logjams. Pendant bars usually show well-developed foreset bedding with well-sorted, subrounded boulders and cobbles and an open-work structure.

Large-scale gravel ripples are coarse, symmetrically spaced bedforms in bedrock streams. Ripple heights range from 0.5 to 10 m, and wavelengths range from 10 to 150 m. The ripples consist of well-sorted, open-framework gravel foresets alternating with discrete layers of granules and pebbles.

The sediments in any flood deposit or bedform need to be described as a columnar section in the same manner as the stratigraphy of an ancient alluvial sequence is recorded. Unfortunately, there are not enough good descriptions of modern flood sediments like this (e.g., McKee et al., 1967). If possible, the structures and stratigraphy of the old floodplain surface also need to be described. Older, buried deposits similar to the fresh overbank sediments may provide additional information.

Particle Sizes

Facies and particle sizes of the flood sediments should be documented. To accurately represent the particle size distribution for any deposit requires decisions on the number of samples to take and the amount of each sample. Horizontal and vertical variability in particle sizes can be significant. Because this spatial variability will be unique for each flood deposit, no blanket rule can be given for the number of samples to take; however, the number of required samples almost always far exceeds one (Mosley and Tindale, 1985). Unless size distributions at specific sampling points are desired, the most practical approach is to take a large number of samples, combine the lot, and split this composite down to a suitable amount for sieve analysis. Heavy-duty clear plastic sample bags, about 25 by 40 cm or larger, or comparably sized special canvas sample bags with prefastened labels, make good containers.

For sand sizes and smaller, about 0.5–1 kg of sediment need to be collected for sieve analysis (Krumbein and Pettijohn, 1938). For gravel, a sample in which the weight of the largest particle is less than 5% of the weight of the total sample is a minimum acceptable amount (Mosley and Tindale, 1985); in practice, however, many kilograms may be needed to ensure that large particles are not eliminated during sample splitting.

In sieve analyses half-ϕ-size intervals should be used to adequately describe the particle size distribution. If the settling tube method is used, remember that sieve diameter and fall diameter are not the same, and the difference is greater for larger particles with smaller shape factors (Subcommittee on Sedimentation, 1958).

For coarse (≥8 mm) flood sediments, bulk sieving is not practical because of the size of clasts and the weight required of each sample. In this situation bulk sieving can be replaced by a frequency-by-number technique based on area sampling. The easiest and most widely used technique is the "pebble count" procedure developed by Wolman (1954), where 100 particles are randomly selected from a grid system and intermediate axes directly measured in the field (Fig. 3). (The grid can be replaced by other random-sampling techniques, if desired.) Considerable areal variability can occur with surface particles, and the number of particles that need to be counted to adequately represent the deposit can range from about 100 (rarely) to many thousand (Mosley and Tindale, 1985).

According to Kellerhals and Bray (1971), frequency-by-number and bulk sieve analyses are equivalent procedures, and they can be combined into one cumulative frequency graph (see example in Williams and Guy, 1973). However, Mosley and Tindale (1985) found only a very weak relationship between mean grain sizes of surface and subsurface material, so there may be some risk in combining the two distributions.

The largest rocks moved by a flood are of great interest and study value. The largest particles present in flood deposits usually are assumed to represent the maximum competence of the stream during the flood. However, this assumption may not be valid. Boulders larger than those available for transport possibly could have been moved, had they been present; or, large boulders may have been deposited by nonflood processes such as rafting, mass movements, or bank erosion (Rubey, 1938; Gage, 1953). However, Krumbein and Lieblein (1956) showed with extreme value theory that most anomalously large particles in gravel deposits are normal members of the stream particle size population. As a measure of the flood "competence," avoid single, isolated large rocks and use the average of the five largest particles thought to have moved.

The average particle diameter of flood sediments should be studied with distance downstream. Plumley (1948), Scott and Gravlee (1968), and Bradley and others (1972) concluded that 75–90% of measured downstream grain size reductions resulted from selective deposition and sorting, and that abrasion accounted for the remaining small percentage (10–25%).

FIGURE 3. Young geomorphologist investigating textural characteristics of coarse flood deposit, Connecticut Valley, Connecticut, 1955.

Grain Shape and Roundness

Changes in particle shape and roundness with distance of travel in flood sediments are poorly understood; only a few reports contain sufficient data to provide some insight. The two most comprehensive studies are Krumbein's (1940) in the San Gabriel Mountains, California, and Scott's (1967) in the Rubicon River, California. Both studies documented the rapid rounding of large clasts in relatively short transport distances. Grain shape showed no significant change in about 10 km of transport.

For roundness investigations the visual comparison charts of Krumbein (1941) can be used. There are a variety of shape indexes (see Pettijohn, 1975, for a summary); most of them require measuring the three orthogonal axes of individual particles.

PHOTOGRAPHS OF FLOOD EFFECTS

Photographs are an excellent way to show the geomorphologic effects of a flood. In addition, a photograph permits a review and appraisal of certain conditions on subsequent occasions. One example might be the estimation of approximate bed material sizes and the selected value of the roughness coefficient. Also, photography is an essential part of monitoring long-term response of channels and deposits following a flood.

In addition to photographs that you take, good photographs probably will be available from other sources. Many organizations (newspapers, state and county agencies, and others) and individuals (landowners, freelance photographers, and others) take photographs—both aerial and ground photographs—right after a flood. Besides photographs, such information as channel cross sections, rainfall data, and other hydrologic data for before and after the flood might be available from the U.S. Geological Survey (Water Resources Division), U.S. Forest Service, U.S. Army Corps of Engineers, U.S. Soil Conservation Service, U.S. Bureau of Reclamation, state and local highway departments, the U.S. Weather Bureau, historical societies, libraries, and museums. Extensive inquiries will bring to light some spectacular pictures and other valuable information that otherwise would be missed. Permission to use photographs in a scientific publication usually is not hard to get.

Timeliness in taking photographs is critical, as mentioned before in connection with collecting on-site data. For evaluating flood effects, and for present and future field work, a set of small-scale aerial photographs taken soon after the flood will be extremely valuable. If these have not already been taken, inquire with aircraft rental companies or pilots about the cost. The price might be surprisingly affordable.

In addition to general cross-channel views, take ground photographs of stream sites, looking both upstream and downstream. A cross-section location and peak flow level can be indicated by a person standing at the cross section, holding a stadia rod horizontally at the elevation of the peak. If possible, include a sharply defined, readily identifiable, rather permanent object (e.g., rock outcrop in foreground or mountain peak), for repeat photography in subsequent years. Scenes of the streambed, including closeups of particle sizes and bedforms, are always useful, but a low-water condition may not occur for some time after a flood.

One of the most common deficiencies in many published photographs is the scale indicator. For landscape views, if at all possible, always have a *person* in the picture for scale. Place the person against a distinctive background, not too far from the camera. For scenes near a road a vehicle is also a suitable indicator. For closeup views (distances $\lesssim$ 1 or 2 m), measuring tapes, stadia rods, rulers, backpacks, notebooks, pencils, and hammers have varying degrees of usefulness, depending on the viewer's ability to interpret their actual size in the picture.

As soon as each photograph is taken, record what is shown in the picture in a field notebook, including site, date, direction of view, and time of day. For more detailed work, and with future repeat photography in mind, a tripod might be used and various other camera-related information might be recorded, as suggested by Malde (1973).

Ideally, two cameras should be used—one for color slides for use in oral presentations and one for black-and-white film for publication. In practice, however, only one camera usually is carried to the field. It is easier and less expensive to get a black-and-white print made from a color slide, so use color-slide film if only one camera is available. The slower the film, the clearer the picture. A wide-angle lens (28 or 35 mm) can be useful for some shots.

As soon as any slides or prints are received, mark all information on each item. This marking takes one or two hours but is very important; files are full of beautiful old photographs that have no date, no site, and no other identifying information, which makes them less valuable for future investigators.

With monitoring of postflood recovery in mind, the locations of selected photographs should be marked permanently at the time of the initial photography, for subsequent repeat photography. One of the best marking methods is by means of an iron rod (rebar) or long gutter nail driven into the ground, as explained later. Choose a spot that will remain as permanent as possible; for example, the alluvial bank of a stream would not be a good choice.

Useful information on all aspects of field photography is found in the periodical "Repeat Photography Newsletter," which appears twice yearly and is available for $4.00 per year (1985 price). Address: Wayne Lambert, editor; Repeat Photography Newsletter; Department of Geosciences; West Texas State University; Canyon, Texas, 79016.

MONITORING RECOVERY RATES

Rates of natural processes, and time scales in general, deserve much more attention in geomorphologic research. As an integral part of a flood study, establish monumented cross sections and reaches that can be resurveyed and rephotographed at selected future times (see discussion of the Vigil Network, at end of this chapter). The rate and nature of channel response or recovery thereby can be measured. Some of the more important channel features worthy of periodic postflood measurements or observations are the cross section (including special mention of positions of banktops), longitudinal profile, bed sediments, vegetation (species, density of coverage, sizes, and approximate ages), and general morphology.

Some changes might occur very soon after the flood, so establish the measuring sites as soon as possible. Try to choose an easily accessible reach.

With long-term monitoring in mind, assume that other investigators will want to continue observations at some later time. Describe the reach and cross-section locations so clearly that a new person can use your descriptions to find the field sites easily. Prepare a tabulated road log of distances, using permanent roadside features as intermediate points, for getting to the study reach. Include a location sketch map, topographic map, ground photographs (clearly labeled), and aerial photographs. Mark all cross sections on the topographic map and on the aerial photographs. Describe the site, with comments on the geology, soil, vegetation, climate, and topography. Finally, record the name, address, and telephone number of the person to contact for permission to go on the land.

Monuments

Permanent monuments or markers must be installed in the ground at both ends of each cross section. Concrete has been used for this purpose; however, iron rods (rebars) about ⅜–½ in. in diameter, pounded into the ground, seem to be more common. Carry odd lengths of these rebars (up to a little more than one meter) to the field site because soil conditions at a particular spot will be unpredictable. A small metal identification tag fixed to the monument is very useful.

Opinions differ as to the desirable height of the monument above the ground. Typical heights range from about 0 to 2 dm. The higher the monument, the more visible it will be, and this is extremely useful in surveying and refinding. However, a high monument is subject to vandalism and carries a great risk of damage to farm animals, wildlife, and farm machinery. These risks are minimized if the monument is installed flush with the ground, but at this level it can be very difficult and time consuming to find on subsequent trips. A metal detector can help find such monuments and can in fact be indispensable where subsequent deposition has buried a monument. If some of the monument is left exposed, carefully record its height above the ground; later measurements then will give some indication of any local soil erosion or aggradation.

Regardless of the height of the monument, mark its top with a bright color to help find it on later occasions. Paint is one possibility. Or, bright plastic field tape can be wrapped around the top if repeat visits will be made at least every few years; "international orange" is the recommended color for this purpose. Small, yellow plastic caps (plastic surveyors' markers) are also available for fitting over rebars or onto pipe ends. These caps cost only a few cents each and can be heat branded by the factory with some letters and numbers for identification (same branding for all markers).

Monuments, even though clearly marked, often are lost due to various causes. Besides vandals and animals, many tens of meters of bank erosion can occur during a later flood. The goal is to be able to refind the cross section, and this frequently is very difficult. With this goal in mind, include the following seven features for each cross section: backup monuments, triangulation distances, azimuths, sketch maps, topographic maps, aerial photographs, and an exact written description of all pertinent information.

On each side of the stream install several backup monuments behind the principal monument along the same cross section. These backup monuments can be spaced as far as 100 m or more apart, depending on the channel width and local topography.

In addition to noting the horizontal distances between all monuments, describe the location of each monument in terms of its distance from at least two rather permanent features. Large nails driven into big trees and crosses chiseled into large rocks or bedrock commonly serve as such features. Distances from two fixed points and triangulation can be used to locate the monument even with no other monuments. Record azimuths for (1) the compass direction of each monument from each permanent feature and (2) the compass orientation of the cross section.

Complete the local sketch map, with locations and azimuths of all monuments and permanent features clearly labeled, when the cross sections are first established. At the same time mark the cross sections and monument locations on the aerial photographs and topographic maps. These items, used with triangulation and associated azimuths, ought to make it possible to locate at least one of the monuments. Then, even if no other monuments can be found, the general compass orientation of the cross section will indicate, approximately, the location of the entire cross section.

Rephotographing from Monumented Sites

Make repeat photography a standard part of monitoring channel response to floods. The four requirements for repeat photography are:

1. Always place the camera at the same site for each cross section (right over the monument).
2. Always aim the camera in the same direction, with a permanent, easily recognizable feature always at the same place in the picture.
3. Take photographs at the same time of year, to eliminate the effect of seasonality on apparent vegetation changes. (If at all possible, always take the pictures at about the same time of day, also, so that the sun direction and shadow factors will be as constant as possible.)
4. Write the site, date, and approximate direction of view on the back of every photograph.

The Vigil Network

The types of measurements to be made in documenting channel reactions following a flood are the same as those used in the Vigil Network. The Vigil Network (Leopold, 1962; Emmett and Hadley, 1968) is an international network of observational areas on which periodic measurements are made and preserved. The chief purposes of the Vigil Network are to document changes in the landscape with time, especially over a period of many decades, and to make the data available to present and future generations of scientists. Observations are made on stream channels, valley floors, hillslopes, reservoirs, precipitation, and vegetation. An approximate standard format is preferred for submission of data (see Emmett and Hadley, 1968). Data are submitted to either of two repositories, namely, the Vigil Network Repository in the libraries of (1) the U.S. Geological Survey, MS 950 National Center, Reston, VA 22092 or (2) the Geomorphology Laboratory, Uppsala University, Box 554, S 751 22 Uppsala, Sweden.

CONCLUSIONS

Investigations of floods by geomorphologists are common. However, the observations, measurements, and data have been inconsistent. Aspects that deserve more attention are hydraulic characteristics associated with sediment entrainment and deposition, evidence of debris flows (in hilly regions), and well-organized plans for monitoring the landscape's postflood recovery. A standard list of features to observe, measure, or estimate after a flood includes the following:

1. Evidence of a debris flow or water flood
2. Cross-sectional flow area
3. Water surface width
4. Slope of water surface and of channel
5. Peak flow mean velocity and water discharge
6. Roughness coefficient (n value)
7. Spatial distribution and volume of deposits
8. Primary sedimentary structures
9. Particle sizes
10. Grain lithologies, shape, and roundness
11. Vegetation affected by the flood

Photographically document as many of these features as possible, and establish long-term monitoring sections.

ACKNOWLEDGMENTS

We very much appreciate informative discussions with Robert H. Meade, Raymond M. Turner, Harold E. Petsch, Jr., and William W. Emmett. Kenneth L. Wahl, Robert D. Jarrett, and James C. Knox provided many helpful comments on an initial manuscript.

REFERENCES

Baker, V. R. (1984). Flood sedimentation in bedrock fluvial systems. *Mem.—Can. Soc. Pet. Geol.* **10**, 87–98.

Barnes, H. H., Jr. (1967). Roughness characteristics of natural channels. *Geol. Surv. Water-Supply Pap. (U.S.)* **1849**, 1–213.

Barnes, H. H., Jr., and Davidian, J. (1978). Indirect methods. *In* "Hydrometry: Principles and Practices" (R. W. Herschy, ed.), pp. 149–204. Wiley, New York.

Bathurst, J. C. (1985). Flow resistance estimation in mountain rivers. *J. Hydraul. Eng.* **111**(4), 625–643.

Benson, M. A., and Dalrymple, T. (1967). General field and office procedures for indirect discharge measurements. *Tech. Water-Resour. Invest. (U.S. Geol. Surv.)*, Book 3, Ch. A1, pp. 1–30.

Bradley, W. C., Fahnestock, R. K., and Rowekamp, E. T. (1972). Coarse sediment transport by flood flows in the Knik River, Alaska. *Geol. Soc. Am. Bull.* **83**(5), 1261–1284.

Chow, V. T. (1959). "Open-Channel Hydraulics." McGraw-Hill, New York.

Church, M., and Jones, D. (1982). Channel bars in gravel-bed rivers. *In* "Gravel-Bed Rivers" (R. D. Hey, J. C. Bathurst, and C. R. Thorne, eds.), pp. 291–324. Wiley, New York.

Costa, J. E. (1983). Paleohydraulic reconstruction of flash-flood peaks from boulder deposits in the Colorado Front Range. *Geol. Soc. Am. Bull.* **94**, 986–1004.

Costa, J. E. (1984). The physical geomorphology of debris flows. *In* "Developments and Applications of Geomorphology"

(J. E. Costa, and P. J. Fleisher, eds.), pp. 268–317. Springer-Verlag, Berlin and New York.

Costa, J. E., and Baker, V. R. (1981). "Surficial Geology: Building with the Earth." Wiley, New York.

Costa, J. E., and Williams, G. P. (1984). Debris-flow dynamics (videotape). *Geol. Surv. Open-File Rep. (U.S.)* **84-606**, 22½ minutes.

Dalrymple, T., and Benson, M. A. (1967). Measurement of peak discharge by the slope-area method. *Tech. Water-Resour. Invest. (U.S. Geol. Surv.)*, Book 3, Ch. A2, pp. 1–12.

Davidian, J. (1984). Computation of water-surface profiles in open channels. *Tech. Water-Resour. Invest. (U.S. Geol. Surv.)*, Book 3, Ch. A15, pp. 1–48.

Emmett, W. W., and Hadley, R. F. (1968). The Vigil Network—Preservation and access of data. *Geol. Surv. Circ. (U.S.)* **460-C**, 1–21.

Foley, M. G. (1977). Gravel-lens formation in antidune-regime flow—A quantitative hydrodynamic indicator. *J. Sediment. Petrol.* **47**, 738–746.

Gage, M. (1953). Transport and rounding of large boulders in mountain streams. *J. Sediment. Petrol.* **23**(1), 60–61.

Hand, B. M., Wessel, J. M., and Hayes, M. O. (1969). Antidunes in the Mount Toby conglomerate (Triassic), Massachusetts. *J. Sediment. Petrol.* **39**, 1310–1316.

Harms, J. C., and Fahnestock, R. K. (1965). Stratification, bed forms, and flow phenomena (with an example from the Rio Grande). *Spec. Publ.—Soc. Econ. Paleontol. Mineral.* **12**, 84–115.

Harrison, S. S., and Reid, J. R. (1967). A flood frequency curve based on tree-scar data. *Proc. N. D. Acad. Sci.*. **21**, 23–33.

Jarrett, R. D. (1984). Hydraulics of high-gradient streams. *J. Hydraul. Eng.* **110**(11), 1519–1539.

Jarrett, R. D. (1985). Determination of roughness coefficients for streams in Colorado. *Water-Resour. Invest. (U.S. Geol. Surv.)* **85-4004**, 1–54.

Jarrett, R. D., and Costa, J. E. (1986). Hydrology, geomorphology, and dam-break modeling of the July 15, 1982, Lawn Lake Dam and Cascade Lake Dam failures, Larimer County, Colorado. *Geol. Surv. Prof. Pap.* (U.S.) **1369**, 1–78.

Jopling, A. V. (1963). Hydraulic studies on the origin of bedding. *Sedimentology* **2**, 115–121.

Jopling, A. V. (1966). Some principles and techniques used in reconstructing the hydraulic parameters of a paleo-flow regime. *J. Sediment. Petrol.* **36**, 5–49.

Kellerhals, R., and Bray, D. I. (1971). Sampling procedures for coarse fluvial sediments. *J. Hydraul. Div. Am. Soc. Civ. Eng.* **97**(HY8), 1165–1180.

Koster, E. H. (1978). Transverse ribs: Their characteristics, origin, and paleohydraulic significance. *Mem.—Can. Soc. Pet. Geol.* **5**, 161–186.

Krumbein, W. C. (1940). Flood gravel of San Gabriel Canyon, California. *Geol. Soc. Am. Bull.* **51**, 639–676.

Krumbein, W. C. (1941). Measurement and geologic significance of shape and roundness of sedimentary particles. *J. Sediment. Petrol.* **11**(2), 64–72.

Krumbein, W. C. (1942). Flood deposits of the Arroyo Seco, Los Angeles County, California. *Geol. Soc. Am. Bull.* **53**, 1355–1402.

Krumbein, W. C., and Lieblein, J. (1956). Geological application of extreme-value methods to interpretation of cobbles and boulders in gravel deposits. *Trans. Am. Geophy. Union* **37**(3), 313–319.

Krumbein, W. C., and Pettijohn, F. J. (1938). "Manual of Sedimentary Petrography." Appleton-Century-Crofts, New York.

Leopold, L. B. (1962). The man and the hill. *Geol. Surv. Circ. (U.S.)* **460-A**, 1–5.

Limerinos, J. T. (1970). Determination of the Manning coefficient from measured bed roughness in natural channels. *Geol. Surv. Water-Supply Pap. (U.S.)* **1898-B**, 1–47.

Malde, H. E. (1973). Geologic bench marks by terrestrial photography. *U.S. Geol. Surv. J. Res.* **1**, 193–206.

McDonald, B. C., and Banerjee, I. (1971). Sediments and bed forms on a braided outwash plain. *Can. J. Earth Sci.* **8**, 1282–1301.

McKee, E. D., Crosby, E. J., and Berryhill, H. L. (1967). Flood deposits, Bijou Creek, Colorado, June 1965. *J. Sediment. Petrol.* **37**(3), 829–851.

Mosley, M. P., and Tindale, D. S. (1985). Sediment variability and bed material sampling in gravel-bed rivers. *Earth Surf. Processes Landforms* **10**(5), 465–482.

Pettijohn, F. J. (1975). "Sedimentary Rocks," 3rd ed., Harper & Row, New York.

Plumley, W. J. (1948). Black Hills terrace gravels: A study in sediment transport. *J. Geol.* **56**, 526–577.

Riggs, H. C. (1976). A simplified slope-area method for estimating flood discharges in channels. *J. Res. U.S. Geol. Surv.* **4**, 285–291.

Rubey, W. W. (1938). The force required to move particles on a stream bed: *Geol. Surv. Prof. Pap. (U.S.)* **189-E**, 121–141.

Scott, K. M. (1967). Downstream changes in sedimentological parameters illustrated by particle distribution from a breached rockfill dam. *In* "Symposium on River Morphology," Publ. No. 75, pp. 309–318. Int. Assoc. Sci. Hydrol., Bern.

Scott, K. M., and Gravlee, G. C., Jr. (1968). Flood surge on the Rubicon River, California—hydrology, hydraulics, and boulder transport. *Geol. Surv. Prof. Pap. (U.S.)* **422-M**, 1–40.

Shaw, J., and Kellerhals, R. (1977). Paleohydraulic interpretation of antidune bedforms with applications to antidunes in gravel. *J. Sediment. Petrol.* **47**, 257–266.

Simons, D. B., and Richardson, E. V. (1966). Resistance to flow in alluvial channels. *Geol. Surv. Prof. Pap. (U.S.)* **422-J**, 1–61.

Smith, D. G. (1973). Aggradation of the Alexandria—North Saskatchewan River, Banff Park, Alberta. *In* "Fluvial Geomorphology" (M. Morisawa, ed.), SUNY Pub. Geomorphol., pp. 201–219. Binghamton, New York.

Subcommittee on Sedimentation (1958). Some fundamentals of particle-size analysis. *In* "Measurement and Analysis of Sediment Loads in Streams," Rep. No. 12. Inter-Agency Committee on Water Resources, St. Anthony Falls Hydraulic Laboratory, Minneapolis, Minnesota.

Thorne, C. R., and Zevenbergen, L. W. (1985). Estimating mean velocity in mountain rivers. *J. Hydraul. Eng.* **111**(4), 612–624.

Williams, G. P. (1978). Bankfull discharge of rivers. *Water Resour. Res.* **14**(6), 1141–1154.

Williams, G. P. (1984). Paleohydrologic equations for rivers. *In* "Developments and Applications of Geomorphology" (J. E. Costa and P. J. Fleisher, eds.), pp. 343–367. Springer-Verlag, Berlin and New York.

Williams, G. P., and Guy, H. P. (1973). Erosional and depositional aspects of Hurricane Camille in Virginia, 1969. *Geol. Surv. Prof. Pap. (U.S.)* **804**, 1–80.

Wolman, M. G. (1954). A method of sampling coarse river-bed material. *Trans. Am. Geophy. Union* **35**(6), 951–956.

PART II

FLOOD PROCESSES

> Nothing under heaven is softer and more yielding than water, but when it attacks things hard and resistant there is not one of them that can prevail.
>
> —*Daodejing* ("*The Way and Its Power*") by Lao Zi

Humankind has always been mystified by the processes operating during great floods. However, because some people have had to live close to flooding rivers, they learned by observation. Consider the words of the Indian riverman in Rudyard Kipling's *In Flood Time*:

> Listen, sahib! The river has changed its voice. It is going to sleep before the dawn, to which there is yet one hour. With the light it will come down afresh. How do I know? Have I been here thirty years without knowing the voice of the river as a father knows the voice of his son?

As scientists we learn "the voice of the river" by measuring important parameters that describe fluvial processes. This empirical approach serves us well when the processes recur and operate on temporal and spatial scales appropriate to measurement. The riverman quoted by Kipling knew the daily and seasonal changes for a 30-year time sample. However, for knowledge of events operating on time scales of centuries, occurring over broad regions in places unobserved, he would have probably relied on legends and folk tales.

The importance of processes in geomorphology was emphasized in the influential text of Leopold and others (1964). In that book the existing wealth of process measurements was related to the form and change of riverscapes. However, a quality of streamflow gauging, sediment sampling, and the like is that such measurements are biased toward processes that occur frequently enough and are small enough to be easily studied. Processes of great rarity and immense intensity may defy direct measurement. Such processes, which include intense floods, must be studied indirectly, through theory or by the study of their effects on the landscape.

Leopold and others (1964, p. 7) observed that geomorphologists have a much better understanding of depositional rather than erosional landforms. This is because deposits can be easily viewed in the field, and interpretations can be made of the mechanics of their emplacement. With erosion much of the evidence of formation is removed by the erosional process itself. In Chapter 5, Victor R. Baker introduces the topic of flood erosion. His essay emphasizes the importance of resistance factors as well as dynamic ones in establishing the erosive effects of floods. Especially important is the distinction between alluvial channels and nonalluvial (resistant boundary) channels. Chapter 5 concentrates on nonalluvial channels, since alluvial channels are inextricably linked to concerns with sediment erosion and transport. The latter topics are treated in Chapter 6 by Paul D. Komar.

Komar's approach to sediment transport by extraordinary floods emphasizes the theoretical mechanics of the problem. Because most sediment transport formulae are totally or in part empirical and because their observational basis comes from flume experiments and/or normal river flows, Komar is skeptical of their applicability in scientific studies of sedimentary phenomena in cataclysmic floods. It should be remembered that much of the existing wealth of sediment transport methodology derives from the efforts of hydraulic engineers concerned with channel stability or structural design under noncataclysmic flow conditions. Komar considers the first principles of grain thresholds, flow competence modes, and quantities of sediment transport as applicable up to the scales of the great Pleistocene glacial lake bursts of the northwestern United States.

Small, steep drainage basins in regions of abundant sediment supply are subject to flooding that may mobilize immense concentrations of sediment. The resulting flow phenomena may differ fundamentally from water flows. Because standard hydraulic and sediment-transport equations are strictly applicable only to water flows, serious errors of flow prediction and remedial design may occur if these equations are applied to debris flows or hyperconcentrated flows. Since such analyses are often made without observing a cataclysmic flow event, such errors can be easily made by field engineers. Chapter 7 by John E. Costa takes a major step toward the avoidance of this problem. The rheological character of past flow events can be generally distinguished by an in-depth study of the geomorphic and sedimentologic features that they produce. Clearly, where the consequences of misdesign or poor prediction are grave, a geomorphic-sedimentologic analysis of past flow events is an essential component of flood studies for small, mountainous drainage basins.

Until recently, flood sedimentation in bedrock fluvial systems has received relatively little attention from both geomorphologists and sedimentologists. Therefore, Chapter 8 on this topic by Baker and R. Craig Kochel describes some phenomena that may be unfamiliar to many readers. Chapter 8 focuses on resistant-boundary rivers that display narrow, deep cross sections. Flood flows in such rivers produce phenomenally high flow velocity, shear stress, and stream power per unit area (Baker and Pickup, 1987). Indeed, these values may be orders of magnitude greater than experienced even in great floods for alluvial channels (Baker and Costa, 1987). The sedimentary phenomena include distinctive bedload deposits, including various gravel bars and giant gravel waves.

A most important aspect of intense flood flows in narrow, deep bedrock channels is their capability of transporting sand in suspension or even as washload. Even the central Chang Jiang (Yangtze River) of China has reaches that can achieve the necessary flow intensity. When this coarse suspended load settles in channel-margin areas of velocity reduction, it can build up slackwater deposits that subsequently prove useful in flow reconstruction (see Chapter 21 of this volume). Tributary mouths are especially susceptible to this form of deposition when backflooding from the mainstem channel carries suspended sand upstream into the tributary.

Flood flow phenomena on bedrock streams have proven very difficult to measure in the field. This may be one reason, in this era of empirical process geomorphology, that these phenomena have been underappreciated by modern fluvial geomorphologists. In Chapter 8, Baker and Kochel document considerable field experience and observations of bedrock channel sedimentation in China, northern Australia, and the southwestern United States. In addition, recent flume experiments on the flood slackwater sedimentation are described.

Chapter 9 by G. Robert Brakenridge discusses the complex architecture of floodplain stratigraphic sequences. Brakenridge argues that various magnitude–frequency relationships of flow stages lead to various facies geometries for floodplains. Examples from the southwestern United States, southern Missouri, central Tennessee, northern Vermont, Poland, Germany, and Austria show that episodic deposition along meandering rivers causes stratigraphic sequences containing paleosols and varying facies relationships. Studies of the complex internal stratigraphy of floodplains can be used to retrodict past river flood regimes and the interplay of internal and external variables in controlling river history.

William B. Bull in Chapter 10, "Floods—Degradation and Aggradation," emphasizes the important concept of threshold exceedence by floods for stream subsystems that are in equilibrium. Because equilibrium (or graded) conditions for streams derive from responses over a period of years, this chapter provides a long-term perspective on flood response. The important concept of stream power (Bagnold, 1966, 1977, 1980) is used to establish a criterion for net degradation or aggradation. This criterion, the ratio of stream power to resisting power, can indicate departures from equilibrium conditions which influence how a given reach of stream will respond to flood events. Temporal variations in the criterion are dictated in part by feedback mechanisms involving hillslope and streambed armoring. Climatic change and tectonism have major long-term influences on the stream power criterion.

REFERENCES

Bagnold, R. A. (1966). An approach to the sediment transport problem from general physics. *Geol. Sur. Prof. Pap. (U.S.)* **422-I**, 1–37.

Bagnold, R. A. (1977). Bed-load transport by natural rivers. *Water Resour. Res.* **13**, 303–312.

Bagnold, R. A. (1980). An empirical correlation of bedload transport rates in flumes and natural rivers. *Proc. Roy. Soc.* **372A**, 453–473.

Baker, V. R., and Costa, J. E. (1987). Flood power. *In* "Catastrophic Flooding" (L. Mayer and D. Nash, eds.), pp. 1–22. Allen and Unwin, London.

Baker, V. R., and Pickup, G. (1987). Flood geomorphology of the Katherine Gorge, Northern Territory, Australia. *Geol. Soc. Am. Bull.* **98**, 635–646.

Leopold, L. B., Wolman, M. G., and Miller, J. P. (1964). "Fluvial Processes in Geomorphology." W. H. Freeman, San Francisco.

5

FLOOD EROSION

VICTOR R. BAKER
Department of Geosciences, University of Arizona, Tucson, Arizona

INTRODUCTION

Geomorphic erosion can be analyzed as the dynamic action of forces on resistant geologic materials. Indeed, Strahler (1952) proposed that the science of geomorphology be organized specifically to study forces, both gravitational and molecular, and material properties, especially resistance to deformation by imposed forces. Because floods impose higher than average forces on stream bed and bank materials, their study is critical to all work in fluvial geomorphology.

In studying flood erosion an important distinction must be made between alluvial and nonalluvial river channels. The former have their channel characteristics established by variations in the discharge of water and sediment, as modified by the growth of vegetation along the channel banks (Maddock, 1976a). Such channels are self-formed through the interdependent adjustment of morphologic variables. These systems are characterized by a tendency to adjust toward equilibrium or regime. Critical to this adjustment is the erosion and transport of the individual sedimentary particles that make up the stream bed. Thus, erosion in alluvial channels is intimately tied to the question of sediment transport, as discussed more fully by Komar (this volume).

Nonalluvial river channels are characterized by sufficiently high boundary resistance that an equilibrium or regime adjustment among sediment capacity, water discharge, and morphologic variables is not possible for most flood stages. As will be discussed, extremely intense, rare floods may overcome the high threshold for erosion in such channels, producing boundary roughness adjustments that tend to re-establish regime. However, such systems are sediment limited, and true equilibrium cannot be achieved.

RESISTANCE FACTORS

The forces affecting flood erosion have been well documented in terms of flow physics and hydraulics. Much less attention has been paid to resistance factors in fluvial geomorphology. Resistance involves properties of both intact and modified materials comprising channel beds and banks. For example, intact granite on bare slopes affords immense resistance to fluvial erosion. Only flows of exceptional magnitude (and low frequency) will be of sufficient strength to pluck or scour such a resistant rock. In contrast, however, granite buried beneath an alluvial fill will disintegrate by grusification to sand-sized particles that are easily transported by flows of moderate size (and high frequency).

The action of ice wedging provides another example of variable resistance. In cold climatic environments this process preconditions stream banks that otherwise would be resistant to erosion by melt-season streamflow (Walker and Arnborg, 1966). An extreme example of this preconditioning is the "Eisrinde effect" described by Büdel (1982). The *Eisrinde* (translation: ice rind) is the uppermost zone of permafrost that extends beneath braided meltwater channels in subpolar areas such as Svalbard (Spitzbergen). Seasonal melting and freezing beneath the stream bed concentrates mechanical weathering of bedrock at precisely the point where annual melt-season floods can remove the generated detritus. The result can be extremely rapid degradation, as much as 3000 mm/1000 yr. By contrast modern rates of terrestrial landscape degradation range from 50 to 500 mm/1000 yr (Ollier, 1981).

In alluvial channels with noncohesive beds and banks, resistance to erosion can be minimal, related only to the capability of the flow to entrain the sedimentary particles. Cohesive sediments offer greater resistance, generally precluding grain-by-grain dislodgement. This greater resistance is associated with narrow, deep-channel cross sections, as opposed to the wide, shallow-channel cross sections of noncohesive-sediment streams (Schumm, 1977). However, the association is even more general than in the well-known sequence for alluvial channels. Shepherd (1975, 1979) observed that rock resistance exerted a major influence on bedrock channel morphology in central Texas. For limestone, which yields only large blocks by plucking erosion of joints and bedding planes, channels display pronounced pool-and-riffle development in rock, relatively high sinuosity, and relatively low width-to-depth ratios. In contrast, granite, which yields abundant sand-sized grus fragments through weathering, favors the development of low-sinuosity braided patterns with high width-to-depth ratios.

An excellent example of bedrock control of channel morphology occurs at Wiggley Gorge (Fig. 1) near Alice Springs, central Australia. A resistant ridge of Charles River Gneiss results in an upstream reach (U in Fig. 1) characterized by riffles of locally derived coarse gravel and boulders (R in Fig. 1) separated by scour pools (P in Fig. 1) along the main channel. Thick sand deposits (S in Fig. 1) are derived from weathering of upstream outcrops of grussified Alice Springs Granite. The sand supply is limited, but it flushes through the reach as washload and suspended load during rare large floods. Sand accumulates locally in slackwater deposits, as described by Baker and Kochel (this volume).

Figure 2 shows a bedrock pool and riffle reach in Heavitree Quartzite, another resistant rock unit in the vicinity of Alice Springs. In contrast to such narrow, deep bedrock channels, less resistant rocks allow the development of relatively wide, shallow channels. Reach W in Figure 1 is developed in weathered granite. Active sandbars occur

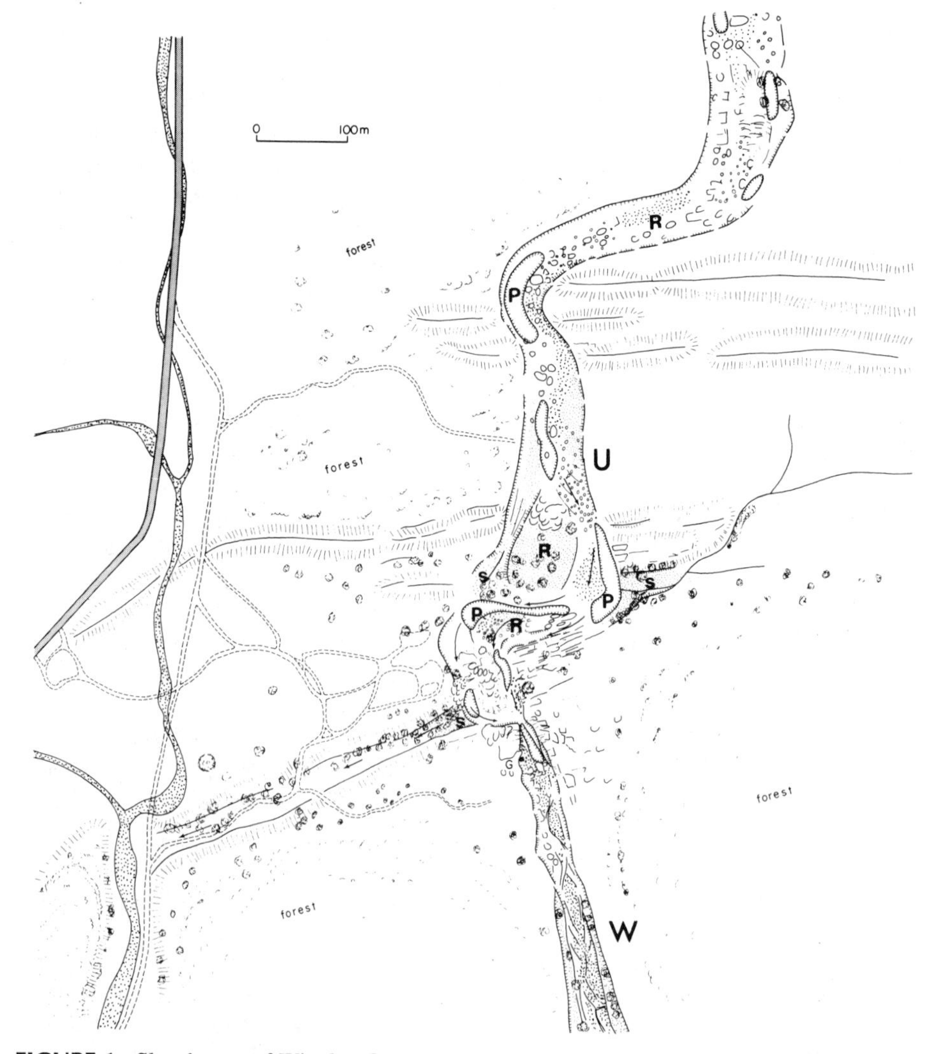

FIGURE 1. Sketch map of Wiggley Gorge near Alice Springs, central Australia. Scour pools and depositional riffles of gravel and boulders characterize reach U, while braided sand bed channel characterizes downstream reach W.

FIGURE 2. Bedrock pool-and-riffle reach at Redbank Gorge, central Australia. This reach is morphologically similar to reach U in Figure 1.

in the braided channel thalweg; and vegetated sand berms, 1–2 m high, line the channel margins (Fig. 3). The latter constitute a floodplain, confining relatively frequent flood flows to a cross-sectional area that is adjusted to those flows. Rarer, large flows inundate these vegetated berms and deposit suspended sand by vertical accretion.

Vegetation plays a critical role in flow resistance, particularly in relation to relaxation times and thresholds for flood erosional effects. Because of the resistance of vegetation-mantled banks and floodplains, there may be progressive, long-term encroachment of vegetation into the channel, especially on small streams. This vegetation increases roughness and reduces flow conveyance. A major, rare flood may be required to overcome the resistance threshold provided by the vegetated valley bottom. Spectacular threshold exceedences occur when intense floods impact bedrock-confined channels and locally rip up vegetation, exposing underlying floodplain sediments to massive scour (Fig. 4). In the absence of subsequent major floods, this flood-widened channelway will recover to its original dimensions at varying rates depending on climatic-geomorphic factors (Wolman and Gerson, 1978).

ALLUVIAL CHANNELS

Flood erosional effects have been especially well documented on rivers with alluvial beds. Where sediments can be easily entrained by a range of flood flows, rivers will scour during high flow (Leopold et al., 1964, pp. 227–241). Detailed monitoring of alluvial channels shows that initial stage increases during a flood produce net filling (Leopold and Maddock, 1953). However, continued rise to the flood peak generates net scour to depths below the preflood level. Filling occurs again during late hydrograph recession. The result is that scour and fill may not be strictly correlated to discharge. A river channel may be filling at a given discharge (or mean flow velocity) on rising stage but then be scouring at the same mean flow velocity when it is reached during falling stage (Leopold and Maddock, 1953, pp. 30–35).

Regime Expressions

The fact that erosional effects in alluvial channels do not correlate directly with hydraulics, for example, mean velocity

FIGURE 3. Photograph of reach W in Wiggley Gorge (Fig. 1). Note the vegetated floodplain and sandy channel.

FIGURE 4. Scour of vegetated floodplain surface, Medina River, near Bandera, central Texas. Flooding occurred during tropical storm Amelia in August 1978.

or discharge, is a consequence of regime behavior. Adjustments in sediment concentration and bed roughness are the variables that are most difficult to evaluate in predicting alluvial channel erosion. The only effective procedure derives from the tendency of alluvial rivers (a) to conserve adjustments that lead to equilibrium and (b) to dissipate adjustments that do not. This observation can be expressed as regime equations (Maddock, 1976b) such as

$$\frac{V^2}{D} \cong c_s d_s \quad (1)$$

where V = mean flow velocity
D = depth
c_s = sediment concentration
d_s = mean sediment size (all in consistent units).

Since hydraulic geometry (Leopold and Maddock, 1953) defines

$$W \propto Q^{1/2} \quad (2)$$

$$D \propto Q^{1/3} \quad (3)$$

$$V \propto Q^{1/6} \quad (4)$$

$$S \propto Q^{-1/6} \quad (5)$$

where Q = discharge
W = width
S = slope

Then it is also possible to write the following forms of Equation (1)

$$VS \cong c_s d_s \quad (6)$$

and

$$DS^2 \cong c_s d_s \quad (7)$$

Note that velocity is inversely proportional to slope as long as c_s and d_s remain constant. Also note that the major role of width [Eq. (2)] leads to prominent channel widening during discharge increases by floods in alluvial channels.

Effects of Rising Stage

In a meandering alluvial stream with a floodplain, the rising stages of flood flow yield a sequence of adjustments. Within the channel, width does not markedly increase as depth increases rapidly up to bankfull stage. The resulting increase in velocity and shear stress greatly increases sediment transport capability. The stream erodes sediment from its bed to keep pace with an increasing equilibrium rate, but there may be limits on sediment availability, either because of supply or resistance. If the sediment concentration increases faster than the equilibrium rate, the overbank flows will deposit the excess sediment as natural levees, thereby restoring equilibrium (Maddock, 1976a). If sediment transport does not keep pace with the equilibrium rate, the flow will attack the banks, widening the channel (Fig. 5). Equilibrium is thereby restored since sediment is added to the flow, and the enlarged channel reduces velocity for a given discharge thereby producing a lower transport rate.

When sediment-deficient water moves onto a floodplain, it can be quite erosive. As stage continues to increase in large overbank floods, the flow may be more efficiently conveyed in a straighter pattern than along the original sinuous channel. The combination of sediment-deficient water and greater discharge leads to chute erosion (Fig. 6) as the overbank flows are carried in a relatively straight, high-gradient flood channel. Such erosion of floodplain surfaces also depends on resistance factors. Vegetated floodplains may have a high threshold for initiating the scour. This results in the need for an exceptionally large, rare event to produce the erosion.

Bank erosion on rising stage occurs by direct shearing through hydraulic action (Knighton, 1984). However, erosion can also occur on falling stage through liquefaction of floodplain sediments. An excellent example of this process occurred during flooding of the Guadalupe River, near Sisterdale, Texas, in August 1978. Preflood banks consisted of calcareous silt-sand alluvium with sufficient cohesion

FIGURE 5. Pedernales River, near Fredericksburg, central Texas, showing flood widening of alluvial banks (right side of photograph). Overbank flow is occurring through a levee breach. Flooding occurred during tropical storm Amelia in August 1978.

FIGURE 6. Spectacular chute developed on the Medina River near Bandera, Texas. Scour occurred during Amelia flooding in August 1978.

to stand for years in vertical cuts (Fig. 7). Flood stages of over 10 m saturated the banks, which flowed into the channel after hydrograph recession (Fig. 8). Spectacular examples of such bank erosion occur on major rivers, such as the Mississippi, Brahmaputra (Coleman, 1969), and Amazon (Sternberg, 1975).

Example: Southern Arizona

The larger streams of desert lowlands in the southwestern United States generally flow on valley floors underlain by fine-grained unconsolidated alluvial fills. Their channel banks, unless stabilized by riparian vegetation, are exceptionally sensitive to erosion during rare, large floods (Cooke and Reeves, 1976; Graf, 1985; Kresan, this volume). The sequence of flow events plus their water and sediment characteristics determine the nature of erosion and channel change.

An especially well-documented flood geomorphic history characterizes an 88-km reach of the Gila River near Safford, Arizona. Burkham (1972) demonstrated that the period 1905 through 1917 was marked by a series of eight relatively large winter floods characterized by low sediment loads. These flows were especially erosive, widening channels from a preflood width of 90 m to a postflood width of 600 m. In contrast, from 1918 to 1965, floods were relatively small, generally consisting of summer flows characterized by high sediment loads. During this period channels generally recovered to a width of about 100 m. These changes, in turn, led to major changes in the mag-

FIGURE 7. Road cut of loamy alluvium at low-water crossing of Guadalupe River near Sisterdale, Texas. Photograph taken in October 1974.

FIGURE 8. Same road cut as shown in Figure 7, photographed in August 1978, shortly after flooding by tropical storm Amelia caused the river rise to top of the alluvial surface. Bank failure occurred by liquefaction.

FIGURE 7.

FIGURE 8.

nitude, timing, and conveyance of flood waves through the reach (Burkham, 1976).

Channel recovery processes in southern Arizona display a pronounced association between riparian vegetation and sedimentation. For example, during the 1941–1965 phase of low-magnitude, sediment-charged summer floods, Rillito Creek, near Tucson, Arizona, narrowed as the riparian vegetation growth increased the hydraulic roughness of channels, thereby trapping sediment (Pearthree and Baker, 1987). However, in December 1965, a prolonged, sediment-deficient winter flood induced major channel enlargement. During the subsequent decade relatively moderate flow events characterized a period of above-average precipitation. Channels again narrowed markedly as riparian vegetation was locally re-established, but urbanization effects are now modifying this vegetative response. Beginning in 1978 another series of large flow events resulted in extensive channel widening, culminating in the spectacular effects of the flood of October 1983 (Kresan, this volume). Similar sequences of channel recovery have been documented for streams in Kansas (Schumm and Lichty, 1963) and Oklahoma (Wolman and Gerson, 1978).

NONALLUVIAL CHANNELS

In channels with exceptionally resistant flow boundaries the equilibria states discussed for alluvial channels cannot be achieved at relatively frequent, low-magnitude discharges. As stage rises, erosive capability and sediment transport capacity increase immensely, but sediment availability is severely constrained. In bedrock channels with narrow, deep cross sections the disparity between increasing erosive capability and sediment deficiency is most extreme. The excess energy of the flow can neither be dissipated in roughness adjustment of the resistant flow boundaries nor can it be damped by sediment transport. This results in the appearance of macroturbulence (Matthes, 1947). Especially intense energy dissipation by large-scale vortex action is the most important manifestation of macroturbulence. Conditions necessary for its appearance include (1) a steep energy gradient, (2) a low ratio of actual sediment transported to potential sediment transport, and (3) an irregular, rough boundary capable of generating flow separation (Matthes, 1947).

As flow depth continues to rise in a narrow, deep bedrock channel, flow velocity, shear stress, stream power, and macroturbulence all increase to immense values (Baker, 1984). If the flood is large enough to achieve values that exceed boundary resistance, spectacular channel enlargement will occur (Baker, 1977). Bedrock can be eroded by plucking, cavitation, and other processes requiring immense flow intensity (Baker, 1978a, 1979). Such erosion cannot achieve adequate sediment loading of the disequilibrium flood flows, but it can modify boundary roughness, thereby allowing more efficient energy dissipation. However, the short duration of extreme events and the great resistance of the boundary generally limits the completeness of this roughness adjustment.

The nature of bedrock boundary roughness response to intense floods can be idealized through flume experiments (Shepherd and Schumm, 1974). An evolutionary sequence occurs during prolonged flows, beginning with longitudinal lineations and potholes, proceeding to longitudinal grooves, and finally to the development of a single pronounced narrow, deep inner channel. The latter is highly efficient as generating the intense flow conditions necessary to erode intact rock. Baker (1973a, 1978b) demonstrated that a similar evolutionary sequence characterizes cataclysmic Pleistocene flood erosion of basalt bedrock in the Channeled Scabland of Washington State. Excess energy in exceptionally deep, high-velocity flood flows is expended in producing form roughness on the bedrock channel boundaries. The characteristic morphology is a prominent inner channel, known as a Dalles-type fluvial channel (Bretz, 1924).

Excellent examples of inner channels occur in the Three Gorges section of the central Chang Jiang (Yangtze River) of China. Ledges of resistant sandstone are exposed at moderate discharge (Fig. 9) but become submerged beneath 10 m or more of floodwater at high stage. The hydraulics of floods in the Three Gorges is such that shear stress and unit stream powers are 2 to 3 orders of magnitude greater than achieved during comparable floods on alluvial rivers (Baker and Kochel, this volume).

Erosion of floodwater has been attributed classically to corrosion, corrasion, plucking (or quarrying), and cavitation. Corrosion effects involve chemical processes, such as solution of limestone. Corrasion is the mechanical process of erosion through contact with the sediment load with the bed. Foley (1980) presents one of the few attempts to analyze the mechanics of abrasion by fluvial sediment. Plucking is most effective under macroturbulent conditions where bedded or jointed rock yields to local hydraulic forces. Spectacular examples occur in the Channeled Scabland (Baker, 1973a, 1978a).

The critical conditions necessary to initiate cavitation in river flow have been evaluated by several investigators (Hjulström, 1935; Barnes, 1956; Baker, 1979). Their calculations show that mean flow velocities necessary to initiate the process are generally higher than about 10 m/s for flow depths greater than about 4 m. Such high mean flow velocities are rarely achieved, except in very pronounced bedrock channel constrictions, such as occurred for Pleistocene cataclysmic flooding in the Channeled Scabland (Baker, 1973b). Even where cavitation occurs, the phenomenon is probably short-lived (Barnes, 1956). A small depression in rock initiated by cavitation would likely trap particles of bedload and continue to enlarge by abrasion

FIGURE 9. Inner channel development on the Chang Jiang (Yangtze River) near Fengdu, Sichuan Province, central China.

through the rotary flow described by Alexander (1932). The end result would be a pothole (Fig. 10). Spiral grooves on pothole walls indicate vortex flow of sediment-charged water (Lugt, 1983).

A variety of small-scale erosional forms develop on rock surfaces in narrow, deep canyons (Fig. 11). Most common are flute marks (Fig. 12), which have been described extensively by Allen (1971, 1982). Excellent examples of fluting and polished rock surfaces occur in the Katherine Gorge (Fig. 13), Northern Territory, Australia (Baker and Pickup, 1987). These occur on a very resistant sandstone, the Kombolgie formation. The polished surfaces locally form facets (Fig. 14), analogous to the morphology of ventifacts (Greeley and Iversen, 1985). Faceted rock surfaces formed by fluvial action are rare, but they have been described as forming in analogous fashion to eolian faceting (Maxson, 1940). Maxson (1940) noted their occurrence in canyon reaches characterized by steady and rapid sediment-charged water flow not split by boulders. Faceted boulders and associated flutes occur at rapids in the Grand Canyon of Arizona (Maxson and Campbell, 1935). Fine-scale erosional pits and grooves occur on the Katherine Gorge facets (Fig. 15).

DISCUSSION

The role of floods in the erosion of stream channels has been one of the most controversial topics in fluvial geomorphology. Nearly every experienced fluvial geomorphologist can cite examples where major floods have produced surprisingly little channel change (Dury, 1973), very temporary effects (Costa, 1974), or where spectacular adjustment has occurred (Baker, 1977; Gupta, 1975, 1983; Stewart and LaMarche, 1967). Indeed, the famous Spokane flood debate, concerning the effects of the greatest known fresh-water floods on the planet (Bretz, 1969; Baker, 1973b; Baker and Bunker, 1985), centered on the issue of the erosive capability of running water (Baker, 1978c). Those who disbelieved the flood theory of J Harlen Bretz did so out of their experience that rivers did not behave as Bretz proposed. Subsequent work showed that their experience, not Bretz's theory, was inadequate. Bretz, not his critics, turned out to be the true uniformitarian (Baker, 1981).

Part of the debate concerning flood effects on river channels centers on the issue of the magnitude and frequency of the input forces (Wolman and Miller, 1960; Baker, 1977; Wolman and Gerson, 1978; Kochel, this volume).

FIGURE 10.

FIGURE 11.

FIGURE 12. Irregular flute marks developed on Buldiva Sandstone, Umbrawarra Gorge, northern Australia.

However, another issue of importance is that of the resistance of the system experiencing those forces. Ultimately the effectiveness of flood erosion depends on the exceedence of a resistance threshold in bed or bank materials, including vegetation, by the stream power per unit area generated during flood flow. Much of the confusion over this issue arises because there is a great range in resistance factors, such as that between sand bed and bedrock channels. Moreover, the response times and relaxation times for the adjustments are highly variable, as are the ranges of discharge at which thresholds are exceeded. For example, only rare, great floods may be capable of eroding highly resistant rock. Moreover, the short duration of such floods generally precludes an adequate response time for boundary roughness to be adjusted to the flow intensity. Instead, incomplete disequilibrium effects will be preserved in the spectacular erosion following an exceptionally intense flood. These may persist for long time periods in some geomorphic settings (Baker, 1977). In contrast, low-threshold systems are extremely sensitive to floods of lower intensity and high frequency. Changes in sediment and water discharge during a large event generated very rapid responses in channel geometry and bed roughness according to regime relationships. Recovery following a major flood may also be rapid, depending on the subsequent sequence of flow events and vegetative response.

This chapter has provided but a brief overview of the complexities of flood erosion. Other chapters in this volume

FIGURE 10. Pothole developed on a rock bar surface in Katherine Gorge, northern Australia. Note the spiral grooving of the pothole walls.

FIGURE 11. Small pits developed a fluted sandstone surface, Umbrawarra Gorge, northern Australia (note especially below pocket knife). Englargement of a prominent flute has resulted in the incipient, elongate pothole at right.

FIGURE 13.

FIGURE 14.

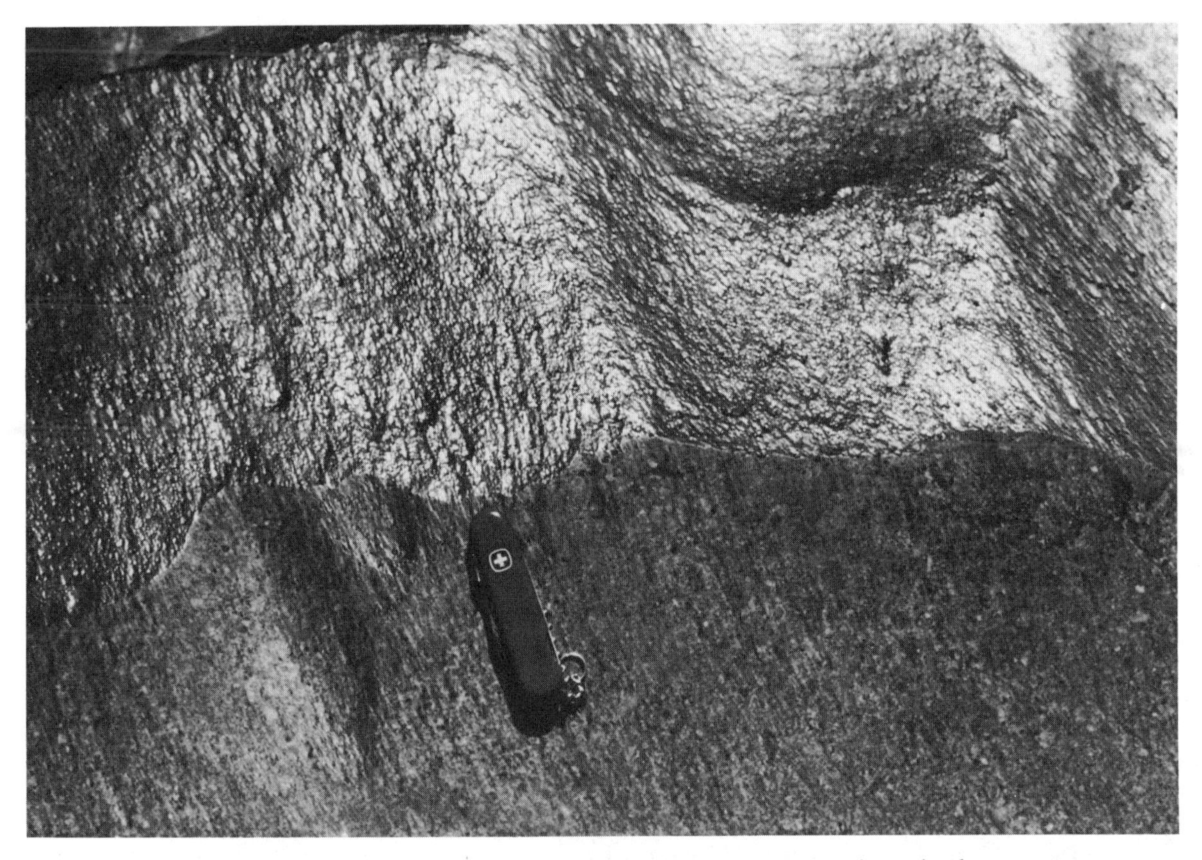

FIGURE 15. Polished sandstone surface in Katherine Gorge. The fine-scale erosional grooves are conjugate spindle-shaped flute marks (terminology of Allen, 1982). The flutes open in a downstream direction to widths of 0.2–0.5 cm, and they extend parallel to flow for lengths of 0.5–2.5 cm. Their proximal (upstream) termini are generally located at individual mineral grains, which may have served as material defects that facilitated their origin (Allen, 1971).

describe interesting details of flood erosion in a variety of geomorphic settings. However, there remains a critical research need to compare and contrast quantitatively the significant factors of flood erosion in a variety of geomorphic settings, including flow strengths, boundary resistances, response times, relaxation times, and morphologic change. Especially crucial is the development of data on effects of rare, large-magnitude events that are inadequately documented by numerous short-term systematic stream gauge records but that occur repeatedly during the long-term geomorphic evolution of a river system.

REFERENCES

Alexander, H. S. (1932). Pothole erosion. *J. Geol.* **40**, 305–337.

Allen, J. R. L. (1971). Transverse erosional marks of mud and rock: Their physical basis and geological significance. *Sediment. Geol.* **5**, 167–385.

Allen, J. R. L. (1982). "Sedimentary Structures, Their Character and Physical Basis," Vol. 2. Elsevier, Amsterdam.

Baker, V. R. (1973a). Erosional forms and processes for the catastrophic Pleistocene Missoula floods in eastern Washington.

FIGURE 13. Oblique aerial view of the Katherine Gorge, Northern Territory, Australia. Note the slotlike canyons produced by stream incision into vertical joints and faults in very resistant sandstone.

FIGURE 14. Polished sandstone produced by abrasive action of sand carried by flood flows in Katherine Gorge. Polish occurs on two facets of the rock surface. Flow direction is from right to left.

In "Fluvial Geomorphology" (M. Morisawa, ed.), Publ. Geomorphol., pp. 123–148. State University of New York, Binghamton.

Baker, V. R. (1973b). Paleohydrology and sedimentology of Lake Missoula flooding in eastern Washington. *Spec. Pap.—Geol. Soc. Am.* **144**, 1–79.

Baker, V. R. (1977). Stream-channel response to floods with examples from central Texas. *Geol. Soc. Am. Bull.* **88**, 1057–1071.

Baker, V. R. (1978a). Paleohydraulics and hydrodynamics of scabland floods. *In* "The Channeled Scabland" (V. R. Baker and D. Nummedal, eds.), Nat. Aeronaut. Space Admin., pp. 59–79. Washington, D.C.

Baker, V. R. (1978b). Large-scale erosional and depositional features of the Channeled Scabland. *In* "The Channeled Scabland" (V. R. Baker and D. Nummedal, eds.), Nat. Aeronaut. Space Admin., pp. 81–115. Washington, D.C.

Baker, V. R. (1978c). The Spokane Flood controversy and the Martian outflow channels. *Science* **202**, 1249–1256.

Baker, V. R. (1979). Erosional processes in channelized water flows on Mars. *JGR, J. Geophys. Res.* **84**, 7985–7993.

Baker, V. R. (1981). "Catastrophic Flooding: The Origin of the Channeled Scabland." Dowden, Hutchinson & Ross, Stroudsburg, Pennsylvania.

Baker, V. R. (1984). Flood sedimentation in bedrock fluvial systems. *Mem.—Can. Soc. Pet. Geol.* **10**, 87–98.

Baker, V. R., and Bunker, R. C. (1985). Cataclysmic late Pleistocene flooding from Glacial Lake Missoula: A review. *Quat. Sci. Rev.* **4**, 1–41.

Baker, V. R., and Pickup, G. (1987). Flood geomorphology of the Katherine Gorge, Northern Territory, Australia. *Geol. Soc. Am. Bull.* **98**, 635–646.

Barnes, H. L. (1956). Cavitation as a geological agent. *Am. J. Sci.* **254**, 493–505.

Bretz, J H. (1924). The Dalles type of river channel. *J. Geol.* **32**, 139–149.

Bretz, J H. (1969). The Lake Missoula floods and the Channeled Scabland. *J. Geol.* **77**, 505–543.

Büdel, J. (1982). "Climatic Geomorphology." Princeton Univ. Press, Princeton, New Jersey.

Burkham, D. E. (1972). Channel changes of the Gila River in Safford Valley, Arizona, 1846–1970. *Geol. Surv. Water-Supply Pap. (U.S.)* **655-G**, 1–24.

Burkham, D. E. (1976). Effects of changes in an alluvial channel on the timing, magnitude, and transformation of flood waves, southeastern Arizona. *Geol. Surv. Prof. Pap. (U.S.)* **655-K**, 1–25.

Coleman, J. M. (1969). Brahmaputra River: Channel processes and sedimentation. *Sediment. Geol.* **3**, 129–239.

Cooke, R. U., and Reeves, R. W. (1976). "Arroyos and Environmental Change in the American Southwest." Oxford Univ. Press (Clarendon), London and New York.

Costa, J. E. (1974). Response and recovery of a piedmont watershed from tropical storm Agnes, June 1972. *Water Resour. Res.* **10**, 106–112.

Dury, G. H. (1973). Magnitude-frequency analysis and channel morphometry. *In* "Fluvial Geomorphology" (M. Morisawa, ed.), Publ. Geomorphol., pp. 91–121. State University of New York, Binghamton.

Foley, M. G. (1980). Bedrock incision by streams. *Geol. Soc. Am. Bull.* **91**, 2189–2213.

Graf, W. L. (1985). "The Colorado River: Instability and Basin Management." Assoc. Am. Geogr., Washington, D.C.

Greeley, R., and Iversen, J. D. (1985). "Wind as a Geological Process on Earth, Mars, Venus and Titan." Cambridge Univ. Press, London and New York.

Gupta, A. (1975). Stream characteristics in eastern Jamaica, an environment of seasonal flow and large floods. *Am. J. Sci.* **275**, 825–847.

Gupta, A. (1983). High-magnitude floods and stream channel response. *Spec. Publ. Int. Assoc. Sedimentol.* **6**, 219–227.

Hjulström, F. (1935). Studies in the morphological activity of rivers as illustrated by the River Fyris. *Bull. Geol. Inst. Univ. Uppsala* **25**, 221–528.

Knighton, D. (1984). "Fluvial Forms and Processes." Arnold, London.

Leopold, L. B., and Maddock, T., Jr. (1953). The hydraulic geometry of stream channels and some physiographic implications. *Geol. Surv. Prof. Pap. (U.S.)* **252**, 1–57.

Leopold, L. B., Wolman, M. G., and Miller, J. P. (1964). "Fluvial Processes in Geomorphology." Freeman, San Francisco, California.

Lugt, H. J. (1983). "Vortex Flow in Nature and Technology." Wiley, New York.

Maddock, T., Jr. (1976a). A primer on floodplain dynamics. *J. Soil Water Conserv.* **31**(2), 44–47.

Maddock, T., Jr. (1976b). Equations for resistance to flow and sediment transport in alluvial channels. *Water Resour. Res.* **12**, 11–21.

Matthes, G. H. (1947). Macroturbulence in natural stream flow. *Trans., Am. Geophys. Union* **28**, 255–262.

Maxson, J. H. (1940). Fluting and faceting of rock fragments. *J. Geol.* **48**, 717–751.

Maxson, J. H., and Campbell, I. (1935). Stream fluting and stream erosion. *J. Geol.* **43**, 729–744.

Ollier, C. D. (1981). "Tectonics and Landforms." Longman, London.

Pearthree, M. S., and Baker, V. R. (1987). "Channel Change Along the Rillito Creek System of Southeastern Arizona, Pre-1941 to 1983: Implications for Floodplain Management." University of Arizona, Bureau of Geology and Mineral Technology, Tucson.

Schumm, S. A. (1977). "The Fluvial System." Wiley, New York.

Schumm, S. A., and Lichty, R. W. (1963). Channel widening and flood-plain construction along Cimarron River in southwestern Kansas. *Geol. Surv. Prof. Pap. (U.S.)* **352-D**, 71–88.

Shepherd, R. G. (1975). Geomorphic operation, evolution, and equilibria, Sandy Creek Drainage, Llano Region, central Texas. Ph.D. Dissertation, University of Texas, Austin.

Shepherd, R. G. (1979). River channel and sediment responses to bedrock lithology and stream capture, Sandy Creek drainage, central Texas. *In* "Adjustments of the Fluvial System" (D. D. Rhodes and G. P. Williams, eds.), pp. 255–275. Kendall/Hunt Publ. Co., Dubuque, Iowa.

Shepherd, R. G., and Schumm, S. A. (1974). An experimental study of river incision. *Geol. Soc. Am. Bull.* **85**, 257–268.

Sternberg, H. O. (1975). "The Amazon River of Brazil." Franz Steiner Verlag, Wiesbaden.

Stewart, J. H., and LaMarche, V. C. (1967). Erosion and deposition produced by the flood of December 1964 on Coffee Creek, Trinity County, California. *Geol. Surv. Prof. Pap. (U.S.)* **422-K**, 1–22.

Strahler, A. N. (1952). Dynamic basis of geomorphology. *Geol. Soc. Am. Bull.* **63**, 923–938.

Walker, H. J., and Arnborg, L. (1966). Permafrost and ice-wedge effect on riverbank erosion. *Proc. Int. Permafrost Conf., 1963*, pp. 164–171.

Wolman, M. G., and Gerson, R. (1978). Relative scales of time and effectiveness of climate in watershed geomorphology. *Earth Surf. Processes* **3**, 189–208.

Wolman, M. G., and Miller, J. P. (1960). Magnitude and frequency of forces in geomorphic processes. *J. Geol.* **68**, 54–74.

6

SEDIMENT TRANSPORT BY FLOODS

PAUL D. KOMAR
College of Oceanography, Oregon State University, Corvallis, Oregon

INTRODUCTION

Exceptional floods can erode and transport vast quantities of sediments, resulting in major modifications of the landscape. Understanding the processes of sediment transport and evaluating the rates and quantities offer difficult enough problems under normal flow conditions; with floods the task is a true challenge. Our sediment transport formulas are totally or in part empirical, based on flume experiments with some supporting measurements from rivers having modest discharges. Applications of these formulas to floods can require extrapolations by factors of 100 or more beyond the original data base. When the equation is purely empirical, such extrapolations are of course unwarranted; if the relationship has a reasonable theoretical foundation based on the mechanics of grain and fluid motions, then there may be more confidence in the application to floods. Such limitations to our observations of sediment transport processes should be kept firmly in mind throughout this chapter.

The present review will begin with the related topics of grain threshold and flow competence, relationships between the flow's magnitude and the sediment size it is capable of transporting. We shall then turn to an examination of the sediment transport processes, considering the several modes of transport as well as the potential for evaluating the quantities of that transport.

GRAIN THRESHOLD AND FLOW COMPETENCE

Standard Threshold of Uniform Grains

The first stage in sediment transport is the grain threshold condition, the flow strength required for the initial movement of grains of a specific size and density. This relationship has been established mainly by experiments, nearly all of them conducted in the controlled conditions of laboratory flumes. A typical experiment consists of employing nearly uniform size grains, commonly a single sieve fraction, placing the sediment in the flume, carefully smoothing the bed, and then progressively increasing the flow discharge and accompanying bottom stress until grains begin to move. Although seemingly straightforward, there are difficulties with such experiments, especially in the subjectivity of defining when grains "begin to move," and this has led to a considerable scatter in the resulting threshold data (Miller et al., 1977). In spite of such problems, the resulting measurements have served to establish the standard threshold curves. The Shields threshold diagram as updated by Miller et al. is shown in Figure 1*a*, a plot of the Shields parameter:

$$\theta_t = \frac{\tau_t}{(\rho_s - \rho)gD} = \frac{\rho u_{*t}^2}{(\rho_s - \rho)gD} \tag{1}$$

versus the grain Reynolds number

$$\mathrm{Re}_* = u_{*t}D/\nu \tag{2}$$

where τ_t = threshold flow stress
ρ_s = grain density
ρ = fluid density
g = acceleration of gravity
D = grain diameter
$u_{*t} = \sqrt{\tau_t/\rho}$, the threshold shear "velocity"
ν = fluid's kinematic viscosity

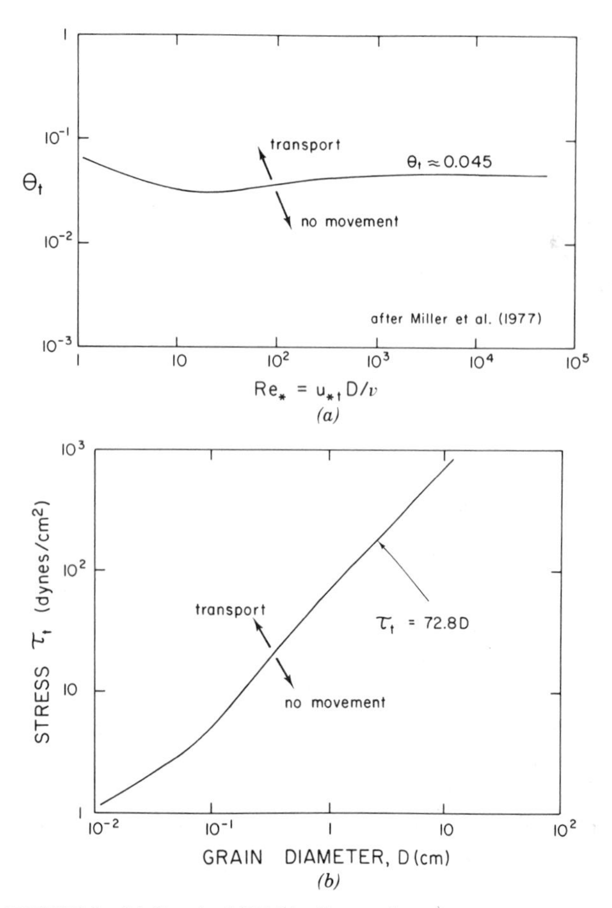

FIGURE 1. (*a*) Standard Shields diagram based on measurements compiled by Miller and coworkers (1977) for the threshold of uniform grains. (*b*) The threshold flow stress τ_t required for the entrainment of grains of diameter D from a deposit of uniform sizes.

Although difficult to employ, the advantage of the Shields diagram over simpler threshold relationships is that it is nearly universal, being applicable to any combination of grain and liquid compositions (but not to gases) and can even be employed for different gravity fields (such as to analyses of floods on Mars). Simpler threshold curves such as the τ_t versus D diagram of Figure 1*b* can apply only to a single combination of grains–liquid–gravity, in this case to quartz density grains in water at 20° C on Earth.

Of particular interest here is the threshold of coarse-grained sediments by floods, and for such materials the Shields diagram of Figure 1*a* yields

$$\theta_t = \text{constant} \approx 0.045 \tag{3}$$

so that from Equation (1),

$$\tau_t = 0.045(\rho_s - \rho)gD \tag{4a}$$

The threshold of quartz density grains in water ($\rho_s = 2.65$ g/cm^3 and $\rho = 1.00$ g/cm^3) is then given by the simple proportionality

$$\tau_t = 72.8D \tag{4b}$$

where τ_t has units of dynes per centimeters squared and D is in centimeters, seen in Figure 1*b* to hold approximately for $D > 0.1$ cm. Various investigators have reported values ranging from 0.03 to 0.06 for this constant θ_t portion of the Shields curve, differences resulting mainly from the subjectivity of the threshold assessment [see Neill and Yalin (1969) and Miller et al. (1977) for discussions of such problems in the evaluation of gravel threshold]. The data reviews of both Miller et al. and Yalin and Karahan (1979) demonstrate that the 0.045 value used here is reasonable.

The relationship of Equation (4) would seem to offer a simple evaluation of the flow stress required to entrain and transport gravel. Even though the threshold data employed to establish this relationship are at maximum for small gravel, the trend of θ_t = constant appears well enough established to provide confidence in extrapolating the application to still coarser material. In addition, there is some theoretical justification for expecting that the threshold condition for uniform grains would be expressed by Equations (3) and (4). For such reasons a number of investigators have employed these relationships for the evaluation of the threshold of gravel, at times for estimating paleoflood τ_t values responsible for the entrainment and deposition of a coarse-grained deposit. However, recent gravel entrainment measurements in rivers and tidal currents raise the question of the general applicability of the standard threshold curves of Figure 1, empirical curves derived from experiments with uniform grain sizes but generally applied to deposits consisting of mixtures of sizes.

Selective Entrainment from Deposits of Mixed Sizes

The studies by Milhous (1973), Carling (1983), and Hammond et al. (1984) of gravel entrainment from natural deposits of mixed sizes all obtained measurements that demonstrate systematic departures from the standard threshold curves of Figure 1. Carling conducted his study in streams of the upland Pennines of England, deriving the threshold estimates from the largest clast sizes caught in a bedload trap under different flow discharges and bottom stresses. His data are shown plotted in the Shields diagram of Figure 2 and as the stress versus diameter in Figure 3. The measurements are seen to systematically depart from the standard Shields curve for uniform grains, θ_t now decreasing with increasing Re_* rather than remaining constant at approximately 0.045. The stress required for entrainment

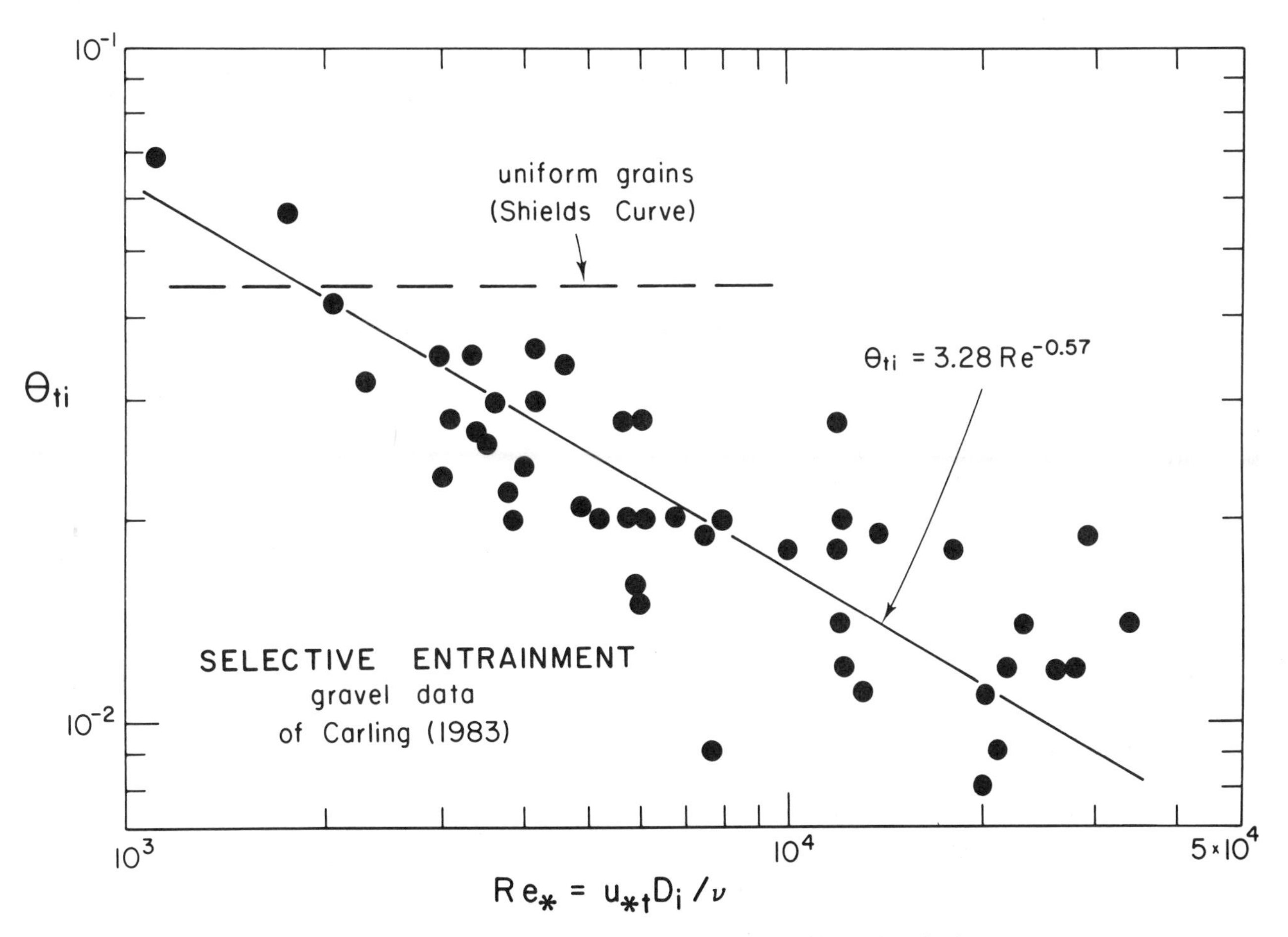

FIGURE 2. The systematic departure from the standard Shields curve (dashed line) by the measurements of Carling (1983) for the entrainment of gravel from a deposit of mixed sizes.

does increase with increasing grain diameter, yielding the relationship

$$\tau_{t,i} = 110D_i^{0.38} \tag{5}$$

where the units are the same as in Equation (4a). Here the subscript i serves as a reminder that the relationship is for selective entrainment, the movement of individual grains of diameter D_i from a bed of mixed sizes. This is conceptually different from the case of threshold of grains from a bed of uniform sizes, and it is seen in Figure 3 that the $\tau_{t,i}$ versus D_i data of Carling again show a marked departure from the threshold curve for uniform grains.

Comparable results were found for the gravel threshold data of Milhous (1973) and Hammond et al. (1984). The measurements of Milhous were similar to those of Carling (1983), obtained by employing a bedload trap in a gravel bed stream, the largest particle size caught ranging from 0.8 to 11.7 cm depending on the flow discharge and bottom stress. The resulting empirical relationship is

$$\tau_{t,i} = 108D_i^{0.57} \tag{6}$$

Hammond et al. obtained their measurements under tidal currents in the West Solent, the strait separating the Isle of Wight from the mainland of England. Their data yield the equation

$$\tau_{t,i} = 55D_i^{0.42} \tag{7}$$

for grain diameters in the range 0.5–4 cm. Each of these three studies, therefore, has obtained data that demonstrate systematic departures from the standard threshold curves of Figure 1.

This pattern of departures is summarized in Figure 4 where the curves based on the flume experiments of Day (1980) and the analyses of Komar and Wang (1984) are also included. All of these curves for the selective entrainment of grains from beds of mixed sizes form similar trends, obliquely crossing the standard threshold curve based on flume experiments with uniform grains. In each

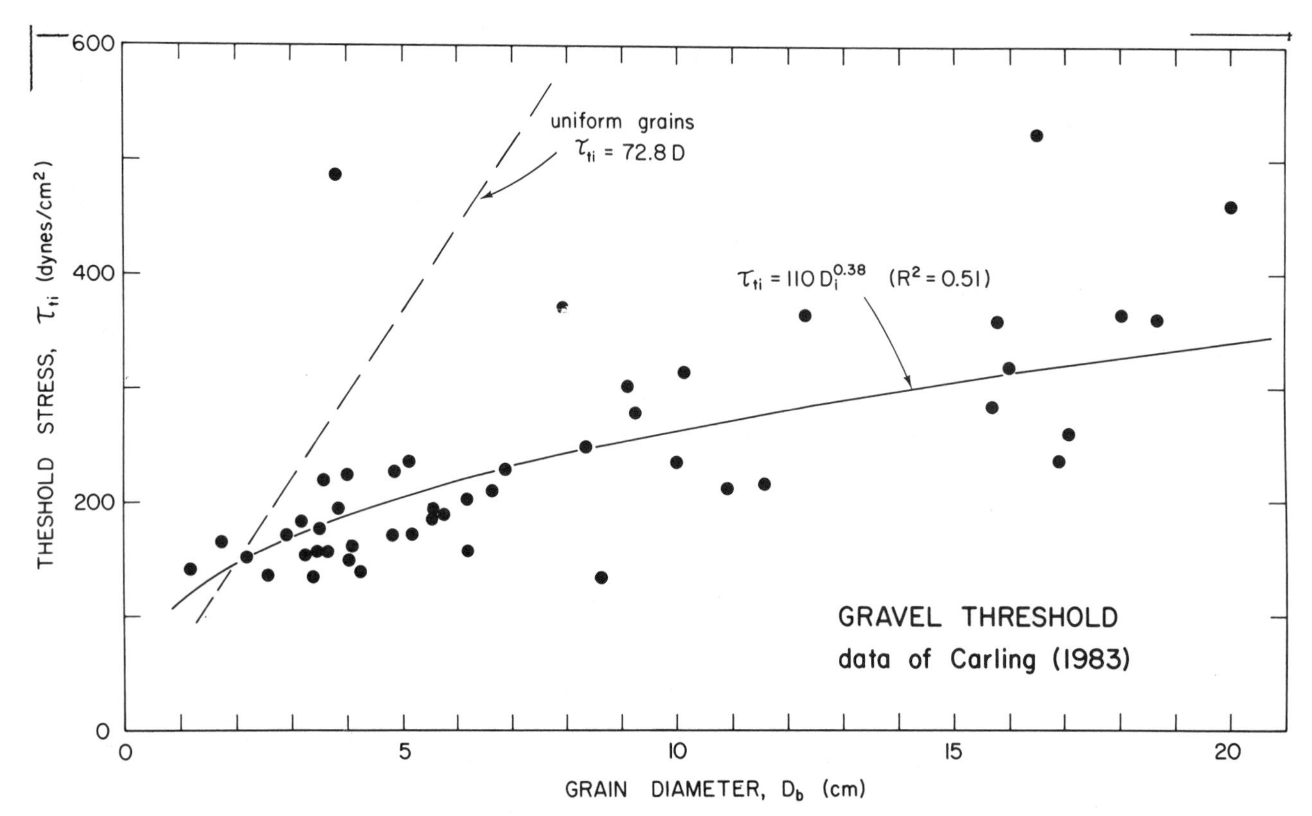

FIGURE 3. The gravel threshold data of Carling (1983) for the selective entrainment from a deposit of mixed sizes, systematically departing from the standard threshold curve (dashed) based on experiments with uniform sizes.

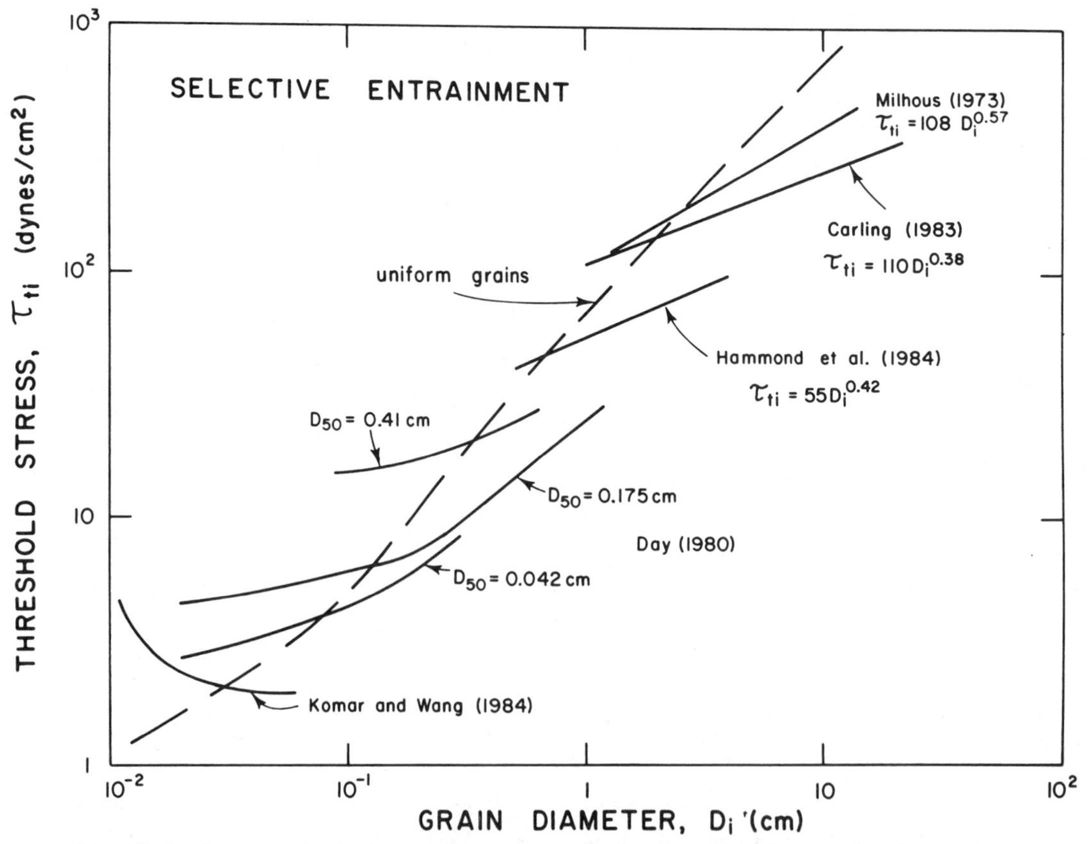

FIGURE 4. The series of selective entrainment curves for deposits of mixed sizes obliquely crossing the standard curve (dashed) for the threshold of uniform grains (Komar, 1987).

data set the coarser grains in the deposit are entrained at lower flow stresses than if they formed beds of uniform sizes, whereas the finer size fractions are more difficult to entrain than their uniform-bed equivalents.

The crossover point of a selective entrainment curve and the standard curve for uniform grains is roughly at the median diameter of the mixed-size deposit, best shown by the controlled measurements of Day (1980), but this needs to be better established by additional experiments (Komar, 1987). We have seen that the proportionality coefficients of Equations (5)–(7) differ due to the contrasting grain sizes involved in the various studies, also apparent in the vertical order of the selective entrainment curves in Figure 4. This suggests that the measurements of the several studies might converge if normalized to their respective D_{50} median diameters according to the dimensionless relationship

$$\theta_{t,i} = a(D_i/D_{50})^b \tag{8}$$

where a and b are empirical coefficients, an equation employed by Parker, Klingeman, and McLean (1982), Andrews (1983), and Komar (1987) in examinations of gravel entrainment. The data of Carling (1983) are analyzed according to Equation (8) in Figure 5 where it is seen that reasonable agreement occurs when $a = 0.045$ and $b = -0.68$. In this analysis D_{50} was taken as 2 cm, obtained from the crossing point with the threshold curve for uniform grains (Fig. 4). That selection sets the a coefficient in Equation (8) to the Shields curve value since $\theta_{t,i} = a$ when $D_i/D_{50} = 1$. With such a selection for the proportionality coefficient in Equation (8), the relationship yields the departure from the Shields curve, $\theta_{t,i}$ decreasing from the 0.045 Shields curve value as D_i/D_{50} increases. Due to the considerable variability in the grain sizes observed in the bed of the river studied by Carling, it is not possible to determine whether this 2-cm diameter is actually a reasonable choice for D_{50}; representative grain size distributions presented by Carling (Fig. 3) do offer some support. Similar analyses of the measurements of Hammond et al. and Milhous yield comparable results (Komar, 1987). The data of Hammond et al. give almost the same coefficients ($a = 0.045$ and $b = -0.71$) as obtained for the measurements of Carling. The data of Milhous are more scattered, but again basically agree with Equation (8) when the a and b values from Carling are employed. On the other hand the studies of Parker et al. (1982) and Andrews (1983) obtained markedly different coefficients, Parker et al. finding $a = 0.0876$ and $b = -0.982$ while Andrews obtained $a = 0.0834$ and $b = -0.872$. The cause of these differences is explored more fully in Komar (1987), the main factors involving contrasting procedures in their respective analyses. Of particular interest is that Parker et al. also reanalyzed the data of Milhous, yet their $b = -0.982$ exponent indicates essentially no selective entrainment [Equation (8) reduces to $\tau_{t,i}$ = constant when $b = -1$, indicating that all grain size fractions are entrained at the same flow stress]. This contrasts with Equation (6), which shows substantial selective entrainment for the measurements of Milhous. Parker et al. based their application of Equation (8) on measured transport rates of the different size fractions, extrapolated back to a very small transport rate to serve as an effective threshold criterion. However, such an approach largely eliminates the possibility for evaluating selective entrainment of the different size fractions. The analysis procedures used here cannot strictly be compared with the results obtained by Parker et al. and Andrews.

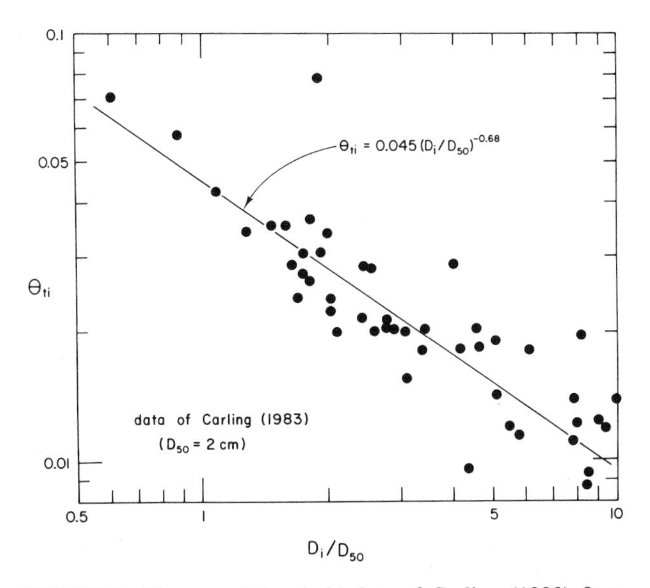

FIGURE 5. The gravel threshold data of Carling (1983) fit to Equation (8).

A general understanding of these observed patterns of selective entrainment by size can be gained from the diagram of Figure 6. With mixed grain sizes the larger gravel particles will project higher above the bed into greater flow velocities, and this increased exposure will aid in their entrainment in spite of their greater weights (Fenton and Abbott, 1977). It is also apparent in Figure 6 that the larger grains will have smaller pivoting angles and therefore can rotate more easily out of their resting positions on the bed. This dependence of threshold on the pivoting angle has been shown by the entrainment analyses of Komar and Li (1986), based on direct measurements of grain pivoting by Miller and Byrne (1966) and Li and Komar (1986). Such pivoting analyses are also the origin for the curve in Figure 4 obtained by the study of Komar and Wang (1984) in their investigation of selective entrainment leading to the formation of placers. Their results indicate that in the fine sand range the sorting will be opposite to that seen above for gravels and coarse sands, $\tau_{t,i}$ now increasing with decreasing D_i. This reversal from that seen for gravels results in the selective removal of the coarser size fractions by a current, tending to con-

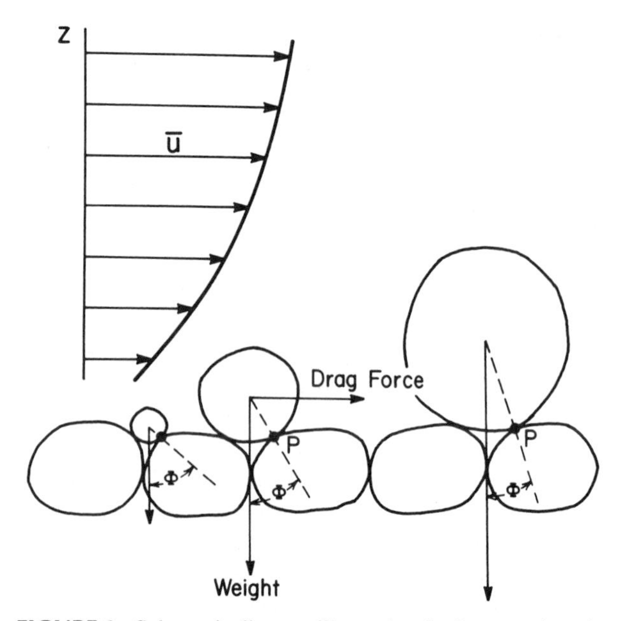

FIGURE 6. Schematic diagram illustrating the factors enhancing the entrainment of larger grains in spite of their greater weights, factors of increased exposure to the flow and a decreased Φ pivoting angle that permits the grain to more easily roll out of position.

centrate the heavy minerals due to both their smaller sizes and higher densities.

The compilation of curves in Figure 4 demonstrates that entrainment from beds of mixed sizes differs substantially from our traditional flume experiments utilizing uniform grains. Applications of the standard threshold curves can lead to serious errors in estimates of the flow stresses required for entrainment. For example, the largest particle diameter trapped during the study of Carling (1983) was 20 cm (Fig. 2). According to the standard threshold curve of Figure 1*b* or the equivalent equation (4b), a stress τ_t = 1500 dynes/cm^2 would be required for its entrainment and transport; Carling's measurements actually indicate that the stress at the time of its movement was only 560 dynes/cm^2. Had this been an attempt to estimate the hydraulics of a paleoflood, it is apparent that use of the standard threshold curve would have yielded a substantial over-evaluation of the flow stress. In such evaluations it is necessary to consider selective entrainment from deposits of mixed sizes.

Flow Competence

The concept of flow competence is specific to the application of evaluating flood strengths (stresses, velocities, and discharges) from the maximum size material transported. The principal application has been to the evaluation of extreme, nonhistoric floods of an otherwise still active river system. For example, Bradley and Mears (1980) applied flow competence evaluations to boulders up to 230 cm diameter found adjacent to Boulder Creek where it flows through Boulder, Colorado, apparently representing some extreme flood stage for this watershed, the potential return of which represents an obvious hazard to the community. Particularly interesting are applications of this approach to analyses of catastrophic paleofloods such as carved the Channeled Scabland in eastern Washington (Bretz, 1923, 1969; Baker, 1973). The evaluation of flow competence can equally be applied to ancient river flood deposits preserved in the geologic rock record.

A variety of approaches have been employed for the evaluation of flow competence: (1) empirical correlations based on material transported by historic floods; (2) theoretical formulas for the initiation of grain movement by pivoting, sliding, or rolling; and (3) empirical engineering formulas for the evaluation of the stability of large rocks used in river control works. Reviews as well as applications of these separate approaches can be found in Novak (1973), Baker and Ritter (1975), Bradley and Mears (1980), and Costa (1983).

A compilation of coarse-particle transport is given in Figure 7, based on the data summary of Costa (1983). These measurements come from a wide variety of sources but generally represent the movement of clasts by flood stage flows. In Figure 7*a* the size of material is related to the mean flow velocity, while Figure 7*b* provides an evaluation of the flow stress, this latter correlation yielding

$$\tau_c = 26.6D^{1.21} \qquad (9)$$

Empirical relationships such as these potentially offer a direct and simple evaluation of flow competence. For example, the 230-cm boulder found by Bradley and Mears (1980) in Boulder, Colorado, implies a competence flow velocity $\bar{u}_c$ = 670 cm/s and stress $\tau_c = 1.9 \times 10^4$ dynes/cm^2. However, in many cases the reasonableness of such estimates is questionable.

The relationships of Equations (5)–(7), respectively, from the studies of Carling (1983), Milhous (1973), and Hammond et al. (1984), all correlate the flow stress to the largest gravel sizes transported. Therefore, those equations are in effect flow competence relationships, but limited in application to the specific environments of those respective studies. These equations are seen to differ markedly from the standard flow competence relationship of Equation (9), most significantly in their *D* exponents, indicating that the competence equation (9) would yield poor results when compared to the basic measurements of those studies. For example, returning to the 20-cm diameter clast captured in Carling's bedload trap, Equation (9) yields an evaluated flow stress τ_c = 1000 dynes/cm^2, substantially higher than the 560 dynes/cm^2 measured stress; again, had this been

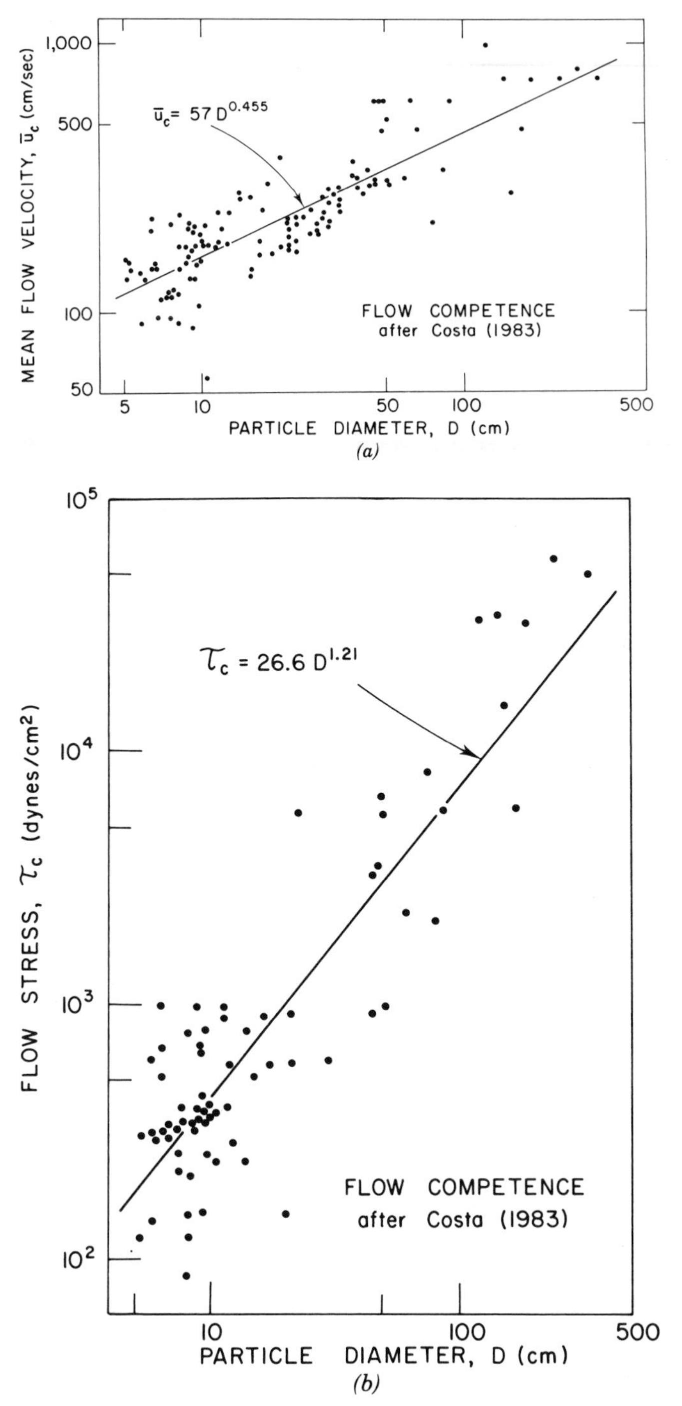

FIGURE 7. The flow competence velocity and stress required for the transport of maximum size clasts by a flood, according to the data compilation of Costa (1983).

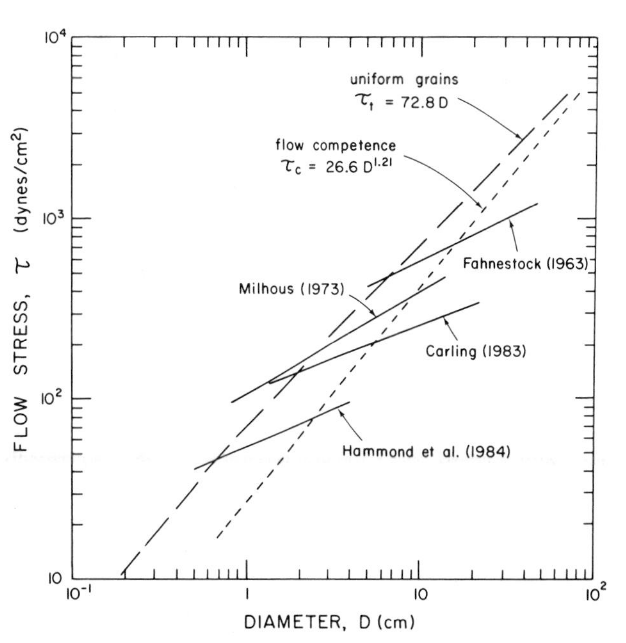

FIGURE 8. A comparison between the flow competence relationship from Figure 7*b* (short-dashed line), the standard curve for the threshold of uniform grains (long-dashed line), and the series of selective entrainment curves from Figure 4 (solid lines) (Komar, 1987).

a paleoflood estimate, the competence equation (9) would have provided a substantial overevaluation of the flow hydraulics.

This basic disagreement is more apparent in Figure 8 where the three types of relationships are compared, Equation (4b) for deposits of uniform grains, the series of curves for selective entrainment from beds of mixed grain sizes, and the empirical equation (9) for the evaluation of flow competence. With additional studies like those of Milhous (1973), Carling (1983), and Hammond et al. (1984), one can envision many more selective entrainment curves obliquely crossing both the standard curve for uniform grains and the flow competence curve. With some reflection it becomes apparent that, for the most part, the flow competence curve is the product of selective grain entrainment, combining data from a number of rivers individually affected by selective entrainment just as seen above for the data of Carling and others. However, it can also be seen in Figure 8 that the competence curve is not shifted far enough toward higher *D* diameters to provide approximate agreement with the coarse-grained ends of the selective entrainment curves, the conditions representing the most extreme floods in those environments. This accounts for the poor comparison of Equation (9) with the measured entrainment of 20-cm diameter clasts as determined by Carling, and it is seen in Figure 8 that similar disagreements would be obtained in comparisons with the maximum sized material transported as measured by Milhous, Hammond et al., and Fahnestock (1963). Due to the contrasting trends of the series of selective entrainment curves in Figure 8 in comparison with the steep slope of the flow competence relationship, had the studies of Carling and others experienced greater discharges causing the transport of still larger clasts, the discrepancy with the flow competence equation (9) would have been still greater.

From this comparison it would appear that in many applications the empirical flow competence relationships such as Equation (9) do not provide satisfactory evaluations of flood hydraulics. Although only Equation (9) was included in this comparison, it can be expected that other similarly derived competence relationships will be defective in the same way, whether the relationships are for the velocity, stress, or other measure of the flow.

One alternative method for evaluating flow competence is to employ Equation (8), its use also recognizing that in many applications flow competence and selective entrainment are conceptually one and the same. Such an approach would also acknowledge that the movement of an individual clast of diameter D_i will not generally depend on only the flow stress as assumed in the flow competence equation (9) but will also depend on the general size of the bed material, that is, on the ratio D_i/D_{50} as given by Equation (8). For the transport of coarse-grained materials, the coefficients $a = 0.045$ and $b = -0.7$ can be employed since these agree with the data of Carling, Milhous, and Hammond et al., utilized at least until additional studies are available that might determine how these coefficients depend on tightness of grain packing, imbrication, and so on. Turning one final time to the 20-cm clast trapped by Carling, $D_i/D_{50} = 20/2 = 10$ and Equation (8) yields $\theta_{t,i} = 0.0095$, which in turn corresponds to the entrainment stress $\tau_{t,i} = 320$ dynes/cm^2, essentially the same value (340 dynes/cm^2) obtained from Equation (5) and the curve in Figure 3 fit to Carling's data.

At present, Equation (8) has been tested for D_i/D_{50} ranging up to about 20; many flow competence evaluations are likely to involve higher ratios so that further tests are required. It is clear that the relationship is unlikely to apply to extreme D_i/D_{50} ratios, for example, a 200-cm boulder resting on 2-cm gravel. Additional complicating factors are then involved, illustrated by the experiments of Fahnestock and Haushild (1962) with cobbles on a sand bed. They found that the cobbles moved downstream only during the upper-flow regime where the bed is flat or covered by low-amplitude antidunes, the cobbles tipping upstream into scour pockets during low-regime flow. From this it can be seen that grain movement at high D_i/D_{50} will depend on the fluid forces tending to roll or slide the particle downstream while scouring at the same time acts to bury the particle.

Another potential approach to flow competence evaluations is through applications of theoretical analyses of the forces involved in the process. White (1940) developed a force moments analysis for grain entrainment, balancing the force of fluid drag against the grain's weight. Helley (1969) undertook a similar analysis, obtaining a relationship that Bradley and Mears (1980) applied to the evaluation of flow competence in Boulder Creek. One version of such a relationship is

$$\theta_{t,i} = A\left[\log\left(\frac{14D_i}{\zeta D_{50}}\right)\right]^{-2} \tan\Phi \qquad (10)$$

obtained by balancing the moments of the fluid drag and grain weight as diagramed in Figure 6. The term in brackets accounts for the grain's exposure due to its projection above the bed, similar to that measured by Fenton and Abbott (1977). The ζ parameter is the ratio of the bottom roughness k_s to D_{50}, and would have values on the order of 1 to 5 as found by a number of studies of boundary layers. The proportionality coefficient is $A = 0.040/C_o$ where C_o is the drag coefficient for the flow around the grain. As seen geometrically in Figure 6 and demonstrated by the measurements of Miller and Byrne (1966) and Li and Komar (1986), the pivoting angle Φ is also a function of D_i/D_{50}, decreasing with an increase in that size ratio. Therefore, Equation (10) becomes a relationship predicting the decrease of $\theta_{t,i}$ with increasing D_i/D_{50}, just as seen in the empirical equation (8). Although somewhat more difficult to apply than that empirical relationship, Equation (10) can serve as the basis for an improved understanding of the processes of grain threshold and selective entrainment, perhaps eventually leading to improved evaluations of flow competence. The model can also be altered to consider grain sliding or rolling rather than pivoting, $\tan\Phi$ in Equation (10) simply being replaced by the appropriate friction coefficient. Unfortunately, at present we have little information available to guide us in selecting values for the friction coefficient, and such models of pivoting, sliding and rolling are still largely untested.

MODES OF SEDIMENT TRANSPORT

Sediment transport in rivers occurs as a combination of bedload, suspended load, and as a fine-grained component termed washload. The coarsest sediments are found in the bedload, rolling and bouncing (saltating) along the river bottom. Finer sediments are lifted well above the bed by the flow's turbulence and comprise the suspended load. The washload is the finest-grained fraction of the suspension transport, sufficiently fine grained that the river is able to transport it in nearly unlimited quantities and at rates equal to the flow of the water itself. The ranges of sediment grain sizes that comprise these three transport modes are governed by the flow strength of the river—its velocity, stress, or power. Much of the increase in sediment transport during floods results from shifts in the grain sizes found within these three modes; grain sizes that normally move as bedload are transported at high rates during a flood as part of the suspended load, and coarser grain sizes than usual are found in the washload. It is therefore of interest to deduce hydraulic criteria that evaluate the approximate divisions between these modes of sediment transport as a function of the river's strength. Most attention has centered on bedload versus suspension transport because this determination is important to subsequent evaluations of transport rates and quantities. The division between suspension transport and the washload has received comparatively little attention and, as will be seen, is inherently ill-defined.

Bagnold (1966) and Francis (1973) make clear distinctions between bedload and suspension transport; grains in the bedload make frequent contact with the bottom or with other grains as they slide, roll, and bounce, whereas grains transported in suspension are supported above the bottom by the turbulence field of the flowing water. In principle this distinction is clear whereas in practice it is somewhat arbitrary. This is shown by the experiments of Francis (1973) who photographed the paths of individual grains moving over a bed of fixed grains of the same size. With the flow stress τ just above that necessary for threshold of grain motion, the grains were observed to roll in continuous contact with the bed. With a somewhat higher τ the grains began to saltate, taking jumps with a ballistic trajectory (the maximum rise above the bed was only about 2–4 grain diameters, however, so that saltation in water follows much lower paths than in aeolian transport). At still higher flow velocities and stresses, Francis found that the grains began to transport in suspension, making much longer and higher trajectories, the flat upper parts of which often appeared wavy due to the action of turbulent eddies on the grain's motion. With full suspension such irregularities in the grain's path become dominant.

The distinction between bedload movement and suspension transport is in the role played by turbulence in carrying the grains above the bottom, opposed by the grain's settling velocity, which acts to return it to the bed. Bedload versus suspension therefore depends on the relative magnitudes of the grain's settling velocity w_s versus the w' vertical velocity component of the turbulent eddies. Theoretical and experimental work on boundary layer turbulence demonstrates that w' is determined by the flow stress τ or its surrogate, the shear velocity $u_* = \sqrt{\tau/\rho}$. A criterion for bedload versus suspension transport therefore is provided by the dimensionless ratio

$$w_s/u_* = k \tag{11}$$

where k is its value (determined experimentally) at this division. Reflecting the subjectivity of distinguishing between these modes of transport, a range of k values has been employed: $k = 1$ (Lane and Kalinske, 1939; Inman, 1949; Francis, 1973), $k = 1.20$ (Einstein, 1950), $k = 1.25$ (Bagnold, 1966), and $k = 1.79$ (McCave, 1971).

According to Equation (11), for a given flow τ or u_*, grains having settling velocities greater than w_s will be in the bedload whereas those with lower settling rates will be transported predominantly in suspension. This relationship can be plotted as in Figure 9 directly onto a graph of τ or u_* versus the grain diameter D for a fixed sediment composition, having employed standard settling velocity equations to convert w_s to D. The grain threshold curve from Figure 1*b* is included for comparison. The suspension criterion of Equation (11) separates regions of τ and u_* versus D in which bedload and suspension prevail. For coarse sediment of a specific diameter, as τ progressively increases the grains first become part of the bedload until τ and u_* exceed the critical level given by Equation (11), at which point the sediment is transported in suspension. Grain diameters less than about 0.1 mm immediately go into suspension once threshold is achieved, there being no bedload phase for these fine grain sizes.

Although the division between bedload and suspension transport is somewhat subjective, the relationship of Equation (11) appears to offer a reasonable criterion for their separation

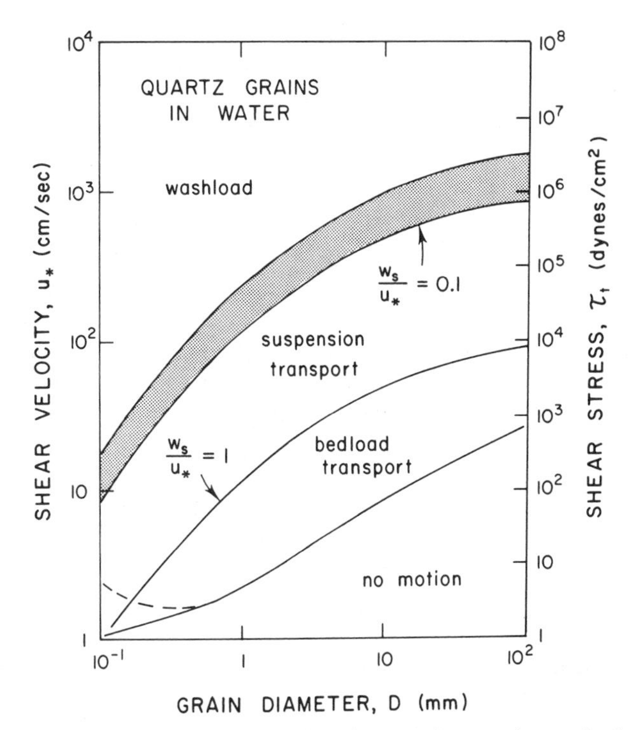

FIGURE 9. Curves based on Equation (11) for u_* and τ required for the transition of a grain diameter D from bedload to suspension and then to washload.

and has served that purpose in evaluations of sediment transport rates.

The washload of a river consists of the finest-grained portion of the total sediment load, grains having settling velocities that are so low they are carried passively along by the flowing water and essentially never reside on the bed. In that this washload is vertically mixed by the flow turbulence, technically it can be considered as the finest-grained portion of the suspended load. It was first distinguished as a separate component by Einstein et al. (1942) who divided the total sediment transport into a bed material load and washload, the former being coarse enough to be found in abundance on the river bed whereas the finer grain sizes of the washload are nearly absent in a bed sample. This differs from the bedload versus suspension division in that most of the suspended load increases in concentration downward toward the bed and is found "in abundance" in the bottom sediments. In contrast, the fine-grained washload is vertically homogeneous in concentration and comparatively little occurs in the bottom sediments. Einstein et al. were particularly interested in making the distinction that the quantities and transport rates of the washload component are basically unpredictable, showing little or no relationship to the hydraulic parameters of the river flow (Colby, 1963; Guy, 1964). The quantity of washload is governed instead by its availability in the river's watershed. In contrast, the "bed material load" of Einstein et al., comprising the bedload and coarser sizes of the suspension transport, are controlled by the river's flow strength and therefore have the potential for being predicted from the hydraulic parameters.

Unlike the division between bedload and suspension transport, little consideration has been given to the grain size cutoff between coarser-grained suspension and washload. From the above discussion it is apparent that this cutoff is highly subjective so a washload counterpart to Equation (11) will be more arbitrary. In 1964 W. Kresser (see discussion in Graf, 1971, p. 204) proposed the relationship $\bar{u}^2/gD = 360$ where $\bar{u}$ is the mean velocity of the river and D is the cutoff diameter, the coarsest grain size of the washload. However, comparisons of this formula with data from the Mississippi River indicate that it yields D values that are too coarse (Komar, 1980). The observation that a river can transport nearly unlimited concentrations of washload suggests that its transport mechanism may be by autosuspension, a concept introduced independently by Bagnold (1962) and Nordin (1963). This approach compares the energetics of the river flow and sediment transport system, finding that when $w_s < \bar{u}S$ where S is the channel slope, the sediments are "self-suspending" in the sense that they supply more power to the flow than extracted in becoming suspended. This relationship appears to yield grain diameters that are too small for the onset of washload (Komar, 1980), but the concept of autosuspension does account for many of the properties of the washload and therefore warrants further study as a potential mechanism.

In my analyses of washload (Komar, 1980), I applied Equation (11) but with a much lower value for k than used in the bedload versus suspension cutoff. Based primarily on the extensive engineering literature for the pumping of suspensions through pipes, I employed the range $k = 0.05–0.10$ for the washload cutoff, using this range as a reminder of the subjectivity of the determination. This criterion is shown plotted in Figure 9, providing a transition from suspension transport to washload as τ and u_* increase.

Equation (11) and its graphical representation in Figure 9 can serve to assess grain size shifts in the sediment transport modes between normal flow conditions and those during a flood. The flood stress is approximately proportional to the square of the mean flow velocity, so that doubling u during a flood will increase τ by a factor of 4, and during major floods τ could be increased by an order of magnitude. It is apparent from the curves in Figure 9 that such an increase in τ would result in a marked coarsening of the grain size ranges found in the different modes of transport. This approach can also serve to analyze the transport mode for extreme flood events such as those that carved the Channeled Scabland in eastern Washington (Bretz, 1923, 1969). In his study of the hydraulics of these flows, Baker (1973) obtained estimated flows ranging up to $u_* \approx 100$ cm/s or $\tau \approx 10^4$ dynes/cm^2; according to the graph of Figure 9, sediments as coarse as 10–30 cm could have been transported in suspension, diameters as large as 0.4–0.8 mm having been in the washload. My analyses of floods on Mars, which carved the large outflow channels, similarly indicate that the flows may have been capable of transporting cobbles in suspension with nearly all sand size material and finer having been in the washload (Komar, 1980).

QUANTITIES OF SEDIMENT TRANSPORT

A great number of formulas have been developed over the years to evaluate quantities of sediment transport. Textbooks such as that by Graf (1971) provide useful summaries of this vast literature. Although many of the formulas have at least some theoretical basis, even those are partly empirical in being calibrated with sediment transport measurements obtained in laboratory flumes and rivers of modest discharges. Even when evaluating the total sediment transport for rivers under normal flow conditions, the estimates based on these many formulas commonly spread over nearly an order of magnitude. When extrapolated to flood conditions, the results are lacking in any credibility.

The earliest of the bedload transport models is that developed by DuBoys in 1879, a model that assumes that sediment moves along the bottom in layers of progressively decreasing rate with depth within the bedload carpet. This model yields a relationship of the form

$$q_s = \xi\, \tau(\tau - \tau_t) \quad (12)$$

where q_s = volume transport rate of sediment per unit channel width (cm^3/s cm)

τ = actual flow stress that exceeds the τ_t threshold value required for initiating sediment movement

ξ = proportionality coefficient that must be based on actual measurements of sediment transport.

The form of Equation (12) is seen to be intuitively reasonable in that q_s depends on the flow strength as measured by the degree to which τ exceeds τ_t. Although DuBoys' model has been criticized for its basic assumptions, and subsequent measurements over the years have established that the ξ proportionality coefficient depends on factors such as grain size and perhaps bottom slope, many bedload formulas still used today are similar in form to Equation (12).

During flood conditions $\tau >> \tau_t$, and since $\tau \propto \overline{u}^2$, from Equation (12) we have approximately $q_s \propto \overline{u}^4$ for the relationship of the bedload transport rate to the mean flow velocity. This suggests a very strong dependence on $\overline{u}$ so that during a flood there should be a considerable increase in quantities of bedload movement. For example, doubling $\overline{u}$ produces a $2^4 = 16$ factor increase in q_s. This is the transport rate per unit channel width, and since the river will widen during a flood the increase in total bedload transport would be still greater.

Dependencies of q_s on τ and $\overline{u}$ shown by the DuBoys equation are typical for bedload transport formulas. A summary of such relationships is given in Table 1, a representative though far from exhaustive list of available equations. These formulas variously yield $q_s \propto \overline{u}^3$ to $\propto \overline{u}^5$. The available sediment transport measurements are sufficiently scattered that it is not possible to conclusively establish the relationship between q_s and τ or $\overline{u}$. With normal flow conditions and modest τ and u magnitudes, the exponents of these parameters are not too critical, but when extrapolated to flood conditions the value of the exponent makes a great deal of difference in the resulting evaluated bedload transport. For this reason at present we can have little confidence in making such extrapolations to estimate bedload movement during floods.

Other factors may also be involved that depend on the nature of the watershed and the conditions of the flood. Bedload formulas are derived to predict the maximum bedload that a stream can transport in equilibrium for a specific flow strength and sediment grain sizes. The actual quantity of bedload transport may be much less than this maximum, especially during floods when erosion rates are insufficient to keep pace with the river's demand for additional bedload. The actual load depends then on geologic factors and on such hydrologic parameters as the intensity and duration of rainfall.

Quantities of suspension transport depend on the relative availability of grains that are sufficiently fine to be lifted above the bed by the flow's turbulence. The sediment concentration C at an elevation z above the bed is found to show reasonable agreement with a relationship of the form

$$\frac{C}{C_a} = \left(\frac{h - z}{z}\,\frac{z_a}{h - z_a}\right)^{w_s/\kappa u_*} \quad (13)$$

TABLE 1 Velocity Relationships for Typical Bedload Formulas

Source and Date[a]	Transport Dependence	Velocity Proportionality
DuBoys, 1879	$\tau(\tau - \tau_t)$	$\overline{u}^4$
Donat, 1929	$\overline{u}^2(\overline{u}^2 - \overline{u}_t^2)$	$\overline{u}^4$
Schoklitsch, 1930	$S^{1.4}q^{0.6}(q^{0.6} - q_t^{0.6})$	$\overline{u}^4$
Schoklitsch, 1934	$S^{3/2}(q - q_t)$	$\overline{u}^3$
O'Brien, 1934	$(\tau - \tau_t)^m$ with $1.5 < m < 1.8$	$\overline{u}^3$ to $\overline{u}^{3.6}$
Straub, 1935	$\frac{1}{D^{3/4}}\tau(\tau - \tau_t)$	$\overline{u}^4$
Shields, 1936	$\frac{qS}{D}(\tau - \tau_t)$	$\overline{u}^5$
Meyer-Peter, 1948	$\tau^{3/2}$	$\overline{u}^3$
Einstein, 1950	—	$\overline{u}^3$
Bagnold, 1966	$\tau\overline{u}$	$\overline{u}^3$

[a] References can be found in Graf (1971).

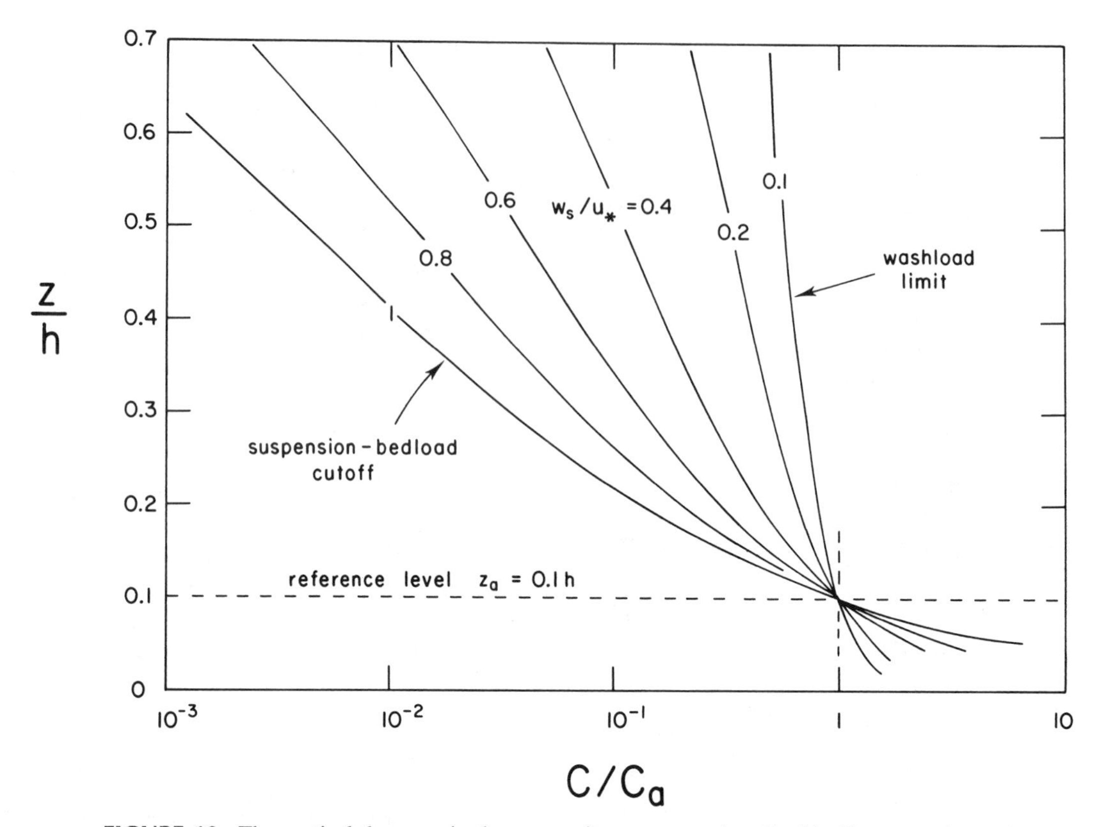

FIGURE 10. The vertical decrease in the suspension concentration C with distance z above the bed according to Equation (13), C_a and z_a being the reference concentration and elevation, and h is the water depth.

where C_a = "reference" concentration at an elevation z_a

w_s = grain's settling velocity

h = water depth

$\kappa = 0.4$ = von Karman's constant

This relationship is shown plotted in Figure 10 for a range of w_s/u_* ratios, demonstrating the expected upward decrease in C away from the bottom. The higher the w_s/u_* ratio, the lower the concentration of grains above the bed so that with a given u_*, most of the coarser grains would be found near the bottom whereas the finer grains with lower w_s occur in higher concentrations throughout the river's depth. Note that when $w_s/u_* = 1$ little sediment is found above the bed, this value being the suspension-bedload cutoff [Equation (11)].

During a flood u_* increases so that for a fixed grain size and w_s, the ratio w_s/u_* decreases and that grain size is carried on average to greater distances above the bottom. In that the water flow velocity increases with distance above the bed, the suspension transport rate will thereby increase. However, this is only one factor in enhancing the suspension transport during a flood. The reference concentrations C_a for the various grain sizes, generally evaluated at the bed, may also increase; for example, doubling C_a will double the C concentrations of that grain size at every level in the flow.

Evaluations of suspension transport rates involve coupling the C vertical profiles with the flow velocity profile, integrating their product over the flow depth. This must be done for each grain size (or w_s) and then summed to yield the total suspended load. The major uncertainty to this approach is in determining the reference C_a concentrations for each of the grain sizes present in suspension. According to the approach of Einstein (1950), sediments within the bed layer become the source of suspended load so that C_a corresponds to concentrations as found in a bottom sediment sample; Einstein also takes $z_a = 2D$ where D is the grain diameter.

A more direct indication of changing quantities of suspension transport with flow discharge are actual measured loads obtained at river gauging stations. Figure 11 shows an example from the Powder River, Wyoming, analyzed by Leopold and Maddock (1953). Although highly scattered, there is a clear trend of increasing total suspended load with increasing discharge. At the highest discharges represented by the data, the suspended load is some 1000

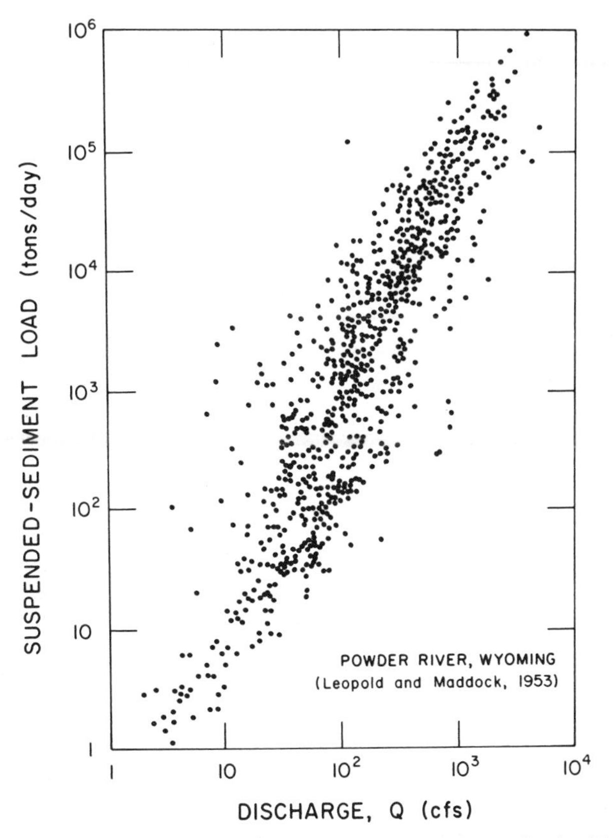

FIGURE 11. Variations in the suspended-sediment load with discharge in the Powder River, Wyoming, demonstrating a 1000-factor increase during floods (Leopold and Maddock, 1953).

times that for average flow discharges. Although the graph of Figure 11 empirically applies only to Powder River, it is typical in its indication of vast increases in the quantities of suspension transport during floods.

As discussed earlier, the washload of a river consists of the finest-grained portion of the total sediment load, grains having settling velocities so low that there is essentially no vertical concentration gradient through the flow depth (seen by the $w_s/u_* = 0.1$ curve in Figure 10). Einstein et al. (1942) distinguished washload as a separate transport mode in that, unlike the bedload and suspension transport already discussed, the quantity of washload typically is not controlled by the river flow parameters. This is seen in Figure 12 from Colby (1963) where the concentration of washload bears no relationship to the river's discharge, contrasting with the coarser-grained suspension transport graphed in Figure 11. Instead, the washload concentration and total quantity depend more on the geology of the river's drainage basin and climate. Concentrations tend to be high if there is a ready source of fine-grained sediments in the drainage basin, either derived from geological formations or where chemical weathering promotes clay development. Extremely high concentrations of washload are a common occurrence during flash floods in the semi-arid southwestern United States. Beverage and Culbertson (1964) used the term *hyperconcentrations* for streamflow samples that exceed 40% sediment by weight (normal rivers generally have less than about 10%). They tabulated the hyperconcentration measurements obtained by the U.S. Geological Survey river-sampling program, finding measured values as high as 65% sediment by weight. That weight concentration corresponds to a volume concentration of 41% (volume of sediment per unit total volume), so it is seen that an appreciable portion of the stream discharge actually constitutes sediment rather than water. Mud flow concentrations have been placed at 79–85% sediment by weight (Sharp and Nobles, 1953), so it is seen that some of the hyperconcentration measurements in rivers are approaching a mud flow consistency.

All of the data of Beverage and Culbertson (1964) came from relatively small rivers (discharges 1.1–6270 cfs). Large rivers seldom reach hyperconcentration levels because their drainage basins do not supply adequate quantities of fine-grained material in proportion with the increase in available water. However, Todd and Eliassen (1938) mention that the Yellow River in China once carried a concentration of 40% sediment at a discharge of 812,000 cfs, indicating that high concentrations can occur even in large rivers if an adequate source is available (extensive loess deposits in the case of the Yellow River).

Nordin (1963) discusses the flow of streams with hyperconcentration sediment loads, basing his observations on the Rio Puerco in New Mexico. He makes the important point that at the concentrations of fine material involved, the settling velocities of still coarser grains would be so

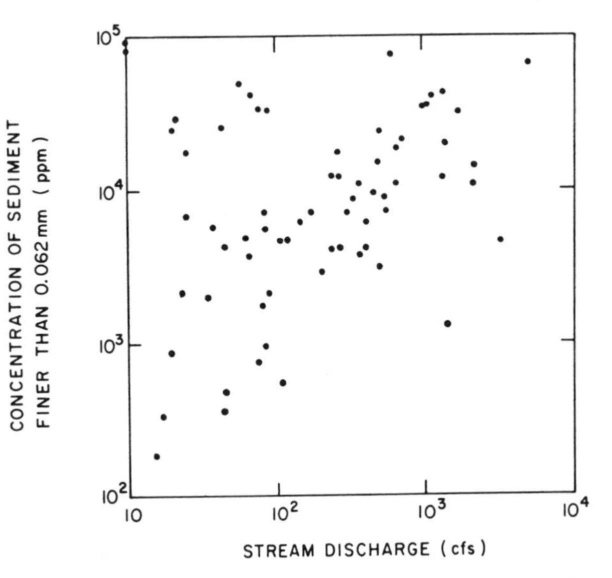

FIGURE 12. A lack of dependence of the washload concentration on the flow discharge for the Powder River, Wyoming (Colby, 1963).

reduced that they would also become part of the washload. It is therefore probable that in floodwaters containing high concentrations of fine-grained washload, much of the sediment that would normally be present in the bedload is now found in the suspended load due to its reduced settling rates, while that normally in the suspended load is present in the washload. This of course would greatly enhance the total sediment transport rate. Furthermore, if Bagnold's (1962) concept of autosuspension is correct, then this washload actually contributes to the power of the flow, enhancing its ability to cause erosion and to transport large diameter gravel and boulders.

Floods have the potential for carrying considerable quantities of washload even if hyperconcentration levels are not achieved. In flowing over floodplains and weathered soils, a flood may have a ready source of clay and silt that will become part of the washload. During catastrophic floods that flow over "fresh" terrain, the washload may be particularly significant. This would have been especially the case for the Lake Missoula floods that eroded the Channeled Scabland (Baker, 1973); those floods removed thick deposits of loess (the Palouse formation), which would have become part of the washload and could have been transported away at high concentrations and at rapid rates. Sand-sized grains likely also became part of the washload due to their reduced settling velocities in the concentrated suspension, perhaps explaining the almost total absence of sand deposits laid down by the floods (at least until they reached the Pacific Ocean). In a similar fashion ancient floods on Mars likely transported all sand-size and finer grains as washload, due both to the flood strengths and reduced gravity of that planet, leading to the rapid erosion of the large outflow channels (Komar, 1980).

CONCLUSION

Due to the complexities of sediment transport processes and the inability to obtain satisfactory measurements during extreme flood events, attempts to make estimates of transport rates of sediments carried by a flood can lead only to questionable results. The trends are clear, however. We have seen that the transport rate of bedload is proportional to the flow velocity with an exponent on the order of 3–5 so that floods can result in dramatic increases. The increase in the suspended load may be still greater, direct measurements in a typical river showing a 1000-fold increase during floods. Although the quantity of the fine-grained washload is inherently unpredictable, it undoubtedly can play an extremely significant role in determining the total quantity of sediment transported during a flood and thus in the amount of erosion. The greater the strength of the flood flow, the larger the size gravel and boulders entrained and transported. Attempts at evaluating the flood hydraulics by the largest transported particles is fraught with problems, however, due to processes of selective entrainment from a bed of mixed sizes and because of size restrictions of available clasts.

Major topographic changes produced by floods attest to their ability to entrain and transport huge quantities of sediments. However, due to the scale of such events, especially in comparison with our laboratory and field studies of sediment transport processes, quantitative assessments of sediment transport rates or flow competence evaluations require considerable extrapolations and therefore can yield only very rough estimates.

ACKNOWLEDGMENTS

This review was undertaken while supported by the Planetary Geology Program of the National Aeronautics and Space Administration through Grant NSG-7178 to Oregon State University.

REFERENCES

Andrews, E. D. (1983). Entrainment of gravel from naturally sorted riverbed material. *Geol. Soc. Am. Bull.* **94**, 1225–1231.

Bagnold, R. A. (1962). Auto-suspension of transported sediment; turbidity currents. *Proc. R. Soc. London, Ser. A* **265**, 315–319.

Bagnold, R. A. (1966). An approach to the sediment transport problem from general physics. *Geol. Surv. Prof. Pap. (U.S.)* **422-I**, 1–37.

Baker, V. R. (1973). Paleohydrology and sedimentology of Lake Missoula flooding in eastern Washington. *Spec. Pap.—Geol. Soc. Am.* **144**, 1–79.

Baker, V. R., and Ritter, D. F. (1975). Competence of rivers to transport coarse bedload material. *Geol. Soc. Am. Bull.* **86**, 975–978.

Beverage, J. P., and Culbertson, J. K. (1964). Hyperconcentrations of suspended sediment. *J. Hydraul. Div., Am. Soc. Civ. Eng.* **90**(HY6), 117–128.

Bradley, W. C., and Mears, A. I. (1980). Calculations of flows needed to transport coarse fraction of Boulder Creek alluvium at Boulder, Colorado. *Geol. Soc. Am. Bull.* **91**, 135–138.

Bretz, J H. (1923). The channeled scablands of the Colorado Plateau. *J. Geol.* **31**, 617–649.

Bretz, J H. (1969). The Lake Missoula floods and the channeled scabland. *J. Geol.* **77**, 505–543.

Carling, P. A. (1983). Threshold of coarse sediment transport in broad and narrow streams. *Earth Surf. Processes* **8**, 1–18.

Colby, B. R. (1963). Fluvial sediments—A summary of source, transportation, deposition, and measurement of sediment discharge. *Geol. Surv. Bull. (U.S.)* **1181-A**, 1–47.

Costa, J. E. (1983). Paleohydraulic reconstruction of flash-flood peaks from boulder deposits in the Colorado Front Range. *Geol. Soc. Am. Bull.* **94**, 986–1004.

Day, T. J. (1980). A study of initial motion characteristics of particles in graded bed material. *Geol. Surv. Can., Curr. Res., Part A, Pap.* **80-1A**, 281–286.

Einstein, H. A. (1950). The bed-load function for sediment transportation in open channel flows. *U.S., Dep. Agric., Tech. Bull.* **1026**, 1–78.

Einstein, H. A., Anderson, A. G., and Johnson, J. W. (1942). A distinction between bed-load and suspended load in natural streams. *Trans., Am. Geophys. Union* **19**, Part 2, 628–633.

Fahnestock, R. K. (1963). Morphology and hydrology of a glacial stream—White River, Mount Rainier, Washington. *Geol. Surv. Prof. Pap. (U.S.)* **442-A**, 1–70.

Fahnestock, R. K., and Haushild, W. L. (1962). Flume studies on the transport of pebbles and cobbles on a sand bed. *Geol. Soc. Am. Bull.* **73**, 1431–1436.

Fenton, J. D., and Abbott, J. E. (1977). Initial movement of grains on a stream bed: The effect of relative protrusion. *Proc. R. Soc. London, Ser. A* **352**, 523–527.

Francis, J. R. D. (1973). Experiments on the motion of solitary grains along the bed of a water stream. *Proc. R. Soc. London, Ser. A* **332**, 443–471.

Graf, W. H. (1971). "Hydraulics of Sediment Transport." McGraw-Hill, New York.

Guy, W. H. (1964). An analysis of some storm-period variables affecting stream sediment transport. *Geol. Surv. Prof. Pap. (U.S.)* **462-E**, 1–46.

Hammond, F. D. C., Heathershaw, A. D., and Langhorne, D. N. (1984). A comparison between Shields' threshold criterion and the movement of loosely packed gravel in a tidal channel. *Sedimentology* **31**, 51–62.

Helley, E. J. (1969). Field measurements of the initiation of large bed particle motion in Blue Creek near Klamath, California. *Geol. Surv. Prof. Pap. (U.S.)* **562-G**, 1–19.

Inman, D. L. (1949). Sorting of sediments in the light of fluid mechanics. *J. Sediment. Petrol.* **19**, 51–70.

Komar, P. D. (1980). Modes of sediment transport in channelized water flows with ramifications to the erosion of the Martian outflow channels. *Icarus* **42**, 317–329.

Komar, P. D. (1987). Selective grain entrainment by a current from a bed of mixed sizes: A reanalysis. *J. Sediment. Petrol.* **57**, 203–211.

Komar, P. D., and Li, Z. (1986). Pivoting analyses of the selective entrainment of sediments by shape and size with application to gravel threshold. *Sedimentology* **33**, 425–436.

Komar, P. D., and Wang, C. (1984). Processes of selective grain transport and the formation of placers on beaches. *J. Geol.* **92**, 637–655.

Lane, E. W., and Kalinske, A. A. (1939). The relation of suspended to bed material in rivers. *Trans., Am. Geophys. Union* **20**, 637–641.

Leopold, L. B., and Maddock, T. (1953). The hydraulic geometry of stream channels and some physiographic implications. *Geol. Surv. Prof. Pap. (U.S.)* **252**, 1–57.

Li, Z., and Komar, P. D. (1986). Laboratory measurements of pivoting angles for applications to selective entrainment of gravel in a current. *Sedimentology* **33**, 413–423.

McCave, I. N. (1971). Sand waves in the North Sea off the coast of Holland. *Mar. Geol.* **10**, 199–225.

Milhous, R. T. (1973). Sediment transport in a gravel-bottomed stream. Ph.D. Thesis, Oregon State University, Corvallis (unpublished).

Miller, M. C., McCave, I. N., and Komar, P. D. (1977). Threshold of sediment motion under unidirectional currents. *Sedimentology* **24**, 507–527.

Miller, R. L., and Byrne, R. J. (1966). The angle of repose for a single grain on a fixed rough bed. *Sedimentology* **6**, 303–314.

Neill, C. R., and Yalin, M. S. (1969). Quantitative definition of beginning bed movement. *J. Hydraul. Div., Am. Soc. Civ. Eng.* **95**(HY1), 585–588.

Nordin, C. F. (1963). Sediment transport in alluvial channels: A preliminary study of sediment transport parameters, Rio Puerco near Bernardo, New Mexico. *Geol. Surv. Prof. Pap. (U.S.)* **462-C**, 1–21.

Novak, I. D. (1973). Predicting coarse sediment transport: The Hjulström curve revisited. *In* "Fluvial Geomorphology" (M. Morisawa, ed.), Publ. Geomorphol., pp. 13–25. State University of New York, Binghamton.

Parker, G., Klingeman, P. C., and McLean, D. G. (1982). Bedload and size distribution in paved gravel-bed streams. *J. Hydraul. Div., Am. Soc. Civ. Eng.* **108**(HY4), 544–571.

Sharp, R. P., and Nobles, L. H. (1953). Mudflows of 1941 at Wrightwood, Southern California. *Geol. Soc. Am. Bull.* **64**, 547–560.

Todd, O. J., and Eliassen, C. E. (1938). The Yellow River problem. *Proc. Am. Soc. Civ. Eng.* **64**, 1921–1991.

White, C. M. (1940). The equilibrium of grains on the bed of a stream. *Proc. R. Soc. London, Ser. A* **174**, 332–338.

Yalin, M. S., and Karahan, E. (1979). Inception of sediment transport. *J. Hydraul. Div., Am. Soc. Civ. Eng.* **105**(HY11), 1433–1443.

7

RHEOLOGIC, GEOMORPHIC, AND SEDIMENTOLOGIC DIFFERENTIATION OF WATER FLOODS, HYPERCONCENTRATED FLOWS, AND DEBRIS FLOWS

JOHN E. COSTA

U.S. Geological Survey, David A. Johnston Cascades Volcano Observatory, Vancouver, Washington

INTRODUCTION

When investigating floods, especially in small, mountainous basins, one of the most important tasks is to properly identify the flow process that occurred in the basin. Variation in flow processes are caused mainly by the character and content of sediment entrained in the flow. Failure to properly identify the type of flow has lead to significant scientific misunderstanding and erroneous remedial practices in many areas. Mitigating procedures for normal floods, such as channelization and damming, may be ineffective for other types of flows. For example, channelization for debris flows is ineffective because channels can quickly become blocked, causing subsequent surges to flow in new directions. Channel improvements during the 1964 dry season in the Rio Reventado channel in Costa Rica proved unsuccessful. The first storm of the rainy season promptly filled the enlarged channel with mud and rock debris (Waldron, 1967). Peak discharge values calculated using evidence from debris flows usually lead to excessive estimates of floods in small basins, and because of sparse rainfall data in mountainous regions, indirect-discharge estimates have been used by some hydrologists and engineers to determine the rainfall that occurred during a storm. This usage may lead to inaccurate estimates of rainfall and flood discharges for the design of flood control structures (Costa and Jarrett, 1981).

The purpose of this chapter is to outline and describe the rheologic, geomorphic, and sedimentologic evidence for the most common types of flows that can occur in the channels of small, steep basins. The possible flow processes can be divided conveniently into three categories: (1) water floods, (2) hyperconcentrated flows, and (3) debris flows. Each category has some unique and diagnostic effects and products. However, in nature, there exists a continuum of flow conditions and sediment concentrations.

RHEOLOGY OF DIFFERENT TYPES OF FLOWS

Water Floods

Water floods, which are Newtonian fluids, have viscosities that are unique to a particular fluid composition at a specified temperature, and have essentially no yield or shear strength. Newtonian fluids have a linear relation between shear stress and rate of strain in which the slope of the line is the dynamic viscosity of the fluid:

$$\tau = \mu \frac{dv}{dy} \tag{1}$$

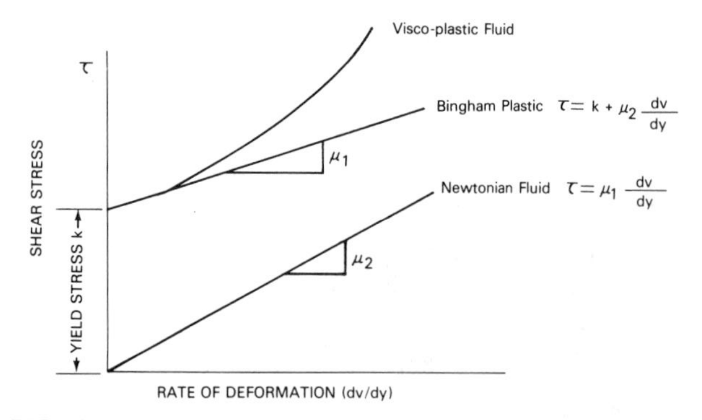

FIGURE 1. Stress–strain curves for idealized Newtonian fluid and plastic rheologic bodies. Newtonian fluid has no shear strength. Plastic bodies begin to deform only after yield stress, k, is exceeded (from O'Brien and Julien, 1985).

where τ = shear stress
μ = dynamic viscosity
v = velocity
y = depth (Fig. 1)

For pure water μ is a constant at constant temperature. Increasing concentrations of fine sediment begin to contribute shear strength to the flow, which then must be exceeded before any deformation or flow occurs. Water floods are unable to resist shear stress without deformation or to have any appreciable shear strength. Although pure Newtonian fluids have no shear strength, a small finite shear strength can exist in water floods with large sediment loads. These flows are beginning to lose their Newtonian character, and small shear strengths (less than 100 dynes/cm^2) are estimated to be possible (Kang and Zhang, 1980; Mingfu et al., 1983).

During water floods, sediment and water are two distinct and separate phases. Sediment moves by suspension and by rolling and saltating along the channel by transfer of energy from moving water to the sediment particles. Sediment support mechanisms primarily are electrostatic charges on small particles in slow-moving water, and turbulence. Water floods are turbulent flows that transport relatively small amounts of sediment (about 1–40% by weight or 0.4 to 20% by volume), have nonuniform sediment concentration profiles, are undergoing shear throughout the flow, and have bulk densities in the general range of 1.01–1.33 g/cm^3, viscosities between 0.01 and about 20 poises, and particle fall velocity between 100 and about 33% of clear-water fall velocities (Bradley and McCutcheon, 1985) (Table 1, Fig. 2*a*). These conditions represent the normal flood flows in most channels (*water flows* of Bull, 1964; *flash floods* of Gagoshidze, 1969; and *clear-water floods* of the Committee on Methodologies for Predicting Mudflow Areas, 1982). With water floods standard hydraulic and sediment transport equations, such as Manning's or Einstein's equations (Graf, 1971), generally describe the flow characteristics.

Hyperconcentrated Flows

With increasing sediment concentrations, flow and fluid properties gradually begin to change, fall velocity of particles decreases, and fluid density and viscosity increase. The decrease in fall velocity enables fine sediment to remain in suspension longer and sediment transport rates to increase substantially (Nordin, 1963).

At sediment concentrations of 40% by weight (20% by volume), Beverage and Culbertson (1964) define the initiation of hyperconcentrated flows. Sediment concentrations less than these percentages are very common, but greater percentages are unusual. Hyperconcentrated flows are streamflows containing large quantities of sediment (40 to about 70% by weight, 20 to about 47% by volume) and bulk densities in the general range of 1.33–1.80 g/cm^3 (Table 1, Fig. 2*b*). Hyperconcentrated flows possess a small, but measurable shear strength, probably about 100–400 dynes/cm^2. Shear strength was not used in originally defining hyperconcentrated flow (Beverage and Culbertson, 1964).

Rheologic and fluid characteristics are greatly controlled by the amount of fines (silt and clay) in the flow. Flows with neutrally buoyant coarse particles at low strain rates can maintain Newtonian behavior up to sediment concen-

TABLE 1 General Rheologic Classification of Water and Sediment Flows in Channels[a]

Flow	Sediment Concentration	Bulk Density (g/cm^3)	Shear Strength (dyne/cm^2)	Fluid Type
Water flood	1–40% by wt. 0.4–20% by vol.	1.01–1.33	0–100	Newtonian
Hyperconcentrated flow	40–70% by wt. 20–47% by vol.	1.33–1.80	100–400	non-Newtonian (?)
Debris flow	70–90% by wt. 47–77% by vol.	1.80–2.30	>400	Viscoplastic (?)

[a] Assumes silt and clay content <10%

FIGURE 2. Photographs of (*a*) Newtonian water flood, Parish Creek, Utah. Note the turbulent nature of the flow. (*b*) Hyperconcentrated flow, Kanab Creek, Utah. Note the smooth flow surface and greatly decreased turbulence. (*c*) Debris flow, Slate Creek, Colorado. Note the absence of turbulence and the very coarse material being moved.

TABLE 1 *(Continued)*

Major Sediment-Support Mechanism	Viscosity (poise)	Fall Velocity (% of Clear Water)	Sediment Concentration Profile	Predominant Flow Type
Electrostatic forces, turbulence	0.01–20	100–33	Nonuniform	Turbulent
Buoyancy, dispersive stress, turbulence	20–≥200	33–0	Nonuniform to uniform	Turbulent to laminar
Cohesion, buoyancy, dispersive stress, structural support	>>200	0	Uniform	Laminar

trations of about 73% by weight (50% by volume) (Bagnold, 1954; Rodine, 1974). Flows with more poorly sorted sediments sustain Newtonian characteristics with sediment concentrations as much as 59% by weight (35% by volume) (Fei, 1983). Fluid suspensions of montmorillonite and kaolinite clays acquire shear strength at concentrations of 3 to 13% by volume (Hampton, 1972). Hyperconcentrated flows with maximum suspended sediment concentrations of 70% by weight have been measured in Rudd Creek, Utah (Pierson, 1985), 68% by weight in the Rio Puerco, New Mexico (Nordin, 1963), 67% by weight in the Toutle River near Mount St. Helens, Washington (Pierson and Scott, 1985), 65% by weight in the Paria River, Arizona (Beverage and Culbertson, 1964), 62% by weight in the Little Colorado River, Arizona (Beverage and Culbertson, 1964), and 47% by weight in Kanab Creek, Utah (J. E. Costa, U.S. Geological Survey, unpublished data, 1975). Criterion for defining the range of hyperconcentrated flows originally was based on perceived natural boundaries in actual measured sediment concentrations, not on rheologic characteristics. Hyperconcentrated flows also are called *noncohesive mudflows* (Kurdin, 1973), *turbulent mudflows* (Gagoshidze, 1969), *intermediate flows* (Bull, 1964), *type III and IV sediment flows* (Lawson 1982), and *mud floods* (Committee on Methodologies for Predicting Mudflow Areas, 1982).

In hyperconcentrated flows, as in water floods, the solids and water are separate components of the flow. Vertical turbulence fluctuations keep sediment in suspension through viscous drag on the particles. This process smooths out large-scale fluctuations and decreases turbulence. Particle fall velocities can be only one-third or less of their values in clear water, and sediment concentration profiles begin to become more uniform. With larger sediment concentrations, sediment support mechanisms begin to change. Particle collisions begin to extract as much energy from the flow as turbulence. Buoyancy and dispersive stress, in addition to turbulence, are primary sediment support mechanisms (Costa, 1984). When the flows slow, coarser solids are deposited. Sediment transport capacity equations are questionable because empirical constants for such equations are based on clear-water conditions.

In the hyperconcentrated flow range, logarithmic velocity profiles have been measured in the Yellow River, China (Zhou et al., 1983), and in the Rio Puerco, New Mexico (Nordin, 1963). Resistance of hyperconcentrated flows in channels and pipes have been found similar to clear-water values (Fan and Dou, 1980), and flows generally are undergoing shear throughout (Bradley and McCutcheon, 1985). Thus, hyperconcentrated flows seem to be approximated as non-Newtonian fluids where shear strength is about 100–400 dynes/cm^2 (Fig. 2*b*).

The concentration at which a sediment–water mixture acquires sufficient shear strength to produce diagnostic landforms and deposits (about 200–400 dynes/cm^2) depends on the particle size distribution of the mixture. Flows with large amounts of silt and clay acquire shear strength when sediment concentrations are 53–59% by weight (30–35% by volume) (Lane, 1940; Qian et al., 1980). Flows with a large quantity of clay have shear strength at weight concentrations as small as 23% (10% by volume) or less (Hampton, 1975).

Debris Flows

There is a definite distinction between water floods and hyperconcentrated flows and debris flows. In debris flows solid particles and water move together as a single viscoplastic body (Johnson, 1970). Sediment entrainment is irreversible, water and solids move at the same velocity, and debris flows cannot deposit any but the coarsest particles as flow velocities decrease. Solids may constitute 70–90% by weight (47–77% by volume) of the flow mass, and bulk densities generally are 1.80–2.30 g/cm^3 for typical poorly sorted sediments (Table 1, Fig. 2*c*). Shear is concentrated in a thin zone at flow boundaries. When deposited, there is no separation of debris flows into solid and liquid components as in water floods and hyperconcentrated flows. However, some dewatering of coarse debris flow deposits may occur shortly after deposition, and the very coarsest particles may settle in the fluid (Rodine and Johnson, 1976; Pierson, 1980).

The Coulomb-viscous and Bingham-plastic models for debris flows generally are known as viscoplastic rheologic models (Fig. 1). They originated from rheologic investigations by Bingham and Green (1919) on oil paints and seem to characterize the features of debris flows and their deposits better than Newtonian models (Johnson, 1970).

In debris flows resistance to flow (or deformation) results from shear strength originating from cohesion and internal friction and viscosity. Cohesion and internal friction constitute the shear strength of the debris that must be exceeded before any flow occurs. Viscosity only affects flow resistance in moving debris flows. The model combines the Coulomb equation

$$\tau = c + \sigma \tan \alpha \tag{2}$$

where τ = shear stress
c = cohesion
σ = normal stress
α = angle of internal friction

and the Newtonian viscous-flow equation

$$\tau = \mu \frac{dv}{dy} \tag{3}$$

and is called the Coulomb-viscous model and has the form

$$\tau = c + \sigma \tan \alpha + \mu \frac{dv}{dy} \qquad (4)$$

Takahashi (1981) modeled debris flows as dilatant fluids, based on the experimental results of Bagnold (1954). Using Bagnold's concept of dispersive pressure P, the shear stress of debris flows is

$$\tau = P \tan \theta \qquad (5)$$

where τ = shear stress
P = dispersive pressure
θ = the dynamic angle of internal friction

Dominant sediment-support mechanisms in debris flows are profoundly different from those in water floods. Cohesion is controlled by the amount of clay in the debris. Slurries with 8–10% clay can support sand-sized particles indefinitely. Buoyancy, controlled by the density difference of submerged solids and transporting fluid, is a major particle support mechanism in debris flows and could support 75–90% of the particle weight in debris flows (Costa, 1984). Dispersive stress (Bagnold, 1954) results from lift produced when forces are transmitted between particles in collision or near collision as one is sheared over another. Where sediment concentrations are large, dispersive stress is a dominant process in dynamic sediment flows.

Structural support, or grain-to-grain contacts providing a framework of particles in contact with the bed and each other, occurs at sediment volume concentrations greater than 35% and supports about one-third of the weight of coarse particles (Pierson, 1981). The efficacy of turbulence in debris flows is questionable because of the substantial viscosity and cohesion, as well as the laminar appearance of most debris flows (Enos, 1977).

Shear strength, and its effect on landforms and sediment characteristics, can be used to separate flow processes. During flood flows when sediment concentration is relatively small, shear strength increases slowly with increasing sediment loads, but the fluid can be considered to be approximately Newtonian. However, at some critical value, shear strength increases rapidly with increasing sediment concentration (Costa, 1984). Differences in this shear strength result in different sedimentology and landforms that are diagnostic of the different flow processes. These differences have been directly measured in active subaerial sediment flows originating at Matanuska Glacier, Alaska, studied by Lawson (1982).

A classification of sediment-transporting flows using measured or calculated sediment concentrations and shear strengths from laboratory slurries and natural flows is shown in Figure 3. The critical values of shear strength and sediment concentration between hyperconcentrated flows and debris flows vary with composition, texture, and sorting of sediments, and no single value can differentiate all situations. Water floods and hyperconcentrated flows trans-

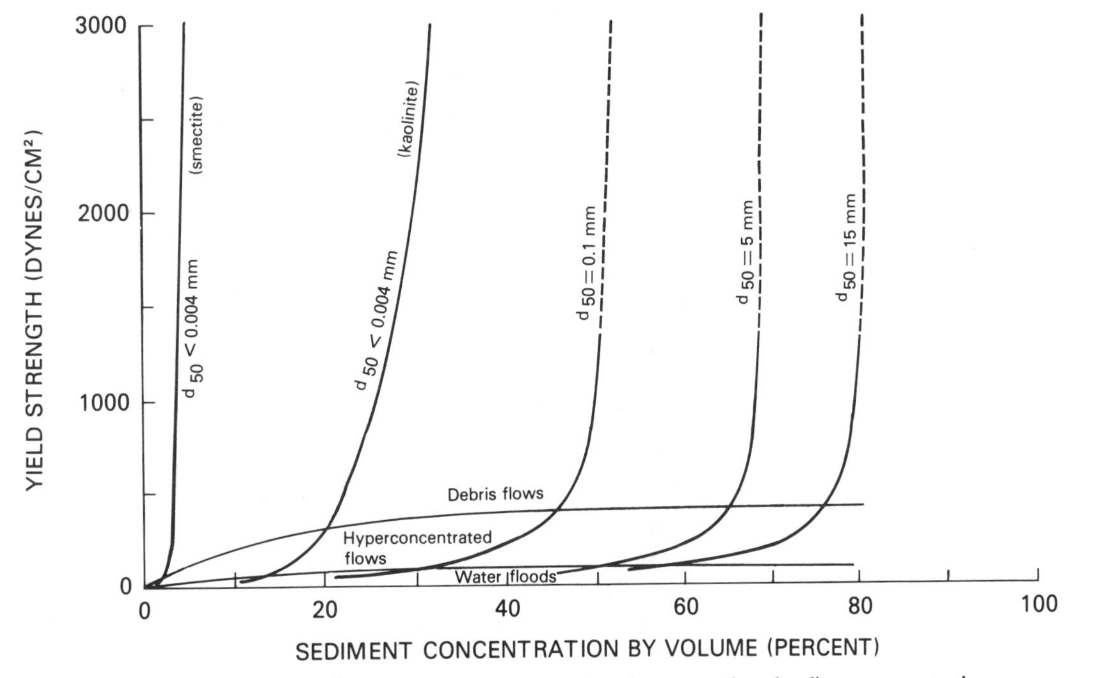

FIGURE 3. Classification of sediment-laden flows based on shear strength and sediment concentration. Data from Hampton (1972) and Kang and Zhang (1980).

port sediment by turbulence, shear, lift, drag, and dispersive stresses, and conventional hydraulic formulas generally apply because the fluids still are nearly Newtonian. Once a sediment concentration threshold is exceeded, shear strength of the flow increases exponentially. The flow boundary between hyperconcentrated flows and debris flows is considered to occur at the marked increase in shear strength that occurs at about 400 dynes/cm^2 for natural, poorly sorted sediments (Fig. 3). In debris flows sediment is transported by cohesive strength, buoyant forces, grain interactions, structural support, and perhaps turbulence (Costa, 1984).

GEOMORPHIC EVIDENCE OF FLOW TYPES

There are many reports of the different landforms associated with water floods in small, steep basins (Wolman and Eiler, 1958; Costa, 1974; Knox, 1980). Depositional landforms can be generalized as different types of bars that form at locations of energy dissipation (Baker, 1984), fans at the mouths of small tributaries, and sheets and splays of coarse sediments that have flat bases and convex tops on relatively fine-grained floodplain surfaces (Costa, 1974; Ritter, 1975). These water flood deposits usually will have primary sedimentary structures, including horizontal stratification, cross-bedding, and imbrication. Rapidly deposited, well-sorted sediments may appear massive. Stream channels eroded by water flows have a different morphology than debris flow channels. Wide, shallow channels that have width-to-depth ratios greater than 12 are characteristic of bedload-dominated water flood channels (Schumm, 1960; DeSloges and Gardner, 1984), while debris flow channels tend to have a semicircular to "U"-shape (Johnson, 1970; Costa and Jarrett, 1981).

Varying degrees of shear strength result in distinctive landforms and deposits for each of the different flow processes (Table 2). Because of the rapid increases in shear strength of sediment–water flows above a critical sediment concentration (Fig. 3), Costa and Jarrett (1981) suggested that in the field, the geomorphic (landforms) and sedimentologic (deposits) evidence resulting from a large flow should be unequivocal as to whether a water or hyperconcentrated or debris flow had occurred. Debris flows in open channels have the following dynamic characteristics different from normal flood flows: (1) a steep, lobate snout at the flow front, commonly containing a large concentration of large boulders; (2) lateral levees of very coarse deposits; (3) a tendency to flow in obvious pulses or surges; (4) a concentration of the largest particles toward the surface and edges of the flow; and (5) formation of a nondeforming rigid plug in the center of the flow, and maximum shear concentrated at flow boundaries (Costa and Williams, 1984; Pierson and Scott, 1985).

Geomorphic evidence for hyperconcentrated flows is very difficult to separate from water floods; but debris flow evidence usually is diagnostic: (1) marginal levees of coarse, poorly sorted clasts bordering channels; (2) steep-fronted terminal lobes of coarse, poorly sorted sediments on fans and in channels; (3) great damage to vegetation, including large trees, in the direct flow path, but little or no damage to vegetation, except burial, at the edges of the flow and on gentle gradients. Dried, gravelly mud (similar to dried concrete) may coat branches and trees at the flow margins; (4) the presence of a wide, trapozoidal-shapped stream channel that has a relatively small width-to-depth ratio of less than 10. This form results from the passage of a nondeforming, rigid-plug flow (Johnson, 1970; Costa, 1984; DeSloges and Gardner, 1984) (Fig. 4); and (5) where flows turn in bends, the strength of the slurry straightens stream-

TABLE 2 Geomorphic and Sedimentologic Characteristics of Water and Sediment Flows in Channels

Flow	Landforms and Deposits	Sedimentary Structures	Sediment Characteristics
Water flood	Bars, fans, sheets, splays; channels have large width-to-depth ratio	Horizontal or inclined stratification to massive; weak to strong imbrication; cut-and-fill structures; ungraded to graded	Average Trask sorting coefficient 1.8–2.7; clast supported; normally distributed; rounded clasts; wide range of particle sizes
Hyperconcentrated flow	Similar to water flood	Weak horizontal stratification to massive; weak imbrication; thin gravel lenses; normal and reverse grading	ϕ graphic sorting 1.1–1.6 (poor); clast-supported open-work texture; predominantly coarse sand
Debris flow	Marginal levees, terminal lobes, trapozidal to U-shaped channel	No stratification; weak to no imbrication; inverse grading at base; normal grading near top	Average Trask sorting coefficient 3.6–12.3; ϕ graphic sorting 3.0–5.0 (very poor to extremely poor); matrix supported; negatively skewed; extreme range of particle sizes; may contain megaclasts

FIGURE 4. Photograph of debris flow channel on Mount Massive, Colorado. View is toward source area. Note well-formed levees bordering the channel and the extraordinary large boulders in the right levee. Two people standing in the channel for scale.

lines, restricts the formation of secondary circulation, and pushes sediment to the outside of meanders where it commonly is deposited. This is the opposite of water flood where secondary currents in bends form points bars on the inside of meanders.

SEDIMENTOLOGIC EVIDENCE OF FLOW TYPES

Sedimentologic differentiation of different types of subaerial water and sediment flows primarily is based on sorting and sedimentary structures (Table 2). In poorly sorted sediments water flood deposits usually produce sedimentary structures, including horizontal and inclined laminations or stratification, imbrication, cut-and-fill structures, and cross-bedding (Allen, 1982) (Fig. 5*a*). Some well-sorted flood sediments can appear massive. Sorting characteristics seem to be valuable clues to processes in mountain channels. Water flood sediments, commonly poorly sorted, generally are better sorted than debris flow deposits. Average Trask sorting coefficients (actually a limited measure of sorting because it takes into account only the middle 50% of the distribution) for water floods range from 1.8 to 2.7 (Costa, 1984). Textural plots of median grain size against quartile deviation (Pe and Piper, 1975) and median grain size versus the grain size of the coarsest one percent (CM diagrams) (Bull, 1962) seem to differentiate water flood sediments and debris flow sediments.

Hyperconcentrated sediments and their characteristics are poorly understood. At Mount St. Helens, Washington, hyperconcentrated flows in large rivers have suspended-sediment concentrations of 57–67% by weight (35–45% by volume). These deposits have a coarse, sandy texture with distinctly less fines than debris flow deposits and are more poorly sorted than most water flood deposits of similar median size, with graphic sorting values of 1.1–1.6 ϕ. These deposits have a generally massive or poorly developed horizontal stratification with thin gravel lenses, a clast-supported noncohesive open-work structure, and reverse-graded subunits (Scott, 1985) (Fig. 5*b*).

Starkel (1972) studied debris flows in the Darjeeling Hills of northern India. Field evidence for the transition from debris flows to hyperconcentrated flows along the lower valley bottoms consists of (1) a decrease in the relative amounts of fine-grained sediments, (2) numerous percussion marks on large boulders indicating turbulent transport, (3) imbrication of coarse-gravel clasts, (4) extensive erosion of valley fill and bedrock, and (5) extensive deposition of open-framework boulders on valley floors (Starkel, 1972).

Debris flow deposits consist of a uniform distribution of sizes from clay to boulders. The largest clasts are supported by a matrix of sand, silt, and clay, but some debris flow deposits can be clast supported if the matrix drains or is washed away. Despite this complication, the distinguishing feature of undisturbed debris flows is a mud matrix surrounding larger particles (Blackwelder, 1928; Crandell, 1971) (Fig. 5*c*). A debris flow matrix also may contain lightweight materials such as wood and bark fragments, pine needles and cone chips, and animal droppings that should have floated away if water or hyperconcentrated flows were responsible for the deposits. Numerous bubble holes (vesicles) also are more common in the fine matrix material of debris flows than in water-deposited fine sediments (Sharp and Nobles, 1953; Crandell, 1971).

Sedimentary structures, including stratification, are virtually nonexistent in debris flow deposits; however, contacts between different flows tend to be distinct. Debris flow sediments are very poorly to extremely poorly sorted. Average Trask sorting coefficients range from 3.6 to 12.3 (Costa, 1984), and at Mount St. Helens, graphic sorting of debris flow deposits ranges from 3.0 to 5.0 ϕ (Pierson and Scott, 1985). Other textural characteristics that may be helpful in identifying debris flow deposits are positive skewness and bimodal size distributions (Sharp and Nobles, 1953; Scott, 1971).

Because of the small difference in density between boulders and fluid material in debris flows, buoyant forces and dispersive pressures may concentrate boulders at the top of the deposit, forming reverse grading (Fisher, 1971; Naylor, 1980). However some debris flow deposits are normally graded. Clast fabric can also be used to identify debris flow deposits. In thick, viscous flows that have a relatively small water content, the larger clasts have a random orientation throughout the deposit (Lawson, 1982). In more fluid flows that are less viscous, particles may have a poorly preferred orientation parallel or perpendicular to the flow direction (Lindsay, 1968; Mills, 1984).

FIGURE 5. Photographs of (*a*) stratified and cross-bedded flood gravel, upper Wabash River Valley, Indiana; (*b*) weakly stratified to massive, reverse-graded, hyperconcentrated flow deposits, Mount St. Helens, Washington (photograph courtesy of Kevin M. Scott, U. S. Geological Survey); (*c*) unstratified gravel and boulders in fine-grained matrix; debris flow deposits, Ophir Creek, Nevada.

CONCLUSIONS

A variety of different flow processes may occur in the channels of small, steep watersheds. The flow processes in these channels can be divided into water floods, hyperconcentrated flows, and debris flows, depending on the composition, texture, and sorting of sediments. The criteria proposed here assumes the source materials are natural, poorly sorted sediments.

Water floods are Newtonian fluids with turbulent flow, nonuniform sediment concentration profiles, sediment concentrations less than about 40% by weight (20% by volume), and shear strengths less than 100 dynes/cm^2. Landforms diagnostic of water floods include various types of bars formed on the inside of meander bends and at points of flow separation, and sheets and splays of coarse sediments. These flood deposits usually will contain primary sedimentary structures such as stratification, cross-bedding, cut-and-fill structures, and imbrication, and overlie relatively fine-grained floodplain surfaces. Hyperconcentrated flows are approximated as moderately turbulent to laminar non-Newtonian fluids with nonuniform to uniform sediment concentration profiles, sediment concentrations ranging from 40 to 70% by weight (20 to 47% by volume), and shear strengths ranging from 100 to 400 dynes/cm^2. Landforms deposited from hyperconcentrated flows are poorly understood. Deposits are massive to crudely stratified, and sorting is intermediate between that of debris flow and most water flood deposits. Debris flows are non-Newtonian viscoplastic or dilatent fluids having laminar flow and uniform sediment concentration profiles, sediment concentrations ranging from 70 to 90% by weight (47 to 77% by volume), and shear strengths greater than about 400 dynes/cm^2. Landforms consist of marginal levees bordering channels and steep-fronted terminal lobes of coarse, poorly sorted clasts on fans and in channels. Deposits consist of matrix-supported coarse clasts in massive beds with extremely poor sorting and reverse grading.

Because of the fundamental rheologic differences among water floods, hyperconcentrated flows, and debris flows, it is possible to accurately reconstruct the correct flow process from geomorphic and sedimentologic evidence. Landforms and sediment characteristics are diagnostic of the different flow processes because of the rapid increase of shear strength with increasing sediment concentrations. The onset of this strength results in different types of flow and sediment support mechanisms, which in turn produce different landforms and sedimentary structures. Hyperconcentrated flow landforms and deposits rarely have been studied and are poorly understood. These flows are transitional between more normal water floods and debris flows, and they retain characteristics of both. They have been difficult to identify as causing unique landforms and deposits because such flows are relatively rare and transitional. Difficulties in interpretation of process also may occur when a single rainstorm produces multiple types of flows in the same basin. However, the correct identification of flow processes is essential for understanding sediment mobilization, routing, and storage, proper design of mitigation measures, and accurate delineation of hazard zones. The solution is accurate landform and sediment interpretation in the field.

ACKNOWLEDGMENTS

I would like to thank Patrick A. Glancy, R. Craig Kochel, Donald E. Hillier, and Carol Anderson for helpful and constructive reviews and Thomas C. Pierson and Kevin M. Scott for support and inspiration.

REFERENCES

Allen, J. R. L. (1982). "Sedimentary Structures: Their Character and Physical Basis; Vols. 1 and 2. Elsevier, Amsterdam.

Bagnold, R. A. (1954). Experiments on a gravity-free dispersion of large solid spheres in a Newtonian fluid under shear. *Proc. R. Soc. London, Ser. A* **225**, 49–63.

Baker, V. R. (1984). Flood sedimentation in bedrock fluvial systems. *Mem.—Can. Soc. Pet. Geol.* **10**, 87–98.

Beverage, J. P., and Culbertson, J. K. (1964). Hyperconcentrations of suspended sediment. *J. Hydraul. Div. Am. Soc. Civ. Eng.* **90** (HY6), 117–128.

Bingham, E. C., and Green, H. (1919). Paint, a plastic material and not a viscous liquid; the measurement of its mobility and yield value. *Proc. Am. Soc. Test. Mater.* **19**, Part II, 640–664.

Blackwelder, E. (1928). Mudflow as a geologic agent in semiarid mountains. *Geol. Soc. Am. Bull.* **39**, 465–484.

Bradley, J. B., and McCutcheon, S. C. (1985). The effects of high sediment concentration on transport processes and flow phenomena. *In* "Erosion, Debris Flow, and Disaster Prevention" (T. Aritsune, ed.), pp. 219–226. Toshindo Printers, Tokyo.

Bull, W. B. (1962). Relation of textural (CM) patterns to depositional environment of alluvial fan deposits. *J. Sediment. Petrol.* **32**, 211–216.

Bull, W. B. (1964). Alluvial fans and near-surface subsidence in western Fresno County, California. *Geol. Surv. Prof. Pap. (U.S.)* **437-A**, 1–71.

Committee on Methodologies for Predicting Mudflow Areas (1982). "Selecting a Methodology for Delineating Mudslide Hazard Areas for the National Flood Insurance Program." Nat. Res. Counc., Nat. Acad. Press, Washington, D. C.

Costa, J. E. (1974). Stratigraphic, morphologic, and pedologic evidence of large floods in humid environments. *Geology* **2**, 301–303.

Costa, J. E. (1984). The physical geomorphology of debris flows. *In* "Developments and Applications of Geomorphology" (J. E. Costa and P. J. Fleisher, eds.), pp. 268–317. Springer-Verlag, Berlin and New York.

Costa, J. E., and Jarrett, R. D. (1981). Debris flows in small mountain stream channels of Colorado, and their hydrologic implications. *Bull. Assoc. Eng. Geol.* **18**, 309–322.

Costa, J. E., and Williams, G. P. (1984). Debris-flow dynamics (videotape). *Geol. Surv. Open-File Rep. (U.S.)* **84-606**, 22-½ minutes.

Crandell, D. R. (1971). Postglacial lahars from Mount Rainier volcano, Washington. *Geol. Surv. Prof. Pap. (U.S.)* **677**, 1–75.

DeSloges, J. R., and Gardner, J. S. (1984). Process and discharge estimation in ephemeral channels, Canadian Rocky Mountains. *Can. J. Earth Sci.* **21**, 1050–1060.

Enos, P. (1977). Flow regimes in debris flows. *Sedimentology* **24**, 133–142.

Fan, J., and Dou, G. (1980). Sediment transport mechanics. *In* "Proceedings of the International Symposium on River Sedimentation," pp. 1167–1177. Guanghua Press, Beijing, China.

Fei, X. (1983). Grain composition and flow properties of heavily concentrated suspensions. *In* "Proceedings of the Second International Symposium on River Sedimentation," pp. 296–308. Water Resources and Electrical Power Press, Beijing, China.

Fisher, R. V. (1971). Features of coarse-grained, high concentration fluids and their deposits. *J. Sediment. Petrol.* **41**, 916–927.

Gagoshidze, M. S. (1969). Mudflows and floods and their control. *Sov. Hydrol.* **4**, 410–422.

Graf, W. H. (1971). "Hydraulics of Sediment Transport." McGraw-Hill, New York.

Hampton, M. A. (1972). The role of subaqueous debris flow in generating turbidity currents. *J. Sediment. Petrol.* **42**, 775–793.

Hampton, M. A. (1975). Competence of fine-grained debris flows. *J. Sediment. Petrol.* **45**, 834–844.

Johnson, A. M. (1970). "Physical Processes in Geology." Freeman & Cooper, San Francisco, California.

Kang, Z., and Zhang, S. (1980). A preliminary analysis of the characteristics of debris flows. *In* "Proceedings of the International Symposium on River Sedimentation," pp. 225–226. Chinese Society of Hydraulic Engineering, Beijing, China.

Knox, J. C. (1980). Geomorphic evidence of frequent and extreme floods. *In* Proc. Eng. Found. Conf., "Improved Hydrologic

Forecasting—Why and How," pp. 220–238. Am. Soc. Civ. Eng., New York.

Kurdin, R. D. (1973). Classification of mudflows. *Sov. Hydrol.* **4**, 310–316.

Lane, E. W. (1940). Notes on limit of sediment concentration. *J. Sediment. Petrol.* **10**, 94–95.

Lawson, D. E. (1982). Mobilization, movement, and deposition of active subaerial sediment flows, Matanuska Glacier, Alaska. *J. Geol.* **90**, 279–300.

Lindsay, J. F. (1968). The development of clast fabric in mudflows. *J. Sediment. Petrol.* **38**, 1242–1253.

Mills, H. H. (1984). Clast orientation in Mount St. Helens debris-flow deposits, North Fork Toutle River, Washington. *J. Sediment. Petrol.* **54**, 626–634.

Mingfu, W., Yizheng, Z., Jianjun, L., Wenzhong, D., and Weimin, W. (1983). An experimental study on turbulence characteristics of flow with hyperconcentration. *In* "Proceedings of the Second International Symposium on River Sedimentation," pp. 45–53. Water Resources and Electrical Power Press, Beijing, China.

Naylor, M. A. (1980). The origin of inverse grading in muddy debris flow deposits—a review. *J. Sediment. Petrol.* **50**, 1111–1116.

Nordin, C. F. (1963). A preliminary study of sediment transport parameters, Rio Puerco near Bernardo, New Mexico. *Geol. Surv. Prof. Pap. (U.S.)* **462-C**, 1–21.

O'Brien, J. S., and Julien, P. Y. (1985). Physical properties and mechanics of hyperconcentrated sediment flows. *In* "Delineation of Landslide, Flash-flood, and Debris Flow Hazards in Utah" (D. S. Bowles, ed.), pp. 260–279. Utah Water Res. Lab., Logan.

Pe, G. G., and Piper, D. F. W. (1975). Textural recognition of mudflow deposits. *Sedimentology* **13**, 303–306.

Pierson, T. C. (1980). Erosion and deposition by debris flows at Mount Thomas, North Canterbury, New Zealand. *Earth Surf. Processes* **5**, 227–247.

Pierson, T. C. (1981). Dominant particle-support mechanisms in debris flows at Mount Thomas, New Zealand, and implications for flow mobility. *Sedimentology* **28**, 49–60.

Pierson, T. C. (1985). Effects of slurry composition on debris flow dynamics, Rudd Canyon, Utah. *In* "Delineation of Landslide, Flash-flood, and Debris Flow Hazards in Utah" (D. S. Bowles, ed.), pp. 132–152. Utah Water Res. Lab., Logan.

Pierson, T. C., and Scott, K. M. (1985). Downstream dilution of a lahar: Transition from debris flow to hyperconcentrated stream flow. *Water Resour. Res.* **21**, 1511–1524.

Qian, Y., Yang, W., Zhao, W., Chang, X., Zhang, L., and Xu, W. (1980). Basic characteristics of flow with hyperconcentration of sediment. *In* "Proceedings of the International Symposium on River Sedimentation," pp. 175–184. Guanghua Press, Beijing, China.

Ritter, D. F. (1975). Stratigraphic implications of coarse-grained gravel deposited as overbank sediment, southern Illinois. *J. Geol.* **83**, 645–650.

Rodine, J. D. (1974). Analysis of mobilization of debris flows. Ph.D. Thesis, Stanford University, Stanford, California (unpublished).

Rodine, J. D., and Johnson, A. M. (1976). The ability of debris, heavily freighted with coarse clastic materials, to flow on gentle slopes. *Sedimentology* **23**, 213–234.

Schumm, S. A. (1960). The shape of alluvial channels in relation to sediment type. *Geol. Surv. Prof. Pap. (U.S.)* **352-B**, 17–30.

Scott, K. M. (1971). Origin and sedimentology of 1969 debris flows near Glendora, California. *Geol. Surv. Prof. Pap. (U.S.)* **750-C**, C242–C247.

Scott, K. M. (1985). Lahars and lahar-runout flows in the Toutle-Cowlitz River System, Mount St. Helens, Washington—origins, behavior and sedimentology. *Geol. Surv. Open-File Rept. (U.S.)* **85-500**, 1–202.

Sharp, R. H., and Nobles, L. H. (1953). Mudflow of 1941 at Wrightwood, southern California. *Geol. Soc. Am. Bull.* **64**, 547–560.

Starkel, L. (1972). The role of catastrophic rainfall in the shaping of the relief of the lower Himalaya (Darjeeling Hills). *Geogr. Pol.* **21**, 103–147.

Takahashi, T. (1981). Debris flows. *Annu. Rev. Fluid Mech.* **13**, 57–77.

Waldron, H. H. (1967). Debris flow and erosion control problems caused by the ash eruptions of Irazu Volcano, Costa Rica. *Geol. Surv. Bull. (U.S.)* **1241-I**, 1–37.

Wolman, M. G., and Eiler, J. P. (1958). Reconnaissance study of erosion and deposition produced by the flood of August 1955 in Connecticut. *Trans. Am. Geophys. Union* **39**, 1–14.

Zhou, W., Zeng, Q., Fang, Z., Pan, G., and Fan, Z. (1983). Characteristics of fluvial processes for the flow with hyperconcentration in the Yellow River. *In* "Proceedings of the Second International Symposium on River Sedimentation," pp. 608–619. Water Resources and Electrical Power Press, Beijing, China.

8

FLOOD SEDIMENTATION IN BEDROCK FLUVIAL SYSTEMS

VICTOR R. BAKER
Department of Geosciences, University of Arizona, Tucson, Arizona

and

R. CRAIG KOCHEL
Department of Geology, Southern Illinois University, Carbondale, Illinois

INTRODUCTION

The distinction between alluvial and nonalluvial river channels is critical for the analysis of sedimentation by large, rare flood events (Baker, 1984). Where bedrock or indurated regolith has sufficient erosional resistance, narrow, deep confined channels may result (Fig. 1). Such channels are subject to extremely high flow velocities, shear stresses, and stream powers during extreme floods (Baker, this volume). The high discharges of extraordinary floods are accommodated by exceptionally great flow depths (Tinkler, 1971; Baker, 1977). The sedimentary phenomena of these narrow, deep bedrock channels will be described in this chapter.

It is somewhat ironic that the dramatic flood flow phenomena that defy direct measurement in bedrock channels also provide unique opportunities for paleohydraulic flow reconstruction. Two methods are applied. The first involves the fact that, given a suitable supply of sediment, the extreme competence of bedrock channel flows produces a broad range of transported boulder sizes that can be related to responsible flow hydraulics (Baker, 1973; Bradley and Mears, 1980; Costa, 1983). The second arises from the rapid fall in stage that follows from the flood peak through narrow, deep bedrock channels. Suspended sand and silt are deposited at high flood levels in slackwater areas, and preserved as the flood rapidly recedes. Such deposits allow the reconstruction of past floodwater surface profiles, in turn allowing the calculation of the flood magnitude. Stratigraphic sequences of such deposits, as preserved in certain ideal settings, allow the generation of long-term records of magnitudes and ages for the largest ancient flow events (Kochel and Baker, this volume). Because of their importance for flow reconstruction, this chapter will emphasize the fine-grained deposits of bedrock channels. The discussion of coarse-grained deposits will supplement a published review (Baker, 1984).

Bedrock channels draining relatively small, mountainous basins may be subject to extremely high sediment concentrations. The resulting hyperconcentrated flows or debris flows may be distinguished from water floods by their distinctive resulting landforms and sediment characteristics (Costa, this volume). This chapter will exclusively treat water floods in which turbulence is sufficient to separate entrained sediment into the three primary modes of transport: bedload, suspended load, and washload (Komar, this volume). This distinction generally results in stratified-to-massive, relatively well-sorted bedload materials comprising bars, fans, sheets, and splays (Baker, 1984; Costa, this volume).

MODES OF SEDIMENT TRANSPORT

Narrow, deep bedrock canyons in the southwestern United States typically show a sorting of flood channel sediments as shown in Figure 2. Coarse gravel and boulders cover channel bottoms, while slackwater deposits of sand and

FIGURE 1. Bedrock fluvial channels with narrow, deep cross sections, Aravaipa Creek, southeastern Arizona. (*a*) Survey crew standing on alternate bar of gravel and boulders, typical of channel floor. (*b*) Constricted inner gorge of Aravaipa Creek. Debris from flood of October 1983 can be seen adjacent to saguaro cactus at left center. Slackwater deposit of sand (Fig. 3) occurs at mouth of Painted Cave Creek, which enters Aravaipa Creek from top left.

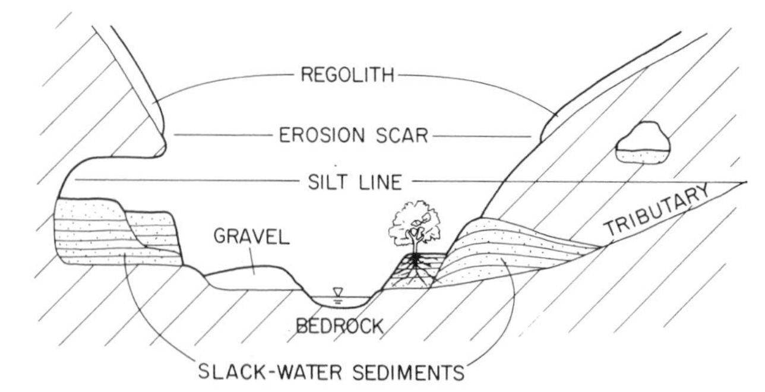

FIGURE 2. Schematic cross section of a bedrock canyon river, typical of the southwestern United States. Gravel comprises bars on channel floor. Slackwater sediments, mainly sand, comprise terraces and mounds in areas of velocity reduction for rare, great floods (tributary mouths, caves, alcoves, etc.). Silt lines and erosion scars may also serve as useful paleoflood stage indicators (Kochel and Baker, this volume).

silt occur in channel margin areas, where flow velocities during rare, large floods were reduced by flow separation or eddies. This sorting develops because of hydraulic factors for extreme flow events (Table 1).

As shown by Komar (this volume) a bed shear stress of 10 N/m^3 (10^2 dynes/cm^2), as typical of flood flows on low-gradient alluvial rivers (Table 1), can entrain and transport pebble gravel (1-cm diameter) as bedload. Sediment finer than 1 mm (coarse sand) can be in suspension, while washload must be finer than about 0.1 mm (very fine sand and silt). This distinction very nicely corresponds to usual experience in observing river flow. However, that experience may be misleading for bedrock streams.

Table 1 shows that extraordinary floods in narrow, deep canyon settings, such as the Pecos Canyon and the Katherine Gorge, may have bed shear stresses as high as 10^3 N/m^3 (10^4 dynes/cm^2). Komar's analysis shows that such flows could transport cobbles and small boulders (10–30 cm diameters) in suspension. Particles 0.4–0.8 mm (medium sand) and finer would move as washload. Bedload transport is completely off the scale of Komar's diagram (Komar, this volume, Fig. 9), but clearly boulders a few meters in diameter could be easily moved.

For the bed shear stresses of 10^4 N/m^3 achieved by Missoula flood flows (Baker, 1973), washload might include granule gravel. Boulders could be moved in temporary suspension, and immense blocks tens of meters in diameter might move as bedload. These sizes are confirmed by studies of Missoula flood sedimentation (Baker, 1973).

Figure 3 illustrates the striking contrasts in sediment sorting near the mouth of Painted Cave Creek, at its confluence with Aravaipa Canyon (Fig. 1) in southeastern Arizona. The coarse gravel layers comprise bedload transported by flash flooding on Painted Cave Creek. During those tributary floods, sand was flushed through the system as washload and no sand accumulated on the bed. The intercalated sand layers are washload and suspended load from Aravaipa Creek, which were deposited upvalley in Painted Cave Creek by backflooding from Aravaipa. The sands contain organic debris that also settled in this slackwater zone for the mainstem flooding of Aravaipa Canyon.

Bed Material Deposits

Material comprising the streambed is transported as bedload or in temporary suspension. The flood bedload may be organized into complex assemblages of bedforms. In narrow, deep bedrock systems, somewhat distinctive bars and giant

TABLE 1 Comparative Flow Dynamics of Great River Floods

River	Channel Type	Peak Discharge (m^3/s)	Slope	Depth (m)	Velocity (m/s)	Shear Stress (N/m^2)	Power per Unit Area (W/m^2)
Amazon	Alluvial	3×10^5	1×10^{-5}	60	2	6	12
Mississippi	Alluvial	3×10^4	5×10^{-5}	12	2	6	12
East Fork (Wyoming)	Alluvial	20	1×10^{-3}	1.2	1.2	10	12
Chang Jiang (Three Gorges)	Bedrock	1×10^5	5×10^{-4}	150	9	3×10^2	3×10^3
Pecos River (Texas)	Bedrock	2.7×10^4	3×10^{-3}	30	12	9×10^2	1×10^4
Katherine Gorge (Australia)	Bedrock	6×10^3	3×10^{-3}	45	7.5	1.5×10^3	1×10^4
Elm Creek (Texas)	Bedrock	1×10^3	4.5×10^{-3}	7	6.4	3×10^2	2×10^3
Missoula flood (Average)	Bedrock	1×10^7	3×10^{-3}	70	10	2×10^3	2×10^4
Missoula flood (Constriction)	Bedrock	5×10^6	1×10^{-2}	100	30	1×10^4	3×10^5

FIGURE 3. Slackwater sand from Aravaipa Creek intercalated with tributary gravel of Painted Cave Creek, southeastern Arizona. The deposit occurs at the mouth of Painted Cave Creek (Fig. 1*b*). The field investigator is sampling a layer of leaf litter dominated by shrub live oak (*Quercus turbinella*) and desert hackberry (*Celtis pallida*). The former was transported a considerable distance to the site by flooding since it only grows at elevations 600–1600 m higher than the site.

ripple forms develop as a response to very high relative roughness, flow velocities, and stream power (Baker, 1984). Bar forms occur where flood bed shear, flow velocity, and stream power were locally reduced. Examples occur downstream of flow obstructions, at eddy locations, and at expansions. Such large-scale flood-generated bedforms are especially well documented for the Channeled Scabland of Washington State, where cataclysmic Pleistocene flooding has produced a phenomenal assemblage of forms (Baker, 1978).

During major floods along vegetated stream channels small-scale pendant bars commonly develop where trees in the streambed serve as obstructions to the transport of bedload (Fig. 4). In unobstructed reaches the bedload may be organized into large-scale transverse bedforms. These are especially common in the Channeled Scabland, where both the bedform size and the included sediment sizes correlate well to shear stress and stream power (Baker, 1973). Observations of somewhat smaller gravel waves in central Texas (Fig. 5) and Australia (Fig. 6) seem consistent with these trends (Baker, 1984). In general, the development of bars and giant ripple forms is well described in the scientific literature (Allen, 1982).

FIGURE 4. River red gum trees and associated flood deposition, Finke River, central Australia. (*a*) Pendant bar of gravel developed downstream (center to left of photograph) from flood-damaged tree. Note flood debris on upstream sides of trees. (*b*) Scour hole developed on upstream side of tree trunk. A small gravel bar extends downstream from the trunk (left center).

More difficult to envision is the suspension transport of coarse gravel and boulders. There are numerous observations of high-magnitude floods moving bed material load from channel scour holes up on to adjacent highs of the valley floor surface, where the material may organize into broad, lobate sheets. In some cases these coarse bed material particles are carried on to floodplain surfaces, where they may subsequently become intercalated with fine-grained overbank sediments (Ritter, 1975). Boulders may be scoured from rock outcrops or channel lag and be deposited high on adjacent surfaces (Baker, 1977).

Suspension of boulders is possible at exceptionally high velocity or shear stress. Where field criteria are used to exclude the possibility of debris flow transport (Costa, 1984), suspension probably can be attributed to macroturbulent lift forces (Matthes, 1947; Foley et al., 1978). Unfortunately, macroturbulent processes in streamflow have defied measurement and quantification. This important mechanism of sediment transport during intense floods in bedrock channels requires further study.

Washload and Suspended-Load Deposits

As shown above, during bedrock channel flood flows of exceptionally high velocity, shear stress, and power, unusually coarse sediment can be transported as washload. Sand, which would comprise bedload at low flows, can be flushed through narrow, deep channels either in suspension or, for more extreme shear stresses, as washload. Because this coarse sediment has a high fall velocity, it will rapidly settle in channel margin areas where mean flow velocities decrease. The requisite conditions are achieved at tributary mouths, downstream of flow obstructions, in channel margin caves, and other areas of slow eddying and slackwater. Sedimentation at such sites results in slackwater deposits (Kochel and Baker, this volume).

Washload transport is limited by sediment supply. Thus, ideal conditions for slackwater sediment deposition occur where an abundant source of sand-sized sediment occurs immediately upstream of a narrow, deep confined channel reach (most commonly developed in bedrock) subject to occasional, exceptionally intense flood flows. An excellent example is Umbrawarra Gorge, Northern Territory, Australia (Fig. 7). The gorge developed in the Proterozoic Buldiva Sandstone, a very resistant pink quartz sandstone with

FIGURE 5. Large-scale gravel wave that developed transverse to Medina River flood flow (right to left) during tropical storm Amelia flooding in central Texas, August 1978. The ripple chord is about 80 m and the maximum intermediate-axis grain diameter of the gravel armor on its surface averages about 10 cm. The peak flood flow over the ripples had a depth of 10 m, a velocity of 3.5 m/s, and a bed shear stress of 300 N/m^2.

local pebble conglomerates. It drains an extensive lowland area underlain by Proterozoic Cullen Granite. Grusification in the savanna climate yields abundant particles of sand-sized grus that are subject to extensive erosion in floods that follow the annual dry season. This grus is flushed through the gorge as washload, settling in channel margin crevices to mark the high-water surface of past flow events (Fig. 7).

The most comprehensive study of slackwater sedimentation in bedrock channels is that of Kochel (1980) and Kochel and others (1982) in west Texas (Fig. 8). Figure 9 illustrates typical settings of slackwater sediment accumulation in the west Texas flood channels. If slackwater deposition occurs in sites that are relatively well shielded from erosion by subsequent mainstream and tributary flows, then long-term, thick accumulations may provide a continuous record of large-magnitude floods by the mainstream (Kochel and Baker, 1982).

SLACKWATER SEDIMENTATION: CENTRAL CHANG JIANG

Because most narrow, deep bedrock canyons occur close to headwater areas, slackwater sedimentation studies generally have been conducted on relatively small rivers. However, slackwater deposition also characterizes large rivers, including the Mississippi and the Huang He (Yellow River) of China (Shih Fucheng et al., 1985). This section provides some observations on the geomorphic setting for slackwater sedimentation on the Chang Jiang (Yangtze River) of central China (Fig. 10). The general principles apply to slackwater sedimentation in numerous settings.

The largest historic flood of the Chang Jiang occurred during July 1870. In the Three Gorges area of central China (Fig. 11) this flood had a peak discharge of approximately 105,000 m^3/s (Shih Winshing, 1985). The flood flow was

FIGURE 6. Large-scale gravel waves produced by the flood of March 1983, Finke River Gorge, central Australia. These bedforms have an average spacing of 10 m and heights of 0.5 m. The maximum clast size averages 0.4 m in intermediate diameter. Nearby high-water marks indicate a flow depth of approximately 6–7 m.

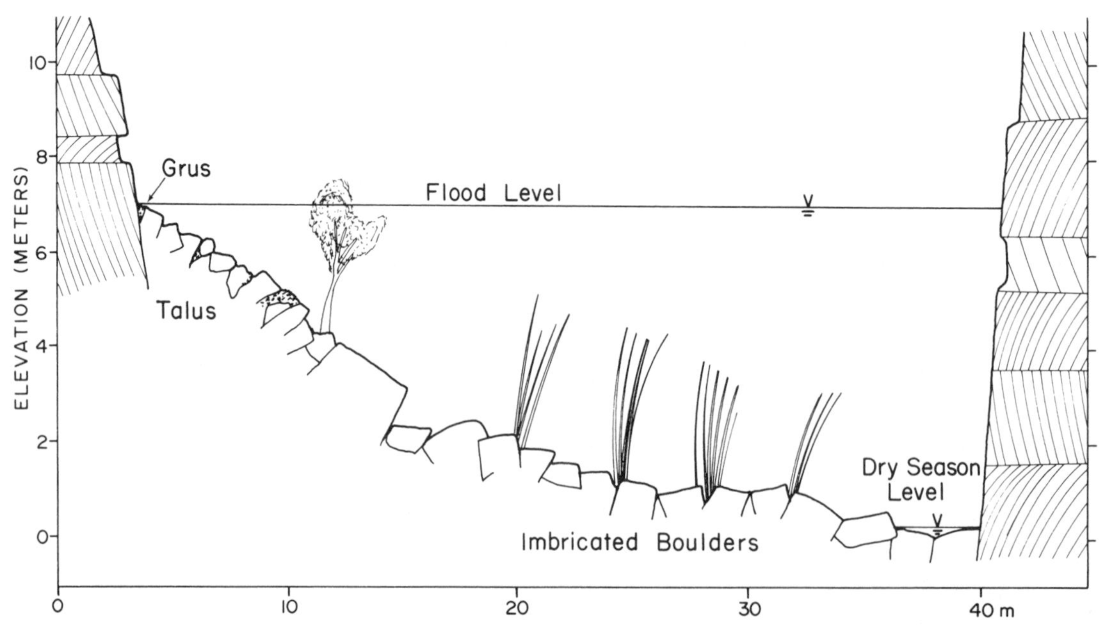

FIGURE 7. Cross section in Umbrawarra Gorge, Northern Territory, Australia. Note that the width-to-depth ratio is approximately 3 for the effective cross-sectional area of the indicated extreme flow event. The gorge is developed in sandstone. Grus (granitic sand) transported by rare large floods settles on the sandstone talus up to the highest stage reached by flooding.

FIGURE 8a. Bedrock channel of Seminole Canyon near the confluence of the Pecos River and Rio Grande. The cross section is nearly rectangular, with canyon depth just slightly less than canyon width. Virtually no coarse or fine sediments are visible in this straight reach extending for about 1 km.

FIGURE 8b. Slackwater sediments in the mouth of Zixto Canyon along the Pecos River. See Kochel and Baker (this volume) for a detailed description of slackwater sedimentation in west Texas.

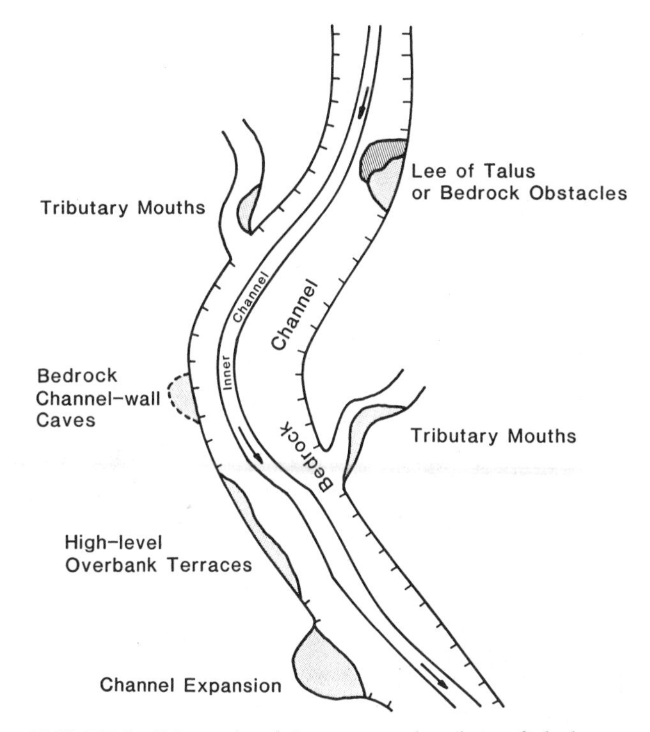

FIGURE 9. Schematic of the common locations of slackwater deposits. See also discussion in Kochel and Baker (this volume).

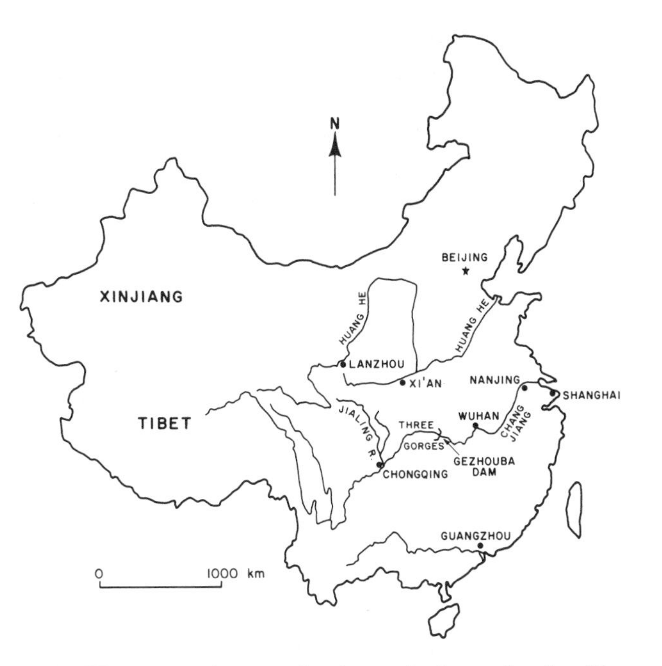

FIGURE 10. Location map showing study sites and major cities of China.

FIGURE 11. Xiling Gorge on Chang Jiang (Yangtze River) of east-central China.

reconstructed by the use of 400 flood marks and 90 rock inscriptions that documented the great calamity (Shih Winshing, 1985). Additional flood marks and records indicate that other extraordinary floods occurred in the years 1153, 1227, 1560, 1788, 1796, and 1860 (Luo Chengzheng, 1985). From such data Chinese hydrologists developed water surface profiles for ancient floods (Fig. 12). These were then utilized in hydraulic calculations of paleodischarges for the appropriate cross sections (Shih Winshing, 1985).

The gradient of Chang Jiang is steep for such a large river. Through the Three Gorges' section (Fig. 12) the river falls 100 m in about 200 km. This slope of 5×10^{-4} is an order of magnitude steeper than comparable-sized alluvial rivers, such as the Mississippi and Amazon (Table 1). The steep gradients reflect flow constrictions. At Three Gorges the Chang Jiang transects a series of resistant limestone mountains. Upstream of the gorges, between Fengdu and Fenje, the river mainly flows in broad strike valleys, resulting in very gentle slopes (Fig. 12). Transverse ranges are also encountered near Chongqing, where gradients are steeper.

Cross sections in the Three Gorges are narrow and deep (Fig. 11). At Qutang Gorge the river is only 100 m wide. During the 1870 flood over 80 m of stage change occurred in this remarkably narrow reach. Floods in the Three Gorges' section have left huge river steamers stranded on rock bars 20 m above normal water levels. The flow properties of these immense floods can be documented on the Jialing River, a major tributary, just upstream of Chongqing.

The gauging station on the Jialing River at North Hot Springs, Beipei District (Fig. 13), was established in 1939. Drainage area of the Jialing River upstream of the gauge is 160,000 km^2. The maximum recorded discharge at this site was 44,000 m^3/s, measured in 1981. This was associated with a depth of 62 m, a width of 220 m, and an average velocity of 7.1 m/s. The minimum recorded discharge at this site was 249 m^3/s, and the long-term average discharge is 2138 m^3/s. Historic flood marks show that the July 1870 flood had a depth of 67 m and a discharge of 60,000 m^3/s. The Jialing River contributed most of the flow to the great July 1870 Chang Jiang flood.

Table 2 compares the approximate flood flow dynamics for Qutang Gorge versus the lower Amazon River. Note

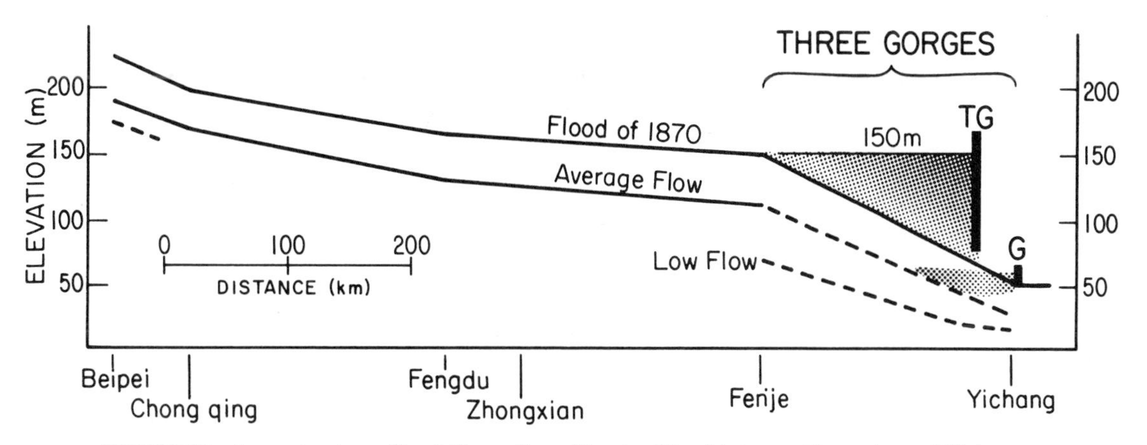

FIGURE 12. Approximate profile of Chang Jiang (Yangtze River) between Chongqing and Yichang. Major dam projects at Gezhouba (G) and Three Gorges (TG) will modify profiles as shown. Profiles are drawn from approximate notes (Shih Winshing, oral communication, 1985).

FIGURE 13. Stream gauge site on Jialing River at North Hot Springs, Beipei District, Chongqing, Sichuan Province, China. The Jialing River drains 160,000 km^3 at this point. This station was used to measure a flow of 44,000 m^3/s in June 1981.

TABLE 2 Comparison of Approximate Flood Flow Dynamics, Chang Jiang and Amazon

Properties	Chang Jiang Qutang Gorge	Lower Amazon
Flow boundaries	Resistant Bedrock	Alluvial
Width (m)	100	3000
Depth (m)	150	60
Discharge (m^3/s)	1×10^5	3×10^5
Stage change in flood (m)	80	15
Velocity (m/s)	7–9	2
Shear stress (N/m^2)	300	6
Power/unit area (W/m^2)	3000	12

that shear stresses and unit stream powers are 2 to 3 orders of magnitude greater for the Chang Jiang flood flow. Sand moves as flood washload, settling as slackwater deposits at tributary mouths (Fig. 14). Backflooding may extend for many kilometers up tributaries (Fig. 15) during extraordinary flow events, and fine-grained washload is easily flushed upstream during major floods on the mainstem river.

FLUME EXPERIMENTS

There are relatively few direct field observations of slackwater sedimentation during rare, large flow events. This arises from the infrequency of the flows, the usual short duration of the events, and the difficulty of access to flood-choked, steep-walled canyons. Therefore, to study the process, Kochel and Ritter (1987) conducted a series of ex-

FIGURE 14. Slackwater deposit of sand transported by moderate flood flow of Chang Jiang (Yangtze River) in Xiling Gorge, east-central China.

FIGURE 15. Win River approximately 5 km upstream from its junction with the Jialing River near Beipei, Chongqing, Sichuan Province, China. Backflooding from the Jialing River during the great floods of 1870 and 1981 was nearly 40 m above low water levels for this river. Terraced rice fields have reworked Jialing River flood sediments emplaced up the Win River valley.

periments utilizing an artificial bedrock stream in a flume. The goal of their study was to determine the influences of variable flood hydrographs and channel geometries on slackwater sedimentation.

Experimental Design

A cement channel was constructed to simulate a nonerodible bedrock channel reach similar to the west Texas channel shown in Figure 8*a*. Channel walls were constructed so that they were nearly vertical, resulting in a semirectangular cross section with an average width of 30 cm and average depth of about 25 cm. Because the most common field setting for stratigraphic accumulations of slackwater sediments is in the mouths of tributaries, the flume model was constructed with 20 tributaries entering the main channel at various junction angles. Adjustments were also made so that the system would exhibit a range of tributary and mainstream channel gradients (Fig. 16). The lateral extent of the tributaries was limited by the width of the 2.5-m-wide flume, causing their distal margin to be constructed as a vertical headwall. This headwall ultimately resulted in interference with slackwater backflooding and sedimentation processes at some sites, particularly during runs made at the highest discharges.

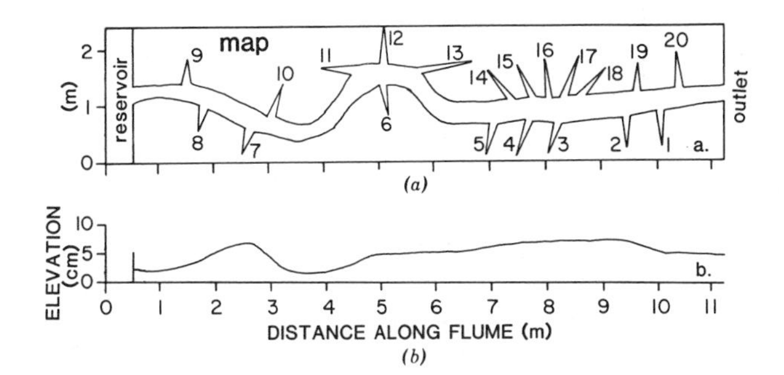

FIGURE 16. Schematic diagram of the cement channel used in the flume to study slackwater sedimentation. (*a*) Map view of the channel and tributaries. (*b*) Profile of main channel measured along the channel axis. The irregularities were emplaced in order to study the effects of varying mainstream gradient on backflooding.

TABLE 3 Summary of Flume Experiments

Run Number	Discharge Q^a (W) (m)	(D) (m)	(V) (m/s)	=	Q^b (m³/s)	Slope	Duration of Run (hr)	Type of Run
1	(0.4)	(0.05)	(0.5)	=	0.01	0.020	1.15	Low flow, low slope
2	(0.49)	(0.1)	(0.6)	=	0.03	0.020	1.15	High flow, low slope
3	(0.38)	(0.035)	(0.72)	=	0.01	0.030	1.15	Low flow, high slope
4	(0.45)	(0.09)	(1.0)	=	0.04	0.030	1.15	High flow, high slope

[a] Measured at fixed point each time.
[b] W = width, D = depth, V = velocity, and Q = discharge.

Numerous test runs, conducted in two separate experimental periods in 1984 and 1985, revealed that it was possible to achieve significant backflooding into tributary mouths in this kind of model. Moreover, the number of sedimentation units deposited in these sites corresponded exactly to the number of floods sent down the channel. Table 3 summarizes the conditions used in four subsequent runs where detailed measurements were made of discharge and sedimentology of the resulting slackwater deposits (Kochel and Ritter, 1987). The four runs were designed to compare slackwater sedimentation under conditions of relatively high and low discharge and under the same discharges with varying mainstream slope achieved by tilting the flume with its hydraulic jacks. The slopes used in the experiments were somewhat higher than the slopes experienced along the Pecos River, which are between 0.0015 and 0.003 (Kochel, 1980). Sediment was introduced into the channel by mixing about 10 kg of poorly sorted sand and mud into a mixing reservoir at the upper end of the flume, with successive sediment injections every 10 min throughout each 70-min flood. Samples of suspended sediment measured near the mouth of the flume showed that variations between runs were within 25% of the mean (Kochel and Ritter, 1987).

Geometry of Slackwater Deposition

For all flume runs sedimentation occurred at tributary mouths because the artificial tributaries (Fig. 16) did not flow. Although this corresponds to one end member in the range of possible tributary/main-channel interactions (Best, 1986), it is the common situation for slackwater sedimentation utilized in paleoflood reconstruction (Kochel and Baker, 1982). Deposits became thinner and finer grained uptributary, away from the main channel as observed in the field (Kochel et al., 1982). Most of the deposits were asymmetrical in section with the thickest portion of the slackwater sediment wedge located along the downstream side of the tributary channel (Fig. 17). This can be explained from observations of the pattern of backflooding into these sites. Dye tracers injected into the flow showed that backflood eddies developed in the lower few centimeters of tributary mouths. Eddy circulation was such that main-channel water entered the tributaries along their downstream walls, whereas return flow to the mainstream exited along the upstream wall of the tributaries. Therefore, the bulk of the slackwater sediments settled rapidly from suspension upon entering

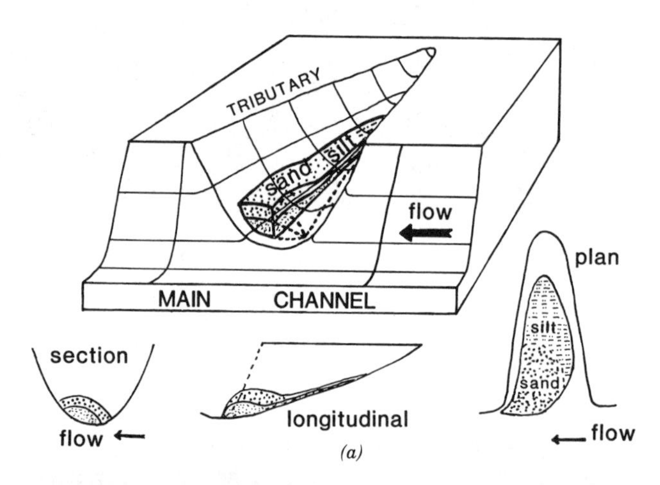

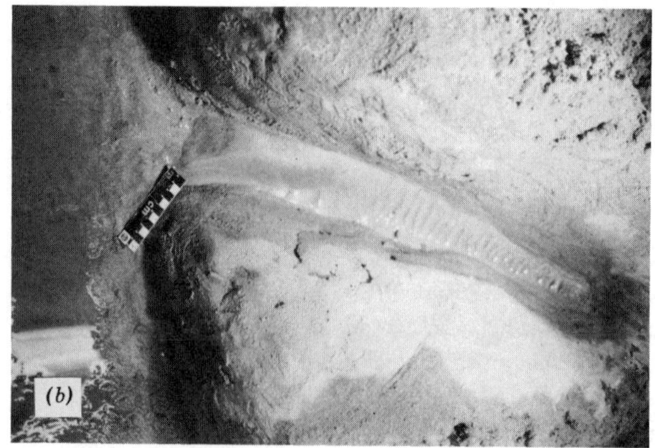

FIGURE 17. Tributary mouth slackwater deposits formed in the flume. (*a*) Schematic diagram illustrating the typical geometry of deposition. (*b*) Deposit at site 2 formed during run 2. Note the marked thinning of the sediment wedge uptributary (to the right). The deposit is also thicker along the downstream tributary wall (toward the top).

the channel along its downstream margin. Return flows probably contained less suspended sediment than backflows and may also have carried some potential slackwater sediment back into the mainstream channel. Deposit asymmetry was most pronounced where tributary junction angles were nearly normal to the mainstream (Kochel and Ritter, 1987).

Variations in Deposit Thickness

A major influence on the efficiency of backflooding is the geometry of the junction angle between the tributary and the mainstream. Maximum backflooding and slackwater sedimentation occurs when junction angles are between 50° and 130°. Below 50°, mainstream flood flows tend to bypass the tributary mouth, resulting in nominal backflooding. Sites with high junction angles (above 130°) are extensively backflooded, resulting in significant slackwater deposition. However, subsequent floods may severely alter the pre-existing deposits because backflood velocities are very high in these high-angle situations. Hence, the continuous stratigraphic sequences required for paleoflood studies are unlikely to be preserved where junction angles are high. These observations have been confirmed both in the flume (Kochel and Ritter, 1987) and in the field (Kochel and Baker, this volume). Moreover, variation in the size of bedforms on the surfaces of flume slackwater deposits further document the range of backflood discharges experienced in tributaries with different junction angles. Ripple chord was greatest for sites having nearly normal junction angles at least for sites with particularly acute junction angles (Fig. 18).

Mainstream gradient also affected the efficiency of backflooding in the flume such that thicknesses of slackwater sediments varied between sites of similar junction angle geometry. Unusually thin slackwater deposits were produced during all flume runs where mainstream gradient was oversteepened (Figs. 16 and 19, sites 7 and 10). Backflooding in these steep reaches was limited because, at constant discharge, mainstream flows responded to the increased gradient by increasing their velocity. These high-velocity mainstream flows bypassed the tributary mouth sites, resulting in reduced backflow stage and associated slackwater sedimentation. The same phenomenon occurred at all sites when the effects of similar discharges were compared at different flume tilt. This relationship can also be observed in field data collected along the Pecos River (Kochel and Baker, this volume).

Slackwater deposit thickness also increased directly with increasing flood discharge when flume slope was held constant. Between-site variations in deposit thickness occur during the same flood due to local variations in channel morphology, junction angle, and slope. At-a-site variations in thickness among deposits of different floods seem to depend on the peak discharge and duration of the flood.

FIGURE 18. Photographs of ripples on slackwater deposits in the flume formed during run 2. (*a*) Large ripples with chord averaging 35 mm at site 17, where the junction angle was 90°. (*b*) Smaller ripples with chord averaging 20 mm at site 11, where the junction angle is about 45°.

Flood Stage and Slackwater Sediment Elevation

The flume experiments provided a means of testing a major assumption in slackwater paleoflood hydrology: The highest elevation of slackwater sedimentation closely corresponds with the associated peak flood stage. Kochel and Ritter (1987) demonstrated that the assumption is probably valid. Figure 20 shows that, in almost all cases, slackwater sediments extended to within a few millimeters of the peak flood stage measured at the mouth of the tributary. The only exceptions occurred at two sites with exceptionally steep tributary slopes. Flume-generated slackwater sediments probably slid downslope from their depositional position at these sites because of increased gravitational stress exerted on saturated sediments (Fig. 20, sites 8 and 9).

During the high discharge runs, however, the above relationships were not so consistent (Fig. 19). At high flow, backflooding inundated tributaries to their artificial

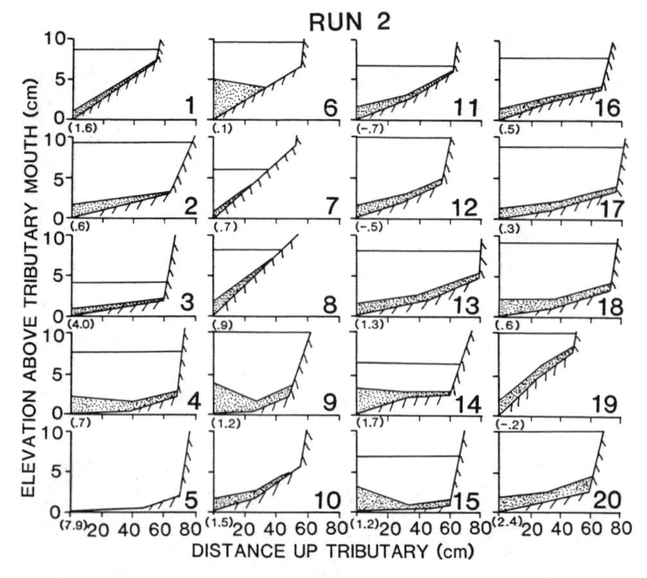

FIGURE 19. Cross section of flume tributaries and slackwater sediments resulting from run 2, characterized by high discharge and low slope. Tributary cross sections show the tributary bedrock profile, depth of backflooding, and the extent of slackwater sedimentation (dotted pattern). The elevation of tributary mouths above the main channel floor are shown in parentheses (cm).

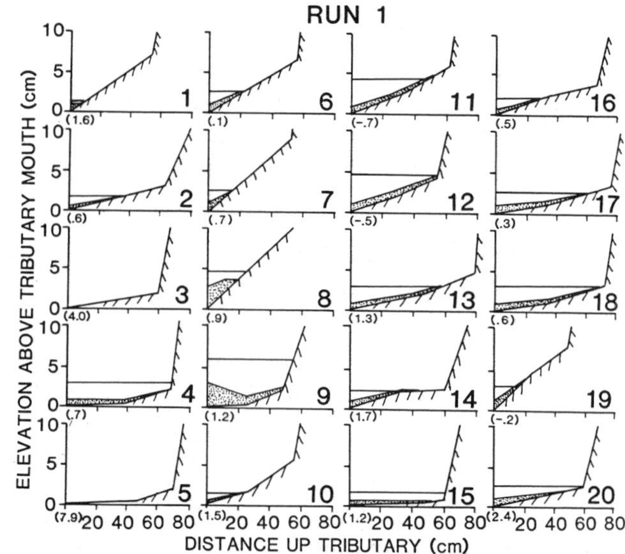

FIGURE 20. Cross section of flume tributaries and slackwater sediments resulting from run 1, characterized by low discharge and low slope.

headwalls. In these cases sediments piled up in the distal portions of the tributaries, rather than extending farther upstream because of the interference of the vertical headwall. Thus, design problems of the flume experiment probably explain this lack of correspondence (Kochel and Ritter, 1987). For many of the sites (Fig. 19), the normal distal thinning trend reversed for the high discharge runs, indicating that slackwater sediments would have extended well upstream had they not been inhibited by the constraints of the flume width.

DISCUSSION

Because resistant-boundary bedrock channels cannot easily adjust their geometry to rare, large flow events, sedimentation is dominated by the influence of complex, three-dimensional flow characteristics dictated by channel boundaries. Important influences occur at channel confluences (Best, 1986). In cases where tributary flow is negligible, backflooding from the mainstem results in slackwater sedimentation. Modes of sediment transport for various size fractions of the load are dictated by the bed shear stress and unit stream power of the flood flow. For narrow, deep bedrock channels the flows are especially intense, resulting in sand transport as washload.

Because of the dynamics of rare, large floods that affect them, the bedrock channel types described in this chapter constitute a distinct fluvial sedimentary environment. Flood sedimentation in such streams consists of gravel and boulder deposits. However, channel margin areas may accumulate slackwater deposits of relatively coarse suspended load and washload. Even very large rivers, such as the Chang Jiang, may display such sedimentary phenomena in reaches with appropriate geometry and erosional resistance.

Processes of tributary backflooding and associated slackwater sedimentation may be studied by simulation in flume experiments. Kochel and Ritter (1987) found that the degree of backflooding varies greatly among tributary mouth sites, depending on their junction angles, mainstream gradients, and the elevations of the tributary above the mainstream. Perhaps the most significant results of the flume studies and their comparisons with field studies (Kochel and Baker, this volume) are in the development of guidelines for the selection of optimum field sites for slackwater paleoflood hydrology studies. Table 4 summarizes this field and laboratory experience in order to increase the chances of selecting the best sites for slackwater paleoflood studies along bedrock rivers.

TABLE 4 Criteria for Selection of Optimum Slackwater Sites for Paleoflood Analysis

Criteria likely to have stable channel cross sections[a]
1. Bedrock channel reaches
2. Reaches where scour and fill during floods are likely to be minimal

Criteria likely to yield thick and continuous slackwater sequence[b]
1. Tributaries with junction angles between 50° and 130°
2. Tributaries along low-gradient reaches of the mainstream
3. Tributaries along fairly straight reaches of the mainstream
4. Low-gradient tributaries
5. Tributaries with basins having morphometries indicative of low flash-flood potential

[a]From Kochel et al. (1982).
[b]From Kochel (1980), Kochel et al. (1982), and this study.

REFERENCES

Allen, J. R. L. (1982). "Sedimentary Structures: Their Character and Physical Basis, vols. 1 and 2." Elsevier, Amsterdam.

Baker, V. R. (1973). Paleohydrology and sedimentology of Lake Missoula flooding in eastern Washington. *Spec. Pap.—Geol. Soc. Am.* **144**, 1–79.

Baker, V. R. (1977). Stream-channel response to floods with examples from central Texas. *Geol. Soc. Am. Bull.* **88**, 1057–1071.

Baker, V. R. (1978). Large-scale erosional and depositional features of the Channeled Scabland. *In* "The Channeled Scabland" (V. R. Baker and D. Nummedal, eds.), pp. 81–115. Nat. Aeronaut. Space Admin., Washington, D.C.

Baker, V. R. (1984). Flood sedimentation in bedrock fluvial systems. *Mem.—Can. Soc. Pet. Geol.* **10**, 87–98.

Best, J. L. (1986). The morphology of river channel confluences. *Prog. Phys. Geogr.* **10** (2), 157–174.

Bradley, W. C., and Mears, A. I. (1980). Calculations of flows needed to transport coarse fraction of Boulder Creek alluvium at Boulder, Colorado. *Geol. Soc. Am. Bull.* **91**, Part II, 1057–1090.

Costa, J. E. (1983). Paleohydraulic reconstruction of flash-flood peaks from boulder deposits in the Colorado Front Range. *Geol. Soc. Am. Bull.* **94**, 986–1004.

Costa, J. E. (1984). The physical geomorphology of debris flows. *In* "Developments and Applications of Geomorphology" (J. E. Costa and P. J. Fleischer, eds.), pp. 268–317. Springer-Verlag, Berlin and New York.

Foley, M. G., Vessell, R. K., Davies, D. K., and Bonis, S. B. (1978). Bedload transport mechanisms during flash floods. *In* "Conference on Flash Floods: Hydrometeorological Aspects and Human Aspects," pp. 109–116. Am. Prepr., Meteorol. Soc., Boston, Massachusetts.

Kochel, R. C. (1980). Interpretation of flood paleohydrology using slackwater deposits, lower Pecos and Devils Rivers, southwest Texas. Ph.D. Dissertation, University of Texas, Austin.

Kochel, R. C., and Baker, V. R. (1982). Paleoflood hydrology. *Science* **215**, 353–361.

Kochel, R. C., and Ritter, D. F. (1987). Implications of flume experiments on the interpretation of slackwater paleoflood sediments. *In* "Regional Flood Frequency Analysis" (V. J. Singh, ed.), pp. 365–384. D. Reidel, Boston.

Kochel, R. C., Baker, V. R., and Patton, P. C. (1982). Paleohydrology of southwestern Texas. *Water Resour. Res.* **18**, 1165–1183.

Luo Chengzheng (1985). A survey of historical flood and its regionalization in China. Paper presented at the U.S.-China Bilateral Symposium on the Analysis of Extraordinary Flood Events, Nanjing, China.

Matthes, G. H. (1947). Macroturbulence in natural stream flow. *Trans. Am. Geophys. Union* **28**, 255–262.

Ritter, D. F. (1975). Stratigraphic implications of coarse-grained gravel deposited as overbank sediment, southern Illinois. *J. Geol.* **83**, 645–650.

Shih Fucheng, Yi Yuanjun, and Han Manhua (1985). Investigation and verification of extraordinarily large floods of the Yellow River. Paper presented at the U.S.-China Bilateral Symposium on the Analysis of Extraordinary Flood Events, Nanjing, China.

Shih Winshing (1985). Application of historic flood in the design of Three Gorge Project. Paper presented at the U.S.-China Bilateral Symposium on the Analysis of Extraordinary Flood Events, Nanjing, China.

Tinkler, K. J. (1971). Active valley meanders in south-central Texas and their wider implications. *Geol. Soc. Am. Bull.* **82**, 1783–1800.

9

RIVER FLOOD REGIME AND FLOODPLAIN STRATIGRAPHY

G. ROBERT BRAKENRIDGE
Department of Geological Sciences, Wright State University, Dayton, Ohio

INTRODUCTION

How do river flood regimes affect floodplain stratigraphic and facies development? In order to address this interesting but complex question, the fundamental processes involved in floodplain genesis are described. Then, relationships between flow stages of various frequencies and facies geometries are used to infer the production of varying floodplain facies architecture. Finally, the story of floodplain stratigraphic development is completed by considering the effects of alternating periods of fluvial stability and activity on floodplain sedimentation.

It will be demonstrated that fluvial history can be described from studying floodplain sequences and that floodplain studies provide valuable perspectives on flood-related fluvial processes at time scales of 100–10,000 yr. Floodplain stratigraphy is shown to be a complex, but decipherable, record of present and past river flood regimes, of other aspects of river history, and of the internal and external geomorphic variables that control such history.

FORMATION OF FLOODPLAINS

Meandering rivers create floodplains by the combined operation of (1) lateral channel migration, with accompanying in-channel deposition, and (2) suspended-load fallout from slowly moving or still water during higher than normal river stages (Fig. 1). The two floodplain-forming processes are independent, although both may occur simultaneously.

In regard to the first process, lateral channel movement broadens valley floors by concave bank erosion. At the same time bedload (relatively coarse) sediments accumulate by point bar deposition on the channel bed, and suspended-load (relatively fine) sediments are deposited on the river's convex bank. The resulting sedimentary sequence may thus be dominated by lateral and not vertical accretion.

The initiation of lateral movement (meandering) may be related to the longitudinal fluid compression that occurs due to channel bed friction and to gradient reduction in the downstream direction (Bejan, 1982). Such a compressed, semiconfined flow responds by exerting lateral forces on its channel in the same way that a partially plugged garden hose bends. The mechanics, geometric patterns, and evolution of meandering, once meander initiation has occurred, are described by an abundant literature (e.g., Leopold et al., 1964; Brice, 1974). Note especially that river meandering can occur during constant flow and need not be related to floods. "Floodplains" formed solely by this lateral accretion process may therefore also form without floods (Fenneman, 1906).

In regard to suspended-load deposition during floods, high discharges cause increased rates of sediment transport along river channels: Enhanced erosion and not deposition should be favored. However, high discharge events still do result in deposition of fine-grained sediment because of (1) reductions in flow velocity caused by channel overtopping and the resulting large increase in flow width and (2) greatly increased roughness and decreased flow velocities related to terrestrial vegetation present on the flooded surfaces, including the upper banks. Thus, deposition of suspended-load sediment is a fundamental result of flow variability along many rivers: Many river channels are, at certain times, too small, floodplains are flooded, and suspended-load skimmed from the top of the flow column is deposited.

FIGURE 1. Overbank flood sedimentation occurring along the Sevier River, southern Utah, during June 1984. The channel is marked near the center of the right side of the photograph by two lines of partially submerged vegetation.

FLOODPLAIN FACIES

As a result of both channel lateral movement and suspended-load deposition, meandering river channels and floodplains include the following two assemblages of sedimentary faces:

1. *Channel bed deposits*: channel lag, point bar, side bar, longitudinal bar, and low-stage slackwater facies.
2. *Bank and overbank deposits*: bank, levee, back-levee swale, terrace veneer, and crevasse splay facies.

Bottom stratum and *top stratum*, respectively, are facies assemblage terms that are equivalent to the two assemblage categories and are especially useful when mapping sediments in cross section (Fig. 2). Bottom stratum refers to the relatively coarse sediments deposited on the channel bed and preserved by burial during lateral migration of the channel. Top stratum refers to the relatively fine sediments deposited on upper channel banks, on levees, and on floodplain and terrace surfaces during infrequent high river stages. Friedman and Sanders (1978) conclude that the presence of both top stratum and bottom stratum facies is the best criterion for the identification of meandering river deposits in the ancient sedimentary record.

Exceptions to the fine-grained (top stratum) and coarse-grained (bottom stratum) grouping exist and can be important for paleohydrologic studies. Thus, relatively coarse-grained crevasse splays are deposited on floodplain surfaces during floods that breach natural levees. They may consist of coarse sands or even (along large rivers) fine pebble gravels. When preserved, they serve as important records of the individual floods that produced them. Also, low-stage slackwater facies are fine-grained (silty) deposits that accumulate on channel beds in local slackwater areas during low river discharges. Such deposition occurs due to the lack of river competence during very low flows and also from the effects of flow separation and other secondary current-related processes (e.g., Thorne et al., 1985) during low flows. Such deposits frequently form at the downstream ends of point bars during low flows and may be superposed directly over the coarse-grained bedload sediments moved during higher stages. Subsequent channel migration may then bury these facies by lower bank accretion and incorporate these fine sediments into the lower portions of the floodplain sequence.

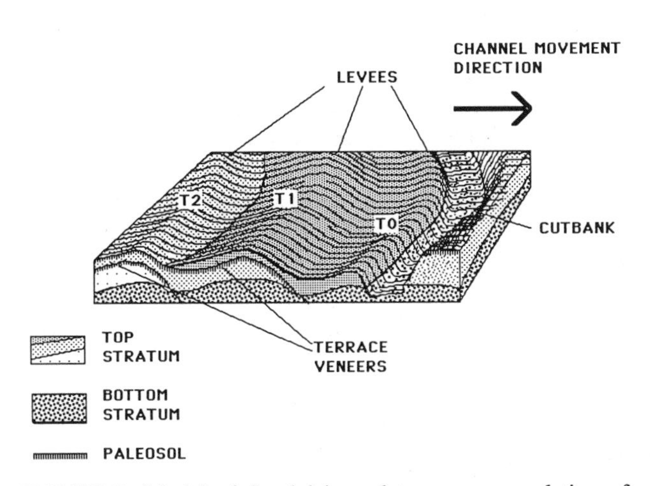

FIGURE 2. Model of floodplain and terrace accumulation of bottom stratum and top stratum sediments by a meandering river. Note the conformable deposition, buried soils marking past channel positions, and terrace veneer facies burying the primary portions of the two older floodplain accumulations and thus creating raised fill terraces T1 and T2.

FIGURE 3. Buried subfossil log, dating from 6400 14C yr B.P., retrieved from low-stage slackwater facies of the Missisquoi River in northern Vermont (Brakenridge, et al., 1987).

Along rivers in Vermont, and perhaps other locations as well, low-stage slackwater facies are a common site of lodgment and burial of large organic debris, such as tree logs (Fig. 3). These materials can be radiocarbon dated and thus can help establish an absolute chronology of river sedimentary history (Brakenridge et al., 1987).

A third type of facies may be locally significant in floodplain environments but is not directly related to river action: alluvial fans (Fig. 4). Such fans occur in both arid

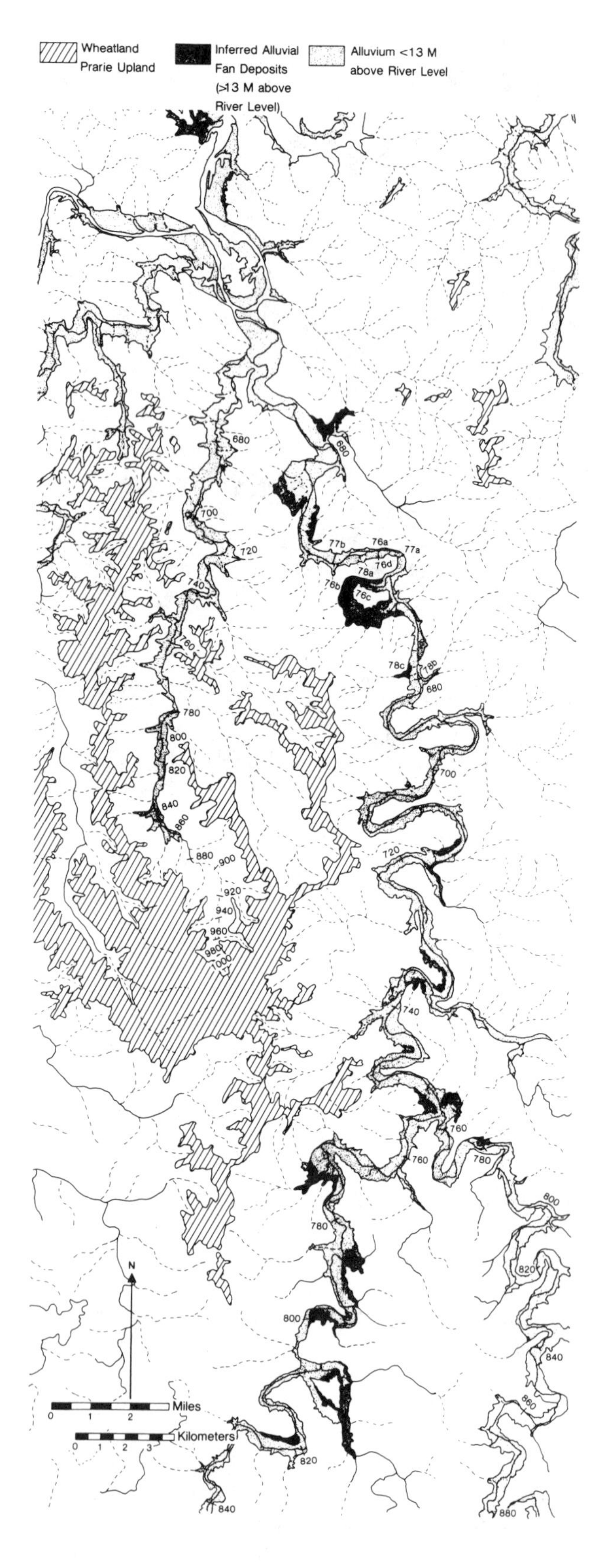

and humid climatic regimes wherever river valley floors are bounded by higher bedrock slopes. Episodic flooding and sedimentation on such fans may occur coevally with major channel floods or, at other times, be out of phase with trunk stream floods. In many upland areas bedrock valley meander cutoffs have created local sedimentary basins of different ages, and these abandoned meanders are today slowly filling with Quaternary alluvial fan material deposited during local flood events (e.g., see cross sections in Haynes, 1985, and Brakenridge, 1981).

FLOOD REGIMES AND FLOODPLAIN FACIES ARCHITECTURE

Many "self-formed" alluvial channels increase their flow capacity during high flows by deepening their channels and altering their bedforms. Given sufficient response times during floods, river channels may simply enlarge to accommodate high flows, and no overbank discharge may occur. Aside from flow variability, therefore, another fundamental requirement for the occurrence of overbank floods is a significant time lag associated with channel adjustment: The channel does not adjust instantaneously.

Time lags for channel response to high flows are normally significant even in many sandy bed, relatively noncohesive river channels. There are examples, however, of quite rapid enlargements of major channels resulting from high flow events (e.g., see Kresan, this volume, and Baker, 1984). Along such rivers, overbank flooding does not occur (Fig. 5).

Aside from erosion, another possibility exists for channel enlargment and high-flow accommodation. Successive increments of top stratum deposited during each overbank flood will slowly increase the height of any given channel's banks, and thus the channel's flow capacity (Fig. 6). Therefore, a marked slowing of top stratum accumulation should occur, through time, after a series of floodplain flood events. The channel capacity will be increased, and progressively larger floods will be required for bank overtopping to occur (Fenneman, 1906). The critical point is that a tendency toward an equilibrium channel size and top stratum thickness must exist, even though this equilibrium may be repeatedly interrupted by lateral movement of the channel.

FIGURE 4. Geomorphic map of the Pomme de Terre River Valley in the Ozark Highlands of southern Missouri. See Haynes (1985) and Brakenridge (1981, 1983) for stratigraphic information concerning the lower valley reach (near top). The dark shading represents areas of alluvial fan sedimentation, which has partially filled two abandoned valley meanders.

FIGURE 5. An entrenched tributary to the Santa Cruz River near Tucson, Arizona.

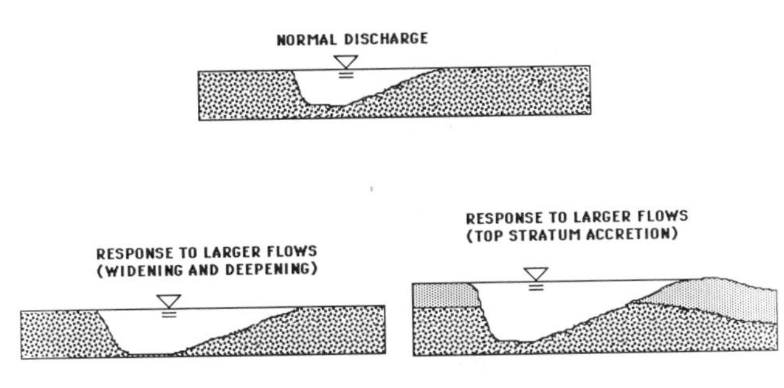

FIGURE 6. Two responses of self-formed river channels to infrequent, higher-than-normal, discharges. Option 1 is an erosional response, and option 2 is a depositional response; both act to enlarge channel capacity. The erosional response may occur quickly during large floods, whereas the depositional response occurs at time scales of decades to centuries as channel enlargement occurs due to progressive floodplain sedimentation.

To illustrate this concept, Figure 7 shows four different river flow regimes (plotted here as frequency of various daily mean river stages) compared to inferred thicknesses of the local floodplain's top stratum and bottom stratum facies. These are shown as sections in Figure 7*a*–7*d* and are discussed in this order below.

Floodplains without Top Stratum Facies

The upper cross section (Fig. 7*a*) shows a river with a constant flow regime, and thus one without top stratum deposition. This is the hypothetical case of a river that does not experience floods, even in-channel ones. The flow frequency series mode is also the bankfull stage along such a river. The floodplain in this case is an alluvial accumulation formed entirely by in-channel processes (mainly point bar sedimentation): It is never flooded and no overbank sedimentation occurs. Such rivers are uncommon in nature.

Rivers that do not experience discharge variation have reduced opportunities for fine-grained deposition, because the flowing current normally occupies its entire channel. However, as noted, other mechanisms exist for limited, in-channel deposition of fine-grained sediments. These include secondary current-related sedimentation and effects

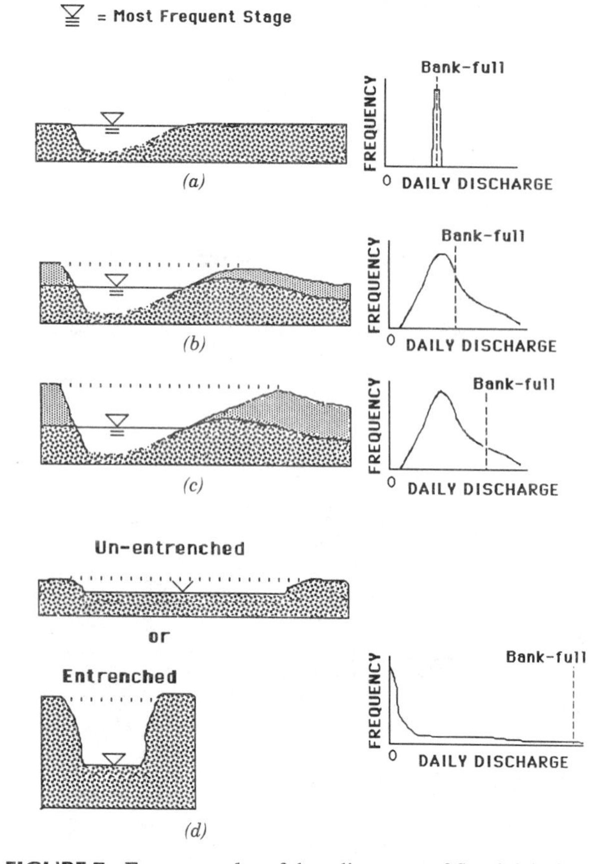

FIGURE 7. Four examples of the adjustment of floodplain facies geometry to differing flow regimes. The hydrographs show the modal (most frequent) daily discharges that are illustrated also in the cross sections. The discharge represented by the bank-full stage increases from (*a*) to (*c*), but the modal discharge remains approximately constant. (*a*) illustrates a river without flood events: discharge never varies. (*b*) and (*d*) illustrate typical meandering rivers with bankfull discharges progressively larger in comparison to the modal discharge. (*d*) illustrates a river whose channel and floodplain geometry oscillates between entrenched and unentrenched configurations, and infrequent large discharge events control channel dimensions.

related to channel position changes. Flow complications as well as geomorphologic changes in the depositing system thus act to affect sedimentation patterns even when the overall flow regime is extremely simple.

Floodplains with Varying Top Stratum Thicknesses

Figure 7*b* shows a perennial river whose modal flows are not greatly less than that needed for the bankfull state. The sandy and gravelly bottom stratum comprises most but not all of the floodplain, and extends nearly to the level of the most frequent daily stage. The top stratum of fine sand, silt, and clay forms a relatively thin cover. Many meandering rivers exhibit such an overall sedimentary architecture. However, another common architecture is that shown by Figure 7*c*, which exhibits a considerably thicker top stratum. What determines the equilibrium thicknesses attained by the top stratum along such rivers (compare Fig. 7*b* and 7*c*)? Do such thicknesses provide information of value in interpreting flood hydrology?

Flood Regime Controls over Top Stratum Thicknesses. Referring again to Figure 7*b*, note that the channel capacity is sufficient to contain flows that are greater than the most frequent discharge. Thus, even more rare high discharges are necessary for overbank floods. This hydraulic geometry adjustment is common in many humid regions: The channel is large enough to contain those high discharges having a recurrence interval of ca. 1.58 yr (Dury, 1973).

The balance between frequency and magnitude of stream power in carving the channel is an important theme in fluvial geomorphology (see review by Richards, 1982, pp. 122–145). Due to the variety of the processes involved, the adjustment of natural channels to higher than normal discharges is a complex compromise between the frequency, duration, competence, and capacity of river flows, the nature of the river's bed and bank materials, and other factors (such as ice jams). Therefore, empirical relationships derived in one region (or along one reach) may not apply to another. The important principle for the present purpose is that the capacity of many self-formed river channels is commonly set by higher than normal discharges, but not, necessarily, by extremely rare, high-magnitude events. Also, in a system at equilibrium, the local river stage associated with this bankfull discharge determines, by definition, the maximum elevation reached by the top stratum facies.

The bankfull stage may, in turn, be either much higher than, or only slightly higher than, the "most frequent stage" which defines the channel bed environment (compare Fig. 7*b* and 7*c*). In the zone between the most frequent river stage and the top of the channel banks, rooted vegetation strongly retards flow competence. Although terrestrial plants capable of surviving periodic inundation are common, such plants do not become established on the mobile bed of the channel itself. As a result, the sloping, vegetated, lower channel banks are much different, in roughness and other hydraulic parameters, from the channel bed (see, e.g., Klimek and Starkel, 1974), and they are important sites of top stratum (but in-channel), suspended-load deposition during intermittent high flows.

The most frequent stage (often perceived as the normal stage) of perennial rivers is thus an important hydrogeomorphic parameter controlling the location of the active, largely vegetation-free, channel bed environment. The difference in local stage between bankfull and most frequent stages may be one important factor affecting equilibrium thicknesses obtained by top stratum sedimentation (as shown in Fig. 7*b* and 7*c*).

Other Factors Affecting Top Stratum Thickness. Slow lateral migration may allow top stratum equilibrium thickness to be quite closely approached. In contrast, river channels experiencing fast lateral migration effectively prohibit floodplains from aggrading to the level permitted by prevailing flood regime. Studies of floodplain sediments along such rapidly meandering rivers often conclude (correctly) that point bar sedimentation is the most important factor in local floodplain development because such floodplains are underlain mainly by point bar and other bottom stratum facies. However, such results do not apply to all meandering rivers. Thus, floodplains along a river in Tennessee are underlain by relatively large thicknesses of top stratum (Fig. 8), and this river experienced low rates of channel migration during the Holocene (0.5–2 m/100 yr; Brakenridge, 1985). The rate of lateral movement along this river was apparently sufficently slow to allow floodplain vertical aggradation to adjust to flood hydrology (Fig. 9).

Flood-Dominated Systems

The last example, Figure 7*d*, shows a quite different facies pattern: The top stratum is absent, the channel is comparatively large in relation to mean daily discharge, and the channel cross section is unusually symmetrical and exhibits steep, almost vertical, banks. This is an entrenched river channel such as is common today in populated semiarid regions in the United States: for example, the Santa Cruz River (Kresan, this volume; also see Fig. 5). The channel is shaped mainly by rare, very high discharge events that tend to cause bank cutting on meander bend insides as well as outsides. The much smaller flows, or no flows, occurring between these events are much less important in determining channel morphology.

Such river channels are extremely large in relation to their average daily discharges. Overbank floods are often essentially absent along entrenched channels because the channel enlarges quickly in response to high discharge events. These fluvial morphologies are favored by at least two important environmental factors: (1) the occurrence of high discharge events that are very large in relation to

FIGURE 8. Large (6–7 m) top stratum thicknesses along the Duck River, Tennessee (Brakenridge, 1984) were exposed by this combined bulldozer-backhoe excavation, with 2–3 m of sandy and gravelly bottom stratum sediments exposed at the base of the excavation.

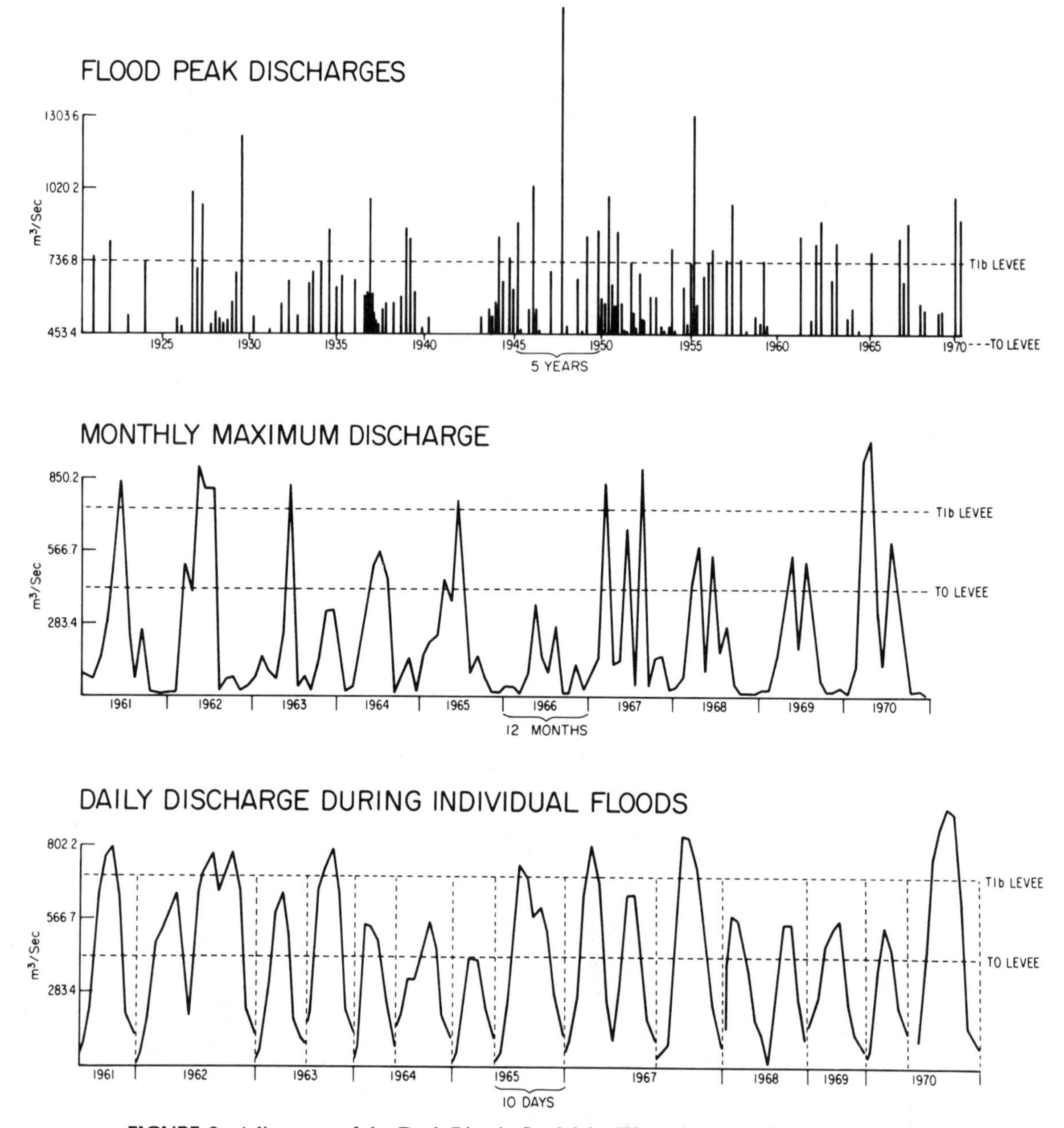

FIGURE 9. Adjustment of the Duck River's floodplain (T0) and lowest alluvial terrace (T1) to river hydrologic regime. Estimated discharges at which the T0 and T1 levees are overtopped are shown. The upper graph shows all recorded overbank discharges; the middle graph shows monthly maximum discharges for 1961–1970 (as an example), and the lower graph shows the duration of overbank discharge for individual floods during the same time interval. The upper graph shows that T0 (the modern, rapidly aggrading floodplain) is being innundated by floods about twice per year, whereas T1b (the lowest well-marked alluvial terrace) is being flooded, on the average, only once every 1.6 yr and has developed upon it a mollic undifluvent soil profile. See Figure 13*c* for a comparison cross section.

normal discharge and (2) low local water tables, which permit erosion and preservation of deeply cut channels.

Most entrenched rivers and streams are not entrenched for very long distances. In the up and downstream directions from entrenched reaches, the channel is instead unusually broad and shallow; floods carve the entrenched reaches while they spread out to occupy the entire (in some cases, several kilometers) widths of the unentrenched channels (see again Fig. 7*d*). Abrupt geomorphic thresholds (headcuts) separate such aggrading reaches from the degrading, entrenched reaches (Fig. 10; also see Patton and Schumm, 1981), and the headcuts may move through time in the upstream direction. Such a fluvial system is dominated by the geomorphic and sedimentologic effects of floods, and a cut-and-fill alluvial stratigraphy results from the alternation, through time, of both entrenched and unentrenched configurations along a given reach.

Summary

Figure 7 shows a progression from floodplain (*a*), a perennial stream with no high flows and no overbank floods, to floodplains (*b*) and (*c*), perennial streams with cross-sectional areas adjusted to high discharges, but not to catastrophic events, to floodplain (*d*), a discontinuous ephemeral stream in which sedimentary and geomorphic processes are adjusted to rare, high-magnitude events occurring in either very broad and shallow or very deep and narrow channels. The models shown are largely time independent: No sense of sedimentary *history* is conveyed. Now to be considered is how sedimentation proceeds though time, and thus how sequences of stratigraphically distinct units, each including the above-discussed sedimentary facies, are produced.

FLOODPLAIN SYSTEMS THROUGH TIME

Causes for Floodplain Stratigraphy

Floodplain stratigraphy is the science of mapping and interpreting floodplain deposits of different age. The basic goal is to reconstruct the sequence of depositional events. The existence of floodplain stratigraphy itself indicates that noncontinuous sedimentation has occurred: The boundaries of stratigraphic units are, by definition, based on a lack

FIGURE 10. Headcut separating unentrenched (to the right) and entrenched (to the left) ephemeral stream reaches in a semiarid portion of southern Arizona.

of continuity. What causes such noncontinuous sedimentation?

Figure 7*a*–7*c* cross sections were discussed as laterally migrating stream channels that may cut continuously outward on the meander bends while slowly accreting sediments on convex point bars and channel banks. As the channels move, they may maintain their equilibrium morphology. The production of a local alluvial unconformity by cut bank erosion thus proceeds simultaneously with conformable point bar, bank, and overbank aggradation on the opposite side of the river. Slow, unidirectional channel migration must eventually be interrupted by (1) meander cutoffs or (2) movement of the apex of the meander in the downstream direction, thus causing one or both "legs" of the meander to begin moving in the opposite direction (Fig. 11).

In either case the cut bank location subsequently become the scene of point bar and convex bank sedimentation. This younger alluvium is separated from the earlier deposits by the (cut-bank-produced) erosional unconformity. This example illustrates a common system-internal mechanism for producing floodplain stratigraphy: No changes in flow regimes or environmental perturbations are necessary. Along meandering rivers, floodplain sedimentation is thus unlikely to be continuous through time even given constant environmental conditions.

When we combine this situation with the documented histories of late Quaternary climatic and other environmental changes in many regions, and their predicted effects on watershed hydrology (Schumm, 1965), then it appears probable that floodplain sedimentation in nature must be *episodic* (marked by periods of much slower and much faster net sedimentation rates) on both a local and regional basis. Discrimination of the two types of episodicity requires information as to whether observed contacts and unconformities are restricted to a specific site or are widespread in the valley or along different valleys.

As a result, two states of channel–floodplain morphologies can be defined for meandering rivers: (1) *stable configurations*, during which channel capacity and floodplain height are both adjusted to prevailing flow regimes, and net top stratum deposition is occurring very slowly and (2) *active configurations*, which occur following a change in the sense of river channel movement or following some environmental perturbation, and during which floodplain sedimentation and cut bank erosion may occur rapidly. These two configurations for meandering river types b and c (Fig.7) are illustrated in Figure 12. The alternation between these two modes of sedimentation is an important cause of floodplain stratigraphy along perennial meandering rivers. In contrast, along river systems in semi-arid and arid regions, where comparatively rare and large flow events are more important (Fig. 7*d*), it is not clear that a relatively stable configuration ever exists even on a local (bend or reach) basis. Instead, as noted above, floodplain stratigraphy is

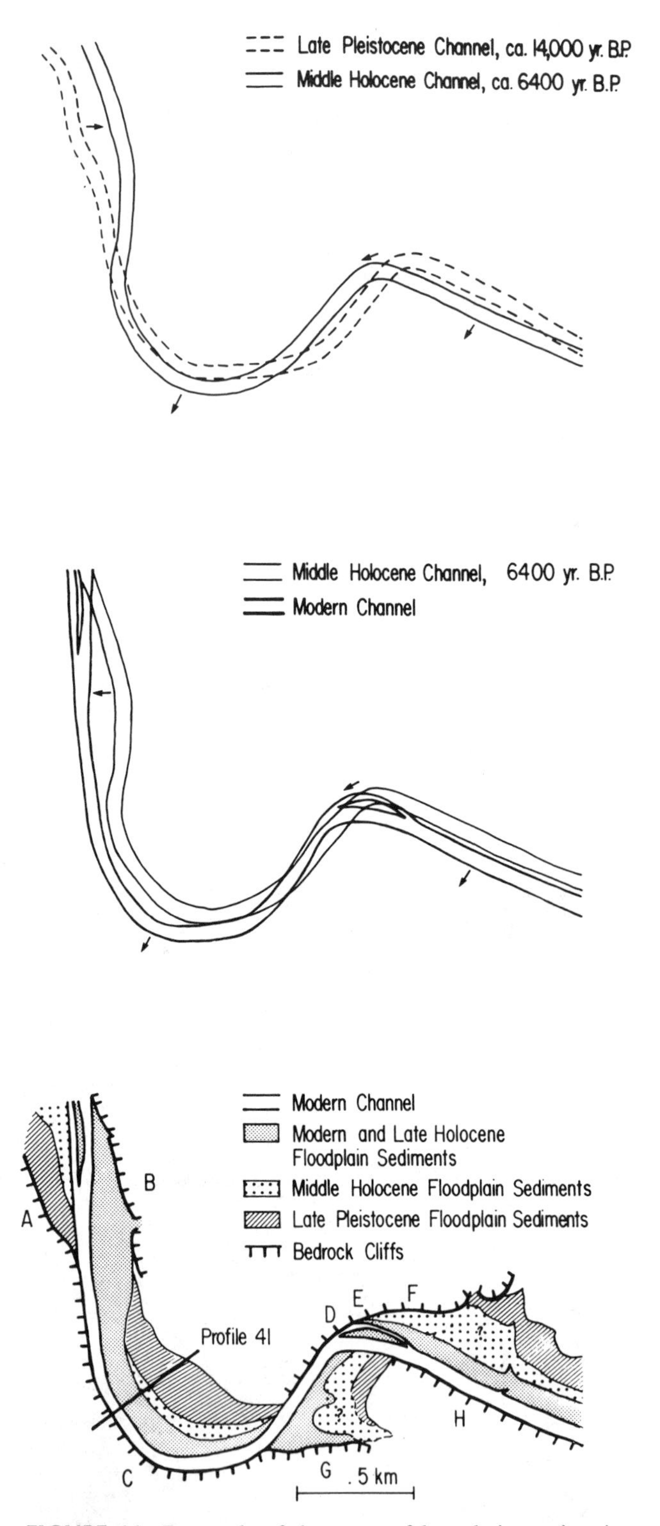

FIGURE 11. Reversals of the sense of lateral river migration along the Duck River in Tennessee. Such reversals truncate previously deposited floodplain alluvium by progressive cut-bank retreat; a second reversal then emplaces a new, adjacent floodplain fill (locations A and B). In this manner, local "cut-and-fills" and inset alluvial terraces are created. At location C, however, channel migration has been mainly uni-directional and a series of conformable fill terraces resulted. From Brakenridge (1985).

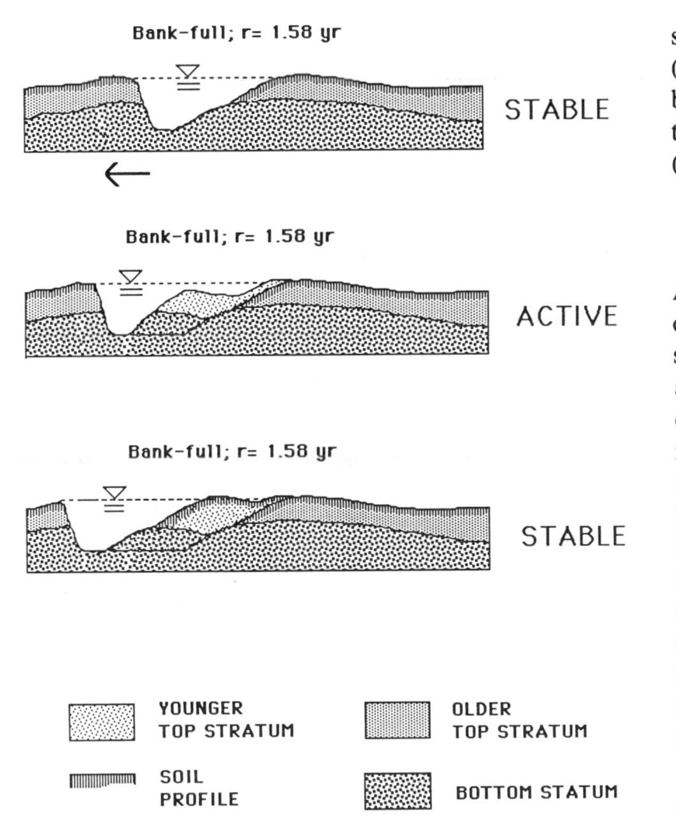

FIGURE 12. Accumulation of a new floodplain during a period of fluvial activity. New top stratum is accumulated as both lateral bank accretion and vertical floodplain accretion during the active phase. During an ensuing stable phase (slow-channel migration), relatively slow deposition rates on the new floodplain surface and channel banks occur, and soil profiles develop. The events shown could occur at a particular site due to local events, such as the reversal of channel migration direction, or throughout a valley or valley reach, in response to environmental changes in the watershed.

produced by large changes in channel geometry, driven by channel cutting and headcut migration in response to individual flow events.

Examples of Floodplain Stratigraphy

Field studies of floodplain sediments show that a strong episodic pattern in Holocene floodplain sedimentation is in fact commonly observed (Antevs, 1955; Brakenridge, 1980, 1981, 1984; Brakenridge et al., 1987; Fink, 1977; Haynes, 1968, 1976, 1985; Knox, 1977; Kozarski and Rotnicki, 1977; Miller and Wendorf, 1954; Patton and Schumm, 1981; Schirmer, 1983; Starkel, 1982). Figure 13*a*–13*g* shows summaries of Holocene floodplain stratigraphies mapped along seven rivers in different regions of North America and Europe. Note especially the presence of buried soils that mark old terrace surfaces, the presence, side by side, of floodplain deposits of quite different ages (*row terraces* of Schirmer, 1983), and the existence of both conformable contacts (marked by dipping paleosols that mark former bank positions) and unconformable contacts (marked by truncation of stratification) at terrace boundaries.

Floodplain Stratigraphic Techniques

Absolute dating methods such as radiocarbon analysis are of much utility in floodplain stratigraphic work. However, such analyses are expensive, and the needed sample materials are not present in every exposure. Moreover, absolute dating methods are all subject to various possible errors, making it desirable for stratigraphic mapping to be conducted first so that samples submitted for dating can be placed in their correct, relative age, order. Each date then obtained becomes much more informative. Floodplain stratigraphers also rely on several traditional stratigraphic tools, including soil stratigraphy (using buried soils as stratigraphic markers), morphostratigraphy (using surface topography to define terraces), and lithostratigraphy (defining sedimentary units on the basis of physical characteristics and stratigraphic relations).

Soil Stratigraphy. Pedogenesis (soil profile development) occurs on subaerial floodplain and terrace surfaces by a variety of physical, biological, and chemical weathering processes, including shrink–swell, clay eluviation and illuviation, bioturbation, organic material decay, oxidation, and hydrolysis. Master soil horizons can often be mapped in field exposures, and include O (organic), A (humified organic), E (leached), Bh, Bk, Bq, Bs, Bt, Bw, By, and Bz (zones of accumulation of various substances such as clay, iron sesquioxides, and calcium carbonate), and K (petrocalcic or caliche) carbonate horizons (Birkeland, 1983, pp. 7–8). Common soil types cover a very broad range, and include entisols, aridisols, mollisols, alfisols, spodosols, ultisols, and vertisols.

Time (hundreds of years for O and A horizons, thousands of years for most other horizons) is required for master horizon formation under most climatic regimes, so that sites of active deposition along rivers are not characterized by the presence of such horizons, and the sediment itself is relatively fresh and unweathered. In contrast, if a stable floodplain configuration is attained by the river, and yearly net sediment additions are very small, then pedogenesis dominates and soil profiles develop.

Several of the master soil horizons can be preserved upon burial by younger sediment from an ensuing active phase, and such buried paleosol horizons can be traced in the subsurface for considerable distances in floodplain excavations. Each such buried soil is thus a useful stratigraphic marker and serves to separate two floodplain sedimentary accumulations, one older and one younger than the time

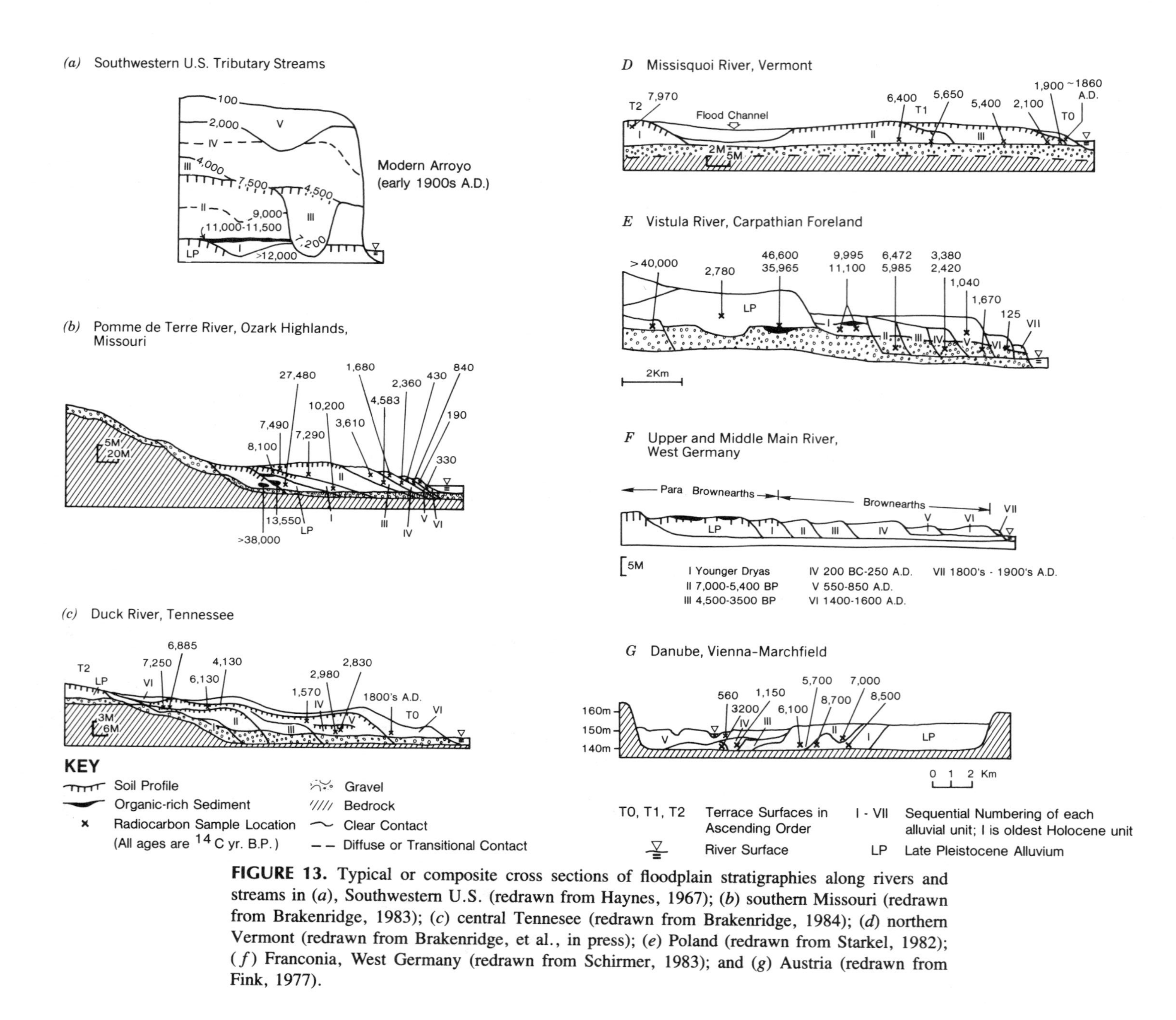

FIGURE 13. Typical or composite cross sections of floodplain stratigraphies along rivers and streams in (*a*), Southwestern U.S. (redrawn from Haynes, 1967); (*b*) southern Missouri (redrawn from Brakenridge, 1983); (*c*) central Tennesee (redrawn from Brakenridge, 1984); (*d*) northern Vermont (redrawn from Brakenridge, et al., in press); (*e*) Poland (redrawn from Starkel, 1982); (*f*) Franconia, West Germany (redrawn from Schirmer, 1983); and (*g*) Austria (redrawn from Fink, 1977).

of soil profile development. Such soil stratigraphic units are shown in cross sections in Figure 13*a*–13*g*.

The erection of a valleywide soil stratigraphy consists of mapping exposures at various locations along a floodplain reach, and dating the buried soil profiles, both in relative age terms (on what sedimentary unit is the soil developed and by what unit is it buried?), and by absolute dating methods, such as radiocarbon. Note that the soil itself need not be dated if reliable age information is available from the sedimentary sequence in which it is entombed.

Because soil profiles are mainly the result of weathering of already deposited sediment, and not the addition of new sediment, soil profiles belong stratigraphically to the sedimentary accumulation on which they are developed. The upper boundaries of buried soil profiles are commonly relatively sharp, whereas their lower boundaries, and the boundaries between horizons, are commonly transitional. Thus, the abruptness of any transitions noted is a powerful field tool for the differentiation of soil horizon boundaries from sedimentary unit boundaries (see illustrated example in Brakenridge, 1983).

Regional Environmental Changes and Soil Stratigraphy. Because of the considerations already outlined, an important question in conducting soil stratigraphic studies is whether the particular buried soils being mapped at different sites are actually of similar ages. As noted, floodplain sedimentary responses to external changes may be complex, and buried soils may exist for local, system-internal reasons, such as a change in the sense of lateral channel migration, with or without any external environmental change. However, when buried soils found along an entire river reach, or on a regional basis, can be demonstrated to be coeval, then the case for external causation is strong.

For example, many rivers in the northcentral, southeastern, and northeastern United States are immediately bordered by low alluvial surfaces, without soil profiles, that are the scene of active floodplain sedimentation (e.g., Fig. 13 and 14, Duck River, Tennessee). Such low active floodplains are flooded one or more times per year, whereas the next higher surfaces are flooded only once every 2–3 yr (see, e.g., Fig. 9). Soil profiles are commonly developed on the higher surfaces but are now being thinly buried by

FIGURE 14. The T0 (right side) and the T1 (left side) surfaces along the Duck River in central Tennessee. The T0 surface can be considered an incipient floodplain (Brakenridge, 1984): it is frequently being flooded and is still actively aggrading.

"terrace veneer" facies of the younger alluvium (Fig. 13, cross sections *c* and *d*; also see Fig. 2).

Radiocarbon and archeological age assessments of the younger sediment in Tennessee, Vermont, and Wisconsin testify to a maximum age of about 120–180 yr. Deposition of these units thus began immediately after the radical alteration of watershed characteristics by row cropping, grazing, logging, small river impoundments for hydroelectricity generation, and other human impacts (Knox, 1977; Brakenridge, 1984; Brakenridge et al., 1987). It appears that external causation of this sedimentary change is a reasonable inference, and that the buried soil profiles on the older surfaces record the older and relatively stable floodplain landscape present before the land-use changes took place. The resulting soil stratigraphy is well developed over a broad region of the eastern and central United States and has proved useful for floodplain stratigraphers, geomorphologists, and archeologists (see Knox, 1977, for a well-described Wisconsin example).

Morphostratigraphy. Episodicity of floodplain sedimentation is also related to the development of floodplain and terrace surfaces. A simple morphostratigraphic nomenclature has proved useful in many alluvial stratigraphic investigations. In this system T0 is the lowest alluvial surface bordering the river along a particular reach and is also the "modern floodplain" as defined by Gary et al. (1972). T1 is the next higher alluvial surface (the lowest geomorphic terrace), T2 is the next higher terrace, and so on. There need be no one-to-one correlation between surfaces and soil stratigraphic or lithostratigraphic units (e.g., examine Fig. 2 and 13), although establishing which lithostratigraphic formations underlie which surfaces is clearly desirable.

Note also from Figures 2 and 13 that conformable deposition associated with lateral channel movement can create the stepped alluvial surfaces. Thus, terraces do not necessarily indicate cut-and-fill cycles. Cut-and-fill terraces do exist as well, however, and are, by definition, stepped surfaces, separated by erosional unconformities in which the younger alluvium is said to be "inset" against the older alluvium through the erosional unconformity. Note that a cut-and-fill stratigraphic relationship may also exist between two adjacent fills whose surfaces are at similiar levels: No geomorphic terrace may be evident even though a cut-and-fill event occurred (see cross sections in Fig. 13*a*, 13*b*, 13*e*, and 13*g*).

In summary, geomorphic terraces formed by river alluvium may be either conformable or nonconformable, and the scarps formed between terraces may thus represent either fossil river banks or erosional unconformities. Good field examples of both types of contacts separating morphostratigraphic surfaces are shown in Figure 13.

As is the case for soil stratigraphic studies, an important question that arises when mapping either type of geomorphic terrace concerns the correlation of surfaces from one site to another. For example, the T0 and T2 surfaces shown along the Duck River (Fig. 13) are distinctive and easily mapped geomorphic landforms everywhere along the studied river (Brakenridge, 1984). However, the T1 terrace complex that was also defined occurs along the middle and lower reaches of the river as two side-by-side row terraces (Schirmer, 1983). These surfaces were designated as T1a and T1b (Fig. 15, top). In contrast, the terrace morphology at upstream localities (or along many small tributaries) in this same watershed is "condensed": The surface morphology is deceptively simple compared to the underlying lithostratigraphy (Fig. 15, bottom).

As a result, mapping of terrace surfaces is best done in conjunction with subsurface data. Where such data is not obtainable, levee-swale morphology, or truncation thereof, can be a useful key to help differentiate conformable terrace scarps from nonconformable ones. Soil profile development may also help to differentiate surfaces of different age (see, e.g., Fig. 13*b* and 13*f*). However, surface soil profiles (or lack thereof) can also deceive: Terrace veneer facies are common along meandering rivers, and older terraces with associated well-developed soil profiles may be present, at shallow depths, below thin covers of younger alluvium without soil profiles (Fig. 2).

Lithostratigraphy. Noncontinuous floodplain sedimentation also results in recognizable lithologic contrasts between alluvia of different age. These contrasts may develop from the effects of progressive in situ weathering, from changes in sediment source area, or from changes in the dominant fluvial processes occurring (e.g., river metamorphosis), and are discussed in this order below.

First, consider the effects of sediment age. Following deposition, floodplain sediments begin to weather even if no soil profile develops. In humid regions oxygenated water moves downward to the water table and may promote a

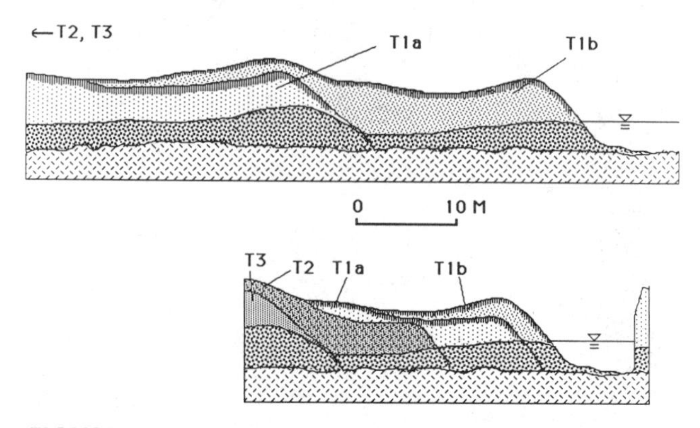

FIGURE 15. Schematic comparison of laterally extensive (top) and condensed (bottom) floodplain stratigraphies based on comparisons of the upper and lower Duck River Valley (see Brakenridge, 1984). Note that the upstream terrace morphology is less expressive of floodplain stratigraphy and history, and slower lateral migration rates also prevail.

variety of chemical processes in this interval. Even in semi-arid regions, occasional rains penetrate deeply, although K (plugged) soil horizons may halt such flow. Aside from wetting of the section, other weathering processes may occur, including compaction, bioturbation, and the decay of organic matter in the presence of oxygen. All of these processes are included by some sedimentologists as an aspect of *early diagenesis* of sediments; they are complex, and they cause older deposits to exhibit different physical characteristics than younger deposits.

Table 1 summarizes lithologic characteristics useful for field differentiation of four late Quaternary alluvial formations mapped in central Tennessee. For this river's fine-grained top stratum facies, with increasing age:

1. Sediment color becomes redder or yellower.
2. Sediment texture becomes increasingly a bimodal mixture of sand and clay. Study of petrographic thin sections suggest that this is due to progressive in situ chemical weathering of silt size grains.
3. Iron and magnesium staining and concretions become more abundant throughout the vertical thicknesses of the top stratum.
4. Charcoal becomes less abundant, apparently due to bacterial digestion (as suggested by C. V. Haynes, personal communication).
5. Pedogenic structure also becomes more pronounced, throughout the top stratum, as shrink–swell processes associated with wetting and drying produce "pressure cutans" on ped surfaces and fractures that allow deep downward penetration of illuvial clay.

TABLE 1 Summarized Physical Characteristics of Top Stratum Facies of Four Lithostratigraphic Units Defined for the Central Duck River Valley in Tennessee. (All Colors Are for Moist Sediment).

UNIT	MUNSELL COLOR	TEXTURES	CHARCOAL	Fe-Mn*
Sowell Mill Formation	Brown 10YR4/3	Silt Loams, Silt Clay Loams	Abundant	Absent
Leftwich Formation	Dark Brown 7.5YR3/4	Silt Clay Loams, Clay Loams	Common	Rare
Cannon Bend Formation	Brown 7.5YR4/4	Silt Clay Loams, Clay Loams, Silty Clays	Rare	Common
Cheek Bend Formation	Brownish-Yellow and Brown (mottled) 10YR6/6, 7.5YR4/6	Clay Loams, Silty Clays, Clays	Absent	Abundant

* Concretions, clots, or coatings

Table 1 thus illustrates the variety of tools the field worker has available for mapping floodplain sediments in this and similar regions. Laboratory data can add to and quantify such observations, but it is difficult to map field stratigraphy using laboratory data unless the field area can be frequently revisited.

In Tennessee the warm, humid climate and the clay-rich nature of the top stratum probably enhance rapid early diagenesis of the top stratum. Along other valleys other factors may be more important than age in producing a mappable lithostratigraphy. Thus, changing source areas for the deposited sediment can be important even at relatively short time scales. In southcentral Missouri, the early Holocene top stratum sediments are composed of a brown silt that appears to be reworked loess, which has now been nearly completely stripped from the watershed's hillslopes (Haynes, 1985; Brakenridge, 1983). Thus, during the late Quaternary, eolian activity added sediment to this watershed's upland. In early Holocene time this loess was apparently washed off the uplands and added significantly to the fine-grained suspended-load carried by this river. At present, only very thin (less than 1 m) loess deposits are present in the basin, and the T0 alluvium of this river is different in color and texture from the brown, silty, early Holocene (T1) alluvium (Haynes, 1985) composed of reworked loess.

In the preceding example an input in sediment from outside the basin was partially responsible for the observed floodplain lithostratigraphy. However, other studies have shown that glacial-interglacial environmental changes had even more severe effects on rivers, effects that directly changed the lithology of deposited sediments because of an overall change in river morphology. Along the Main River in Bavaria, differences in the texture and fabric of the bottom stratum deposits, related to fluctuations between braided and meandering channel patterns, allow the mapping of different-aged bottom stratum sands and gravels (Schirmer, 1983). Also, in regions as diverse as Texas, Finland, and Poland (Baker and Penteado-Orellana, 1977; Koutaniemi, 1979; Starkel, 1982), late Quaternary paleohydrologic changes have altered rivers from braided to meandering patterns, or changed greatly the geomorphic dimensions of meandering river features such as channel sizes and meander wavelengths.

The pre-Holocene fluvial histories of many rivers have thus included river metamorphosis: drastic changes in overall river morphology. The changes were as large as those commonly attributed, in the older geologic record, to tectonism, progressive landscape peneplanation, and other long-term variables. It is clear, also, that river metamorphosis, changes in sediment source area, and postdepositional alteration through early diagenesis can act in concert to produce floodplain lithostratigraphy.

At-a-site and Valleywide Floodplain Stratigraphies. It must be demonstrated (not assumed) that any defined stratigraphic units are discrete entities useful for mapping purposes within the specified study area. If each unit could be physically traced to the different exposures examined, would such correlation lines agree with the correlations inferred on the basis of lithologic, pedogenic, or morphologic characteristics, on the stratigraphic relations, and on the age criteria?

For example, we have noted already that stable and active configurations along a laterally migrating river may occur as a basic characteristic of the meandering process. Thus, the ages of buried soils, surfaces, and lithostratigraphic sedimentary units may be variable between two different bends of the river; no one-to-one correlation necessarily exists for the stratigraphy between the two sites. Such a stratigraphy is "at a site" only, and probably exists along most meandering river floodplains (for reasons already outlined) whether or not a valleywide stratigraphy also exists. Thus, it is important to identify which units are clearly and unambiguously present everywhere in a particular studied valley or valley reach and which units (and associated conformable and erosional contacts) are present only at specific sites.

In this respect the use of type sections is important in mapping valleywide stratigraphy. Defining such reference sections forces important questions to be asked during the course of the field work: Is the unit currently being described actually the same unit as that mapped at the type section and can this be demonstrated? Is the stratigraphy, as defined, of predictive value; that is, will the measured properties at a new exposure fall within the range already specified for the unit at the type section? The best type sections include as much of the unit's internal variability as possible, and exhibit the unit's stratigraphic relations to other units older and younger than itself. Where possible, stratigraphic units should be named for their type sections, again to encourage such questions.

Type sections are less appropriate for at-a-site (only) floodplain investigations. For example, in some microstratigraphic studies, such as those conducted locally at archeological sites, valleywide stratigraphy may not be the objective. Instead, what is needed is as precise a determination of local floodplain sedimentary history as possible. In such cases it is normally possible to perform many necessary correlations directly, by physical tracing of contacts or buried soils in continuous exposures. Only for broader studies examining a whole valley reach is it necessary to use type sections, test for the accuracies of correlations between sections, and limit defined units to those most easily mapped throughout the study area.

CONCLUSION

This chapter offers several conclusions concerning how floodplain stratigraphy develops, how it is studied, and how high-stage hydrology and sedimentary history both affect floodplain sedimentary facies patterns and floodplain stratigraphies. Channel meandering, in-channel high flows, and overbank high flows are all natural processes that occur together to produce river floodplains. Lateral movement of the channel bed is associated with bottom stratum (mainly channel bar) accretion, whereas high-stage deposition causes top stratum (bank, levee, back-levee swale, terrace veneer, and other facies) deposition of relatively finer sediment. Such deposition occurs rapidly following episodes of rapid channel movement, and, after a time lag, slows. Sequential periods of deposition give rise to floodplain stratigraphy, which includes soil stratigraphic, morphostratigraphic, and lithostratigraphic components.

Seven cross sections of late Quaternary floodplain alluvium, from Arizona, Missouri, Tennesee, Vermont, Poland, Bavaria, and Austria are given as examples of floodplain stratigraphy and facies. River metamorphosis is one cause of floodplain stratigraphy and may result, for example, in braided river and meandering river facies to be recognized in the same valley. However, this chapter has focused on episodic deposition along meandering rivers (without such drastic overall system changes) as a cause for Holocene floodplain stratigraphy. In this regard meandering river floodplain stratigraphy may be either at a site (due to local, perhaps system-internal, causes) or valleywide (most likely due to external environmental changes). The latter may be caused by any perturbing factor capable of affecting a whole river reach, including climatic change, human land use, tectonism, base-level changes, or a major catastrophic change such as a very large flood. Because such events are independently documented for many regions, it appears reasonable to suggest that most floodplain stratigraphies will reflect both types of genetic processes.

Late Quaternary floodplain stratigraphies have much to tell us regarding how meandering river systems work. This analysis has defined bottom stratum sediments as formed only in the channel bed environment and top stratum deposits as including upper bank (in-channel) as well as overbank facies. This stems from field observations indicating that meandering river deposits cannot be easily categorized into "channel"and "overbank" facies because there is no abrupt sedimentological process discontinuity between vegetated upper bank surfaces and the immediately adjacent overbank environments along many rivers. The sharpest sedimentological discontinuity is instead at the upper boundaries of the active (mobile) river bed (the surfaces of active point and other bars), which are inundated during frequent flows and below which terrestrial vegetation does not grow. These channel bed sediments are transported by traction and saltation and do not form stable substrates for rooted plants. Above such river stages suspended-load sedimentation dominates, but it may be intermittently interrupted by levee breeching (crevasse splays).

The model of floodplain stratigraphy thus presented has two other process implications. First, periods of enhanced fluvial activity are capable of producing river terraces; valleywide cut-and-fill episodes should not automatically be invoked. Many terrace scarps are actually fossil river banks that exhibit conformable contacts with the younger alluvium. Second, older floodplain surfaces may stand somewhat higher than younger ones, or the modern floodplain, without downcutting by the river because, during rare high floods, not only are the active floodplain surfaces flooded but also the adjacent low terraces. The terrace veneer facies thus deposited raise terrace surfaces, bury paleosols, and obscure the underlying lithology. The benefit of these sediments to the floodplain stratigrapher is that they record, where bedding is preserved, the low-frequency, high-magnitude portion of river flow regimes. The drawback is that they make correct interpretation of river history from terrace sequences almost impossible without subsurface data.

In conclusion there are other topics that need to be more fully addressed. One is how floodplain sedimentation proceeds on a flow-by-flow basis: How and why are the resulting beds either subsequently reworked or, instead, preserved, and how can the number of flow events best be estimated from the number of preserved beds? Also, point bar sedimentation appears to be an oversimplification as a model for the accumulation of even the bottom stratum portion of a floodplain deposit. Witt (1979) illustrates some of the other mechanisms, including the evolution of channel side bars into vegetated floodplains when a reversal of channel migration sense occurs, but these known processes have yet to be incorporated into a widely accepted facies model for meandering rivers. Last, the effects of infrequent, very large floods (catastrophic floods) on floodplain stratigraphy also remain understudied. In this respect this chapter has noted the differences in floodplain sedimentation between river systems that are dominated by frequent flow events and those that are dominated by infrequent flows. It may be that very rare flow events are well recorded by the frequent-flow-dominated (perennial) rivers. The frequent small floods of such rivers may provide abundant opportunities for burial and preservation of the sedimentary traces of large floods. If so, then future meandering river facies models should include the likely locations of these traces in meandering river deposits.

ACKNOWLEDGMENTS

I wish to thank Professor Wolfgang Schirmer of the Universitaet Duesseldorf for showing me the floodplain stratigraphy of a portion of the Main Valley in Franconia (Federal Republic of Germany) and Dr. Andrzej Witt and Professor Stefan Kozarski, both at the Institute for Quaternary Studies of Adam Mickiewicz University, Poznan, for field trips to sites along the Warta and other rivers in western Poland. A research fellowship provided by the German Academic Exchange Service (1982) and visiting scientist support by the Adam Mickiewicz University (1984), as well as travel and other research assistance from Wright State University (1984), helped make my work possible and are gratefully acknowledged.

REFERENCES

Antevs, E. A. (1955). Geologic-climatic dating in the West. *Am. Antiq.* **20**, 317-335.

Baker, V. R. (1984). Questions raised by the Tucson flood of 1983: "Hydrology and water resources of the Southwest." *Proc. Ariz. Sect. Am. Water Resour. Assoc. Hydrol. Sec. Ariz.-Nev. Acad. Sci.* **14**, 211–219.

Baker, V. R., and Penteado-Orellana, M. M. (1977). Adjustment to Quaternary climatic change by the Colorado River in central Texas. *J. Geol.* **85**, 395-422.

Bejan, A. (1982). Theoretical explanation for the incipient formation of meanders in straight rivers. *Geophys. Res. Let.* **9**, 831–834.

Birkeland, P. W. (1983). "Soils and Geomorphology." Oxford Univ. Press, London and New York.

Brakenridge, G. R. (1980). Widespread episodes of stream erosion during the Holocene and their climatic cause. *Nature* **283**, 655–656.

Brakenridge, G. R. (1981). Late Quaternary floodplain sedimentation along the Pomme de Terre River, southern Missouri. *Quat. Res.* **15**, 62-76.

Brakenridge, G. R. (1983). Late Quaternary floodplain sedimentation along the Pomme de Terre River, southern Missouri. Part II. Notes on sedimentology and pedogenesis. *Geol. Jahrb., Reihe A* **71**, 265-283.

Brakenridge, G. R. (1984). Alluvial stratigraphy and radiocarbon dating along the Duck River, Tennessee: Implications regarding floodplain origin. *Geol. Soc. Am. Bull.* **95**, 9-25.

Brakenridge, G. R. (1985). Rate estimates for lateral bedrock erosion based on radiocarbon ages, Duck River, Tennessee. *Geology.* **13**, 111-114.

Brakenridge, G. R., Thomas, P. A., Schiferle, J. C., and Conkey, L. E. (1987). Fluvial stratigraphy as an indicator of river channel migration and floodplain sedimentation rates, Missisquoi River, Vermont. *Quat. Res.*, in press.

Brice, J. C. (1974). Evolution of meander loops. *Geol. Soc. Am. Bull.* **85**, 581-586.

Dury, G. H . (1973). Magnitude-frequency analysis and channel morphology. *In* "Fluvial Geomorphology" (M. Morisawa, ed.), Pub. Geomorphol., pp. 91-121. State University of New York, Binghamton.

Fenneman, N. M. (1906). Floodplains produced without floods. *Am. Geogr. Soc. Bull.* **38**, 89-91.

Fink, J. H. (1977). Jungste Schotterakkumulatione im Oesterreicheische Donauabschnitt. *In* "Dendrochronologie und postglaziale Klimaschwankungen in Europa" (B. Frenzel, ed.), pp. 191-211. Steiner Verlag, Wiesbaden.

Friedman, G. M., and Sanders, J. E. (1978). "Principles of Sedimentology." Wiley, New York.

Gary, M., McAfee, R., Jr., and Wolf, C. L. (1972). "Glossary of Geology." Am. Geol. Inst., Washington, D.C.

Haynes, C. V. (1968). Geochronology of late Quaternary alluvium. *In* "Means of Correlation of Quaternary Successions" (R. B. Morrison and H. E. Wright, eds.), pp. 591-631. Univ. of Utah Press, Salt Lake City.

Haynes, C. V. (1976). Late Quaternary geology of the lower Pomme de Terre River, Missouri. *In* "Prehistoric Man and His Paleoenvironments" (W. R. Wood and R. B. McMillan, eds.), pp. 47-61. Academic Press, New York.

Haynes, C. V. (1985). Mastodon-bearing springs and late Quaternary geochronology of the lower Pomme de Terre Valley, Missouri. *Spec. Paper 204 Geol. Soc. Am.*, Boulder, Colorado, 1–45.

Klimek, K., and Starkel, L. C. (1974). History and actual tendency of flood-plain development at the border of the Polish Carpathians. *In* "Geomorphologische Prozesse und Prozesskombinationene in der Gegenwart unter verschiedene Klimabedingungen" (H. Poser, ed.), pp. 185-196. Vandenhoeck & Ruprecht, Goettingen.

Knox, J. C. (1977). Human impacts on Wisconsin stream channels. *Ann. Assoc. Am. Geogr.* **76**, 323-342.

Koutaniemi, L. (1979). Late-glacial and post-glacial development of the valleys of the Oulanka River Basin, northeastern Finland. *Fennia* **157**, 13-73.

Kozarski, S., and Rotnicki, K. (1977). Valley floors and changes of river channel patterns in the north Polish Plain during the late Wuerm and Holocene. *Quaest. Geogr.* **4**, 51-93.

Leopold, L. B., Wolman, M. G., and Miller, J. P. (1964). "Fluvial Processes in Geomorphology." Freeman, San Francisco, California.

Miller, J. P., and Wendorf, F. (1954). Alluvial chronology of the Tesuque Valley, New Mexico. *J. Geol.* **66**, 177-194.

Patton, P., and Schumm, S. A. (1981). Emphemeral stream processes: Implications for studies of Quaternary valley fills. *Quat. Res.* **15**, 24-43.

Richards, K. S. (1982). "Rivers: Form and Process in Alluvial Channels." Methuen, New York.

Schirmer, W. (1983). Criteria for the differentiation of late Quaternary river terraces. *Quat. Stud. Poland* **4**, 199-204.

Schumm, S. A. (1965). Quaternary paleohydrology. *In* "The Quaternary of the United States" (H. E. Wright and D. G. Frey, eds.), pp. 783-794. Princeton Univ. Press, Princeton, New Jersey.

Starkel, L., ed. (1982). Evolution of the Vistula River Valley during the last 15,000 years. *Pol. Acad. Sci. Geogr. Stud. Spec. Issue*, pp. 9-20.

Thorne, C. B., Zevenbergen, L. W., Pitlick, J. C., Rais, S., Bradley, J. B., and Julien, P. Y. (1985). Direct measurements of secondary currents in a meandering sand-bed river. *Nature (London)* **315**, 746-747.

Witt, A. (1979). Present-day mechanism of floodplain lateral accretion in the middle course of the Warta River. *Quaest. Geogr.* **5**, 153-167.

10

FLOODS; DEGRADATION AND AGGRADATION

WILLIAM B. BULL
Department of Geosciences, University of Arizona, Tucson, Arizona

INTRODUCTION

Floods are important output processes of fluvial systems. Flood durations typically range from minutes to days; such brief time spans encourage earth scientists to focus on the complex and highly interesting interrelations of rainfall–runoff events. However, the processes and landforms of a given fluvial system are also the result of interactions between variables over much longer time spans. Rates of biomass production, rock weathering, sediment yield, and water yield from the hillslope subsystem may change with time and can be affected by runoffs of the flood magnitude.

The effects of floods on the modes of operation of stream subsystems range from minimal to profound. The occurrence of infrequent but extremely large flow events also raises intriguing questions about the operation of fluvial systems over time spans of centuries or millennia.

1. What effects do floods have on reaches of stream subsystems in equilibrium?
2. Do floods result in net aggradation or net degradation?
3. Are different reaches of stream subsystems affected in similar or dissimilar ways by flood events?
4. Do climatic and tectonic events of the geologic past exert a recognizable and important influence on the generation of floods and the consequences of these large streamflows?

The scope of this chapter centers on these questions. An underlying theme is that variations in spatial or temporal rates of aggradation or degradation of streams during floods are a function of whether or not a given reach is close to or distant from equilibrium or threshold conditions. Past equilibrium or threshold conditions are represented by terraces, which generally may be considered as time lines in stream systems.

MODES OF STREAM OPERATION

Over a period of years reaches of streams may undergo either net aggradation or degradation, or they may remain in equilibrium (a condition of grade). "A graded stream is one in which, over a period of years, slope, velocity, depth, width, roughness, pattern, and channel morphology delicately and mutually adjust to provide the power and efficiency necessary to transport the load supplied from the drainage basin without aggradation or degradation of the channels" (Leopold and Bull, 1979). Thus, reaches of streams in equilibrium attain stable longitudinal profiles through a variety of interactions between independent and dependent variables. Aggradational and degradational modes of operation occur along those reaches where insufficient time has elapsed for one or several interactions between variables to result in equilibrium conditions. Prior longitudinal profiles of valley floors may be represented by the treads of paired fill, cut, and strath terraces (Leopold and Miller, 1954). Treads of fill terraces represent the passage of a stream from aggradational to degradational modes of operation—a change that commonly occurs during a flood. Floods that erode laterally usually result in beveled surfaces indicative of equilibrium conditions, such as cut (in alluvium) and strath (in bedrock) longitudinal profiles. Subsequent channel downcutting leaves remnants as cut and strath terraces. These modes of stream operation generally occur in a roughly synchronous manner along reaches of substantial length.

It has also been recognized that aggradation may be occurring in one part of the stream subsystem and degradation in another part (Schumm, 1973; Patton and Schumm, 1975). Such streamflow behavior is referred to as a complex response and typically is a spatial adjustment to a change in one or more variables. An example of a complex response would be a slug of bedload sediment that works its way downstream in a kinematic fashion; aggradation occurring along one reach may be concurrent with degradation in the upstream and downstream reaches. Complex responses typically result in low terraces that generally are unpaired. In contrast, major episodes of aggradation or degradation are the result of changes in independent variables such as climate and uplift, occur over a much longer time spans, and normally result in paired stream terraces. Perturbations such as stream capture also can initiate aggradation and degradation (Ritter, 1967).

Aggradational and degradational modes of operation for a reach of a stream are separated by the threshold of critical power (Bull, 1979). Magnitudes of streamflow events are a highly important, but not the sole, factor in determining whether or not this erosional-depositional threshold is crossed. The critical-power threshold is defined as a ratio where the numerator consists of those variables that if increased favor degradation and the denominator consists of those variables that if increased favor aggradation:

$$\frac{\text{stream power (driving factors)}}{\text{resisting power (resisting factors)}} = 1.0 \tag{1}$$

The available stream power to transport bedload, and thus tend to prevent valley floor aggradation, has been defined by Bagnold (1973, 1977) as

$$\Omega = \gamma Q S \tag{2}$$

where Ω is total kinetic power, or in terms of power per unit area of streambed ω.

$$\omega = \frac{\gamma Q S}{w} = \gamma d S v = \tau v \tag{3}$$

where γ = specific weight of the sediment-water fluid
Q = stream discharge
S = the energy slope
w = streambed width
d = streamflow depth
v = mean flow velocity
τ = shear stress exerted on the streambed

Flood discharges have a major influence on stream power because discharge changes rapidly relative to the slope of the energy gradient, which roughly parallels the longitudinal profile of the stream. Flood discharges greatly affect resisting power when they decrease hydraulic roughness by destroying riparian vegetation and streambed armor, and by straightening channels.

Stream power is a most useful element of the threshold definition because sediment transport is highly sensitive to changes in discharge and slope of streamflow (e.g. see Baker, 1973, Fig. 54). Shear stresses needed to entrain bedload are also a function of channel geometry (Baker and Ritter, 1975). The importance of flood discharges on stream power is underscored by the marked increases in entrainment of suspended sediment that occur with increasing streamflow at a stream gauge station. Suspended-sediment transport rates, G_s, increase exponentially with increases in discharge, Q (Leopold et al., 1964). Rapid increases in sediment concentration are indicated by positive exponents of 1.8–3.

$$G_s = pQ^{2.5} \tag{4}$$

Fewer studies have been made of transport of bedload, which comprises most valley floor aggradation materials, but bedload transport rates also increase exponentially with increases in bed shear stress (Leopold and Emmett, 1976; Emmett, 1976; Andrews, 1981; Lekach and Schick, 1983).

The denominator of Equation (1) is resisting power, which is the stream power needed to transport the bedload, and consists of those variables that if increased favor deposition of bedload. Resisting power increases with increases in hydraulic roughness, amount of bedload, and bedload size. Thus the threshold of critical power is influenced by fully as many variables as the equilibrium stream as defined by Leopold and Bull. Resisting power varies primarily with changes in amount and size of sediment yielded from hillslope subsystems and with changes of hydraulic roughness for alluvium- and bedrock-floored stream channels.

Processes of bedrock erosion depend largely on how far removed a given reach of a stream is from threshold or equilibrium conditions. Degradational reaches undergo maximum net downcutting during floods, together with concurrent decreases in valley floor slope. Equilibrium reaches undergo degradation into bedrock less frequently because stream power may exceed resisting power only temporarily during times of peak discharges. Instead, lateral erosion predominates and straths attest to the permanent nature of the lateral erosion. Although valley floor width is increased, slope remains about the same.

Aggradation by a broad, shifting pattern of braided channels may be terminated by a flood event if the reach is close to the threshold of critical power. Subsequent streamflows in the resulting shallow, but incised, channel may be sufficiently increased in efficiency to maintain the degradational mode of operation in the recently deposited valley fill. A parallel case can be made for floods as a prime factor in the initiation of arroyo cutting in the American West.

EFFECTS OF FLOODS ON MODES OF OPERATION

Both spatial and temporal variations of stream power and resisting power can determine whether or not a given reach of a stream is close or far from equilibrium or threshold conditions.

Intuitively, it would seem that floods should degrade valley floors. An exponent far in excess of 1 for the power function of Equation (4) would seem to make it likely for streambeds to undergo degradation during floods. Indeed scour to, and exposure of, bedrock is commonly associated with flood flows. Small, frequent streamflows with lesser amounts of stream power usually cause aggradation.

Whether or not net aggradation or degradation occurs is determined by the summations of the effects of all variables that influence aggradation and degradation. The following factors tend to increase the likelihood of net degradation of a reach:

1. Stable hillslopes that yield minimal sediment by fluvial erosion or mass movement processes.
2. Increases in vegetation density on hillslopes.
3. Climatic changes that increase the frequency of high-intensity prolonged rainfalls, increase annual rainfall, or decrease the proportion of precipitation occurring as snow.
4. Tectonic uplift that increases relief, drainage density, and longitudinal slopes of valley floors. Even if all of the other variables remain constant, continuing uplift will cause tectonically induced channel downcutting over long time spans with the maximum amounts of degradation occurring at times of major floods.

Whether or not equilibrium is attained for a given reach is also a function of tectonic uplift rate relative to frequency of flood flows. For a given uplift rate and bedrock erodibility, equilibrium is most likely to be attained along streams with frequent large discharges. The relative amounts of uplift, u, and channel downcutting, cd, can have three relations:

$$\Delta u \lesseqgtr \Delta(\mathrm{cd}) \quad (5)$$

Channel downcutting tends to exceed long-term uplift rates after episodic fault movements, which tends to steepen gradients. The two factors are equal in equilibrium reaches. Uplift tends to exceed downcutting where vertical differential uplift is rapid, available stream power is low, and bedrock resistance to erosion is high.

Flood-prone fluvial systems that achieve equilibrium despite uplift rates of 1–2 m/ky (1 ky = 1000 yr) are common in flood-prone mountains such as the San Gabriel Mountains of southern California and in the Southern Alps of New Zealand.

Floods may be associated with massive aggradation even in tectonically active terrains if the increases of resisting power are sufficiently large to overwhelm the stream subsystem with sediment. The following factors tend to promote net aggradation:

1. An abundance of stored sediment on hillslopes as a result of either abundant soft rock types and/or rates of rock weathering that exceed denudation rates.
2. Climatic changes that greatly decrease vegetative cover on hillslopes or increase rainfall intensity.
3. Fires that remove the vegetative cover and expose hillslope soils to greatly accelerated erosion by rillwash and sheetflow.
4. Unstable slopes subject to landslides that introduce large volumes of earth materials directly into the stream channels.
5. Lack of relative vertical uplift, which leads to attainment of equilibrium conditions and development of straths as valley gradients are decreased to minimal values associated with efficient transport of the imposed sediment load. Once equilibrium is attained, additional net downcutting cannot occur because lesser slopes would result in stream power being less than that necessary to transport the imposed sediment load. The equilibrium conditions associated with tectonic inactivity and/or frequent large streamflows define reaches of streams that may require only moderate changes in variables in order to change to an aggradational mode of operation. Again, the independent variable of tectonic uplift directly affects the consequences of floods on stream subsystems.

Stream systems that aggrade during floods include watersheds with abundant landslides in the Coast Ranges of California. Climatically induced decreases of plant cover and concurrent increases of hillslope sediment yield have favored aggradation of valley floors in climatic settings that range from extremely arid to extremely humid (Bull, 1987). Abundant examples occur in the western United States, the Middle East, and New Zealand. The Appalachian Mountains of the eastern United States are representative of tectonically inactive watersheds where floods caused by tropical storms may cause aggradation.

SPATIAL VARIATIONS OF THE RATIO OF STREAM POWER/RESISTING POWER

The relative strengths of stream power and resisting power, and the consequences of floods, vary greatly within most drainage basins. Consider the hypothetical longitudinal profile of a tectonically active semi-arid, watershed depicted in Figure 1. Uplift along the range-bounding fault zone is a tectonic perturbation that occurs in a highly restricted linear zone crossing the canyon mouth. This tectonic input tends to increase stream power through increases in both slope and increases of orographically induced precipitation, but such increases occur only during long response times of 10^5-10^7 yr. In marked contrast perturbations caused by late Quaternary climatic changes affect the entire watershed in a response time of 10^2-10^4 yr, and change both stream power and resisting power.

Basin area and therefore magnitudes of flood events generated by basinwide rainfalls increase markedly in the downstream direction (Wolman and Gerson, 1978), but erodibility of materials (for a given lithology) remains roughly the same. Therefore, floods will tend to downcut valley floors to equilibrium slopes in the downstream reaches of mountains much sooner than will streamflows of the same frequency in upstream reaches. Even several million years after cessation of tectonic activity, headwater reaches of semi-arid streams will be far removed from equilibrium conditions and will continue to downcut at slow rates. A semi-arid example is the Santa Catalina Mountains of southern Arizona, and a humid-region example is the Appalachian Mountains of the eastern United States. Virtually all mountains have degradational reaches near their headwaters where stream power exceeds resisting power during floods.

The opposite situation is depicted in the aggradational reach of Figure 1, where deposition has been continuous because resisting power has exceeded stream power. Such reaches are particularly obvious in basins of internal drainage where accumulation of playa and associated alluvial fan deposits constitutes a base-level rise that gradually tends to decrease slope and thereby stream power. Examples include the depositional basins of the Great Salt Lake in Utah and Death Valley in California.

Floods may have opposite effects in different parts of the same watershed. Large discharges in headwater reaches cause rapid degradation, whereas large discharges in the basin cause rapid aggradation. Clearly one has to estimate

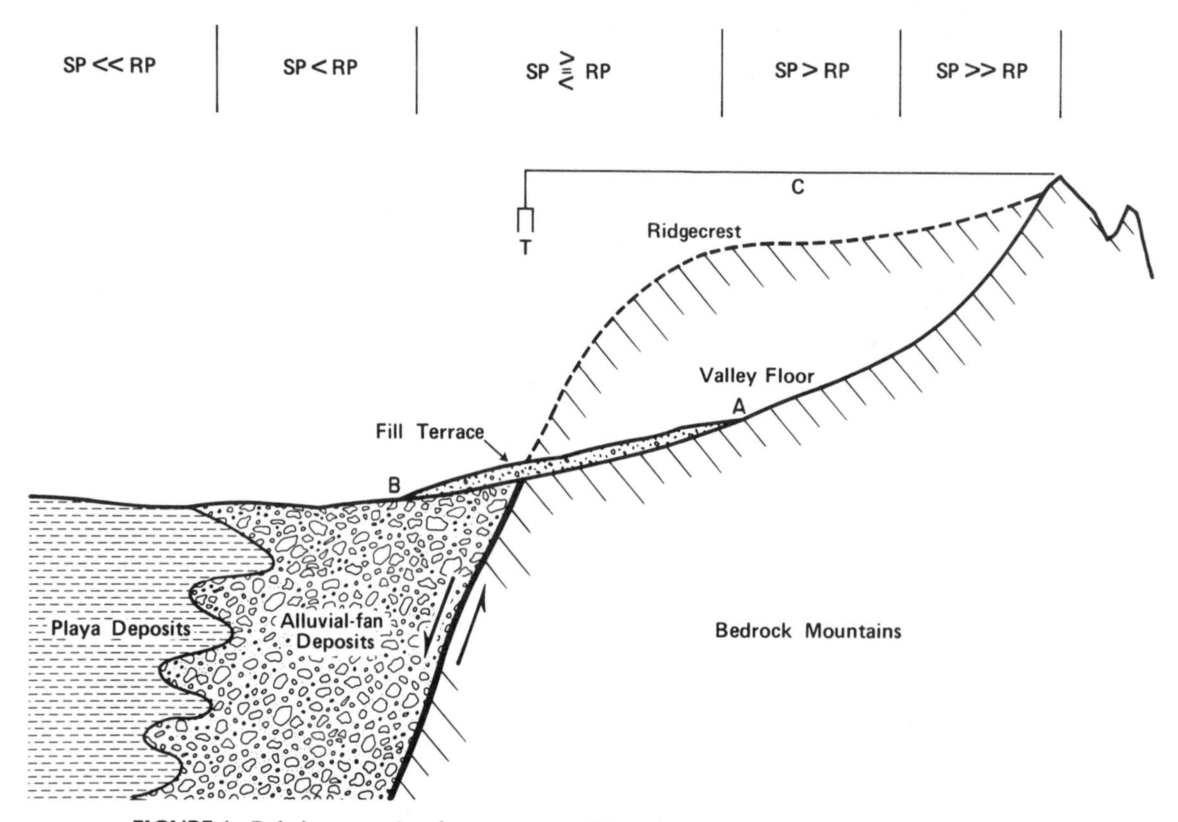

FIGURE 1. Relative strengths of stream power (SP) and resisting power (RP) along a hypothetical fluvial system in an arid closed basin. Tectonic perturbations are initiated in reach T and climatic perturbations are initiated in reach C. The reach at the mountain front will be the most likely to change mode of operation as a consequence of flood discharges. From Bull (1987).

how far removed from equilibrium or threshold conditions a given reach is before predictions can be made regarding potential reversals of modes of stream subsystem behavior.

The fluctuating reach of Figure 1 (SP ⋚ RP) is most sensitive to future floods because it is the reach where stream power may exceed, equal, or be less than resisting power for a given flow event. In such reaches moderate increases or decreases in resisting power may initiate aggradation or degradation, respectively. Relative discharges of water and bedload are especially important for this sensitive reach. Common examples are the discontinuous ephemeral streams that have been downcut into arroyos during the last century as a result of grazing and other human impacts and/or climatic variations that affected the plant cover and discharge of water and sediment.

Threshold intersection points occur along longitudinal profiles of valleys where the threshold of critical power is crossed in a spatial sense. For example, a downvalley change from degradation to aggradation defines the location of a threshold intersection point. In Figure 1 aggradation of a reach that formerly was downcutting into bedrock moved the threshold intersection point upstream to A at the time of maximum aggradation. Subsequent downcutting through the valley fill and new fan deposits then moved the point downstream to B. Threshold intersection points may shift in a great variety of time scales. For the hypothetical shift of Figure 1, 10^3 yr may be needed. Sediment discharge commonly peaks before water discharge during floods, resulting in an upstream shift of the threshold intersection point followed by a downstream movement of the point during the later stages of the same flow event as resisting power decreases more rapidly than does stream power.

TEMPORAL VARIATIONS OF THE RATIO OF STREAM POWER/RESISTING POWER

Interrelations between floods and modes of stream behavior can be highly dependent on self-enhancing and/or self-arresting feedback mechanisms in systems with long response times to major climatic perturbations.

Arid-Region Example

Consider the streams in the presently barren, rocky mountains of the deserts of the Middle East and the southwestern United States. Comparatively higher vegetative densities during the cooler and/or wetter times of the late Pleistocene favored colluviation and partial burial of hillslope bedrock outcrops. Subsequent changes to drier and/or warmer climates markedly decreased the vegetative protection for the accumulated hillslope colluvium. Hillslope sediment yields were greatly increased, and valleys aggraded in a manner first described by Huntington (1907) in central Asia. The sequence of changes in the operation of the hillslope subsystem and its impact on the stream subsystem is diagramed in Figure 2. Reduction in plant cover decreased infiltration rates and exposed soil to erosion. Both runoff of water and concentration of entrained sediment increased for a given precipitation event. Accelerated hillslope erosion also exposed sources of coarser-grained detritus. Increases in both sediment load and sediment size increased resisting power so much that the threshold of critical power was not exceeded despite presumed increases in stream power caused by flashier runoff and larger flood peak discharges associated with (1) increases in area of outcrops and (2) valley-aggradation-induced increases in gradient. Continued decreases in hillslope soil thickness and concurrent increases in areas of exposed bedrock resulted in more rapid runoff of water and more frequent floods. But sediment concentration progressively decreased as more bare rock was exposed and areas of colluvium decreased.

Self-enhancing feedback mechanisms (Fig. 2) were important in perpetuating net removal of soil from the slopes. Decrease in soil thickness and increase in outcrop area both resulted in continued decrease in soil moisture and vegetation density. Such a process response model for the operation of hillslope subsystems does not tend toward

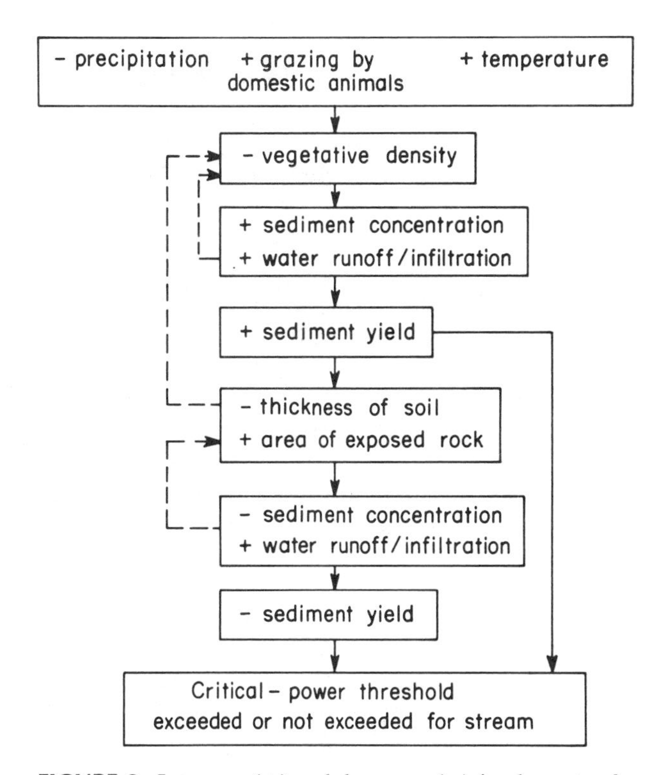

FIGURE 2. Increases (+) and decreases (−) in elements of a hypothetical arid, rocky hillslope subsystem. Self-enhancing feedback mechanisms are shown by dashed lines with arrows. Figure 7 of Bull (1979).

equilibrium conditions, so such arid mountainsides are continuing to become more barren and the stream subsystem is being driven progressively farther to the degradational side of the threshold of critical power.

The response of arid fluvial systems to flood discharges depends in large part on the amount of time that has passed since the major climatic perturbation switched the mode of operation of hillslopes (midslopes and footslopes) from net aggradation to net degradation. In early stages of hillslopes stripping, the amount of available sediment was so large that intense rainfall–runoff events caused debris flows and accelerated valley floor aggradation. An opposite result occurred during later stages when major runoff events accelerated removal of remaining sediment on hillslopes and thereby caused still larger increases of stream power relative to resisting power. The resulting degradation cut through the valley fill to bedrock. Additional downcutting depends partly on base-level controls such as tectonism.

Humid-Region Example

Floods can also disrupt equilibrium conditions along some streams. Straths are indicative of equilibrium conditions, but permanent degradation resulting from flood events can only occur where tectonic or other base-level fall can potentially increase slope, and stream power, sufficiently to include net channel downcutting that converts a strath into a strath terrace.

Attainment of equilibrium is the result of interactions between many factors, including slope. Variables such as hydraulic roughness are important and may increase resisting power sufficiently to create equilibrium conditions, even where the potential for tectonically induced channel downcutting remains. Excellent examples of this type of attainment of equilibrium conditions can be illustrated by the self-arresting feedback mechanisms and subsequent disruption by floods that occur along the Charwell River of New Zealand.

The Charwell River drains 40 km^2 of the Seaward Kaikoura Range in the northeastern part of the South Island of New Zealand. The drainage basin is underlain by sheared graywacke and rises from an altitude of 450–1605 m. The weakly seasonal climate is humid, and mesic to frigid. There is no evidence of Pleistocene glaciation. During full-glacial climates, South Island snowlines were about 800 m lower than at present (Porter, 1975; Soons, 1979) and treeline—presently at 1150 m—was depressed a similar amount (McGlone and Bathgate, 1983; M.S. McGlone, Botany Division, New Zealand DSIR, written communication, Jan. 24, 1986). A period of transitional climate, and geomorphic processes, prevailed between 14 and 10 ka between cold-dry full-glacial and warm-wetter early Holocene times. Full-glacial climates favored periglacial processes, high sediment yields, and valley floor aggradation (Soons, 1962; Stevens, 1974). Subsequent degradation was accompanied by the formation of 12 fill, cut, and strath terraces whose ages have been estimated by the relative-absolute dating (calibration of a relative dating technique by radiocarbon dating) done by Knuepfer (1984) (Table 1). Knuepfer's approach to dating provided age estimates for the terrace treads, which is preferable to ^{14}C ages of the underlying deposits.

In contrast to the early Holocene aggradation described for the arid, thermic drainage basins of Figure 2, the Charwell River underwent 36 m of aggradation during times of late Pleistocene full-glacial climates, which was followed by 75 m of intermittent degradation during the latest Pleistocene and Holocene.

Holocene channel downcutting continued far below the pre-aggradation event strath as the powerful stream caught up with uplift that had continued during times of aggradation. Studies of remnants of uplifted marine terraces infer uplift rates upstream from the range-bounding active Hope fault of 3.8 ± 0.2 m/ky and in the multiple-terrace reach downstream from the fault of 1.2 ± 0.2 m/ky.

Each of the 11 cut-and-strath terraces formed since the time of the latest Pleistocene fill terrace represents an equilibrium period of a few hundred years and is the result of a self-arresting feedback mechanism. Winnowing of alluvial gravels caused armoring of the streambed of the Charwell River with boulders, greatly affecting bedload transport rates and channel degradation. The effects of progressive armoring as described by Gomez (1983) suggest periodic major increases in resisting power caused by increase in size of material to be entrained from the streambed and by concurrent increases in hydraulic roughness. These increases in resisting power continued in a self-arresting manner until resisting power equaled stream power for the majority of discharges. Each time of terrace formation was followed by subsequent degradation that was initiated by a flood. Each major flood event disrupted the streambed

TABLE 1 Age Estimates of the Charwell River Terraces[a]

Terrace	Age (ka)
1	11.1 ± 3.4
2	8.3 ± 2.3
3	6.8 ± 1.8
5	5.8 ± 0.8
6	5.1 ± 0.7
9	4.4 ± 0.6
11	3.7 ± 0.5

SOURCE: Knuepfer, 1984, p. 185.

[a]Terrace ages are based on rates of surficial cobble weathering-rind development, calibrated to radiocarbon-dated sites.

armor and riparian vegetation, thereby allowing renewed degradation to occur along with concurrent winnowing of a new streambed armor. Thus, without invoking either tectonic perturbations or secular climatic changes, a flight of degradation terraces (cut-and-strath terraces) were formed. Each degradation terrace is capped by remnants of a boulder bed that reflects a much different fluvial regime than that of the sandy medium-grained gravels of the latest Pleistocene aggradation event.

The relative magnitudes of stream power and resisting power are estimated for the Charwell River since 28–30 ka ago, which is the time of the pre-Holocene sea-level highstand (Chappell, 1983). The plots of Figure 3 assume similar climates at times of interglacial conditions; these were also periods of prolonged attainment of equilibrium when major straths were formed. These constraints dictate that stream power and resisting power plots are horizontal and identical at roughly 30 ky ago and at the present. The vertical separation of the stream power and resisting power plots is a measure of how far removed the terraced reach of the stream subsystem was from equilibrium or threshold conditions. A single threshold is indicated where the two plots cross but are not horizontal at 14 ka (1 ka = 1 ky before present).

The late Pleistocene climate was relatively cold and dry; it reduced stream power while resisting power was being greatly increased by major increases in sediment yield caused by active periglacial processes on treeless hillslopes. Prior to the 14-ka threshold, aggradation produced a 36-m-thick valley fill and piedmont alluvial fans. The pronounced separation between the stream power and resisting power plots was reversed with the onset of climatic amelioration, which encouraged bushes and then dense forests to recolonize the hillslopes. The time of full-glacial climate was also a time of maximum aggradation rate, which is not an easily recognized parameter in the stratigraphic record.

The post-threshold domain is characterized by multiple variations in resisting power associated with streambed armoring and riparian plant growth during each time of terrace formation. Major increases in sediment load were responsible for the major increase in resisting power prior to 14 ky ago, whereas marked fluctuations in hydraulic roughness and size of streambed material to be entrained dominated the post-14-ka domain. The inferred mid-Holocene increase in stream power reflects a period of greater runoff—more frequent floods—during which streambed armoring was less likely to occur. The mid-Holocene climatic optimum was about 1°C warmer than present (Moar, 1980) and was a period when minor glacial re-advances did not occur in the Southern Alps (Burrows, 1979; Birkeland, 1982). About 32 m of channel downcutting by the Charwell River occurred during this 1-ky period of maximum separation of stream and resisting power.

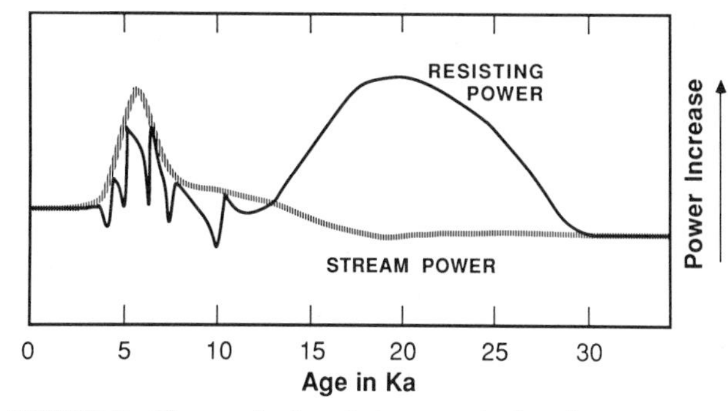

FIGURE 3. Changes in the relative magnitudes of stream power and resisting power during the past 30 ka, Charwell River, South Island, New Zealand. From Bull (1987).

Floods probably had minimal importance during the pre-14-ka domain because snow was a more important type of precipitation, and occasional snowmelt floods may have merely accelerated aggradation. In marked contrast the importance of floods since 14 ky ago is underscored by the necessity of flood flows to disrupt streambed armoring in order to renew tectonically induced valley downcutting. Rainfalls for these flood events may have been derived from tropical moisture sources. A modern example was cyclone Alison whose floods devastated much of the Kaikoura coast in 1975.

CONCLUSION

Every drainage basin is affected differently by floods of a given recurrence interval, but general classes of responses to floods can be outlined that are functions of tectonic uplift, climatic change, and erodibility of surficial materials. These independent variables control output responses of aggradation and degradation through the intermediate hillslope variables of vegetation and water and sediment yield. Despite an inadequate understanding of causal links between variables that affect aggradation and degradation, some general conclusions can be made regarding floods:

1. Net degradation cannot occur over the long term in absence of base-level fall, even where floods are frequent. Rising drainage basins upstream from highly active faults will be more likely to degrade during floods than streams in slowly uplifting, or tectonically inactive, watersheds.
2. Other factors being equal, humid-region streams will tend to degrade more rapidly than will arid-region streams because their greater annual stream power (and frequency of large streamflows) tends to be more than sufficient to transport sediment yielded from hillslope subsystems.

3. Massive influxes of coarse-grained sediment into streams after landslides, fires, or major climatic changes increase resisting power so much that aggradation overwhelms stream subsystems, even during floods in tectonically active drainage basins of humid regions.
4. The tendency of streams to attain equilibrium conditions is a vital concept, but it is equally important to recognize departures from equilibrium conditions and how different sizes of departures influence responses to flood events.
5. Headwater reaches have the greatest tendency to degrade, and depositional basins have an increasing tendency to aggrade downslope from where streams emerge from mountains. Intermediate reaches—upstream and downstream from the mountain–piedmont junction—fluctuate between aggradation and degradation or attain an equilibrium mode of operation; modes of operation that are partly a function of flood frequency and magnitude. These sensitive reaches are shorter in tectonically active than in inactive settings.
6. Prediction of the effects of floods should take into account temporal variations of the ratio of stream power/resisting power that are controlled in part by feedback mechanisms. One example is the self-enhancing feedback mechanism that promotes progressive stripping of soil from arid rocky hillsides, where increases in outcrop area cause large, flashy floods. A second example involves self-regulating feedback mechanisms in alluvial streams where long-term tectonically induced degradation is interrupted by brief periods of equilibrium. Equilibrium is attained by progressive streambed armoring and growth of riparian vegetation, both of which increase hydraulic roughness and shear stresses needed to entrain streambed materials. Degradation resumes only when floods remove the armor and plants.

REFERENCES

Andrews, E. D. (1981). Measurement and computation of bed-material discharge in a shallow sand-bed stream, Muddy Creek, Wyoming. *Water Resour. Res.* **17**, 131–141.

Bagnold, R. A. (1973). The nature of saltation and bed-load transport in water. *Proc. R. Soc. London, Ser. A* **332A**, 473–504.

Bagnold, R. A. (1977). Bed-load transport by natural rivers. *Water Resour. Res.* **13**, 303–312.

Baker, V. R. (1973). Paleohydrology and sedimentology of Lake Missoula flooding in eastern Washington. *Spec. Pap.—Geol. Soc. Am.* **144**, 1–79.

Baker, V. R., and Ritter, D. F. (1975). Competence of rivers to transport coarse bedload material. *Geol. Soc. Am. Bull.* **86**, 975–978.

Birkeland, P. W. (1982). Subdivision of Holocene glacial deposits, Ben Ohau Range, New Zealand, using relative-dating methods. *Geol. Soc. Am. Bull.* **93**, 433–449.

Bull, W. B. (1979). Threshold of critical power in streams. *Geol. Soc. Am. Bull.* **90**, 453–464.

Bull, W. B. (1987, in press). "Geomorphic Responses to Climatic Change," Oxford University Press, New York.

Burrows, C. J. (1979). A chronology for cool-climate episodes in the Southern Hemisphere 12,000-1,000 years B.P. *Palaeogeogr., Palaeoclimatol., Palaeoecol.* **27**, 287–347.

Chappell, J. (1983). A revised sea-level record for the last 300,000 years on Papua New Guinea. *Search* **14**, 99–101.

Emmett, W. W. (1976). Bedload transport in two large gravel-bed streams, Idaho and Washington. *Proc. Fed. Inter-Agency Sediment. Conf., 3rd, 1976,* pp. 4.101–4.113.

Gomez, B. (1983). Temporal variations in bedload transport rates, the effect of progressive armoring. *Earth Surf. Processes Landforms* **8**, 41–54.

Huntington, E. (1907). Some characteristics of the glacial period in non-glaciated regions. *Geol. Soc. Am. Bull.* **18**, 351–388.

Knuepfer, P. L. K. (1984). Tectonic geomorphology and present-day tectonics of the Alpine shear system, South Island, New Zealand. Ph.D. Dissertation, University of Arizona, Tucson.

Lekach, J., and Shick, A. P. (1983). Evidence for transport of bedload in waves; analysis of fluvial sediment samples in a sample upland stream channel. *Catena (Cremlingen-Destedt, Ger.)* **10**, 268–279.

Leopold, L. B., and Bull, W. B. (1979). Base level, aggradation, and grade. *Proc. Am. Philos. Soc.* **123**, 168–202.

Leopold, L. B., and Emmett, W. W. (1976). Bedload measurements, East Fork River, Wyoming. *Proc. Nat. Acad. Sci. U.S.A.* **74**, 2644–2448.

Leopold, L. B., and Miller, J. P. (1954). A post-glacial chronology for some alluvial valleys in Wyoming. *Geol. Surv. Water-Supply Pap. (U.S.)* **1261**. 1–90.

Leopold, L. B., Wolman, M. G., and Miller, J. P. (1964). "Fluvial Processes in Geomorphology." Freeman, San Francisco, California.

McGlone, M. S., and Bathgate, J. L. (1983). Vegetation and climate history of the Longwood Range, South Island, New Zealand, 12,000 B.P. to the present. *N. Z. J. Bot.* **21**, 293–315.

Moar, N. T. (1980). Late Otiran and early Aranuian grassland in central South Island. *N. Z. J. Geol.* **3**, 4–12.

Patton, P. C., and Schumm, S. A. (1975). Gully erosion, northwestern Colorado; a threshold problem. *Geology* **3**, 88–90.

Porter, S. C. (1975). Equilibrium line of late Quaternary glaciers in the Southern Alps, New Zealand. *Quat. Res.* **5**, 27–48.

Ritter, D. F. (1967). Terrace development along the front of the Beartooth Mountains, northern Montana. *Geol. Soc. Am. Bull.* **78**, 467–484.

Schumm, S. A. (1973). Geomorphic thresholds and complex response of drainage systems. *In* "Fluvial Geomorphology" (M. Morisawa, ed.), Publ. Geomorphol., pp. 229–310. State University of New York, Binghamton.

Soons, J. M. (1962). A survey of periglacial features in New Zealand. *In* "Land and Livelihood; Geographical Essays in Honor of George Jobberns." New Zealand Geographical Society, Christchurch.

Soons, J. M. (1979). Late Quaternary environment in the central South Island of New Zealand. *N. Z. Geogr.* **35**, 16–23.

Stevens, G. R. (1974). "Rugged Landscape, the Geology of Central New Zealand." Reed, Wellington.

Wolman, M. G., and Gerson, R. (1978). Relative scales of time and effectiveness of climate in watershed geomorphology. *Earth Surf. Processes* **3**, 189–208.

PART III

FLOODS, CLIMATE, LANDSCAPE

The interaction of climate and landscape dictates the impact that floods have on the morphology and operation of fluvial systems. The combined effect of these two variables controls the sensitivity of fluvial landforms to change during floods and the stability and permanence of flood-created fluvial morphology, In Chapter 11, R. Craig Kochel notes that the permanence of flood-generated landforms is one criterion that can be used to measure the geomorphic significance of a flood event. Climate and landscape control the amount of sediment transported during large floods, another measure of geomorphic work. The magnitude of flood-transported sediment compared to the mean annual sediment yield, is a measure of the relative effectiveness of rare geomorphic events in accomplishing denudation. Finally, at the broadest scale, climate and landscape determine the relative magnitude of the flood event itself.

Part III provides an overview of flood hydrology and geomorphology across a spectrum of climatic and physiographic regions. The contributions demonstrate the considerable variability in flood hydrology and geomorphic response of rivers as a function of climate. However, quantitative comparisons of the flood response from different climatic regions is prohibited by a lack of observational data on the fluvial geomorphology and hydrology of several climatic regions; specifically the tropical, periglacial, and hyperarid. In this part, three of the seven case histories focus on stream systems in humid–temperate regions, which is a statement about the global bias of our geomorphic knowledge. Nevertheless, several general observations can be made from the available data.

The climate-controlled precipitation regime is, of course, strongly linked to the frequency and magnitude of individual runoff events. One measure of the geomorphic significance of a rainfall event is the ratio of the runoff-producing precipitation to the mean annual precipitation. In Chapter 12, for example, Asher P. Shick shows that in hyperarid environments rainfall events necessary to generate large super floods commonly exceed the mean annual precipitation by several orders of magnitude, a statistic that in itself describes the long recurrence interval associated with these flood events. Infiltration losses, in the absence of any antecedent moisture within the ground or base flow in channels, represent a major loss to runoff and severely curtail the magnitude of floods in larger basins. These factors combine to produce the characteristic high variability of the annual peak series in arid regions.

In humid and tropical environments, even catastrophic rainfalls do not exceed the mean annual precipitation. In addition, antecedent conditions can greatly increase the flood response and magnitude of the flood peak. Although there may be significant variability to the annual flood series in these environments, it is not nearly as great as in arid regions. The uniformity of the hydrologic response in humid regions results in a strong correlation between climate and high frequency flood magnitudes.

In contrast to rainfall-generated floods, floods on periglacial rivers result from the melt out of a stored snowpack. The magnitude of the floods depends on the flux of solar energy to the snowpack, and the transmission of the water through the snow. In Chapter 13, Michael Church demonstrates that these are conservative processes that limit the magnitude of the flood peak. Thus, the annual peak series of pure snowmelt periglacial floods has exceptionally low variability. Floods on periglacial rivers are also caused by other processes, including rainfall on snowpack. The result is a combined hydrologic record characterized by mixed populations of floods with increased hydrologic variability.

Floods create distinctive fluvial deposits and landforms; but the effectiveness of floods in land sculpture is related to the permanence of these features. Landscape recovery (Wolman and Gerson, 1978), the return to the preflood fluvial morphology, is most rapid in those environments where the ratio of the flood magnitude to the mean annual flood is lowest. Conversely, the persistence of flood-related landforms is greatest in those fluvial environments where the flood ratio is large and significant erosional thresholds must be exceeded in order to cause morphologic change (Baker, 1977). Finally, when geomorphic effectiveness and work accomplished are defined by the amount of sediment transported, the relative flow duration is the important hydrologic variable. Typically, high-frequency flood events are the dominant geomorphic agent of denudation (Wolman and Miller, 1960).

In arid environments, the erosional stress applied to the fluvial system during a catastrophic flood can be several orders of magnitude greater than that experienced on a more frequent basis. These large floods cause arid and semi-arid streams to cross geomorphic thresholds, change stream morphology, and cause long-term or even permanent change to the river system. Thus, in the hyperarid Negev Desert of Israel, fluvial landforms created by large floods may persist until a larger flow occurs. Given the long waiting time between even moderate flood events, the recovery time of these streams cannot be adequately assessed from the historic record. But, recovery of the system, in the sense that the stream returns to its preflood morphology, may not occur.

In Chapter 14, William L. Graf proposes that one theoretical approach to analyzing geomorphic change on arid-region rivers may be through the use of catastrophe theory. This concept is appropriate because morphologic change on these river systems is not progressive, but instead proceeds through time in a discontinuous, but perhaps, predictable pattern. In addition, threshold changes in channel pattern can produce transitional channel forms that cause catastrophic sediment yields during subsequent high-frequency flows.

In humid region river systems, significant morphologic change can occur during large floods. However, in these environments, subsequent moderate flows are capable of reforming an equilibrium channel adjusted to the lower end of the discharge spectrum. Thus, flood-generated channel change is rarely permanent and stream recovery time can be expressed as a small percentage of the annual flow of the system.

Moreover, in spite of the scale of flood-generated landforms, humid-region vegetation is capable of masking the erosional and depositional morphology related to the flood events. As shown by Wolman and Miller (1960), the bulk of suspended sediment transport for such rivers occurs during moderate flow events. Ultimately the fluvial landforms represent the average hydrologic conditions, except for local evidence of large-scale flood events preserved in the stratigraphic record and as distinct landforms. Circumstances leading to this caveat are illustrated in Chapter 15 by Dale F. Ritter. Ritter illustrates one type of floodplain alteration caused by localized scour and the resulting deposition of a large lobe of sand and gravel during a severe flood on the Gasconade River in Missouri.

Peter C. Patton in Chapter 16 and James C. Knox in Chapter 17 show that the sedimentary deposits of large-scale floods can be recognized in the stratigraphy of humid-region floodplains. Such deposits provide useful paleohydrologic information (Patton, 1987), especially in relation to probable changes in the flood-producing climate. Knox demonstrates that historical changes in the flood frequency spectrum for the upper Mississippi River Valley can be attributed to historical climate change. Such a documentation of nonstationarity for the high-magnitude flood events can be an important geomorphic contribution to flood risk analysis.

Avijit Gupta in Chapter 18 provides an analysis of the variety of fluvial landforms along tropical river systems. Tropical fluvial landforms include channel features and floodplains related to low-flow conditions and flood terraces related to the more severe, but frequent, monsoonal runoff events that create a suite of erosional and depositional features along the outer margin of the valley.

Periglacial regions provide yet another model of the role of climate in flood geomorphology. The mixed populations of floods are reflected in the geomorphic work accomplished within the fluvial system. The relatively uniform magnitude of nival floods is such that they represent a moderate bankfull flood that is probably the single most important hydrologic event in terms of transporting suspended load on these streams as well as determining the channel characteristics. However, major morphologic change may be related to the rarer more catastrophic rainfall–runoff floods that are capable of transporting the gravel that makes up the channel boundaries of these rivers.

For the future, paleoflood hydrology offers the most promising avenue of research for increased understanding of the geomorphic response of rivers to different flood regimes. To date, studies in paleoflood hydrology have generated long-term flood records for fluvial systems under relatively stable climatic conditions and have documented the influence of climate change on long-term flood history. Such information provides the the potential to understand better the linkage between climate and the flood response. The next phase of this research must be to correlate changes in flood response with specific fluvial landforms to understand better the linkages among climate change, flood hydrology, and the geomorphic development of the fluvial landscape. Although modern process studies will continue to improve our understanding of the hydrologic and hydraulic processes during super floods; only studies of the stratigraphy of flood events will provide the long-term view necessary to place isolated hydrologic events into a broader geomorphic context.

REFERENCES

Baker, V. R. (1977). Stream-channel response to floods with examples from central Texas. *Geol. Soc. Am. Bull.* **88**, 1057–1071.

Patton, P. C. (1987). Measuring the rivers of the past: A history of fluvial paleohydrology *In* "The History of Hydrology" (E. R. Landa and S. Ince, eds.), pp. 55–67. *Am. Geophy. Un.*, History of Geophysics, v. 3.

Wolman, M. G., and Gerson, R. (1978). Relative scales of time and effectiveness of climate in watershed geomorphology. *Earth Surf. Processes* **3**, 189–203.

Wolman, M. G., and Miller, J. P. (1960). Magnitude and frequency of forces in geomophric processes. *J. Geol.* **68**, 54–74.

11

GEOMORPHIC IMPACT OF LARGE FLOODS: REVIEW AND NEW PERSPECTIVES ON MAGNITUDE AND FREQUENCY

R. CRAIG KOCHEL

Department of Geology, Southern Illinois University, Carbondale, Illinois

INTRODUCTION

The relative role of infrequent, large-magnitude floods versus the cumulative effect of frequent, small-magnitude floods on stream channels and floodplains has been the focus of considerable debate for many years. The concept of geomorphic work has different connotations to various workers in geomorphology. Presently, two perceptions of work are maintained. Geomorphic work is frequently defined as the amount of sediment transported by a flood (Wolman and Miller, 1960). Alternatively, geomorphic effectiveness is defined as the modification of landforms (Wolman and Gerson, 1978). This chapter will adopt an approach that contains both views and discusses flood effects from both perspectives.

Wolman and Miller (1960) first quantified this debate when they defined geomorphic work in terms of the amount of suspended sediment transported by a river. Using this working definition, they analyzed the suspended-load data for rivers throughout the United States and showed that large floods transported only a minor percentage of the annual suspended-sediment load. They concluded that most geomorphic work is accomplished by frequent events of low magnitude such as bankfull discharge, which, on the average, occurs every 1–2 yr. They cautioned, however, that this conclusion should be tempered with the understanding that suspended-sediment transport is not the only measure of the geomorphic effectiveness of floods. Wolman and Miller (1960) noted that large floods can also alter floodplain and channel morphology, but they had little data to draw conclusions about this type of flood effect.

Subsequent observations allowed Wolman and Gerson (1978) to present an alternative definition of geomorphic effectiveness, which is the ability of floods to alter the shape of landforms (Wolman and Gerson, 1978). Major floods can conceivably produce significant geomorphic effects on floodplain and channel morphology without transporting extremely large quantities of suspended sediment. Wolman and Gerson (1978) defined geomorphic work as the amount of erosion produced by a flood and also stressed that effectiveness was additionally dependent on the rate of recovery of landforms to their normal character that prevails between successive floods. Landform modification occurs only when critical competence values are exceeded such that sediment is entrained (Wolman and Gerson, 1978). Threshold conditions are exceeded only during rare, large floods on many rivers. These thresholds may not be passed during low-magnitude floods, regardless of how frequently they occur. Many investigators have documented large-scale landscape modification caused by rare, threshold-passing, large floods in various parts of the United States (Hack and Goodlett, 1960; Schumm and Lichty, 1963; Stewart and LaMarche, 1967; Scott and Gravlee, 1968; Williams and Guy, 1973; Baker, 1977; Gupta, 1983; Nolan and Marron, 1985).

Assessment of the relative importance of large and small floods is a complicated task. Part of the complexity stems from the various ways geomorphic work is defined. Considerably different conclusions can be drawn from observations of floods on a given river, depending on whether one looks only at the amount of sediment transported (geomorphic work, Wolman and Miller, 1960) or the impact of floods upon the landscape (geomorphic effectiveness, Wolman and Gerson, 1978). Additional complications arise when rivers with exceedingly different geomorphic characteristics and flood regimes are compared; for example, comparing the effects of large floods in semi-arid, bedrock rivers with the effects of large floods in humid, alluvial rivers. Baker (1977) presented a detailed discussion that illustrated how sediment load, rainfall–runoff regime, and channel cross section were important in causing only the large, infrequent floods to have significant effects on stream channels in semi-arid regions of central Texas. These parameters were shown to be distinctly different in semi-arid channels than in most humid regions. Clearly, climate and hydrology are important parameters influencing the role of floods of differing magnitude and frequency.

Striking contrasts in river response to large floods also exist within similar climates. Dramatic channel and floodplain erosion and deposition resulted from large-magnitude floods in Virginia (Hack and Goodlett, 1960; Williams and Guy, 1973), while insignificant changes were produced by the large Hurricane Agnes flood in Pennsylvania (Moss and Kochel, 1978). These studies illustrate the effect of topography and lithology on the ability of rivers to exceed thresholds necessary to cause significant landscape modification.

In this chapter I will attempt to discuss the major geomorphic variables that affect the response of rivers to large-magnitude, infrequent floods. In doing so, I will provide a brief review of studies that focused on this issue but will attempt to synthesize information from these studies with the goal of determining which of a long list of variables may be most important in predicting geomorphic response of stream channels during large floods. In addition, observations from selected regions will be included as examples to support my conclusions. This review is not meant to be comprehensive, but it may hopefully stimulate a new approach to this problem that could ultimately result in quantitative multivariate analyses to resolve some of these questions.

Large floods will be defined as those that equal or exceed the discharge expected by the 50-yr flood. The large floods addressed by many studies have been the largest historical flood recorded at a particular location, and recurrence intervals commonly exceed the 50-yr flood by more than an order of magnitude.

FACTORS CONTROLLING GEOMORPHIC RESPONSE TO LARGE FLOODS

Geomorphic response to large floods is controlled by numerous factors. In order to discuss these factors, I have chosen to divide them into two broad categories based on their relationship to the channel–floodplain system. Any grouping of processes or variables involves overlap, hence, the categories proposed here are used only to expedite discussion. Drainage basin factors are controls on the river that are external to the channel and floodplain, including: (1) climatic (mainly rainfall regime), (2) hydrology (chiefly peak discharge, contributing drainage area, and hydrograph response), (3) basin morphometry, (4) sediment load, (5) vegetation, and (6) soils. River channel factors are internal controls on the river resulting from physical characteristics of the channel and flow, including: (1) channel gradient, (2) channel and floodplain geometry, (3) channel morphology, and (4) bank cohesion. Table 1 summarizes observations made in studies of the effects of large floods largely in the United States. Table 1 should be referred to throughout for specific details of the floods discussed. In addition to the basin and river channel factors already mentioned, the temporal ordering of floods will also be treated, that is, the recovery time between events.

DRAINAGE BASIN FACTORS

Rainfall Regime and Flood Generation Processes

Major floods usually result from one of four kinds of storms: (1) tropical storms, (2) convective storms associated with cold fronts, (3) easterly waves, and (4) amplified precipitation due to orographic effects (see Hirschboeck, this volume, for a related discussion of flood hydroclimatology). Floods due to snowmelt typically do not exceed the 50-yr recurrence interval on most rivers and are not considered in this analysis. Many areas of the world experience arid or semi-arid conditions except when hurricanes (tropical storms) move inland over these areas. Hayden (this volume) shows the locations of these regions throughout the world. The area between Baja California and southwest Texas is a good example of this kind of flood regime in North America. The western portion of the area receives major flooding from the influences of Pacific tropical storms such as the one that caused severe flooding in the Tucson, Arizona, area in 1983 (Kresan, this volume). Major floods in west Texas, like the catastrophic Pecos River flood of 1954, are caused by tropical storms that move northwestward from the Gulf of Mexico, such as Hurricane Alice in 1954 (Kochel et al., 1982) and tropical storm Amelia in 1978 (Baker and Kochel, this volume). Tropical storms have also caused many catastrophic floods in humid regions where annual rainfall is not dominated by tropical influences such as the 1969 Hurricane Camille flood in Virginia and mid-Atlantic floods in 1972 (Hurricane Agnes) and 1985 (Hurricane Juan).

Major floods can also result from the passage of extratropical cyclones associated with cold fronts. These floods are most common in midlatitude regions where the interaction between cold polar air and warm tropical air masses occurs frequently (see maps in Hayden, this volume). Major convective storms often develop along these cold fronts, which can result in significant rainfall amounts and severe flooding, particularly when cold fronts become stationary for periods of several days. An example of this type of flood was the 1982 flood in south-central Missouri discussed by Ritter (this volume).

Exceedingly large, short-duration rainfalls are common near orographic barriers, particularly where moist marine air masses are lifted as they move inland. The Balcones Escarpment of central Texas is a good example of an orographic barrier. Although relief at the escarpment between the inner coastal plain of Texas and the Edwards Plateau is less than 400 m, Baker (1977) showed that many major flood-producing rainfall events have occurred along the Balcones Escarpment.

In fact, these high-intensity rainfall events approach the United States and the world records for durations up to 24 hr (Caracena and Fritsch, 1983). Warm, moist air moving out of the easterly trade winds may also result in major rainfall and flooding in low midlatitude regions of the world. In Texas, for example, these phenomena are known as easterly waves. Easterly waves are low-pressure systems and are similar to extratropical cyclones, except that they are warm and typically saturated with Gulf moisture. When these air masses are orographically lifted over escarpments like the Balcones in Texas, exceedingly large, intense rainfalls can occur.

The effect of orography sometimes combines with tropical storm systems to produce extremely intense rainfall such as the Hurricane Camille flood along the Blue Ridge of central Virginia (Williams and Guy, 1973). Most major Appalachian rainfalls have been associated with the movement of tropical moisture over the mountains (Kochel, 1987).

Hydrologic Factors

Contributing Drainage Area. Contributing area for large floods is extremely variable. Semi-arid-and arid-region floods usually result from rainfalls that were experienced over only a minor portion of the drainage area. The 1954 Pecos River flood in west Texas, for example, one of the largest floods in the United States, had a peak discharge of over 27,000 m^3/s, produced by rainfall from Hurricane Alice in only the lower 9300 km^2 of the more than 91,000 km^2 drainage of the basin. Similarly, the third largest flood in over 200 yr on the James River at Richmond, Virginia, was produced by rainfall from Hurricane Camille localized in an area of approximately 4000 km^2 of the 17,500-km^2 drainage area above Richmond. However, most large humid-region floods result from less intense, widespread rainfall over the entire drainage basin, like the 1972 Hurricane Agnes and 1985 Hurricane Juan floods in the mid-Atlantic region of the United States.

In its purest sense contributing area of a flood would appear to exert little control on the ability of a flood to result in significant geomorphic change. However, most of the floods noted in Table 1 that resulted in significant geomorphic response resulted from floods whose contributing area was small in comparison to the area of the entire basin, regardless of climate. The rainfall required to produce a large flood with only a small portion of the drainage basin contributing runoff must be large and of short duration. Therefore, floods from small subbasin contributing areas are likely to be flashy and more likely to exceed the competence threshold for channel and floodplain erosion.

Peak Discharge and Flash-Flood Magnitude Index. Peak discharge of large floods varies greatly, depending on drainage area and climate. The absolute flood discharge of a river is not as important with regard to geomorphic change as the ratio between peak discharge and mean annual discharge. Floods likely to result in significant geomorphic change are those that produce discharges many times above that normally experienced by the river, that is, those with high maximum peak discharge to mean annual discharge ratios. The morphology of many rivers is adjusted to the flows most commonly experienced, that is, bankfull or smaller magnitude flows, as exemplified by rivers in humid regions (Wolman and Miller, 1960). In contrast, rivers in semi-arid regions commonly adjust channel and floodplain morphologies to less frequent large floods because these are the only flows competent enough to alter river morphology (Baker, 1977). However, if the time between successive large floods is long, even semi-arid rivers may adjust their morphologies to lower magnitude flows of higher frequency. Kochel (1980) found evidence for this phenomenon in studies of the effects of two major floods along the Pecos River in Texas.

Beard (1975) mapped the United States according to its potential for flash flooding, a parameter called the Flash-Flood Magnitude Index (FFMI) (Fig. 1). The FFMI is calculated from the standard deviation of the logarithms of annual maximum discharge as illustrated:

$$\mathrm{FFMI} = X^2/N - 1$$

where

$X = X_m - M$
X_m = annual maximum discharge
M = mean annual discharge
N = number of years of record
X, X_m, and M = logarithms.

TABLE 1 Geomorphic Effects of Catastrophic Floods

River/State	Data Source[a]	Date	Flood Hydrology: Peak Q (m^3s^{-1})	$\overline{D}$ (m)	$\overline{V}$ (m/s)	FFMI	Sediment Characteristics: R.I. (yr)	Bedrock Lithology	Channel Sediment
Pecos/Texas	2, 3	1954 (6/31)	27,000	30	11–13	0.7–0.8	2,000	Limestone	Gravel
Pecos/Texas	2, 3	1974	16,000	20	8–9	0.7–0.8	700	Limestone	Gravel
Devils/Texas	2, 4	1932	17,000	9	6–7	0.7–0.8	2,500	Limestone	Gravel
Devils/Texas	2, 4	1954 (6/31)	16,000	8	5–6	0.7–0.8	1,600	Limestone	Gravel
Medina/Texas	5, 1	(1978) (8/2)	6,800	10–15	3–4	0.8	500	Limestone, granite	Gravel
Elm Creek/Texas	6, 7	1972 (5/2)	1,130	7	6–7	0.8	400	Limestone	Gravel
Big Thompson/ Colorado	8, 9	1976 (7/31)	884	3.2	7–8	0.5–0.6	500–5,000	Mostly crystaline rocks	Gravel
Rubicon/ California	10	1964 (12/23)	7,000	20	6–7	0.5	>100	Granite, gneiss	Gravel
Coffee Creek/ California	11	1964 (12/23)	500		4–5	0.3	100	Granite, gneiss, schist	Gravel
Blieders Creek/ Texas	6	1972 (5/2)	1,370	9	2–4	0.7–0.8	400	Limestone	Gravel
Santa Cruz/ Arizona	12	1983 (10/2)	1,490			0.6–0.7	<1,000	Alluvium volcanics-headwaters	Minor gravel
Shoal Creek/ Texas	13	1981 (5/24)	450	7		0.7–0.8	100	Limestone	Gravel
Davis Creek/ Virginia	1	1969 (8/19)	400	4–5	5–7	0.4–0.5	3,000	Granite, gneiss	Sand gravel
Conestoga/ Pennsylvania	1, 14	1972 (6/22)	2,500	6	2–3	0.2–0.3	200	Limestone, shales	Silt, minor sand, gravel
Pequea Creek/ Pennsylvania	1	1984 (7/1)				0.2–0.3	>100	Limestone, schist	Silt, sand minor gravel
Fishing Creek/ Pennsylvania	1	1984 (7/1)				0.2–0.3	>100	Schist	Sand and gravel
Patuxent/ Maryland	15	1971 (9/11)	600	5		0.3–0.4	>100	Schist	Sand and gravel
Western Run/ Maryland	16	1972 (6/22)	1,100	8		0.3–0.4	>200	Gneiss, schist	Sand and gravel
Yallahs/Jamaica	17, 18	1970 (11/9)	1,400				>30	Conglomerate sandstone, volcanics	Gravel and sand
Sexton Creek/ Illinois	19	1973 (5)		3–4		0.2–0.3	>100	Limestone, chert	Gravel and sand
Gasconade/ Missouri	20	1982 (12/3–5)	3,966	10	1–2	0.3	>100	Limestone, chert	Mixed gravel and silt
James/Virginia	1	1985 (11)				0.3–0.4	>100	Mixed sedimentary, igneous and meta sediments	Gravel and sand
Shenandoah Valley— Mountain front streams/ Virginia	1	1985 (11)	Various streams			0.3–0.4	>50	Sandstone, igneous rocks	Gravel, some sand

[a] Data sources: (1) this study, (2) Kochel (1980), (3) Kochel and Baker (1982), (4) Kochel et al. (1982), (5) Baker (1984), (6) Baker (1977), (7) Patton and Baker (1977), (8) Grozier et al. (1976), (9) Costa (1978), (10) Scott and Gravlee (1968), (11) Stewart and LaMarche (1967), (12) National Research Council (1984), (13) National Research Council (1982), (14) Moss and Kochel (1978), (15) Gupta and Fox (1974), (16) Costa (1974), (17) Gupta (1975), (18) Gupta (1983), (19) Ritter (1975), (20) Ritter, this volume.

Table 1 *(Continued)*

Sediment Characteristics		Basin Characteristics					
Sediment Availability	Bank Cohesion	Channel Gradient†	Climate	Other Basin	Trigger	Years Since Last Flood	Summary Flood Effects[b]
Moderate to high	High: bedrock; low: gravel and sand	0.002–0.003	Semi-arid	Flashy	Hurricane Alice	500–700	Extreme a, b, c, d
Moderate	Bedrock: high	0.002–0.003	Semi-arid	Flashy as above		20	Minor l
High	Bedrock: high; gravel: low	0.002–0.003	Semi-arid	Flashy as above		>1,200	Extreme a, b, c, d, e, f, k
High	As above	0.002–0.003	Semi-arid	Flashy as above	Hurricane Alice	22	Minor l
Moderate to high	Low: gravel and sand	0.002–0.003	Semi-arid, subhumid	Flashy high HD, shallow soil	Tropical storm Amelia		Extreme a, b, c, d, e, f, h, k
High	Low: gravel and sand	0.0045	Semi-arid	Flashy as above			Extreme a, b, c, d, e, f, h, k
High	Low: gravel and sand	0.02–0.04	Semi-arid	Flashy			Extreme a, b, c, d, e, f, k
Very high	Low to moderate gravel	0.01–0.02	Subhumid				Extreme a, b, c, d, e, f, j, k
High	As above	0.02–0.04	Subhumid, humid				Extreme a, b, c, d, e, f, k
High	As above	0.005	Semi-arid	Flashy as above			Extreme a, b, c, d, e, f, h, k
Sand: high; gravel: moderate	Moderate, some caliche	0.003	Arid	Flashy	Tropical storm related	>70	Moderate a, b, c
Moderate	Moderate	0.006	Semi-arid	Flashy	Escarpment and tropical storm	>50	Extreme a, b, c
High	Moderate	0.03–0.04	Humid temperate	Flashy, high ruggedness no., debris avalanches	Hurricane Camille	>200	Extreme a, b, c, d, e, f, h, j, k
Low: gravel; high: silt	High	0.006–0.004 HW 0.001–0.004 DS	Humid tempeate	Moderately flashy	Hurricane Agnes	>50	Minor g
Low: gravel; high: sand, silt	High	0.002 HW 0.004 DS	Humid temperate		Thunderstorm	>50	Minor g
Moderate gravel, high: sand	High to moderate	0.014	Humid temperate	Debris avalanches	Thunderstorm	>50	Extreme a, b, c, d, e, f, h, i, k
Moderate to high	High to moderate	0.007–0.02	Humid temperate				Extreme a, b, c, d, e, f, h
Moderate to high	High to moderate	0.002	Humid temperate		Hurricane Agnes		Moderate a, d, f, l
High	High to moderate		Humid tropical	Debris avalanches			Extreme a, b, c, d, e, f, h, i, k
High	Moderate to high	0.01	Humid temperate	Flashy			Moderate e, h, l
Mixed		0.0003	Humid temperate				Moderate h, k, l
Moderate to high	High	0.001	Humid temperate		Hurricane Juan	13	Minor d, l
High	Moderate to low	0.04–0.09	Humid temperate	Flashy	Hurricane Juan		Extreme a, b, c, d, e, f, h, k, l

[b] Flood effects: (a) bank erosion, (b) channel widening, (c) channel erosion, (d) floodplain erosion, (e) channel deposition, (f) floodplain deposition, (g) obstacle lee deposition only, (h) overbank gravels, (i) mass wasting in basin, (j) boulder levees, terraces, (k) large-scale gravel bedforms, (l) bar reorganization.

† HW = headwater.

DS = downstream.

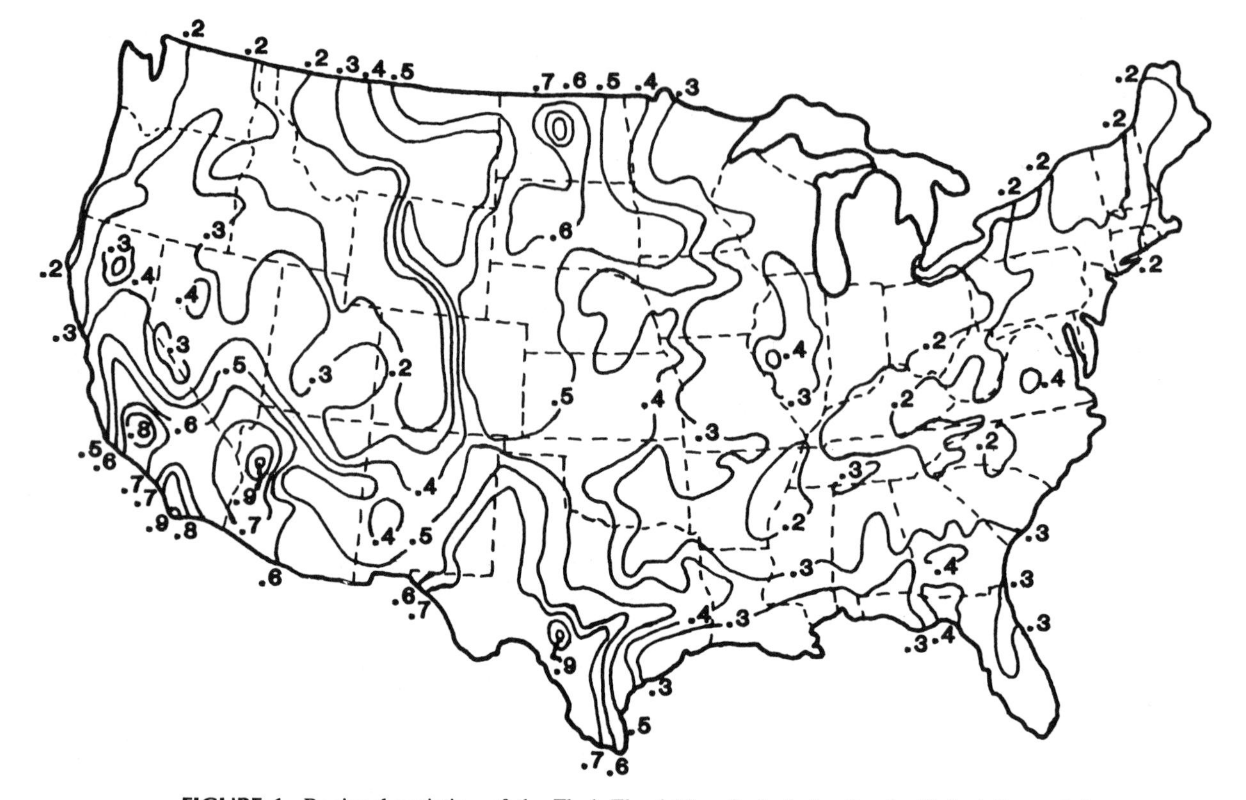

FIGURE 1. Regional variation of the Flash-Flood Magnitude Index for the United States (after Beard, 1975, Center for Research in Water Resources, CRWR-124, Fig. 20). Map is based on gauging records from 2900 stations that had records exceeding 20 yr. Data shown represents only basins whose area is less than 2590 km^2.

Streams with extremely high FFMI generally occur in the semi-arid southwestern United States and in mountainous regions of the western United States. Most studies of the geomorphic effects of floods in these regions have described major response to large floods (e.g., Patton and Baker, 1976; Baker, 1977; Gustavson, 1978). The FFMI, therefore, is a good measure of the variability of flood frequency measured as an index of flood flashiness. Areas of the United States where FFMI values are low include large areas of the low-relief Midwest, Appalachian piedmont, and southeastern coastal plain, where geomorphic effects of large floods have been described as minor (e.g., Costa, 1974; Moss and Kochel, 1978). Therefore, high FFMI may serve as a useful predictor of the likelihood of streams to experience major geomorphic change during large floods.

The FFMI values are not uniformly high in all arid regions. Baker (1977) noted that low FFMI values occur for extremely arid areas of the Colorado Plateau, eastern Oregon, and eastern Washington. This suggests that climatic factors alone cannot account for the response of rivers to major floods. Other factors, such as topography, vegetation, lithology, and basin morphometry are also important controls on flood response.

A study of channel response to the 1969 Hurricane Camille flood in Virginia suggested that peak discharge was much less important compared to other variables in explaining the spatial pattern of channel erosion (Johnson, 1983). The Camille flood, with an estimated recurrence interval between 3000 and 4000 yr, was characterized by numerous debris avalanches on steep slopes in the Blue Ridge province of Nelson County, Virginia (Kochel and Johnson, 1984; Kochel, 1987). Extreme erosion and deposition resulted in downstream areas along stream channels draining the mountainous area affected by catastrophic rains in 1969 (Williams and Guy, 1973). Johnson (1983) compared pre- and postflood aerial photographs to document channel widening produced by the flood. Stepwise multiple regression of 17 morphometric parameters of the channel and upstream drainage basin revealed that the amount of channel widening depended primarily on the percentage of upstream area affected by debris avalanching (Fig. 2). Secondary factors affecting the amount of channel widening included channel gradient and basin shape according to the following equation (Johnson, 1983):

$$CW = 8.2DA + 4592CG - 126K + 93$$

where

CW = channel widening in percent

DA = percentage of drainage area affected by debris avalanching

CG = channel gradient

K = the basin shape expressed as a lemniscate

Significant channel erosion only occurred where there was a supply of coarse gravel bedload and where steep channel slopes existed on which this material could be transported. Peak discharge ranked 14th out of the 17 variables Johnson (1983) used in his study, accounting for less than 4% of the variance. These results demonstrate the importance of the process linkage between basin characteristics and slope systems upstream with downstream channel processes during large floods in determining the geomorphic effects of a given event.

Basin Morphometry

The preceding discussion of channel widening in Virginia streams during the Camille flood showed that the morphometry of the drainage basin can influence river response to large floods. In that case basin shape was an important variable leading to downstream channel widening during the flood. Johnson (1983) found that basins with equant shape (low lemniscate value) occurred where channel widening was greatest. Surface runoff tends to arrive simultaneously from all distal parts of equidimensional basins, providing other morphometric variables affecting lag time are similar, resulting in exceedingly flashy hydrographs with high peak discharge. In contrast, elongate basins may

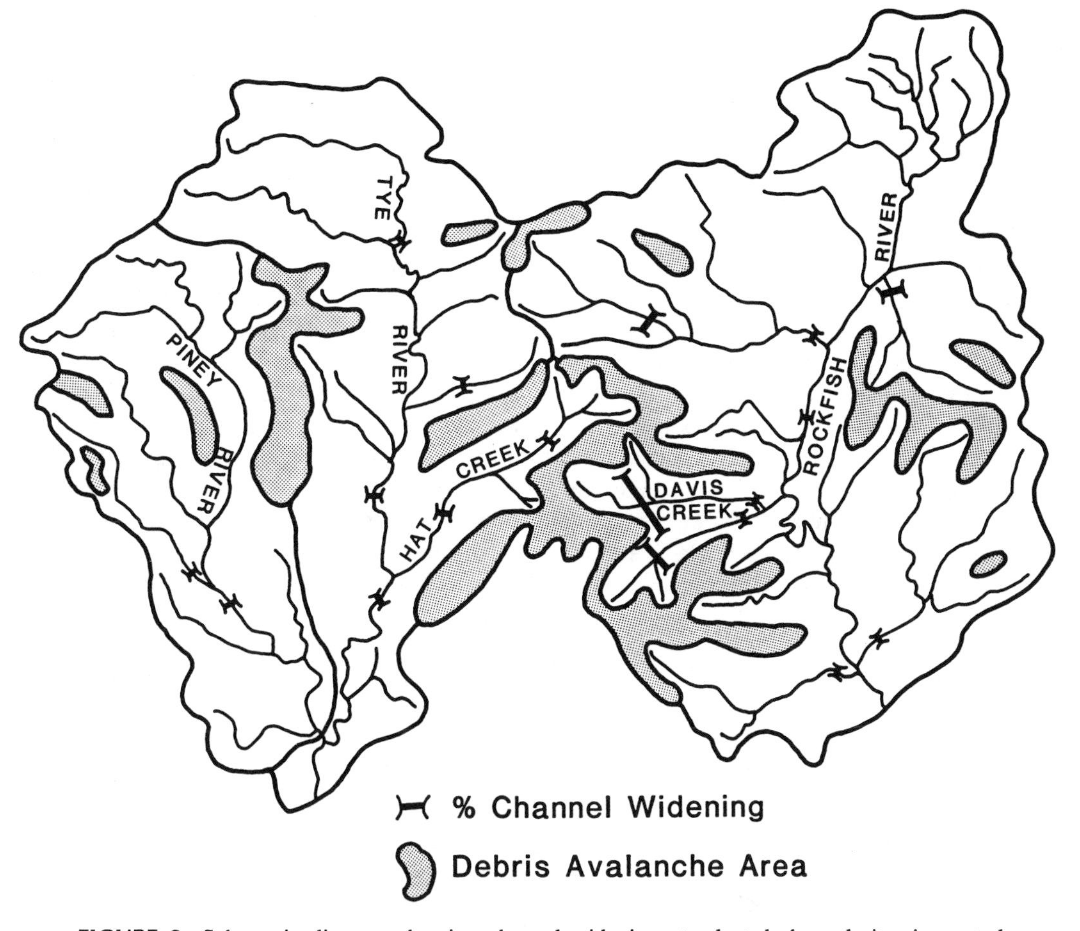

FIGURE 2. Schematic diagram showing channel widening at selected channel sites in central Virginia affected by the Hurricane Camille flood (after Johnson, 1983). The length of the bars shows relative percent of channel widening. The shaded area shows the relative extent of debris avalanching in basin headwaters.

have less flashy hydrographs because runoff is attenuated and arrives at a given point on the mainstream at different times because of the wide range of travel times from various parts of the basin. Thus, circular-shaped basins are more efficient in concentrating runoff than elongate basins (Gregory and Walling, 1973).

Patton and Baker (1976) showed that peak discharge from watersheds of similar size vary in accord with differing basin morphometry. In particular, they showed that ruggedness number, the product of drainage density and relief, in combination with basin magnitude (Shreve magnitude) and first-order channel frequency were most important in explaining the variation in peak discharge from drainage basins within similar climatic and topographic regions. Patton (this volume) presents a review of the effect of basin morphometry on flood hydrographs. The influence of basin morphometry to fluvial response to large floods is dominated by its effect on peak discharge and flash-flood potential.

Vegetative cover and soil characteristics can affect the flood response of the drainage basin. In general, runoff per unit area declines with increasing vegetation density for a given rainfall intensity and duration. Langbein and Schumm (1958) demonstrated that sediment yield also varies directly with vegetative cover, with peak values in semi-arid regions near the transition between desert scrub and grassland vegetation types. Soil thickness also has considerable effect on flood hydrographs. Basins with thin soils tend to produce flashier floods with higher peak discharges, other morphometric factors being equal, for given rainfalls than those with thick soil cover. When thick soils are present, significant quantities of potential runoff enter the subsurface system and may be temporarily lost from the surface system in groundwater flow or near-surface interflow. Once water enters the subsurface system, hydrograph peaks are reduced as the water is delayed in its path to the main channel. Although variations in vegetation and soil cover can have significant effect on hydrographs, the magnitude of these effects are probably not large enough to dramatically affect channel response to flooding.

RIVER CHANNEL FACTORS

Channel Gradient and Bedload

Channel, bank, and floodplain erosion is facilitated by high-velocity flows carrying coarse bedload material that can be used as abrasive tools. Lithology, in combination with climate, largely controls the size of sediments delivered to channels from the drainage basin. Hack (1957) showed that channel gradient is affected by lithology for basins of similar size. Steep gradients generally occur in areas of resistant bedrock, while gentle gradients dominate in regions of weaker rocks. High channel gradients are required to efficiently transport coarse bedload delivered to channels in areas of resistant bedrock; hence gradient should be an important channel factor influencing the geomorphic effects of large floods. The earlier discussion of channel response to the Hurricane Camille flood in Virginia demonstrated the importance of steep gradient permitting streams to deliver coarse material from debris flows on alluvial fans in the headwaters of basins to the downstream areas where catastrophic channel widening occurred. Channel widening in areas of steep gradients exceeded 100% at several sites (Johnson, 1983).

Inspection of the gradient values in Table 1 demonstrates the importance of this variable in affecting flood response in a wide variety of settings. Channel gradient was an order of magnitude higher on rivers that experienced extreme erosion and deposition during floods than on rivers where only minor changes occurred. Where extreme changes occurred, mean gradient was 0.016 (standard deviation = 0.017), while mean gradient for rivers that experienced little change was 0.003 (standard deviation = 0.003).

Moss and Kochel (1978) observed distinctly different responses to the Hurricane Agnes flood in the Conestoga River basin of Pennsylvania, largely depending on channel gradient and bedload. The headwaters of the Conestoga basin are underlain by resistant sandstones and metamorphic rocks, while downstream areas are underlain by weak carbonate rocks (Fig. 3). The Agnes flood caused significant bank erosion, channel scour, and floodplain alterations in headwater streams where coarse bedload was transported. In contrast, the remainder of the basin (over 80% of the drainage area) experienced little effect from the flood. Pre- and postflood channel cross sections at downstream gauging stations showed insignificant channel modification. In addition, floodplains were rarely scoured and received insignificant deposition even though 6 m of water occupied floodplains for over 24 hr (Moss and Kochel, 1978).

Downstream reaches of the river dominantly transported suspended sediment during the Agnes flood. Table 2 summarizes the suspended-sediment yield from 10 gauging stations on the Conestoga River and major tributaries during 1972, including the Agnes flood. The main station (station 4) shows over a million tons of suspended sediment were exported from the Conestoga River during 1972. Approximately one-third of the year's total sediment yield occurred during the 3-day period of Agnes flooding in June. If one defines geomorphic work in terms of flood effectiveness (like Wolman and Gerson, 1978), the Agnes flood had very little impact except in localized headwater areas. Even if one defines work in terms of suspended sediment transported, the Agnes flood only accomplished as much work as is normally done in over 180 days. Given the rarity of

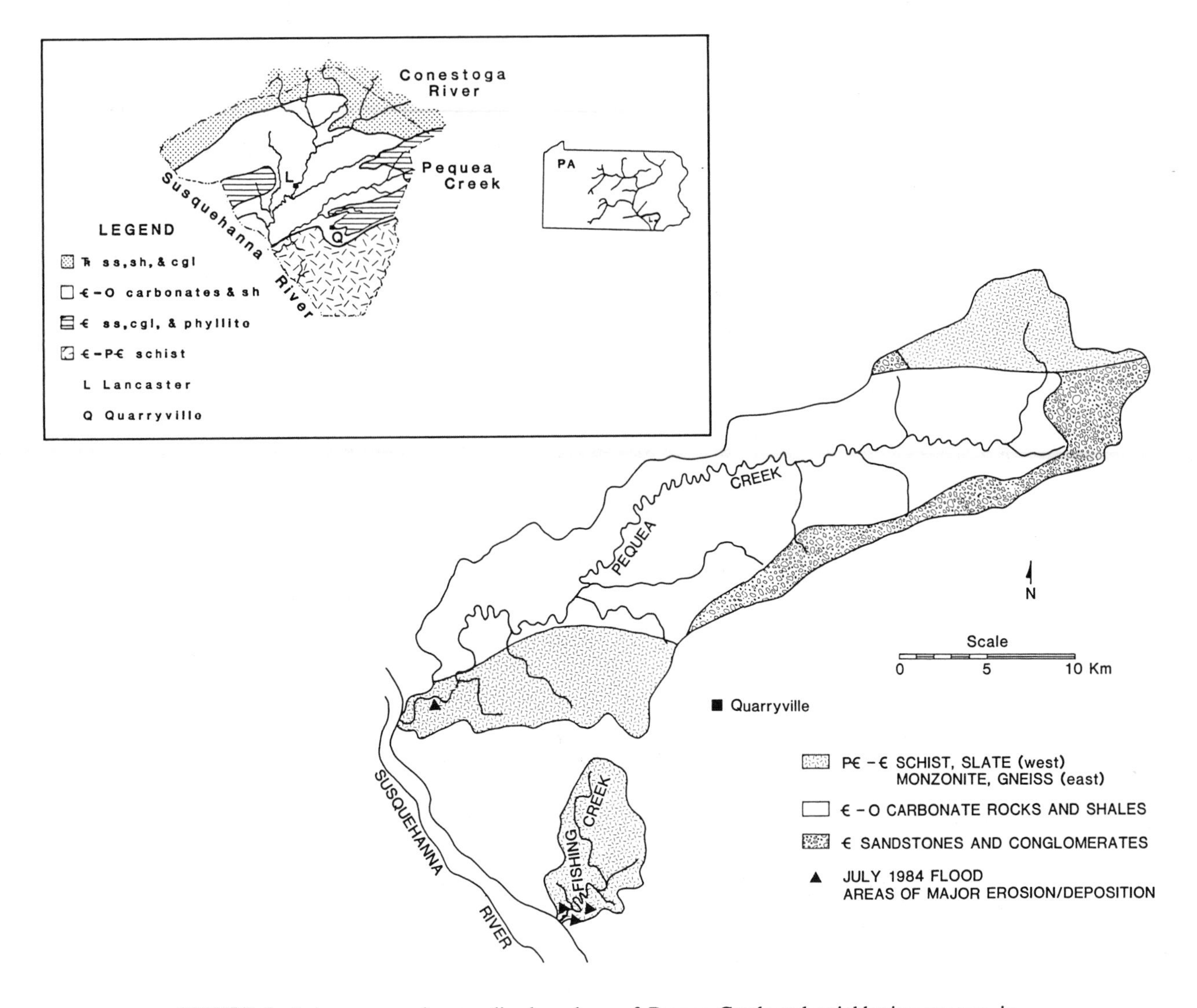

FIGURE 3. Index map and generalized geology of Pequea Creek and neighboring streams in Lancaster County, Pennsylvania. The inset shows the Conestoga River basin and generalized geology.

the Agnes flood, this amount of sediment transport is still insignificant.

Reconnaissance observations of a major flood on July 1, 1984, on a neighboring stream in Pennsylvania also underscore the importance of bedload and gradient in controlling channel response to flooding. An intense convective storm produced flood stages that were higher than the Agnes flood in 1972 along the lower Pequea Creek and adjacent smaller streams (Fig. 3). Inspection of the channel and floodplain less than a week after the flood revealed negligible evidence of channel widening and floodplain scour and deposition. A few isolated sites along the lower 10 km of the creek experienced some bank erosion and floodplain deposition. Coarse bedload is locally abundant in the channel along its lower reaches because that area is underlain by relatively resistant schists in contrast to the carbonate rocks that underlie most of the basin. In dramatic contrast, extensive channel scour, channel widening, and floodplain deposition occurred on a neighboring small tributary to the Susquehanna River called Fishing Creek (Fig.

TABLE 2 Summary of Total Suspended Sediment in the Conestoga River Basin, 1972

Stream	Drainage Area (mi^2)	Total Suspended Sediment (tons/yr)		Denudation ($tons/mi^2/yr$)		Rate in/1000 yr	
		+Agnes	−Agnes	+Agnes	−Agnes	+Agnes	−Agnes
Conestoga River	56.5	504,000	39,000	8,135	635	42.3	3.3
Muddy Creek	51.6	467,959	21,979	8,964	404	46.6	2.1
Cocalico Creek	140.0	490,033	225,033	3,500	1,607	18.2	8.4
Conestoga River	324.0	474,337	225,336	1,473	758	7.7	3.9
Mill Creek	56.0	442,293	110,203	7,772	1,937	40.4	10.1
Little Conestoga Creek	65.1	76,203	31,029	1,163	473	6.0	2.5
Hammer Creek	35.0	70,868	24,720	2,013	702	10.5	3.7
Middle Creek	32.1	66,592	31,073	2,024	968	10.5	5.0
Little Muddy Creek	15.3	51,565	40,603	3,144	2,475	16.3	12.9
W. Br. Little Conestoga Creek	11.5	6,900	1,195	543	95	2.8	0.5
Total Conestoga	477.0	1,092,832	729,904	2,291	1,530	12.8	8.5

+Agnes = values including Hurricane Agnes flood, June, 1972; −Agnes = values for the year without Agnes flood data.

3). Debris avalanches contributed abundant load of coarse sediment to the channel of Fishing Creek, whose entire basin is underlain by the more resistant schist.

The effects of the Hurricane Camille flood in central Virginia in 1969 (Williams and Guy, 1973) and a similar flood in western Virginia in 1949 (Hack and Goodlett, 1960) are in dramatic contrast with the observations made of the Agnes flood (Costa, 1974; Moss and Kochel, 1978), which occurred in a similar climatic regime. Thus, climate was not an important factor controlling the different responses of these major floods. The main differences between the two areas are in topography, bedload, and channel gradient. The Pennsylvania and Maryland floods occurred in moderate-relief areas of the Appalachian piedmont where channel gradients were relatively low and bedload was relatively fine grained (Table 1). In contrast, the areas affected by floods in Virginia occurred in the high-relief areas of the Blue Ridge and the Ridge and Valley provinces where channel gradient was an order of magnitude higher and bedload was dominated by coarse gravel. The Virginia areas were influenced by upstream debris avalanching that provided additional coarse debris that was transported on high gradients to downstream areas. These examples demonstrate the extreme variability that can occur in fluvial response to floods even within the same climatic region.

The Hurricane Juan flood of November 1985 in central Virginia also provides some interesting observations relating to gradient and bedload variations in the same climatic region. Juan resulted in the second highest discharge on the James River at Richmond since records began over 200 yr ago. An aerial and ground reconnaissance survey made in November 1985 showed that very insignificant changes resulted from this flood along the James river (Fig. 4) in the Virginia piedmont. Channel gradient is relatively low along this reach of the James River, and the bedload is a mixture of gravel and sand (Table 1). In sharp contrast Hurricane Juan flooding resulted in extensive channel scour, channel widening, and transport of large boulders along the flanks of the Blue Ridge Mountains and Massanutten Mountain in west-central Virginia (Fig. 5). Mountain streams there have steep channel gradients and an abundant supply of coarse gravel bedload (Table 1).

Many other cases where extreme geomorphic response to flooding was reported (Table 1) occurred in mountainous areas where rivers have steep channel gradient and abundant coarse bedload, regardless of climate and vegetation (Stewart and LaMarche, 1967; Scott and Gravlee, 1968; Baker, 1977, 1984; Grozier et al., 1976; Costa, 1978; Nolan and Marron, 1985).

Channel Geometry

The geometry of stream channels appears to play an important role in determining the geomorphic effect of large floods. Two major categories of rivers will be considered as end members of the fluvial spectrum in terms of flood processes. These are bedrock rivers and alluvial rivers (see Baker and Kochel, this volume; Brakenridge, this volume). Distinctly different flow processes usually result in these types of channel systems that can affect the kinds of sediment

FIGURE 4. Aerial view showing the absence of scour and deposition in the channel and floodplain of the James River near Scottsville, Virginia. Photograph was taken 2 weeks after the November 1985 flood.

transported and kinds of geomorphic work done during large floods.

Bedrock channels are common in uplifted plateau areas, especially in semi-arid and arid regions like central and west Texas (Baker, 1984; Baker and Kochel, this volume). Bedrock channels, like those in Texas, usually have deep, narrow cross sections and are characterized by extremely deep flows during floods. Increased flood discharges are accommodated by dramatic increases in depth and velocity. For example, the catastrophic flood on the lower Pecos River in west Texas in 1954 had an average flow depth of more than 25 m and flow velocities that averaged 12 m/s.

Deep, high-gradient flood flows in bedrock channels usually result in macroturbulent flow phenomena characterized by the birth and decay of vorticity around obstacles and along irregular channel boundaries (Baker, 1984). Matthes (1947) described macroturbulent phenomena that include kolks and other forms of secondary flow circulation. Baker (1978, 1984) described the intense upward vortex action developed when kolks are formed along irregular flow boundaries such as in the lee of bedrock protrusions, downstream of large boulders, and along irregular bedrock channel floors and walls. Because of macroturbulent phenomena, bedrock river floods can readily transport large boulders in suspension for considerable distance. Kolks and other macroturbulent flow phenomena such as cavitation can be particularly erosive and may result in significant erosion of channel walls and floors during a major flood. Baker (1984) and Baker and Kochel (this volume) describe the distinctive suite of gravel-dominated sedimentary deposits produced by these bedrock river floods.

Large volumes of coarse bedload are normally moved during bedrock river floods, and large-scale modification occurs on channel bars and floodplain areas. Examples of the dramatic effects of these bedrock channels were described by Baker (1977, 1978, 1984) and Baker and Kochel (this volume). In spite of the dramatic sediment redistribution that can occur in bedrock channels during major floods, the resistance of bedrock channel walls is generally sufficient to prevent significant change. Thus, the development of intensely energetic macroturbulent flow phenomena occurs in these narrow, deep bedrock channels during floods. If

FIGURE 5. Extensive erosion and deposition caused by the Hurricane Juan flood along the western flank of the Blue Ridge Mountains in the Shenandoah Valley, Virginia (photographs taken in November 1985). (*a*) Aerial view of coarse gravel deposited on a fan on Gap Run along the base of the Blue Ridge. (*b*) View of extensive channel scour and widening on Stony Run along the western flank to the Blue Ridge near Waynesboro. (*c*) Same site as (*b*), photographed in July 1985. Tractive force studies in progress at this and other similar sites along the Virginia Blue Ridge showed that boulders up to 1.5 m were moved (Simmons, 1988). These represent the largest boulders available in these channels.

such deep, high-velocity flood flows were to occur in alluvial river systems, channels would be enlarged by extensive channel widening and scour.

Baker (1977) discussed the relative effects of large-magnitude, infrequent floods and small-magnitude, frequent floods in bedrock channels in central Texas in terms of their ability to transport sediment and modify landforms. Because the sediment load of these bedrock rivers was dominated by coarse gravel, low flows failed to achieve the threshold competence level required to entrain the sediment. Significant geomorphic change occurred only when macroturbulent flow phenomena were established during rare, large-magnitude floods. Baker (1977) explained these observations by shifting the geomorphic work curve of Wolman and Miller (1960), developed primarily for alluvial channels, to reflect the higher competence levels required to entrain coarse sediments when applied to semi-arid, bedrock channels.

Once geomorphic changes occur in bedrock channels during large floods, subsequent floods of lower magnitude are unable to redistribute sediments and readily alter landforms. Therefore, recovery of the landscape to preflood conditions is exceedingly slow in semi-arid bedrock channels. In addition, the slow rate of vegetative re-establishment in arid regions prohibits rapid recovery (Wolman and Gerson, 1978). In contrast, alluvial rivers in humid regions are normally quick to recover from changes incurred during major floods. For example, localized channel scour and widening produced during Hurricane Agnes in the Appalachian piedmont was masked by recovery within a few weeks to a year (Costa, 1974; Gupta and Fox, 1974; Moss and Kochel, 1978).

TEMPORAL ORDERING OF GEOMORPHIC EVENTS

The discussion so far has focused on explaining how variations in the physical setting and dominant geomorphic processes produce differences in channel response to large floods. This section will address the effect of temporal variations in the occurrence of large floods on the same river. In addition, I will examine how varying responses can occur during floods having similar magnitude and frequency. The importance of temporal ordering of floods has only been addressed by a few studies of channel response to floods. Wolman and Gerson (1978) suggested that the time sequence of floods is an important factor in controlling the varying rates of recovery from flood-produced changes in different climates. They concluded that rivers in humid regions experienced relatively minor landform modifications during large floods and that alluvial channels recovered to preflood morphologies within a few weeks to a year or so. Local channel scour and widening in the Appalachian piedmont during Hurricane Agnes, for example, was masked by recovery within a few weeks to a year (Costa, 1974; Gupta and Fox, 1974; Moss and Kochel, 1978). At the opposite end of the spectrum, Wolman and Gerson (1978) discussed the dramatic changes experienced by arid channels during major floods and noted that recovery on these rivers required as long as hundreds of years. Their observations, however, compared flood effects on different rivers in varying climates, rather than varying response by the same river to different floods of large magnitude.

Kochel (1980) studied the effects of two catastrophic floods on channel morphology in the Pecos River of west Texas using aerial photographs (Fig. 6). Major floods occurred on the Pecos River in 1954 and 1974. Paleohydrologic studies have shown that the recurrence intervals for these floods are approximately 2000 yr and 500–800 yr, respectively (Kochel et al., 1982). The 1954 flood resulted in dramatic redistribution of channel gravel and erosion of fine-grained terraces along the channel margin (Figs. 6 and 7). Radiocarbon-dated material in the fine-grained terraces indicated that it had been at least 500 yr since the terraces were last eroded. Additional effects of the 1954 flood included erosion of extensive gravel fans deposited in the Pecos River channel at the mouths of tributaries

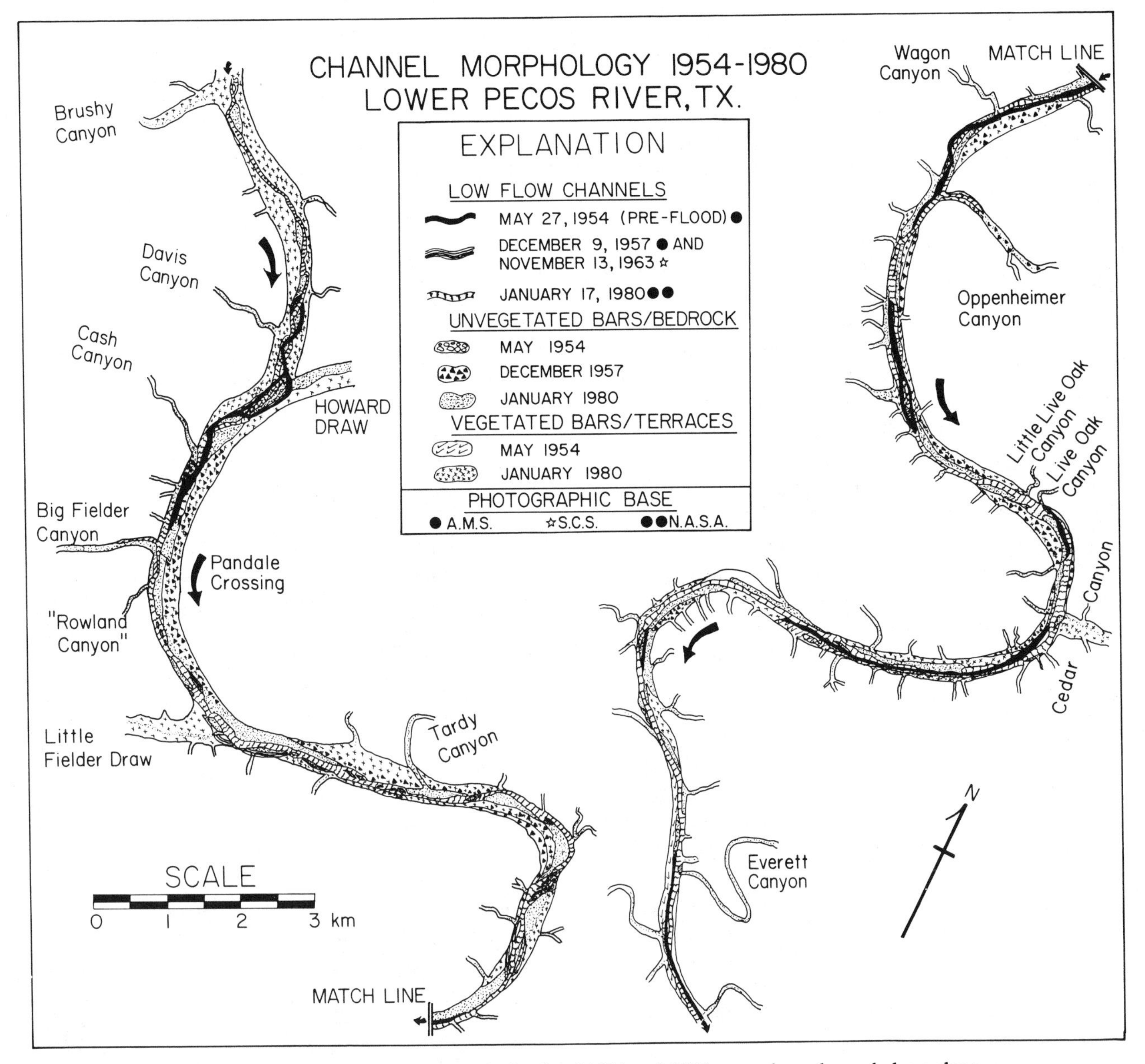

FIGURE 6. Effects of the catastrophic floods of 1954 and 1974 upon channel morphology along the lower Pecos River. Maps were made from aerial photographs.

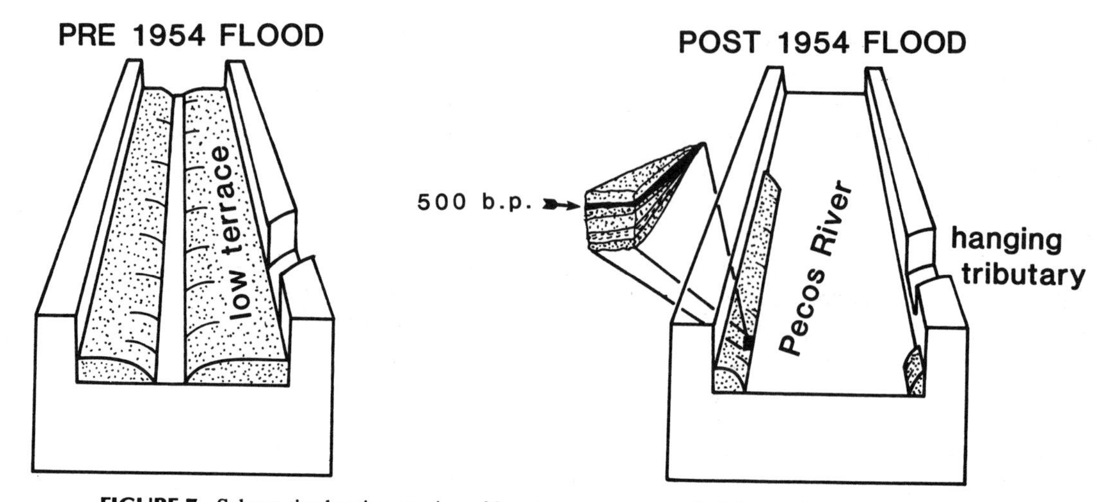

FIGURE 7. Schematic showing erosion of low terraces composed of fine-grained overbank sediments deposited during floods exceeding 100-yr recurrence interval stage. The 1954 flood eroded these terraces in many areas like the one depicted here, scouring the channel bedrock, and leaving some tributaries hanging above the Pecos River floor. Radiocarbon dates from fine-grained organic detritus in the terrace sediments suggest that it had been at least 500 yr since the last major episode of similar erosion had occurred.

during a long period of occasional tributary flood discharges into the mainstream.

In sharp contrast the effects of the 1974 flood on the Pecos River were minor. Only minor redistribution of channel bar gravel occurred in 1974. Apparently, recovery rates in this arid region were slow enough that the 1974 preflood channel morphology was still largely adjusted to the high flows of 1954. Insufficient time had elapsed in the 20 yr between the two floods for tributaries to debouch new sediment into the mainstream and for new overbank terrace sediments to accumulate so that they could be reworked by flood discharges in 1974. Clearly, significant adjustment of channel morphology to frequent, low discharges must be achieved before notable channel modification can occur during succeeding floods. Along the Pecos River, this recovery period seems to be at least 500 yr (Fig. 8).

Newson (1980) reported similar variability in flood response to successive floods in Wales. A large rainfall caused numerous mass movements in the Wye and Severn river basins in 1973 but resulted in little effect on downstream channel systems. In contrast, a similar rainfall in 1977 in the Severn watershed resulted in significant erosion and deposition in downstream channels while causing only one slope failure. Newson (1980) interpreted these observations as follows. The 1973 rainfall triggered extensive slope failures because slopes were primed for failure as a long period of time had elapsed since the last major rainfall. Hence, weathering processes probably caused the slopes to be near their threshold of failure, which was exceeded during the 1973 storm. When a similar rainfall event followed in 1977, there was little unstable material on the slopes because they were denuded only 4 yr earlier. However, the slope material eroded in 1973 had been delivered to downstream reaches where it could then be extensively reworked by the 1977 flood. Therefore, Newson's study illustrates yet another example of dramatically different response within the fluvial system to large floods due to the ordering of events. Kochel et al. (1987) reported a similar constrast in slope and channel response in Little River, Virginia, due to extreme floods in 1949 and 1985. Major rainfalls produce extensive slope failures only after a period of time has elapsed for new colluvial and soil material to accumulate on bedrock slopes again. This recovery time will vary depending on climate and lithology, but should be most rapid in humid climates where rates of soil formation and colluvial production would be highest. However, studies of repeated Holocene debris avalanching in central Virginia indicate that hundreds of years are probably required for slope recovery even in humid regions (Kochel, 1987).

Finally, Bevan (1981) addressed the issue of event ordering viewed in terms of a threshold model fitted to bank erosion data on the River Exe in England. His model combined observed and simulated data and showed that responses to floods of equal magnitude and frequency would vary greatly depending on the temporal ordering of events. The importance of temporal ordering of events should be considered in any analysis of geomorphic response to floods. This factor is likely to be exceedingly important for fluvial systems that are characterized by long recovery periods that may exceed the recurrence intervals of large floods

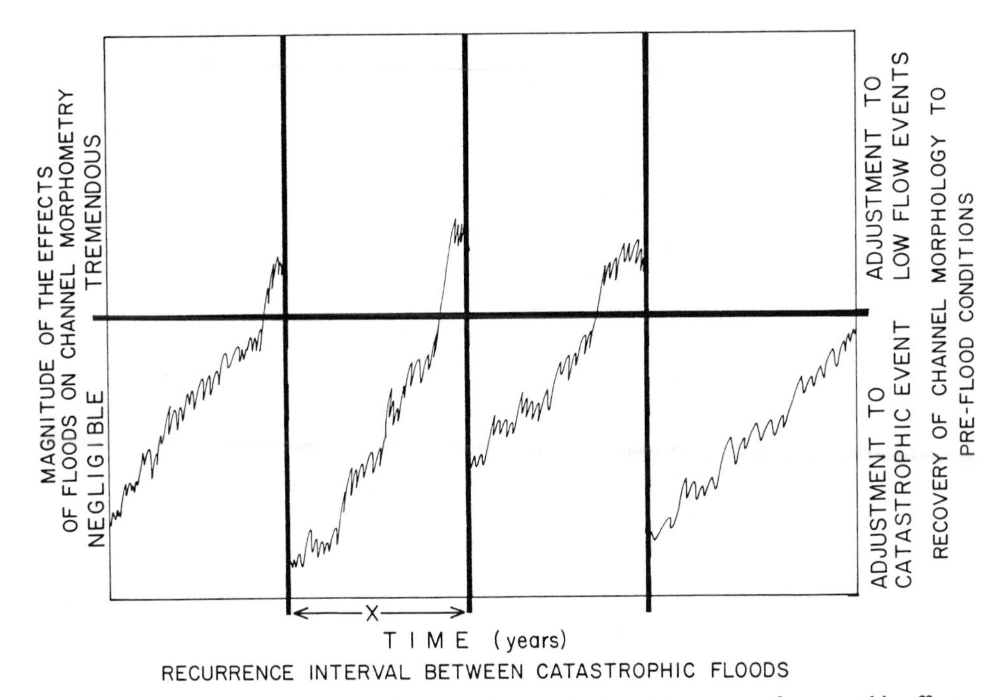

FIGURE 8. Theoretical relationship between the magnitude and frequency of geomorphic effects of large floods on stream channels and recovery time for channel morphology to readjust to low-flow conditions. Some time period must elapse between successive large floods before considerable changes will occur. Solid vertical bars represent major floods, separated by *X* years. After a major flood, channel and floodplain morphology is adjusted to high-flow configurations. The steady climb on the curves between floods represents slow modification of the channel to low-flow conditions, that is, formation of bars, overbank sediments, and tributary fans. If successive floods occur before channel morphology has recovered to low-flow conditions (which may require hundreds of years in arid areas), little change occurs from the succeeding flood. The length of the recovery period depends on climate and the frequency of low-magnitude floods. Along the Pecos River, this recovery time appears to be greater than 500 yr. When a subsequent flood occurs after recovery to low-flow morphology is complete, catastrophic changes are likely to occur as the morphology is suddenly forced to adjust to high flows.

and for geomorphic systems, in general, where change in response to rainfall is strongly influenced by intrinsic thresholds (Schumm, 1973), which depend on time.

CONCLUSION

Making predictions on the expected channel response of a particular flood are exceedingly difficult. The interdependency of variables in fluvial systems causes flood response to vary considerably among rivers in different climates. Even within the same climate, variability in river or basin characteristics negate generalizations about response of rivers to major floods. In addition, the complication of event ordering causes flood responses to vary for a given river. Furthermore, this discussion has purposely excluded the effects that human activities may have had on the system. It is likely that man's activities that affect basin variables (such as sediment load and discharge) as well as channel variables (such as changes in gradient and geometry) will influence river response to flooding. Ritter and Blakley (1986) discuss an example of how anthropogenic modifications have influenced flood response along the Gasconade River in Missouri.

Figure 9 attempts to summarize the trends that can be drawn from this analysis and detailed study of Table 1. Basin variables important in affecting fluvial geomorphic response to large floods include (1) climate, particularly the intensity–duration relationship for rainfall and the frequency of intense rainfall inputs into the system; (2) river hydrology, particularly factors that relate to the flashiness of the hydrograph such as the portion of the basin contributing runoff to the mainstream and the peak discharge; and (3) basin morphometry, particularly factors that affect the

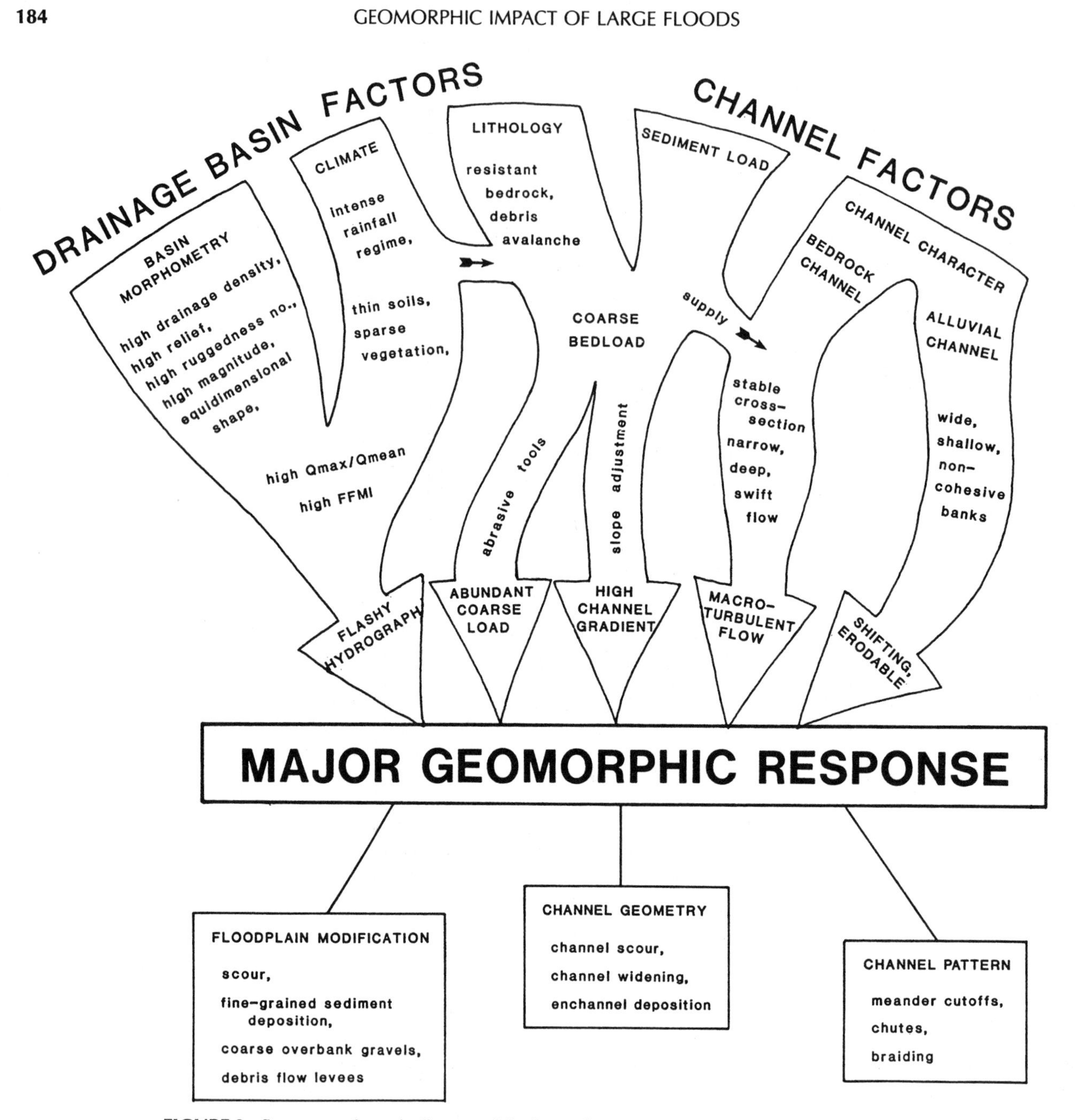

FIGURE 9. Summary schematic diagram of the factors important in controlling channel and floodplain response to large-magnitude floods.

flashiness of the mainstream hydrograph such as basin relief measures, drainage density, basin shape, vegetative cover, and soil cover.

Streams that experience major geomorphic changes during large floods appear to be characterized by (1) flashy hydrographs (recognized by having high FFMI, high Q_{max}/Q_{mean} ratio); (2) high channel gradient; (3) abundant coarse bedload; (4) relatively low bank cohesion; and (5) channel cross-sections that enable flood discharges to be accompanied by deep, high velocity flows where macroturbulent flow phenomena can be initiated and maintained. Flashy streams occur under many climatic regimes, but they are most common in semi-arid and arid regions where intense, short-duration, local rainfalls are common. Intense rainfalls are also common in mountainous areas and other regions where orographic influences predominate in humid climates. Within these climatic zones where catastrophic rainfalls are common, rivers exhibit particularly flashy characteristics where drainage basin morphometry increases the rate and quantity of runoff delivered to the main channel. Basin parameters likely to produce flashy conditions include (1) high relief, (2) high drainage density, (3) equant basin shape, (4) high first-order stream frequency, (5) high basin magnitude, (6) sparse vegetation, and (7) thin soils.

The factors coincident with streams showing dramatic response to floods but having less dependence on climate are (1) high channel gradient, (2) coarse bedload, and (3) channel geometry. These parameters are influenced by lithology and tectonics. Large quantities of coarse bedload are typically available to channels draining resistant lithologies. Therefore, gradients are normally higher because the rivers adjust their slope according to the load they are required to transport. High-gradient reaches can occur anywhere along a river profile, but they are most common near the headwaters. Therefore, during large floods on many streams, considerable geomorphic change occurs in headwater regions of high-gradient channels while minor response occurs downstream where gradient and bedload availability are lower in spite of higher discharge (e.g., Moss and Kochel, 1978). Regions where dramatic channel response should be expected from large floods would be in mountainous basins of low stream order and/or in regions where recent tectonism has created high-gradient channels. These conditions may be independent of climate, although there would be a greater tendency for the production of coarse sediment in arid climates because of the reduced rate of soil production compared to humid regions of similar lithology.

Dramatic channel response to large floods occurs primarily when peak flood velocity and depth exceed threshold values necessary for the development and maintenance of macroturbulence. These conditions result in exceedingly high values of unit stream power on the channel bed. Such conditions are most common in bedrock channels whose relatively stable cross sections allow increases in discharge to be accommodated by dramatic increases in flow depth and velocity, rather than increased channel width that might occur in alluvial rivers. Bedrock channels are not confined to a particular climatic regime but tend to be best developed in semi-arid to arid regions. Channels with this kind of geometry can better exceed competence levels required for the transportation of coarse bedload during major floods than ones that may adjust rapidly to increased discharges through channel widening.

I wish to close with a brief discussion of where I believe we should focus future research efforts on this topic and address a few of the problems that have plagued studies of the effects of floods of differing magnitude and frequency. First, we need to agree on how geomorphic work and/or response is to be defined. At present, we still maintain two different perceptions of geomorphic work, one measured as the amount of transported suspended load (Wolman and Miller, 1960) and another called geomorphic effectiveness, defined as the modification of landforms (Wolman and Gerson, 1978). This requires that each study must indicate how flood response is being defined. Alternatively, we could adopt a new definition, containing both views and discuss the effects of each flood from both perspectives. The approach of measuring stream power, as advocated by Baker and Costa (1987), would help quantify indices of flood effectiveness.

Second, we need to increase our efforts of determining the true recurrence interval of large floods so that they can more properly be placed into their relative position of magnitude and frequency. Recently developed paleohydrologic techniques discussed in Part IV "Paleofloods" of this volume provide methods of refining these estimates of both magnitude and frequency. This problem is especially important for considerations of the response of fluvial systems to large floods because large floods often have recurrence intervals too long to be accurately assessed using standard hydrologic engineering practices.

Finally, the complex interaction of variables important in controlling the channel response to large floods suggests that a systematic multivariate approach may yield useful information. In particular, future reports of flood effects should attempt to be as comprehensive as possible and provide the kinds of data summarized in Table 1. When a larger number of cases are documented, multivariate statistical techniques such as stepwise multiple regression and principal components analysis can be applied to this question.

We shall probably gain the best insights into understanding the factors controlling flood response from comparing numerous studies where certain variables can be held constant in the analysis. For example, systematic observations along a river that exhibits different gradient, sediment load, or channel geometry along its length would permit isolation of the effects of individual variables on geomorphic response to a flood. Establishing a checklist of parameters to be measured after a flood, similar to those in Table 1, would be an important step in providing the data base necessary for quantitative analysis of this problem.

ACKNOWLEDGMENTS

This research was partially supported by the Department of Geology and Office of Research Development Administration, Southern Illinois University. The research in west Texas was supported by the National Science Foundation, grant EAR 77-23025. I thank Andy Mason, DeAnn Kirk, Sara Zimmerman, and Betty Atwood for help with figures, tables, and word processing. I also thank

Dale F. Ritter, Thomas W. Gardner, and Peter C. Patton for their constructive reviews of this manuscript.

REFERENCES

Baker, V. R. (1977). Stream-channel response to floods with examples from central Texas. *Geol. Soc. Am. Bull.* **88**, 1057–1071.

Baker, V. R. (1978). Large-scale erosional and depositional features of the Channeled Scabland. *In* "The Channeled Scabland" (V. R. Baker and D. Nummedal, eds.), pp. 81–115. Nat. Aeronaut. Space Admin., Washington, D.C.

Baker, V. R. (1984). Flood sedimentation in bedrock fluvial systems. *Mem.—Can. Soc. Pet. Geol.* **10**, 87–98.

Baker, V. R., and Costa, J. E. (1987). Flood power. *In* "Catastrophic Flooding" (L. Mayer and D. Nash, eds.), pp. 1–22. Allen and Unwin, London.

Beard, L. R. (1975). Generalized evaluation of flash-flood potential. *Tech. Rep.—Univ. Tex. Austin, Cent. Res. Water Resour.* **CRWR-124**, 1–27.

Bevan, K. (1981). The effect of ordering on the geomorphic effectiveness of hydrologic events. *IAHS-IASH* **132**.

Caracena, F., and Fritsch, J. M. (1983). Focusing mechanisms in the Texas Hill Country flash floods of 1978. *Mon. Weather Rev.* **111**, 2319–2332.

Costa, J. E. (1974). Response and recovery of a Piedmont watershed from tropical storm Agnes, June 1972. *Water Resour. Res.* **10**, 106–112.

Costa, J. E. (1978). Colorado Big Thompson flood: Geologic evidence of a rare hydrologic event. *Geology* **6**, 617–620.

Gregory, K. J., and Walling, D. E. (1973). "Drainage Basin Form and Process." Wiley, New York.

Grozier, R. U., McCain, J. F., Lang, L. F., and Merriman, D. C. (1976). "The Big Thompson River flood of July 31-August 1, 1976, Larimer County, Colorado." Water Conservation Board, Denver, Colorado.

Gupta, A. (1975). Stream characteristics in Eastern Jamaica, an environment of seasonal flow and large floods. *Am. J. Sci.* **275**, 825–847.

Gupta, A. (1983). High magnitude floods and stream channel response. *Spec. Publ. Int. Assoc. Sedimentol.* **6**, 219–227.

Gupta, A., and Fox, H. (1974). Effects of high-magnitude floods on channel form: A case study in the Maryland piedmont. *Water Resour. Res.* **10**, 499–509.

Gustavson, T. C. (1978). Bedforms and stratification types of modern gravel meander lobes, Nueces River, Texas. *Sedimentology* **25**, 401–426.

Hack, J. T. (1957). Studies of longitudinal stream profiles in Virginia and Maryland. *Geol. Surv. Prof. Pap. (U.S.)* **294-B**, 45–97.

Hack, J. T., and Goodlett, J. C. (1960). Geomorphology and forest ecology of a mountain region in the central Appalachians. *Geol. Surv. Prof. Pap. (U.S.)* **347**, 1–66.

Johnson, R. A. (1983). Stream channel response to extreme rainfall events: The Hurricane Camille storm in central Nelson County, Virginia. M.S. Thesis, University of Virginia, Charlottesville.

Kochel, R. C. (1980). Investigation of flood paleohydrology using slackwater deposits, lower Pecos and Devils Rivers, southwest Texas. Ph.D. Dissertation, University of Texas, Austin.

Kochel, R. C. (1987). Holocene debris flows in central Virginia. *In* (J. E. Costa and G. F. Wieczorek, eds.),"Debris Flows/Avalanches: Process, Recognition, and Mitigation," pp. 139–155. Rev. Eng. Geol. vol. VII, Geol. Soc. Am., Boulder, Colorado.

Kochel, R. C., and Baker, V. R. (1982). Paleoflood hydrology. *Science* **215**, 353–361.

Kochel, R. C., and Johnson, R. A. (1984). Geomorphology and sedimentology of humid temperate alluvial fans, central Virginia. *Mem.—Can. Soc. Pet. Geol.* **10**, 109–122.

Kochel, R. C., Baker, V. R., and Patton, P. C. (1982). Paleohydrology of southwest Texas. *Water Resour. Res.* **18**, 1165–1183.

Kochel, R. C., Ritter, D. F., and Miller, J. (1987). Role of tree dams on the construction of pseudo-terraces and variable geomorphic response to floods in the Little River valley, Virginia. *Geology* **15**, 718–721.

Langbein, W. B., and Schumm, S. A. (1958). Yield of sediment in relation to mean annual precipitation. *Trans., Am. Geophys. Union* **39**, 1076–1084.

Matthes, G. H. (1947). Macroturbulence in natural stream flow. *Trans., Am. Geophys. Union* **28**, 255–262.

Moss, J. H., and Kochel, R. C. (1978). Unexpected geomorphic effects of the Hurricane Agnes storm and flood, Conestoga drainage basin, south-eastern Pennsylvania. *J. Geol.* **86**, 1–11.

National Research Council (1982). "The Austin, Texas Flood of May 24-25, 1981," Report of Committee on Natural Disasters. Natl. Res. Counc., Washington, D.C.

National Research Council (1984). "The Tucson, Arizona Flood of October 1983," Report of Committee on Natural Disasters. Natl. Res. Counc., Washington, D.C.

Newson, M. (1980). The geomorphological effectiveness of floods—a contribution stimulated by two recent events in mid-Wales. *Earth Surf. Processes* **5**, 1–16.

Nolan, K. M., and Marron, D. C. (1985). Contrast in stream channel response to major storms in two mountainous areas of California. *Geology* **13**, 135–138.

Patton, P. C., and Baker, V. R. (1976). Morphometry and floods in similar drainage basins subject to diverse hydrogeomorphic controls. *Water Resourc. Res.* **12**, 941–952.

Patton, P. C., and Baker, V. R. (1977). Geomorphic response of central Texas stream channels to catastrophic rainfall and runoff. *In* "Geomorphology in Arid Regions" (D. O. Doehring, ed.), pp. 189–217. Allen & Unwin, Winchester, Massachusetts.

Ritter, D. F. (1975). Stratigraphic implications of coarse-grained gravel deposited as overbank sediment, southern Illinois. *J. Geol.* **83**, 645–650.

Ritter, D. F., and Blakley, D. S. (1986). Localized catastrophic disruption of the Gasconade River floodplain during the December, 1982 flood, southwest Missouri. *Geology* **14**, 472–476.

Schumm, S. A. (1973). Geomorphic thresholds and complex response of drainage systems. *In* "Fluvial Geomorphology" (M. Morisawa, ed.), Publ. Geomorphol. pp. 299–310. State University of New York, Binghamton.

Schumm, S. A., and Lichty, R. W. (1963). Channel widening and floodplain construction along Cimarron River in southwestern Kansas. *Geol. Surv. Prof. Pap. (U.S.)* **352-D**, 71–88.

Scott, K. M., and Gravlee, G. C., Jr. (1968). Flood surge on the Rubicon River, California—Hydrology, hydraulics and boulder transport. *Geol. Surv. Prof. Pap. (U.S.)* **422-M**, 1–38.

Simmons, D. W. (1988). Sedimentology and geomorphology of humid-temperate alluvial fans along the west flank of the Blue Ridge Mountains, Shenandoah Valley, Virginia. M.S. Thesis, Southern Illinois University, Carbondale.

Stewart, J. H., and LaMarche, V. C. (1967). Erosion and deposition produced by the flood of December, 1964, on Coffee Creek, Trinity County, California. *Geol. Surv. Prof. Pap. (U.S.)* **422-K**, 1–22.

Williams, G. P., and Guy, H. P. (1973). Erosional and depositional aspects of Hurricane Camille, in Virginia, 1969. *Geol. Surv. Prof. Pap. (U.S.)* **804**, 1–80.

Wolman, M. G., and Gerson, R. (1978). Relative scales of time and effectiveness of climate in watershed geomorphology. *Earth Surf. Processes* **3**, 189–203.

Wolman, M. G., and Miller, J. P. (1960). Magnitude and frequency of forces in geomorphic processes. *J. Geol.* **68**, 54–74.

12

HYDROLOGIC ASPECTS OF FLOODS IN EXTREME ARID ENVIRONMENTS

ASHER P. SCHICK
Hebrew University, Jerusalem, Israel

INTRODUCTION

In contrast to the North American continent, which has only a few spots with mean annual precipitation less than 50 mm, other parts of the globe are considerably better endowed with extreme arid areas so defined. Africa possesses by far the largest share, with a continuous strip between latitudes 20° and 30° North, that is, an area of roughly 5.5×10^6 km^2 (Dubief, 1953a). Within this vast expanse of the Sahara Desert, even the high mountains of Tibesti, Ahaggar, and Sinai barely pierce, orographically, the isohyet of 50 mm (Fig. 1). On the other hand, about one-half of it is designated as having less than 5–10 mm per year.

Well over one-half of the extreme arid area of the globe is not covered by sand dunes. Being barren of vegetation, it is, therefore, susceptible to runoff generation and, given sufficient rainfall intensities, to flooding. Over most of the extreme arid areas, fluvial forms are indelibly imprinted on the landscape. Contrary to widespread beliefs in the last century but long refuted since, most of these forms are the product of contemporary processes or, at least, are sustained by contemporary processes. Some of the present-day events, witnessed by experienced observers, had a traumatic influence on them because of their abruptness and sheer force. Vivid descriptions of these events abound, but the body of scientific studies on them is still meager.

Are the floods of the extreme desert basically different from those in somewhat wetter arid areas? Or is it only a matter of degree, with basically identical processes operating and landforms resulting? This question is as yet unanswerable; too few data, long periods of idle waiting, too little applicative interest, and logistical difficulties are some of the reasons. In recent years the rapid development of extreme arid areas in some of the oil countries has confronted the planners, for the first time on a grand scale, with catastrophic floods and their geomorphic implications. However, it will take some years, at least, until this information and the studies that emanate from it will be incorporated into science. In the following discussion of floods in extreme arid environments, we are limited to the results of studies done in the southern Negev and in the Sinai.

RAINFALL

Although extremely high point rainfalls have been reported—for example, 142 mm in 24 hr measured at Themed, Sinai (mean annual rainfall 30 mm) (Ashbel, 1951, p. 124)—they are exceedingly rare and reports on them must be regarded with caution (Dubief, 1953a). A 35-yr composite annual rainfall series for Elat, Israel, mean annual rainfall 25.3 mm, included 5 yr with an annual amount in excess of double the mean, 1 yr in excess of three times the mean, and 9 yr with an annual amount less than one-quarter of the mean. Most of the 50.5 mm that fell within the rainiest consecutive period of 24 hr, was highly concentrated in time, and amounted to nearly double the annual mean (Schick, 1971a, pp. 131–133). One year in the series was absolutely rainless and another had only 1.9 mm of total rainfall.

This immense temporal variability seems to be replicated in other hyperarid areas of the world as well. The unique hydrometeorological information assembled in North Africa during the first half of this century still provides the most instructive areal data bank for hyperarid environments. At

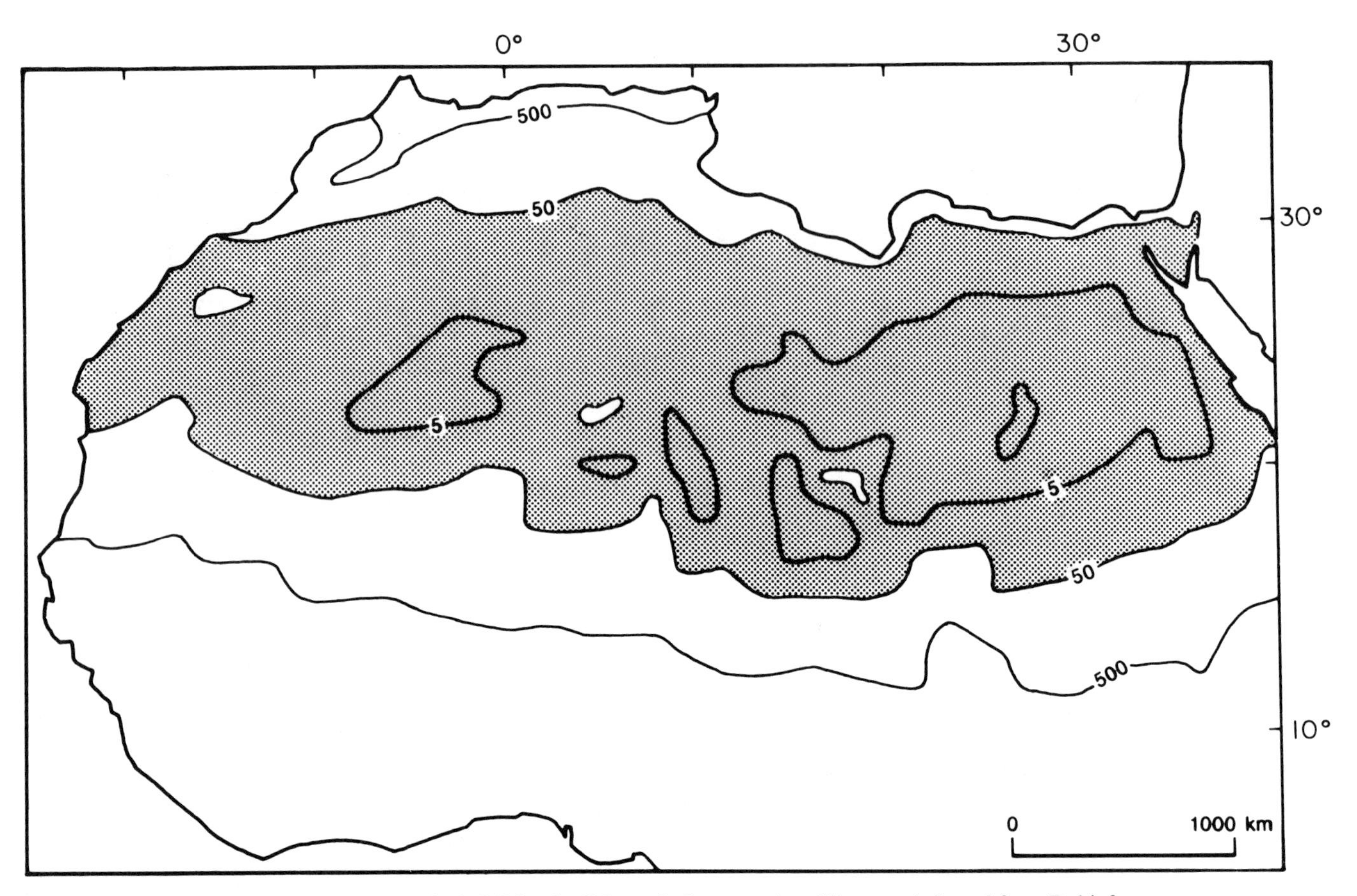

FIGURE 1. Mean annual rainfall for the Sahara. Isohyets are in millimeters (adapted from Dubief, 1953a).

Adrar, Central Sahara (mean annual rainfall 12.7 mm), over a period of 40 yr, 3 yr had rainfall in excess of four times the mean (Dubief, 1953a, p. 32). Table 1 compares the mean annual rainfall of several hyperarid Saharan stations, most of which have been in operation for a period of 20–25 yr, with their 24-hr rainfall maximum. In seven out of nine cases, this 24-hr amount exceeds the mean annual value; in four cases the difference is more than twofold. A few of the largest annual totals seem to indicate frequencies of 50–100 yr, but the bulk of Saharan data show that, for extreme hyperarid areas, a 24-hr rainfall can be expected to exceed the mean annual rainfall by a factor of 3 to 4 once in a 20 to 30-yr period (Fig. 2). During a period of similar duration, nearly all such areas will experience at least one completely rainless year. In regions even deeper inside the extreme arid zone, periods of several rainless years in a row have been reported. The hyperarid core of the Atacama Desert is presumed to have some areas in which rain has never been recorded.

The erratic character of the time series of point rainfall in extremely arid areas finds its counterpart in the areal patterns for a given event. In a climatically uniform area in southern Israel, measuring 80 km^2 and characterized by a mean annual rainfall of 30–35 mm, an extraordinarily large spatial variability of short-term rainfall has been documented (Sharon, 1972). For example, out of point rainfall amounts recorded on 21 separate days at three stations within 15 km of each other, only 4 days were of the same order of magnitude. For 11 of the rain days the ratio between the highest and least receiving station exceeded 20:1. "Spotty" rainfall in this area accounts for 60% of the total. The proportion may be higher when only hydrologically significant rainfall is considered, and seems to characterize extreme arid areas elsewhere (Sharon and Kutiel, 1986).

Intensities of what are popularly described as desert cloudbursts can be very high. In Tamanrasset in the Ahaggar Mountains, Sahara, these come to around 1 mm/min, and if partial periods are considered, approach in extreme cases 2 mm/min (Table 2). In typical cloudbursts of the extreme desert the transition between total dryness and full-blast rain is near instantaneous, with the first few minutes of the rain event recording intensities over 1 mm/min. A typical example is a rainfall event in Nahal Yael, southern Israel, which registered a sharp instantaneous rise in its

TABLE 1 Saharan Rainfall: Mean Annual and Maximum 24-hr Intensity

Station	Mean Annual Rainfall (mm)	Maximum Recorded 24-hr Intensity (mm/day)
Adrar	12.7	32.7
Beni Abbès	32.1	38.5
Djanet	19.1	42.4
El Goléa	38.7	34.4
Ft. Flattres	18.9	48.0
Ft. de Polignac	16.8	29.7
In Salah	14.1	31.0
Ouargla	40.0	26.1
Tamanrasset	42.6	47.4

SOURCE: Dubief, 1953a.

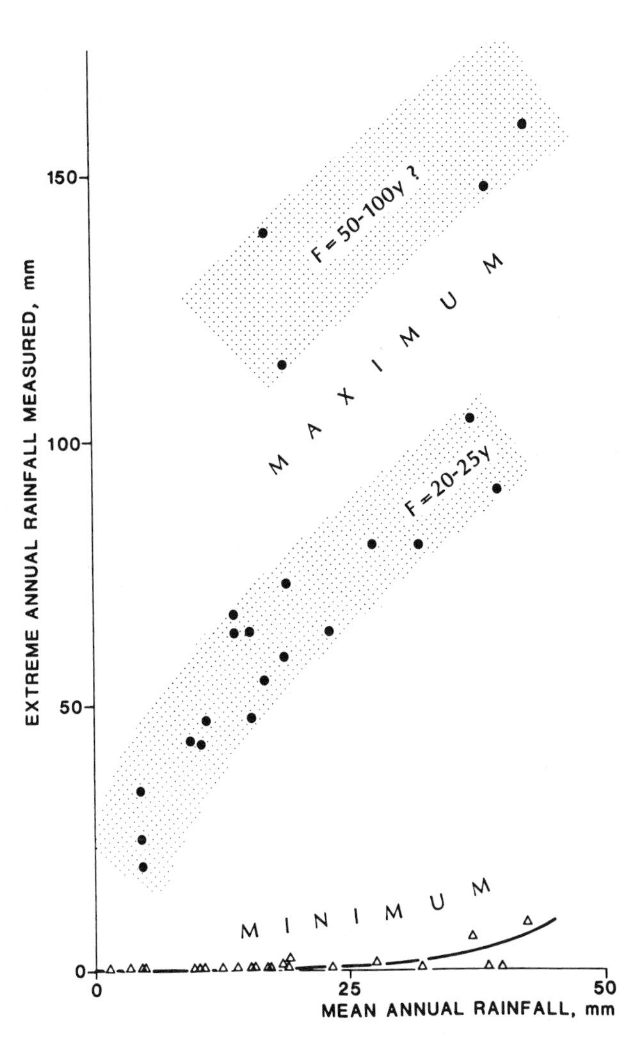

FIGURE 2. Extreme annual rainfall in the extremely arid parts of the Sahara Desert (mean annual rainfall <50 mm) for stations with a record of 20 yr or more. Broad dotted bands indicate possible rainfall frequencies. Data from Dubief (1953a, pp. 55–59).

TABLE 2 High-Intensity Rainfall at Tamanrasset

Date	Rainfall Amount (mm)	Rain Duration (min)	Intensity (mm/min)
4/15/48	12.3	10	1.23
9/13/50	37.0	40	0.92
5/22/33	20.0	30	0.66
5/24/33	47.1	85	0.55[a]
08/05/32	12.5	45	0.28[b]

SOURCE: Dubief, 1953a, p. 204.

[a] Nearly all the rain fell in 20 min, an intensity of nearly 2 mm/min.

[b] This duration includes spells of light rain that preceded and followed the cloudburst.

rain gauges (Fig. 3A). Each rainfall record has a tail of a low-intensity after-rain, while the pre-rain, though discernible in both, was insignificant in amount as well as in duration. Over a period of 17 yr at Nahal Yael (1968–1985), intensities exceeding 14 mm/hr accounted for nearly one-half of the total rain—223.4 out of 448.8 mm. Out of this intense rain (>14 mm/hr), 37% fell in intensities exceeding 2 mm/min (Fig. 4). The highest intensity recorded was 200 mm/hr but accounted for only 0.8% of the total (1.7 mm) (Greenbaum, 1986).

In the Mount Sodom area near the Dead Sea, where the mean annual rainfall is 50 mm, the maximum intensities measured are 50 mm in 30 min. For small watersheds an intensity of 30 mm in 10 min has been used by Gerson (1972) as a base for estimating the maximum probable flood in the region at 40 $m^3/s/km^2$.

Most of the excessive intensities in extreme arid environments seem to be associated with relatively high temperatures, and are, therefore, the result of convective processes. Averaged for five stations in the Negev Desert, 65% of the high-intensity rains (>30 mm/hr) occur between

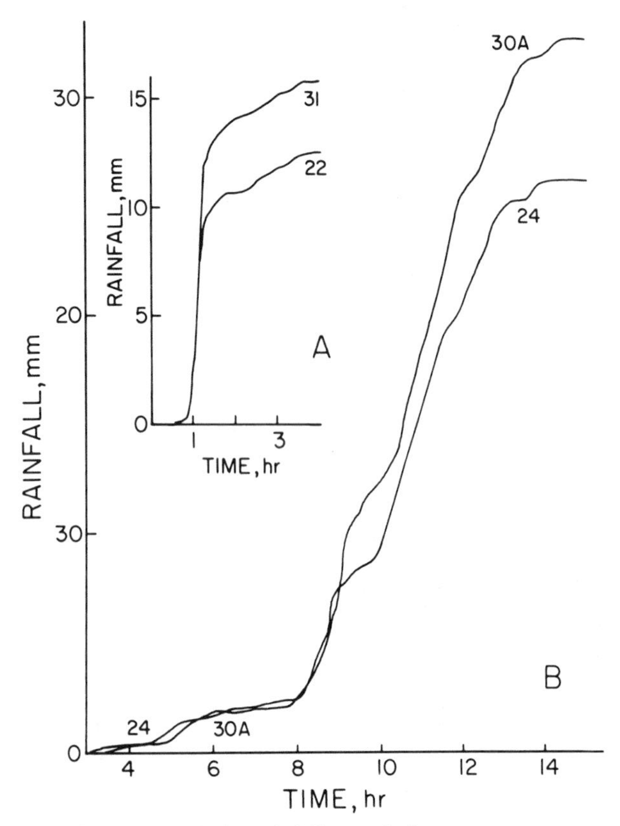

FIGURE 3. Cumulative rainfall records for two storm events at Nahal Yael, Israel. (*A*) Event 12A, 2/20/1975; (*B*) event 4, 5/24/1968. Stations 22 and 24 are on mountaintop sites, elevation about 300 m above sea level. Stations 30A and 31 are located in the valley, roughly 100 m lower. All four stations are within several hundred meters of each other.

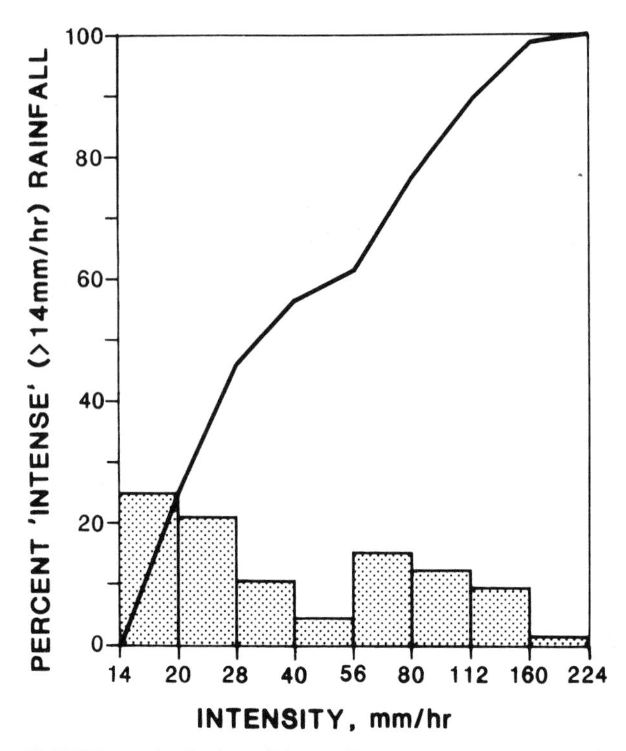

FIGURE 4. Distribution of "intense" (>14 mm/hr, time resolution 0.05 hr) rainfall, by intensity, Nahal Yael, Israel, 1968–1985 (after Greenbaum, 1986).

12 and 21 hr local time, when surface temperature is high, while less than 15% occur in an identical span of 9 hr in the early morning, 00–09 hr local time (Sharon and Kutiel, 1986). These convective storms are not randomly scattered in space but tend to form at preferred distances from each other. In a study in the extremely arid Namib Desert these preferred distances were found to be around 40–50 km and 80–100 km (Sharon, 1981). The rain front is often very sharply defined both in the direction of cell movement as well as laterally. In one fortuitous case where the frontal advance was monitored with the aid of telemetry, the front of runoff-generating rain was found to have proceeded over the 2-km-long Nahal Yael watershed in 18 min (Poráth and Schick, 1974). In other cases the velocity of rain cells in the extreme desert has been found to vary from near zero (stationary cloudburst) to several tens of kilometers per hour (Sharon, 1972). In moving cells lateral boundaries tend to be sharp as well. In the Ma'an flood of 1966 (Fig. 5), event rainfall amounts as recorded near the crest of the 400-km-long storm trajectory halved themselves laterally every 5–15 km (Schick, 1971a). Widespread rains, covering

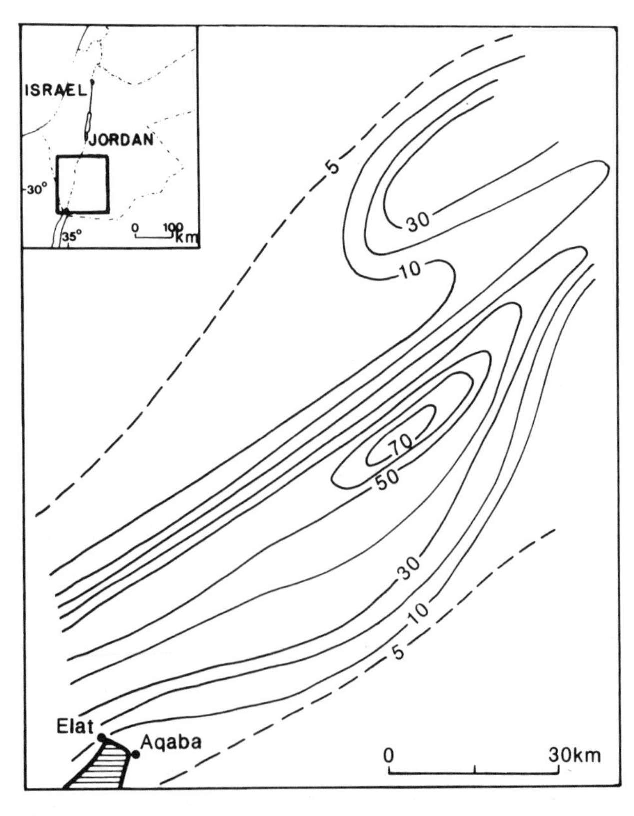

FIGURE 5. Rainfall pattern of the storm of 3/11/1966 (the Ma'an flood) in southern Israel and southern Jordan (after Schick, 1971a).

vast desert areas with lower intensity but relatively high-quantity rains, such as the El Arish flood in 1975, also exist. During the El Arish flood event several parts of the Sinai peninsula and the southern Negev recorded 60–80 mm in 72 hr. This event was, meteorologically, a rare combination of a Red Sea low-pressure trough and an extreme southerly deviation in the trajectory of a Mediterranean frontal system (Levi, 1975).

FLOOD GENERATION

Obviously, storm rainfall such as is characteristic of desert cloudbursts, if converted entirely into runoff, would create floods of enormous magnitudes. Although complete abstraction by infiltration of any conceivable peak rainfall intensity is virtually certain in all sand deserts, this geomorphic land surface type covers only a minority of the global extreme arid terrains, for example, 15% of the Sahara (Mabbutt, 1977). On the other hand bare rock makes up far more of the surface composition of the extreme desert, and its surface water yield in response to rainfall approaches, in some cases, very high values. The detailed mosaic of terrain surface properties is thus crucial to the evaluation of the flood potential of extremely arid watersheds.

The response of nearly all terrains is markedly affected by rainfall intensity to which especially surfaces with intermediate infiltration characteristics show great sensitivity. Other terrain types, such as the *Schaumboden*—a clayey, vesicular layer found around and under detrital coarse-grained alluvial surfaces (Jaekel, 1980)—conceivably go through an abrupt threshold from total absorption to nearly complete runoff when a narrowly definable rainfall amount is exceeded.

Another major factor to be considered in the context of flood generation are the transmission losses. Evaporation in extreme deserts may be around 10 mm per day as indicated, for example, by characteristic values for Elat, southern Israel, varying from 4.7 to 16.0 mm/day for class A pans (Av-Ron, 1970) and from 110 to 330 mm/month for large water surfaces (Atlas of Israel, 1956, Maps IV/6/21-23). For flood trajectories on the order of up to several tens of kilometers, that is, less than one day in total travel time, this factor may be negligible. But for larger and longer runoff events evaporation may be important. Evaporation also minimizes the effect of antecedent moisture, an important factor in many non-arid-region floods. With rainfall events broadly separated in time in the hyperarid environment, it is virtually guaranteed that there is no significant antecedent moisture. Similarly, though exceptions may exist, the coarse channel alluvium has a large available volume for floodwater infiltration practically at all times. This situation takes away much of the pique inherent in the runoff generation controversy—Hortonian overland flow versus variable source contribution—so dominant in nonarid and less arid environments: The proximity of easily saturated areas close to channels does not ensure, in itself, subsequent surface flow through the network.

A series of point infiltration tests were conducted in and around the Nahal Yael watershed in southern Israel (mean annual rainfall 30 mm). The tests used a portable rainulator that effectively covered an area of 0.25 m^2 and produced intensities of 30–35 mm/hr (Salmon and Schick, 1980). These intensities are median to "intense" rains (Fig. 4) and represent the boundary between 75% of the total rainfall, which occurs at lower intensities, and 25% of it, which falls at higher intensities. The infiltration experiments were conducted on different surfaces with different slopes. On coarse-grained, sparsely jointed granite, on a hillslope of 34°, runoff was initiated 72 s after the onset of rainfall. A total input of 1.05 mm of rainfall was needed to stabilize the water losses, primarily due to infiltration, at a rate of 2.8 mm/hr (Fig. 6). On colluvium, in this case underlain by schist and dykes, with a 95% cover of angular stones size 10–40 mm, runoff started only after 340 s. Here, a total input of 6.4 mm was needed before stabilization of infiltration occurred, at a rate exactly 10 times over the rate for the granite miniplot—10.5 mm/hr (Fig. 6) (Salmon and Schick, 1980, pp. 64, 68).

In a recent and more detailed series of tests, using the same method but at intensities ranging from 67.2 to 90.0 mm/hr, terminal runoff–rainfall ratio on three types of magmatic and metamorphic rocks, varying in slope from

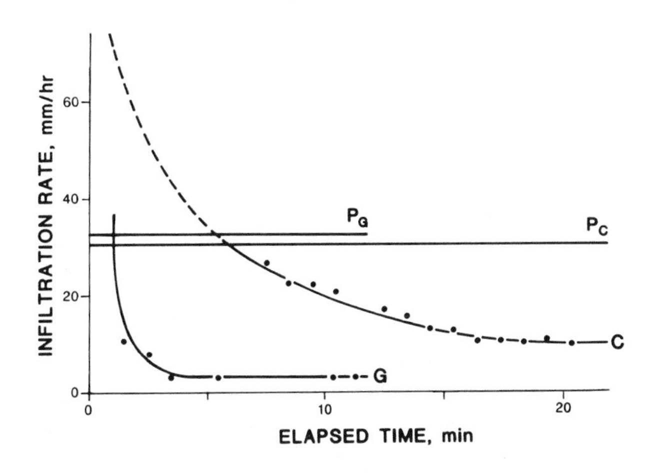

FIGURE 6. Simulated rainfall infiltration tests on small plots underlain by granite (G) and colluvium (C) at Nahal Yael, Israel. The constant rate and duration of the precipitation applied is indicated by horizontal lines marked PG and PC (from Salmon and Schick, 1980, p. 65).

10° to 35° and in degree of jointing from 0.05 to 0.78 joints per linear centimeter, ranged between 95.2 and 98.5% (Greenbaum, 1986). For colluvial surfaces the runoff–rainfall ratio at these high intensities was found to be 60–87%. A large variation in the hydrologic response was found for alluvial terrains. Old, Pleistocene surfaces characterised by a 95% cover of desert pavement, typical stone size 25 mm, and silt-clay content of 60% in the A horizon, revealed a pattern of infiltration losses not unlike certain rock surfaces: a terminal infiltration rate of 4.8 mm/hr and a terminal runoff–rainfall ratio of 93.8%. Younger, Holocene alluvial surfaces yield values of 20–55 mm/hr and rates of 75–40%, respectively. On recent alluvium, size 10–15 mm, slope 3°, silt-clay content of top layer about 10%, 12 min of simulated rainfall applied at an intensity of 87.4 mm/hr failed to produce any surface flow (Greenbaum, 1986).

Simulated rainfall experiments conducted on larger (70–140 m^2) talus-mantled slopes in the southeastern Sinai (mean annual rainfall 30–40 mm), with initially high (43–73 mm/hr) but gradually diminishing intensities, yielded minimum infiltration rates ranging from 8.4 mm/hr for a densely jointed, step-profiled schist and granite slope to 56 mm/hr for a coarse sandstone slope with small, patchy bedrock outcrops (Yair and Lavee, 1976). Initial losses varied from 0.95 to 9.9 mm, and the overall runoff–rainfall ratios ranged from 19.5–40.9% for plots underlain by schist and granite to 0.0–17.4% for the sandstone plot. Interpreted as compounding the effect of contributing and noncontributing areas, resulting from the fine morphologic differentiation in each plot, including gullies, the actual area contributing to the runoff is shown to range from roughly one-third to one-half of the total area (Yair and Lavee, 1974).

When the results of the miniplots are compared with those from slope areas that are two orders of magnitude larger, substantial losses due to overland flow conveyance are evident. These losses increase at an even greater rate at the slope–channel transition. For example, Nahal Yael Station 05 gauges an upstream, rocky channel that drains a 50,000-m^2 mountainous, isometric catchment from which a total runoff of 3130 m^3 was produced during 8 yr (1966/67–1973/74). The aggregate rainfall during the same period was 12,100 m^3, indicating a runoff–rainfall ratio of 26% (Schick, 1977). This average ratio is composed of several events as well as nonevents, which recorded some rainfall but no runoff. The actual runoff–rainfall ratio for specific events shows a wide variation, ranging from a low of 5% for a minor storm of 5.2 mm (Yair and Klein, 1973) to a near 100% runoff yield for a high-intensity rainfall that followed a significant antecedent rainfall event by a few hours (Schick, 1971b, Fig. 6). This particular storm produced a runoff event that peaked at 0.62 m^3/s (Schick and Sharon, 1974, p. 23), a peak flow that is the third highest over a 19-yr period (1967–1986) and is, therefore, a 6- to 7-yr flood. The highest peak flow measured during this period reached 0.83 m^3/s, or 16.6 $m^3/s/km^2$, a figure that compares well with Gerson's estimate of the maximum probable flood in catchments of a similar size (Gerson, 1972). In Mount Sodom, the region where Gerson worked, the lithologic conditions are markedly different from the Elat Mountains. This indicates that, at least for small catchments, differences in terrain properties—so important in normal and even large events—tend to be blurred during truly high-magnitude low-frequency events.

UPPER ALLUVIAL REACHES

Small mountainous upstream tributaries of extremely arid catchments tend, as a rule, to unite into larger channels with a continuous alluvial cover somewhere during their growth from third to fifth order. These alluvial fills, several kilometers long, 3–20 m wide, and of varying depth to bedrock, form an infiltration trap for floodwaters that flow into them either through the orderly tributary system or directly from adjoining slopes. Composed mostly of coarse material, this fill must be effectively, if not totally, saturated if the flood wave is to be conveyed downstream. Flood magnitudes and frequencies in large, extremely arid systems depend considerably, therefore, on the contiguity, three-dimensional size and sedimentary composition of these fills.

The central part of the Nahal Yael channel may serve as an example of these upstream floodwater sinks. Based on a 10-yr water budget, its upper end receives more than 20% of the total rainfall as surface floodwaters, but it delivers downstream less than 5% (Schick, 1979). The 1000-m-long reach is bounded at its upper end by the point of alluviation—the site of gauging station 04 (drainage area 0.10 km^2)—and at its lower end by a 10-m-high dry waterfall on whose lip gauging station 02 (drainage area 0.50 km^2) has been constructed (Fig. 7). Out of 19 flow events that cover the period from 1967 to 1985 and for which both stations yielded dependable data, 9 had a higher peak discharge at the downstream end, 7 at the upstream end, and 2 events had approximately equal flow at both stations. Considering the fivefold increase in drainage area from station 04 to station 02, only massive losses by infiltration between the two stations can explain this relation between peak flows.

Detailed water budget computations for the central alluvial reach of Nahal Yael, based on data from the hydrometric network, seismic refraction survey of the alluvium body, and grain size and moisture analyses of its material, determine the maximum capacity of the reach to be 4000 m^3 [i.e., 4 m^3 per linear meter of alluvial length, or 8 liters per each square meter of drainage area (Schwartz, 1986)].

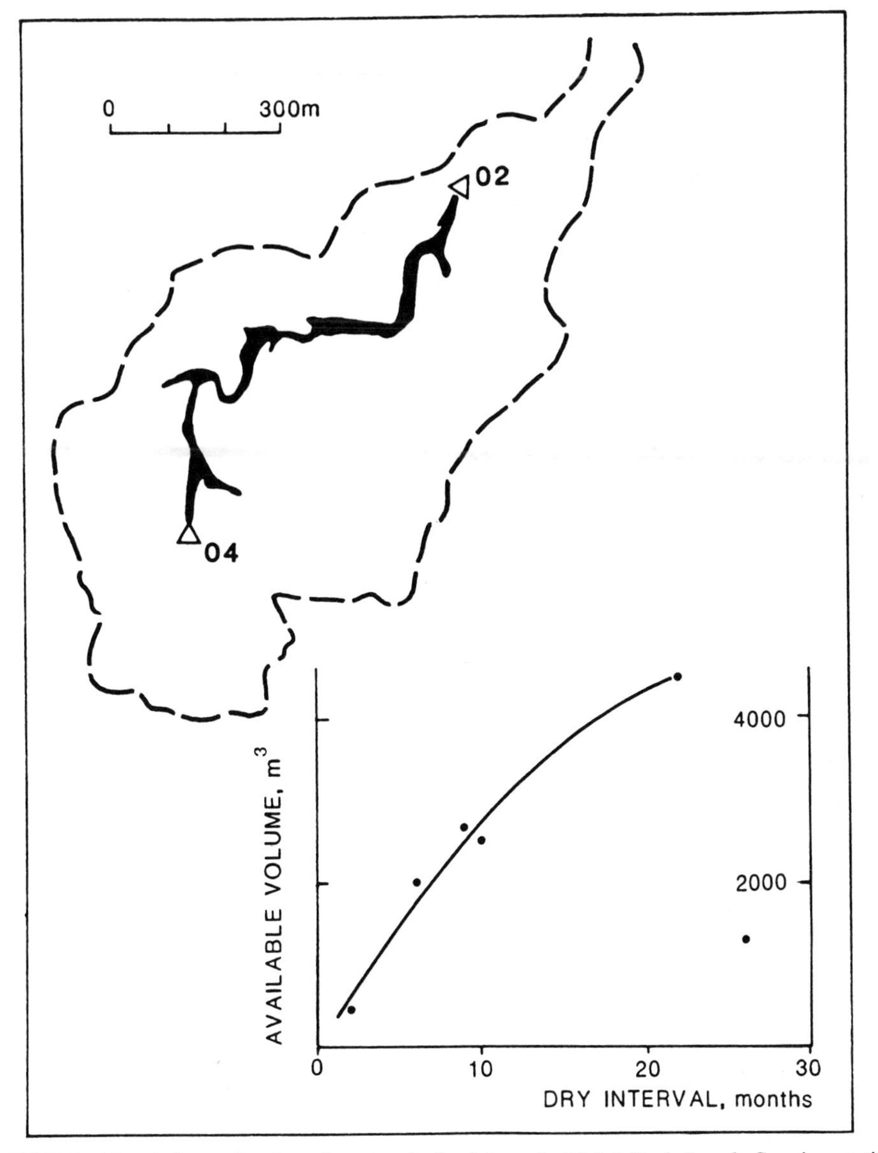

FIGURE 7. Flood absorption into the central alluvial reach, Nahal Yael, Israel. Gauging stations 04 and 02 are at the upper and lower ends of the reach, respectively. The curve shows the relation between the interflood drying time and the absorptive capacity of the alluvium (after Schwartz, 1986).

During floods infiltration is rapid, the wetting front reaches the suballuvial bedrock, and the flood wave proceeds on top of a saturated medium. There is only negligible infiltration into the bedrock, as evidenced by the generally consistent relation between water input into the alluvium and the preceding period of drying (Fig. 7). After-flood gravitative drainage lasts for about 3 days; thereafter the remaining water—about 11% by volume of the alluvial fill—evaporates and transpires at a relatively slow rate.

The state of moisture in the upper alluvial fills can be regarded as playing a similar role in flood generation in extreme arid environments as does antecedent moisture in nonarid areas. The drying curve (Fig. 7)—an essentially deterministic feature—must be juxtaposed with some stochastic distribution of interflood intervals, such as shown in Figure 8, to obtain a true probability distribution of flooding at the downstream end of the system. Thus, the upper, mountainous catchment 05 (drainage area 0.05 km^2) experiences more flow events than the downstream station 02 (drainage area 0.5 km^2), but the difference is exclusively in minor flows (Fig. 8A: events 5, 8, 9, 13, and 14). If the lengths of the "waiting period" between successive floods are arranged by rank (Fig. 8B), it can be seen that, over a period of 21 yr (about 7670 days), three such periods

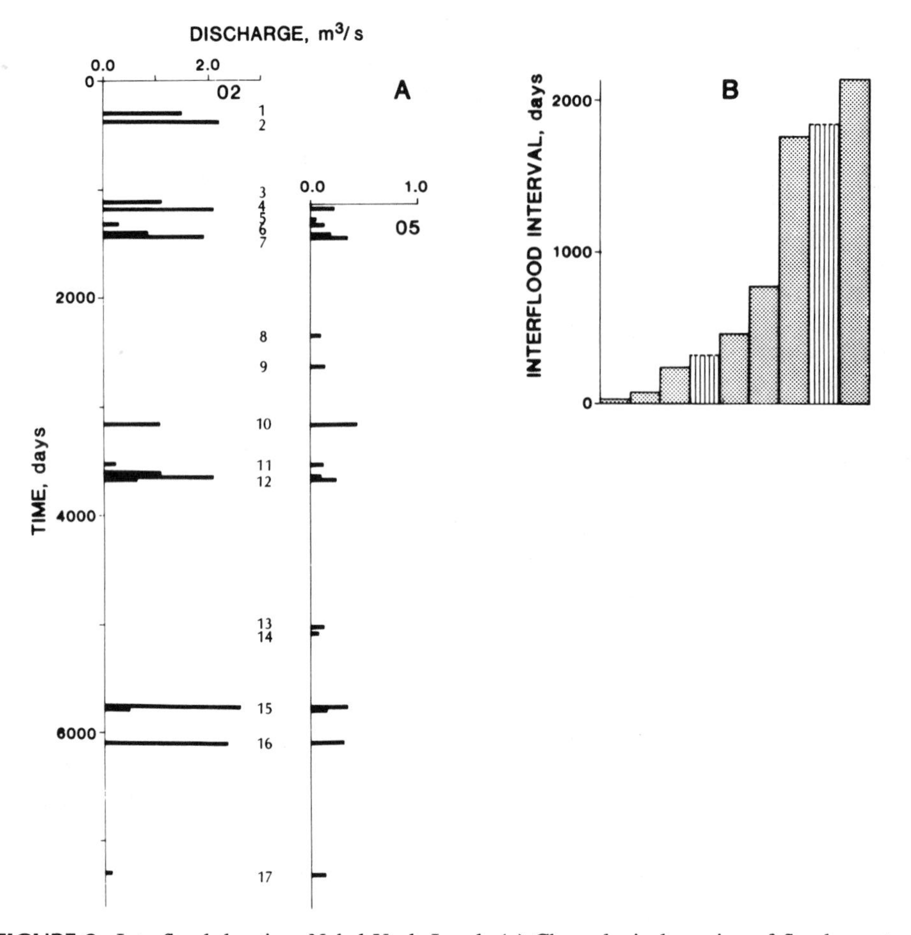

FIGURE 8. Interflood duration, Nahal Yael, Israel. (*a*) Chronological spacing of flood events at an upstream station (No. 05, drainage area 0.05 km^2, 1967–1986) and at a downstream station (No. 02, drainage area 0.50 km^2, 1965–1986). (*b*) Ranked array of durations between successive floods at station 02. Each column represents one event, irrespective of magnitude. Right-hand column shows longest recorded dry interval (more than 2000 days). Left-hand column is the shortest dry interval. The two stippled columns indicate unbounded time intervals at the beginning and end of the observation period.

at station 02 ranged between 1750 and 2250 days. The average waiting period is 365 days (one year), but the range of variation of this parameter is very large.

FLOOD HYDROGRAPHS

A comparison of two hydrographs from different catchments illustrates the importance of infiltration into the channel alluvium. Flood event 12A in Nahal Yael was caused by a 30-mm protracted storm of medium intensity (Fig. 3B). It yielded a 5-hr-long hydrograph of relatively even discharge at a small, upstream gauging station (Fig. 9A). The flow peaked at 0.12 m^3/s and remained above the discharge level of 0.06 m^3/s for 40% of the flow duration. This particular catchment is drained by a third-order channel with negligible alluvial storage and therefore losses must be attributed principally to terrain infiltration, mostly in the colluvium that covers roughly one-third of the drainage area. The rainfall on the catchment amounted to 1500 m^3, of which 600 m^3 infiltrated. This relatively large infiltration was enabled, in part, by the long duration of the event. The remaining 900 m^3—60% of the rainfall—became runoff, even though the peak discharge of the event was so low as to rank only 12th among 17 events in 18 yr for which data are available.

In stark contrast, flood event 4 in Nahal Yael was generated by half the rainfall amount of flood event 12A, but because the intensities were very high (Fig. 3A), the resulting hydrograph at station 04 (drainage area 0.10 km^2, i.e., double the size of the catchment defined by station 05, which is a tributary of catchment 04) was a very sharp

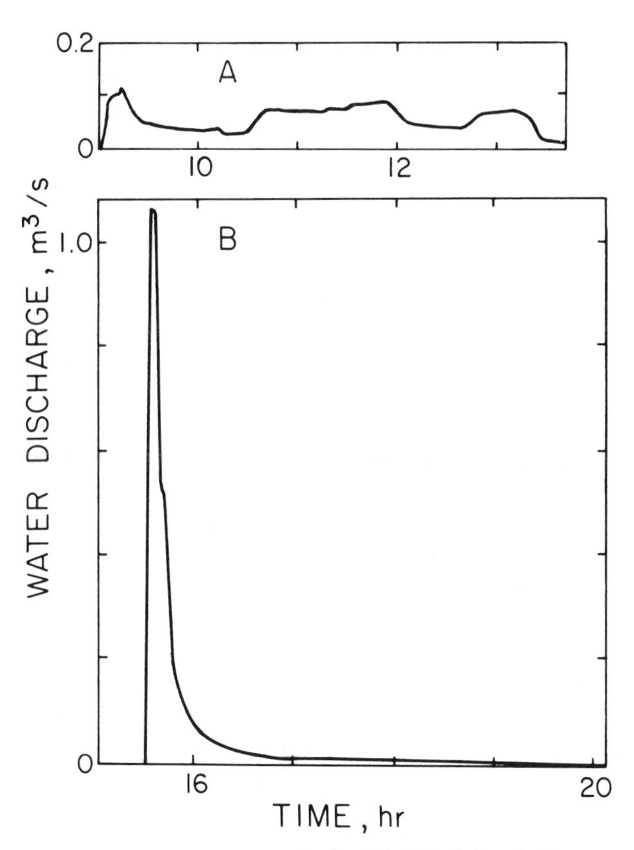

FIGURE 9. Typical hydrographs for Nahal Yael, Israel: (*A*) event 12A, 2/20/1975, at station 05, drainage area 0.05 km^2; (*B*) event 4, 5/24/1968, at station 04, drainage area 0.10 km^2. Rainfall charts for both events are shown in Figure 3.

and short peak (Fig. 9B). The flow remained above the level of one-half of the peak discharge of 1.07 m^3/s for only 8 min. In spite of the fact that the drainage area was twice as large and the storm rainfall only half that of the previous example, similar rainfall volumes (1590 versus 1500 m^3), runoff volumes (860 versus 900 m^3) and runoff coefficients (54 versus 60%) resulted. Still the routing characteristics and hydrograph shape were so different, that a 9:1 relation in peak discharges resulted.

In the southern Negev region, roughly two-thirds of the runoff-producing events behave in a similar way to flood event 4. Floods of this type generate, in small and medium-size catchments, the extremely flashy flows whose geomorphic work often assumes spectacular proportions. About one-third of the events are caused by long-term, medium-to-low rainfall intensity patterns, sometimes lasting several days or more. The famous Tunisian floods of 1969, which lasted for many weeks (Stuckmann, 1969), are an example, though apparently they affected mostly the semi- to subarid parts of the country. In such protracted rainfall events it is the volume that produces the flows, and only rarely do the peak discharges rival those obtained from the intense cloudburst-type rain. Geomorphically these events tend to modify the more drastic changes in landforms caused by the flashy events. Their ability to do so depends on the relative proportion and efficiency of both types of events. In most extreme desert areas, if not in all of them, the advantage is on the side of the flash floods (Wolman and Gerson, 1978).

The flood in Wadi Mikeimin in 1971 is typical of a geomorphically efficient flood (Lekach, 1974). Wadi Mikeimin is a 12.9 km^2 tributary of the major drainage artery of southeastern Sinai—Wadi Watir. As a result of an intense and highly localized rainstorm, Wadi Mikeimin deposited overnight a 6200 m^3 fan that completely obstructed the channel of Wadi Watir at their confluence. A flood reconstruction estimated the peak flow at 91.9 m^3/s, which dissipated completely into the sandy bed of the large Watir, whose catchment was unaffected by the event. A detailed analysis of high-water marks within the Mikeimin catchment for the January 1971 event indicated that one of its two main tributaries, drainage area 5.5 km^3, had a peak discharge of 83 m^3/s (Lekach, 1974). Clearly, the storm did not cover the catchment uniformly, and the flows from the two main tributaries were not synchronous. Wadi Watir remained impounded by the Mikeimin fan for a floodless period of 22 months, until the complete obliteration of the fan by a flood in November 1972 (Schick and Lekach, 1987).

In the 1966 Ma'an flood in southern Jordan (Fig. 5), several catchments ranging in area between 170 and 500 km^2, had a mean storm rainfall of 30–55 mm (Schick, 1971a, p. 121). Resulting peak flows ranged from 325 to 540 m^3/s, with total flow volumes between 1.5 and 7 $\times$ 10^6 m^3. Observations on one of these, Wadi Yutum, indicated supercritical flow, with surface velocities as determined by floats of about 5 m/s. Surges of 12 m/s occurred about once a minute during high flow (Central Water Authority, Hashemite Kingdom of Jordan, 1966).

Similar velocities were observed during the November 1972 flood in Wadi Watir, southeastern Sinai. Standing waves, in rows of up to 8 and reaching at least 1 m in amplitude above the average water surface, were quite common; sometimes they disappeared for a while, only to reform again. Distances between neighboring crests were 4–5 m, and they sometimes moved very slowly downstream (Schick and Sharon, 1974, p. 43).

The Watir catchment covers 3100 km^2 of very rugged mountainous terrain draining through a 20-km-long, canyonlike trunk stream to the Gulf of 'Aqaba (Fig. 10). The hydrograph of the event (Fig. 11), shows a series of very short flood peaks, rising within 1–3 min from insignificant flow levels to peaks of 80–320 m^3/s, and a near-immediate abrupt recession. These "walls of water" can be visualized as racing down the canyon, with the earlier ones decimated by infiltration into the not yet fully saturated channel bed

FIGURE 10. Wadi Watir near its mouth at recession of the 1972 flood. (Photographer, Hanna Magen).

alluvium (e.g., the "preflood" registered on the 23rd between 17 and 22 hr), and the later ones, being larger and somewhat faster, overtaking the smaller ones preceding them, to unite into an even larger flood wave. Some of the flood peaks in the Watir must have been also due to nonsynchronous tributary input, an effect accentuated by the irregular pattern in time and space of the storm that generated the flood (Department of Geography, Hebrew University, 1972). The only rainfall records available are all on the eastern and northern margins of the catchment, and they indicate storm totals ranging from 14.3 to 28.5 mm, without extreme intensities. It is probable that storm rainfall in the central parts of the catchment exceeded these figures. The total flow volume of the Watir was 5.5×10^6 m^3—an amount equivalent to less than 2 mm of runoff if equally distributed over the entire catchment.

Floods in even larger extremely arid catchments preserve most of the characteristics exhibited by the Watir flood. A good example is the El Arish flood of 1975, which resulted from a major 48-hr storm that enveloped most of the Sinai peninsula, with total rainfall depths of up to 73 mm at the Santa Katharina Monastery (Gilead, 1975). A single-peak hydrograph, culminating in a discharge of 1650 m^3/s, was recorded 30 km south of the town of El Arish (Fig. 12). The flood wave needed 39 hr to traverse 250 km from a point in midbasin where it was first identified. Further downstream, a 1 m high wall of water as well as nonsynchronous tributary inputs were observed. Much of the lower part of Wadi El Arish is sandy and was already outside the main rainstorm area; this caused considerable infiltration losses and perhaps also the coalescing of the flood waves. Still, the floodwaters were powerful enough to destroy a railway bridge built 50 yr previously and to deposit a delta 500 by 300 m on the Mediterranean coastline (Ben Zvi and Kornic, 1976). However, much like in the 1972 Watir event, the roughly 150 million m^3 discharged into the Mediterranean in this event amount to only 3–4 mm of catchment rainfall.

The Gulf-Saoura-Messaoud channel system in Algeria and Morocco drains 44,000 km^2 of semi-arid, mountainous and other terrain southward into the extremely arid central Sahara. Its floods have been studied over several decades (Dubief, 1953b; Vanney, 1960). Close to the headwaters

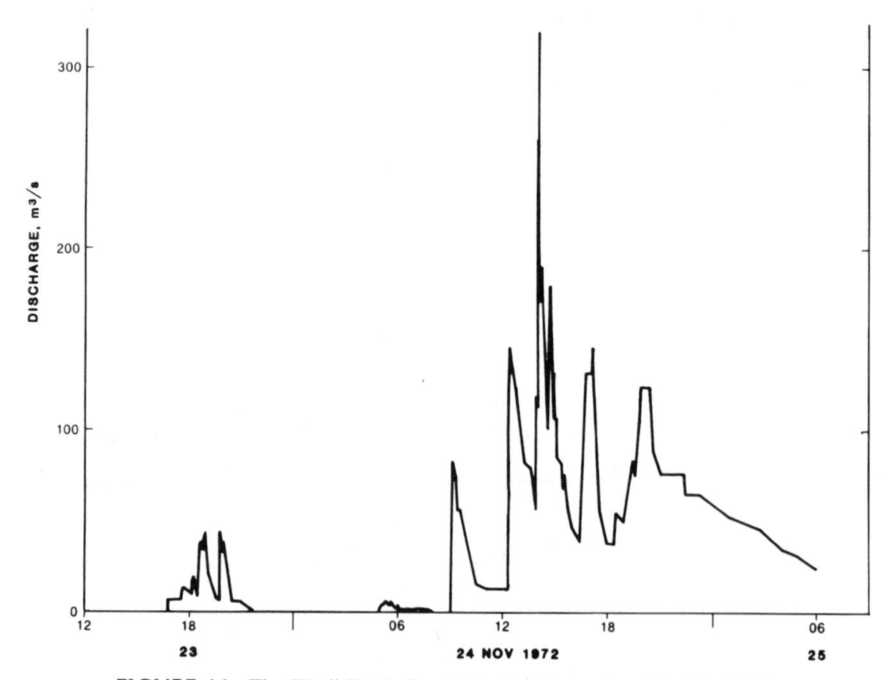

FIGURE 11. The Wadi Watir flood (Sinai), November 23–25, 1972.

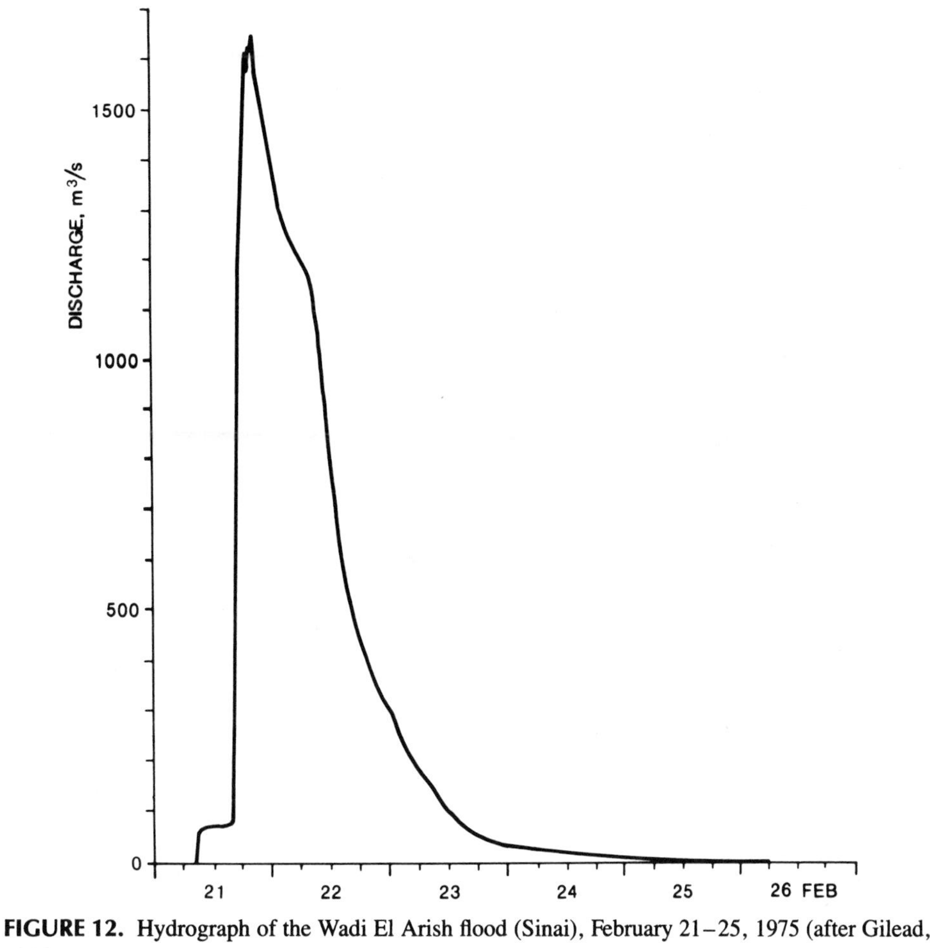

FIGURE 12. Hydrograph of the Wadi El Arish flood (Sinai), February 21–25, 1975 (after Gilead, 1975).

the regime in winter is episodic, with floods superimposed on a quasi-constant base flow. In summer this pattern is replaced by discrete, sharp-peaked ephemeral flows (Fig. 13*a*). The downstream diminution in discharge, caused mostly by infiltration but also by overflow into playas and subsequent evaporation, is shown in Figures 13*b* and 13*c*. Observations made during more than 60 yr reveal a longitudinal channel range of 310 km over which the point of flood extinction is distributed (Fig. 13*d*). The last 230 km receive some floodwaters only during events whose frequency exceeds 10 yr (Vanney, 1960). The advance rate of the flood front was determined, at one instance, to have been 0.6 m/s. The velocity of the flood crests ranged from 1.2 to 1.6 m/s (Teissier, 1965).

FREQUENCY

If completely random in time of occurrence, 20 events over 5 yr are more likely than not to include 2 within 1 week; the probability for this to happen is .53. Since the probability distribution in time is in reality not totally random, even for an extremely arid catchment—certain effects of seasonality and persistence in weather pattern and surface moisture always play a role—a larger degree of event clustering should result. Therefore, a decade or two of measurements cannot be considered, for the purpose of studying floods in extremely arid areas, a sufficiently representative period. However, even such series of observations are rare, and often incomplete or inaccurate. The logistical difficulties of monitoring floods in the extreme desert combine with the above-mentioned factors to make responsible statements as to magnitude and frequency of flooding difficult and risky.

A lognormal probability analysis of an 18 yr long flood series at Nahal Yael station 02, drainage area 0.50 km^2, suggests, by way of a theoretically illicit extrapolation, a 100-yr peak discharge of 5 m^3/s (Fig. 14). The 1971 Mikeimin flood, evaluated as being, for a 5.5 km^2 basin, a 100-yr event, peaked at 83 m^3/s (Lekach, 1974). The discrepancy between the two values is evident, given that peak discharges in catchments of similar terrain relate one to another roughly as a square root of their area ratio.

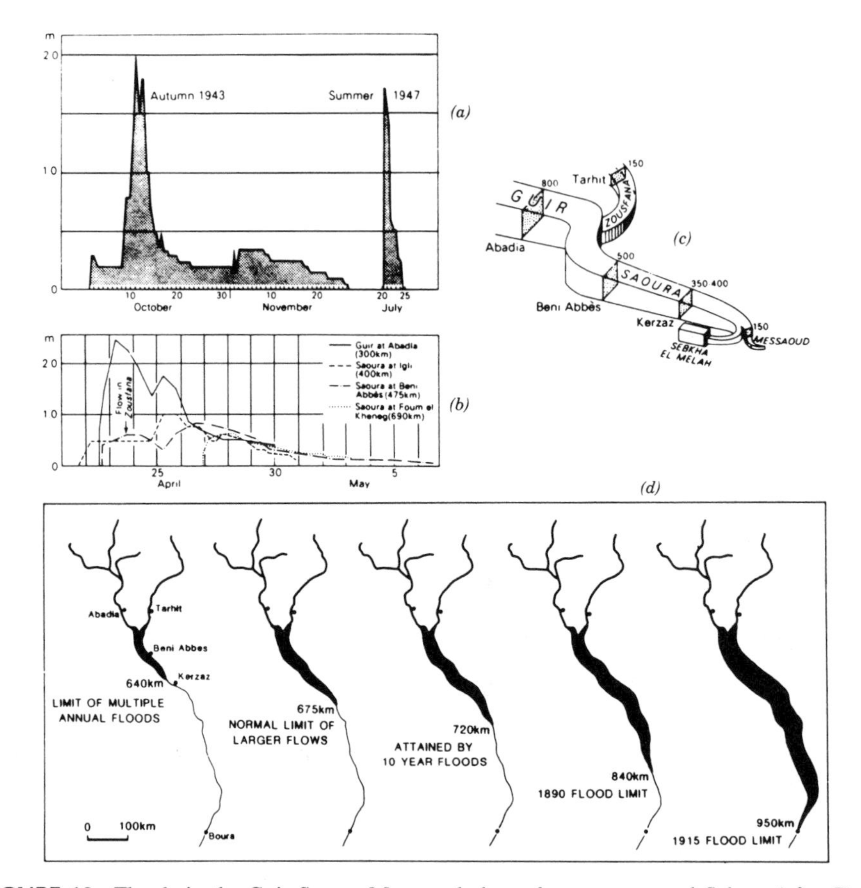

FIGURE 13. Floods in the Guir-Saoura-Messaoud channel system, central Sahara (after Dubief, 1953b; Vanney, 1960; Mabbutt, 1977; Cooke and et al., 1982). (*a*) Typical winter-season and summer-season hydrographs of the Guir at Abadla; (*b*) hydrographs of a single flood in May 1952 in the Guir-Saoura system; (*c*) diagrammatic representation of down-channel diminution of flood discharges (in million m^3) in the Guir-Saoura and Zousfana, March 1959; (*d*) downstream limits of flooding in the Guir-Saoura-Messaoud catchment.

The 320 m^3/s peak discharge of the 1972 Watir flood (Fig. 11) was evaluated as a 5 yr event (Department of Geography, Hebrew University, 1972), though this estimate seems, in retrospect, to have been on the low side. That event was followed 2 yr later by another flood, caused by the same storm that generated the El Arish flood, which peaked to more than 1000 m^3/s (Ben Zvi and Kornic, 1976). Assuming a slope of the probability curve similar to that of the Nahal Yael data (Fig. 14), a frequency of over 1000 yr is suggested for this flood. An event of such magnitude, were it deduced from some past sedimentary record, would certainly lead to conclusions as to major tectonic or climatic changes.

Another approach, a deterministic one, attempts to evaluate the maximum probable flood on the basis of detailed studies of environmental and terrain characteristics. By way of an example, Gerson (1972) found, for several watersheds near the Dead Sea whose size is small enough to ensure equilibrium flow, a maximum probable discharge of 40 $m^3/s\ km^2$. Using this figure as a base for extrapolating peak flow for larger catchments—with the inherent assumption that the square root area ratio takes care also of the decreasing probability of hydrologic input–output equilibrium with increasing area—a maximum probable flood of 145 m^3/s is indicated for the entire 12.9 km^2 Mikeimin catchment; for its above-mentioned 5.5 km^2 tributary a 94 m^3/s peak flow is suggested. For the 3100-km^2 drainage area of Wadi Watir, the maximum probable flood as determined by this procedure amounts to 2000–2500 m^3/s.

With the exception of Nahal Yael, these estimates for the maximum probable flood, spanning three orders of magnitude in drainage area, are 1.5–2 times higher than peak flows that have actually occurred over the last 20 yr. As more data on the magnitude of floods in arid environments become available, this gap between the theoretical maximum flood and actually recorded maximum flows is likely to close.

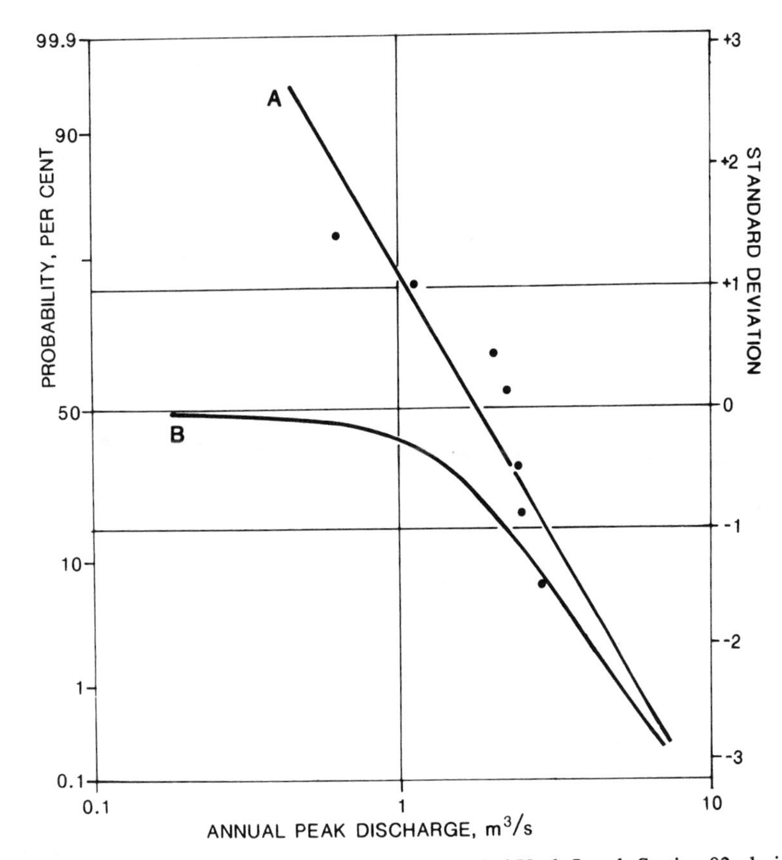

FIGURE 14. Lognormal analysis of peak discharges. Nahal Yael, Israel, Station 02, drainage area 0.50 km^2, with (*A*) dry years excluded and, (*B*) all years considered.

CONCLUSION

Despite the meagerness of data from extremely arid environments, it is of interest to compare their measured or indirectly estimated maximum flood values with those from nonarid or less arid regions. In comparison with values for nonarid environments such as obtained by Baker (1977, Fig. 9) and Costa (1987), most of which rely on estimates by the slope–area method or other means, the maximum probable floods for hyperarid environments as estimated above are clearly lower. The difference is about one-half order of magnitude for small catchments to one order of magnitude for the large ones.

The Finke River catchment in central Australia extends from its headwaters west and south of Alice Springs to the Simpson Desert north of Lake Eyre. The 1967 floods in the Finke River were caused by widespread rains, with many stations receiving more than 100 mm overnight (Mabbutt, 1977). Mean annual rainfall at Alice Springs is 275 mm, and therefore the region must be considered as semi-arid or, at the most, subarid. Nevertheless, the comparison with the hyperarid floods of the Sinai (Watir and El Arish) is of interest because of the similarity in size and in the character of the flood waves (Williams, 1970). Based on bedform analysis, the peak discharge in the middle sector of the Finke was estimated at 1200 m^3/s (Williams, 1971). This value is close to the peak flow of the El Arish flood of 1975, and may represent a similar frequency of 50–100 yr.

A true interregional comparison of floods in hyperarid environments is not possible at this stage of our knowledge. Floods in deserts like the Atacama, the Namib, and those of central Asia are virtually unknown to science on the level required for a modern analysis. Dependable comparative data from neighboring and other arid to subarid areas (<100 mm mean annual rainfall) are also not available or very difficult to obtain. The two exceptions to this generalization are the work of the French scientists in North Africa during the first half of this century and the concentrated effort of the Nahal Yael Project in the southern Negev, Israel, during the last 25 yr. Clearly this lack of knowledge represents one of the challenges to flood geomorphologists in future.

Nahal Yael, as indicated by the proximity in peak flow value of the five largest floods (Fig. 14), may be in a waiting period for a major event; however, there is no way of telling when it will occur. The exceedingly thin cross-combination of erratic distributions both in time and in space, coupled with environmental conditions that can be fully satisfied only on rare occasions, can make the waiting period for a truly exceptional flood very long. Indirect methods for magnitude and frequency determination are bound to play an increasingly important role in the understanding of floods in hyperarid environments.

REFERENCES

Ashbel, D. (1951). "Regional Climatology of Israel" (in Hebrew). Hebrew University, Jerusalem.

Atlas of Israel (1956). Dept. of Surveys, Ministry of Labour and the Bialik Institute of the Jewish Agency, Jerusalem.

Av-Ron, M. (1970). "Evaporation from Class A Pans in Israel" (in Hebrew), HG/70/053. Tahal Water Planning for Israel, Tel Aviv.

Baker, V. R. (1977). Stream channel response to floods, with examples from central Texas. *Geol. Soc. Am. Bull.* **88**, 1057–1071.

Ben Zvi, A., and Kornic, D. (1976). "The Floods of February 1975 in the Sinai and in the 'Arava" (in Hebrew); Rep. Hydro/3/1976. Ministry of Agriculture, Hydrological Service, Jerusalem.

Central Water Authority, Hashemite Kingdom of Jordan (1966). "Floods in Southern Jordan on 11 March 1966." Hydrology Division, Amman.

Cooke, R. U., Brunsden, D., Doornkamp, J. C., and Jones, D. K. C. (1982). "Urban Geomorphology in Drylands." Oxford Univ. Press, London and New York.

Costa, J. E. (1987). Hydraulics and basin morphometry of the largest flash floods in the conterminous United States. *J. Hydrol.* **93**, 313–338.

Department of Geography, Hebrew University (1972). "Floods of 23-25.11.72 in Relation to the Elat—Sharm-el-Sheikh Road" (in Hebrew), No. 3. Floods in Eastern Sinai, Jerusalem.

Dubief, J. (1953a). "Essai sur l'hydrologie superficielle au Sahara." Service des études scientifiques, Gouvernement général de l'Algérie, Birmandreis.

Dubief, J. (1953b). Ruissellement supeficiel au Sahara. *Colloq. Int. C.N.R.S.* **35**, 303–314.

Gerson, R. (1972). Geomorphic processes of Mount Sdom (in Hebrew). Ph.D. Thesis, Hebrew University, Jerusalem (Engl. abstr.).

Gilead, D. (1975). "A Preliminary Hydrological Appraisal of the Wadi El-Arish Flood, 1975" (in Hebrew), Mimeogr. Rep. Israel Hydrological Service, Jerusalem.

Greenbaum, N. (1986). Infiltration and runoff in an extremely arid region: Infiltration experiments in small plots in the Southern Arava Valley and their hydrological, pedological and paleomorphological implications (in Hebrew) M.Sc. Thesis. Department of Physical Geography, Hebrew University, Jerusalem.

Jaekel, D. (1980). Current weathering and fluvio-geomorphological processes in the area of Jabal as Sawda'. *In* "The Geology of Lybia" (M. J. Salem and M. T. Busrewil, eds.), Vol. 3, pp. 861–875. Academic Press, London.

Lekach, J. (1974). Reconstruction of the January 1971 event in Wadi Mikeimin, eastern Sinai (in Hebrew). M. Sc. Thesis, Department of Geography, Hebrew University, Jerusalem.

Levi, M. (1975). The synoptic situation of 20.2.1975 which brought the floods in Sinai (in Hebrew). *Meteorol. Be Isr., Q. J. Isr. Meteorol. Soc.* **11**(1), 2–6.

Mabbutt, J. A. (1977). "Desert Landforms." Aust. Nat. Univ. Press, Canberra.

Poráth, A., and Schick, A. P. (1974). The use of remote sensing systems in monitoring desert floods. *IAHS-AISH Publ.* **112**, 133–139.

Salmon, O., and Schick, A. P. (1980). Infiltration tests. *In* "Arid Zone Geosystems" pp. 55–115.(A. P. Schick, ed.), Division of Physical Geography, Hebrew University, Jerusalem.

Schick, A. P. (1971a). A desert flood: physical characteristics, effects on Man, geomorphic significance, human adaptation—a case study in the southern 'Arava watershed. *Jerusalem Stud. Geogr.* **2**, 91–155.

Schick, A. P. (1971b). Desert floods—interim results of obsevations in the Nahal Yael research watershed, 1965-1970. *IASH-AISH Publ.* **96**, 478–493.

Schick, A. P. (1977). A tentative sediment budget for an extremely arid watershed in the southern Negev. *In* "Geomorphology in Arid Regions" (D. O. Doehring, ed.), pp. 139–163. Allen & Unwin, Winchester, Massachusetts.

Schick, A. P. (1979). Fluvial processes and settlement in arid environments. *GeoJournal* **3**, 351–360.

Schick, A. P., and Lekach, J. (1987). A high magnitude flood in the Sinai desert. *In* "Catastrophic Flooding" (L. Mayer and D. Nash, eds.), pp. 381–410. Allen and Unwin, Winchester, Massachussetts.

Schick, A. P., and Sharon, D. (1974). "Geomorphology and Climatology of Arid Watersheds, Tech. Rep. Department of Geography, Hebrew University, Jerusalem.

Schwartz, U. (1986). Water in the alluvial fill of stream channels in arid environments—availability, distribution and fluctuations (in Hebrew). M.Sc. Thesis, Department of Physical Geography, Hebrew University, Jerusalem.

Sharon, D. (1972). The spottiness of rainfall in a desert area. *J. Hydrol.* **17**, 161–175.

Sharon, D. (1981). The distribution in space of local rainfall in the Namib desert. *J. Climatol.* **1**, 69–75.

Sharon, D., and Kutiel, H. (1986). The distribution of rainfall intensity in Israel, its regional and seasonal variations and its climatological evaluation. *J. Climatol.* **6**, 277–291.

Stuckmann, G. (1969). "Les inondations de septembre-octobre 1969 en Tunisie. Partie II. Effets morphologiques." UNESCO, Paris.

Teissier, M. (1965). Les crues d'oueds au Sahara algérien de 1950 a 1961. *Trav. Inst. Rech. Sahariens* **24**, 7–29.

Vanney, J. R. (1960). "Pluie et crue dans le Sahara nord-occidental," Vol. 4. Mémoirs Régionaux de l'Institute des Recherches Sahariens, Algiers.

Williams, G. E. (1970). The central Australian stream floods of February-March 1967. *J. Hydrol.* **11**, 185–200.

Williams, G. E. (1971). Flood deposits of the sand-bed ephemeral streams of central Australia. *Sedimentology* **17**, 1–40.

Wolman, M. G., and Gerson, R. (1978). Relative scales of time and effectiveness of climate in watershed geomorphology. *Earth Surf. Processes* **3**, 189–208.

Yair, A., and Klein, M. (1973). The influence of surface properties on flow and erosion processes on debris covered slopes in an arid area. *Catena (Cremlingen-Destedt, Ger.)* **1**, 1–18.

Yair, A., and Lavee, H. (1974). Areal contribution to runoff on scree slopes in an extreme arid environment—a simulated rainstorm experiment. *Z. Geomorphol., Suppl.* **21**, 106–121.

Yair, A., and Lavee, H. (1976). Runoff generative process and runoff yield from arid talus mantled slopes. *Earth Surf. Processes* **1**, 235–247.

13

FLOODS IN COLD CLIMATES

MICHAEL CHURCH
Department of Geography, The University of British Columbia, Vancouver, British Columbia, Canada

INTRODUCTION

Cold climates, in the present context, are ones in which the hydrologic cycle is influenced by the seasonal (or perennial) accumulation of snow and ice. The period of accumulation must be sufficiently long to alter the pattern of runoff from that which would otherwise be determined by the distribution of precipitation. At the minimum, then, snow must persist on the ground for periods longer than that typical of synoptic weather events. From a climatological point of view, a practical criterion is the persistence of snow cover for longer than one month in 50% of the years. In such an environment engineering designs and water management procedures must begin to be modified in view of snow accumulation and snowmelt.

Figure 1 illustrates the approximate extent of this *nival region* in the Northern Hemisphere. The map depicts the region within which snow cover occurred in 50% or more of the weeks in January during the 15-winter period 1966–1982, as interpreted from satellite images (Matson et al., 1986). The compilation is based on weekly data so that it does not conform with the definition for nival region given above. However, it matches very closely the same compilation for February, indicating that midwinter snow cover is persistent and that the climatological boundary is relatively stable. The dashed line in Figure 1, indicating the 50% probability for at least 1 in. of snow cover at the end of January (Dickson and Posey, 1967) from an old compilation based on surface data, also confirms the stability of the nival region. The definition of nival region, which incorporates a specific persistence criterion, has not been mapped on a global scale. The geography of the region is, of course, more complicated than can be illustrated on the global scale of Figure 1 because of the effect on climate of elevation: high mountains throughout the world experience nival conditions. One should also bear in mind that runoff generated by melting of snow and ice may be experienced downstream beyond the limit of the nival environment as we have defined it. This is an important aspect of hydrologic regime in alpine forelands.

Snow and ice effect a temporal redistribution of liquid water occurrence at the earth's surface from that which would be set by the regime of precipitation, hence affecting the timing and character of flood runoff. Figure 2 illustrates a modified terrestrial hydrologic cycle in cold regions. In extremely cold environments seasonal or perennial ground-frost may influence the runoff regime by drastically altering infiltration properties of soils and changing runoff pathways. In permafrost regions of low relief, bog or muskeg terrain develops above the frost-perched summer water table and has a further influence on the runoff. High water in rivers need not be associated with extreme flows in cold environments because ice in the river channel may form jams that produce large stage rise by ponding. In small channels winter snowdrifts may have similar effects. Table 1 presents a summary of flood types in the nival regime.

The range of normal hydrometeorologic mechanisms that may produce flood flows in nival regions complicates analysis and projection of extreme flows. Conventional methods for extreme-event analysis suppose that the sequence of extremes is homogeneous in the sense that the events are all drawn from the same population. The various antecedent mechanisms for extreme flow undermine this assumption. The effect of this varies regionally according to climate and according to drainage basin size. The geomorphologic effects of extreme flows may, to some degree, be affected by this situation as well. For the purpose of flow frequency analysis, a similar circumstance affects extreme low flows, for which there usually are distinct late summer and late winter occurrences in most nival regions.

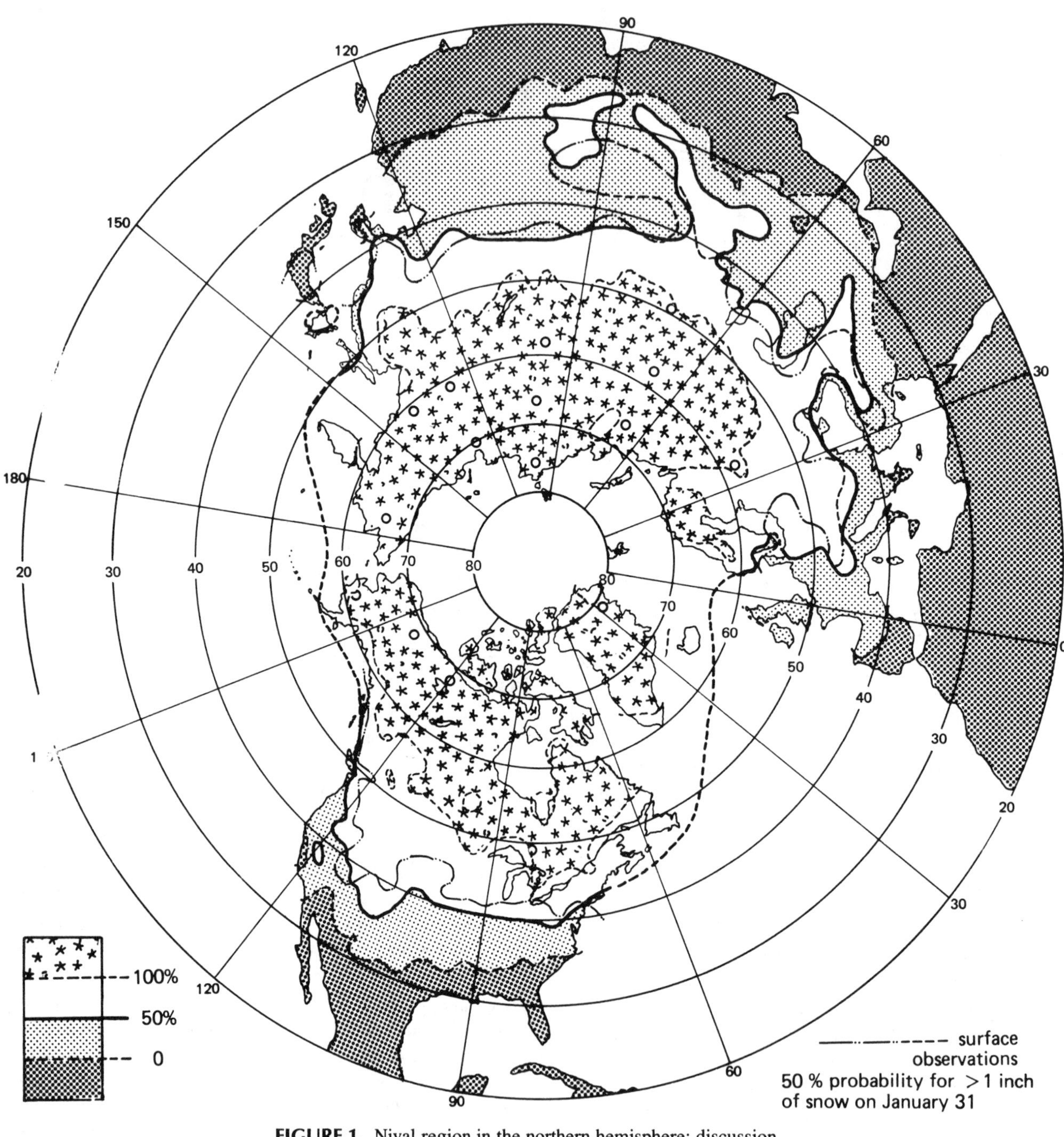

FIGURE 1. Nival region in the northern hemisphere: discussion in text.

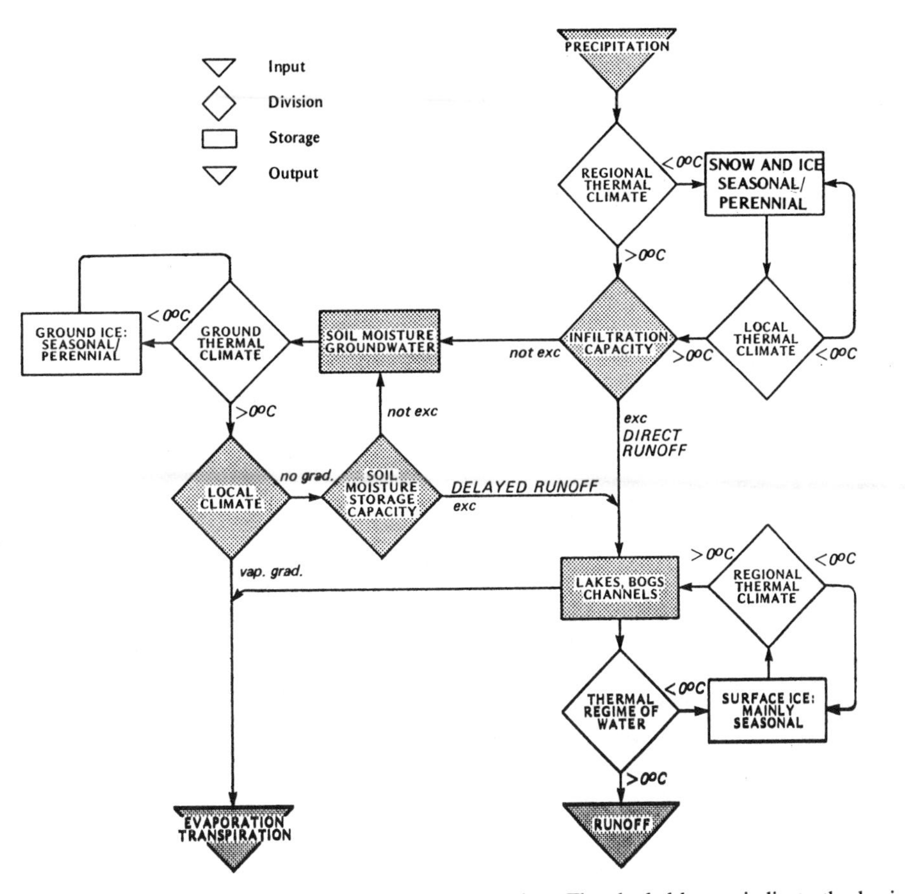

FIGURE 2. Terrestrial hydrologic cycle in the nival region. The shaded boxes indicate the basic elements of the hydrologic cycle not dependent on extreme thermal conditions.

The following sections are organized to follow the main entries in Table 1. In the penultimate section we return to the question of mixed flood populations. The chapter concludes with some discussion of the morphologic effects of nival floods. Examples are drawn from Cordilleran and arctic North America because these are the regions in which the writer's experience has been gained.

TABLE 1 Classification of Floods in Cold Regions

I. Hydrometeorological floods
  a. Nival
    i. Snowmelt
    ii. Glacier ice melt
  b. Rainstorm
  c. Mixed (rain on snow)
II. Channel blockage by snow and ice
  a. Channel blockage by ice (snow)
    i. Winter jams
    ii. Breakup jams
  b. Icing (aufeis)
  c. Jökulhlaups
III. Azonal floods (moraine or landslide failures)

NIVAL FLOODS

Snowmelt and Runoff Formation

Snowmelt runoff occurs when energy is added to a snowpack that is at 0°C. The principal sources of energy for snowmelt are direct solar radiation or, under tree cover, long-wave radiation or sensible heat transfer, which are related to ambient temperature. Hence, a strong diurnal rhythm dominates snowmelt. The form of the diurnal melt flood is controlled by the pattern of melt at the surface, by the passage of water through the snow, and by the saturated flow at the base of the snow to an unimpeded channel.

Passage of water through homogeneous snow produces a shock wave front and long recession because of the dependence of water transmission rate on liquid water content (cf. Colbeck, 1977). Most snowpacks, however, present strata of varying texture and permeability, including perforated ice layers, which guide water movement along discrete pathways, hence serve to diffuse the melt wave and reduce peak flow somewhat (cf. Marsh and Woo, 1985). Melt-wave propagation through snow proceeds typically at rates of 0.1–0.4 m/hr, so that this phase dominates

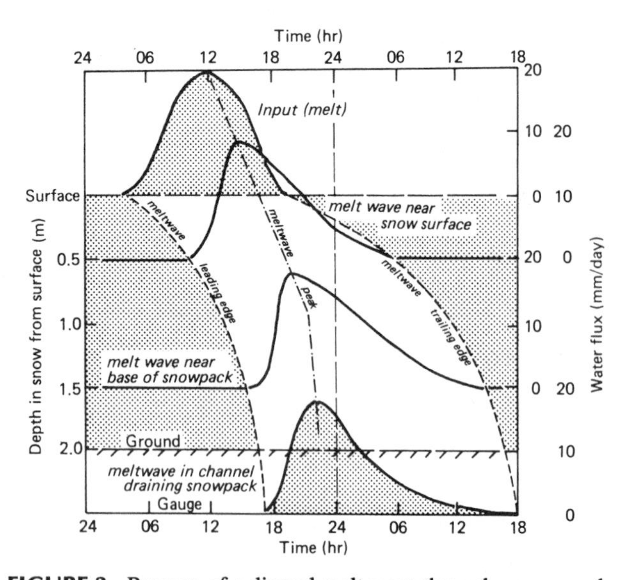

FIGURE 3. Passage of a diurnal melt wave through a snowpack and small drainage basin, constructed schematically from observations reported by Jordan (1979, 1983). The gradients of the dashed line joining the hydrograph extrema give the melt wave celerity: their divergence indicates diffusion of the melt wave while convergence on the rising limb indicates shock development within the snowpack. Marsh and Woo (1985) illustrate the local spatial variability of the process.

runoff timing from deep snow until the pack is thoroughly wetted or "ripe." Hydrograph recessions are not simply exponential because of the nonlinear behavior of the draining snow and because of the variable behavior of the saturated reservoirs at the base of the pack (Jordan, 1979). Flow rapidly concentrates in preferred pathways at the base of the snow, but these change as the melt proceeds. Observed melt-wave shape is further affected by the variation in travel times from distributed sources to a gauging point. Figure 3 illustrates the propagation of a diurnal melt wave through a deep snowpack, and Figure 4 illustrates fair weather control of nival floods in a small basin.

These phenomena have several consequences for the form of diurnal or compound nival flood hydrographs:

1. In shallow snow or late in the season, runoff delay due to the passage through the snow becomes less dominant, while flow paths at the base of the snow become better defined and less impeded. Hence stream hydrographs become more peaked.
2. The recession becomes quite steep and variable in shallow or late snow in sensitive response to the pattern of melt.

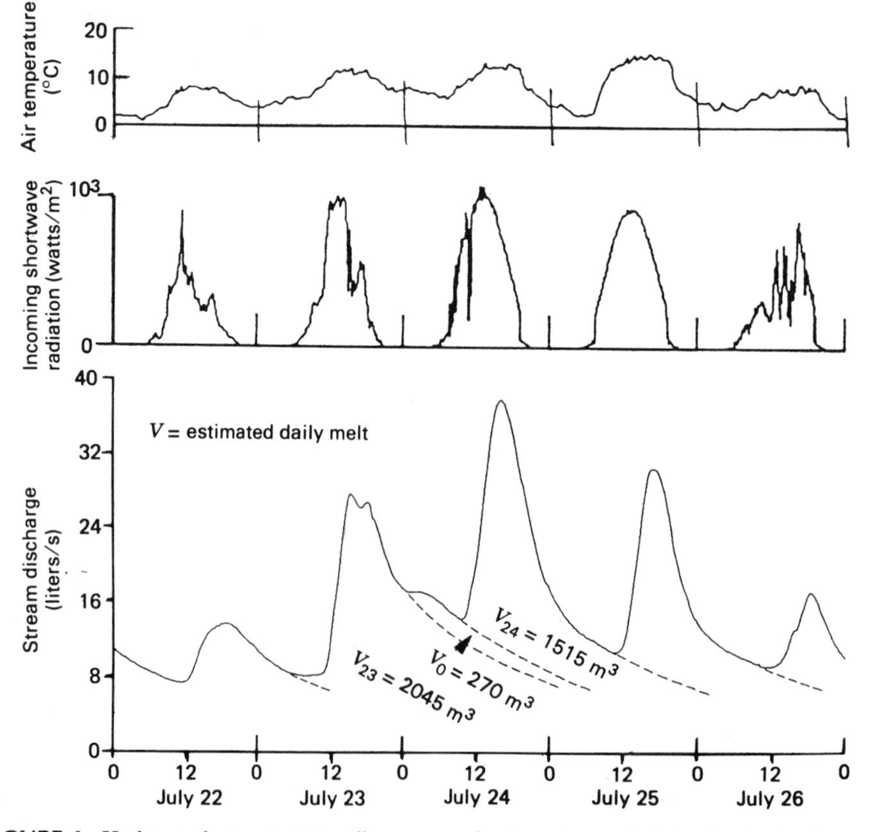

FIGURE 4. Hydrograph response to radiant energy forcing of a small alpine basin of 3.9-ha area. The dashed lines indicate inferred daily recession flows: The additional contribution, V_0, on July 24 was due to high overnight temperatures on July 23–24 (unpublished observations of R. P. Jordan).

3. The phase lag between melt generation and downstream runoff declines as the snow becomes shallower and flow paths better integrated.

Flow routing through drainage systems attenuates diurnal flow pulses; however, once the snowpack becomes discontinuous, the sources of flow become more local and the basin may behave like a much smaller one, which, in effect, it has become with respect to effective contributing area. Whereas the diurnal rhythm of runoff in small drainage areas is a direct response to energy inputs for melt, large basins exhibit a seasonal nival flow peak in which the details of the hydrograph are dominated by synoptic weather periods. Figure 5 illustrates several features of the regime type.

Fraser River at Hope, British Columbia (Fig. 5*a*), illustrates the classic nival regime of a basin so large that only snowmelt generates runoff over most of it at one time. Rainfall events (October and November on the illustrated hydrograph) are entirely subordinate, though they may generate substantial spikes in suspended (washload) sediment transport. Beatton River near Fort St. John, British Columbia, illustrates the mixed flood regime of rivers that are an order of magnitude smaller. In 1974 (Fig. 5*b*), the peak flow observed during the nival melt period (1026 m^3/s on April 30) was actually a substantial rain-on-snow event only 4 days after ice clearance. The annual peak flow (1270 m^3/s on July 18) resulted from heavy summer rain. Synoptic weather control is evident in small basin runoff, illustrated in Figure 5*c* and 5*d*. Boundary Creek is located in semi-arid interior British Columbia, so that in most years the nival flood—with synoptic modulation—dominates runoff. Jamieson Creek is a small mountain watershed at Vancouver, British Columbia. The highest elevation in this basin is near 1500 m and the winter snow load at higher elevations may be as great as 1.5 m water equivalent. Melt continues over several months, but individual peak flows are always rain-on-snow or autumn rain events.

In semi-arid regions (which includes much of the arctic), the nival flood may comprise well over one-half and as much as 90% of the annual runoff in a particular year (cf. Fig. 5*c*), and the balance may be nivally recharged baseflow. Nonetheless, even in the high arctic, the most extreme runoff eventually is contributed by summer rain. Figure 5*e* illustrates nival regime in an arctic drainage of moderate size for years with no notable summer rain (1975) versus those with such rain (1976). In Figure 5*f* strong diurnal control is evident in a small nival basin on Cornwallis Island (75° north latitude). Here, the effect of seasonal weather in determining the timing of the main nival runoff is strongly evident.

Pure nival runoff occurs as a well-defined event during a few days or weeks (according to basin size and initial snowpack depth) in spring. If the short-term fluctuations introduced by synoptic weather are smoothed, the event hydrograph can be analyzed (Fig. 6). In large basins this smoothed performance is apt to be consistent from year to year; nonetheless, the synoptic variations yield the absolute flood peaks.

In glacial regime peak runoff is displaced into mid or late summer when seasonal snow is much reduced, glacial drainage paths are well integrated, and—most important—glacier ice with relatively low albedo is exposed to melt. Figure 7*a* illustrates these features by example of the 1963 melt season hydrograph of Lewis Glacier in eastern Baffin Island. Despite the significance of glacier ice exposure in promoting the greatest melt rate, the volumetric contribution from ice melt to total runoff in each year at most sites remains relatively small (cf. Young, 1982), on the order of 10% of total melt. In drainage basins containing glaciers, that is, with persistent melt in a more or less restricted part of the basin, diurnal rhythm may be sustained in basins on the order of 10^3 km^2 area with about 10% glacier cover or more. Figure 7*c* illustrates an annual hydrograph for Lillooet River near Pemberton, British Columbia. Here, 22% glacier cover sustains high nival flows through the summer, but the most severe floods are autumn storms, as illustrated.

Because of the contrast in customary runoff timing, Church (1974) distinguished nival from glacier runoff regime. In the arctic, however, weather during the short summer season and, particularly, the persistence of drifted snow in gullies and hollows may modify the nival regime so that it appears similar to glacial regime (cf. field examples in Marsh and Woo, 1981, and Fig. 5*f*).

Energy Limits to Nival Flood Generation

It is interesting to ask, How large may a nival event become? Relatively strict limits are set by the energy available for snowmelt, Q_m. The energy balance at the snow surface is

$$Q_m = Q^* + Q_H + Q_E + Q_P \tag{1}$$

where

$$Q^* = (1 - a)K\downarrow + L^* \tag{2}$$

and Q^* = net radiant energy balance

a = surface albedo, 0.8 or higher for clean, new snow, declining to 0.4 or less for old, granular snow

$K\downarrow$ = incoming shortwave radiation

L^* = net long-wave radiant energy balance

Q_H = turbulent sensible heat flux

Q_E = latent energy flux

Q_P = sensible heat added to the snow by rainfall

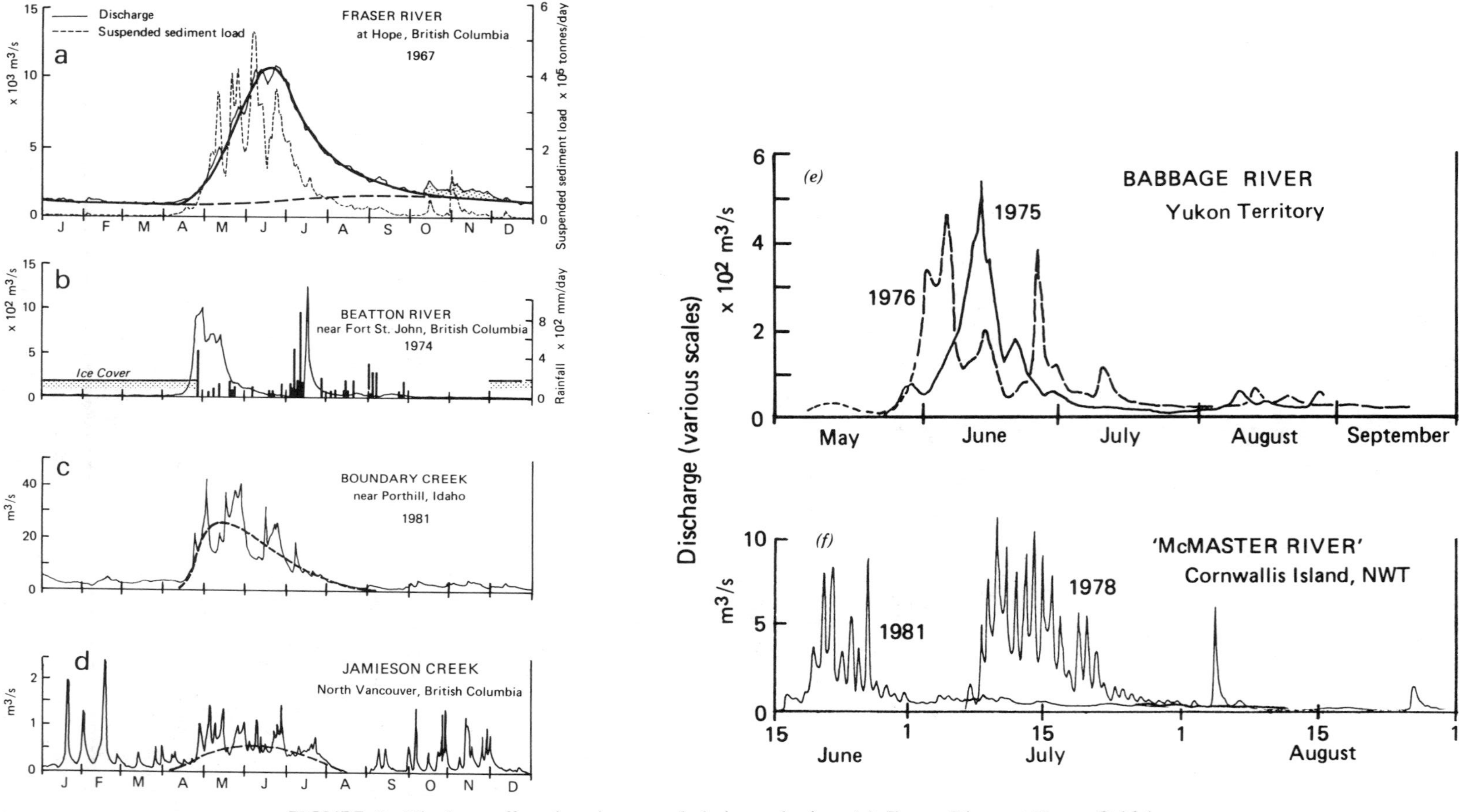

FIGURE 5. Nival runoff regime in several drainage basins: (*a*) Fraser River at Hope, British Columbia (WSC gauge 08MF005), 1967; A = 217,000 km^2. Discharge and suspended-sediment transport. The heavy solid line is a smoothed approximation of the annual runoff regime; cross-hatched area is autumn rainstorm runoff, ignored in the smoothing; heavy dashed line is an estimated base flow component (see Fig. 6 for event analysis). (*b*) Beatton River near Fort St. John, British Columbia (WSC gauge 07FC001), 1974; A = 15,600 km^2. Rainfall at Fort St. John is included to illustrate the rainstorm augmentation of the nival flow peak and rainstorm origin of the annual peak flow. Horizontal line and stipple indicates the occurrence of ice. (*c*) Boundary Creek near Porthill, British Columbia (WSC gauge 08NH032), 1981; A = 251 km^2. The nival period is emphasized by a generalized, dashed line. (*d*) Jamieson Creek, North Vancouver, British Columbia (WSC gauge 08GA062), 1971; A = 2.85 km^2. The nival period is emphasized by a generalized, dashed line. (*e*) Babbage River, Yukon north slope (1975); A = 5000 km^2 (unpublished date of D. L. Forbes). (*f*) "McMaster River," Cornwallis Island, Northwest Territories; A = 33 km^2. An arctic, semi-arid nival regime (data of Woo, 1983).

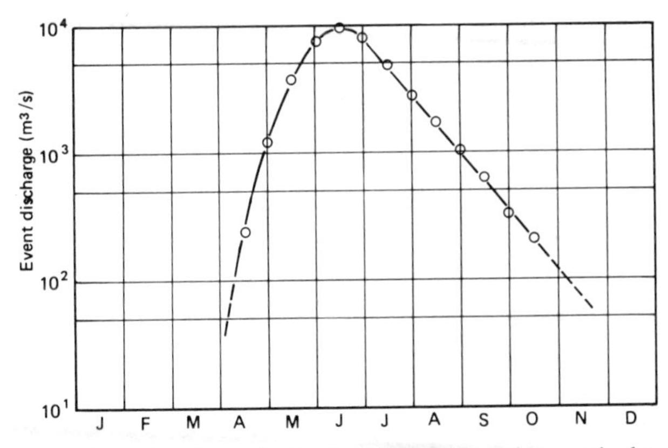

FIGURE 6. Semilogarithmic plot of the snowmelt event hydrograph of Fraser River at Hope, 1967 (see Fig. 5*a* for the complete hydrograph and its separation), to illustrate the exponential recession. The principal reservoirs controlling the recession are large lakes.

In comparison with Q_m the ground heat flux at the base of the snow is negligible. Snowmelt (millimeters of water) is

$$S = 10^3 \, Q_m/\rho\mathscr{L}_f \tag{3}$$

where

ρ = density of water

$\mathscr{L}_f$ = 0.33 MJ/kg is the latent energy of fusion of ice

Q_m is given in SI units (megajoules)

If we suppose that we have a relatively deep snowpack and that relatively warm, clear weather persists for several days so that water storage places in the drainage system become filled, then the maximum nival flood is given by the amount of snow that may be melted. This can be approximated for an unshaded, continuous snow cover on a clear day by $Q^*/\rho\mathscr{L}_f$ since Equation (1) reduces to $Q_m \approx Q^*$ in these circumstances and in the absence of strong, regional advection. Figure 8 gives some results based on estimates of $K\downarrow$ in various latitudes. The following remarkable facts emerge (which would not be much affected by the presence of numerical bias in the calculations):

1. Near summer solstice there is no major difference in potential melt between latitudes. This typically is the time when arctic snow disappears, so it may (potentially) be melted as effectively as snow in more temperate latitudes (cf. comments in Woo, 1983);
2. In midlatitudes, particularly below 50°, there is a period of two or three months about summer solstice when potential energy inputs do not vary drastically, so the variation in timing of spring melt from year to year will not have a large effect on maximum potential melt rate of deep (alpine) snowpacks nor, consequently, on the potential maximum purely nival flood.

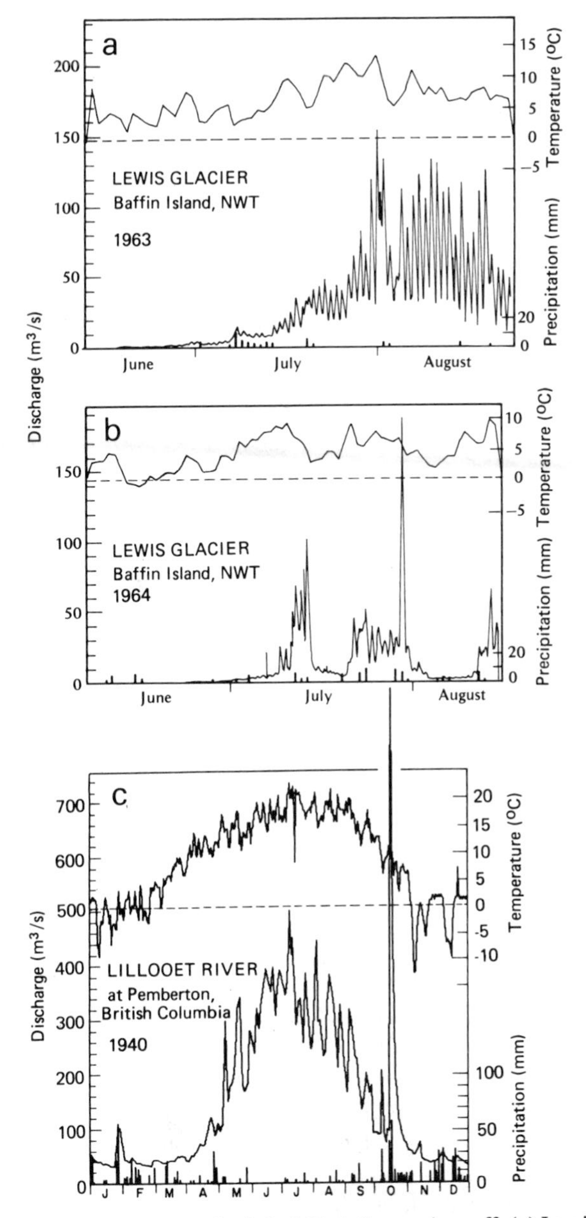

FIGURE 7. Hydrograph of glacially influenced runoff: (*a*) Lewis Glacier, Baffin Island, 97% glacierized, 1963; A = 204 km². Mean daily air temperature at the streamgauge (i.e., off glacier) is shown for comparison. (*b*) Lewis Glacier, 1964. This season was dominated by summer storms: The major runoff peaks are due to summer rain (see Fig. 11). (*c*) Lillooet River, near Pemberton, British Columbia, 22% glacierized (WSC gauge 08MG005), 1940; A = 2160 km². Temperature and precipitation data are for Pemberton Meadows, a valley station near the gauge (data analyzed by R. Gilbert).

Quite different conditions may control melt. When forest cover is present, long-wave radiation from the trees is

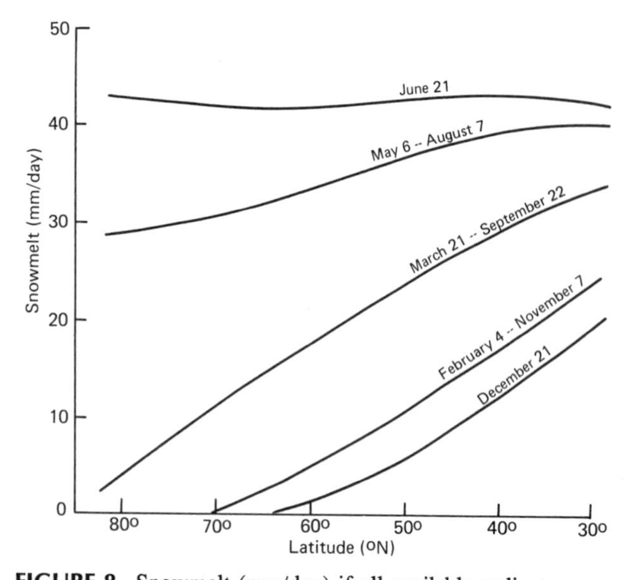

FIGURE 8. Snowmelt (mm/day) if all available radiant energy is utilized. The calculations assume the Davies-Hay (1980) atmospheric transmission coefficients for global solar radiation (from List, 1966) through a standard atmosphere including aerosol and 2.9 mm precipitable water. Snow albedo is assumed to be 0.5 and available energy $Q^* = 0.8(1 - \alpha)K\downarrow$ to account for longwave energy losses. At specific sites, more complex relations between Q^* and $K\downarrow$ have been obtained, based on empirical regressions (cf. Dunne and Price, 1976; Aguado, 1985). The displayed results are considered to be conservative (i.e., likely to yield overestimates) for open sites.

dominant. However, the magnitude of $L^*(T)$ for ambient temperature (T) is unlikely to approach $(1 - a)K\downarrow$ for the same climate. The turbulent fluxes may contribute significant energy: Q_E may be effective when atmospheric humidity is high so that there is condensation onto the snow surface (the latent energy of condensation $\mathcal{L}_v = 7.5\mathcal{L}_f$, so that much more snow is melted for each unit of water condensed), but this most commonly is the case in stormy weather when radiation-driven melt is low. Q_H may become important locally when the snow becomes patchy, but now only restricted areas will contribute melt, and so the averaged melt rate over a drainage basin is not apt to outstrip the potential rate we have established for open terrain. Rain on snow is discussed separately below: for the energy content of the precipitation to be significant, the water input of the rain must dominate runoff formation.

Regional advection may complicate the situation since Q_H may then be important (cf. review in Male and Granger, 1981). In most cases this does not yield melt rates that approach those displayed in Figure 8. For example, in the absence of frontal activity, regional air mass temperature, as measured aloft at the 85-kPa level, may account for on the order of 10-mm melt per day (Granger and Male, 1977). However, föhn or chinook-type winds or vigorous frontal advection may generate much stronger melt, which may approach or exceed radiation melt limits.

It has nonetheless been noted in studies of alpine, glacial, and prairie snowpacks that radiant energy accounts for a high proportion of all the melt (cf. Miller, 1955; Granger and Male, 1977; Jordan, 1979). Storr (1979), using the energy balance parameterization of the U.S. Army Corps

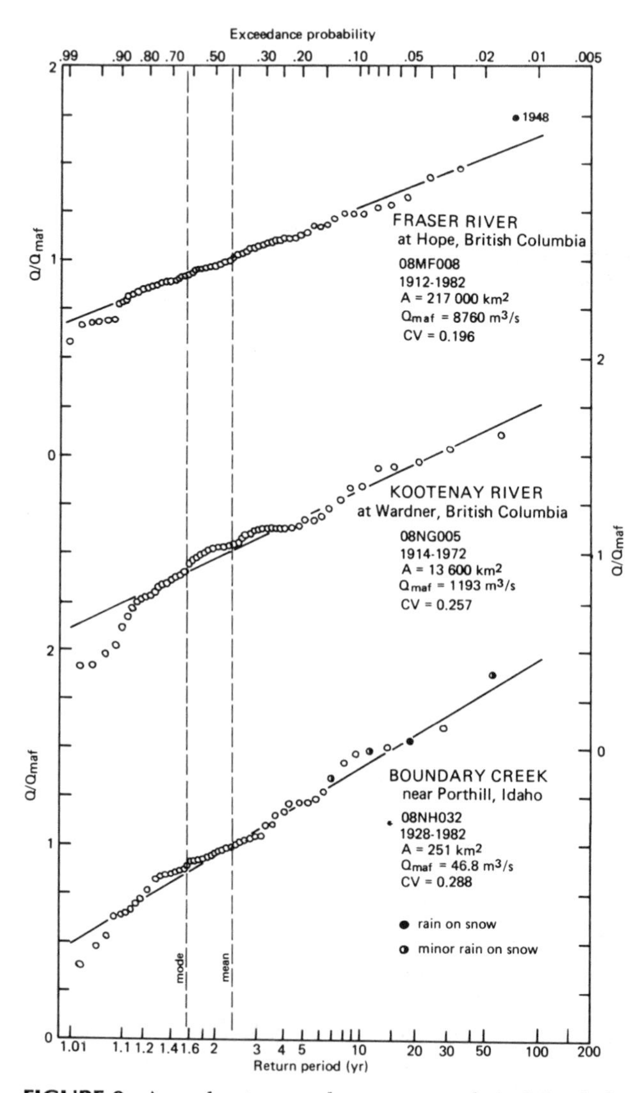

FIGURE 9. Annual extreme value sequence of nival floods for several rivers. Annual peaks on the ordinates are scaled as Q/Q_{maf} (mean annual flood) so that the slope of each sequence is a direct, comparable measure of flood variability. The smallest floods fall off the general trend, most notably for Kootenay River: this results from exhaustion of the snowpack in dry years at low elevations before higher areas are contributing flow. CV is the coefficient of variation of the flood sequence. Data are plotted on Gumbel paper. In a sequence unconstrained as to type, two autumn floods—both smaller than MAF—would enter the Boundary Creek population. Otherwise these are absolute extreme sequences and coincide with the partial duration sequences for the same number of events.

of Engineers (USACE) (1960), determined that in the subalpine Marmot Creek watershed, Alberta—a subhumid, continental region—about 80% of the melt energy was derived from shortwave energy receipt and 20% from longwave energy. The division was related to forest cover and snow distribution.

In summary, it appears that the radiation-driven "potential melt" rate does constitute a fairly strict upper limit for the generation of purely nival floods. The potential rate is apt not to be reached in most circumstances. Furthermore, snowpack and drainage network effects will alter the runoff peak at downstream points. However, the limit is apt to be approached, easily, to within a factor of 2 or so in small watersheds, so that the range of nival floods remains fairly strictly constrained.

Frequency Distribution of Nival Floods

Figure 9 gives some examples of nival annual flood sequences. In each case the sequence assuredly consists of snowmelt-generated flows or of snowmelt with minor addition of rainfall (rain on snow is considered below). In practice, records are checked by noting the date of the peak flow in each year and by investigating the synoptic antecedents. The records are drawn from regions with relatively heavy snowfall so that "snow exhaustion" does not in most years limit the record before the runoff pathways become saturated. The records do, indeed, give the appearance of a fairly strict limit and consequent relatively low variability predicted in the last section.

In regions with shallow snow cover the nival flood is limited by snow exhaustion over a substantial portion of the basin (*not* necessarily by complete disappearance) and the flood magnitude may be well correlated with snow volume. Figure 10 illustrates this for a small basin in the semi-arid Okanagan Valley of British Columbia.

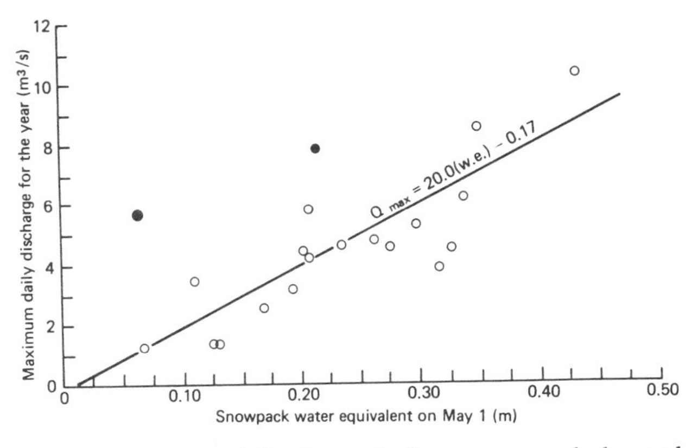

FIGURE 10. Annual flood magnitude versus annual observed maximum snow accumulation for Shatford Creek, British Columbia (WSC gauge 08NM037); A = 101 km^2. Snow accumulation is indexed by the Nickel Plate snow course (2G02) at 1890-m elevation in the basin headwaters. The flood sequence is purely nival except for rain on snow in 2 yr (closed circles). The indicated regression (excluding rain on snow) has $r^2 = 0.68$; standard error of estimate = ± 1.3 m^3/s.

RAINSTORM FLOODS

Event Magnitude

The largest streamflows in small and medium-size basins are generated by rainstorms. Rainstorm floods are considered in several chapters in this volume and will not be reviewed systematically. This discussion is to emphasize the size and timing of rainstorm floods in comparison with nival floods.

The timing of storms depends on synoptic weather, which exhibits little or no diurnal regularity. The effect of this on runoff timing can be seen by comparing the phase lag for runoff in a glacierized basin between melt-dominated and storm-dominated seasons (Fig. 11). Of course, regional climate determines the seasonal pattern of occurrence of major storms (e.g., the autumn/winter preponderance of rainstorm floods on the west coast of North America).

The magnitude of rainstorms depends on the specific humidity of the convergent air mass and the efficiency of the lift and condensation mechanisms that induce precipitation. Realized precipitation intensity exceeds feasible snowmelt rates—about 40 mm in 24 hr—even in far northern regions (Table 2). Two examples emphasize the synoptic conditions that generate extreme rain.

On July 19–21, 1970 a severe rainstorm occurred over about 80,000 km^2 of the Mackenzie Mountains in northwestern Canada (Fig. 12). Rain was generated by a cyclonic storm upon which was superimposed convective instability in the warm sector. The convection cells intruded into the air mass of a cold low trailing the surface pressure pattern aloft triggering intense rainfall (MacKay et al., 1973; Burns, 1974). A high proportion of the estimated available precipitable water was released. The maximum runoff rates were calculated to be equivalent to 50 mm in 24 hr (but no gauges in the region operated throughout the storm) compared with interpolated maximum 24-hr precipitation of 62 mm on July 20 in the headwaters of the Keele, Mountain, and Arctic Red rivers. Severe flooding occurred in these major left-bank tributaries of the Mackenzie River and increased the flow of the Mackenzie River at Norman Wells by 70% over that immediately before the storm. Meteorological estimates of the recurrence interval for the storm, based on station analyses in the region, were about 10 yr. Dendrochronological determinations on storm-destroyed floodplain vegetation (see Fig. 22) indicated that no flood had approached this magnitude in 100 yr. This inference is corroborated by historical chronicles and il-

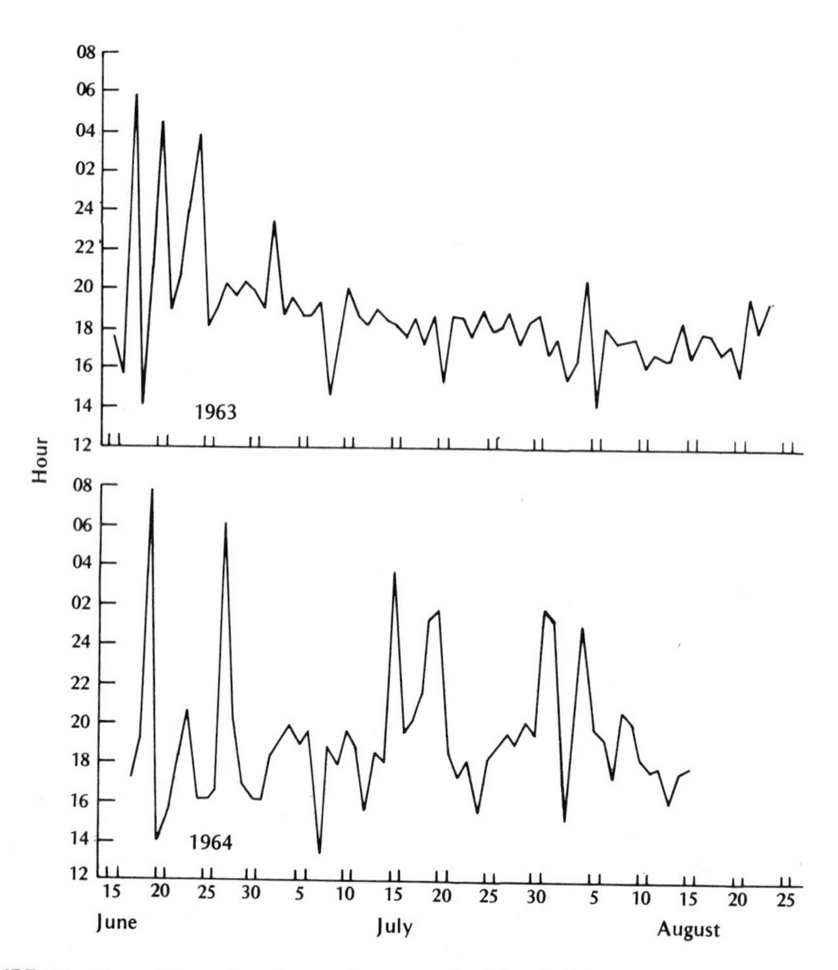

FIGURE 11. Phase (time of peak runoff at gauge) of Lewis River, Baffin Island, Northwest Territories. A = 205 km^2; 85% glacier covered. 1963 was a fair weather season dominated by diurnal melt floods (Fig. 7*a*); 1964 was a storm-dominated season (Fig. 7*b*). The diurnal distribution of peak flow times is far more variable in the latter year.

TABLE 2 Extreme 24-hr Rainfalls in Northern Regions[a]

Station	Length of Record (yr)	24-hr Rainfall (mm)
A. Arctic stations		
Brevoort Island	16	68.3
Cape Dyer	22	51.0
Coppermine	46	63.5
Coral Harbour	37	77.0
Hall Beach	24	52.6
Holman	25	50.5
Mould Bay	33	47.8
Nottingham Island	41	56.4
Pond Inlet	26	58.4
Resolution Island	14	69.6
Eureka	33	41.7
B. Subarctic		
Old Crow	15	58.4
Dawson	78	52.8
Aklavik	35	50.8
Baker Lake	33	52.1
Chesterfield Inlet	50	57.9
Fort Good Hope	65	69.3
Fort Norman	50	73.7
Fort Simpson	62	86.4
Yellowknife	38	82.8

[a] Data pertain to climatic day reporting periods. It is conceivable that greater rainfall has occurred in 24 hr falling partly in each of 2 days.

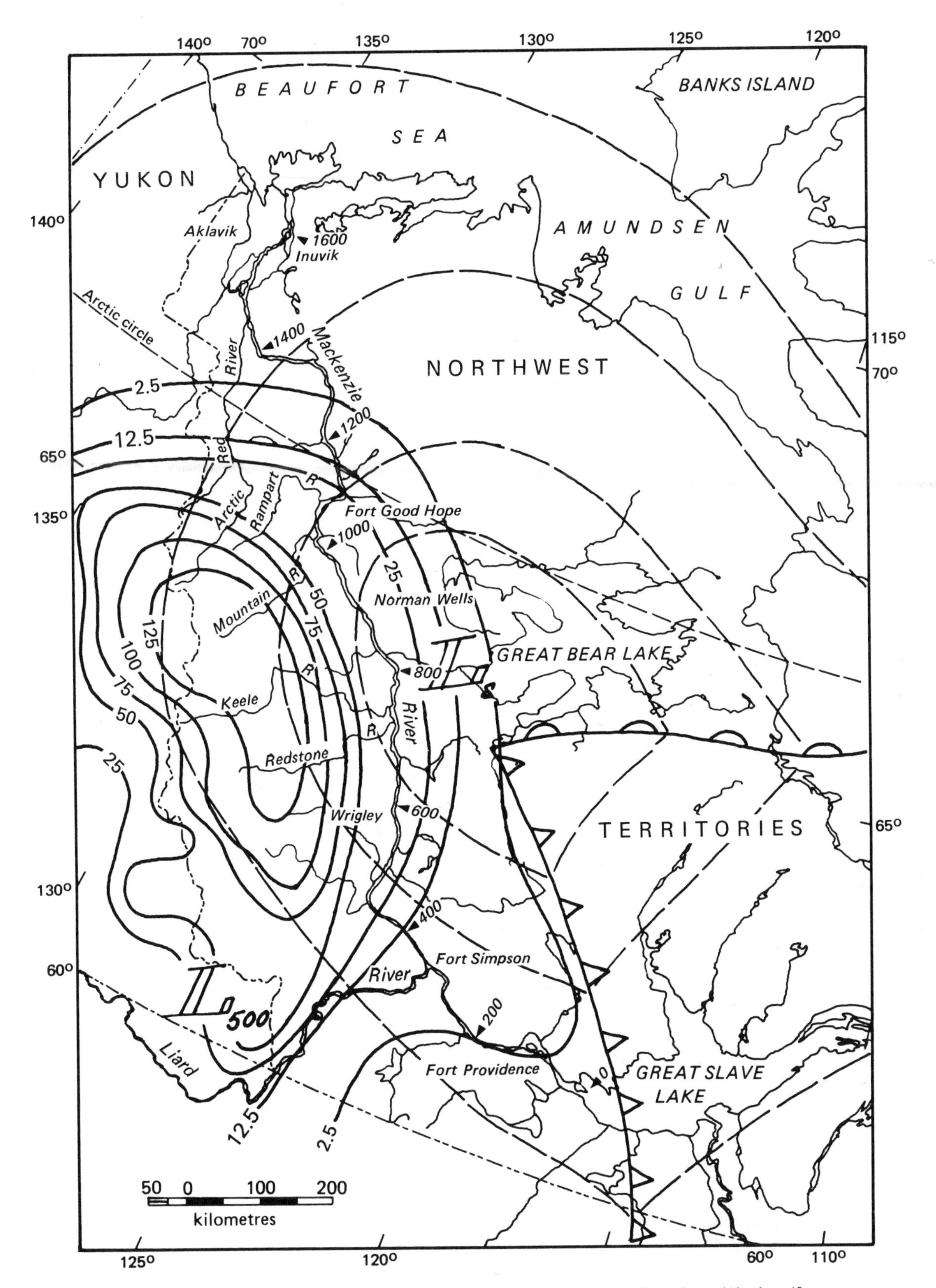

FIGURE 12. Arctic Red River storm of July 18–21, 1970: estimated total precipitation (from Burns, 1974; Fig. 10.11), and synoptic analysis of July 20, 1200Z. L_s and L_{500} indicate centers of surface and 500 mb low pressure. Further discussion in text.

lustrates the problem of determining recurrence intervals for areally extensive events in regions with few stations.

A notable storm centered over southern Ellesmere Island in the Canadian high arctic (75°–80°N) in 1973 was described by Cogley and McCann (1976). The pattern of precipitation is relatively well known from observations in several research camps. At Vendom Fjord, near the apparent storm center, 50 mm of rain fell on July 22 (total precipitation: 55 mm on July 21–23). A northeasterly moving cyclonic storm was responsible for the rain (and some wet snow). Such extreme rains are not frequently experienced in the Arctic so that, in most years, the nival flood provides the annual maximum flow.

In very large basins nival flood dominance persists (e.g., in the Fraser River) because individual rainstorm events never deliver sufficient precipitation over a substantial portion of the entire basin to exceed total snowmelt runoff rates. As a rough estimate of the limit size for basins with storm flood dominance, one may use the approximate dimensions of intense cyclonic storm activity, on the order of 3×10^2 km, to yield the order of 10^5 km^2 area. However, the real limit will depend on regional factors (cf. Figs. 5*a* and 5*b*).

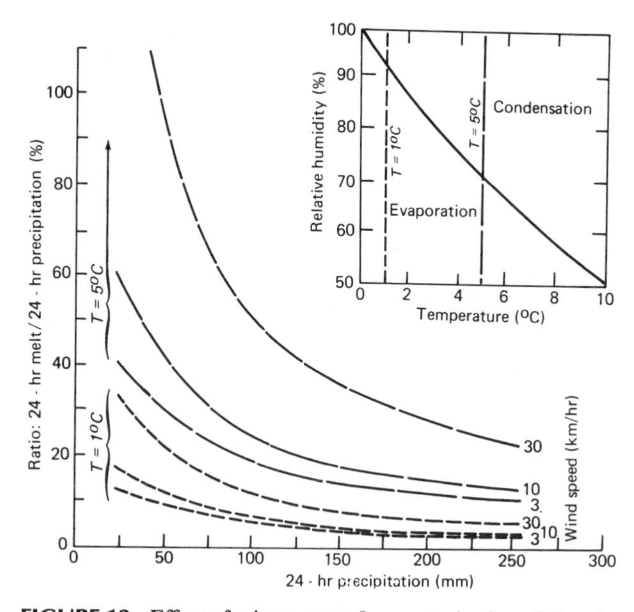

FIGURE 13. Effect of rain on snow [parameterization of Equation (4)] for various combinations of meteorologic elements and relative humidity 100%. Inset: temperature and humidity control of the occurrence of condensation onto the snow surface.

Rain on Snow

Rain on snow may induce the most extreme floods recorded in nival regions, particularly in temperate, coastal mountains where autumn storms are the most vigorous of the year and frequently bring heavy rain onto shallow snow. The sources of energy for snowmelt are quite different than in fair weather. USACE (1960) provided a parameterization of Equation (1) that serves as a basis for discussion. Terms are given as millimeters of water equivalent melted (M) by the energy supplied:

$M(K\downarrow)$ becomes small, say of order 0.1 mm/hr

$M(L^*) = 1.33T_t$, where T_t is mean air temperature for the period, t hours

$M(Q_H + Q_E) = 0.86V_tT_t$, where V_t is average wind speed in meters per second, at 15 m above the surface, and the air is saturated

$M(Q_P) = 0.013P_tT_t$, where P_t is total precipitation in millimeters

$$M_t \approx 0.1t + (1.33 + 0.86V_t + 0.013P_t)T_t \cdot (t/24) \tag{4}$$

For the sensible heat contained in the rain to become significant, P must approach 100 mm in 24 hr. This occurs moderately frequently (with return period less than 1 in 5 yr, say) only on exposed, high mountain slopes. Figure 13 illustrates the sensitivity of Equation (4) to the other underlying weather parameters. In general, temperature has by far the largest impact of any individual parameter and obviously becomes relatively more important as precipitation declines. The effect of wind becomes greater as temperature rises, expressing the increasing importance of turbulent energy exchange in these circumstances. "Condensation melt" plus turbulent sensible energy exchange at the snow surface [collectively, the third term on the right-hand side of Equation (4)] is important if $V > 1$ m/s, or about 3.5 km/hr, which occurs frequently. Hydrologists have come to regard this as the principal source of energy for snowmelt during rain. The effect is most important with moderate rainfall and relatively high winds, a common occurrence around mountains.

In a study of rain-on-snow events over 25 yr in a 60-ha, forested, mountain watershed in western Oregon, where snow persists for relatively short periods, Harr (1981) found that snowmelt increased peak runoff by up to 30%. He obtained separate frequency analyses for rainfall and rain-on-snow events, shown here as Figure 14, which emphasize the importance of snowmelt in contributing to the highest flows in most years. In the adjacent Willamette River, draining 19,000 km^2, 18 of the 24 greatest floods since 1814 have been rain-on-snow events (and the antecedents of three of the remaining floods are unknown).

A further example illustrates the effects of snow on flood magnitude and flood formation. Major storms in the southern Cascade Mountains of British Columbia in December 1980 and January 1984 yielded data of Table 3

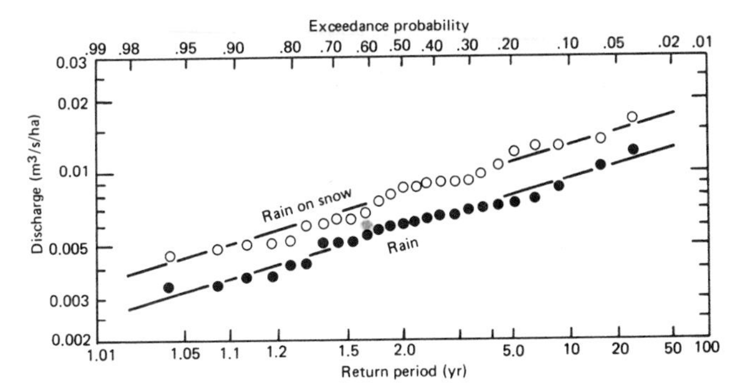

FIGURE 14. Partial duration plots for rain on snow and for rainstorm floods (25 events) for watershed 2, H. J. Andrews Experimental Forest, Oregon; A = 60 ha. Data from Harr (1981, Table 2). The plot is on lognormal probability scales, and the lines are fitted by eye and deliberately made parallel. Waylen (1982) analyzed the corresponding annual extreme value sequences and showed that Gumbel's distribution provides a reasonable fit.

for a small watershed instrumented with a snow pillow. There was a substantial snowpack in 1980 but very little snow in 1984: however, in 1984 the ground was frozen. Details of the storms are given by Church and Miles (1987). The 1980 event produced maximum streamflow well in excess of that expected for the storm, to judge from return period criteria for the rainfall. The total proportion of runoff from the two storms was similar, but the 1980 storm experienced a relatively more peaked hydrograph even though rainfall intensity seems to have been greatest in 1984. The effect of drainage through the snow appears to be the most plausible reason for this distinctive behavior. Harr also noted that the rise time of headwater rain on snow floods was shorter than that of rain floods. One might speculate that previously established passages through the snow present less resistance to flow than rough ground, that these preferred passages concentrate runoff more rapidly, or that depression storage under a snowpack is greatly reduced by the occurrence of ice or slush.

RIVER BLOCKAGE

Snow

The highest stage observed in rivers in nival regime may not be associated with meteorological events at all or with high flows. Ice and, in small streams, snow may block channels and pond water during winter or early spring.

Snow will retain ponded water for only short periods and is notable only in small, headward tributaries. It is particularly effective in incised channels on prairie or tundra, where snow drifts frequently. A. C. Wankiewicz (personal communication, 1986) has studied annually recurrent flooding of prairie fields produced by blockage of drainage ditches (Fig. 15). Woo and Sauriol (1980; cf. Woo, 1983) described spring breakup in a gullied drainage system in the Canadian arctic where miniature "damburst" floods followed the rupturing of snow barriers to increase downstream discharge by factors of up to five times in a few hours. Snow also impedes channel re-establishment in winter-dry channels, creating further irregularities in the runoff early in the nival flood.

Ice

Ice jams affect large rivers. There are some geographic and climatic regularities in their occurrence. Ice jams can be divided into two general types: winter jams and breakup jams. The former comprise slush and frazil ice accumulations that tend to occur regularly at the downstream end of steep, turbulent reaches. They form "hanging dams" where frazil ice is drawn below surface ice cover (cf. Ashton, 1978) and have some effect on upstream stage because of the channel constriction they create.

Breakup jams of moving river ice may be massive and destructive, producing large stage rises upstream and destructive floods downstream. They recur at constricted places along channels (MacKay and Mackay, 1973); however,

TABLE 3 Data of Two Winter Storms in Upper Coquihalla Valley, British Columbia

Datum	Dec. 24–27, 1980	Jan. 1–4, 1984
4-day precipitation (P), Ottomite gauge	218 mm	175 mm
4-day snowmelt (M) Ottomite snow pillow	21 mm	—
Storm runoff (R), Coquihalla River at Needle Creek	142 mm	110 mm
$R/(P + M)$	0.59	0.63
Greatest day precipitation (day)	81 mm (26)	112 mm (4)
Greatest 2-day precipitation	155 mm (25/26)	152 mm (3/4)
Instantaneous peak flow	65.3 m^3/s (26th)	47.7 m^3/s (4th)
Return period of instantaneous peak flow	27 yr	19 yr
Daily peak flow	46.0 m^3/s	33.5 m^3/s
Return period of daily peak flow	38 yr	11 yr

FIGURE 15. Flooded fields 60 km southwest of Winnipeg, Manitoba, produced by blockage of drainage channels and culverts by snow. Once drainage lines are established, the fields are cleared in hours (photograph by A. C. Wankiewicz, Environment Canada).

their occurrence from year to year depends on the weather and streamflow preceding and during breakup. Persistent fair weather and gradual warming in spring may not induce major jams because river ice will be weakened by warm valley conditions and insolation before considerable runoff appears from the uplands. On the other hand, an abrupt warm spell with rain may create serious ice jams as substantial runoff encounters relatively strong ice.

Rivers that flow long distances to the north or have major, early breaking tributaries are prone to major jams. Mackenzie River experiences both of these circumstances (Mackay, 1963). It is interesting to note that the regulation of northern rivers for hydroelectric power generation and their use for peaking power has in recent years created the phenomenon of winter "breakup" and ice jam flooding when flows are abruptly increased.

The irregular occurrence of ice jams, their restriction to northern, continental rivers mainly in sparsely settled regions, and the difficulty of maintaining hydrological observations during their occurrence has resulted in a dearth of information about their effect on flood levels. One expects, from their characteristics and occurrence, that they may produce a highly variable frequency distribution. Gerard and Karpuk (1979) constructed a 120-yr record of ice jam floods for Peace River at Fort Vermilion, Alberta, using historical sources. Their result (Fig. 16) confirms the expectation. For return periods greater than about 10 yr, ice jam stage, rather than river discharge, creates the most severe floods at this site.

The effect of ice may vary substantially along a river because some reaches are more prone to jams than others. MacKay and Mackay (1973) have determined the elevation of ice push limits along Mackenzie River (Fig. 17). In places ice is shoved to substantial heights above the water level so that the effect is not simply one of stage increase.

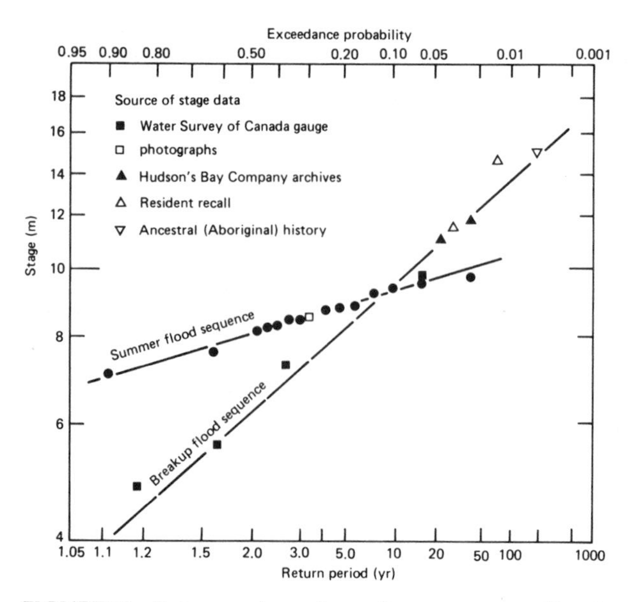

FIGURE 16. Extreme values of stage for open water and breakup floods on Peace River at Fort Vermilion, Alberta (WSC gauge 07HF001). Data from Gerard and Karpuk (1979, their Fig. 6). The plot is on lognormal probability scales.

Icing

In northern continental regions an additional winter ice phenomenon may influence nival flood behavior: that is, the occurrence of "icing" or *aufeis* (Russian: *naled*) in river channels. Icing occurs where the river channel freezes to the bed along part of the channel, but there continues to be a source of water discharge into the channel. The water is forced to the surface and freezes there. Over the winter the channel may largely or entirely fill with ice. Situations where persistent water seepage leads to extensive icing development (Fig. 18*a*) appear to be of two types:

1. Riverbed seepages, where flow persists in riverbed gravels after the surface has frozen. When advancing seasonal frost affects closure of the passage against an impermeable stratum or against permafrost below the channel, hydrostatic pressure develops upstream and may force water to the surface at a weak point in the frozen gravels. Lake outlet channels commonly exhibit this effect.
2. Perennial springs tapping deep groundwater (in permafrost regions, subpermafrost water flowing to the surface via bedrock passages or karst resurgences).

In spring the first meltwater that encounters an ice-full channel must flow over or around the ice, often on the floodplain of the river (Fig. 18*b*). Sediments frequently are found deposited on the ice. After some days, a channel normally is eroded through the ice, but the river may remain in part ice-bound for much of the summer. Redirection of water by ice may create unusual flood conditions (Fig. 18*c*) and increase river instability. Carey (1973) and Kane (1981) have described icing in greater detail.

GLACIAL FLOODS

Proglacial drainage is subject to daily nival floods or, for a sufficiently large basin, to a distinct seasonal flood modulated by synoptic weather. The characteristics of such floods have already been described. It bears emphasis, however, that even on a glacier, by far the largest proportion of water is derived from the melting of seasonal snow.

Singular events occur as the result of more or less sudden drainage of water stored on the surface, within, or beneath the glacier, trapped at the margin of the ice, or dammed where the glacier blocks normal drainage lines. Often, drainage of water trapped in or around the glacier produces unexceptional floods because the reservoirs are not large or because they do not drain completely. However, they do often drain during heavy summer rain as the result of erosive activity of the storm runoff, so they may further increase relatively high floods. They may also be accompanied by ice or slush runs that increase stage. Because considerable water volumes may be stored within or beneath glaciers, no visible lake need be present for the possibility to exist of drainage flows.

Drainage of ice-dammed lakes may yield catastrophic floods. Nye (1976) and Clarke (1982) have developed a hydrostatic-thermodynamic model to describe the drainage of an ice-dammed lake by the development of a tunnel under the ice. The principal postulates of the model are as follows:

1. Outflow commences when the hydrostatic pressure of the water equals the weight of the ice at the critical point on the potential outflow route; that is, when $g\rho_w h_w = g\rho_i h_{ic}$, or $h_w/h_{ic} \approx 0.9$.
2. Thereafter, the circular tunnel enlarges at a rate proportional to discharge and temperature of the water since the enlargement is supposed to be accomplished by melt of the tunnel walls using advected sensible heat and friction losses of the water. This entails an estimation of the frictional resistance to flow in the drainage passages.

In summary,

$$Q = f[\mathrm{h_i}, z_0, \mathrm{L_0}, \mathrm{Q_i}, \mathrm{n}, \theta_l, V(h_w)] \qquad (5)$$

where

Q = the outflow
z_0 = initial lake elevation above the tunnel outlet
$\mathrm{L_0}$ = drainage tunnel plan length
$\mathrm{Q_i}$ = water inflow to the lake
n = Manning's resistance number in the tunnel
θ_l = lake water temperature
$V(h_w)$ = lake hypsometry, or volume distribution with depth

The italicized quantities are treated as fixed parameters of the problem even though they may not all in fact be constant throughout the event. When $Q > \mathrm{Q_i}$ the lake begins to drain.

The principal functional dependences in the model lead to:

$$Q_{\max} \propto \frac{V_0\langle -\partial\phi/\partial s\rangle^{11/16}}{\mathrm{n}} \qquad \theta_l \to 0 \qquad (6)$$

in which $\langle -\partial\phi/\partial s\rangle$ is the average fluid potential gradient along the tunnel and V_0 is the initial volume of water in the lake:

$$Q_{\max} \propto (\theta_l V_0)^{4/5}\frac{\langle -\partial\phi/\partial s\rangle^{21/50}}{\mathrm{n}^2} \qquad \theta_l > 0 \qquad (7)$$

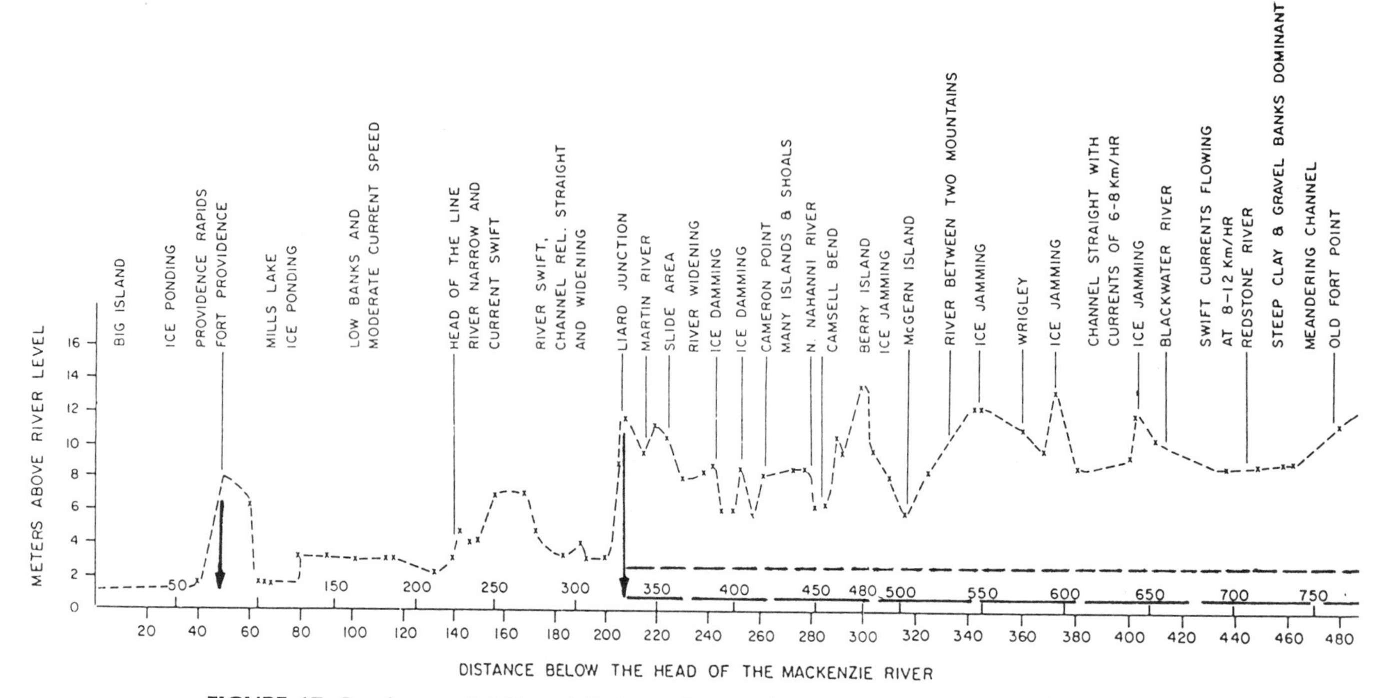

FIGURE 17. Ice-shove and driftwood limits on Mackenzie River, Northwest Territories (from MacKay and Mackay, 1973, their Figs. 6–8). The reference datum is the waterline of June 30, 1970, which was the date of maximum flood at Fort Providence. However, at Fort Good Hope,

FIGURE 18. Icing. (*a*) Large river icing, Marsh Fork of Canning River, Alaska north slope. Photograph taken June 25, 1971 (view upstream). (*b*) River flowing around an icing in nival flood (view upstream) (photograph by C. G. Seagel). (*c*) Stream in flood in an ice-choked channel (view upstream) (photograph by C. G. Seagel).

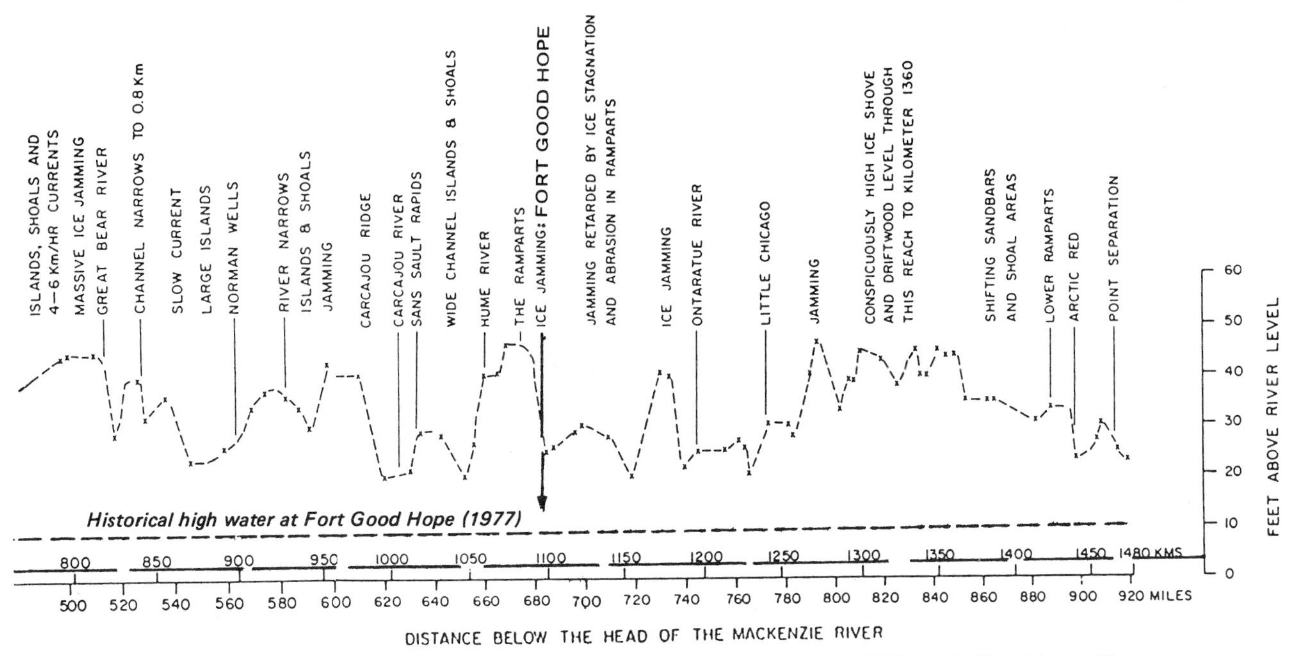

high water (June 14) was 0.87 m higher, and the historical high (1977) was 2.89 m higher. These levels are drawn in for the river below the Liard River confluence: They are precisely relevant only at Fort Good Hope. Figure 12 shows Mackenzie River, including river distances in kilometers.

(See Clarke, 1982, for the complete development.) The difference between the two results reflects the relative dominance of frictional energy losses or advected heat in enlarging the tunnel. Results from Equation (7) are as much as an order of magnitude larger than those from Equation (6) when $\theta_l > 5°C$. Figure 19*a* illustrates the sensitivity of the model to the parameters n and θ_l. Although several parameters must normally be guessed, it seems important to measure θ_l.

The Clarke model may not describe all jökulhlaups ("glacier burst" floods). Figure 19*b* is a hydrograph showing the typically increasing flow until exhaustion of the reservoir: However, it resulted from drainage over rather than through ice. Exceedingly arbitrary parameter adjustments would be necessary to force the model to fit.

An alternative approach is to obtain an empirical scale correlation of Q_{max} and V_0. This was first accomplished by Clague and Mathews (1973) as:

$$Q_{max} = 0.0075 V_0^{2/3}$$

The result has been reworked by several investigators (see, for example, Costa, this volume). The exponent remains well below 1.0, which is at variance with the rational results in Equations (6) and (7), but no explanation is available.

Even more catastrophic floods may be associated, directly or indirectly, with glaciers in the cases of volcanic eruptions and moraine dam failures (which may be sudden). The former are altogether exceptional geophysical phenomena. The latter class of events is considered by Costa (this volume). Neither is pursued here.

MIXED POPULATIONS OF FLOODS

In Figures 14 and 16 flood populations derived from two quite distinct phenomena are illustrated for each of two rivers. More generally, it was noted in the introduction that mixed populations of nival and rainstorm floods occur on many rivers. Both basin size and geographic factors influence the relative importance of each sequence.

In humid regions small drainage basins (say, less than 10^2 km^2 area) tend to yield pure populations of rainstorm floods or mixed sequences of rain on snow. Heretofore, only Harr (1981) has distinguished the latter in frequency analysis. The distinction requires careful retrospective synoptic analysis or field observations. It was suggested above that basins larger than 10^5 km^2 in nival regions are dominated by snowmelt floods because of storm size limitations. Even in this instance, however, there may be significant complications. The modern flood of record on the Fraser River (Fig. 9)—the 1948 event—was the consequence of sudden

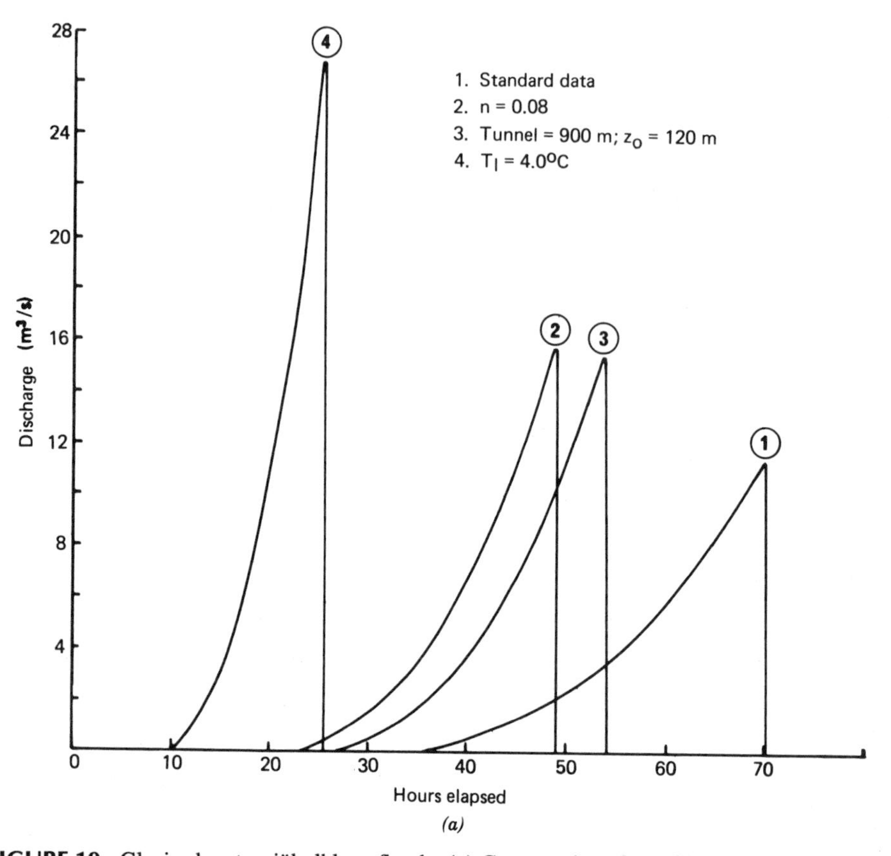

FIGURE 19. Glacier-burst or jökulhlaup floods. (*a*) Computations from Clarke's (1982) model for jökulhlaups from a small lake in the Canadian Coast Mountains (Homathko River drainage). The input data [Equation (5)] were, for model 1, h_i = 100 m; z_0 = 150 m; L_0 = 1800 m; Q_i = 1.0 m/s; n = 0.12; θ_l = 1.0°C; $V_0 = 0.5 \times 10^6 m^3$. Model 2 has n = 0.08, otherwise as model 1. Model 3 has z_0 = 120 m, L_0 = 900 m (i.e., part of the tunnel is assumed to exist in stagnant ice before the drainage), otherwise as model 1. Model 4 has θ_l = 4.0°C, otherwise as model 1.

warming over most of the basin at the beginning of May after an unusually cold spring, accompanied by above normal precipitation during that month. The maximum flow occurred on May 31 and may be regarded as a snowmelt plus rainfall induced flood.

Drainage basins with areas between 10^2 km^2 and 10^5 km^2 tend to exhibit well-developed mixed populations, but whether one type or another is most prominent depends on regional climate and drainage basin size. Figure 20 illustrates a transect in southern British Columbia from the humid coast to the semi-arid interior east of the Coast and Cascade Mountains. On the coast rainstorm floods are dominant: in the interior snowmelt floods dominate, although exceptional rainstorm floods undoubtedly occur, and rain on snow may be significant (Fig. 10). Mixed populations are common within the mountains.

Figure 21 illustrates mixed populations of floods for Chilliwack and Bella Coola rivers, in the British Columbia Coast Mountains. Four sequences are recognized in the Chilliwack River—rain, nival, autumn–winter rain on snow, and spring rain on snow. Spring rain on snow yields a flood sequence with similar variability but somewhat greater mean flow than snowmelt. In comparison, autumn–winter storms produce much more variable and larger floods. Rainstorms produce a sequence of intermediate characteristics.

Bella Coola River, several hundred kilometres north of Chilliwack River, drains an area about 3.5 times as large, with much of the difference represented by drainage of the interior plateau east of the mountains. It is about 9% glacier covered (there are also small glaciers in Chilliwack basin): accordingly a fifth "summer nival" sequence is recognized, when water is derived from ice and from restricted patches of alpine snow. Again, autumn rain on snow yields the greatest floods, but that sequence is not so absolutely dominant as in Chilliwack River. Rain is relatively prominent and yields a sequence with about the same variance as spring melt. Spring rain on snow is not at all prominent

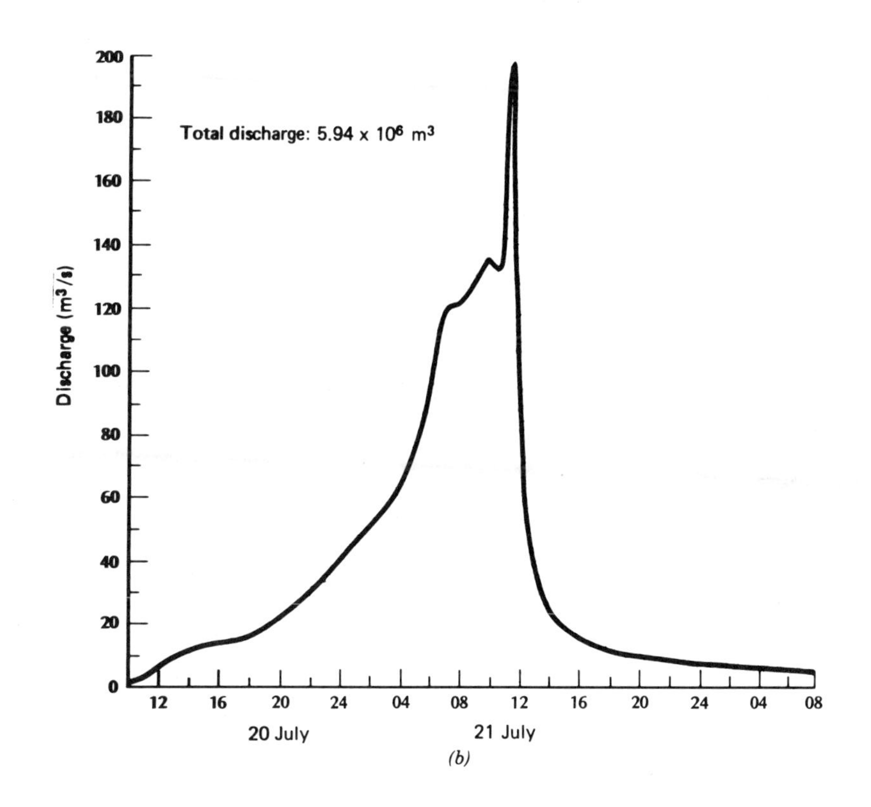

FIGURE 19. (*Continued*). (*b*) Jökulhlaup at Ekalugad Fjord, Baffin Island, July 1967, generated by overflow of an ice dam.

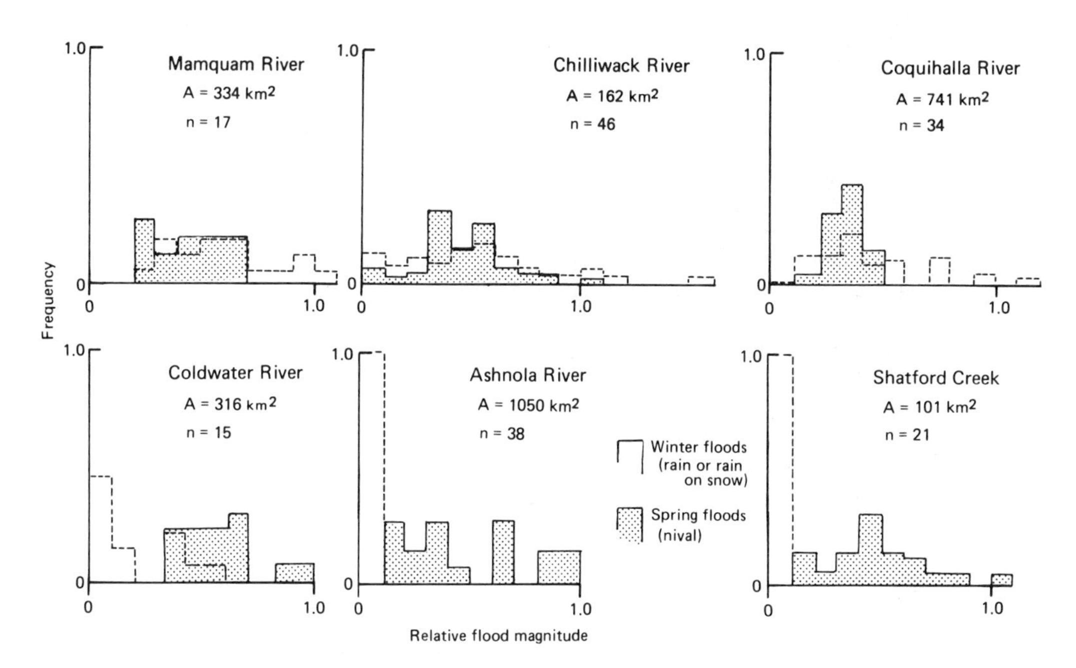

FIGURE 20. Frequency distribution of (spring) nival and (autumn) rainstorm floods in drainage basins of intermediate size on a transect from the coast (Mamquam River) to the interior (Shatford Creek) across the Coast and Cascade Mountains, southern British Columbia. The rivers all drain high mountains. The flows have been normalized by a selected flow that is near the maximum flow observed in each river in order to make the plots more directly comparable.

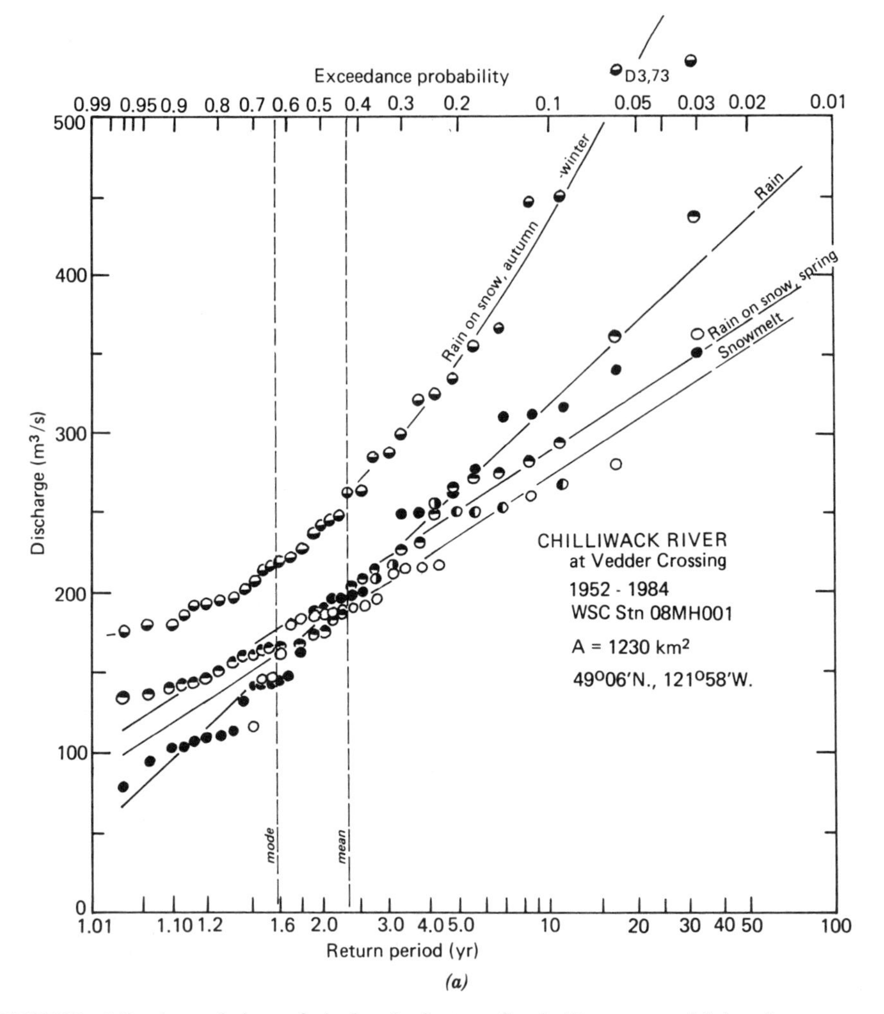

FIGURE 21. Mixed populations of nival and rainstorm floods. Data are partial duration sequences with n = number of years of record, plotted on Gumbel paper. Autumn–winter rain on snow is divided from spring rain on snow at March 1: Before that date rain is apt to fall on fresh or cold snow, after onto a deep, partly or entirely isothermal pack. In Bella Coola basin, there are no floods in the January–April period, so the chance for misclassification is restricted to Chilliwack basin. "Summer nival" floods in Bella Coola basin are those occurring after July 1. Rain floods are divided from autumn rain on snow by antecedent temperature records: there may remain some misclassification. Some sequences have fewer than n events identifiable in the record. Lesser events are confounded with other conditions and are not separated: the plotted floods are the largest of each type to have occurred and include all the largest flows in the record. Some events of one type have minor runoff contribution from another type (i.e., minor rain during a melt flood), and these are marked. Spring nival floods (snowmelt) are restricted to one event per season. (*a*) Chilliwack River, British Columbia: four populations. The autumn–winter rain-on-snow sequence is notably skewed. (In Fig. 20 the spring sequences and the autumn–winter sequences have been combined.)

here. The summer nival sequence has notably low variability and does not produce very large floods.

Synthesis of mixed flood sequences has been considered by Waylen and Woo (1982; Woo and Waylen, 1985) and will not be pursued here. The occurrence of mixed flood populations in rivers of nival regions raises an important cautionary note in respect of extension of flood records from geologic evidence. How can one be sure what were the antecedents of a particular flood? Hence, to what sequence should it be assigned? Can one be sure that an outstanding prehistoric flood does not represent an altogether exceptional event that should not be considered in any

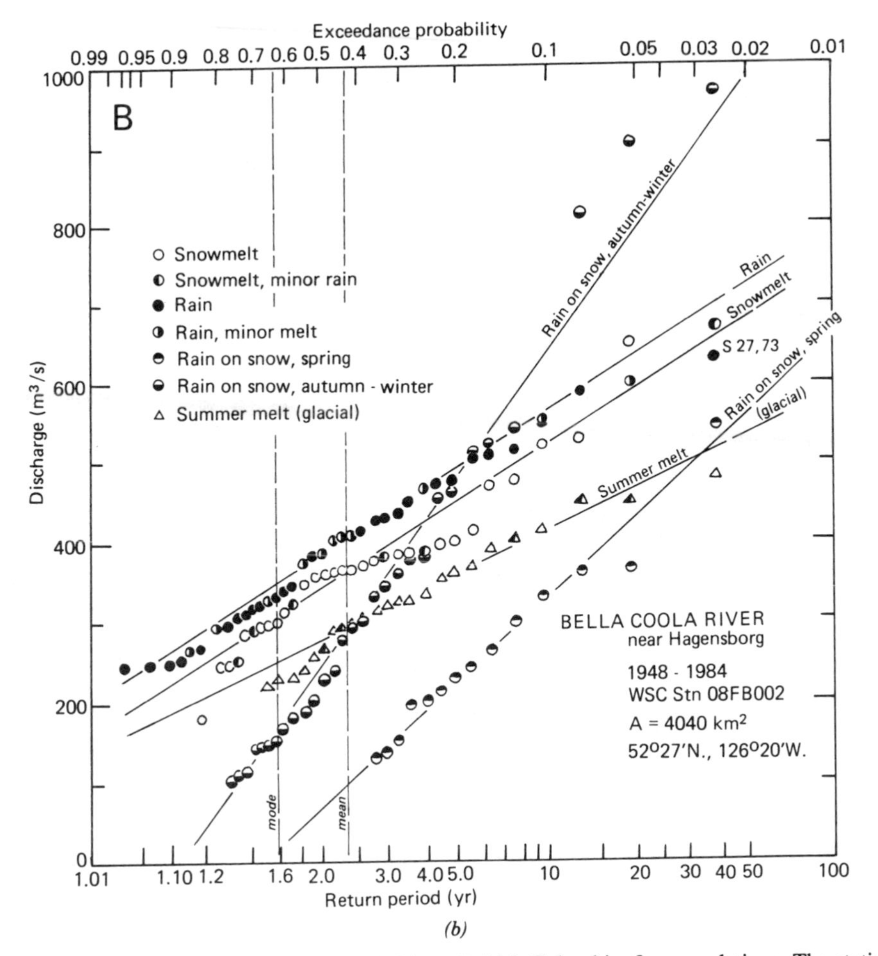

FIGURE 21. *(Continued).* (*b*) Bella Coola River, British Columbia: five populations. The station has been moved: flows after 1968 are increased by 17% in accordance with the comparison of flows between the two sites during the period of combined record.

sequence analysis at all? Sometimes the geologic evidence will permit discrimination of causes, but there appears to be no general procedure to answer these questions.

MORPHOLOGIC EFFECTS OF NIVAL FLOODS

So far as geomorphologic work is concerned, all flows that transport sediments are significant. Hence, we should be interested in the total flow duration or in the partial duration sequence of flows that exceed the threshold for sediment transport. For frequency analysis nival floods introduce extraordinary complications for the selection of successive independent events since snow conditions over the watershed may effectively impart an unusually long "memory" into the flow record, as inspection of the hydrographs in Figures 4, 5, and 7 will show. For the large basins for which the frequency distribution of nival floods is displayed in Figure 9, the annual sequence and the partial duration sequence for the same number of events are identical.

The relatively flat frequency plots exhibited by such populations of nival floods suggest that they perform relatively limited work to change river channel form. Nival floods frequently flow near to just over bankfull. Severe floods sometimes occur when a rapid thaw encounters ice or snow-choked channels, so that blockage of drainage ways causes ponding of water in channels or on the land. In fact, the appearances may be misleading. For many rivers the nival flood is the major sediment-transporting event of the year. On Fraser River (Fig. 5*a*), the bulk of the annual sediment yield of about 20 million tonnes is delivered during the May–July period. Much of the sediment is picked up along the river banks as the river rises in spring. Nanson (1974) demonstrated a similar phenomenon on a very small stream in the Canadian Rocky Mountains.

In both cases a substantial hysteresis occurs in sediment concentration between the spring (rising) and summer (falling) limbs of the nival flood.

Abrupt changes in river morphology more often are associated with major rainstorm floods. On Chilliwack River the December 1975 flood, with return period about 20 yr, deposited some 200,000 m^3 of bed material caliber sediment on the alluvial fan where the river enters Fraser lowland. This is about four times the estimated average annual bed material charge. In the 7-km reach downstream from Vedder Crossing, bed elevation increased by up to 0.6 m, averaging 0.3 m. Substantial widening of the channel occurred in the main depositional reach (unpublished data of D. G. McLean). The channel of Bella Coola River consists of a sequence of relatively stable reaches divided by braided or laterally shifting "sedimentation zones," where episodically mobile bed material is stored. The flood of September 1973 prompted a major cutoff of the most distal sedimentation reach, the "big bend" near Bella Coola village, when sediment filled the former channel which had persisted for nearly a century (Church, 1983).

Even in the high arctic, summer rainstorm floods carrying runoff from the thawed surface soil appear to deliver the highest sediment loads and to produce the greatest changes along channels (Rudberg, 1963; Cogley and McCann, 1976).

Nonetheless, the Arctic Red River storm of 1970 (Fig. 12) produced deep, fast-moving water over the floodplain, so that the forest of 100-yr class spruce was flattened (Fig. 22), yet fluvial erosion was restricted to incremental activity on channel bends and around the heads of channel islands. No major pattern reorganization of the channel occurred. It is possible that the presence of permafrost in the floodplain limited the erodibility of the sediments. In this situation erosion in the short run is limited by thaw rate. This appears not to produce any notable peculiarities of channel morphology in the long run but probably exaggerates the uniformitarian nature of fluvial processes.

The effects observed on Arctic Red River are not unique and not restricted to frozen soils. Gardner (1977) reported very minor effects of an estimated 500-yr flood on Grand River in Ontario and reviewed similar findings in other regions. He supposed that the Grand River valley is "geomorphologically well-adjusted" to carry high flows. It is hard to see what that means: the effect may signify that the floodplain is largely clear of forest, well turfed, and able to convey substantial runoff without suffering major morphologic changes. This was effectively the outcome of the forest destruction on Arctic Red River.

The morphologic effects of ice jams are controversial (Smith, 1979; Kellerhals and Church, 1980). Direct observations are limited to scoured grooves and ice-pushed ridges found locally on bars and along the river bank, shearing of earth projections—usually slumped banks—and sometimes extensive damage to woody vegetation.

FIGURE 22. Morphologic effects of the Arctic Red River flood of July 1970. The channel course has not notably been altered although floodplain forest has been flattened by the force of the floodwater.

Mackay and MacKay (1977) described boulder pavements, rhythmically spaced boulder ridges, and ice-pushed "island buttresses" (in effect natural boulder riprap) along Mackenzie River. All of the features are much smaller than channel scale, although they extend well above normal high water (Fig. 17). Winter frazil dams induce scour holes in the river bed at sites where they recur regularly.

On the other hand the increased frequency of floodplain inundation because of channel blockage by ice (including icing; Fig. 18*c*) may lead to more rapid accumulation of the floodplain, possibly to ultimately higher levels. Evidence is not definitive in this matter either. Smith (1979, 1980) illustrated a low-level bench along rivers in Alberta subject to frequent ice jams and drives upon which extensively ice-damaged shrubs occur. Trees along the edge of a higher bench—corresponding with an approximately 10-yr open-water flood level—are ice scarred. Smith speculates that this morphology is maintained by ice drives, but the particular combination of erosional and sediment depositional work involved remains unclear.

Icing may have major effects. Where channels are blocked for a significant part of the summer, water often is deflected against the banks of the channel (where melt often proceeds most quickly), so that bank erosion and development of a greatly widened channel proceed. In the absence of the icing the river gives the appearance of being extensively braided. Sometimes, icing forms beneath the channel bed as freezeback occurs in autumn. The bed is then heaved and, in the following spring, the bed material is unstable and prone to be entrained in the nival flood. At least locally, this greatly reduces riverbed stability.

The energy of jökulhlaups often is dissipated over an extensive proglacial braidplain. However, where they are confined, they produce major changes. Dramatic indication of this is provided by the increasing realization that the most consistent explanation for the remarkable dimensions of many prairie valleys in the northern United States and southern Canada is that they are jökulhlaup channels (cf. Kehew and Lord, 1986). For at least some examples this may eventually lead to a reassessment of the antecedents for the underfit streams of Dury (1965).

Taken together, these results suggest an elaboration of the concept of what constitutes the most significant class of events accomplishing sediment-transporting work in a river. It is claimed often that relatively high frequency events of moderate magnitude, approximating the mean annual flood, are most effective in accomplishing geomorphologic work. Nival rivers appear to belong in this class, at least in terms of transport of sediments found along the channel. Figure 23*a* illustrates this for Fraser River. Baker (1977) introduced the notion that rivers with relatively high thresholds for fluvial work—particularly,

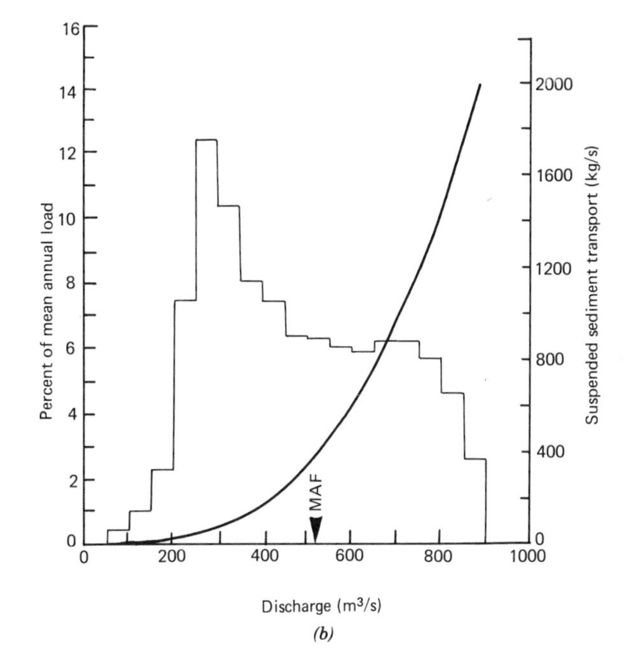

FIGURE 23. Distribution of morphologic work in rivers. MAF = mean annual flood. (*a*) Bedload and suspended-sand transport in Fraser River at Agassiz, British Columbia. Bedload transport rating curve is superimposed. Based on flow and sediment transport records for 1966–1983 (data analyzed by D. G. McLean). (*b*) Total suspended-sediment transport in Lillooet River at Pemberton, British Columbia, a river with major autumn rainstorm events (see Fig. 7*c*) superimposed upon a nival-glacial summer regime. The suspended-sediment rating curve is superimposed. Plot is based on averaged flow duration for 1914–1970. (*c*) Conceptual Wolman diagram for occasional extreme floods with relatively resistant river bed and banks.

for moving channel bed sediments—and under the influence of strongly skewed flow distributions (cf. Fig. 21), may exhibit a "catastrophic" response: most of the work is accomplished by relatively rare floods. In nival regions this may be an apt characterization of the effect of major rain or rain-on-snow floods described above. In the long term most of these rivers may be affected in this way; even ones characterized in this chapter—from rather short flow records—as "purely nival." Figure 23*b* illustrates this for Lillooet River near Pemberton, British Columbia, a nival-glacial river that experiences major autumn rain and rain-on-snow floods. The Wolman diagram for many rivers in nival regions tends to exhibit two peaks, as illustrated in Figure 23*c*—the outer peak often regrettably poorly defined because of record limitations. The difficulty of defining properly the distribution of extreme flows makes the form of Figure 23*b*, for Lillooet River, very tenuous.

The hypothesis advanced here may be taken one step further. The work peak associated with frequently recurring flows entails, mainly, the movement of fine sediment—much of it initially the product of frost weathering or frost heaving of soils—which is moved relatively quickly through the drainage system or stored in floodplain deposits without effecting major changes in the channel. Translation of this material through the drainage system represents *geomorphologic* work in the landscape. The peak associated with rarer flows reflects the mobilization of channel bed sediments, accomplishing *morphologic* work along the channel. In Fraser River at Agassiz (Fig. 23*a*), bedload is largely coincident with the gravel bed material. The bed material transport distribution shows a prominent maximum at high flows that is only weakly expressed in suspended sand load. Lillooet River (Fig. 23*b*) transports mainly volcanic sand (much of it pumiceous) so that the total sediment transport is morphologically significant along the river.

The distinction of geomorphologic work (pertaining to the entire sediment load) and morphologic work along the river (pertaining to bed material) is not restricted to nival rivers. It may, however, be particularly significant in many rivers of the nival region in which relatively rare rainstorms or rain-on-snow events eventually dominate channel modification, whereas the annual nival floods largely move the annual washload of streams.

ACKNOWLEDGMENTS

John E. Hay, Olav Slaymaker, Peter Waylen, and Ming-ko Woo reviewed and made helpful comments on a draft of this chapter.

REFERENCES

Aguado, E. (1985). Radiation balances of melting snow covers at an open site in the central Sierra Nevada, California. *Water Resour. Res.* **21**, 1649–1654.

Ashton, G. (1978). River ice. *Annu. Rev. Fluid Mech.* **10**, 369–392.

Baker, V. R. (1977). Stream-channel response to floods, with examples from central Texas. *Geol. Soc. Am. Bull.* **88**, 1057–1071.

Burns, B. M. (1974). "The Climate of the Mackenzie Valley-Beaufort Sea," 2 vols. Environment Canada, Atmospheric Environment Service, Ottawa.

Carey, K. L. (1973). "Icings Developed from Surface Water and Ground Water," Monogr. III-D3. U.S. Army Corps of Engineers, Cold Regions Research and Engineering Laboratory, Hanover, New Hampshire.

Church, M. (1974). "Hydrology and Permafrost with Reference to Northern North America," Proc. Workshop Semin. Permafrost Hydrol., pp. 7–20. National Committee for the International Hydrological Decade, Canada.

Church, M. (1983). Pattern of instability in a wandering gravel bed channel. *Spec. Publ. Int. Assoc. Sedimentol.* **6**, 169–180.

Church, M., and Miles, M. J. (1987). Meteorological antecedents to debris flow in coastal British Columbia: Some case histories. *Rev. Eng. Geol.* **7**, 63–79.

Clague, J. J., and Mathews, W. H. (1973). The magnitude of jökulhlaups. *J. Glaciol.* **12**, 501–504.

Clarke, G. K. C. (1982). Glacier outburst floods from "Hazard Lake", Yukon Territory, and the problem of flood magnitude prediction. *J. Glaciol.* **28**, 3–21.

Cogley, J. G., and McCann, S. B. (1976). An exceptional storm and its effects in the Canadian high arctic. *Arct. Alp. Res.* **8**, 105–110.

Colbeck, S. C. (1977). Short-term forecasting of water runoff from snow and ice. *J. Glaciol.* **19**, 571–587.

Davies, J. A., and Hay, J. E. (1980). Calculation of the solar radiation incident on a horizontal surface. *In* "First Canadian Solar Radiation Data Workshop" (J. E. Hay and T. K. Won, eds.), pp. 32–58. Atmospheric Environment Service/National Research Council of Canada.

Dickson, R. R., and Posey, J. (1967). Maps of snow-cover probability for the northern hemisphere. *Mon. Weather Rev.* **95**, 347–353.

Dunne, T., and Price, A. G. (1976). Estimating net radiation over a snowpack. *Climatol. Bull. (McGill Univ.)* **18**, 40–48.

Dury, G. H. (1965). Theoretical implications of underfit streams. *Geol. Surv. Prof. Pap. (U.S.)* **452-C**, 1–43.

Gardner, J. S. (1977). Some geomorphic effects of a catastrophic flood on the Grand River, Ontario. *Can. J. Earth Sci.* **14**, 2294–2300.

Gerard, R., and Karpuk, E. W. (1979). Probability analysis of historical flood data. *J. Hydraul. Div. Am. Soc. Civ. Eng.* **105**, 1153–1166.

Granger, R. J., and Male, D. H. (1977). Melting of a prairie snowpack. *J. Appl. Meteorol.* **17**, 1833–1842.

Harr, R. D. (1981). Some characteristics and consequences of snowmelt during rainfall in western Oregon. *J. Hydrol.* **53**, 277–304.

Jordan, R. P. (1979). Response of an alpine watershed to snowmelt. *In* "Canadian Hydrology Symposium 79: Proceedings," pp. 324–333. National Research Council of Canada, Associate Committee on Hydrology, Ottawa.

Jordan, R. P. (1983). Meltwater movement in a deep snowpack. I. Field observations. *Water Resour. Res.* **19**, 971–978.

Kane, D. (1981). Physical mechanics of aufeis growth. *Can. J. Civ. Eng.* **8**, 186–195.

Kehew, A. E., and Lord, M. L. (1986). Origin and large-scale erosional features of glacial-lake spillways in the northern Great Plains. *Geol. Soc. Am. Bull.* **97**, 162–177.

Kellerhals, R., and Church, M. (1980). Comment on "Effects of channel enlargement by river ice processes on bankfull discharge in Alberta, Canada" by D. G. Smith. *Water Resour. Res.* **16**, 1131–1134.

Likes, E. H. (1966). Surface-water discharge of Ogotoruk Creek. *In* "Environment of the Cape Thompson Region, Alaska" (N. Wilimovsky, ed.), pp. 115–124. U.S. At. Energy Comm., Div. Tech. Inf., Oak Ridge, Tennessee.

List, R. J. (1966). Smithsonian Meteorological Tables, 6th edition. *Smithson. Misc. Collect.* **114**, Table 132, 417–418.

MacKay, D. K., and Mackay, J. R. (1973). Locations of spring ice jamming on the Mackenzie River, N.W.T. *In* "Hydrologic Aspects of Northern Pipeline Development," Rep. No. 73-3, pp. 233–257. Task Force on Northern Oil Development, Environmental-Social Committee Northern Pipelines, Canada.

MacKay, D. K., Fogarasi, S., and Spitzer, M. (1973). Documentation of an extreme summer storm in the Mackenzie Mountains, Northwest Territories. *In* "Hydrologic Aspects of Northern Pipeline Development," Rep. No. 73-3, pp. 192–221. Task Force on Northern Oil Development, Environmental-Social Committee Northern Pipelines, Canada.

Mackay, J. R. (1963). Progress of break-up and freeze-up along the Mackenzie River. *Geogr. Bull.* **19**, 103–116.

Mackay, J. R., and MacKay, D. K. (1977). The stability of ice-push features, Mackenzie River, Canada. *Can. J. Earth Sci.* **14**, 2213–2225.

Male, D. H., and Granger, R. J. (1981). Snow surface energy exchange. *Water Resour. Res.* **17**, 609–627.

Marsh, P., and Woo, M.-K. (1981). Snowmelt, glacier melt, and high arctic streamflow regimes. *Can. J. Earth Sci.* **18**, 1380–1384.

Marsh, P., and Woo, M.-K. (1985). Meltwater movement in natural heterogeneous snow covers. *Water Resour. Res.* **21**, 1710–1716.

Matson, M., Ropelewski, C. F., and Varnadore, M. S. (1986). "An Atlas of Satellite-Derived Northern Hemispheric Snow Cover Frequency, USGPO No. 1986-151-384. U.S. Dept. of Commerce, National Oceanic and Atmospheric Administration/ National Environmental Satellite Data, and Information Service, Washington, D.C.

Miller, D. H. (1955). Snow cover and climate in the Sierra Nevada, California. *Univ. Calif., Berkeley, Publ. Geogr.* **11**, 1–218.

Nanson, G. C. (1974). Bedload and suspended-load transport in a small, steep mountain stream. *Am. J. Sci.* **274**, 471–486.

Nye, J. F. (1976). Water flow in glaciers, jökulhlaups, tunnels and veins. *Proc. R. Soc. London, Ser. A* **219**, 477–489.

Rudberg, S. (1963). Geomorphological processes in a cold, semi-arid region. *In* "Jacobsen-McGill Arctic Research Expedition, 1959-62: Preliminary Report 1961-62: Axel Heiberg Island Research Reports" (F. Müller, ed.), pp. 140–150. McGill University, Montreal.

Smith, D. G. (1979). Effects of channel enlargement by river ice processes on bankfull discharge in Alberta, Canada. *Water Resour. Res.* **15**, 469–475.

Smith, D. G. (1980). River ice processes: Thresholds and geomorphological effects in northern and mountain rivers. *In* "Thresholds in Geomorphology" (D. R. Coates and J. D. Vitek, eds.), pp. 323–343. Allen & Unwin, Boston, Massachusetts.

Storr, D. (1979). A comparison of daily snowmelt calculated by the U.S. Corps of Engineers theoretical model with measured amounts on a snowpillow. *In* "Canadian Hydrology Symposium 79: Proceedings," pp. 334–345. National Research Council of Canada, Associate Committee on Hydrology, Ottawa.

U.S. Army Corps of Engineers (USACE) (1960), "Runoff from Snowmelt," Engineer Manual EM 1110-2-1406. U.S. Army Corps of Engineers, Washington, D.C.

Waylen, P. (1982). Some characteristics and consequences of snowmelt during rainfall in western Oregon—a comment. *J. Hydrol.* **58**, 185–188.

Waylen, P., and Woo, M.-K. (1982). Prediction of annual floods generated by mixed processes. *Water Resour. Res.* **18**, 1283–1286.

Woo, M.-K. (1983). Hydrology of a drainage basin in the Canadian high arctic. *Ann. Assoc. Am. Geogr.* **73**, 577–596.

Woo, M.-K., and Sauriol, J. (1980). Channel development in snow-filled valleys, Resolute, N.W.T., Canada. *Geogr. Ann.* **62A**, 37–56.

Woo, M.-K., and Waylen, P. R. (1985). Areal prediction of annual floods generated by two distinct processes. *Hydrol. Sci. J.* **29**, 75–88.

Young, G. J. (1982). Hydrological relationships in a glacierized mountain basin. *IAHS Publ.* **138**, 51–59.

14

DEFINITION OF FLOOD PLAINS ALONG ARID-REGION RIVERS

WILLIAM L. GRAF
Department of Geography, Arizona State University, Tempe, Arizona

INTRODUCTION

The general purpose of this chapter is to review the geomorphic features produced by flood processes in arid regions and to explore the various approaches to using the features as evidence for flood hazards. The primary thesis that emerges from this review is that rivers in arid environments have fundamentally different structures for temporal and spatial change than rivers in other regions. Management of flood hazards and interpretation of geomorphic evidence of flood conditions in dry regions are faulty because these differences have yet to be adequately accounted for by geomorphologic theory, engineering applications, and the law, endeavors that do not even share common basic definitions.

This chapter focuses on environments herein referred to as *arid*, meaning those areas generally characterized by Meigs (1953, reviewed by McGinnies et al., 1968) as arid and semi-arid with annual precipitation of less than 500 mm. Examples are drawn primarily from representative streams in Arizona.

TEMPORAL AND SPATIAL CHARACTERISTICS

The concept of a general system operating in a manner that establishes and maintains an equilibrium condition is the most fundamental assumption underlying fluvial geomorphic and hydraulic theory (Tanner, 1968; Shen, 1979a; Dingman, 1984). Usually this assumption takes the form of a presumed balance among the hydraulic forces, material resistances, and morphology of the fluvial system (as in a *graded* river; Knox, 1975). If one of these factors is altered, the others respond through feedback processes of erosion and deposition to re-establish the initial balance. The maintenance of an equilibrium condition is facilitated by the nearly continuous operation of streamflows in major rivers of humid regions. When floods occur in such systems, the discharge may be several times the mean annual discharge, but the excess water is usually distributed over flood plains where its geomorphologic work is accomplished. Streams in humid Pennsylvania, for example, have 50-yr floods that are only about two and one-half times the magnitude of the annual flood (Leopold et al., 1964, p. 66).

Arid-region rivers, on the other hand, may not exhibit long-term (several decades) tendencies toward some equilibrium condition (Stevens et al., 1975; Graf, 1981). In arid regions the climatic forcing functions that control floods are extremely variable on an annual as well as a decadal basis. For small streams (with basins up to several hundred square kilometers), summer thunderstorms provide the input for flood events, so that years with several floods may be separated by years with no flow. For large streams (with basins of over 10,000 km^2), annual spring floods may be an order of magnitude larger than the mean annual flow, while over a century the annual flood series may contain an additional order of magnitude of variation. The Gila River in Arizona has a regime wherein the 50-yr flood is 280 times the mean annual discharge (Osterkamp et al., 1982, p. 46). These wide fluctuations in discharge that can occur over short periods of time imply that fluvial systems in arid environments cannot operate on a continuous basis as they do in humid regions, and that trends toward a process-form equilibrium are unlikely to be completed (as pointed out by Stevens et al., 1975, for the Gila River, Arizona).

The radical short-term variations in discharge result in a system operation that is pseudocatastrophic in that very large discharges drastically alter channel morphology to a new configuration suited to high flow, but that is not in equilibrium with subsequent low flows (Schumm and Lichty, 1963; Burkham, 1972). Conversely, a long period of relatively low flows produces a channel unable to accommodate the occasional very large flood that completely alters the process–morphology relationship (e.g., Hunt et al., 1953). For this reason the order of prior events is critical in understanding the channel configuration observed at any given time. The arid-region channel is a time-dependent system (Pickup and Rieger, 1979). A statistical summary of events that fails to account for order of occurrence is likely to be misleading at best.

Arid-region rivers frequently accomplish radical adjustment to extreme events by complete changes of channel configuration. Although channel configuration may be generally classified according to the designations of straight, meandering, and braided (Leopold and Wolman, 1957) or by designations of bedload dominant, suspended-load dominant, or mixed (Schumm, 1968), a full range across a scale of sinuosity and form ratio is present in the real world (Richards, 1982). At one end of this scale are those rivers that are relatively deep and narrow and that have meandering courses, while at the opposite end of the scale are those that are relatively wide and shallow with braided characteristics. The position of a particular river reach on this scale represents a partial adjustment to varying water and sediment loads and bank stability. Braided streams generally have steep courses and velocity distributions that permit large amounts of bedload transport. Meandering streams generally have more shallow gradients and cross-sectional forms more efficient for suspended-load transport (reviewed by Schumm, 1977, pp. 143–173).

Because arid-region rivers are subject to wide fluctuations in discharges, their channels change configuration to accommodate the variations in input of mass and energy. They may adopt different channel configurations from one place to another or the same reach may change configuration from one time to another. These spatial and temporal changes in form occur abruptly and are not accommodated by general mathematical models of channel behavior.

Many arid-region rivers have both meandering and braided characteristics and fall in the middle of the configuration scale with compound characteristics. At low flows water occupies a meandering single channel nested inside a larger braided channel that is occupied only during infrequent high discharges (compound channels are also referred to as "channel in channel": Richards, 1982, p. 266; Gregory and Park, 1974). The same channel therefore may have two entirely different behavior patterns depending on discharge conditions, with intermediate discharges causing damage through erosion even though overbank flow does not occur outside the braided channel boundaries (Fig. 1). Compound channels frequently occur downstream from dams and diversion works as a result of flow depletion (Williams, 1978; Petts, 1979; Gregory and Park, 1974), but they may also develop under natural circumstances.

Figure 2 is a catastrophe theory representation of river–channel states that include compound channels. The state of a channel reach at any given time is represented by a point on the cusp surface, while change through time is represented by moving the point about on the surface. Gilmore (1980) reviews general catastrophe theory, while Graf (1979) and Thornes (1980) provide discussions specifically related to geomorphology. The paths of changes in channel states are (1) gradual infilling during sustained periods of low flows causing a change from braided to meandering configurations and (2) rapid erosion during

FIGURE 1. Aerial view of an example of a compound channel: the Salt and Gila rivers at their junction west of Phoenix, Arizona (view looking east). The channel of the Salt River through the center of the view has an outer braided configuration and an inner well-defined meandering low or main flow channel marked by dark phreatophytes. Photograph courtesy of the Department of the Army, Corps of Engineers: Maddox Associates Photo #44, October 2, 1962.

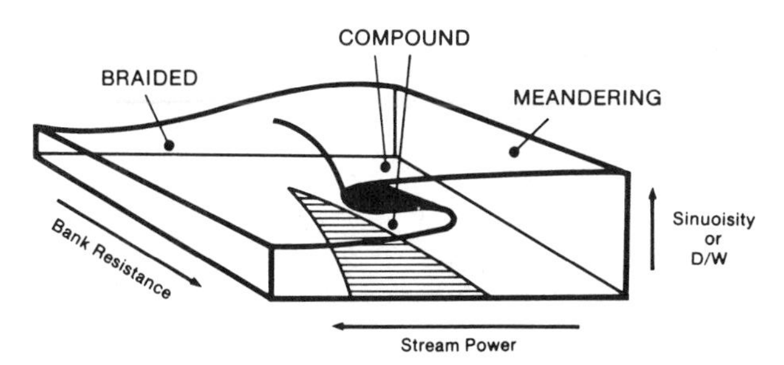

FIGURE 2. A catastrophe theory representation of the interaction of the controlling variables stream power and channel slope with a responding variable, channel configuration as measured by sinuosity. Compound channels occur in the folded portion of the diagram where two possible states exist for given combinations of power and slope.

rare floods causing a change from meandering to braided states. As shown in Figure 2, the nature of the change depends on the interplay between stream power and channel slope (Richards, 1982, p. 215).

The high mobility of bed materials and the general lack of cohesive bank materials in arid-region rivers results in substantial locational instability (Miall, 1977). Lateral shifts in channel location are especially prominent on alluvial fan surfaces (Scott, 1971), but alluvial valley fills also have unstable channel locations. The meandering course of the low-flow thalweg of the channel may exhibit some regularity or locational preference where the channel is constrained by bedrock outcrops or man-made control works, but in other areas lateral shifts may span a distance of several kilometers usually occurring during floods (Graf, 1981).

The Gila River provides a useful example because it has experienced dramatic lateral shifts in channel geometry. Two reaches of specific interest are an upper one in southeast Arizona near Safford and a lower one near Phoenix. In the upper reach the stream had a meandering course in the late 1800s with well-defined tree-lined banks (following account from Swift, 1926, and Burkham, 1972). During the first few years of the twentieth century, drought ruled the region and stockmen watered large herds by driving cattle to the low-flowing river. The cattle trampled the banks, reducing them to gentle inclines, and the trees were either browsed by cattle or removed for lumber.

In 1905 a 25-yr flood occurred, altering the river to a braided and sandy channel 1.5 km wide. A subsequent series of flows did not alter the channel for more than 30 yr, but by about 1940 the wide, sandy channel was overgrown with tamarisk and other phreatophytes that reduced channel capacity and initiated a slow return to meandering channel conditions. By 1980 channel configurations still had not returned to their pre-1905 arrangements. The upper reach of the Gila River therefore appears to have responded mostly to hydrologic changes but also to vegetation influences.

In another reach of the Gila River in southcentral Arizona between Phoenix and Gila Bend (about 80 km downstream from the area already discussed), the river has shown a different series of changes (Fig. 3; following account from Graf, 1981, 1983). Between 1868 and 1929 the channel was braided, and the 1905 flood had no particular geomorphic significance. The sinuosity of the channel thalweg in five 11-km sample reaches ranged from 1.08 to 1.14. In the period 1930–1950, however, tamarisk, willow, and cottonwood trees invaded the braided channel and blocked

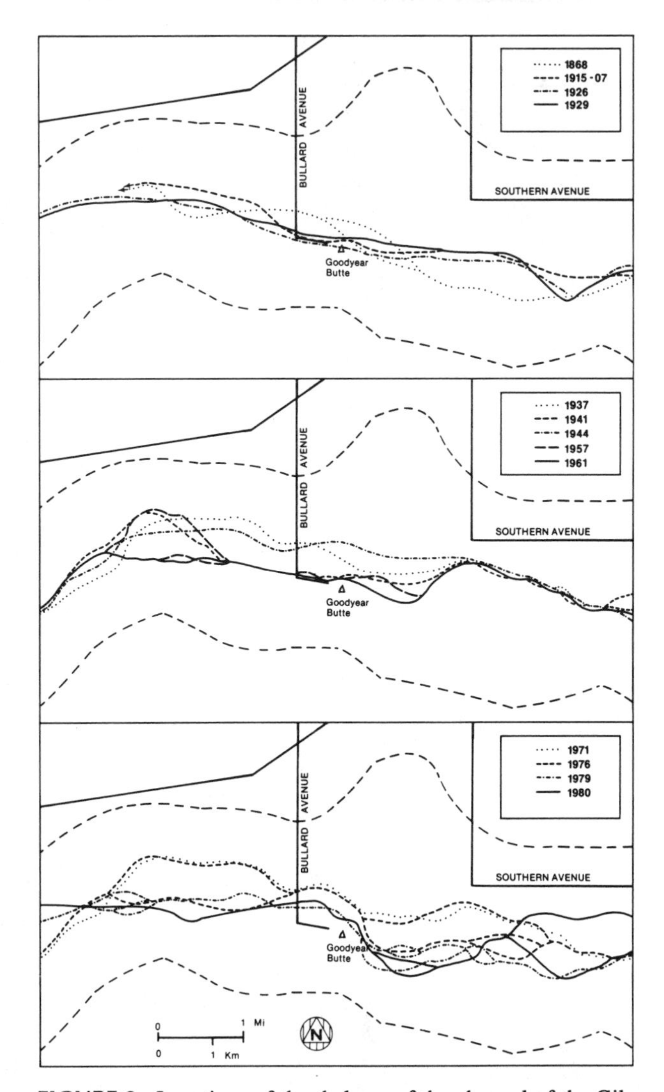

FIGURE 3. Locations of the thalweg of the channel of the Gila River 20 km west of Phoenix, Arizona, 1868–1980. The channel in the period 1868–1929 was relatively stable in its location. In the period 1937–1961 dense phreatophyte growth reduced channel capacity and destablized its location. In the period 1971–1980 several major floods caused further locational changes.

many areas. The results were channel avulsion, the establishment of sinuousities of up to 1.23, and lateral channel migration of up to 1.6 km (Fig. 4). After 1950, groundwater pumping drew down water tables in many parts of the lower reach, and the phreatophytes declined in density. The river channel in the 1960s and 1970s adopted unstable arrangements that were characterized by a meandering low-flow channel and a braided high-flow geometry—a compound channel as discussed previously. The lower reach of the Gila therefore appears to have responded more to vegetation changes than to hydrologic controls.

SPECIAL FLUVIAL FEATURES IN ARID REGIONS

Streams in arid regions develop three features that have implications for their flood characteristics: alluvial fans, alluvial plains, and arroyos. Flooding on alluvial fans is difficult to assess because channel avulsion is common—the plugging of the active channel with sediment and diversion of waters to alternative channels (Cooke, 1984, p. 50). Throughout their recent geomorphologic histories most fans have experienced such changes frequently, usually during floods (Beaty, 1963), resulting in a roughly equal distribution of sediment across their surfaces (Scott, 1971; Weaver, 1984). Present geomorphologic theory does not provide for the calculation of the probability that any particular distributary channel will receive flow. On natural fan surfaces the flood hazard is therefore largely indeterminant. The common engineering and planning solution is to design and maintain a single channel for flow, an approach that usually requires trapping and artificial removal of sediments delivered to the fan apex from the upstream watershed (Cooke, 1984, pp. 168–169).

The discussion heretofore has concentrated on the flood evidence provided by channels, but in some parts of the arid-region landscape, channels may not be present. When floods disgorge onto the nearly flat surfaces of alluvial plains, also known as alluvial aprons (Cooke et al., 1982, pp. 192–193), their waters spread laterally into wide zones of very shallow flow. In southcentral Arizona, for example, floods on some valley floors in 1983 attained widths of several kilometers with depths of less than 2 m. Instead of well-defined channels, alluvial plains frequently demonstrate wide flow zones without clearly marked banks. The zones are not obvious from ground observation, but they are clearly defined on spacecraft imagery (Fig. 5).

FIGURE 5. LANDSAT image showing the broad flow zone across the floor of King Valley, southwest Arizona. North is toward the top of the image, which shows an area approximately 70 by 80 km. The flow zone is the northwest–southeast trending light tone area several kilometers in width. Darker areas are the Castle Dome Mountains on the west and the Kofa, Tank, and Palomas mountains on the east. LANDSAT image 1069-17441-6, September 30, 1972.

The processes by which floods operate over these wide, nearly flat valley floors are not well known. However, the Santa Cruz River north of Tucson, Arizona, provides an instructive example (Aldridge and Eychaner, 1984). The stream flows northward through southern Arizona where it is entrenched into an arroyo, but north of Tucson the stream has poorly defined banks and a channel formed of gentle, interconnected swales on the surface of deep valley fills (generally described by Bryan, 1923).

As floodwaters issued from the entrenched portion of the channel in 1983, they spread laterally into a sheet 4.8

FIGURE 4. The Gila River 10 km upstream from Gila Bend and 60 km southwest of Phoenix, Arizona. Upper view taken in 1949 shows dense growth of phreatophytes and a narrow channel hidden by vegetation but indicated by the original photographer's sketched arrow (photograph courtesy of the U.S. Army Corps of Engineers, Phoenix Urban Studies Office). Lower view in 1982 shows reduced phreatophyte growth and a braided channel in the lower left corner of the frame, a lateral channel displacement of about 1.2 km from the 1949 location as well as a change in channel plan form.

km wide but less than 1.5 m deep. Because of the shallow nature of the flow, agricultural developments on the valley floor, such as lateral canals and roads, influenced the distribution of floodwaters. The total flow zone as observed in the field resembled the dimensions suggested by the evidence from a similar nearby area shown in Figure 5. Eventually the broad flow was collected and concentrated by the Greene Canal, a major cross-valley structure, which experienced significant downcutting as unexpectedly large quantities of water scoured its bed (Haigh and Rydout, 1985).

The relationship between these flow zones and sheetfloods in arid regions may be that they are one and the same (Davis, 1938; Rahn, 1967). During earlier periods of field research only a limited portion of the sheetflood could be observed at any one time. The first scientific report of a sheetflood was an observation by McGee (1897), and a considerable debate ensued about the existence of the phenomenon (Davis, 1938; Rich, 1938). Irrigation survey maps of the Salt River valley in central Arizona show with 2-ft contour intervals and outlines of "alkaline soil" areas the presence of flow zones before the imposition of man-made canals and drains (Salt River Valley Water Users' Association, 1903). After agricultural development the original flow zones are apparent only during periods of extraordinary runoff that exceeds the capacity of the artificial drains. Flooding then occurs through the now obscured flow zone that existed prior to development.

Arroyos also present special problems in the interpretation of the flood hazard from geomorphic evidence in arid regions. Arroyos are entrenched channels that result from excavation of alluvial materials by channel erosion. The arroyo is frequently so deep that floods with recurrence intervals of several centuries would be required for overflow.

The upper alluvial surface near the edge of the arroyo is generally considered to be safe from flood damage. Rillito Creek, an arroyo in urban Tucson, Arizona, demonstrates the containment effect of entrenched channels. Before Anglo-American settlement, the creek was a narrow, indefinite channel lined with trees and interrupted by beaver ponds (Smith, 1910, p. 98). In the 1870s the area was used for hay growing and grazing, and in the 1870s and 1880s there were several floods with wide flow zone characteristics as previously described for the Santa Cruz River. Flows with return intervals of less than 10 yr easily overtopped the banks and spread laterally over the valley floor.

Entrenchment of an arroyo several meters deep began in the 1890s, and once established the arroyo confined floods and perpetuated vertical erosion in a positive feedback process. In the 1980s the arroyo is so deep that it completely contains the 100-yr flow. Areas near the channel that once were inundated frequently are now safe from inundation even in relatively rare events (Fig. 6).

Although these locations are relatively safe from inundation, they are at high risk from erosion damage because

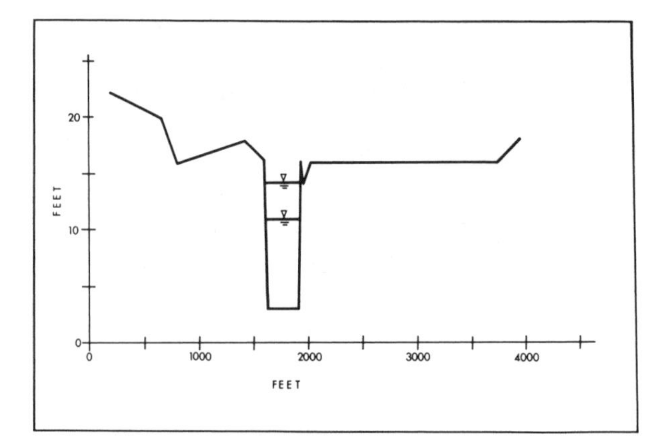

FIGURE 6. Surveyed cross section of the arroyo of Rillito Creek, north of Tucson, Arizona, showing the effect of arroyo development on flood overflows. The lower water surface represents the level of the 50-yr flood; the upper water surface represents the level of the 100 yr flood. Survey data courtesy of the Department of the Army, Corps of Engineers, original data collected by Dames and Moore, Inc., Phoenix, Arizona.

the floor of the arroyo may be frequently occupied by flood flows with the competence necessary to entrain sediments in the bed and banks. Severe erosion and bank collapse result (e.g., Saarinen et al., 1984). For example, along Rillito Creek, in an arroyo in urban Tucson, Arizona, the damages from erosion in a 50-yr simulation period would exceed damages from inundation by a factor of 5 (Graf, 1984). Flood hazard mapping for arroyos therefore must also account for the erosion hazard, which is more severe than the inundation hazard (see also Kresan, this volume).

PERSPECTIVES ON THE MORPHOLOGIC EVIDENCE

Because arid-region channels change their location and configuration, accurate interpretation of the geomorphologic evidence of the floods that cause the changes is critical in hydrologic-geomorphic research, engineering practice, and planning-legal applications. These various approaches can be complementary (Shen, 1979b). Unfortunately practitioners in these various fields rarely exchange information with those outside their home discipline, and a confusing set of perspectives with conflicting basic definitions of the evidence has evolved (Dingman and Platt, 1977). The following outline of the various perspectives demonstrates the usefulness of adopting a combined geomorphologic-hydrologic approach to interpreting evidence of the spatial and temporal behavior of arid-region rivers.

Each perspective is presented through repetitive interpretation of the same channel cross section that is broadly representative of rivers in arid regions. The cross section is a composite of several cross sections along the Agua Fria River, a northern tributary of the Gila River, between Waddell Dam and Calderwood Butte in central Arizona. For a general description of the Agua Fria River see Wildan Associates (1981). Sources of data for the representative cross section include aerial photography by the Soil Conservation Service and U.S. Geological Survey, a 1 : 24,000 scale topographic map (Calderwood Butte 7.5′ Quadrangle, 1974), a soil survey (Hartman, 1984), a geologic map (Wilson et al., 1957), and detailed floodplain maps (Maricopa County Flood Control District, Phoenix, Arizona, unpublished data). The representative cross section has features similar to those of most arid-region rivers of medium size draining about 1000–10,000 km^2.

As shown in the following discussion, the term *flood plain* has several established meanings in arid regions depending on the perspective of the interpreter. The term also has several forms. The term *flood-plain* (the now antiquated hyphenated noun) was introduced into the scientific literature by Geikie (1882, p. 382) as "the level tracts over which a river spreads in flood." The preferred geologic form is as two separate words (Bates and Jackson, 1980, p. 235). This form also appears in geomorphology texts (e.g., Ritter, 1978, p. 258; Bloom, 1978, p. 233), is used by the U.S. Geological Survey and U.S. Soil Conservation Service, and is used in the present chapter. When the two-word form is combined to be used as a modifier, a hyphen is used: "flood-plain deposits," for example (Bishop et al., 1978, p. 235). The single-word form, "floodplain," is common in geography and hydrology textbooks (e.g., Muller and Oberlander, 1984, p. 390), and is the preferred common English form (Merriam Company, 1981, p. 438). In legal parlance the term appears as "flood plane" (*Mader v. Mettenbrink*: Complete legal citations for cases follow the list of references at the end of this chapter).

Geomorphologic Perspective

To the geomorphologist the cross section of the Agua Fria River represents a palimpsest of forms generated by modern flood events superimposed on the effects of Quaternary hydroclimatic and tectonic changes (Fig. 7). In defining the various components of the cross-sectional system, the geomorphologist takes into account surface morphology and subsurface materials. The braided channel is characterized by a single low-flow channel plus secondary channels flanked by terraces, some of which have overflow channels that are occupied during rare, high-magnitude floods. Each set of banks defines a channel, and each channel may be assigned a designation in a hierarchy of bankfull capacities. Williams (1978) analyzed the problem of assessing bankfull flows based on geomorphic evidence.

In modern geomorphology the flood plain is recognized as the surface adjacent to the channel, separated from the

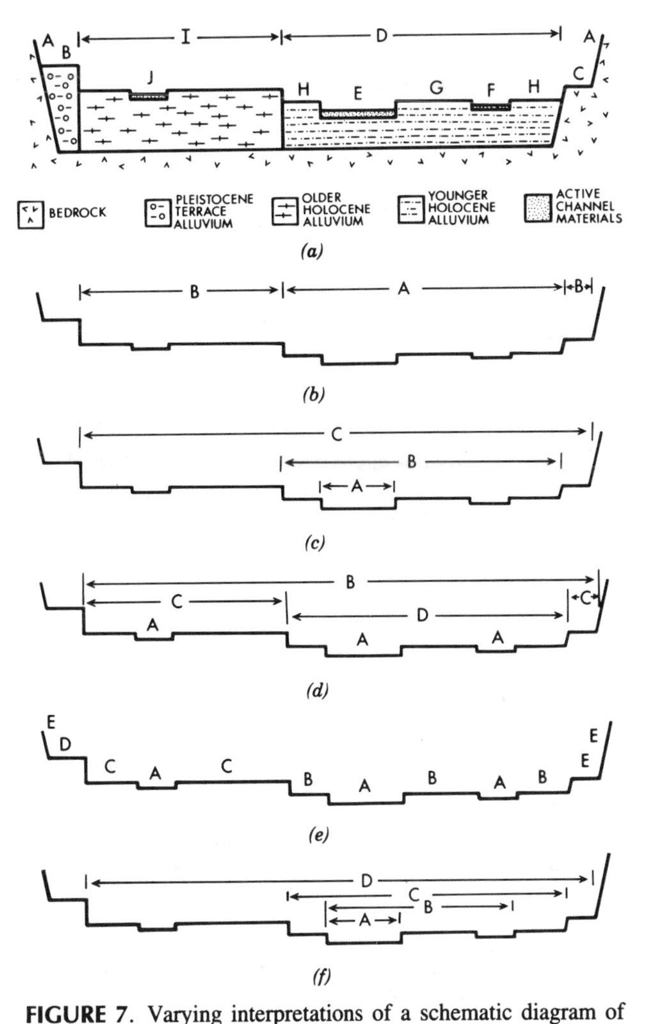

FIGURE 7. Varying interpretations of a schematic diagram of a representative cross section of the lower Agua Fria River, west central Arizona. The diagram is not to scale, but has about 10 : 1 vertical exaggeration; total width is about 800 m. Letters represent surface components as identified using standard definitions in each case. (*a*) Geomorphologic interpretation relying on analysis of surface expression and subsurface materials. (A) bedrock, (B) Pleistocene terrace, (C) erosional terrace or rock bench, (D) braided channel, (E) main or low flow channel, (F) secondary channel, (G) midchannel bar, (H) channel-side bar, (I) recent terrace, (J) overflow channel. (*b*) Topographic interpretation of the representative cross section shown in (*a*). Letters represent geometric interpretations. (A) channel, (B) flood plain. (*c*) Hydrologic interpretation of the representative cross section shown in (*a*). (A) channel, (B) 50-yr flood plain with channel, (C) 100-yr flood plain with channel. (*d*) Engineering interpretation of the representative cross section shown in (*a*). (A) channel, (B) 100-yr flood plain, (C) flood fringe, (D) flood way. (*e*) Pedologic interpretation of the representative cross section shown in (*a*). Letters represent interpretations from the soil survey that includes the lower Agua Fria River (Hartman, 1984). (A) Excessively well-drained flood-plain soils, (B) well-drained low terrace soils, (C) well-drained middle terrace soils, (D) well-drained upper terrace soils, (E) soils developed on bedrock. (*f*) Legal interpretation of the representative cross section shown in (*a*). Letters represent four possible alternative definitions of the "channel" depending on the set of banks used for limits.

channel by banks, and built of materials deposited in the present regime of the river (Wolman and Leopold, 1957; Kellerhals et al., 1976, pp. 820–821). Though no such surface exists in the representative cross section, there is a low terrace next to the channel. Its origin is recognized as an effect of some past hydroclimatic regime or tectonic episode. It is not a flood plain in the sense that the general bulk of its materials are not affected by the modern river regime. Under the geomorphologic interpretation the representative cross section exhibits a braided channel and terraces without flood plains, a situation common in many arid regions.

The primary advantage of the geomorphologic perspective is that it takes into account process as well as form and so is an effective method of hazard assessment. The major disadvantage of the perspective is that it is expensive because it requires large amounts of surface and subsurface information. Without additional hydrologic data the frequency of inundation of various parts of the surface is unknown.

Topographic Perspective

In the search for useful definitions for legal, planning, and management purposes, a strictly geometric set of definitions might be applied to the representative cross section (Ritter, 1978, p. 258). In this application only surface morphology is taken into account (Fig. 7*b*). The "channel" is that portion of the cross section between the banks of the active stream ("active" based on some return interval). If the active stream were to be defined for the Agua Fria example, the 50-yr flow probably would extend across the entire portion labeled by the geomorphologist as a braided channel. In the topographic perspective the flood plain is the relatively flat surface adjacent to the channel and separated from the channel by banks. The overflow channel on the left side of the cross section is difficult to account for in the topographic perspective, but presumably it is part of the flood plain.

Under the topographic definition, the cross section contains a channel and a flood plain, but the flood-plain designation is not useful as an interpretive tool. The relatively flat surfaces adjacent to the channel might be underlain by sediments or might be a bedrock feature, two possibilities that have differing implications for surface management and erosion potential. There is also a problem in selecting which set of banks to use as a definitional base.

Hydrologic Perspective

From the hydrologic perspective the representative cross section has a channel flanked by flood plains that are inundated by water once during a given return interval (Ward, 1978). As increasingly large discharges are considered after the channel capacity has been exceeded, higher discharges inundate more of the near-channel surfaces and the flood plains become progressively larger (Fig. 7*c*). Dalrymple (1964) provided a general review of this process.

The advantage of the hydrologic perspective is that it is directly attached to a specific probability of inundation. Those areas within the 100-yr flood plain have an annual probability of flooding of one percent. The method is also well known and widely applied. The disadvantage of the hydrologic perspective is that it does not account for variation in surface materials and potential stability of the surfaces, so it is most effective when combined with the geomorphologic perspective.

Engineering Perspective

Civil and hydraulic engineering perspectives on the representative cross section are in part based on the concepts associated with the hydrologic approach. In the United States the most common mathematical model used to determine the area inundated by flows of a given return interval is the Water Surface Profiles computer program of the Hydrologic Engineering Center of the U.S. Army Corps of Engineers (Hydrologic Engineering Center, 1982). Sometimes water surface calculations are supplemented by scour and deposition analysis (Hydrologic Engineering Center, 1976). The program for water surface levels, also known as HEC-2, uses as input surveyed cross sections of channels and near-channel areas along with discharge estimates for the flow of a given return interval. The program processes the data using widely accepted hydraulic functions, and produces as output water surface elevations. These elevations may then be compared to detailed topographic maps in order to determine the areas likely to be inundated by the flow.

The resulting designation for the inundated surface is a flood plain (Fig. 7*d*) and corresponds to the topographic channel plus flood plains as defined in Figure 7*b*. Because many property owners fear the financial implications of the designation of flood plain for their property, increasingly common usage is to divide the engineering-defined flood plain into two parts, the floodway and the flood fringe. The floodway is the area of the cross section "that could theoretically convey the 100-yr flood with only a one-foot rise of water level above the height of an unconstricted flood" (Dunne and Leopold, 1978, p. 427). Therefore, if the rest of the cross section were filled by human construction activity, the constriction caused by that infilling would result in an increase in the stage of the 100-yr flood of 1 ft or less. The portion of the cross section within the 100-yr flood plain and outside the floodway is the flood fringe.

The engineering perspective as outlined is commonly employed by planning and land management agencies in the United States. Its major advantages are its wide acceptance and the ability to map probabilities of inundation. Its drawbacks include the difficulty of its application to arid-region rivers that have highly mobile beds and unstable banks that change configuration during flood events and that render suspect the cross-sectional surveys and basic assumptions upon which the program depends.

Pedologic Perspective

The distribution of soils across the representative cross section gives rise to still another perspective because when the soils are mapped, usually through interpretation of aerial photographs with field checks, some judgments are required of the analyst (Buol, et al., 1973). In arid regions significant portions of channels remain dry for most of the year, so even channel floors are mapped with identified soil units, usually xerofluvents (Soil Conservation Service, 1975, pp. 193–194). Whether or not the pedologic perspective on the Agua Fria River is typical remains to be investigated, but it at least provides a useful case example.

The soil survey covering the representative reach of the Agua Fria River contains no soil series that precisely fits the previous definitions of river channel (Hartman, 1984), so analysts designated the main or low-flow channel, secondary channels, and overflow channels as being areas of flood-plain soils. Flood-plain areas are characterized by "excessively drained soils," an apt description because the geomorphologic channels have surface materials dominated by coarse-grained, unconsolidated alluvium. As shown in Figure 7*e*, the soil survey was sensitive enough to discriminate among several surfaces, but the labels assigned (low terrace and terrace) do not correlate with the geomorphologic interpretation shown in Figure 7*a*, and assessment of relative flood hazards is not apparent. Soil surveys are effective aids to geomorphologic interpretations.

Legal Perspective

From the perspective of American jurisprudence, the general fluvial system has three elementary components: surface waters, stream waters, and floodwaters (Clark, 1976). Surface waters are uncollected runoff that is not concentrated into channel flow (*Southern Pacific Company v. Proebstel; Mogle et al. v. Moore et al.*), roughly correlative to what the geomorphologist or hydrologist would refer to as overland flow. Once collected into a watercourse, the flow is designated a stream water (*Martinez v. Cooke*), and if it leaves the channel through overflow, it is designated as floodwater (*Mader v. Mettenbrink, Maricopa County Municipal Water Conservation District No. 1 v. Warford, Southern Pacific Company v. Proebstel*). The general legal assumption is that floodwaters return to the channel at some point downstream.

Because most American water law originated in British law or in applications to the humid eastern United States (Trelease, 1979), several problems develop in its application to the streams of the arid western United States. For example, a watercourse or natural stream is defined by its longitudinal and cross-sectional characteristics, both derived primarily from humid-region experiences. To be designated as a natural stream, the feature in question is usually considered to have a definable beginning (where surface waters collect), and a definable end, such as another stream, a lake, sea, or ocean (*Mogle et al. v. Moore et al.*). In arid regions, however, it is common that a natural stream loses its definition on broad basin floors or sinks where the waters percolate into the subsurface, a circumstance that requires the definition of the watercourse to be one of fact determined by jury or court (*Costello v. Brown*).

The watercourse that conducts moving water is usually defined as a feature that is a definite channel with definite bed and well-marked banks (*State v. Hiber, Southern Pacific Company v. Proebstel, Mogle v. Moore*). Though there may be exceptions to this definition (*State v. Hiber*), the general legal concept is that the channel cross section consists of a single channel with its lateral extent marked by banks, a concept applicable to most humid-region situations. When this concept is applied to the usual arid-region situation, however, the multiple channels of the braided or compound system pose the problem of multiple banks. In the representative cross section discussed previously, there are several alternative legal definitions of "the channel" depending on which set of banks are selected (Fig. 7*f*). The significance of the choice relates in part to defining whether one's property lies inside or outside "the channel."

According to American law, the flood plain of a natural stream consists of the lands adjacent to the channel (and therefore presumably separated from the channel by banks) that are overflowed in times of high water and that return the overflow to the channel at some point downstream (*Le Brun v. Richards, Southern Pacific Company v. Proebstel, Maricopa County Municipal Water Conservation District No. 1 v. Warford*). In humid regions such a definition is usually easy to apply because the natural streams have only one channel and one pair of banks. In the arid-region example of Figure 7 there are several possible choices, each with different financial and legal implications for property owners. For example, if one's property is in the legally defined "channel," he would expect no governmental protection from water damage. If the property is in the legal "flood plain" the damaging waters are considered a "common enemy" produced by natural processes (*Gillespie Land and Irrigation Company v. Gonzales*). If the property is not in the "channel" or "flood plain," water damage might be considered to be the result of actions of other property owners nearby or upstream. As shown in Figure 7*f*, the critical designation for arid-region rivers with multiple banks is problematic.

CONCLUSION

The foregoing review of flood-related processes and forms in arid regions demonstrates that despite some similarities, fundamental differences exist between arid-region and humid-region rivers. The presence of compound channels and the absence of geomorphologic flood plains are significant differences that result from the operations of the arid-region hydroclimatic system. There are several noncongruent perspectives on flood evidence that must be reconciled before progress is possible in interdisciplinary analysis. The presence of unstable alluvial fans, broad flow zones instead of channels on alluvial plains, and unstable arroyos may have muted and smaller counterparts in humid regions, but they are common, imposing features in arid regions.

There are three lines of endeavor likely to benefit the harmonious association between society and flood-prone areas of arid-region landscape. First, fluvial geomorphologic theory must be extended to account for the special properties of arid-region systems. The analysis of the nonequilibrium operations of rivers with compound channel configurations, no flood plains, and sometimes without banks provide important research challenges. Second, engineering applications need to account for the erodible banks and mobile beds often found in arid-region rivers. Simple applications of formulations derived from humid-region experiences need to be avoided. Finally, the administrative and legal system based on case law from humid regions must be receptive to alternative interpretations of evidence provided by geomorphic and hydrologic perspectives that are firmly rooted in actual rather than presupposed conditions.

ACKNOWLEDGMENTS

Discussions with Stan Schumm concerning applications of geomorphology to legal problems were helpful in formulating the theses of this chapter. Comments on an earlier draft of the chapter by Jim Knox and Pete Patton were helpful in clarifying issues. Steve Foote provided valuable research assistance for the legal material. William Lowell-Britt was the cartographic assistant.

REFERENCES

Aldridge, B. N., and Eychaner, J. H. (1984). Floods of October 1977 in southern Arizona and March 1978 in central Arizona. *Geol. Surv. Prof. Pap. (U.S.)* **2223**, 1–143.

Bates, R. L., and Jackson, J. A., eds. (1980). "Glossary of Geology." Am. Geol. Inst., Falls Church, Virginia.

Beaty, C. B. (1963). Origin of alluvial fans, White Mountains, California and Nevada. *Ann. Assoc. Am. Geogr.* **53**, 516–535.

Bishop, E. E. et al. (1978). "Suggestions to Authors of the Reports of the United States Geological Survey," 6th ed. U.S. Govt. Printing Office, Washington, D.C.

Bloom, A. L. (1978). "Geomorphology: A Systematic Analysis of Late Cenezoic Landforms." Prentice-Hall, Englewood Cliffs, New Jersey.

Bryan, K. (1923). Erosion and sedimentation in Pagago country, Arizona. *Geol. Surv. Bull. (U.S.)* **720-B**, 19–90.

Buol, S. W., Hole, F. D., and McCracken, R. J. (1973). "Soil Genesis and Classification." Iowa State Univ. Press, Ames.

Burkham, D. E. (1972). Channel changes of the Gila River, Safford Valley, Arizona, 1846-1970. *Geol. Surv. Prof. Pap. (U.S.)* **655-G**, 1–24.

Clark, R. E., ed. (1976). "Water and Water Rights; A Treatise on the Law of Waters and Allied Problems: Eastern, Western, Federal," Vol. 5. A. Smith Co., Indianapolis, Indiana.

Cooke, R. U. (1984). "Geomorphological Hazards in Los Angeles: A Study of Slope and Sediment Problems in a Metropolitan County." Allen & Unwin, London.

Cooke, R. U., Brunsden, D., Doornkamp, J. C., and Jones, D. K. C. (1982). "Urban Geomorphology in Drylands." Oxford Univ. Press, London and New York.

Dalrymple, T. (1964). Flood characteristics and flow determination. Part I of section 25. Hydrology of flow control. In "Handbook of Hydrology" (V. T. Chow, ed.), pp. 25/1–25/32. McGraw-Hill, New York.

Davis, W. M. (1938). Sheetfloods and streamfloods. *Geol. Soc. Am. Bull.* **49**, 329–339.

Dingman, S. L. (1984). "Fluvial Hydrology." Freeman, New York.

Dingman, S. L., and Platt, R. H. (1977). Flood plain zoning: Implications of hydrologic and legal uncertainty. *Water Resour. Res.* **13**, 519–523.

Dunne, T., and Leopold, L. B. (1978). "Water in Environmental Planning." Freeman, San Francisco, California.

Geikie, A. (1882). "Text-book of Geology." New York, Readex Microprint, New York (1974 reprint).

Gilmore, R. (1980). "Catastrophe Theory for Scientists and Engineers." Wiley, New York.

Graf, W. L. (1979). Catastrophe theory as a model for change in fluvial systems. *In* "Adjustments of the Fluvial System" (D. D. Rhodes and G. P. Williams, eds.), pp. 13–32. Kendall/Hunt Publ. Co., Dubuque, Iowa.

Graf, W. L. (1981). Channel instability in a sand-bed river. *Water Resour. Res.* **17**, 1087–1094.

Graf, W. L. (1983). Flood-related change in an arid region river. *Earth Surf. Processes Landforms* **8**, 125–139.

Graf, W. L. (1984). A probabilistic approach to the spatial assessment of river channel instability. *Water Resour. Res.* **20**, 953–962.

Gregory, K. J., and Park, C. C. (1974). Adjustment of river channel capacity downstream from a reservoir. *Water Resour. Res.* **10**, 870–873.

Haigh, M. J., and Rydout, G. B. (1985). Catastrophic erosion in southern Arizona: The case of Greene's Canal. Mimeographed manuscript of paper presented at the First International Conference on Geomorphology, Manchester, England, September, 1985.

Hartman, G. W. (1984). "Soil Survey of Maricopa County, Arizona, Central Part." U.S. Dept. of Agriculture, Soil Conservation Service, Washington, D. C. and University of Arizona, Agricultural Experiment Station, Tucson.

Hunt, C. B., Averitt, P., and Miller, R. L. (1953). Geology and geography of the Henry Mountains region, Utah. *Geol. Surv. Prof. Pap. (U.S.)* **228**, 1–234.

Hydrologic Engineering Center (1976). "HEC-6: Scour and Deposition in Rivers and Reservoirs: User's Manual," 723-G2-L2470. U.S. Army Corps of Engineers, Davis, California.

Hydrologic Engineering Center (1982). "HEC-2: Water Surface Profiles: Program User's Manual," 723-X6-L202A, U.S. Army Corps of Engineers, Davis, California.

Knox, J. C. (1975). Concept of the graded stream. *In* "Theories of Landform Development" (W. N. Melhorn and R. C. Flemal, eds.), pp. 169-198. State University of New York, Binghamton.

Kellerhals, R., Neil, C. R., and Bray, D. I. (1976). Classification and analysis of river processes. *J. Hydraul. Div., Am. Soc. Civ. Eng.* **107**, 813–829.

Leopold, L. B., and Wolman, M. G. (1957). River channel patterns—braided, meandering and straight. *Geol. Surv. Prof. Pap. (U.S.)* **282B**, 39–85.

Leopold, L. B., Wolman, M. G., and Miller, J. P. (1964). "Fluvial Processes in Geomorphology." Freeman, San Francisco, California.

McGee, W J (1897). Sheetflood erosion. *Geol. Soc. Am. Bull.* **8**, 87–112.

McGinnies, W. G., Goldman, B. J., and Paylore, P. (1968). "Deserts of the World: An Appraisal of Research into Their Physical and Biological Environments." Univ. of Arizona Press, Tucson.

Meigs, P. (1953). "Reviews of Research on Arid Zone Hydrology." United Nations, Paris.

Merriam Company (1981). "Webster's New Collegiate Dictionary." G. and C. Merriam Company, Springfield, Massachusetts.

Miall, A.D. (1977). A review of the braided-river depositional environment. *Earth Sci. Rev.* **13**, 1–62.

Muller, R. A., and Oberlander, T. M. (1984). "Physical Geography Today: A Portrait of a Planet," 3rd ed. Random House, New York.

Petts, G. E. (1979). Complex response of river channel morphology subsequent to reservoir construction. *Prog. Phys. Geogr.* **3**, 329–362.

Pickup, G., and Rieger, W. A. (1979). A conceptual model of the relationship between channel characteristics and discharge. *Earth Surf. Processes* **4**, 37–42.

Rahn, P. H. (1967). Sheetfloods, streamfloods, and the formation of pediments. *Ann. Assoc. Am. Geogr.* **57**, 593–604.

Rich, J. L. (1938). Origin and evolution of rock fans and pediments. *Geol. Soc. Am. Bull.* **46**, 999–1024.

Richards, K. (1982). "Rivers: Form and Process in Alluvial Channels." Methuen, London.

Ritter, D. F. (1978). "Process Geomorphology." W. C. Brown, Dubuque, Iowa.

Saarinen, T. F., Baker, V. R., Durrenberger, R., and Maddock, T. Jr. (1984). "The Tucson, Arizona Flood of October 1983." Natl. Acad. Press, Washington, D. C.

Salt River Valley Water Users' Association (1903). "Irrigation Survey of the Salt River Valley" (map in 4 sheets). Salt River Valley Water Users' Assoc., Phoenix, Arizona.

Schumm, S. A. (1968). River adjustment to altered hydrologic regimen—Murrumbidgee River and paleochannels, Australia. *Geol. Surv. Prof. Pap. (U.S.)* **598**, 1–65.

Schumm, S. A. (1977). "The Fluvial System." Wiley, New York.

Schumm, S. A., and Lichty, R. W. (1963). Channel widening and flood-plain construction along Cimarron river in southwestern Kansas. *Geol. Surv. Prof. Pap. (U.S.)* **352-D**, 71–88.

Scott, K. M. (1971). Origin and sedimentology of 1969 debris flows near Glendora, California. *Geol. Surv. Prof. Pap. (U.S.)* **750-C**, 242–247.

Shen, H. W., ed. (1979a). "Modeling of Rivers." Wiley, New York.

Shen, H. W. (1979b). Additional remarks on extremal floods, basic equations, river channel patterns, modeling techniques and research needs. *In* "Modeling of Rivers" (H. W. Shen, ed.), Wiley, New York.

Smith, G. E. P. (1910). Groundwater supply and irrigation in the Rillito Valley. *Ariz., Agric. Exp. Stn., Bull.* **64**, 81–242.

Soil Conservation Service (1975). Soil taxonomy: A basic system of soil classification for making and interpreting soil surveys. *U.S., Dep. Agric., Agric. Handb.* **436**, 1–754.

Stevens, M. A., Simmons, D. B., and Richardson, E. V. (1975). Non-equilibrium river form. *J. Hydraul. Div., Am. Soc. Civ. Eng.* **101**, 557–566.

Swift, T. T. (1926). Rate of channel trenching in the Southwest. *Science* **63**, 70–71.

Tanner, W. F. (1968). Equilibrium in geomorphology. *In* "The Encyclopedia of Geomorphology" (R. W. Fairbridge, ed.), pp. 315–316. Reinhold, New York.

Thornes, J. B. (1980). Structural instability and ephemeral channel behavior. *Z. Geomorphol.* **36**, 233–244.

Trelease, F. J. (1979). "Cases and Materials on Water Law," 3rd ed. West Publishing Company, St. Paul, Minnesota.

Ward, R. (1978). "Floods: A Geographical Perspective." Wiley, New York.

Weaver, W. E. (1984). Geomorphic thresholds and the evolution of alluvial fans. *Geol. Soc. Am., Abstr. Programs* **16**, 688.

Wildan Associates (1981). "Agua Fria River Study," Maricopa County Flood Control District Contract Rep. FCD 81-2. Wildan Assoc., Phoenix, Arizona.

Williams, G. P. (1978). The case of the shrinking channels—the North Platte and Platte Rivers in Nebraska. *Geol. Surv. Circ. (U.S.)* **781**, 48.

Wilson, E. D., Moore, R. T., and Pierce, W. (1957). "Geologic Map of Maricopa County, Arizona" (map). Arizona Bureau of Mines and University of Arizona, Tucson.

Wolman, M. G., and Leopold, L. B. (1957). River flood plains: Some observations in their formation. *Geol. Surv. Prof. Pap. (U.S.)* **282-C**, 87–107.

COURT CASES

Costello v. Bowen, 80 Cal. App.2d 621, 182 P.2d 615.

Gillespie Land and Irrigation Company v. Gonzales, 93 Ariz. 152, 379 P.2d 135.

Le Brun v. Richards, 210 Cal. 308, 291 Pac. 825, 828, 72 A.L.R. 336.

Mader v. Mettinbrink, 65 N.W.2d 334, 344, 159 Neb. 118.

Maricopa County Municipal Water Conservation District No. 1 v. Warford, 69 Ariz. 1, 206 P.2d 1168.

Martinez v. Cooke, 56 N.M. 343, 348, 244 P.2d 134.

Mogle et al. v. Moore et al., 16 Cal.2d 1, 104 P.2d 785, 789.

Southern Pacific Company v. Proebstel, 150 P.2d 81, 84, 61 Ariz. 412.

State v. Hiber, 44 P.2d 1005, 1010, 48 Wyo. 172.

15

FLOODPLAIN EROSION AND DEPOSITION DURING THE DECEMBER 1982 FLOODS IN SOUTHEAST MISSOURI

DALE F. RITTER

Department of Geology, Southern Illinois University at Carbondale, Carbondale, Illinois

INTRODUCTION

The December 1982 floods on rivers in southeast Missouri received national attention because of the environmental problems created at Times Beach, Missouri (Fig. 1). During overflow of the Meramec River, the entire town of Times Beach was inundated, causing enormous flood damage. In addition, discovery of dangerously high levels of dioxin after the flood made Times Beach unsafe for habitation, and the town is now essentially abandoned.

It is not my intent to diminish the significance of the Times Beach disaster. Its cost in terms of dollars and disruption of human lives was severe, in fact, catastrophically so. The purpose of this chapter, however, is to demonstrate that alteration of floodplains by geomorphic and sedimentologic processes operating during these rare large floods are equally disastrous but receive little public notice.

A limited reconnaissance of the region (Fig. 1) following the floods indicated that floodplain damage was widespread and varied in magnitude from minor changes to significant alteration of the floodplain topography and surface materials. In some cases valuable valley bottom farms were rendered useless and cannot be rehabilitated for agricultural purposes without enormous investment. In addition, supports of highway bridges placed in the floodplain surfaces were partially exposed by vertical erosion caused by turbulent flow around the piers. In several cases this created an extremely dangerous situation. In this report the general erosional and depositional effects will be briefly examined. These will be followed by a more detailed case study involving a single reach of the Gasconade River.

THE OZARK REGION

Geology and Physiography

The dominant geologic feature in Missouri is a low asymmetrical dome known as the Ozark dome. Its core is located in southeast Missouri, but the structure extends regionally into Arkansas, Oklahoma, and southwestern Illinois. The dome is marked by older rocks occupying a central geographic position and younger rocks, flanking the center of the uplift, that dip in all directions away from the core.

The oldest rocks exposed in the region are Precambrian granites and other igneous rocks that exist in the center of the dome. These rocks, underlying the St. Francois Mountains, are exposed in St. Francois, Madison, and Iron Counties about 120 km south of St. Louis and 90 km west of Cape Girardeau.

Surrounding the Precambrian core is a thick Paleozoic rock sequence of Cambrian to Pennsylvanian age. Most rocks exposed in southeastern Missouri, however, are Cambrian and Ordovician carbonates (Stout and Hoffman, 1973). It is important to recognize that many of the lower Paleozoic carbonates in southeastern Missouri are characterized by an enormous chert content. This property is especially significant in the Cambro-Ordovician dolomites, which encircle the core of the Ozark uplift, because these units provide rivers draining the domal area with a notable chert, gravel bedload. For example, the Gasconade formation caps divides in the St. Francois Mountain area and is exposed in most of the deeper valleys draining the uplift area such as the Gasconade, Osage, White, Current, and

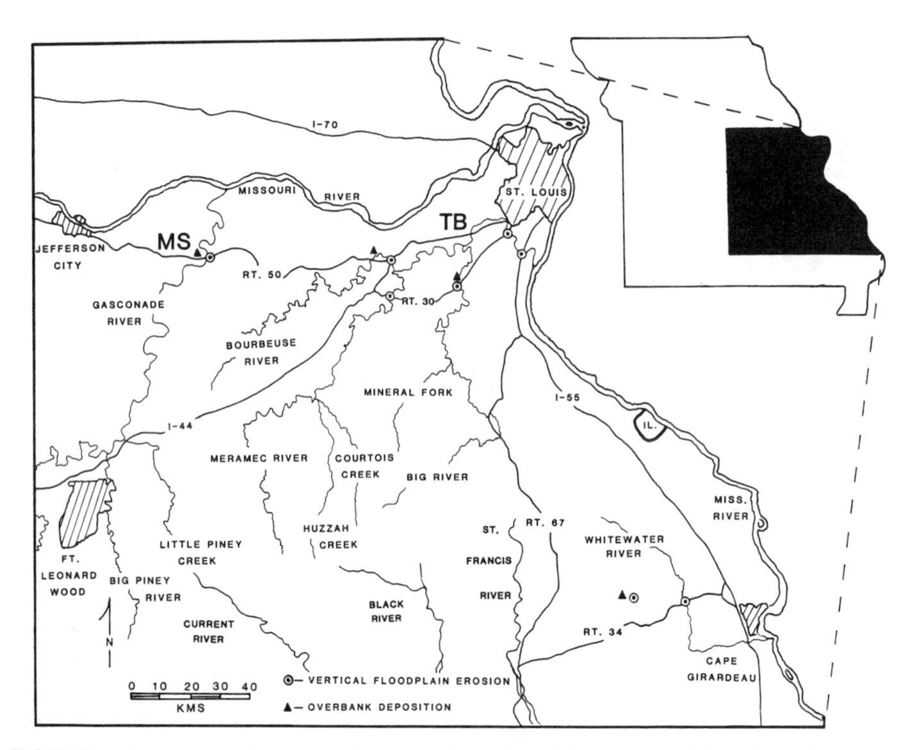

FIGURE 1. Index map showing major rivers in southeast Missouri and locations of erosion and deposition found in limited reconnaissance. TB is Times Beach and MS is Mt. Sterling.

Meramec (Wilson, 1922). The unit is exposed in almost every county in the region under question. Chert in this and other carbonate formations exists as discrete beds, honeycombed masses, or isolated nodules. Dissolution of the carbonate leaves a residiuum of chert that underlies almost every slope and floors the tributary stream channels. The coarse material is continuously introduced to the major rivers and constitutes a significant control on the regional fluvial character and mechanics. A thin loess cap, up to 3 m thick, covers many of the upland surfaces and provides a fine-grained component to the river loads.

Physiographically, the region of the Ozark uplift constitutes what is known as the Ozarks Plateaus Province (Pirkle and Yoho, 1977). It covers an area of approximately 103,600 km^2 and is bounded on the north, west, and east by the Central Lowlands, on the south by the Ouachita Province, and on the southeast by the Gulf Coastal Plain.

The Ozarks Plateaus Province of Missouri is coincident with the highest elevation and greatest relief in the state. The highest point in the state, 540 m, is located in the St. Francois Mountains. The region of the Ozark uplift generally stands above 365 m. Local relief is greatest along the largest rivers, which are deeply incised into the underlying bedrock. In the St. Francois Mountains maximum local relief exceeds 300 m, but generally outside of the St. Francois area local relief is 30–60 m and never is greater than 100 m.

Hydrology of the Region

The hydrologic character of south-central Missouri has been documented by the United States Geological Survey and the Missouri Geological Survey and Water Resources (Gann et al., 1976). Mean annual precipitation varies from 96 cm near St. Louis to approximately 117 cm near the Arkansas border. Nearly 30% of the annual precipitation becomes direct runoff; the remainder is lost by evapotranspiration and by sinking into the underground system, which consists primarily of interconnected solution openings and fractures. Resurgence of groundwater is commonly through some of the largest springs in the world. Major rivers draining the region are shown in Figure 1. Discharge and runoff characteristics for selected stations along major rivers are presented in Table 1.

It is somewhat difficult to estimate the potential for flooding in the Ozark region. For example, the Flash-Flood Magnitude Index (Beard, 1975; Patton and Baker, 1976) is not particularly great for Ozark basins up to 2590 km^2 (1000 mi^2) and is considerably lower than regions such as south-central Texas where flash flooding is common and severe. In contrast, observed floods in the region suggest a high flash-flood potential. For example, the June 17–18 1964 floods in the eastern part of the region (Petersen, 1965) showed very high flash-flood potential. In fact, Plattin Creek (drainage area = 216 km^2), a tributary of the Mis-

TABLE 1 Discharge and Runoff Characteristics at Selected Stations on Major Rivers in Southeast Missouri

River	Station	Drainage Area (km^2)	Discharge (m^3/s) Max.	Min.	Mean	Mean Annual Annual Runoff (cm)	Period of Record
Gasconade	Jerome, MO	7356	2861	7	70	30	1923–1971
Meramec	Eureka, MO	9811	3399	6	84	27	1922–1971
St. Francis	Patterson, MO	2476	2244	0.2	31	39	1921–1971
Current	Doniphan, MO	5278	2674	24	75	45	1919–1971
Black	Annapolis, MO	1254	1286	2	15	38	1940–1971

SOURCE: Data from Gann et al. (1976).

sissippi River, had a peak discharge of 853 m^3/s. This flood plots close to the trend line representing the U.S. national maximum flood discharges as compiled by Hoyt and Langbein (1955). In addition, Leopold and others (1964, pp. 65, 66) show relatively high discharge values in the Ozark region for the mean annual and 10-yr floods in basins less than 780 km^2 (300 mi^2).

It is possible, therefore, that small streams in the Ozarks may have a relatively high flash-flood potential. This might be explained by steep valley-side slopes, thin soil covers, and the intense rainfall regime of the area. In larger basins, however, expanding floodplain area tends to increase storage capacity and retard the downstream peak discharge. Floodplains are normally well defined along the major rivers and commonly exceed 450 m in width. The sedimentary sequence underlying the floodplains is variable, but most have a basal chert-rich gravel representing point bar deposition. These gravels are usually covered by 3—8m of silt and clay deposited as overbank material. Commonly, lenses of gravel are interspersed in the fine-grained overbank sediment. These are probably lobate deposits of overbank gravel lifted onto the floodplain surfaces during high-magnitude floods (Ritter, 1975).

Most floods (about 40%) occur during April and May. Nonetheless, floods have occurred in every month, and late fall and early winter floods cannot be considered as rare events. All major floods in the Ozark region result from intense rainfall, usually associated with frontal storm centers. Rainfall intensities in the area are quite high, being equivalent to those found in flood-prone areas such as central Texas (Hoyt and Langbein, 1955, p. 27). The magnitudes of floods with selected recurrence intervals are given in Table 2 and compared with discharges experienced during the 1982 floods.

THE DECEMBER 1982 FLOOD

The Flood-producing Storm

On December 3 and 4, 1982 southeast Missouri experienced widespread and intense rainfall. Precipitation totals during those two days are shown in the isohyetal map presented as Figure 2. The precipitation was distributed in two distinct centers, one located near Van Buren, Missouri, and the second spreading south from Rolla, Missouri, to Mountain Grove, Missouri. Thus, the storm affected the core of the Ozark region and zones fringing the uplift. Precipitation totals, however, decreased rapidly toward the western part of the state, being over 28 cm (11 in.) at Van Buren and less than 2.5 cm (1 in.) at Joplin, a distance of approximately 300 km. Every stream heading within the Ozarks or the fringing areas experienced significant flooding, although the magnitude and timing of the high flow varied from basin to basin.

The distribution of precipitation on December 2, 1982 (Fig. 3) clearly shows a dominance of rain in the western part of the state, especially near Warrensburg and Springfield where about 10 cm (4 in.) fell during that day. Farther east, however, little precipitation was measured on December

TABLE 2 Discharge Magnitude and Frequency at Selected Stations on Major Rivers Compared with Peak Discharge in 1982 Floods

River	Station	Q_{10}(m^3/s)[a]	Q_{25}(m^3/s)[a]	Q_{50}(m^3/s)[a]	Q–1982(m^3/s)[b]
Gasconade	Jerome, MO	1741	2078	2522	3966
Meramec	Eureka, MO	2311	2838	3484	4108
St. Francis	Patterson, MO	1151	1589	1907	4391
Current	Doniphan, MO	1674	2172	2640	3314
Black	Annapolis, MO	831	1223	1459	960

[a] Values calculated from equations presented in Sandhaus and Skelton (1968).
[b] From Sauer and Fulford (1983).

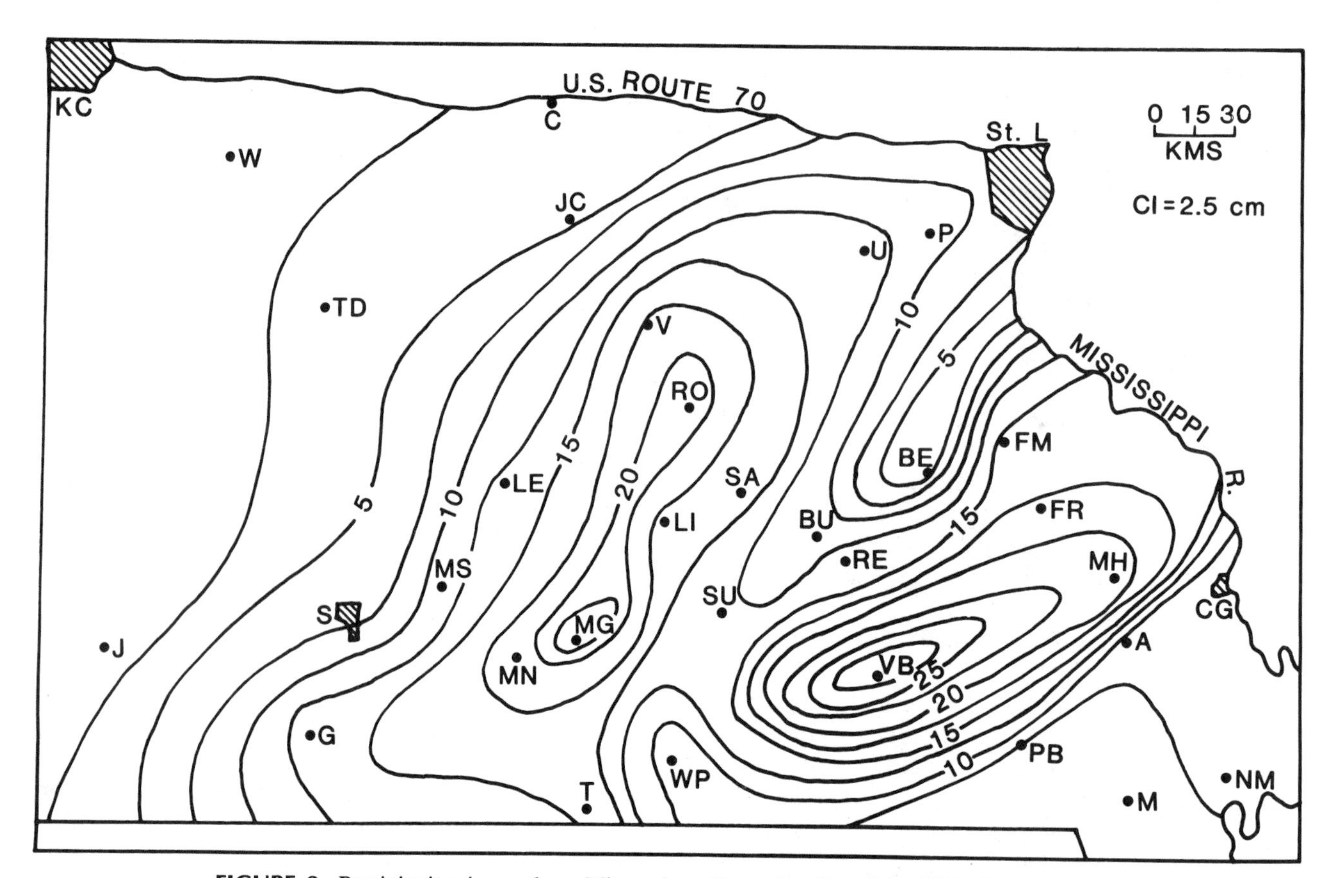

FIGURE 2. Precipitation in southern Missouri on December 3 and 4, 1982. Letters represent rainfall gauge stations: (A) Advance, (BE) Belleview, (BU) Bunker, (C) Columbia, (CG) Cape Girardeau, (FM) Farmington, (FR) Fredricktown, (G) Galena, (J) Joplin, (JC) Jefferson City, (KC) Kansas City, (LE) Lebanon, (LI) Licking, (M) Malden, (MG) Mountain Grove, (MH) Marble Hill, (MN) Mansfield, (MS) Marshfield, (NM) New Madrid, (P) Pacific, (PB) Poplar Bluff, (RE) Reynolds, (RO) Rolla, (S) Springfield, (SA) Salem, (SL) St. Louis, (SU) Summerville, (T) Tecumseh, (TD) Truman Dam, (U) Union, (V) Vienna, (VB) Van Buren, (W) Warrensburg, (WP) West Plains.

2 except for a localized storm at Belleview. Stations surrounding Belleview all recorded less than 0.2 cm of rainfall.

Climate maps for the week November 29–December 5, 1982 published by NOAA show a cold front extending through central Colorado and New Mexico on December 1. At its northern end the front was hooked to a low-pressure area centered in northeastern Utah. On December 2 the leading edge of the front was in western Missouri. By then the front connected two low-pressure centers, one in western South Dakota (which had shifted from Utah) and a developing low in southeastern Texas. The front, therefore, had moved eastward approximately 1000 km between December 1 and December 2. On December 3 the front was directly over eastern Missouri, still joining a northern and southern low-pressure area. By December 4 the northern low-pressure zone and the associated frontal line had moved rapidly to the northeast and east, respectively, but the southern part of the system remained relatively stationary as its eastern migration was blocked by a Bermuda high off the southeast coast of the United States.

Thus, it appears that the major precipitation recorded on December 3 and December 4 resulted when an eastward-moving cold front stalled over the region. Presumably, moisture derived from the Gulf of Mexico was continuously provided to the stalled front by circulation around the Bermuda high-pressure zone and the low pressure centered in western Louisiana. By December 5 the southern limb of the cold front had broken loose and drifted eastward, and the southern low-pressure center had migrated northward into western Illinois.

General Geomorphic Effects

During the 1982 floods many floodplains were affected by deposition and erosion accompanying the high flow. Almost all erosion occurred in the form of vertical cutting into the floodplain surfaces. Little, if any, bank erosion was noted. Most vertical erosion was initiated when overbank flow encountered an isolated obstruction jutting above the floodplain surface. In some cases the obstructions were individual trees, but more commonly, the barriers were bridge supports designed to keep highways off the floodplain surface.

Flow circulating around bridge piers and trees seems to initiate vertical macroturbulence (Matthes, 1947), or kolking, that literally drills holes into the floodplain sediment. Kolking involves an extreme pressure gradient in a vertical column of water, and flow mechanics induced in such a setting is known to create enormous erosional problems (Moore and Masch, 1963; Laursen, 1960; Shen, 1971). The kolks, or macroturbulent eddies, are thought to be created by significant velocity differences between surface and bottom waters and by turbulent energy dissipation as flowing water encounters a barrier such as those described above. The results of kolking action during the 1982 floods were quite variable depending on local conditions. In some cases the effect was severe. For example, near Marble Hill, Missouri (Fig. 2), Crooked Creek created funnel-shaped holes up to 3 m deep and 20 m wide in the floodplain (Fig. 4). The largest holes apparently formed where an earthen dike was breached, creating minor topographic relief on the otherwise flat surface that existed prior to the flood. Turbulence developed in that zone, combined with swirling flow generated around nearby trees lining the channel bank, probably initiated the kolking action. The holes shown in Figure 4 are very similar to the "swirl pits" developed during the 1936 flood in the Connecticut River valley (Collins and Schalk, 1937).

Deposition on the floodplain surfaces was equally problematical. It occasionally occurred as a thin (<0.5 m) sheetlike layer of coarse gravel and sand spread evenly across the floodplain surface (Fig. 5). More commonly, however, sand and gravel were deposited as barlike lobes that traversed the floodplain surface (see Costa, 1974; Ritter, 1975). These were especially prevalent on the inside of meander bends. The coarse debris was invariably derived from load that would normally be contained within the river channels. It was lifted to the floodplain surfaces in suspension or as traction load moving up depositional ramps that build vertically to the level of the floodplain during high-flow conditions. In either case coarse debris was deposited passively on the floodplain surface with no interval of scouring preceding its deposition. Tillable floodplain

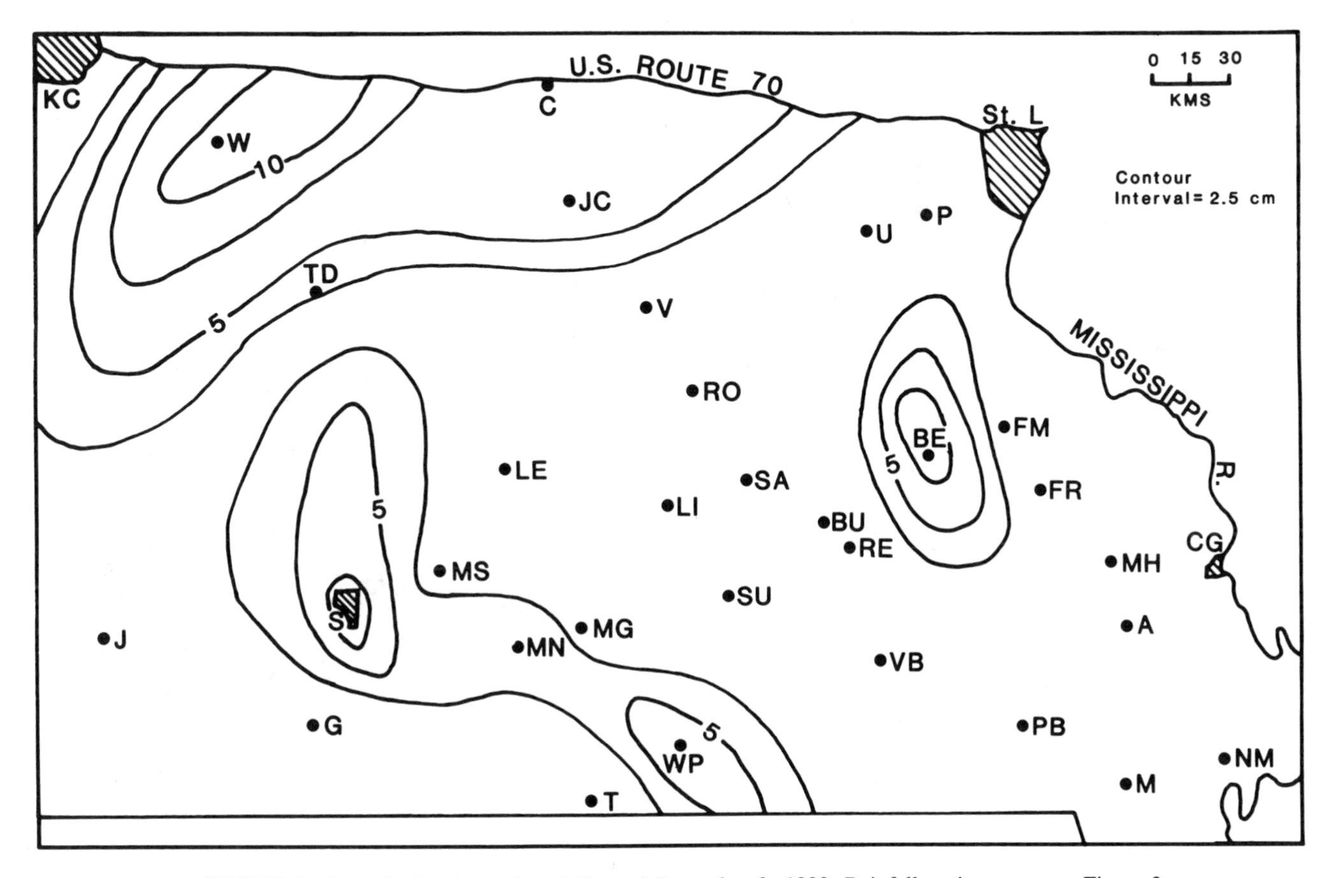

FIGURE 3. Precipitation in southern Missouri December 2, 1982. Rainfall stations same as Figure 2.

FIGURE 4. Eroded floodplain surface of Crooked Creek near Marble Hill, Missouri.

FIGURE 5. Sheet deposit of sand and gravel. Crooked Creek near Marble Hill, Missouri. Upstream from location shown on Figure 4.

soils are, therefore, buried by a variable thickness of coarse sediment that cannot support crops and cannot be reworked into the soil by repeated deep plowing.

THE GASCONADE RIVER: A CASE EXAMPLE

The Gasconade River heads southwest of Rolla, Missouri, and flows to the northeast until it joins the Missouri River about 90 km west of St. Louis. The river drains an area of approximately 9300 km^2 underlain primarily by the chert-rich Cambrian and Ordovician carbonates described earlier. Near Mt. Sterling, Missouri (Fig. 1), the erosional and depositional effects of the flood were particularly dramatic. Here the river exists in a broad meander, the outside of the bend abutting against vertical bluffs of the cherty dolomites. The inside of the meander bend is occupied by a broad floodplain that generally stands about 6 m above the channel bottom. Missouri State Route 50 rests directly on the floodplain surface except in the 75 m closest to the river. In that zone the bridge crossing the river was changed in 1981 from the southeast direction shown in Figure 6 to an almost east–west position, and the roadbed was raised above the floodplain surface. The bridge here is supported by 1.5 m diameter cement piers buried about 2 m into the floodplain sediment and set on iron friction piles that are anchored into the bedrock beneath the floodplain sequence.

During construction of the new bridge, trees north and south of the site were removed (Fig. 6), and ditches were excavated adjacent to both sides of the highway to drain water from the floodplain surface. The channel of the south ditch was directed between the bridge piers, and the two ditches joined just north of the bridge where water was funneled into the river through a large culvert. The west bank of the Gasconade was riprapped to prevent erosion and potential undercutting of bridge supports. Thus, prior to the 1982 flood considerable change in the local environment was brought on by activities associated with construction of the new, elevated bridge.

The December 1982 storm, determined by the U.S. Geological Survey to be the 100-yr storm for a 24-hr period, produced a discharge that generally represents a flood with a recurrence interval of greater than 100 yr (Table 3). High-water marks on bridge piers, trees, and the unsubmerged sides of Route 50 indicated that the water level was 3.9 m (12.9 ft) above the floodplain surface during the flood peak.

Depositional Effect

The floodplain surface near Mt. Sterling was drastically altered by sand and gravel deposition during high-flow conditions. The phenomenon of overbank sand and gravel deposition is not uncommon (Jahns, 1947; Wolman and Eiler, 1958; Costa, 1974; Gupta and Fox, 1974; Stewart and LaMarche, 1967; Ritter, 1975; Teisseyre, 1978, Stene, 1980). However, to my knowledge the Gasconade deposit probably represents the largest of these in a humid region reported to date. Overbank deposition assumed the geometric form of a giant 30–100-m-wide lobe, extending longitudinally for approximately 1 km (Figs. 7 and 8). The lobe deposit is 1–2 m thick and rests conformably on the silts and clays of the floodplain surface (Fig. 9). Like most of these deposits, lateral margins are distinct, changing along a well-defined line from coarse sediment to undisturbed floodplain surface (Fig. 10). This exemplifies the remarkable fluvial mechanics associated with this phenomenon because no laterally confining barriers exist within a sheet of floodwater. Therefore, a thread of deposition must have been present in the moving water, adjacent to which the current was capable of transporting the entire load. The mechanics of the bar formation are probably similar to

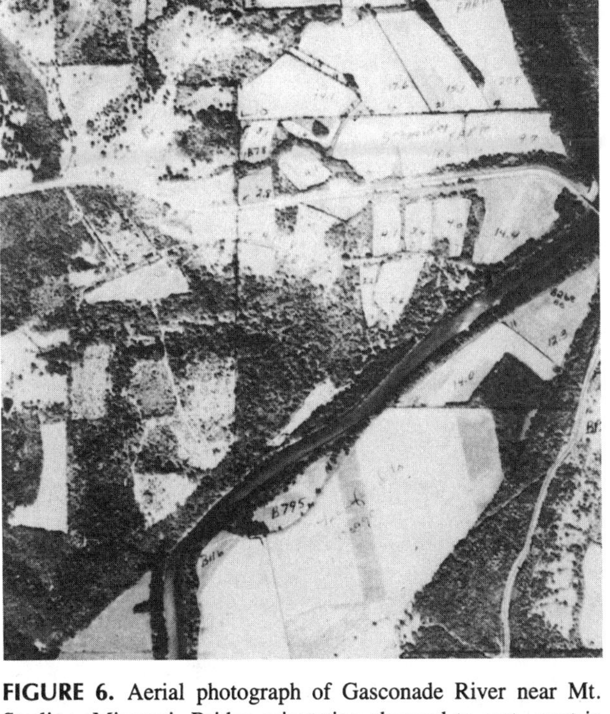

FIGURE 6. Aerial photograph of Gasconade River near Mt. Sterling, Missouri. Bridge orientation changed to east–west in 1981, and trees south and north of the bridge removed. Note that river abuts limestone bluffs on the outside of meander bend at the bridge site.

TABLE 3 Maximum Discharge and Precipitation During December 1982 Flood, Gasconade River

Station	Drainage Area (km^2)	Date	Discharge (m^3/s)	Frequency (yr)
Hazelgreen	3238	12/3/82	2663	40
Jerome	7356	12/5/82	3966	> 100
Rich Fountain[a]	8236	12/6/82	3796	> 100

RAINFALL AT VIENNA WEATHER STATION[b]

Time Period	Rainfall (cm)	Frequency (yr)
4 days (12/2–12/5)	25.3	50
2 days (12/2–12/3)	21.9	100
1 day (12/2–12/3)	19.9	100

[a] 16 km upstream from Mt. Sterling.
[b] 40 km southwest of Mt. Sterling.

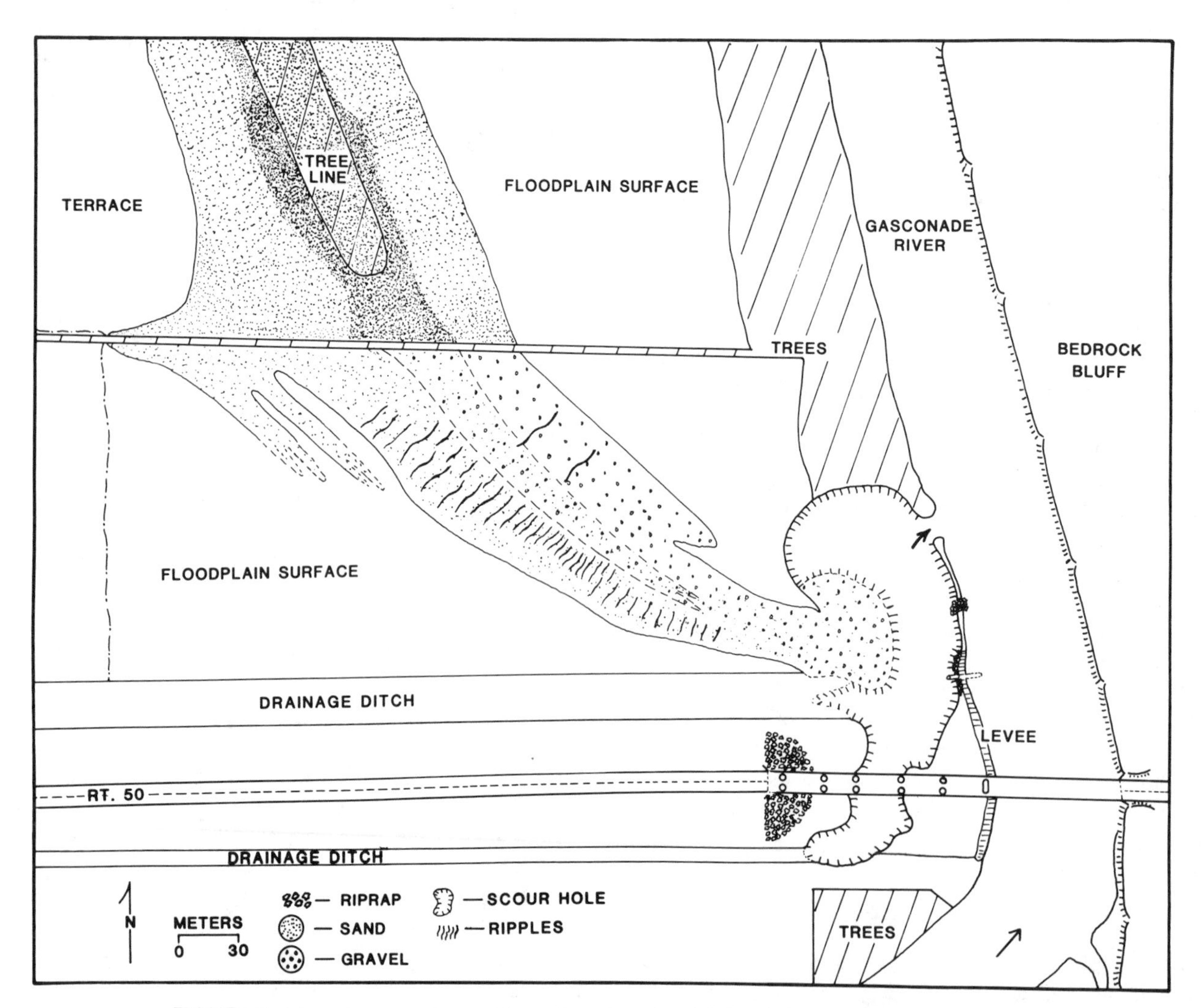

FIGURE 7. Map of sand and gravel lobe deposit on floodplain of the Gasconade River near Mt. Sterling, Missouri.

FIGURE 8. Photograph of sand and gravel lobe deposit shown in Figure 7. Taken from uplands east of the Gasconade River. Note scour hole north of the bridge.

FIGURE 9. Gravel sublobe looking across floodplain surface to Route 50 in background.

FIGURE 10. North margin of gravel sublobe looking east. Note undisturbed floodplain surface and sharp extremity of gravel deposition.

those described in the initiation of braided channels (Leopold and Wolman, 1957), and in flow processes associated with secondary circulations described by Leopold (1982) where bedload will be moving in marginal zones but depositing in the central position. Any chance accumulation of bed debris causes an increase in the velocity gradient over the incipient bar much like a braided channel. The higher the bar gets the steeper the gradient. Therefore, bed shear stress increases, allowing transport of coarse sediment down the bar surface until, at its distal end, velocity decreases and deposition occurs. This allows lobate material to extend longitudinally but, at the same time, have no deposition adjacent to the centrally located bar. Thus, construction of the lobate form instead of a widespread depositional sheet was a function of fluvial mechanics rather than topographic irregularity on the floodplain surface.

The lobate deposit consists of two distinct subzones, one primarily composed of gravel, the other composed of sand. These zones parallel one another but are separated by an intermediate zone that is texturally mixed (Fig. 7). The gravel component extends longitudinally to an east–west stand of trees marking the boundary between adjacent properties (Fig. 7). Downstream from this tree line, no gravel is found in the lobe, and the deposit of sand thickens considerably. Sand accumulation continues down the surface of the floodplain, but it gradually thins until the distinct lobate form is lost several hundred meters from the point of re-entry into the Gasconade River.

The source of the sand and gravel is almost certainly channel sediment of the Gasconade River. A large hole (to be discussed later) eroded into the floodplain surface during the flood, may have provided some of the material included in the lobate deposits. However, volumetric comparison of the hole and the lobate deposit indicates that vertical erosion of the floodplain sequence could not have been the primary source of the lobate material. Total volume of the lobe deposit is estimated to be 112,500 m^3, contrasting sharply to the 43,000 m^3 volume of the hole. Sediment exposed in the hole is almost all overbank silt and clay, debris that is consistently absent in the lobe deposits. Natural levee sands do interfinger with the silt and clay near the river bank, but these are volumetrically inconsequential. In fact, if the entire floodplain stratigraphy was composed of sand, erosion of the hole could only account for 40% of the lobate sand deposit. It is unlikely that gravel was produced from the scour hole as no gravel is exposed within the hole or along the local Gasconade banks. It is, however, possible that some gravel exists lower in the floodplain sequence, but was beneath the lake and river levels at the time of our observations. Nonetheless, its mobilization could not have been a significant contribution to the gravel subzone.

Bedforms in the Flood Deposit. Ripple bedforms of various size were found in both the sand and gravel subzones. The literature concerning current ripples and their relationship to flow dynamics is enormous; therefore, a complete treatment is well beyond the scope of this chapter. In addition, terminology used here to describe bedforms in the Gasconade lobe deposits is not an attempt to provide an encompassing classification, and it is probable that certain terms will have a different connotation than the same terms employed in other studies.

Bed features formed in lower regime flow generally exist as three types: *small ripples*, *megaripples*, and *sand waves* (giant ripples). Megaripples range from 0.6 to 30 m in length and from 0.06 to 1.5 m in height. The ripple index (L/H) is always greater than 15 (Reineck and Singh, 1980). Megaripples are considerably larger than small ripples, which some workers believe are distinct forms that are usually less than 0.3 m in length and are not necessarily transitional into megaripples. Thus, the lower length limit of 0.6 m for megaripples may set them apart geometrically, and presumably genetically, from small ripples.

Two types of megaripples are usually noted. One develops at lower velocities (0.3–0.8 m/s) and tends to be straight crested. These forms have been called type I megaripples (Dalrymple et al., 1978), straight-crested megaripples (Reineck and Singh, 1980), and sand waves (Southard, 1975; Boothroyd, 1978). The second type of megaripple is usually known as a dune, although it has been called a type II megaripple by Dalrymple and others (1978). Dunes tend to be sinuous or cuspate and develop at velocities between 0.7 m/s and 1.5 m/s. They are commonly associated with scour pits or pools that exist in the position of the ripple troughs.

The term *sand wave* has produced some confusion because it is used to describe both low-energy megaripples (Southard, 1975; Boothroyd, 1978) and very large bedforms that develop in coarse sediment under high-energy flow and water depths greater than 4 m (Dalrymple et al., 1978). In this chapter the term *giant ripple* is used to describe very large bedforms, a practice suggested by Reineck and Singh (1980).

In addition to flow, particle size exerts an important control on bedform type. For example, in sands having a mean size smaller than 0.6 mm, the initial forms developed are small ripples that change character to megaripples with increasing current velocity. However, in sand coarser than 0.6 mm no ripples develop until flow velocity exceeds a critical value that produces some type of megaripple as the initial bedform. Some studies have shown that giant ripples are also related to particle size. For example, Dalrymple and others (1978) point out that megaripples and giant ripples can occur under the same water depth and flow velocity, but the giant ripples develop where sediment is coarser than 0.3 mm. Megaripples will form if the sediment is finer than 0.3 mm.

In the Gasconade flood deposit a distinct ripple train extending from the large hole to the east–west tree line was noted in the sandy portion of the lobe. A short train consisting of several large ripples was preserved in the gravel portion of the lobe near the same tree line. A segment of the sandy ripple train is characterized by accumulation of fine gravel (2–8 mm) on the stoss face of successive ripples although the corresponding troughs are formed in sand (Fig. 11). These ripples have shorter chord lengths and smaller heights than the sandy ripples immediately upstream and downstream from the segment containing the gravel caps. It is therefore possible that the gravel caps prevented this segment from developing geometric characteristics similar to the normal sandy ripples. Nonetheless, they are treated below as separate entities.

Ripples in the sand sublobe increase in height downstream, and near the western margin of the floodplain they have disjointed crest lines and scour pools in the troughs (Fig. 12). All sandy ripples have smaller ripple trains superimposed on their stoss limbs. Ripples in the gravel sublobe are much larger than the sandy forms (Table 4). The train is very short and consists of only two well-developed ripples.

All sandy bedforms in the Gasconade case are megaripples based on ripple indices (L/H) and crest orientations (Table 4). Upstream sand ripples and the gravel-capped sand ripples are interpreted as low-energy megaripples. Downstream ripples in the sand sublobe are probably dunes because of the pronounced scour pools associated with the ripple form. Textural analyses show that all sandy ripples are composed of sand smaller than 0.6 mm. Thus, it seems unlikely that geometry of the sand ripples is reflecting particle size rather than flow characteristics.

It also seems unlikely that the 1982 flood produced rapid flow conditions, that is, all bedforms were formed during lower regime, tranquil flows. At a maximum depth of 3.9 m, a Froude number (F) equal to 1 requires a flow velocity of 6.2 m/s across the floodplain surface. Rough calculations suggest that the total cross-sectional area of the floodwater was approximately 2350 m^2. The maximum discharge at Rich Fountain, Missouri, was estimated at 3796 m^3/s. Therefore, at peak flow an average velocity of 1.6 m/s was attained, well below that needed for $F = 1$. This means that bedforms could have formed at any time during the flood and not have been destroyed by flow conditions reaching the upper regime.

FIGURE 11. Bedforms in the Gasconade floodplain lobes. Gravel ripples in upper left of photograph. Dunes in sand sublobe shown in center left. Gravel-capped ripples and sand megaripples in center. North drainage ditch with refrigerator in foreground. All bedforms are found between bridge and east–west treeline.

Bedforms in the gravel sublobe are either megaripples or giant ripples. Figure 13 shows the extension of the best-fit lines derived by Allen (1968) and Baker (1973, 1978), which relate average ripple lengths to average ripple heights. Allen's line represents large asymmetrical ripple trains ranging in height from 0.06 to 21 m and in length from 0.9 to 1800 m. Clearly this broad range of bedform sizes includes a variety of bedform types. In contrast, Baker's line represents measurements on giant ripples formed in the Lake Missoula floods in eastern Washington. Geometrically, ripples from the Gasconade sublobes fit closely to both line extensions and cannot be logically assigned to either. However, extending Baker's (1973, 1978) best-fit lines between depth and both ripple length and ripple height and inserting the Gasconade data (using depth ascertained from the average high-water mark of 3.9 m) shows that only the gravel ripples of the Gasconade plot close to the lines (Fig. 14). Sand ripples and gravel-capped sand ripples plot far from the projected lines.

Based on all the bedform data, the following model is suggested for the development of the Gasconade ripple forms. Dunes in the downstream portion of the sand sublobe probably developed rapidly during the rising stage of the flood, a phenomenon observed elsewhere by Coleman (1969). It is possible that gravel bedforms also developed as dunes during that phase. However, as discussed above, even though dunes and giant ripples can form under similar conditions of current velocity and depth (Dalrymple et al., 1978; Jackson, 1976), only dunes will develop in sediment having a grain size less than 0.3 mm. Sediment coarser than that will tend to develop giant ripples. Importantly, however, both studies show that giant ripples require deep water to develop ($\approx$4 m).

Therefore, considering the size and depth constraints, and the apparent relationship with Baker's analyses, it seems possible that the gravel bedforms did not develop as dunes in the rising phase of the flood but are, in fact, poorly organized giant ripples that began to develop at the peak depth of 3.9 m. This may explain the limited number of bedforms in the gravel train. Megaripples and small ripples in the upstream portion of the sandy sublobe are quite possibly late-phase embellishments on the dune ge-

FIGURE 12. Dunes in downstream portion of sand sublobe. Note prominent scour pools.

ometry (see Fig. 11) as velocity decreased below 0.7 m/s during the recession of the flood. No superimposed megaripples are found on the gravel bedforms because the large particle size probably precluded their formation.

The interpretation that giant gravel ripples formed at peak flow receives support from Jackson's (1976) suggestion that dunelike large-scale ripples are produced from macroturbulent bursting. Bursts seem to occur immediately downstream from the crestlines of large-scale ripples, and their spacing is controlled by the periodicity of the burst cycle. Importantly, Jackson (1976) suggests that the wavelength of these forms is related to the depth of flow in that the ratio of wavelength to depth (λ/d) is approximately 7. Using the wavelength of Gasconade gravel ripples and employing Jackson's ratio suggests that the water depth was approximately 3.4 m when the ripples formed, a value close to the maximum depth of 3.9 m.

In light of the above, however, it is difficult to understand why the sandy dunes did not extend their wavelengths at peak flow. The wavelength–depth relationship would predict a water depth of only 0.7 m at the time of their formation. Thus, it is possible that (1) the sandy dunes

TABLE 4 Bedform Geometry and Deposit Texture in Gasconade River Overbank Lobes[a]

	Ripple Height (m)	Ripple Length (m)	Ripple Index (L/H)	Median Diameter (mm)	Mean Diameter (mm)	Maximum Size (mm)	Bedform Type
Gravel	0.58	23.6	40.7	0.5[b]	0.6[b]	49	Dune or giant ripple
Sand	0.12	5.0	41.7	0.25	0.25	—	Upstream: low-energy megaripple; Downstream: dune
Gravel-capped sand	0.08	2.5	30.9	0.3	0.38	8.7	Low-energy megaripple

[a] Size based on only 10 samples.
[b] Matrix only.

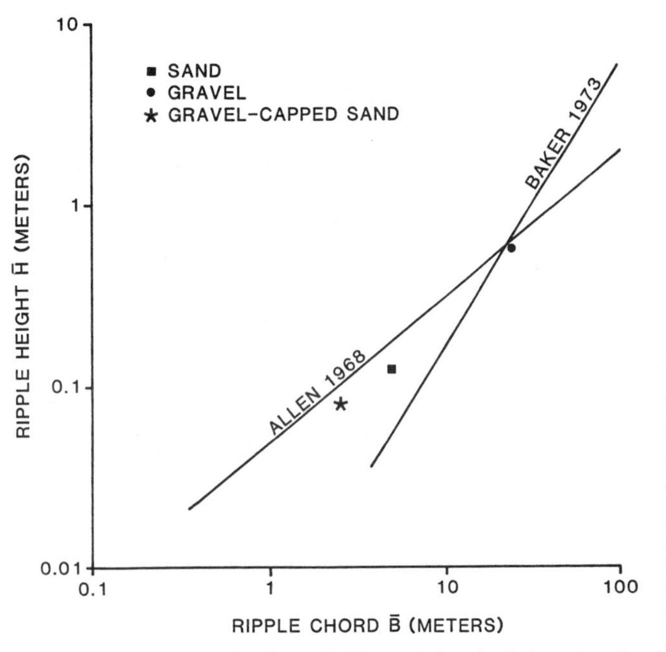

FIGURE 13. Extension of best-fit lines relating ripple length and ripple height with plots of Gasconade bedforms. Based on Baker (1973) and Allen (1968).

were prevented from becoming large-scale forms because of their particle size or (2) the sandy dunes, like the other megaripples, formed during the recessional phase of the flood. The latter alternative almost requires that deposition of the sand lobe occurred after peak flow and is, therefore, not genetically related to the gravel lobe. This hypothesis is not beyond the realm of possibility.

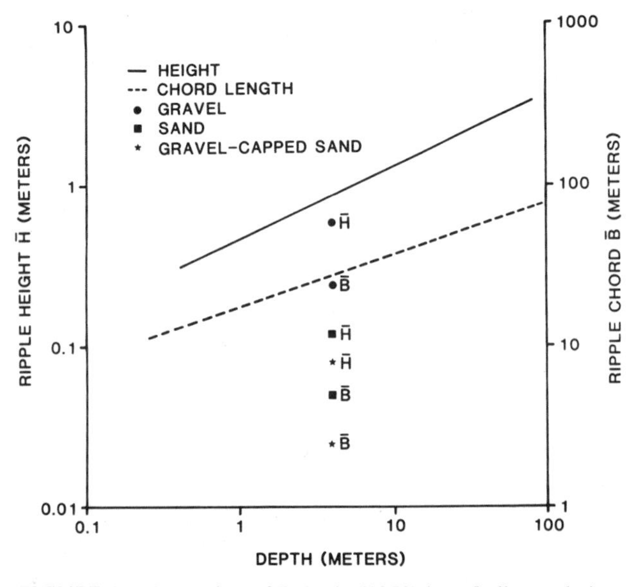

FIGURE 14. Extension of Baker's (1973) best-fit lines relating flow depth to ripple length and ripple height. Gasconade bedforms plotted using average maximum depth (3.9 m).

The Erosional Effect

The upstream end of the lobate deposit is truncated by a large hole scoured into the floodplain surface during the flood (Fig. 7). The hole is approximately 160 m long and 60 m wide, occupying a surface area of 5 ha. Excavation below the floodplain surface is greater than 7 m at its maximum depth. The hole is partially filled with water, creating a small lake, the level of which is being controlled by the elevation of a small outlet channel at its north end.

After the flood the northeast edge of the eroded hole was separated from the Gasconade River by only 3–5 m of floodplain. Recognize that the consumption of material resulting in the thin sliver of floodplain that remains was caused by erosion directed from the inside out. This means that erosion began by circular vertical incision, and that dimensions of the depression expanded outward in all directions during the flood, including toward the river. Dramatic evidence of this phenomenon is shown in Figure 15 where riprap placed on the Gasconade bank was undercut from the inside and fell onto the bank of the hole. Presumably, this arrested further lateral expansion of the hole toward the river.

Vertical erosion was presumably initiated by macroturbulence around the bridge piers and over minor topography associated with the drainage ditch leading to the culvert. Scouring around bride piers develops by two basic vortex flow systems: (1) a *horseshoe vortex* and (2) a *wake vortex* (Shen, 1971). A horseshoe vortex is a flow separation phenomenon and usually results in greatest scouring on the upstream side of the obstruction (Moore and Masch, 1963). In contrast, a wake vortex involves vertical lift associated with kolking and tends to form elliptical scour holes downstream from the obstruction.

Some evidence exists to suggest that both vortex types operated along the Gasconade. Maximum erosion occurred downstream from the bridge supports, and the scour hole attains its greatest depth immediately north of the culvert. It seems certain, therefore, that sucking action of vertically oriented kolks associated with wake vortices produced the major portion of the scour hole. Beneath the bridge, however, scouring around the piers was notably greater on their upstream sides, indicating that horseshoe vortices may have developed (Fig. 16).

It is indeed fortunate that the major incision did not occur immediately adjacent to the piers. The area beneath the bridge represents the zone of maximum flow constriction and greatest depth, factors that are considered to be prime ingredients needed for severe macroturbulent action (Laursen, 1960; Moore and Masch, 1963; Shen, 1971). Even so, the comparatively minor cutting beneath the bridge was still able to expose almost 1.5 m of the friction piles underlying the cement piers (Fig. 16). Additional release of the surrounding pressure by deeper removal of the floodplain sediment could have produced lateral shifting of the

FIGURE 15. East margin of scour hole looking south. Scour hole expanded toward the river (left) until it undercut riprap protecting levee on Gasconade River. Water depth beneath culvert was approximately 2 m.

friction piles, piers and the bridge surface. Such a scenario would have created an extreme hazard and potential loss of the bridge.

Precisely when the scouring occurred during the tenure of the flood is not clear. Field relationships suggest that the rear edge of the gravel lobe was truncated by the scour hole, indicating that scouring followed the gravel deposition. In light of this, it seems most probable that scouring began at or about the same time as the incipient giant ripple development, that is, at or near peak flow. Scouring at this time could also have reworked the natural levee deposits in the floodplain sequence, thereby providing additional sands for the megaripple development during the waning stage of the flood.

CONCLUSION

In December 1982 floods in southeast Missouri caused overbank deposition of sand and gravel on some floodplains and concomitant vertical erosion into other floodplain surfaces. Overbank sand and gravel was deposited as either lobate or sheetlike forms, depending on the mechanism by which channel bedload was carried to the floodplain surface. At the Gasconade site an enormous lobe of sand and gravel was placed on the floodplain surface. The lobe consists of two distinct sublobes composed primarily of sand or gravel. Each sublobe is marked by bedforms that indicate that current velocities probably ranged from 0.3 to 1.5 m/s. Dunes preserved in the downstream portion of the sand sublobe probably developed under rapidly increasing current velocity during the rising phase of the flood, although a waning-stage origin cannot be ruled out. At or near peak stage, incipient giant ripples formed in the gravel sublobe when flow depth approached a critical value. Their development, however, ceased as water levels receded.

When velocity decreased during flood recession, low-energy megaripples and small ripples developed in the upstream portion of the sandy ripple train, partially destroying the original dune geometry. These had insufficient time, however, to totally re-form the downstream dunes. It is possible that the superimposed megaripples and small ripples developed because additional sands entered the system by reworking of natural levee sands in the floodplain stratigraphy.

FIGURE 16. Major scouring on upstream side of bridge pier. Looking north along axis of scour pool. Friction piles exposed by scouring. Depth of water at friction piles is 1.2 m.

Vertical erosison was initiated by kolking or other macroturbulent flow around obstructions and over small topographic irregularities on the floodplain surface. The erosion, combined with sediment deposition on the floodplain surfaces, resulted in a significant agricultural loss. Nonetheless, the greatest erosional hazard rests in the potential of kolking to undermine bridge supports. Reconnaissance after the 1982 floods indicated that bridge supports adjacent to every major river experienced some damage.

At the Gasconade meander the extreme erosion was probably caused by several factors: (1) removal of tree stands that lined the channel prior to construction of the new bridge, (2) elevation of the bridge off the floodplain surface, and (3) the position of the bridge on the inside of a large meander bend that abuts against bedrock bluffs. Vertical erosion here produced a 5-ha (2-acre) hole that presumably developed by wake vorticity and associated kolking initiated around the bridge piers and over artificially produced topography on the floodplain surface. Excavation of the hole probably followed most of the lobe deposition. The deepest part of the hole (7 m) was located downflow of the bridge supports. Even so, enough erosion occurred around the piers to place the bridge in jeopardy of collapse.

ACKNOWLEDGMENTS

I wish to thank D. S. Blakley for his valuable assistance in the field. Sieve analyses were done by J. Marzolf. I also express appreciation to P. C. Patton and R. C. Kochel who reviewed the initial manuscript and made excellent suggestions for its improvement. The Office of Research Development and Administration at Southern Illinois University at Carbondale provided funds to partially support field expenses associated with the study.

REFERENCES

Allen, J. R. L. (1968). "Current Ripples, Their Relation to Patterns of Water and Sediment Motion." North-Holland Publ., Amsterdam.

Baker, V. R. (1973). Paleohydrology and sedimentology of Lake Missoula flooding in eastern Washington. *Spec. Pap.—Geol. Soc. Am.* **144**, 1–79.

Baker, V. R. (1978). Paleohydraulics and hydrodynamics of Scabland floods. *In* V. R. Baker and D. Nummedal, eds., "The Channeled Scabland" pp. 59–115. Natl. Aeronaut. Space Admin., Washington, D.C.

Beard, L. R. (1975). Generalized evaluation of flash-flood potential. *Tech. Rep.—Univ. Tex. Austin, Cent. Res. Water Resour.* **CRWW-124**, 1–27.

Boothroyd, J. C. (1978). Mesotidal inlets and estuaries. *In* "Coastal Sedimentary Environments" (R. A. Davis, Jr., ed.), pp. 287–360. Springer-Verlag, Berlin and New York.

Coleman, J. M. (1969). Brahmaputra River: Channel processes and sedimentation. *Sediment. Geol.* **3**, 129–139.

Collins, R. F., and Schalk, M. (1937). Torrential flood erosion in the Connecticut Valley, March, 1936. *Am. J. Sci.* **234**, 293–307.

Costa, J. E. (1974). Response and recovery of a Piedmont watershed from tropical storm Agnes, June 1972. *Water Resour. Res.* **10**, 106–112.

Dalrymple, R. W., Knight, R. J., and Lambiase, J. J. (1978). Bedforms and their hydraulic stability relationships in a tidal environment, Bay of Fundy, Canada. *Nature (London)* **275**, 100–104.

Gann, E. E., Harvey, E. J., and Miller, D. E. (1976). Water resources of south-central Missouri. *Geol. Surv. Hydrol. Invest. Atlas (U.S.)* **HA-550**.

Gupta, A., and Fox, H. (1974). Effects of high-magnitude floods on channel form: A case study in Maryland Piedmont. *Water Resour. Res.* **10**, 489–509.

Hoyt, W. G., and Langbein, W. B. (1955). "Floods." Princeton Univ. Press, Princeton, New Jersey.

Jackson, R. G. (1976). Sedimentological and fluid-dynamic implications of the turbulent bursting phenomenon in geophysical flows. *J. Fluid Mech.* **77**, 531–560.

Jahns, R. H. (1947). Geologic features in the Connecticut Valley, Massachusetts, as related to recent floods. *Geol. Surv. Water-Supply Pap. (U.S.)* **996**, 1–158.

Laursen, E. M. (1960). Scour at bridge crossings. *J. Hydraul. Div. Am. Soc. Civ. Eng.* **86**(HY2), 39–54.

Leopold, L. B. (1982). Water surface topography in river channels and implications for meander development. *In* "Gravel-Bed Rivers" (R. D. Hey, J. C. Bathurst, and C. R. Thorne, eds.), pp. 359–388. Wiley, New York.

Leopold, L. B., and Wolman, M. G. (1957). River channel patterns; braided meandering and straight. *Geol. Surv. Prof. Pap. (U.S.)* **282-B**, 39–85.

Leopold, L. B., Wolman, M. G., and Miller, J. P. (1964). "Fluvial Processes in Geomorphology." Freeman, San Francisco, California.

Matthes, G. H. (1947). Macroturbulence in natural stream flows. *Trans., Am. Geophys. Union* **28**, 255–262.

Moore, W. L. and Masch, F. D. (1963). Influence of secondary flow on local scour at obstructions in a channel. *Misc. Publ. U.S. Dep. Agric.* **970**, 314–320.

Patton, P. C., and Baker, V. R. (1976). Morphometry and floods in small drainage basins subject to diverse hydrogeomorphic controls. *Water Resour. Res.* **12**, 941–952.

Petersen, M. S. (1965). Floods of June 17–18, 1964 in Jefferson, St. Genevieve and St. Francois Counties, Missouri. *Water Resour. Rep. (Mo., Geol. Land Surv. Div.)* **19**, 1–20.

Pirkle, E. C., and Yoho, W. H. (1977). "Natural Regions of the United States." Kendall-Hunt Publ. Co., Dubuque, Iowa.

Reineck, H. E., and Singh, I. B. (1980). "Depositional Sedimentary Environments." Springer-Verlag, Berlin and New York.

Ritter, D. F. (1975). Stratigraphic implications of coarse-grained gravel deposited as overbank sediment, southern Illinois. *J. Geol.* **83**, 645–650.

Sandhaus, E. H., and Skelton, J. (1968). Magnitude and frequency of Missouri floods. *Water Resour. Rep. (Mo., Geol. Land Surv. Div.)* **23**, 1–276.

Sauer, V. B., and Fulford, J. M. (1983). Floods of December 1982 and January 1983 in central and southern Mississippi River basin. *Geol. Surv. Open-File Rep. (U.S.)* **83-213**, 1–41.

Shen, H. W. (1971). Scour near piers. *In* "River Mechanics" (H. W. Shen, ed.), pp. 23.1–23.25. Water Resources Publications, Fort Collins, Colorado.

Southard, J. B. (1975). Bed configuration. *SEPM Short Course* **2**, 5–44.

Stene, L. P. (1980). Observations on lateral and overbank deposition—Evidence from Holocene terraces, southwestern Alberta. *Geology* **8**, 314–317.

Stewart, J. H., and LaMarche, V. C. (1967). Erosion and deposition produced by the flood of December, 1964, on Coffee Creek, Trinity County, California. *Geol. Surv. Prof. Pap. (U.S.)* **422-K**, 1–22.

Stout, L. N., and Hoffman, D. (1973). An introduction to Missouri's geological environment. *Mo., Geol. Surv. Water Resour., Educ. Ser.* **3**, 1–44.

Teisseyre, A. K. (1978). Physiography of bed-load meandering streams: Imbricated gravels in fine-grained overbank deposits. *Geol. Sudetica* **13**(1), 87–92.

Wilson, M. E. (1922). The occurrence of oil and gas in Missouri. *Mo., Div. Geol. Surv. Water Resour.* [Rep.] **16**, 1–284.

Wolman, M. G., and Eiler, J. P. (1958). Reconnaissance study of erosion and deposition produced by the flood of August, 1955 in Connecticut. *Trans., Am. Geophys. Union* **39**, 1–14.

16

GEOMORPHIC RESPONSE OF STREAMS TO FLOODS IN THE GLACIATED TERRAIN OF SOUTHERN NEW ENGLAND

PETER C. PATTON
Department of Earth and Environmental Sciences, Wesleyan University, Middletown, Connecticut

INTRODUCTION

The variation in geology, physiography, and hydrology make southern New England (Connecticut, Rhode Island, and Massachusetts) an instructive region in which to evaluate geomorphic responses of streams to large floods. Analysis of flood phenomena and generalizations derived from those observations can be applied to other humid temperate regions and to general questions concerning the effects of floods on fluvial landforms.

The region is typical of much of northern North America where the present drainage network morphometry and channel systems were established after the last Wisconsin glaciation. The inherited glacial landscape has had a profound effect on the hydrology and geomorphology of modern drainage basins. For example, basins underlain by low permeability glacial till and fine-grained lacustrine sediment have a more rapid hydrologic response during large flows than basins underlain by high-permeability coarse-grained stratified deltaic and fluvial sediment (Thomas, 1966). Hydrograph characteristics such as lag time are also affected by the deranged drainage and numerous inland wetlands that increase surface detention and storage and are the results of glaciation. Channel morphology and pattern are partly controlled by valley gradient and sediment type, and these are directly related to the glacial topography and sedimentology. Thus, the degree to which floods permanently affect the morphology of channels and flood plains in this region is partly a function of these pre-existing conditions.

The geographic and hydrologic diversity of southern New England permit the comparison of the flood response of bedrock-controlled streams in highland areas with alluvial rivers in adjacent lowlands. Such comparisons are useful in determining the physical factors that control stream response to large floods. Southern New England fluvial systems are also interesting because of the apparent rapid recovery time of the streams to rare great floods. Wolman and Eiler (1958) noted that, 3 yr following the New England floods in 1955, many stream reaches that had suffered scour or deposition were rapidly revegetated and that it was difficult to determine the effect of the flood after only a short time interval. Analysis of the geomorphic effects of floods and their stratigraphic record in this region demonstrates the difficulty encountered in isolating the geomorphic effects of large floods from those related to more frequent flow events. Nevertheless, some geomorphic trace remains of the catastrophic events, and this stratigraphic evidence can be used to evaluate the paleoflood hydrology of streams in this region.

Physical Setting

As noted, the physiography and geology of southern New England constrain the geomorphic response of streams to floods. The region can be subdivided into four geologic provinces (Fig. 1) with similar physiographic characteristics (Denny, 1982). The greatest relief in the region (in places exceeding 300 m) is in the Western Highlands: the Berkshire Hills, the Taconic Mountains, and their foothills.

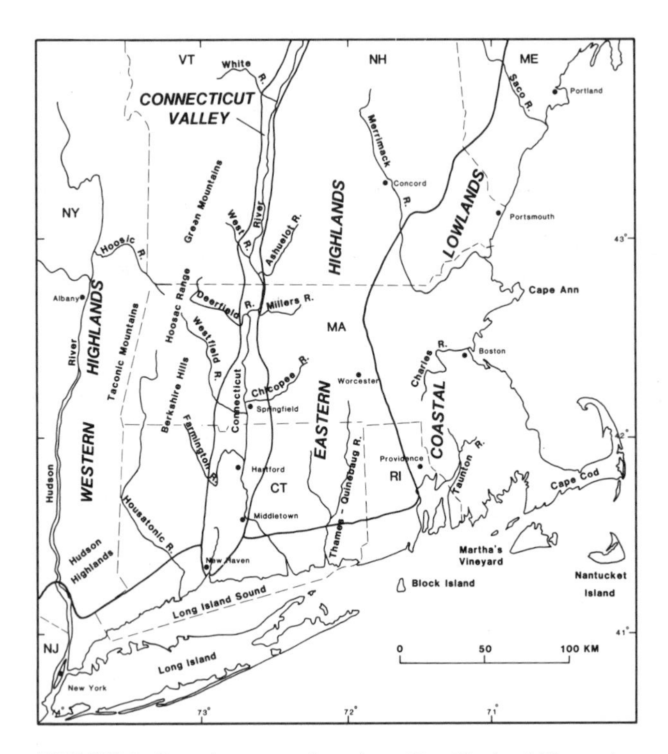

FIGURE 1. Location map of southern New England illustrating the major physiographic provinces and the rivers described in the text.

Here, narrow valleys are cut into the lower Paleozoic and Proterozoic igneous and metamorphic rocks. Only where Cambrian and Ordovician carbonate rocks are exposed have the valleys been significantly widened, such as along the Housatonic River where it crosses the Stockbridge Marble in the vicinity of Kent, Connecticut (Rodgers, 1985). To the east, southern New England is bisected by the Connecticut Valley Lowland, a Mesozoic half-graben that trends north–south (Fig. 1). The Connecticut River, the major drainage of the lowland, flows down the axis of this sedimentary basin until it crosses the eastern border fault and exits the basin at Middletown, Connecticut. Although this is predominantly a region of low relief underlain by nonresistant sandstones, siltstones, and shales, prominent, basalt-capped ridges form a nearly continuous west-facing cuesta that diagonally crosses the basin from northwest to southeast. The Connecticut Valley Lowland is bordered on the east by the metamorphic and igneous rocks of the Eastern Highlands, a region of moderate relief of up to 100 m (Denny, 1982). Finally, southern New England is bordered on the south and east by the Coastal Lowlands that front the Atlantic Ocean and Long Island Sound.

Flood Meteorology

Floods in southern New England are caused by a variety of meteorological conditions including hurricanes, wave cyclones ("northeasters"), frontal systems, and isolated convective storms. In addition, rapid spring snowmelt, often exacerbated by heavy frontal precipitation, has caused some of the largest floods on New England rivers. The following accounts serve to illustrate the meteorological conditions that have produced large floods and the hydrologic response of specific streams.

Major hurricane floods in southern New England occurred in 1927, 1938, and 1955. Figure 2 shows the storm tracks of the hurricanes responsible for these floods. The November 1927 floods resulted from meteorological conditions that funnelled a tropical depression northward between two high-pressure systems, one to the west over the continent and one in the northern Atlantic (Kinnison, 1929). The tropical air mass was orographically lifted over the Western Highlands and additionally lifted over the colder high-pressure systems. Rainfall totals were great because the flow of tropical air was maintained for up to 24 hr. The greatest rainfalls were in the high elevations of northern New England where 230 mm of rainfall fell over a broad area. The areas of maximum precipitation occurred in two north–south belts, one following the Berkshire–Green Mountain trend along the western part of New England, the second extending inland along the Connecticut–Rhode Island border (Fig. 3*a*). The storm rainfall was effectively

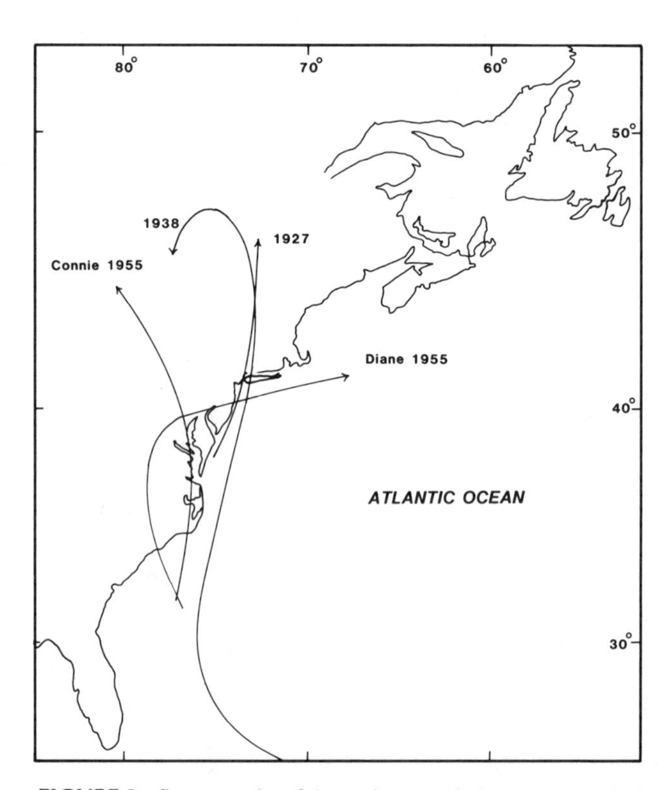

FIGURE 2. Storm tracks of the major twentieth-century tropical hurricanes that have caused catastrophic flooding in southern New England.

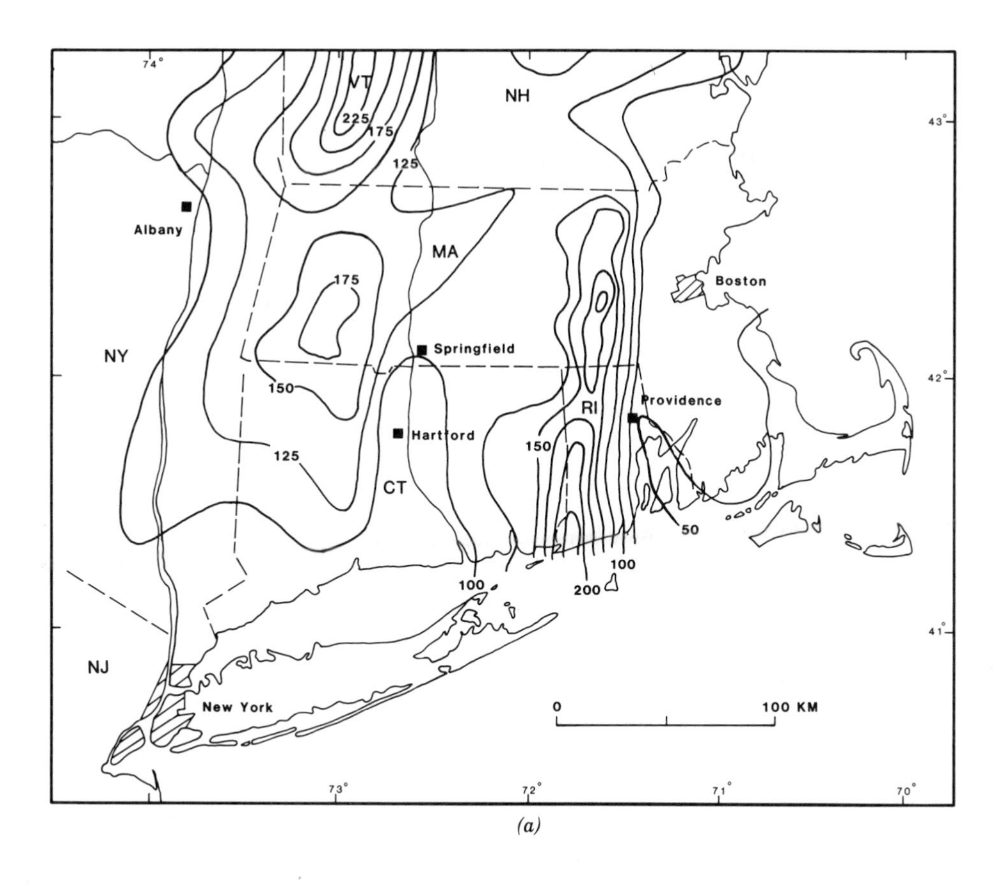

(a)

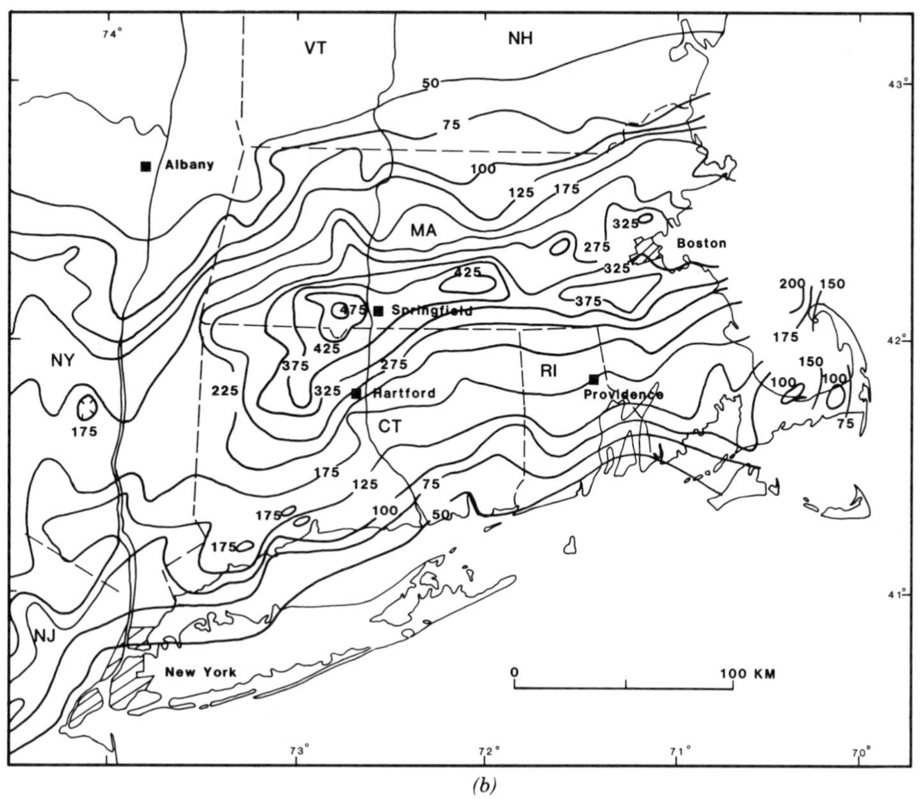

(b)

FIGURE 3. (a) Isohyet map of the rainfall totals for the November 2–5, 1927, hurricane. Note the high rainfall concentrations over the Eastern and Western Highlands (adapted from Kinnison, 1929). (b) Isohyet map of the rainfall totals for Hurricane Diane for the period August 17–20, 1955 (adapted from Bogart, 1960). Rainfall totals are in millimeters.

converted to runoff because the deciduous forests had lost their leaves, and thus interception and surface detention was reduced.

Because the most intense rainfall was in northern New England, most of the record-setting peak discharges occurred there. In southern New England record peak discharges were recorded in streams that drained the Berkshire Hills, and on the Connecticut River as far south as Holyoke, Massachusetts. Although the flood wave on the Connecticut River was attenuated to the south, the 1927 flood on the Connecticut River at Hartford came within 0.25 m of exceeding the 1854 flood of record, which was the largest flood since 1639 (Kinnison, 1929).

The September 1938 hurricane and flood combined to produce a catastrophic natural disaster in southern New England. Antecedent conditions, and the hurricane storm track that focused intense rainfall on the Eastern Highlands, combined to produce a dramatic flood event (Paulson, 1940). Antecedent rainfall during the period of September 12–16 was caused by a low-pressure system that moved eastward across New England and dropped 25–100 mm of rain in the region. This storm saturated the ground and depleted the natural detention storage, but produced only negligible runoff. It was followed by the hurricane rainfall that began on September 17 and culminated with the passing of the storm on the 21st. The storm track of the hurricane was directly across Long Island and up the Connecticut River Valley (Fig. 2). The cyclonic circulation caused the heaviest rainfall to the east of the storm center and resulted in a north–south ridge of high rainfall in the Eastern Highlands. Rainfall totals dropped off rapidly to the east of this ridge. Near Portland, Connecticut, the 4-day rainfall total was 433 mm. Because the ground was saturated, runoff volumes relative to total precipitation were high. An average of 190 mm of rain fell across the Connecticut River basin above Hartford. Of this rainfall 103 mm were converted to runoff, a yield of 54%. In smaller basins the runoff yield approached 75%. Equally important in generating the flood was the rainfall intensity. At Hartford, the total storm rainfall for the 4-day period was 330 mm; but the storm dropped over 225 mm of rainfall in less than 2 days and over 150 mm on September 20th alone (Paulson, 1940).

Along the Connecticut River in southern New England the flood of 1938 was the second largest in 300 yr, exceeded only by the rainfall–snowmelt flood of 1936. At Hartford the Connecticut River crested at 10.8 m, 0.67 m lower than the flood of 1936. However, in many of the smaller tributary basins and in other drainages, the peak discharge exceeded the 1936 flood. Even those basins in the western region, removed from the greatest rainfall intensities, such as the Housatonic River, recorded stages and discharges greater than the 1936 flood. The destruction and loss of life from the 1938 storm was caused by the combination of the floods, hurricane force winds, and the storm surge along the coastline, where tides were more than 3 m above normal high water; it remains the greatest natural disaster ever visited on the region (Paulson, 1940).

In 1955 two late-summer hurricanes, Connie and Diane, produced record-breaking precipitation in the Western Highlands of southern New England. Hurricane Connie moved up Chesapeake Bay, through central Pennsylvania and across Lake Erie between August 12 and 14 (Fig. 2). Rainfall records for western New England between August 11 and 16 indicate that up to 225 mm fell in this high-relief region. Immediately on its heels, Hurricane Diane moved ashore in the mid-Atlantic states, crossed New Jersey, and eventually went out to sea, paralleling the southern New England coastline (Fig. 2). Rainfall during the period August 17–20 from this storm produced record totals with nearly 500 mm of rainfall in the vicinity of Springfield, Massachusetts. In general, the isohyets paralleled the coastline with the most intense rainfall located in a southwest–northeast band approximately 80 km inland from the shore (Fig. 3*b*). This rainfall event was the greatest since systematic records were begun in 1886 and exceeded that generated by the storm of October 3–4, 1869, the previous storm of record (Bogart, 1960).

The first hurricane saturated the ground and caused many rivers to reach bankfull stage. When the second

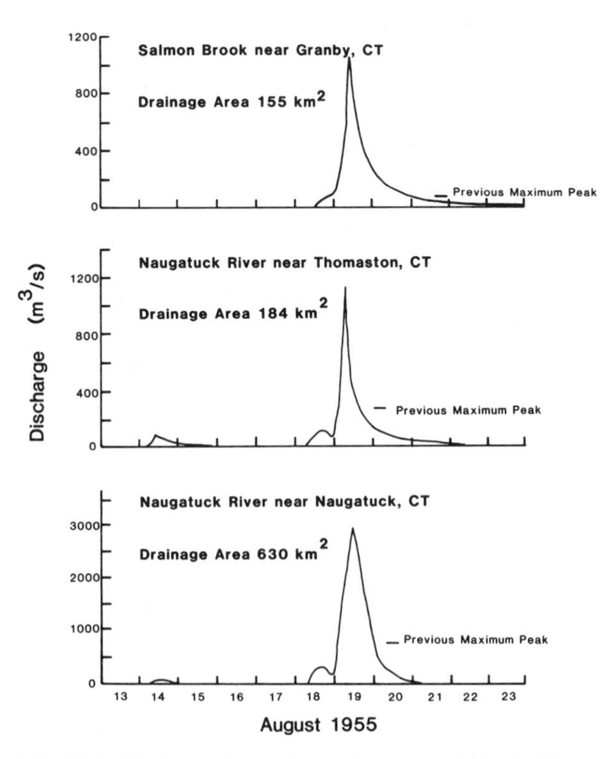

FIGURE 4. Hydrographs for three streams that drain the Western Highlands of Connecticut. Note the sharp rising limb on these hydrographs and the relation of the 1955 flood peak with the previous maximum peak (adapted from Bogart, 1960).

storm hit, the conversion of rainfall to runoff was rapid, creating flash floods along many streams (Fig. 4). For example, the Naugatuck River at Thomaston, Connecticut, rose 5.8 m during a 7-hr period on August 19. In the space of a half hour it rose 1.2 m. The highest unit discharge in the Northeast occurred in the Berkshire foothills where Powdermill Brook near Westfield, Massachusetts, discharged 162 m^3/s from 6.5 km^2, a rate of 24.9 $m^3/s/km^2$. In addition, in Massachusetts and Connecticut alone, 54 gauging stations recorded the maximum flood of record. On many streams the new peak discharge exceeded the previous maximum severalfold, and significantly changed the shape of the flood frequency curves for these streams (Figs. 4 and 5). Whereas most of the severe flooding occurred along small streams and intermediate sized rivers, the mainstem drainages did experience significant flooding in the southern portions of their basins. For example, the combined flood runoff from the Chicopee, Westfield, Scantic, Farmington, and Park rivers was great enough to produce the third greatest flood of record at Hartford (Bogart, 1960).

The greatest flood of record on the Connecticut River was the result of a snowmelt–rainfall flood in March 1936 (Grover, 1937). The flood crest at Hartford was 2.6 m higher than any previous flood during the 300-yr period of record that existed at that time. The March floods resulted from rainfall from four distinct low-pressure systems that passed over New England between March 9–22. The storm of March 16–19 was at the time one of the greatest storms to hit New England and is noteworthy because it occurred earlier in the year than any other previous great storm. The heaviest rainfalls were recorded in northern New England with up to 500 mm recorded in the White Mountains during the period March 9–22. Rainfall in southern New England averaged between 150 and 200 mm with the greater totals in the eastern and western highlands. The rain fell on a snowpack that had a water content ranging from 25 to 125 mm in the Western Highlands of Connecticut to up to 300 mm in the White Mountains of New Hampshire. The combined water content of the rain and snow over the basin was a record. The first storm caused a major breakup of the ice-covered rivers. On the Connecticut River, the ice was 0.4 m thick. The resulting ice breakup produced an ice jam at Windsor Locks, Connecticut, that raised the water level 2.1 m, up to the level of the 1927 flood. The second storm fed the still swollen rivers and produced an

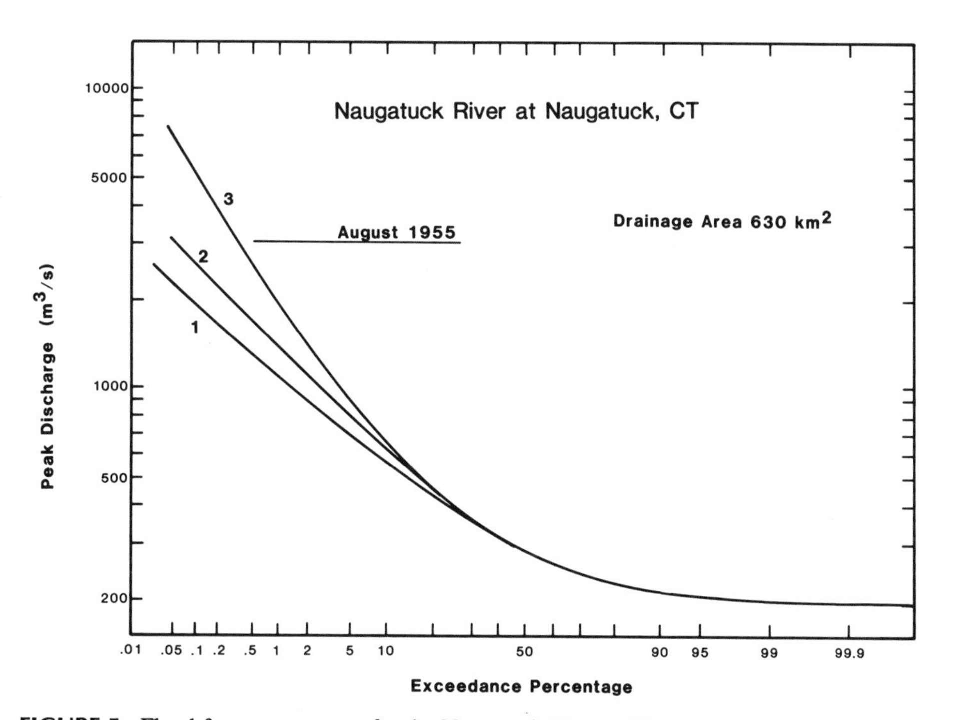

FIGURE 5. Flood frequency curves for the Naugatuck River at Naugatuck, Connecticut. Curve 1 was computed using the regional skewness coefficient of 0.03 and does not include the 1955 flood. Curve 2 is computed with the same skewness but includes the 1955 flood. Curve 3 was calculated with a new regional skewness coefficient of 1.0 proposed for tributaries in southern New England following the 1955 flood. Curves 2 and 3 increase the magnitude of low-probability floods (adapted from U.S. Army Corps of Engineers, 1956).

additional rise of over 2.4 m on the Connecticut River, which crested at Hartford at 11.5 m on March 20 (Grover, 1937).

Based on the above accounts, it is clear that all of the disastrous floods in southern New England have resulted from a combination of hydrologic factors that have served to increase the conversion of rainfall to runoff. The most common factors are antecedent rainfall or ground frost, which decrease infiltration, or a water-saturated snowpack, which increases the total water content on the surface of a basin. Although large floods have occurred in all seasons, the seasonality of the storms can be important in influencing surface detention and infiltration. Finally, storm movement and land surface elevation have been important in controlling the regional distribution of rainfall.

Flood Magnitude and Frequency

By comparison with other regions in North America the absolute magnitude of New England floods is not great. The unit discharge of floods in this region does not approach that of regions like the southern Appalachians or central Texas (Leopold et al., 1964, p. 66). One reason for this is that the flood-producing rainfall is of lower intensity and magnitude (Hoyt and Langbein, 1955).

Within southern New England, however, several regions occur for which the flood potential is somewhat higher than elsewhere in the region. Beard (1975) has suggested that the greater the standard deviation of the annual peak series, the greater the potential for flash floods. This is simply because the higher standard deviation reflects the existence of a large outlier in the flood series. In southern New England the Western Highlands stand out as an area of greater flash-flood potential when compared to the rest of New England. The orographic effect of the highlands increases rainfall during storms and the high-relief and steep slopes rapidly convert rainfall to runoff.

A second aspect of the magnitude and frequency of floods in this region is the length of record and the flood experience. At Hartford, stage records for the Connecticut River were recorded beginning in 1871 (Kinnison et al., 1938). Historical accounts of great floods date to 1639, 3 yr after the city was settled (Kinnison et al., 1938). Stage records prior to 1871 were recorded by individual observers and some of these observations can be tied to the modern datum of the gauge. For example, prior to 1871, flood stages were recorded by spikes driven into the wall of a brewery located near the site of the modern gauge. These elevations were converted to the datum of the gauge by the U.S. Corps of Engineers Survey of 1871 (Kinnison et al., 1938). The earliest reliable stage measurement is for the flood of 1683, which crested at 7.9 m, and is thought to be a lesser flood than that of 1639. Prior to 1936 the largest flood of record was the 1854 flood. The 1936 flood exceeded the stage record by over 2.4 m. Since 1927 the

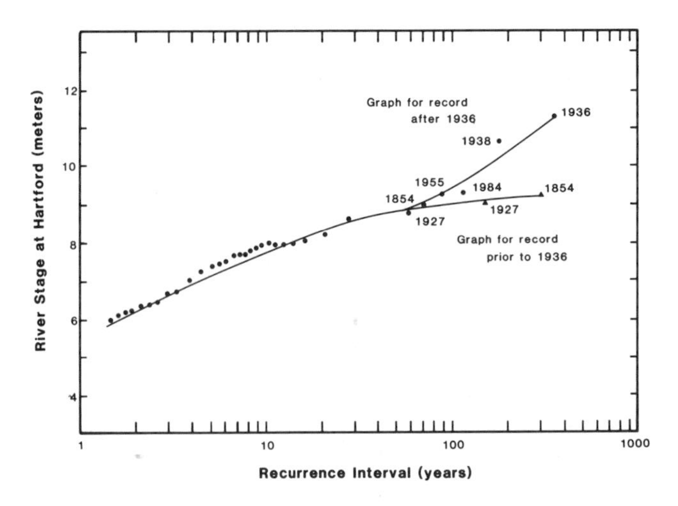

FIGURE 6. Frequency distribution of flood stage at Hartford, Connecticut, since 1639. The lower curve is the record prior to the 1936 flood, the upper curve includes the five large floods of the past 50 yr (modified from Hoyt and Langbein, 1955).

1854 flood has been nearly equaled (1927 flood) or exceeded (1936, 1938, 1955, 1984) five times (Fig. 6). Thus five of the six greatest floods in a 350-yr record have occurred in the past 50 yr (Fig. 6). This illustrates that even in a region characterized by low variability in the annual peak flood series there can be significant departures in the flood magnitude series in spite of a long historic record. The effect of the 1955 flood on the flood frequency curves for smaller streams in the region also supports this view.

GEOMORPHIC EFFECTS OF LARGE FLOODS

The position of streams in the drainage network, as well as the influence of local relief and structure, strongly affect the geomorphic response of streams to large floods. Also, the recovery time of stream systems typically increases with decreasing size of the drainage basin, such that landforms and sedimentary deposits related to flood events are more evident on the smallest streams. Larger grain sizes are moved in the steeper smaller drainages, resulting in less reworking and modification by interflood flows in the small basins.

As a result, southern New England can be broken down into two regions of differing geomorphic response to floods: the highland, strongly flood-affected drainages in the crystalline and metamorphic terrains and the lowland, somewhat less modified drainages of the Connecticut Valley and the Coastal Lowlands.

Flood Effects in Highland Drainages

The river systems in the highlands of eastern and western New England exhibit morphologies that are strongly influenced by large floods. Most of these rivers are gravel bed rivers with relatively steep gradients and bedrock controls on their valley dimensions and shape.

Following the disastrous 1955 floods, Wolman and Eiler (1958) noted that valley width and stream gradient were important local controls on erosion and sedimentation (Fig. 7). Erosion resulted from greater flow depths and greater flow velocities in narrow confined reaches. Comparing streams and valleys of equal cross-valley dimensions, the higher gradient streams suffered more severe erosion for the same reason. Because valley dimensions are important controls on erosion and sedimentation, the spacing of erosional and depositional features was spotty. Wolman and Eiler (1958) also noted the rapid recovery time of these upland streams. They considered that, given the size of the entire drainage system, the degree of erosion and deposition was relatively minor compared to the magnitude of the floods. However, stratigraphic data collected from deep excavations along these floodplains, discussed below, confirms that rare floods are important processes in floodplain construction in this region.

Streams draining the highland regions are characterized by high-gradient tributaries that have created debris fans at their confluences with the main stems. These debris fan deposits have long been associated with severe flooding. For example, in 1902 H. F. Cleland reported the effects of a cloudburst flood on the flanks of Mt. Greylock in western Massachusetts. The flood resulted from a storm that produced up to 86 mm of rainfall in 4 hr. Three ravines on the slopes of Mt. Greylock produced debris avalanches that created long linear scars on the mountainside and debris fans across the agricultural fields at the base of the mountain. The largest scar was 455 m long and 15 m wide at the head and 60 m wide at the base. The stream that filled this scar was estimated to be 7.5 m deep. Where this torrent flowed across the plain at the base of the mountain it was 23–30 m wide and 3 m deep. Cleland described it as being filled with sediment and trees. The deposits from this debris avalanche covered 2–2.5 ha with sand and boulders, some of which were up to 1.2 m in diameter (Cleland, 1902).

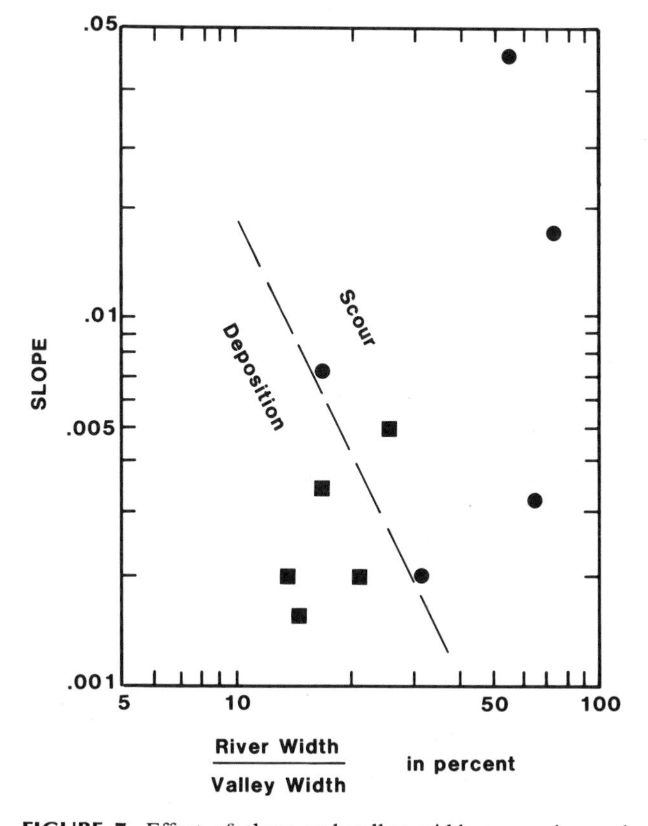

FIGURE 7. Effect of slope and valley width on erosion and deposition during the 1955 flood in Connecticut (from Wolman and Eiler, 1958).

Since Cleland's observation, numerous other investigators have noted the presence of debris avalanches following floods in the high-relief regions of New England. Jahns (1947) described small debris fans that were associated with small active gullies in agricultural fields. He also described large fan-shaped features that occurred at the mouths of larger tributaries where they joined the mainstem. He termed these larger fans, deltas, although they are simply larger scale versions of the debris fans. Jahns noted that along rivers such as the Deerfield River in Massachusetts the fans consist of coarse gravel that must be deposited into the mainstem during large flood events. Similar debris fans have been described following large floods from other regions in the Appalachians (Hack and Goodlett, 1960; Williams and Guy, 1973; Kochel, this volume).

Floods and Floodplain Construction along a Highland Drainage: The Shepaug River in Western Connecticut

The Shepaug River in western Connecticut is typical of highland drainages. Analysis of the effects of the 1955 flood in this basin and detailed studies of the floodplain stratigraphy provide insight into the long-term geomorphic effect of rare great floods on these streams.

The Shepaug River drains an area of 390 km^2 in northwestern Connecticut and is an eastern tributary of the Housatonic River (Fig. 8). The Shepaug's valley is incised into Ordovician metamorphic rocks of the Western Highlands, primarily the Rowe and Ratlum schists (Rodgers, 1985). The valley is structurally controlled and can be divided into high-gradient, narrow, bedrock-controlled constrictions exhibiting narrow floodplain surfaces separated by wider and longer valley reaches. Glaciofluvial and glaciolacustrine deposits are neither thick nor extensive in this valley (Malde, 1967); the Shepaug River has thus been near its present elevation since the beginning of the Holocene (Patton and Handsman, 1983). Finally, the Holocene evo-

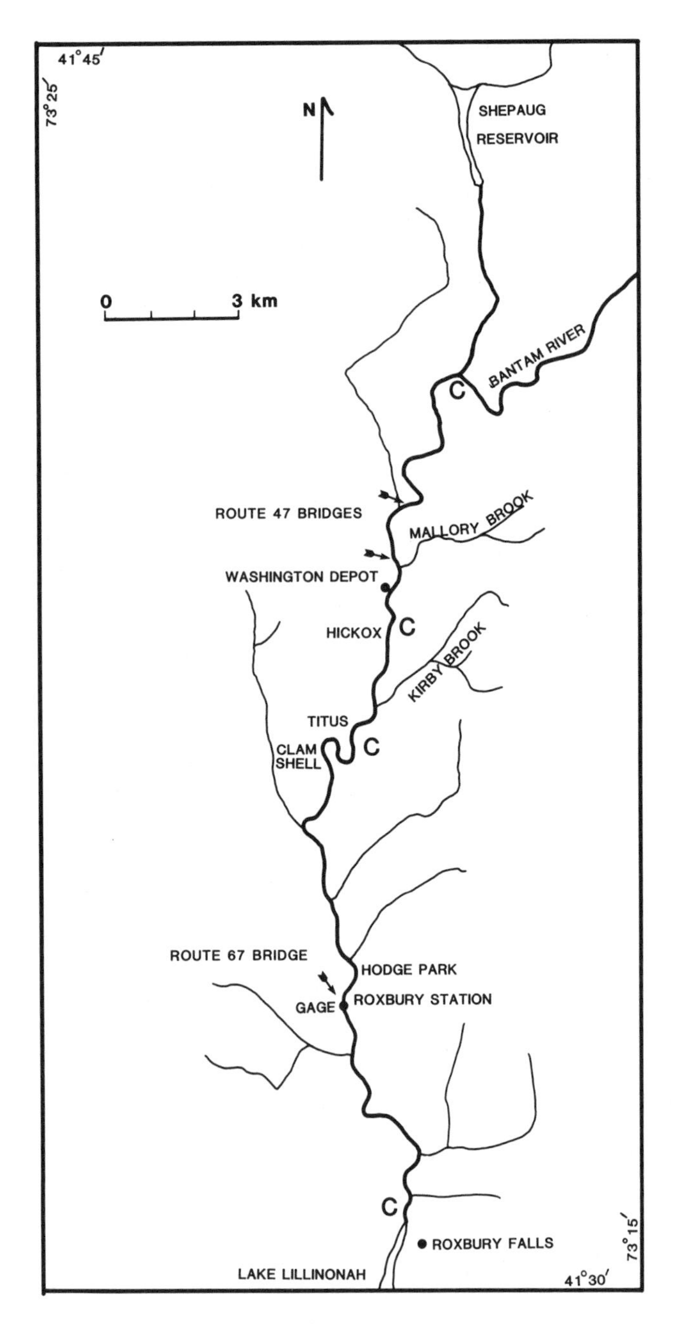

FIGURE 8. Location map of the Shepaug basin in western Connecticut identifying stratigraphic locations mentioned in the text. Bedrock valley constrictions are noted by the letter C. Arrows point to highway bridge locations referred to in Figure 10.

lution of the Shepaug River has also been independent of the downcutting history in the trunk-stream Housatonic drainage because a high-gradient bedrock gorge occurs at Roxbury Falls and forms a steep knickpoint and local base-level.

The 1955 flood in the Shepaug River valley is the largest flood since systematic stream gauging began in 1931, and the flood is thought to have exceeded in magnitude and destruction the previous great flood of November 13, 1853 (Thomson et al., 1964). During Hurricane Diane an estimated 290 mm of rain fell on the basin of which 200 mm was converted to runoff (Bogart, 1960). At the gauging station near Roxbury, the flood peak was estimated at 1423 m^3/s, five times greater than the 1938 hurricane flood, the previous flood of record. The 1955 flood drastically altered the slope of the annual peak frequency curve (Fig. 9), and the true probability of this flood is not known. Water surface elevations of this flood were mapped, and these data illustrate the interrelationship between valley dimensions and the water surface slope. In general, upstream of valley constrictions the water surface slope is reduced compared to the water surface slope in the constrictions (Fig. 10). A similar effect can be seen at bridge crossings where rafted debris dams created low-gradient backwaters upstream of the bridges. The valley dimensions thus partially controlled the energy grade line of the flood, which in turn

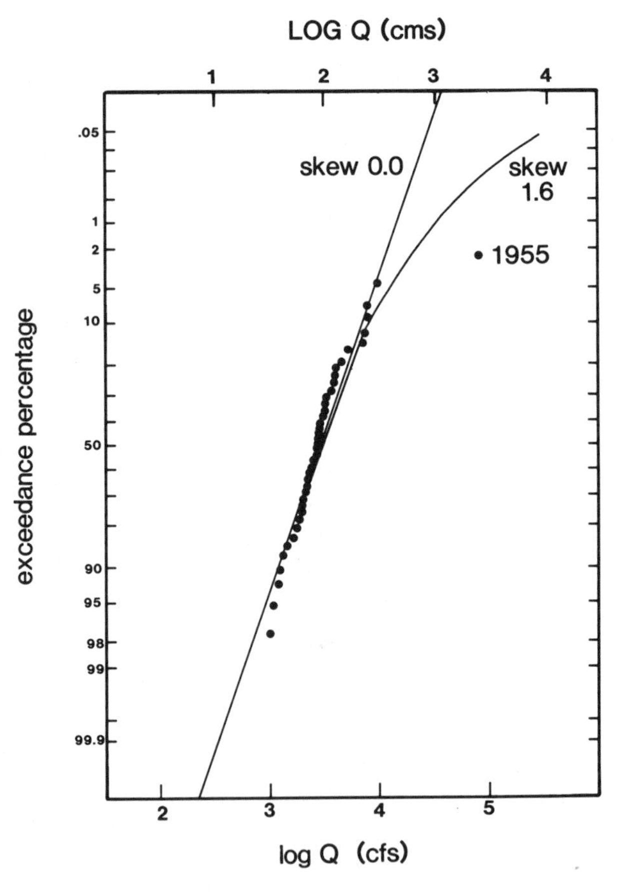

FIGURE 9. Log-Pearson III frequency curve for the Shepaug River near Roxbury, Connecticut. Graph is drawn with zero skewness and with a skewness of 1.6 calculated for the flood series. The 1955 flood is still a significant outlier given the length of record.

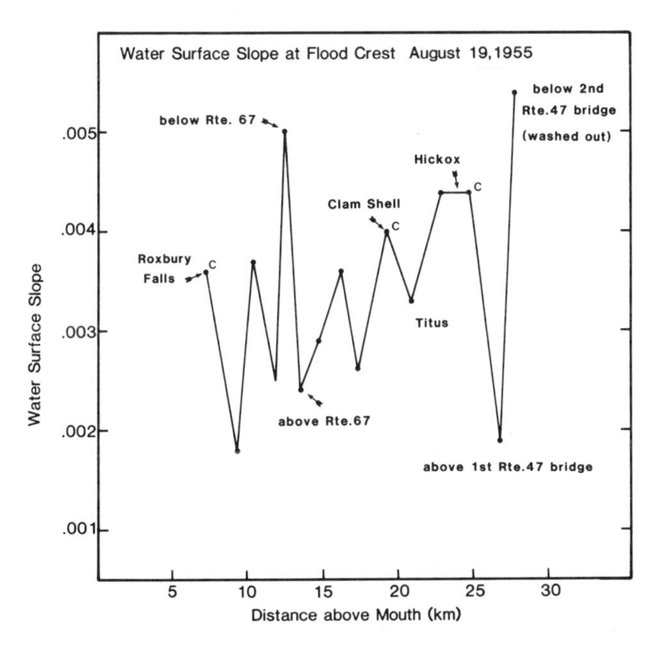

FIGURE 10. Plot of water surface gradient with distance along the Shepaug River for the crest of the 1955 flood. Bedrock valley constrictions at Roxbury Falls, at the Clam Shell, and at Hickox have greater water surface slopes than major channel expansions immediately upstream. Note also the effect of road crossings and bridges on the water surface slope.

affected erosion and deposition on the floodplain surfaces. Analysis of the floodplain morphology and stratigraphy demonstrate that similar processes have been responsible for the development of the floodplains on this river throughout the Holocene.

Floodplain surfaces immediately downstream from bedrock valley constrictions are typically narrow and highly irregular with large scoured overflow channels. Cross sections at two constrictions (Hickox and the Clam Shell) illustrate the scoured character of the floodplains as well as the high stage of the 1955 flood in these narrow reaches (Fig. 11). In these reaches the natural levee separating the main channel from the overflow channel is a boulder berm consisting of imbricated large boulders. In places these boulder berms bury the trunks of trees growing along the banks. Downstream from the constriction, the floodplain surfaces are lower and smoother and the valley wider.

Excavations of the floodplain at the Hickox site, within a valley constriction, revealed an internal stratigraphy of fine-grained overbank alluvium punctuated with layers of gravel deposited on top of and truncating buried soil A horizons (Fig. 12). At the upstream end of the floodplain surface, two buried gravel layers separated by thin partially eroded A horizons cap a gravel layer at least 0.5 m thick. The uppermost surface is capped with sand deposited during the 1955 flood. A radiocarbon date of 3850 ± 610 yr B.P.(TX-3936) was obtained from charcoal recovered from the second buried A horizon 0.6–0.7 m below the floodplain surface (Fig. 12). At the downstream end of the floodplain, the deposit is more fine grained, and the basal gravel was not encountered in a 2-m deep pit. The stratigraphy and radiocarbon date indicate that the floodplain is developing primarily through vertical accretion deposits, and the coarse-grained gravel layers represent deposition during large floods. These floodplain deposits are similar to those described for streams in central Texas (Baker, 1977; Patton and Baker, 1977) and in southern Illinois (Ritter, 1975). The radiocarbon date indicates that at least one flood of similar magnitude to the 1955 flood has occurred on this reach of the Shepaug in the past 3000–4000 yr.

Floodplains that are upstream from channel constrictions are less scoured in appearance. Cross sections of the Titus

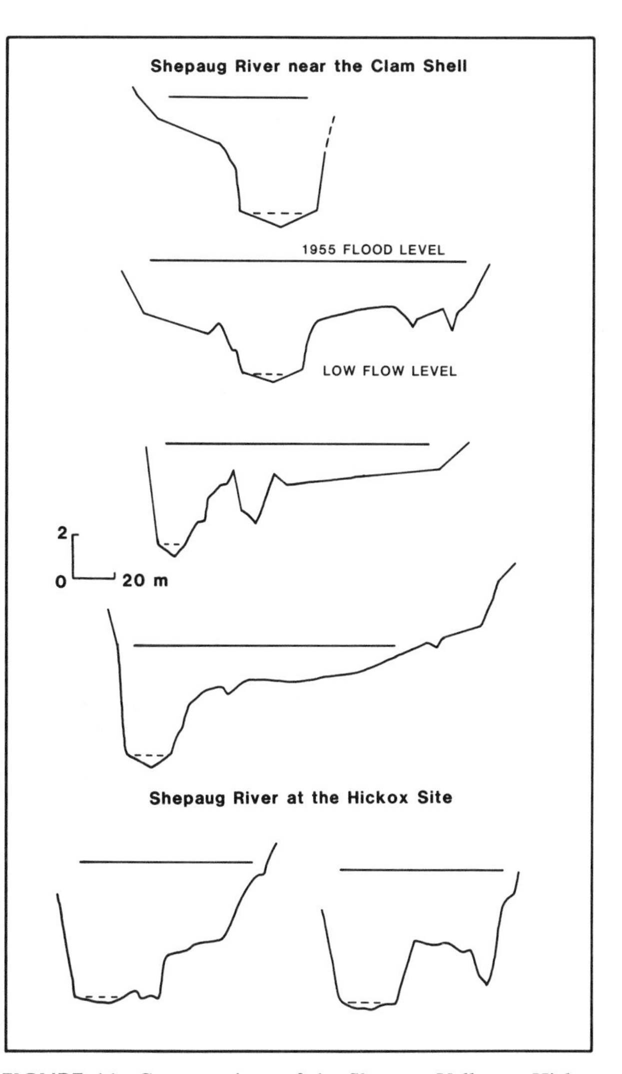

FIGURE 11. Cross sections of the Shepaug Valley at Hickox and at the Clam Shell. Note the stage of the 1955 flood. Levees adjacent to the channel are coarse-grained boulder berms. The floodplain of the Shepaug becomes wider and smoother with distance downstream of the Clam Shell constriction.

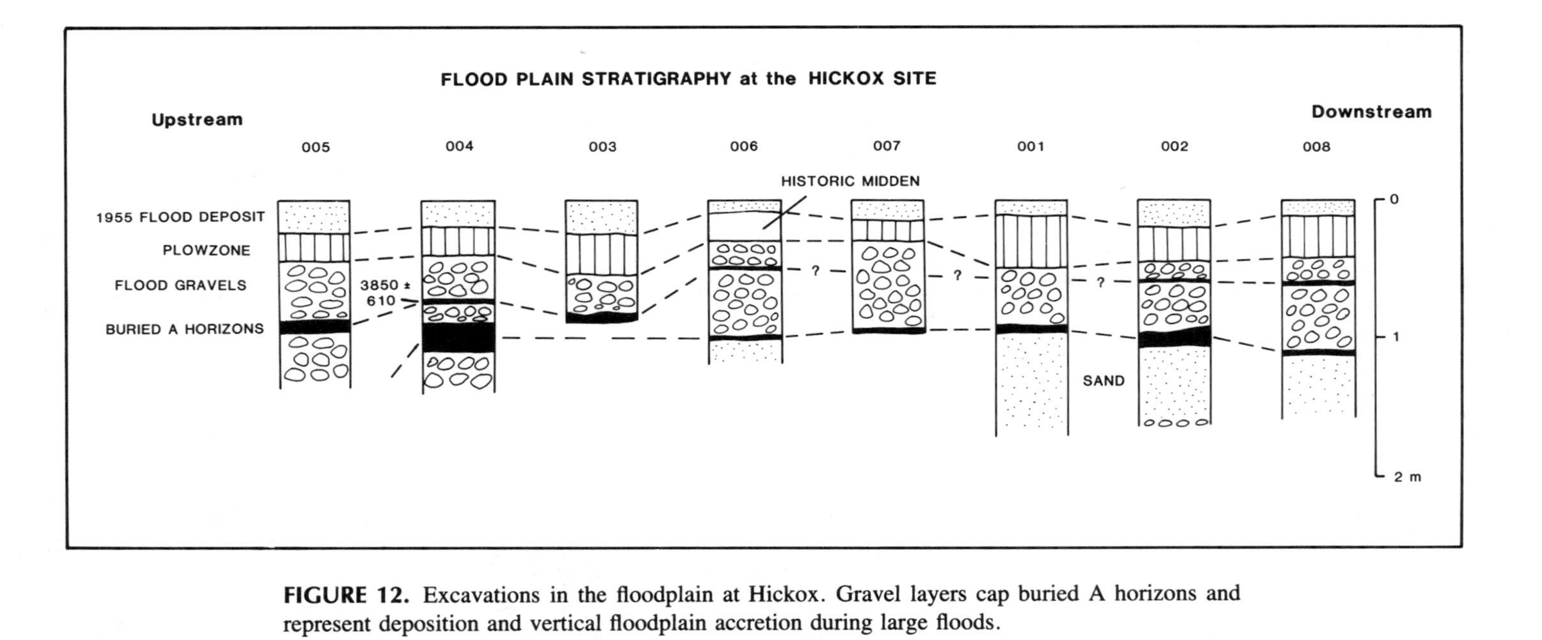

FIGURE 12. Excavations in the floodplain at Hickox. Gravel layers cap buried A horizons and represent deposition and vertical floodplain accretion during large floods.

field upstream from the Clam Shell indicate a gently rolling floodplain surface that gradually increases in elevation away from the river. Excavations at the Titus site and exposures in a cut bank upstream from Roxbury Falls indicate that upstream from prominent valley constrictions the floodplain deposits consists of fine-grained overbank sediment that buries soil horizons.

Three backhoe trenches were cut across the floodplain at the Titus site to better investigate the stratigraphy. The base of each backhoe trench was excavated to a layer of boulder-sized gravel (Fig. 13). This boulder–gravel layer slopes toward the modern river and probably represents the bed material sediment of the Shepaug River deposited as the river migrated across the valley. A radiocarbon age on wood immediately above the basal gravel in trench I indicates that the process of vertical floodplain accretion at this section dates to 12,880 ± 540 (GX-9362) yr B.P.The alluvium that comprises the floodplain at this site is primarily fine-grained sand. Buried soil horizons separate time periods of relative stability and nondeposition from periods of flood deposition. Soils with cambic B horizons occur in the upper part of the depositional sequence. The well-preserved and laterally continuous buried soil horizons suggest that overbank deposition was from suspension and that there has been little erosion at this cross section. However, minor erosion has occurred in the form of an overflow slough that can be traced along the lower portion of the floodplain in the two upstream trenches. This type of floodplain modification is typical of rivers in this region and was noted by Wolman and Eiler (1958) on the Farmington River following the 1955 floods. The overflow slough exposed in the Titus excavations is filled with a sequence of black organic rich sand and silt. This sediment may represent the historic postsettlement alluvium created by accelerated erosion on the hillslopes following the deforestation of the basin in the eighteenth and nineteenth century. The top of the section consists of a historic plowzone that was buried by approximately 30 cm of sand during the 1955 flood. In trench III a layer of unweathered sand is sandwiched between two plowzone horizons and may represent sediment deposited in the 1853 flood.

The lack of buried soils in the lower half of the trenches indicates rapid deposition on a low floodplain surface. As the floodplain built upward through vertical accretion, only larger floods were capable of overtopping the surface. Longer periods of nondeposition allowed soils to develop cambic B horizons. Based on a single radiocarbon date on charcoal found in the bottom of the erosional slough (2585 ± 185, GX-9361), this period of greater floodplain stability dates to before 2500 yr B.P. Since 2500 yr B.P., a minimum of three floods have topped the surface of the floodplain. Prior to the 1955 flood, the previous great flood on this reach of the Shepaug occurred in 1853 and may be represented by the sediment that separates the plowzone in trench III. Previous to this flood a third great flood occurred before the occupation of the floodplain by the Woodland culture, about 1000 yr B.P. and is recorded by flood sediment that underlies Woodland middens in trenches I and III (Fig. 13).

Along the Shepaug River, debris fans can be found on many small tributaries where they enter the mainstem. Where each tributary enters, the Shepaug is "pressed" to the opposite side of the valley due to fan deposition (Fig. 14). Although these fans are predominantly composed of coarse gravel, there is sufficient overbank deposition to mask the gravel deposit morphology with a veneer of fine-grained sediment up to 1 m thick. Archeological excavations on one fan (Mallory Brook) revealed some aspects of the internal stratigraphy of these fans and suggest that deposition of new gravel lobes is relatively infrequent.

Mallory Brook is a small high-gradient tributary of the Shepaug River near the town of Washington Depot. The brook is incised through a stratified glacial deposit and into the Hartland Schist where it has created a narrow steep gorge for approximately 1.5 km before it debouches onto the floodplain of the Shepaug River. Within this gorge the channel is irregular in cross section. It is typified by small gravel levees and gravel benches at various heights above the channel. Organic debris in the form of fallen logs has created small check dams that trap coarse gravel sediment. This stored sediment is probably remobilized during large floods.

Excavations into the downstream portion of this fan indicate that poorly sorted gravels exist 1 m beneath the surface. These gravels exhibit imbrication suggesting sediment transport from Mallory Brook. In addition, the surface of the gravels slopes toward the Shepaug River and is graded to a level about 1 m above the low-flow elevation of the modern stream. A Paleoindian archeological site was uncovered on the gravel surface and charcoal from a hearth was dated at 10,190 ± 300 yr B.P. (Moeller, 1980). This is additional evidence that the Shepaug must have been at or near its present elevation at the beginning of the Holocene and that much of the gravel within the fan had been deposited by the late Pleistocene or early Holocene. The fine-grained overbank sediment that buried the debris fan gravels does not contain any prehistoric buried soils, suggesting that the rate of sediment deposition on this surface was much slower than the rate of pedogenesis and that the thick cambic B horizon was probably cumulic. However, the 1955 flood did inundate this fan surface, and the historic plowzone was buried by up to 30 cm of sand.

Deposition in Mallory Brook during the 1955 flood was confined to a belt adjacent to the active channel. The deposits consist of poorly sorted gravels containing much historical debris; they were deposited in the preflood channel as the brook cut laterally into older debris fan deposits. No fine-

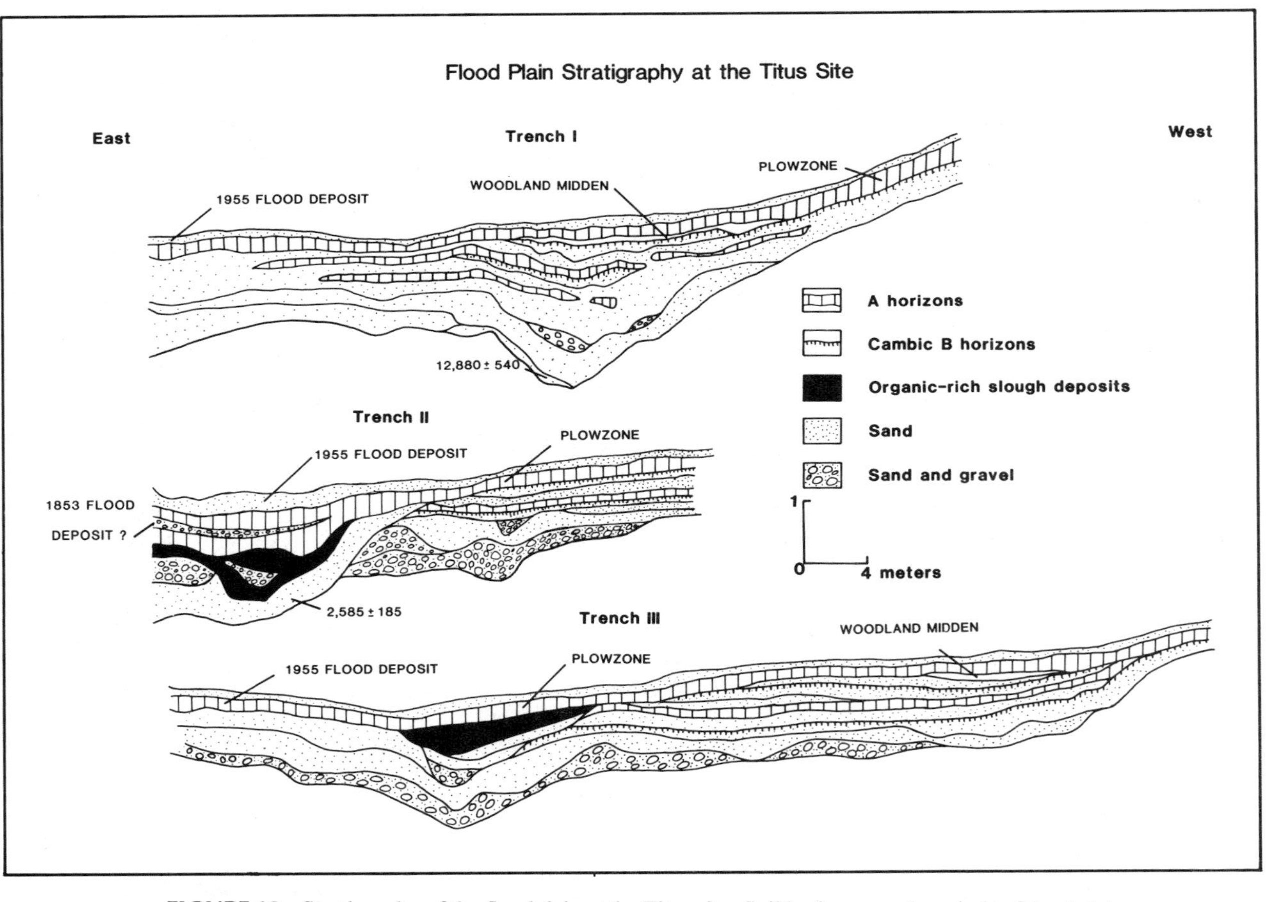

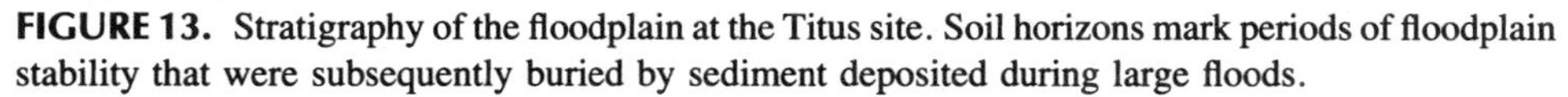

FIGURE 13. Stratigraphy of the floodplain at the Titus site. Soil horizons mark periods of floodplain stability that were subsequently buried by sediment deposited during large floods.

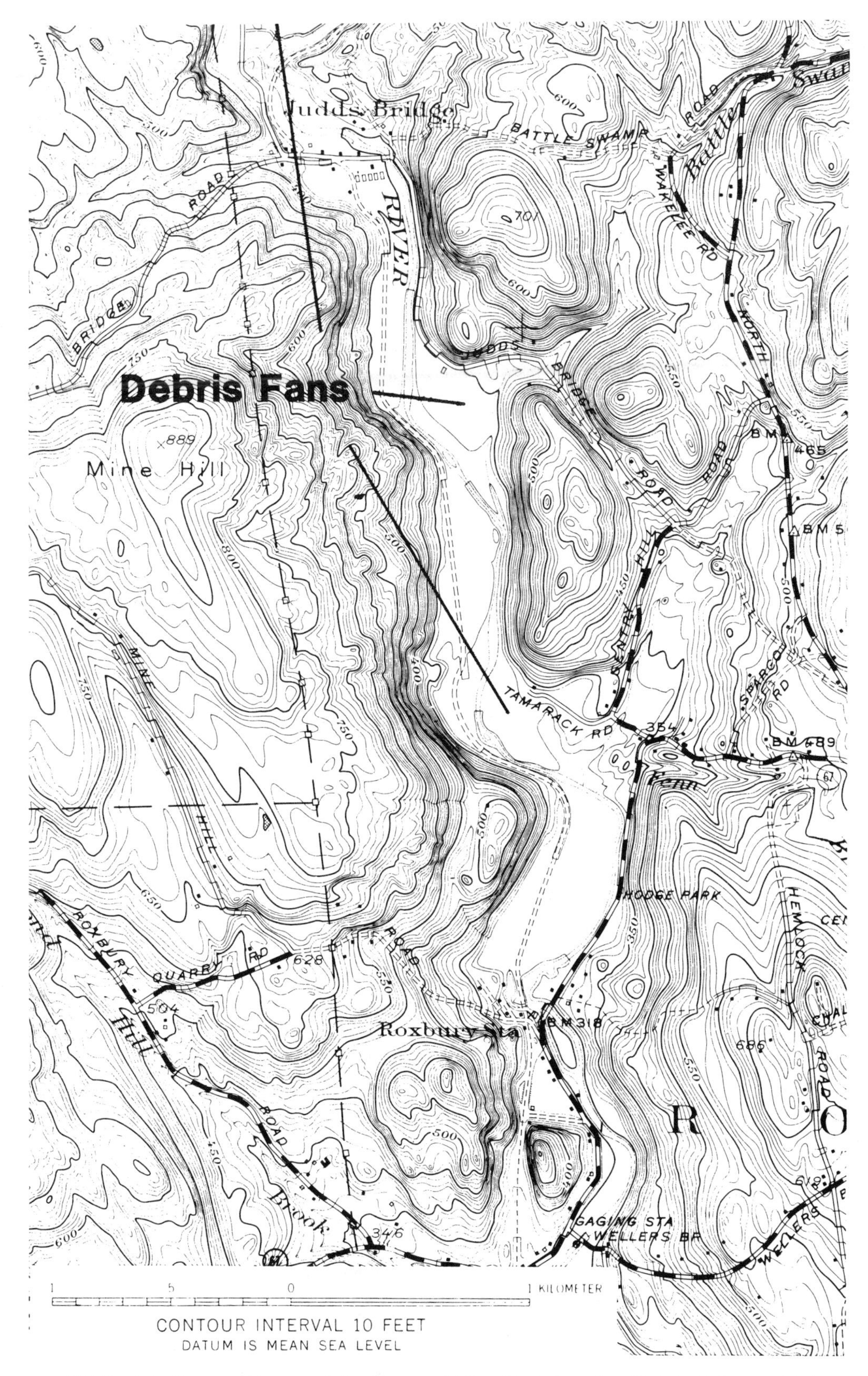

FIGURE 14. Topographic map of the Shepaug Valley illustrating the effect that tributary debris fans have had on the course of the mainstem. Area of map covers the mainstem from just south of the Clam Shell to the Roxbury gauging station (see Fig. 8).

grained deposits cap these gravels, and the surface of the new gravel deposits extend up to the surface of the fan. Based on the mode of deposition of the 1955 flood gravels, one would expect linear gravel bars oriented down the fan to represent past episodes of rapid channel widening and bar deposition during floods. The excavations on the fan covered a large area, but they did not encounter intermediate age deposits. Either the excavations were not located in that portion of the fan affected by late Holocene flooding or the deposition of gravel on these fan surfaces is an exceptionally rare event.

The combined stratigraphic evidence from the Shepaug River valley indicates that large floods are important in constructing the floodplain surfaces bordering the river. Thus, although the cosmetic effects of large flow events may be rapidly obscured, there remains a record of these flow events in the stratigraphy of the Holocene deposits. In addition, because the river has remained in its present position for much of the Holocene, sedimentary sequences in the floodplains represent vertical accretion of flood deposits over this time interval. The stratigraphic evidence also indicates that the 1955 flood was an exceptionally rare event and that the last flood to have topped many of the higher floodplain surfaces may have occurred at least 1000 yr ago at Titus and perhaps as much as 3000–4000 yr ago at Hickox.

Flood Effects in Lowland Drainages

Erosional Processes. Jahns (1947) described the geomorphic effects of the 1936 and 1938 floods in the Connecticut valley. He noted that floodplain erosion was restricted to those reaches where overbank flow depths were unusually great. For example, on the Connecticut River where the flow was contained by an unusually high bank, it was not unusual to find overflow channels scoured across the opposite lower floodplain surface. Other sites of scour occurred where levees were overtopped and eroded and where eddies were created in the wake of trees and buildings. In general, Jahns (1947) noted the general lack of erosion and remarked that in many fields not even the sod cover had been disturbed.

One remarkable example of floodplain erosion occurred during the 1936 flood along the Connecticut River in the vicinity of Hatfield, Massachusetts. Floodwaters spilled through two abandoned meander bends on the west side of the river (Collins and Schalk, 1937) and created large circular erosion pits where the flow spilled back across the floodplain to the main channel. The individual "swirl pits" created in the floodplain were up to 4.5 m deep and tens of meters in diameter. In places the pits coalesced to form long elliptical depressions. Collins and Schalk (1937) hypothesized that the swirl pits were created by large vortices created downstream of a low rise on the floodplain that separated the oxbow channels from the modern river. This ridge ponded water in the oxbow channel above the level of the water in the main channel (M. Schalk, personal communication, 1977). The drop in head across the ridge may have created supercritical flow through the low depressions in the ridge, and the resulting hydraulic jump on the downstream side may have caused the turbulence necessary to erode the swirl pits. In any case it is clear that unusual geomorphic settings are required for extreme erosion on the flat wide lowland floodplains.

Channel erosion consisted primarily of bank erosion on the outside of meander bends. Jahns (1947) noted that the bend erosion was inhibited by the natural vegetation, but that once the banks were cleared by several high flows, even moderate flows could cause significant channel migration. Therefore, it is likely that the cumulative effect of moderate flows on bank erosion is greater than the amount of erosion caused by a single catastrophic flow.

Depositional Processes. Following the 1936 and 1938 floods, Jahns (1947) made detailed observations of the deposition of fine-grained sediment on floodplain surfaces adjacent to the Connecticut River. Jahns (1947) noted the variable thickness of suspended-load deposits on the Connecticut River floodplain in Massachusetts. In places near the river up to 2 m of sediment was deposited during the 1936 and 1938 floods, whereas, near the margins of the floodplain, only a thin veneer of sediment was deposited. Based on 611 measurements of flood deposition, Jahns estimated that an average of about 9 mm was added to the floodplain surface. He also noted that the farming practices on the floodplain soon destroyed much of the stratigraphic record by plowing.

Jahns recognized that tributary ravines that cut through the terraces and floodplain of the Connecticut River were favorable sites for sediment deposition. He noted that "the small ravines that cross the meadows form traps of quiet water during flood stages and retain very complete and well-preserved sections whose beds delicately register the activity of flood waters. In these ravines the depositional sequence for each flood truly reflects the height and duration of each of its different stages" (Jahns, 1947, p. 102). Typically these deposits become thicker and more coarse grained toward the mainstem, although sand transported by the tributary during the recession of the flood could leave a coarse sand cap on top of an individual flood deposit.

Jahns used the stratigraphic relationships observed in these fine-grained deposits to infer the recurrence interval of the 1936 and 1938 floods. Sediment from the 1936 flood occurred at the highest elevation and covered the low terrace (terrace III in Jahn's report) of the Connecticut River. Because the flood sediment buried a fine-grained sedimentary sequence typical of an open floodplain environment, he reasoned that the 1936 flood must have been the largest flood to occur since terrace III had formed. He estimated that terrace III was the active river floodplain approximately 2500–6000 yr B.P. Jahn's work stands as one of the first efforts to utilize geologic evidence to deduce the recurrence interval of large floods. These fine-grained flood deposits in tributary streams have since been termed *slackwater deposits* and are the field evidence on which many paleohydrologic reconstructions are based (Kochel and Baker, this volume).

Costa (1984) has re-excavated the slackwater deposits at the East Deerfield, Massachusetts, locality originally excavated by Jahns 3 yr after the 1938 flood. In 45 yr the boundaries between the sedimentation units have been obliterated by bioturbation and weathering processes. This indicates that in a humid temperate environment, the geologic record of individual flood events can be rapidly destroyed by pedogenesis. Recent studies of slackwater deposits on other New England streams, however, indicate that where the sedimentation units are thick, they can be preserved for longer time intervals.

One example of a thick slackwater deposit representing a prehistoric large flood has been excavated in the Housatonic River valley (Patton and Handsman, 1983, 1984). The slackwater deposit is located in a reach of the Housatonic River where it flows through a broad valley underlain by the Stockbridge Marble in the otherwise high-relief western uplands. This slackwater deposit is located in a tributary that has cut a narrow valley through a sequence of glacial and Holocene terraces (Kelley, 1975).

The slackwater deposit buries the Flynn site, an archeological site situated on a terrace, 7.5 m above the low-water level of the Housatonic River, on the banks of a small ravine, Squash Hollow Brook (Fig. 15 *a*). The upper surface of the terrace deposit consists of a slightly weathered medium to fine sand that in places is up to 90 cm thick (Fig. 15*b*). Soil development consists of a 30-cm-thick plowzone that caps an oxidized C horizon, but the base of this sedimentation unit is unweathered. Beneath this unit is a buried B horizon that is more intensely oxidized and has a slight increase in clay content indicating the presence of an incipient argillic B horizon. The top of the

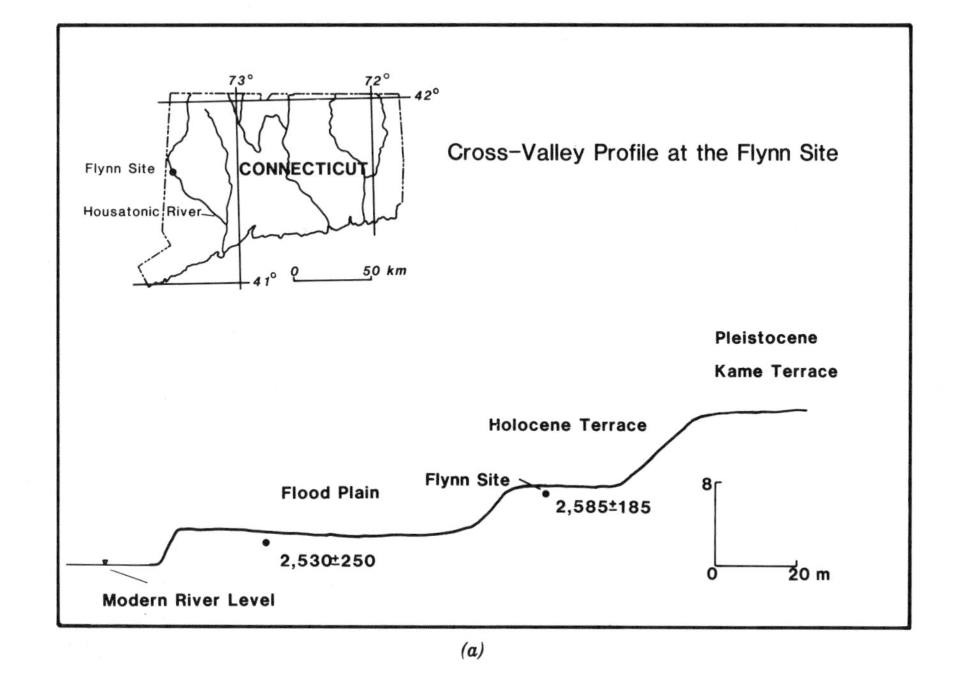

(*a*)

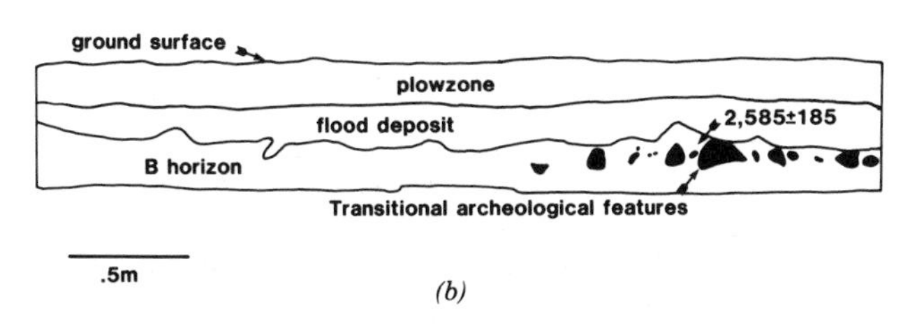

(*b*)

FIGURE 15. (*a*) Cross valley profile of the Housatonic River at the Flynn site. The site is located on the surface of a Holocene terrace adjacent to Squash Hollow Brook. Position of radiocarbon dates in the terrace and floodplain are indicated. (*b*) Stratigraphy of the north wall of the excavation illustrating the stratigraphic relationships between the Transitional archeological site and the overlying flood slackwater deposit.

B horizon is the floor of an archeological site that dates to the transitional period between the late Archaic and the Woodland. Charcoal from a hearth on this surface yielded a radiocarbon age of 2585 ± 185 yr B.P. The deposit that buried the archeological site and protected it from later cultural disturbance is a thick slackwater deposit.

As noted, slackwater deposits can be used to reconstruct the hydrology of past floods (see also Kochel and Baker, this volume). However, one important assumption in this process is that the dimensions of the valley cross section have remained relatively constant since the deposition of the flood sediment. On the Housatonic River, this assumption is difficult to evaluate because the river has downcut through a thick sequence of glacial deposits during the Holocene. Precisely when the river reached its present level is not known. Some evidence that the river was near its present position 2500 yr ago comes from excavations in the modern floodplain surface. On this surface 1.2 m of structureless, unweathered sand overlies coarse gravel. A radiocarbon age of 2530 ± 250 yr B.P. (TX-3686) on charcoal buried 1 m below the surface dates the beginning of overbank sedimentation on this floodplain (Fig. 15*a*). Therefore, approximately 2500 yr ago the river had apparently already downcut below the level of the gravel layer in the modern floodplain. If this assumption is correct, then the flood that deposited the slackwater deposit may have had a stage approaching 7.5 m.

The paleoflood represented by the slackwater deposit may be the greatest flood on this reach of the river in the late Holocene. Based on gauging records collected 4 km upstream at Gaylordsville, Connecticut, the greatest flood of record is that of May 1, 1854, when a gauge height of 6.5 m was recorded (Thomson et al., 1964). The gauging station is in a more narrow reach of the valley, and it is likely that the flow was not more than 6.5 m deep at the Flynn site. Furthermore, the stratigraphic evidence indicates that the flood that deposited the slackwater sediment was prehistoric. Therefore, the slackwater deposit at the Flynn site may represent the greatest flood in the past 2500 yr on this reach of the Housatonic River.

The preliminary work of Jahns (1947) and the more recent investigations of slackwater deposits demonstrates that useful paleohydrologic information can be extracted from the Holocene stratigraphic record of streams in southern New England. Paleoflood analysis of these river systems represents one important avenue for future research. Additional data on the magnitude and frequency of rare great floods and their role in developing fluvial landforms will greatly enhance our understanding of the postglacial fluvial geomorphic history of the streams in this region.

CONCLUSION

The geomorphic significance of floods in southern New England varies as a function of the physical setting of the river systems. In the highland drainages the internal stratigraphy of the floodplain surfaces demonstrates that sedimentation during floods is an important process in constructing the fluvial landforms along these rivers. Observations of erosion and sedimentation during floods indicates that local factors such as valley morphology and stream slope are strong influences on these processes. The result is that erosion and deposition occur in localized reaches while many reaches undergo only minor and indeterminate changes during large floods. Combined with the masking effect of vegetation, which gives the appearance that little macroscopic change has occurred, reworking by moderate flows can rapidly obliterate erosional and depositional morphology created by the biggest floods. Despite this cosmetic alteration, the stratigraphic evidence from floodplains and debris fans along rivers like the Shepaug indicate that periods of stasis and soil formation are interrupted abruptly by major floods that add considerable sediment to the floodplain surfaces. Both scales of process are important and future research is needed to quantify the significance of each. Finally, several field studies have demonstrated that the stratigraphy of flood deposits can be used to evaluate the frequency of rare large floods on these streams.

On streams in lowland regions the response to floods is fundamentally different from that described for the highland drainages. Erosion during rare floods, except where unusual hydraulic conditions exist, is not great; certainly not greater than that achieved by a succession of more moderate flows. Sediment deposited in open floodplain environments is variable in thickness, but again the average thickness of sediment added to the floodplain is minor and is rapidly incorporated into an homogenous floodplain stratigraphy. Rare great floods can add sediment to older, higher alluvial surfaces, but much of the floodplain stratigraphy created by this process is rapidly lost by subsequent natural weathering processes and through land-use practices. However, in selected localities, slackwater deposits may accumulate to enough thickness to remain recognizable for several thousand years. This paleoflood record may be useful in placing the long flood record for this region into a geological perspective. However, paleoflood analysis in glaciated humid temperate environments must consider the changes in valley and channel morphology caused by Holocene entrenchment into Pleistocene glacial deposits and the loss of stratigraphc detail caused by pedogenesis.

ACKNOWLEDGMENTS

Investigations of the stratigraphy of floodplains along streams in western Connecticut would not have been possible without the cooperation and generous support of Dr. R. G. Handsman of the American Indian Archaeological Institute, Washington, Connecticut, who supervised the excavations at the Titus, Hickox, and Flynn sites. Additional support was provided by the Office of Water Research and Technology, Project No. A-078-CONN. A. Burnett, J. Sullivan, and C. Evans assisted in the field. This chapter was significantly improved by G. R. Brakenridge, who reviewed an earlier version.

REFERENCES

Baker, V. R. (1977). Stream-channel response to floods, with examples from central Texas. *Geol. Soc. Am. Bull.* **88**, 1057-1071.

Beard, L. R. (1975). Generalized evaluation of flash-flood potential. *Tech. Rep.—Univ. Tex. Austin, Cent. Res. Water Resour.* **CRWR-124**, 1–27.

Bogart, D. B. (1960). Floods of August-October 1955 New England to North Carolina. *Geol. Sur. Water-Supply Pap. (U.S.)* **1420**, 1–854.

Cleland, H. F. (1902). The landslides of Mt. Greylock and Briggsville, Mass. *J. Geol.* **10**, 513–514.

Collins, R. F., and Schalk, M. (1937). Torrential flood erosion in the Connecticut Valley, March 1936. *Am. J. Sci.* **34**, 293–307.

Costa, J. E. (1984). Fluvial paleoflood hydrology. *Trans. Am. Geophys. Union* **65**(45), 892.

Denny, C. S. (1982). Geomorphology of New England. *Geol. Surv. Prof. Pap. (U.S.)***1208**, 18.

Grover, N. C. (1937). The floods of March 1936. Part 1. New England Rivers. *Geol. Surv. Water-Supply Pap. (U.S.)* **798**, 1–466.

Hack, J. T., and Goodlett, J. C. (1960). Geomorphology and forest ecology of a mountain region in the central Appalachians. *Geol. Surv. Prof. Pap. (U.S.)* **347**, 1–66.

Hoyt, W. S., and Langbein, W. B. (1955). "Floods." Princeton Univ. Press, Princeton, New Jersey.

Jahns, R. H. (1947). Geologic features of the Connecticut Valley, Massachusetts as related to recent floods. *Geol. Surv. Water-Supply Pap. (U.S.)* **996**, 1–158.

Kelley, G. C. (1975). Late Pleistocene and recent geology of the Housatonic River in northwestern Connecticut. Ph.D. Thesis, Syracuse University, Syracuse, New York.

Kinnison, H. B. (1929). The New England flood of November, 1927. *Geol. Surv. Water-Supply Pap. (U.S.)* **636-A**, 45–100.

Kinnison, H. B., Conover, L. F., and Bigwood, B. L. (1938). Stages and flood discharges of the Connecticut River at Hartford Connecticut. *Geol. Surv. Water-Supply Pap. (U.S.)* **836-A**, 1–18.

Leopold, L. B., Wolman, M. G., and Miller, J. P. (1964). "Fluvial Processes in Geomorphology." Freeman, San Francisco, California.

Malde, H. E. (1967). Surficial geologic map of the Roxbury Quadrangle, Litchfield and New Haven Counties, Connecticut. *Geol. Surv. Geol. Quadrangle (U.S.)* **GQ-611**, Scale 1:24,000.

Moeller, R. W. (1980). 6LF21, A Paleo-Indian Site in western Connecticut. *Occas. Pap. Am. Indian Archaeol. Insti.* **2**, 1–160.

Patton, P. C., and Baker, V. R. (1977). Geomorphic response of central Texas stream channels to catastrophic rainfall and runoff. *In* "Geomorphology in Arid Regions" (D. O. Doehring, ed.), Publ. Geomorphol., pp. 189–217. State University of New York, Binghamton.

Patton, P. C., and Handsman, R. G. (1983). Geomorphology and archeology of the Housatonic River basin, western Connecticut. *Geol. Soc. Am. Abstr. Programs* **15**, 179.

Patton, P. C., and Handsman, R. G. (1984). Paleoflood record for the Housatonic River basin, western Connecticut. *Trans. Am. Geophys. Union* **65**(45), 891.

Paulson, C. G. (1940). Hurricane floods of September 1938. *Geol. Surv. Water-Supply Pap. (U.S.)* **867**, 1–562.

Ritter, D. F. (1975). Stratigraphic implications of coarse-grained gravel deposited as overbank sediment, southern Illinois. *J. Geol.* **83**, 645–650.

Rodgers, J. (1985). Bedrock geologic map of Connecticut. *Conn., State Geol. Nat. Hist. Surv.*, Scale 1:125,000, 2 sheets.

Thomas, M. P. (1966). Effect of glacial geology upon the time distribution of streamflow in eastern and southern Connecticut. *Geol. Surv. Prof. Pap. (U.S.)* **550-B**, B209–B212.

Thomson, M. T., Gannon, W. B., Thomas, M. P., Hayes, G. S., et al. (1964). Historical floods in New England. *Geol. Surv. Water-Supply Pap. (U.S.)* **1779-M**, 1–105.

U.S. Army Corps of Engineers (1956). "New England Floods of 1955. Part 1. Storm Data. Part 2. Flood Discharges. U.S. Army, Office of the Division Engineers, New England Division, Boston, Massachusetts.

Williams, G. P., and Guy, H. P. (1973). Erosional and depositional aspects of Hurricane Camille in Virginia. *Geol. Surv. Prof. Pap. (U.S.)* **804**, 1–80.

Wolman, M. G., and Eiler, J. P. (1958). Reconnaissance study of erosion and deposition produced by the flood of August 1955 in Connecticut. *Trans. Am. Geophys. Union* **39**, 1–14.

17

CLIMATIC INFLUENCE ON UPPER MISSISSIPPI VALLEY FLOODS

JAMES C. KNOX
Department of Geography, University of Wisconsin, Madison, Wisconsin

INTRODUCTION

This chapter examines climatic influence on variations in floods of the Upper Mississippi Valley in the north-central part of the central lowlands physiographic province (Fig. 1). A major objective of this chapter is to evaluate the role of climatic change as an influence on magnitudes and frequencies of floods during historical and Holocene (post-glacial) time. Although floods in this region have occurred in all months of the year, the highest frequencies of flooding usually are during the March–April spring snowmelt season and during early summer (June–July) heavy rainfalls. Historical streamflow records show that changes in climate and land use have significantly influenced magnitudes and frequencies of floods and that climatic change is a probable cause of changes in the seasonal concentration of floods during historical time. Prior to the early nineteenth century the level of human alteration of the natural environment was hydrologically insignificant, and floods were primarily related only to variations in direct (temperature and precipitation) and indirect (vegetation) climatic factors. Throughout the Holocene a relatively steep climatic gradient has occurred across the Upper Mississippi River basin and it is associated with a major ecotone separating mixed hardwood forest to the northeast from prairie to the southwest (Bernabo and Webb, 1977). The stratigraphic history of Holocene floods suggests a long-term pattern of variation that is broadly similar to Holocene history of climatic change and ecotone movements that have been reconstructed from fossil pollen (Knox, 1985).

Most of the Upper Mississippi basin is underlain by nearly flat-lying sedimentary rocks of Cambrian, Ordovician, and Silurian age, except in the more northerly parts of the watershed in Wisconsin and Minnesota where Precambrian crystalline rocks may occur at or near the surface. Pleistocene glaciations have either directly or indirectly affected all of the basin with most areas being covered with modest to thick amounts of glacial-derived sediments except for the unglaciated Driftless Area in southwestern Wisconsin and northwestern Illinois. From the standpoint of flood hydrology, the influence of Pleistocene glacial events on the Upper Mississippi basin can be divided into three regional areas, including: (1) the landscape underlain by Wisconsin age glacial deposits, (2) the landscape underlain by older Pleistocene glacial deposits, and (3) the unglaciated Driftless Area of southwestern Wisconsin and northwestern Illinois (Fig. 1). The area of Wisconsin age glacial deposits is characterized by low-relief, gentle slopes, and low drainage density. Loess thicknesses vary from negligible to less than a meter. By comparison the region of pre-Wisconsin Pleistocene deposits is characterized by higher relief, steeper slopes, and higher drainage density. Loess thickness may vary from negligible on eroded hillslopes to several meters thick on interfluves. The Driftless Area is a highly dissected landscape with moderate to steep slopes that separate narrow and gently sloping uplands from narrow bottomlands. In the Driftless Area Cambrian sandstones and Ordovician limestones and dolomites are the dominant rock formations that are thinly covered with late Wisconsin loess. Residuum thicknesses vary from a few centimeters on the steeper slopes to several meters on the wider interfluves. The responsiveness of floods to climatic events increases systematically from region 1 to region 3.

The climate of the Upper Mississippi Valley is distinctly continental with hot summer and cold winters. The Upper Mississippi River Comprehensive Basin Study Coordinating

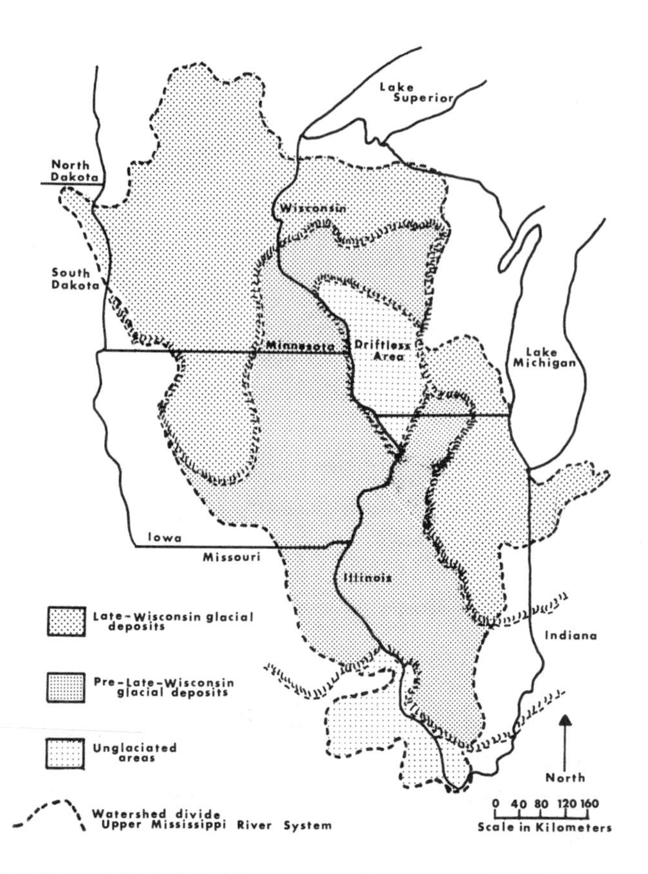

FIGURE 1. The Upper Mississippi River system is represented by three topographically and hydrologically distinct areas caused by the uneven distribution of different age Pleistocene glacial deposits. (Pleistocene deposits after Thornbury, 1965.)

Committee (1970) provides the following climatic generalizations. Average annual precipitation ranges from about 610 mm to about 1015 mm, respectively, between northern and southern parts of the basin. The corresponding variation in average annual snowfall ranges from about 1220 mm to about 510 mm, respectively, between northern and southern parts of the basin. The percentage of the average annual precipitation that falls as snow increases from about 10% in the central part of the basin to 20% or more in the northern part. Much of the basin has an average annual snow cover lasting between 80 and 120 days (Wisler and Brater, 1959, p. 306). Average annual temperature ranges from about 6.1°C in northern areas to about 11.1°C in the southern areas. The corresponding mean daily maximum temperature for July averages about 28.9°C in the north-central basin to about 32.2°C in the south. The mean daily minimum temperature for January averages about −14.4°C in the north-central part of the basin to about −6.7°C in the south-central part of the basin.

TYPES OF UPPER MISSISSIPPI VALLEY FLOODS

Natural floods in the Upper Mississippi Valley result from (1) rainfall, (2) snowmelt, or (3) combined rainfall and snowmelt. Floods resulting from excessive rainfall are primarily restricted to the months of April through November. Snowmelt or snowmelt in combination with rainfall is primarily responsible for floods during the period December through March in the Upper Mississippi Valley, although flooding related to snowmelt is common during April and May on the Mississippi River and its major northern tributaries. Obviously, snowmelt becomes a less important source of floods in the southern part of the watershed.

Seasonality and Causes

Most floods in Upper Mississippi Valley watersheds occur during the months of March through July. Figure 2*a*, for

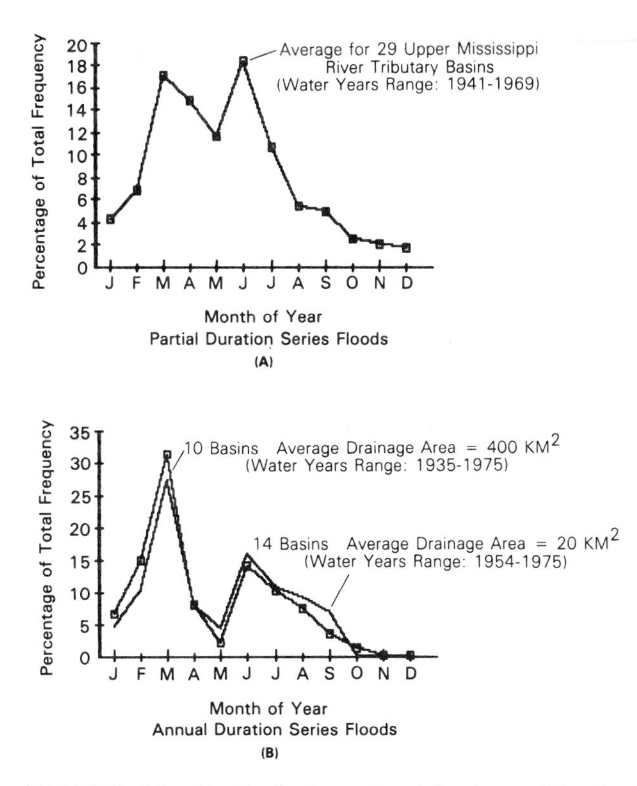

FIGURE 2. Monthly distributions of partial and annual duration series floods in Upper Mississippi Valley watersheds are strongly bimodal with a March peak representing snowmelt floods and a June peak representing rainfall floods. Descriptive data for the 29 watersheds of (*a*) are given in Table 1 and similar data for 24 watersheds of (*b*) are given in Table 2. Data source: U.S. Geological Survey computer files.

example, shows that March through July floods represented approximately 73% of all floods that occurred during the years 1941 through 1969 in a sample of 29 river systems listed in Table 1. The total number of floods exceeding the partial duration series base was 2240 for the period, representing an average of about 77 floods per year within the total 29 watersheds. March and June were the months with the highest frequencies of floods, together accounting for more than one-third of the total of all floods occurring from 1941 through 1969 in the 29 sample watersheds. Figure 2*b* shows that the monthly distribution of annual floods, representing only the largest flood of each year for each river, occurring in 24 sample watersheds located primarily in southwestern Wisconsin, is also bimodal, clustering on March and June. This sample of 24 rivers includes 6 that were also part of the 29 sample watersheds described above (Table 2). The greater importance of March floods over June floods in the annual duration flood series reflects the nearly annual recurrence of spring snowmelt runoff whereas the more annually variable June rainfall often is of insufficient intensity or duration to generate the annual maximum runoff.

Snowmelt Floods. Snowmelt-related floods are the most frequent in the annual flood series for most large and small rivers in the Upper Mississippi River watershed. For example, note that Figure 2*b* contains two distribution curves, with curve 1 showing the monthly distribution of annual floods for watersheds that average about 400 km^2 and curve 2 showing the distribution for watersheds that average about 20 km^2. Initially, one might expect the distribution curves to plot differently because larger basins might have floods dominated by long-duration and widespread runoff events, such as produced by snowmelt or frontal rainfall. Furthermore, one might expect most floods in the smaller basins to result from high-intensity rainfall typical of short-duration and locally concentrated thunderstorm events. Snowmelt and low-intensity frontal rainfall could be viewed as not accumulating runoff rapidly enough to generate significant flooding on the streams of the small watersheds. However, the record shows that even in relatively small basins of the Upper Mississippi Valley, floods that result from isolated high-intensity rainfall events occur too rarely on a given individual watershed to become the most frequent event in the annual flood series. Therefore, the nearly annually occurring March snowmelt event dominates the annual maximum flood series, causing the monthly distribution percentages of floods occurring in large and small tributary watersheds to be remarkably similar (Fig. 2*b*).

Precipitation during the period December through February typically falls as snow and is temporarily stored until the spring snowmelt event. However, winter rains on snowpacks occasionally occur during anomalous brief warm periods. These winter rains often result in moderate to large floods on the larger rivers as runoff from both snowmelt and rainfall combine over extensive areas.

While the month of March is the overall mode for snowmelt floods in the Upper Mississippi Valley, the actual timing tends to occur later in spring with northward position in the watershed and as size of drainage area increases. The modal date of spring snowmelt flood in medium and small size watersheds clearly occurs earlier in southern Iowa and southern Illinois than in northern Minnesota and northern Wisconsin. However, on the large drainages, such as the mainstem Mississippi River, the combined effects of many northern tributaries and long-duration downstream travel time for floods causes the flood date to usually occur in April rather than in March as was true for the aggregate sample of 29 watersheds. For example, April is the most frequent month of flooding on the Mississippi River for both the partial duration series floods at St. Paul, Minnesota, and the annual duration series floods farther downstream at Keokuk, Iowa (Fig. 3).

TABLE 1 29 Upper Mississippi Valley Stream Gauging Sites Used in the Analysis of Partial Duration Series Floods

Stream Gauging Site	Gauge	Period of Record	Drainage Area (km^2)	Base for Partial Flood Series (m^3/s)
Mississippi River, St. Paul, Minnesota	5-3310	1867–1971	95312.0	368.2
Minnesota River, Carver, Minnesota	5-3300	1935–1972	41958.0	184.1
Des Moines River, Tracy, Iowa	5-4485	1920–1972	32320.6	566.4
Cedar River, Conesville, Iowa	5-4650	1940–1972	20163.2	339.8
Cedar River, Cedar Rapids, Iowa	5-4645	1903–1972	16860.9	424.8
Des Moines River, Boone, Iowa	5-4815	1903–1968	14273.5	212.4
Skunk River, Augusta, Iowa	5-4740	1915–1972	11144.8	424.8
Raccoon River, Van Meter, Iowa	5-4845	1915–1972	8912.2	240.7
St. Croix River, Danbury, Wisconsin	5-3335	1914–1971	4112.9	85.0
Turkey River, Garber, Iowa	5-4125	1914–1972	4001.6	226.6
South Raccoon River, Redfield, Iowa	5-4840	1940–1972	2558.9	14.2
Black River, Neillsville, Wisconsin	5-3810	1905–1971	1958.0	141.6
Kickapoo River, Steuben, Wisconsin	5-4105	1934–1971	1787.1	53.8
Root River, Lanesboro, Minnesota	5-2840	1940–1971	1592.8	99.1
Sangamon River, Monticello, Illinois	5-5720	1908–1971	1424.5	51.0
Iowa River, Rowan, Iowa	5-4495	1941–1972	1111.1	34.0
Du Page River, Shorewood, Illinois	5-5405	1941–1971	841.8	36.8
Grant River, Burton, Wisconsin	5-4135	1935–1971	691.5	68.0
Platte River, Rockville, Wisconsin	5-4140	1935–1971	360.0	59.5
West Bureau Creek, Wyanet, Illinois	5-5570	1937–1966	215.7	42.5
Little LaCrosse River, Leon, Wisconsin	5-3825	1934–1971	199.7	14.7
North River, Bethel, Missouri	5-5005	1937–1970	150.2	22.7
Paint Creek, Waterville, Iowa	5-3885	1951–1972	110.9	14.2
Bay Creek, Pittsfield, Illinois	5-5125	1940–1971	102.6	56.6
Indian Creek, Wanda, Illinois	5-5880	1941–1971	95.8	28.3
Big Creek, Wetaug, Illinois	5-6000	1941–1971	83.4	39.6
Canteen Creek, Caseyville, Illinois	5-5895	1939–1971	58.3	22.7
East Fork Galena River, Council Hill, Illinois	5-4155	1949–1969	52.1	19.8
Ralston Creek, Iowa City, Iowa	5-4550	1925–1969	7.8	5.7

SOURCE: U.S. Geological Survey computer files.

TABLE 2 24 Upper Mississippi Valley Stream Gauging Sites Used in the Analysis of Annual Duration Series Floods

Stream Gauging Site	Gauge	Drainage Area (km^2)	Period of Record Used
LARGE WATERSHED TRIBUTARIES			
West Branch Pecatonica River at Darlington, Wisconsin	5-4325	709.66	1940–1975
Grant River at Burton, Wisconsin	5-4135	691.53	1935–1975
Kickapoo River at LaFarge, Wisconsin	5-4080	688.94	1939–1975
East Branch Pecatonica River at Blanchardville, Wisconsin	5-4330	572.39	1940–1975
Platte River near Rockville, Wisconsin	5-4140	360.01	1935–1975
Little Maquoketa River at Durango, Iowa	5-4145	336.70	1935–1975
Galena River at Buncombe, Illinois	5-4150	331.52	1937; 1940–1975
Little LaCrosse River near Leon, Wisconsin	5-3825	199.69	1935–1940; 1954–1975
Paint Creek at Waterville, Iowa	5-3886	145.04	1951; 1953–1975
Yellowstone River near Blanchardville, Wisconsin	5-4335	73.82	1954–1971; 1973; 1975
SMALL WATERSHED TRIBUTARIES			
Richland Creek near Plugtown, Wisconsin	5-4071	49.73	1958–1975
Mount Vernon Creek near Mount Vernon, Wisconsin	5-4360	42.48	1954–1965; 1974–1975
Crooked Creek near Boscobel, Wisconsin	5-4072	33.41	1959–1975
Nederlo Creek near Gays Mills, Wisconsin	5-4098.9	24.35	1968–1975

TABLE 2 *(continued)*

Stream Gauging Site	Gauge	Drainage Area (km^2)	Period of Record Used
SMALL WATERSHED TRIBUTARIES *(continued)*			
Pats Creek near Elk Grove, Wisconsin	5-4149	21.99	1960–1975
Knapp Creek near Bloomingdale, Wisconsin	5-4085	21.94	1954–1968
Bishops Creek near Viroqua, Wisconsin	5-4088	18.34	1959–1969
Pigeon Creek near Lancaster, Wisconsin	5-4134	17.95	1960–1975
Morris Creek near Norwalk, Wisconsin	5-4074	11.89	1960–1975
Rock Branch near Mineral Point, Wisconsin	5-4323	11.66	1959–1975
Bear Branch near Platteville, Wisconsin	5-4142	7.25	1958–1975
North Fork Nederlo Creek near Gays Mills, Wisconsin	5-4098.3	5.70	1968–1975
Rocky Branch near Richland Center, Wisconsin	5-4068	4.35	1954–1975
Skinner Creek Tributary near Monroe, Wisconsin	5-4342	1.24	1959–1975

SOURCE: U.S. Geological Survey computer files.

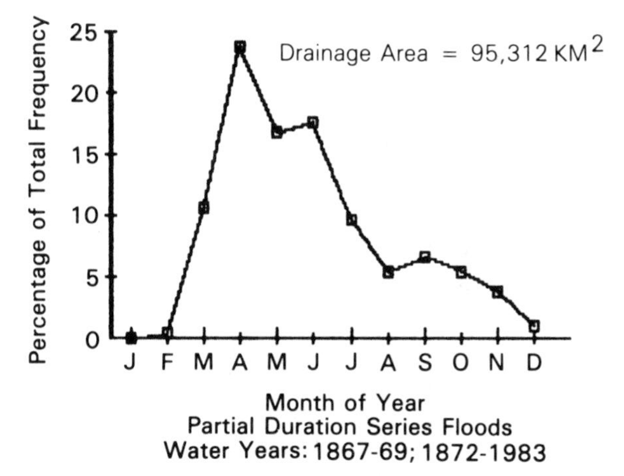

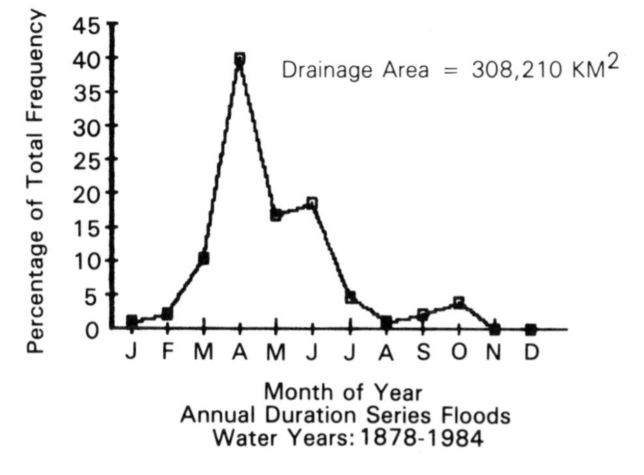

FIGURE 3. Monthly distributions of partial and annual duration series floods for the mainstem Mississippi River show that most floods are snowmelt related and occur during the month of April. Data source: U.S. Geological Survey computer files.

Rainfall Floods. The month of May in the Upper Mississippi Valley has relatively few floods compared to months that immediately precede or immediately follow it (Figs. 2 and 3). The lower frequency of May floods apparently results from its seasonal transition position. Snowmelt is an insignificant contributor to flooding in May, and high-intensity rainfalls of sufficient duration to cause flooding are relatively infrequent. In May, cool dry air masses from the northern Great Plains or Canadian prairies continue to dominate the northern half of the Upper Mississippi Valley. While warm and moist air masses from the Gulf of Mexico may advance into the northern part of the Upper Mississippi Valley, these low-density air masses override the high-density cool dry air and favor low-intensity rainfall that does not generate significant flooding.

Most June floods result directly from excessive or intensive rainfalls, although high levels of antecedent moisture often accelerate flood magnitudes and frequencies. In June warm moist air masses from the Gulf of Mexico dominate most of the Upper Mississippi Valley. The high moisture content of these air masses gives them a high potential for flood generation. Relatively cool dry air masses still are prevalent in nearby southern Canada, and the occasional invasion into the Upper Mississippi Valley by these polar air masses usually triggers heavy rainfalls and flooding when they come into collision with warm moist Gulf air masses.

Although relatively few floods occur in the other summer months, July and August have great potential for occurrences of floods because warm moist Gulf air is frequently dominant over the watershed (Figs. 2 and 3). A principal reason for fewer floods during this period is because the landscape of southern Canada also is relatively warm by July and August. The relatively gentle north–south thermal gradient across the midcontinent region provides only a modest contrast in densities of air masses between the Canadian region and the Upper Mississippi Valley. Hence, intrusion of cool dry polar air masses into the region occurs less frequently in July and August than in June. Nevertheless, when intrusions do occur large floods often follow. There are at least two other significant factors that help explain why July and August have fewer floods than June in the Upper Mississippi Valley. First, many agricultural fields still have significant areas of exposed soil in June, favoring reduced infiltration capacities and increased surface runoff. In July and August much of the land surface is covered by vegetation and more of the rainfall is likely to infiltrate into the subsurface. Second, because July and August usually have less rainfall than June, antecedent moisture levels tend to be lower during July and August compared to June. The low level of antecedent moisture therefore favors high infiltration rates and reduces the magnitudes of surface runoff and flooding.

September floods are relatively few in the Upper Mississippi Valley in spite of frequent occurrences of long-duration frontal rains accompanying the transition to fall season atmospheric circulation (Figs. 2 and 3). The explanation for the relatively small number of floods relates to at least three major factors, including: (1) the rainfall, although often of long duration, usually is widespread and of low intensity; (2) antecedent soil moisture levels often are low before the beginning of this rainy season; and (3) most of the landscape is covered with vegetation. October and November also have relatively few flood events in the Upper Mississippi River watershed, mostly for the same reasons that September has relatively few floods. Precipitation during October and November decreases significantly from September amounts as cool and relatively dry air masses gradually replace warm and moist air masses.

Climatic Associations

Because snowmelt floods are more frequent than rainfall floods in most Upper Mississippi Valley watersheds, statistical relationships between flood magnitudes and standard climatic variables tend to be weak. Graphical portrayal of the relationship between seasonal magnitudes of precipitation and magnitudes of floods in southwestern Wisconsin and adjacent areas shows that winter and summer precipitation correlate best with the magnitude of annual maximum floods (Fig. 4 and Table 3). The relationship confirms the roles of winter snowfall and summer rainfall as two independent sources of magnitudes and frequencies of floods in the Upper Mississippi Valley. On the other hand no statistically significant relationships were found between measures of atmospheric temperature and the magnitudes of annual maximum floods (Table 3). Stronger relationships undoubtedly would be found if flood data were tabulated at a monthly scale, but monthly scale data were not available. Many of the gauging stations used in the present analysis were designed to record only the annual maximum discharge. Climate stations used for precipitation and temperature analyses are listed in Table 4.

Precipitation. A statistically significant relationship between seasonal precipitation and the magnitudes of annual floods in a sample of small southwestern Wisconsin watersheds (average drainage area about 20 km^2) occurs only for summer season precipitation (Fig. 4 and Table 3). Although snowmelt is the source of most floods, it is reasonable that the strongest association occurs with summer precipitation because snowmelt runoff from the limited source areas of small watersheds usually is incapable of generating large-magnitude floods as can be more easily accomplished in large watersheds. In general, the smaller the watershed, the more likely the largest floods will be produced by high-intensity summer rainfalls. Nevertheless, the nearly annual recurrence of a snowmelt flood, often representing the annual maximum, accounts for a modest relationship between winter snowfall and the annual maximum flood. For example, the lack of a statistically significant relationship between winter precipitation and the annual maximum flood magnitude for small watersheds is due mainly to very anomalous floods of 1954 and 1971 (Fig. 4). The 1954 anomaly was a consequence of winter precipitation that was more than one standard deviation below average followed by summer precipitation that was about one and one-quarter standard deviations above average. Most of the summer rainfall occurred as heavy rainfall in June that accounted for a large annual maximum flood in most southwestern Wisconsin watersheds. The 1971 anomaly resulted from winter precipitation that was 2.7 standard deviations above the average followed by a cool dry spring that favored slow snowmelt and minimal flooding. The March average temperature was about one-half of a standard

FIGURE 4. Relationships between the magnitude of the annual maximum flood and seasonal precipitation tend to be weak. The flood data were first transformed to base 10 logarithms and then expressed in standard deviation units from respective mean values. The precipitation data were not similarly transformed before standardization because experimentation showed that logarithmic transformation did not significantly improve statistical relationships between the two data sets. The flood data represent 14 small basins whose average drainage area is about 20 km^2 and 10 larger drainage basins whose average drainage area is about 400 km^2 (Table 2). The relationships that are statistically significant at the 0.05 probability level or better are indicated by an asterisk (*).

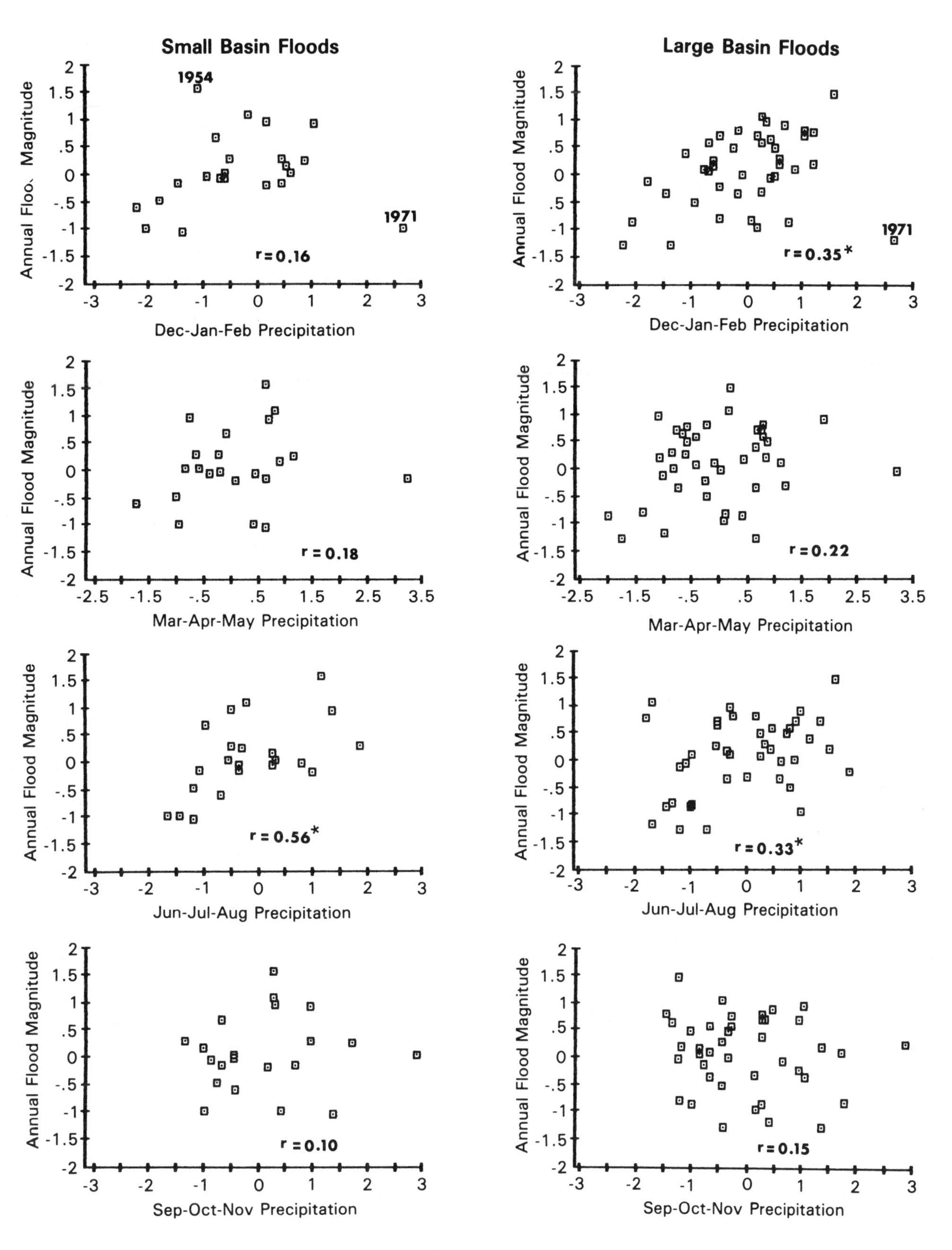

Small Basin Floods
Large Basin Floods
1954
1971
r = 0.16
r = 0.35*
r = 0.18
r = 0.22
r = 0.56*
r = 0.33*
r = 0.10
r = 0.15
Annual Flood Magnitude
Dec-Jan-Feb Precipitation
Mar-Apr-May Precipitation
Jun-Jul-Aug Precipitation
Sep-Oct-Nov Precipitation

FIGURE 4.

TABLE 3 Correlation Matrix Relating Annual Maximum Flood Magnitudes and Climatic Variables for Southwestern Wisconsin Watersheds

	FLD LB (1)	FLD SB (2)	ANN P (3)	WIN P (4)	SPG P (5)	SUM P (6)	FAL P (7)	JAN T (8)	MAR T (9)	JUN T (10)	ANN T (11)	LGSUMP (12)	LGFALP (13)
1	1.00	0.81	0.24	0.35	0.22	0.34	−0.15	0.11	−0.14	−0.14	−0.19	0.09	−0.08
2	0.81	1.00	0.41	0.16	0.18	0.56	0.10	0.19	−0.18	−0.14	−0.04	0.14	−0.13
3	0.24	0.41	1.00	0.31	0.65	0.58	0.67	0.03	0.04	−0.13	−0.05	−0.14	−0.12
4	0.35	0.16	0.31	1.00	0.06	0.11	0.05	−0.10	0.02	0.04	−0.09	−0.10	0.07
5	0.22	0.18	0.65	0.06	1.00	0.19	0.19	0.22	−0.03	−0.05	0.09	0.16	−0.13
6	0.34	0.56	0.58	0.11	0.19	1.00	0.03	0.02	−0.04	−0.12	−0.22	−0.24	−0.13
7	−0.15	0.10	0.67	0.05	0.19	0.03	1.00	−0.09	0.13	−0.12	0.05	−0.15	−0.03
8	0.11	0.19	0.03	−0.10	0.22	0.02	−0.09	1.00	0.02	0.02	0.56	0.21	−0.02
9	−0.14	−0.18	0.04	0.02	−0.03	−0.04	0.13	0.02	1.00	−0.08	0.33	−0.06	0.08
10	−0.14	−0.14	−0.13	0.04	−0.05	−0.12	−0.12	0.02	−0.08	1.00	0.38	−0.17	−0.05
11	−0.19	−0.04	−0.05	−0.09	0.09	−0.22	0.05	0.56	0.33	0.38	1.00	0.07	0.03
12	0.09	0.14	−0.14	−0.10	0.16	−0.24	−0.15	0.21	−0.06	−0.17	0.07	1.00	0.02
13	−0.08	−0.13	−0.12	0.07	−0.13	−0.13	−0.03	−0.02	0.08	−0.05	0.03	0.02	1.00

The flood variables are defined as follows: FLD LB = average flood magnitude standardized for 10 southwestern Wisconsin watersheds with average drainage area 400 km^2 (see Table 2); FLD SB = average flood magnitude standardized for 14 southwestern Wisconsin watersheds with average drainage area 20 km^2 (see Table 2). The flood data were transformed to base 10 logarithmic units before standardization. The climatic variables represent standardized averages of arithmetic (nontransformed) data for 9 southwestern Wisconsin area stations (see Table 4). The climatic variables are defined as follows: ANN P = annual precipitation; WIN P = winter (December, January, February) precipitation; SPG P = spring (March, April, May) precipitation; SUM P = summer (June, July, August) precipitation; FAL P = fall (September, October, November) precipitation; JAN T = average January temperature; MAR T = average March temperature; JUN T = average June temperature; ANN T = average annual temperature; LGSUMP = June, July, and August precipitation of preceding year; and LGFALP = September, October, and November precipitation of preceding year. Data base was 1935–1975 for FLD LB and all climatic variables, requiring a correlation coefficient of at least 0.30 for statistical significance at the 0.05 level of probability. Data base was 1954–1975 for FLD SB, requiring a correlation coefficient of at least 0.40 for statistical significance at the 0.05 level of probability.

deviation below the mean while spring and summer precipitation amounts were, respectively, 1 and 1.7 standard deviations below the long-term average. The deep winter snowfall probably also minimized the depth of ground frost allowing for better infiltration of moisture when snowmelt occurred.

Statistically significant relationships between precipitation and the magnitudes of annual floods in a sample of 10 large tributary watersheds (average drainage area about 400 km^2), centered on southwestern Wisconsin, occurred for both winter and summer seasons (Fig. 4 and Table 3). The dual significance indicates that either snowmelt or large-magnitude summer rainfalls can produce a wide range of flood magnitudes. The spring and fall season precipitation magnitudes have nonsignificant statistical correlations with the magnitudes of annual floods, a condition that is similar in small basins as already noted. The minimal contribution of spring and fall precipitation to magnitudes of the annual maximum flood indicates that the annual maximum floods in these Upper Mississippi Valley watersheds do not have a long memory of antecedent moisture conditions. The generally poor association between the magnitude of the annual maximum flood and antecedent seasonal precipitation was confirmed by correlation and regression analyses that showed that neither precipitation during the preceding fall nor during the preceding summer had a statistically significant relationship with the magnitude of the following year's annual maximum flood (Table 3).

TABLE 4 Climate Stations Used for Evaluating Climatic Contributions to Variations in Southwestern Wisconsin Annual Maximum Floods

Climate Station	Period of Record Used
Blair, Wisconsin	1935–1975
Darlington, Wisconsin	1935–1975
Lancaster, Wisconsin	1935–1975
Prairie du Chien, Wisconsin	1935–1975
Richland Center, Wisconsin	1935–1975
Viroqua, Wisconsin	1935–1975
Maquoketa, Iowa	1935–1975
Lansing, Iowa	1935–1975[a]
Monticello, Iowa	1935–1975[a]

[a] Not included in temperature analyses.

SOURCE: Annual Climatic Summary Reports, National Oceanic and Atmospheric Administration.

Temperature. No statistically significant relationships were found between magnitudes of maximum annual floods and January, March, June, and mean annual temperatures for the region represented by the 24 Upper Mississippi watersheds listed in Table 2 (Table 3). January, typically the coldest month of the year in the region, was selected to best represent the extremes of winter temperatures. March is the month corresponding with the maximum frequency of snowmelt floods and was selected to test for a possible statistical relationship between atmospheric temperature and rate of snowmelt. June is the month corresponding with the maximum frequency of rainfall floods. Therefore, June temperatures might best discriminate between cloudy, cool and wet periods versus sunny, warm and dry periods. Annual temperature was selected to provide an overall synthesis of change in atmospheric temperature from year to year.

There are at least four important reasons for the lack of statistical significance between the temperature variables and magnitudes of the annual maximum flood. First, the time range of investigation is too short for the effects of temperature changes to influence vegetation cover, especially in the north-central Upper Mississippi Valley where agricultural land use limits the ability of vegetation cover to respond to shifts in climatic conditions. Second, responses of floods to temperature changes probably are related indirectly through precipitation, which is more strongly and directly related to the influence of hemispheric patterns of atmospheric temperature to control air mass boundaries and storm tracks. Third, the magnitudes of the annual maximum flood are mostly determined by the rate and magnitude of snowmelt runoff or the rate and magnitude of summer precipitation events, and temperature therefore contributes only modest influence. Fourth, the statistical relationship between climatic events and flood events may be nonlinear and may vary from month to month as variations in ground cover and soil moisture influence rates of infiltration and runoff. However, the monthly differences in ground cover and soil moisture probably have a greater influence on the magnitude and frequency of floods in the partial duration series than in the annual maximum series, recalling that the magnitude of the annual maximum flood is primarily determined by either snowmelt magnitude or summer rainfall magnitude.

Climatic Associations with Flood Anomalies. In light of the weak to statistically insignificant linear relationships between magnitudes of annual maximum floods and seasonal climatic data, further examination of the climatic conditions associated with groups of years having either especially large or especially small annual flood magnitudes was undertaken. Two data groups were identified for analysis based on the average value of the flood magnitude representing 10 large basin tributaries occurring in or near southwestern Wisconsin (Table 2). Only years with floods being equal or greater than 0.5 standard deviations from their respective mean value were selected for analysis. Table 5 presents the climatic associations for the two flood groups. These results show that when annual maximum flood magnitudes are equal or greater than 0.5 standard deviations below the mean, climate tends to be relatively warm and dry, especially during the summer season when the average June, July, and August precipitation total was 0.7 standard deviations below the mean and June temperature was about 0.2 standard deviations above the mean. Climatic conditions of years when annual maximum flood magnitudes are equal or greater than 0.5 standard deviations above the mean are slightly more complex. Not surprisingly, all seasons except fall have precipitation that is above average, but the greatest departure is restricted to winter precipitation. The average annual temperature of years with these large floods was about 0.3 standard deviations below average, although January temperatures were slightly warmer than average. Although the amount of precipitation during the preceding fall and summer was not significantly related to flood magnitude when all data were examined

TABLE 5 Climatic Associations with Flood Anomalies in Southwestern Wisconsin Watersheds[a]

Climate Variable	Years with Floods 0.5 Standard Deviation below the Mean	Years with Floods 0.5 Standard Deviation above the Mean
Annual precipitation	−0.57	+0.13
Winter precipitation	−0.36	+0.45
Spring precipitation	−0.54	+0.14
Summer precipitation	−0.70	+0.14
Fall precipitation	+0.12	−0.17
Annual temperature	+0.39	−0.28
January temperature	−0.01	+0.18
March temperature	+0.10	−0.22
June temperature	+0.19	−0.01
Past year summer precipitation	+0.14	+0.31
Past year fall precipitation	+0.03	+0.00

[a] All data expressed in standardized deviations from mean values. Temperature data represent either mean annual or mean monthly values. Flood data were obtained from U.S. Geological Survey annual reports for the 10 "large" watersheds shown in Table 2. Climatic data were obtained from the National Atmospheric and Oceanic Administration Annual Climatic Summaries for states.

together (Table 3), this analysis shows that years with large annual maximum floods tend to occur when wet summers are followed by above average winter precipitation. On the other hand variations in the amount of precipitation during the preceding fall apparently have little influence on the magnitude of the annual maximum flood (Table 5).

MAGNITUDE AND FREQUENCY OF FLOODS

Modern Floods

It is common practice to assume that the recurrence interval of a flood of a given magnitude will behave in a random fashion such that the probability of recurrence remains the same from year to year (U.S. Water Resources Council, 1981). Close inspection of flood series often shows this assumption is not met because of either changes in climate or land use or both. Understanding the behavior of long-term variations in magnitudes and frequencies of floods is complicated in the Upper Mississippi Valley because snowmelt, rainfall, or combined snowmelt and rainfall all are causes of floods. Most would agree that sample populations of snowmelt and rainfall floods should be separated for magnitude and frequency analyses, but as a matter of practice it is rarely done.

Seasonality. The magnitudes of floods in the Upper Mississippi River watershed reflect the season of occurrence, suggesting that separating flood series into seasonal subpopulations should often be considered in magnitude and frequency analyses. The largest magnitude floods in watersheds that drain less than several hundred square kilometers tend to occur in summer following high-intensity or long-duration rainfalls. The largest magnitude floods on large watersheds such as the Mississippi and its major tributaries tend to occur in spring during snowmelt runoff. The relationship between basin size and seasonality of the largest floods is illustrated for three representative watersheds in Figure 5. Note that most of the largest floods occur during the warm season months of June through August for the two small watersheds represented by Rocky Branch (4.4 km^2) and Platte River (360 km^2). Very large floods in the small tributaries usually result from high-intensity excess rainfall rather than from snowmelt because snowmelt runoff accumulates too slowly from the limited drainage area of small watersheds to generate large floods. The same relationship is true of long-duration, low-intensity rainfalls. In contrast, the Mississippi River at Keokuk, Iowa (308,210 km^2), experiences its largest floods during April, May, and June in response to snowmelt or to the residual effects of snowmelt. Floods on large rivers in the Upper Mississippi Valley typically result from snowmelt or long-duration rainy periods. Even though the rate of

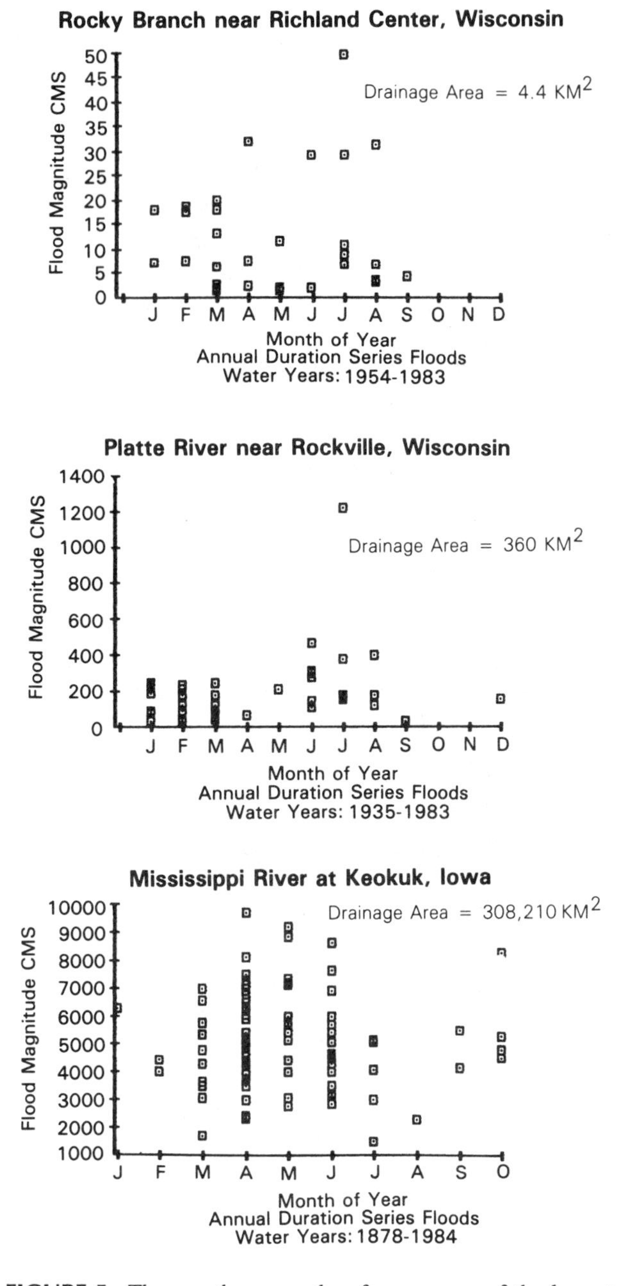

FIGURE 5. The month or months of occurrence of the largest floods is partially dependent on drainage area because drainage area is sensitive to the various types of flood-generating processes. The largest floods on tributary streams tend to occur during summer in response to excess rainfall, but the largest floods on larger rivers, such as the Mississippi River, tend to occur during spring in response to snowmelt. Data source: U.S. Geological Survey computer files.

runoff is relatively low and flooding may not occur in the tributaries, the simultaneous accelerated flow from numerous smaller tributaries often is sufficient to generate large downstream floods.

When the annual variations of floods in the three watersheds are compared to magnitudes of the 50-yr, 10-yr, and mean annual flood in each of the respective basins, nonrandom variation is suggested (Fig. 6). Note that flood magnitudes from Rocky Branch and Platte River have been relatively small in the more recent years compared to the early years of their respective records. For example, on Platte River most annual maximum floods equaled or exceeded the magnitude of the mean annual flood during the period from 1935 to 1954, but the record since 1954 shows conservative variation about the mean. The pattern of variation in annual maximum floods on the large Mississippi River shows that large floods were common in the late 1800s and in recent decades but that magnitudes of floods during the period from about 1895 to about 1950 were small and did not exceed the long-term average magnitude of the expected 10-yr flood (Fig. 6). These systematic patterns of variation in flood magnitudes suggest that flood records might be partitioned to better evaluate the true recurrence probabilities of floods in relation to existing environmental factors. When a flood series represents the combined aggregate of all seasons, it often is unclear whether temporal variations in floods represent a response to change in average annual climatic conditions or a response to shifts in seasonal distributions of temperature and rainfall while average annual conditions have remained nearly constant.

The potential importance of a shift in seasonality of flooding is illustrated by the data of Table 6, which present estimates of flood magnitudes grouped by season and annually for the Rocky Branch and Platte River watersheds cited above. These estimates indicate that summer floods of low-recurrence frequency tend to be substantially larger than winter floods of the same recurrence probability. For example, the 100-yr flood of summer is about three times larger than the winter flood of the same expected return period. The wide discrepancy between seasonal flood magnitudes indicates that magnitudes of low-recurrence floods can be substantially underestimated when the data are not seasonally stratified (Table 6).

Errors associated with estimating probable flood magnitudes can be particularly significant if a climatic shift occurs to cause either an enhancement or suppression of a given season's flooding potential. In the 1940s and early 1950s large summer floods were common throughout much of the Upper Mississippi Valley in response to a strong southerly to southwesterly component in the summer atmospheric circulation that advected warm, moist, and unstable air masses into the region (Knox et al., 1975, pp. 16–18). The average magnitudes of these large floods often exceeded one standard deviation above the long-term

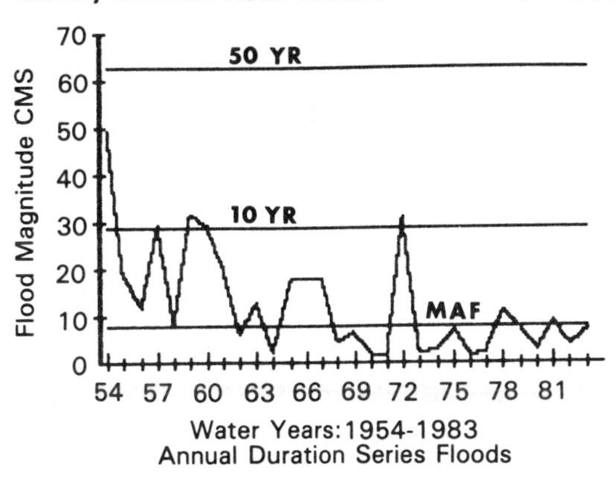

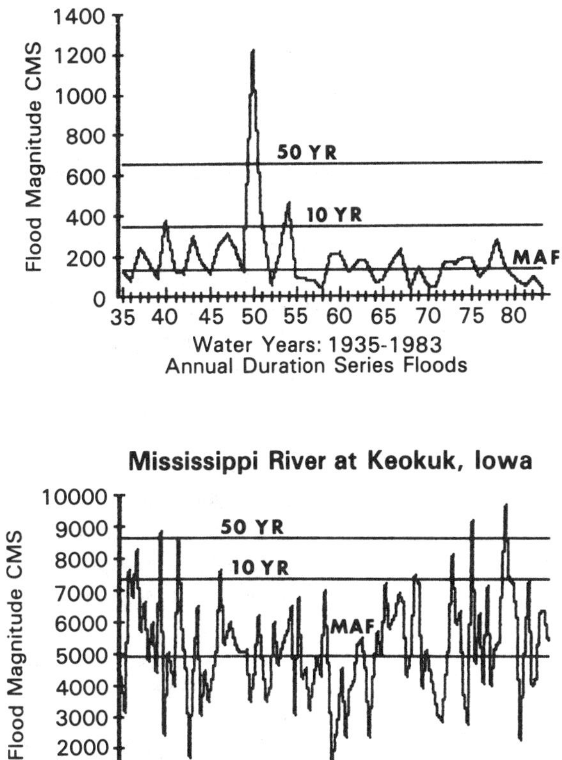

FIGURE 6. Time series of flood magnitudes often show patterns of persistence in departures from the mean annual flood (MAF). The magnitudes of the 50-yr, 10-yr, and mean annual floods were determined by application of the log-Pearson Type III method (U.S. Water Resources Council, 1981). The decreasing relative difference between the mean annual flood and low-frequency floods, as drainage area increases from a few square kilometers in Rocky Branch to a few hundred thousand square kilometers on the Mississippi River at Keokuk, is typical for the region. Data source: U.S. Geological Survey computer files.

TABLE 6 Log-Pearson Type III Estimates of Floods Grouped Annually and Seasonally

Recurrence Probability	All Data Combined m^3/s Discharge	Winter Floods (Dec.–Mar.) m^3/s Discharge	Summer Floods (Apr.–Nov.) m^3/s Discharge
ROCKY BRANCH WATERSHED (1954–1983)			
0.01	81.59	42.72	120.39
0.02	62.83	37.69	84.03
0.04	46.88	32.30	58.00
0.05	41.79	30.19	50.68
0.10	28.98	24.18	32.70
0.63	9.37	6.45	5.07
PLATTE RIVER WATERSHED (1935–1983)			
0.01	831.53	375.95	1257.35
0.02	658.52	343.44	1019.60
0.04	510.81	288.47	813.61
0.05	467.65	273.94	744.56
0.10	346.98	228.63	561.53
0.63	98.87	87.91	155.87

Flood Data Source: U.S. Geological Survey computer files. Respective drainage areas for Rocky Branch and Platte River are 4.35 and 360.01 km^2 (Table 2).

historical mean flood magnitude in watersheds of the north-central Upper Mississippi Valley (Fig. 7). The high frequency of large floods was subsequently suppressed, especially in the 1960s, when the strong southerly to southwesterly atmospheric circulation was replaced by large-scale circulation regimes having a stronger westerly and northwesterly component (Knox et al., 1975, pp. 16–18). Air masses entering the Upper Mississippi Valley from the west and northwest tend to be relatively dry and stable. Consequently, floods become fewer and smaller. Figure 7 shows that the smallest average magnitudes of floods in the large and small Upper Mississippi Valley tributaries listed in Table 2 occurred between 1956 and 1971 when a stronger westerly and northwesterly large-scale atmospheric circulation increased in frequency.

The preceding examples illustrate that the magnitude of the annual maximum flood is sensitive to variations in seasonal climatic conditions. Figure 8 shows that individual monthly frequencies of floods in the Upper Mississippi Valley also are in several instances quite sensitive to seasonal climatic shifts. The monthly flood frequencies shown on Figure 8 were determined by counting the number of times that independent streamflows exceeded the partial duration series flood stage in each of the 29 watersheds listed in Table 1. The plots of Figure 8 confirm the previous assertion that most flooding occurs from March through July. March and June are not only the most prominent with respect to average frequency of floods but they also appear to be the most sensitive to short-term climatic variations. Note, for example, that the number of March snowmelt floods that exceeded the partial duration series base in 29 separate watersheds ranged from 5 or less during the mid-1950s to about 25 in late 1940s and early 1960s. June rainfall floods in the 29 watersheds ranged from an average of about 5 in the late 1950s and early 1960s to commonly more than 20 in the 1940s and early 1950s. By comparison, other months seem to show conservative variation around a more stable long-term historical mean flood frequency. It is im-

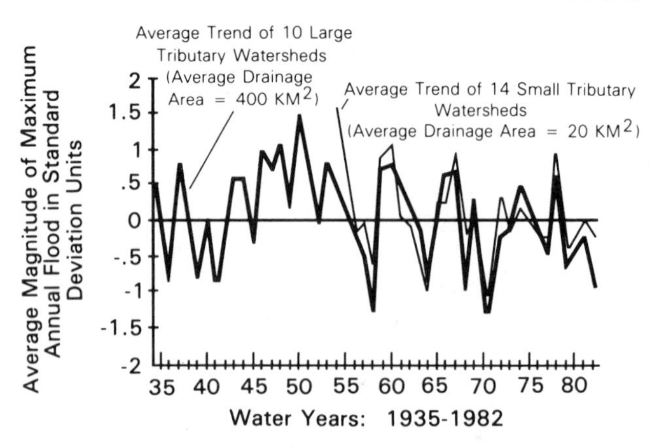

FIGURE 7. Temporal patterns of variation in the average magnitudes of Upper Mississippi Valley annual maximum floods from large tributaries are very similar to those of small tributaries, suggesting dominance by a climatic influence. Group standardized units for flood magnitudes were determined by averaging for each year the individually standardized flood magnitudes representing base 10 logarithmic transformations of flood records for the respective rivers in each category. The flood records used in each group correspond with the listings in Table 2. Data source: U.S. Geological Survey computer files.

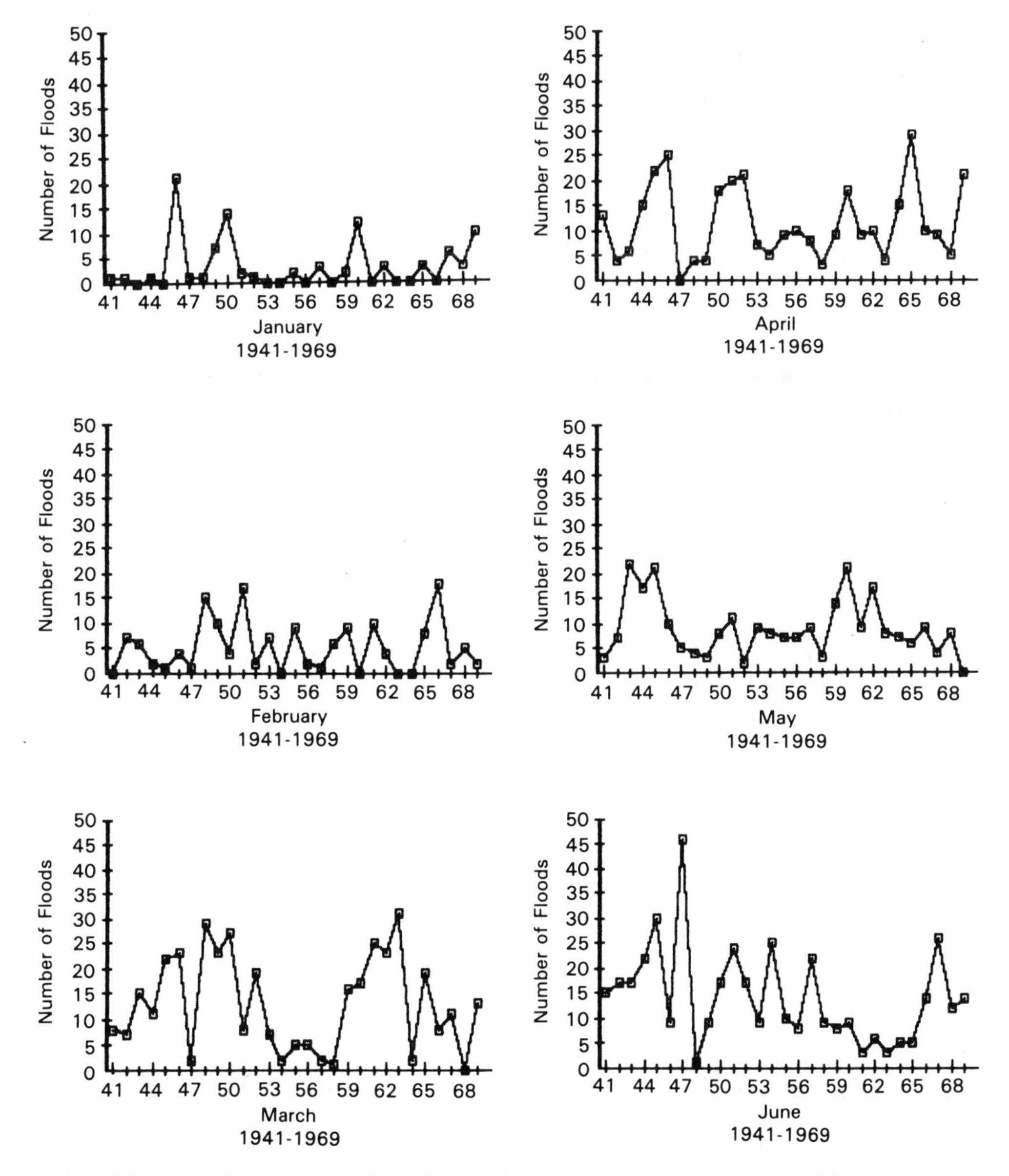

FIGURE 8. Individual monthly flood frequencies for the watersheds listed in Table 1 show that the months of March through June are not only the months with the greatest frequencies of floods, but that each of these months can experience major short-term differences in flood frequencies in response to changes in seasonal climatic conditions. Floods during March and June appear to be the most responsive to seasonal climatic conditions of all the months. Data source: U.S. Geological Survey computer files. (*Figure continues on p. 292.*)

portant to acknowledge the month to month variation in flood frequencies because the summation of these variations can be either cumulative or offsetting on an annual basis. Simple assessment of nonseasonally segregated annual floods may give unrealistic impressions of long-term variations in the magnitude and frequency of floods.

Temporal Change in Magnitude and Frequency. The above examples of short-term changes in flood seasonality suggest that the mean and variance of flood series may not remain stationary as the time scale of focus lengthens. Unfortunately, most historical records are too short to adequately test the hypothesis that contemporary shifts in climate produce significant changes in the mean and variance of flood series. Few records are long enough to estimate the time scale beyond which the assumption of a stationary mean and variance is unrealistic. Additional difficulty in evaluating climatic influence is caused by changing land use that often may greatly influence the magnitude and frequency of floods. Many land uses related to agriculture,

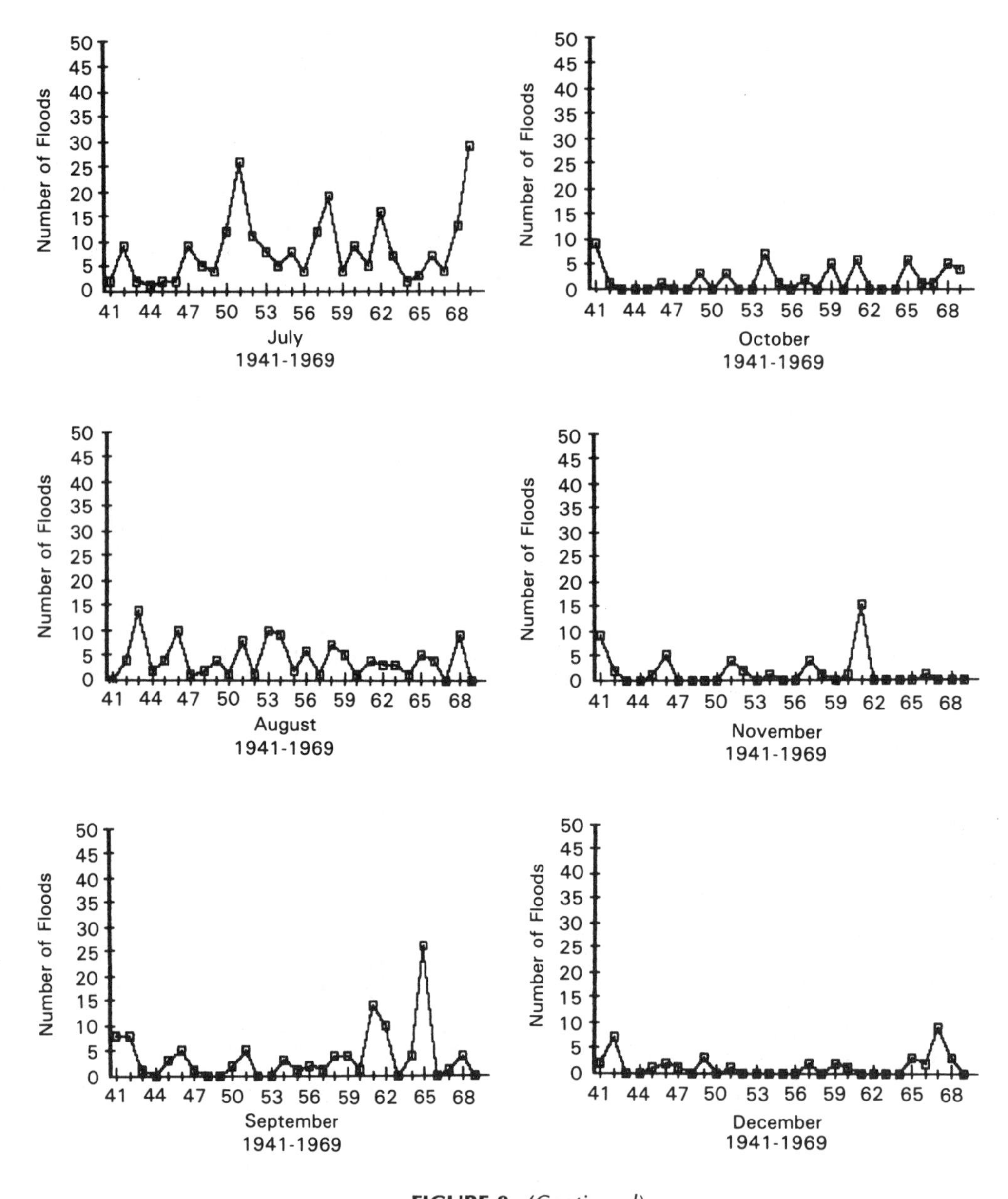

FIGURE 8. *(Continued)*

urbanization, channelization, and other activities increase magnitudes and frequencies of floods, while construction of dams, reservoirs, drainage diversions, and certain types of land use may reduce magnitudes and frequencies of floods. Evaluating the specific hydrologic contributions of land use usually is a difficult exercise and is somewhat subjective.

The longest records of floods in the Upper Mississippi Valley are associated with the Mississippi River where observations began in the late 1860s and 1870s. In an earlier study, I used climatic data to partition the partial duration series floods of the Mississippi River at St. Paul, Minnesota (Knox, 1984). The analysis showed that relatively high frequencies of large floods in the late 1800s before 1895 and during the years since about 1950 were associated with mean annual temperatures that were slightly below the long-term mean and annual precipitation that was above the long-term mean. The tendency for large floods to be clustered in the early and late portions of the record, and infrequent during the middle years of record, also is apparent in the annual flood series record for the Mississippi River at Keokuk, Iowa (Fig. 6). Although flood stages on the Mississippi River have been significantly influenced by human activities (Belt, 1975), the results from the St. Paul record showed that climatic shifts were primarily responsible for statistically significant different episodes of magnitudes and frequencies of floods recorded in the more than century long record. As might be expected, the magnitudes of large floods of low frequency were found to be more sensitive to the climatic episodes than were small floods of high frequency when magnitudes of a given recurrence probability were compared between climatic episodes.

Holocene Floods

Flood records at stream gauging sites for most rivers and streams in the Upper Mississippi Valley date from the 1930s or later. The lack of sufficiently long and reliable flood records has limited efforts to evaluate the influence of climatic change on magnitudes and frequencies of floods. Paleoflood evidence represented in alluvial sediments and relict channels represents an alternative basis for examining responses of floods to long-term environmental change (Baker, 1984; Baker et al., 1983; Costa, 1978, 1983; Knox, 1980, 1985; Kochel and Baker, 1982; Patton et al., 1979; Williams, 1983a, 1984). Whereas many techniques that have been used to extend hydrologic records through mathematical or statistical simulation can only restate the information content of the base data (Bras, 1980), paleoflood data have no such limitation.

I will illustrate how long-term variations in magnitudes and frequencies of Holocene (postglacial) floods can be reconstructed from alluvial stratigraphy by use of examples from the hilly unglaciated Upper Mississippi Valley driftless area (Fig. 1). Driftless Area rivers and streams transport large quantities of bedload and suspended-load sediments. Bedload sediments, derived from the local sedimentary rocks and regolith, range from sand through large boulders. Suspended-load sediments are mainly loess-derived silt and clay.

The Driftless Area is associated with seasonally steep climatic gradients that produce an ecotone separating mixed hardwood forest to the northeast from prairie to the southwest (Borchert, 1950; Bryson, 1966). Fossil pollen from sites in and adjacent to the Driftless Area indicate that the prairie/forest ecotone was established in the region by about 9000 yr B.P. (Davis, 1977; King, 1981; Maher, 1982; Wright et al., 1963). Synthesis of the fossil pollen data indicates that movements of the ecotone have been conservative since 9000 yr B.P., with gradual northeastward migration until about 7000 yr B.P. followed by a general southwestward retreat since then (Bernabo and Webb, 1977). Bartlein and Webb (1982) and Bartlein and others (1984) examined relationships between climate and pollen and suggested that mean annual precipitation decreased between 10 and 20% and mean annual temperature increased between 1 and 2°C when the ecotone moved toward its maximum northeastward position. While there is consensus that the climate during much of the early Holocene before 6000–7000 yr B.P. was effectively more dry than modern conditions in the region, the precise combination of temperature and moisture conditions are still somewhat unclear. More recently, R. M. Forester (personal communication, 1986) and his associates, studied fossil ostracodes and diatoms in sediments from Elk Lake, Minnesota, to the northwest of the Driftless Area, and concluded that the climate during the period of northeastward advance of prairie/forest ecotone was cool/dry rather than warm/dry as indicated by Bartlein and Webb. The principal period of dryness appears to have ended by about 6000 yr B.P. in the central and western part of the Driftless Area, but fossil pollen from a site near the eastern margin of the Driftless Area indicates that relatively dry conditions may have occurred there until about 3000 yr B.P. (Winkler, 1985).

High-Frequency Floods. The magnitudes of floods having an expected recurrence of 1–2 yr can be estimated from the channel capacity for the bankfull stage of relict natural channels. Here, I will only highlight the methodological procedures for this technique because a detailed description has been recently presented elsewhere (Knox, 1985). In general, the procedure is based on the observation that the bankfull stage in a natural alluvial channel is approximated by the upper limit of lateral-accretion deposits and is therefore approximately equal to the tops of point bars associated with the active floodplain (Wolman and Leopold, 1957). This definition of bankfull stage is critical because other definitions of bankfull stage, usually emphasizing morphologic features on channel margins, frequently do not equate well with the flood event of 1–2 yr (Williams, 1978).

Magnitudes of relict floods with recurrence intervals of 1–2 yr therefore can be reconstructed from dimensions of relict channels left behind during meander cutoffs and other channel avulsions. Cross sections of relict channels are determined from borings at intervals of a few meters taken perpendicular across relict channels. Sedimentological analyses of the cores are used to differentiate channel margin from channel fill deposits in the relict channels. The bankfull stage of the relict channel is defined as the top of the fining upward, lateral-accretion deposits in attached buried point bars, following the same criteria used to calibrate the relationship between cross section capacity and bankfull stage discharge in modern channels. Radiocarbon dating is the preferred method of establishing the age of a relict channel, although other types of evidence, including human artifacts, soil development, and fossil plants and animals can provide helpful indicators of relative age. The relative difference between modern and relict floods of 1–2 yr recurrence frequency can be estimated for any given site by comparing the reconstructed bankfull stage discharge with the modern bankfull stage discharge expected for the drainage area associated with the relict channel. The method is most reliable in regions where channel cross section geometry is not susceptible to major metamorphosis when shifts in discharge and sediment load characteristics occur. A major change in channel cross-section shape affects flow resistance and, in turn, flood conveyance.

The record of variations in magnitudes of Holocene floods having recurrence intervals of 1–2 yr is reconstructed for Driftless Area sites in Figure 9*a*. The Holocene flood record probably represents floods resulting from snowmelt because the region's high-frequency, small- to moderate-magnitude flood events in modern annual and partial duration series are dominantly a product of snowmelt (Figs. 2 and 3). The Holocene variations in magnitudes of bankfull stage floods show relatively large and persistent departures from contemporary magnitudes of floods with the same recurrence interval, illustrating that the mean of the series has not remained constant during the Holocene.

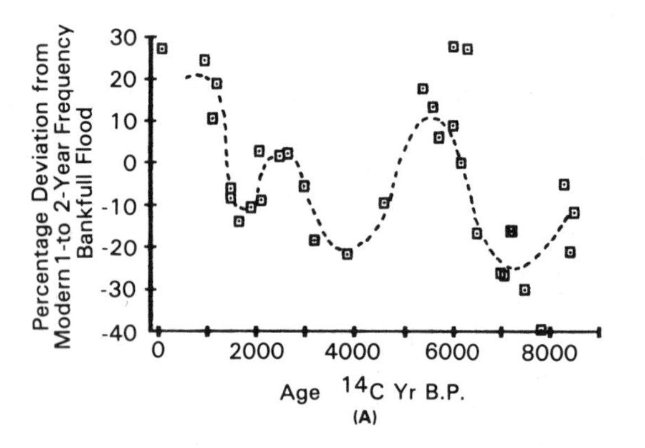

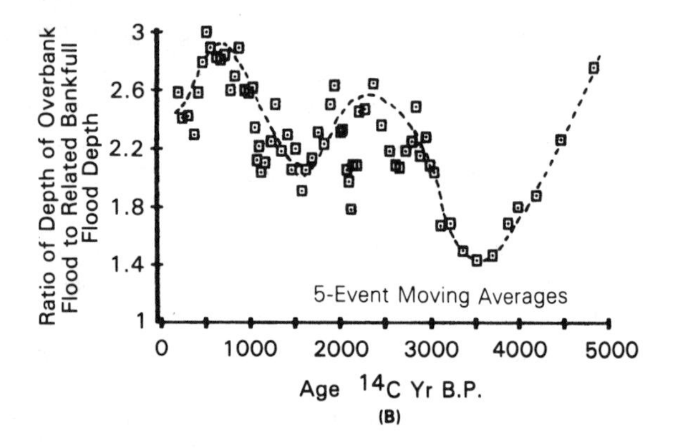

FIGURE 9. (*a*) Snowmelt-dominated Holocene high-frequency, small-magnitude floods ranged from about 10–15% larger to 20–30% smaller than equivalent frequency modern floods. The record of fossil pollen from the region indicates that climatic change was responsible for the pattern of long-term departures from modern flood magnitudes. (*b*) Rainfall-dominated late Holocene low-frequency, high-magnitude floods have varied in a broadly similar fashion to late Holocene high-frequency floods, suggesting that long-term Holocene climatic changes influenced cold and warm season flood-generating mechanisms in a broadly similar way. The flood depth ratio represents the ratio formed by dividing the maximum depth of the overbank flood above a channel bed lag gravel by the bankfull depth of the modern channel at the same site. Although calculated ratios ranged from near 1.0 (bankfull stage) to 4.3 (4.3 times the bankfull depth), a 5-ratio running mean was presented here to give a composite expression of the relative changes in magnitudes of all overbank floods.

During the early Holocene, when mean annual precipitation may have been 10–20% less than present-day amounts (Bartlein and Webb, 1982), magnitudes of bankfull stage floods were 20–30% smaller than comparable frequency modern floods. The small magnitude of the early Holocene floods may have been a consequence of dominance of the Upper Mississippi Valley by air masses derived from the central and northern Great Plains due to a strong westerly circulation that probably prevailed during most of the early Holocene in response to continued presence of the Laurentide ice sheet (Knox, 1983). Maximum dryness and maximum northeastward displacement of the prairie/forest border occurred about 7200 yr B.P. (Wright, 1976), and the prairie/forest border has retreated generally southwestward since then (Bernabo and Webb, 1977).

Fossil pollen indicates that the climate became more moist in association with the southwestward retreat of the ecotone. Increasing moisture is consistent with the relict flood reconstructions that show occurrences of relatively large floods after about 6000 yr B.P. (Fig. 9*a*). The bankfull stage floods occurring after about 6000 yr B.P. averaged 10–15% larger than modern counterparts. The occurrence of these relatively large floods continued until about 4500 yr B.P. when smaller floods became dominant (Fig. 9*a*). Although fossil pollen indicates that climate, in comparison to the early Holocene, remained relatively cool and moist between 4500 and 3000 yr B.P., magnitudes of bankfull stage floods were relatively small (Fig. 9*a*). The difference between the large-magnitude floods occurring between 6000 and 4500 yr B.P. and the small-magnitude floods occurring between 4500 and 3000 yr B.P. apparently is explained by shifts in storms tracks in the large-scale atmospheric circulation. The forest/tundra border in north-central Canada reached its maximum postglacial northern position about 3500 yr B.P. (Sorenson et al., 1971; Sorenson, 1977), indicating the approximate time of maximum postglacial warming there. Prolonged warmer conditions in Canada would lower the incidence of heavy precipitation in the Upper Mississippi Valley, much in the same way that present-day seasonal precipitation is influenced by the atmospheric temperature gradient across central North America.

About 3000 yr B.P. a second sharp rise occurred in the magnitudes of floods with recurrence intervals of 1–2 yr (Fig. 9*a*). Because this rise corresponds with further cooling inferred from fossil pollen and from southwestward movements of the prairie/forest ecotone (Wright, 1976) as well as the forest/tundra ecotone (Sorenson, 1977), it appears to represent a response to large-scale climatic change. A brief period of smaller floods is suggested after about 2000 yr B.P., but larger floods are again suggested after about 1000 yr B.P. (Fig. 9*a*). The evidence in support of large floods about 1000 B.P. is relatively weak, however, because sediment textures and point bar heights did not indicate floods as large as those suggested by the dimensions of the relict channels.

In summary, the history of Holocene floods with recurrence intervals of 1–2 years indicates that the mean value fluctuates with time. Thus, an assumption underlying most techniques for estimating recurrence probabilities of floods (e.g., U.S. Water Resources Council, 1981) is not met in the Upper Mississippi Valley for time scales extending over a few thousand years. The principal cause of nonstationarity is climatic change.

Low-Frequency Floods. Estimating magnitudes of former low-frequency floods that occurred long in the past is difficult because evidence of high-water marks, mud lines, organic debris, and other ephemeral features usually have been destroyed. However, the high energy of large floods often is sufficient to transport coarse sediments from the channel bed onto the adjacent floodplain. Recovery of these overbank deposits provides a basis for estimating depths of former overbank floods of relatively low recurrence frequency. The methodological procedure is based on estimating the depth of flow competent to transport the clastic particles represented by the overbank flood deposits.

One of the more common approaches for estimating competent flow depth is based on the DuBoys equation for boundary shear:

$$\tau = \gamma RS$$

where τ is the critical mean shear stress in newtons per square meters for initiating particle transport, γ is the specific weight of the water and sediment mixture in newtons per cubic meters, R is hydraulic radius in meters, and S is the dimensionless energy slope in meters per meter. Baker and Ritter (1975) summarized the relationship between size of transported particles and channel bottom shear stress by aggregating and standardizing the data from many investigators. They found that critical shear stress explained approximately 85% of the variance in the intermediate axis dimensions of transported particles and that the size of the coarsest particles transported varied with the 0.54 power of shear stress. More recently Williams (1983b), using a smaller subset of the data presented by Baker and Ritter, presented a multivariate regression equation to estimate the critical flow depth directly from the intermediate axis dimension of a particle and the corresponding energy slope associated with its movement. I have followed the procedure of Williams (1983b) by establishing a regression equation based on a limited set of particle diameters (approximate range: 50–3300 mm) that approximate the range of textures represented in the overbank flood gravels in the Driftless Area. The equation is

$$D = 0.0001A^{1.21}S^{-0.57}$$

where D = competent flood depth in meters

A = intermediate axis, in millimeters, representing an approximation of the coarsest particles moved (usually either D_{90} or the mean of the coarsest five or fewer particles)

S = an approximation of the energy slope in meters per meter

The equation assumes a constant of 9800 N/m^3 for the specific weight of the water and sediment load. The equation is based on 52 observations originally reported in selected observations by Lane and Carlson (1953), Wolman and Eiler (1958), Fahnestock (1963), Kellerhals (1967), and Scott and Gravlee (1968). The standard error of estimate of the equation is 0.34 $\log_{10}$ units and the equation is statistically significant at the 0.01 probability level with 74% of the variance in competent flow depth being explained by the two independent variables. D. F. Ritter (personal communication, 1986) suggested that the above competency equation probably underestimates true flood depths because it (1) assumes that all sizes of particles are available in the channel load, and (2) it is primarily derived from data concerning bedload, while suspended load is a more likely mechanism for transportation of overbank flood gravels.

The procedure for estimating magnitudes of former low-frequency overbank floods involved several steps. First, overbank flood gravels were located in the field either in river bank exposures or in floodplain or terrace excavations. During the first half of the field investigation, 300–400 particles were sampled from approximately the top 20–30 cm of each deposit. Because only a few of the overbank flood gravels were associated with gravel deposits thicker than 20–30 cm, the entire vertical section was sampled at most sites. Initially, samples were analyzed to determine the mean intermediate axis dimension of particles coarser than the 90th percentile. However, experimentation showed that equally reliable estimates of the coarse fraction could be determined more efficiently by simply sampling about 30 of the coarsest particles observed and then calculating the mean of the coarsest 5 particles from that subsample. Therefore, the final analysis used the mean of the coarsest five particles for the particle axis dimension in the equation for estimating flow depths of paleofloods. A total of 114 overbank gravel deposits, representing 70 flood events, were sampled in the field. The energy slope corresponding to the flood gravel was estimated by determining the average gradient of the floodplain through the reach containing the flood gravels. After experimenting with several ways of estimating slope, including field surveys of channel bed slope, channel slope at bankfull stage, floodplain slope, and slope determined from 1 : 24,000 scale topographic maps, it was decided that the latter provided the best approximation of the former energy gradients associated with overbank paleofloods. The other measures, which may provide more accurate slope estimates for floods in modern channels, are especially sensitive to local biases of modern meander geometry. Slope estimates determined from topographic maps tend to smooth out local anomalies and appear to better approximate the gradients associated with former deep flows of large overbank paleofloods. Slope usually was computed for valley reaches extending over a distance of several hundred meters, except in steep tributaries where the distance was greatly shortened to reflect significant changes in slope over short distances. Ages of the former overbank floods were determined either directly by radiocarbon dating or indirectly based on the thickness of fine textured vertical accretion sediments between the base of the overbank flood gravel and the top of the underlying buried lateral-accretion sediments. The latter technique has a standard error of estimate of approximately 850 yr (Knox, 1985).

The difference in magnitude between rare extreme floods and the mean annual flood is not as great in humid regions as in semi-arid and arid regions. For example, the largest observed historical floods in the Upper Mississippi Valley usually have been associated with overbank flood depths that do not exceed more than a few meters. The shallowness of these overbank flows causes them to be especially sensitive to the hydraulic influences of vegetation and the presence or absence of even very low terraces. Because the overbank depths of most paleofloods probably were similarly affected by the extent of former terraces and by the density and type of vegetation on the former valley floors, estimates of former paleoflood discharges are subject to considerable error. Because of this unknown error, paleoflood reconstructions were restricted to estimating paleoflood depth in relation to the bankfull stage flood depth at the site in question. Extensive radiocarbon dating of alluvial units and relict channels in the Driftless Area has shown that the channel bed elevations have remained nearly constant during the Holocene while considerable lateral migration has occurred (Knox, 1985). The maintenance of constant bed elevation is associated with resistant lag gravels, and it indicates that the reference base, from which bankfull stage heights and overbank flood stage heights are compared, has remained constant throughout the Holocene.

As noted previously, 114 overbank flood gravels were sampled, but some of the gravels were associated with the same flood event. The ratio of the overbank flood depth to the bankfull stage depth, both measured from the channel bed lag gravel, was computed for all 114 overbank flood gravels. For example, a computed ratio of 3.0 indicates that the depth of the overbank paleoflood was three times the stage height of the bankfull flood at the same site. The ratios were grouped and averaged for those sites representing the same flood event, reducing the sample population of flood ratios to 70 events. While the overbank floods were spread over a time range representing approximately the last 6000 radiocarbon years, most events had occurred since 3000 yr B.P. The bias toward the younger end of the time scale does not mean that overbank floods are increasing with time. Rather, it reflects the fact that most floodplains in the Driftless Area have been extensively reworked by lateral stream migration and that most alluvial units above the resistant channel bed lag gravel date from approximately the last 3000 yr.

The magnitudes of computed depth ratios of overbank floods to bankfull stage floods ranged from slightly greater than 1.0, indicating a flood about equal to the bankfull flood, to slightly over 4.0, indicating an extreme flood of very low recurrence probability. Modern Driftless Area floods having a 2% chance of occurrence in any given year (i.e., the 50-yr flood) are associated with flood depth ratios equal to approximately 2.0, suggesting that ratios of 3.0 and larger probably represent floods that are well beyond the magnitude of the present-day expected 100-yr flood. Unlike paleofloods reconstructed from bankfull stage dimensions of relict channels, where a specific recurrence frequency can be assigned, the reconstruction of water depths associated with overbank flows involves floods from a wide range of expected recurrence intervals.

Because a major objective was to determine how the pattern of temporal variation in magnitudes of overbank floods compared with the pattern of temporal variation in bankfull stage floods, an averaging filter was applied to the flood data to produce a composite expression of the data. Running mean filters, averaged over five-event windows, were computed for the flood depth ratios and their corresponding ages. Results of the filtering show that relative magnitudes of overbank floods in the Driftless Area were relatively small from about 4500 to 3000 yr B.P. but increased considerably in magnitude from about 3000 to 1800 yr B.P. (Fig. 9*b*). Between about 1800 and 1000 yr B.P. magnitudes of overbank floods were slightly reduced from immediately earlier magnitudes, but soon after 1000 yr B.P. and until about 500 year B.P. overbank floods again increased in magnitude. Because the data are not uniformly distributed over time, the smoothed dates must be interpreted with caution due to the tendency for clusters in the base data to overemphasize the importance of certain dates and to distort the timing of a change in trend. Nevertheless, the general similarity of long-term variation in the magnitudes of high-frequency floods (Fig. 9*a*) and low-frequency floods (Fig. 9*b*) suggests a common causal connection related to climatic change. The increased magnitudes of overbank floods after 3000 yr B.P. and again after 1000 yr B.P. are associated with climatic changes toward cooling, a condition that often is followed by an increase in the magnitude and frequency of floods in the Upper Mississippi Valley (Knox et al., 1975). Other traditional factors can be ruled out. The region has not been influenced by either tectonic or eustatic events during the Holocene, and human influence has only been hydrologically significant during approximately the past 150 yr. The indirect effects of climatic change, expressed through vegetation response to changes in temperature and moisture, may have contributed to the long-term variation in magnitudes and frequencies of floods, but the relatively modest and progressively slow changes in Holocene vegetation (Wright, 1976) do not correspond well with the somewhat abrupt and high-amplitude shifts in magnitudes and frequencies of Holocene floods depicted on Figures 9*a* and 9*b*. It appears that Holocene floods responded relatively directly to climatic events that influenced the amounts of winter snowfall and early summer rainfall much in the same way that seasonal concentration and magnitudes and frequencies of modern floods are closely related to these two factors.

The magnitudes of long-term Holocene variations in high-frequency floods of 1–2 yr expected return frequency ranged from about 10–15% larger to 20–30% smaller than contemporary floods of the same recurrence frequency. Estimating between-episode percentage deviations in magnitudes of overbank floods is not possible because reliable discharge estimates cannot be determined at this time. However, a general impression of the relative variation is given by the range limits of flood depth ratios for overbank floods. For example, the largest observed overbank floods during the period 4500–3000 yr B.P. were associated with flood depth ratios of about 2.0, a value that equates with a recurrence interval of approximately once in 50 yr for historically observed floods. However, since about 3000 yr B.P. the largest overbank floods commonly were associated with flood depth ratios of 3.0 or larger on the unsmoothed data. These flood depth ratios represent floods that probably would be expected to recur less than once in 100 yr with respect to modern floods. The changing range limits for flood depth ratios indicates that, as in the case of high-frequency floods of 1–2 yr return frequency, the mean and variance of low-frequency overbank floods have not remained stationary during time intervals spanning many centuries and longer.

CONCLUSION

Natural floods in the Upper Mississippi Valley result from snowmelt, excessive rainfall, and various combinations of snowmelt and rainfall. Nearly 75% of all floods occur during the months of March through July, with March and June being the most important. When only annual maximum flood series are considered, March accounts for about 30% of the total, with June being the next most important month accounting for about 15% of the total. Relatively few floods occur during the winter months of December through February when moisture is temporarily stored in snow cover. Late summer July and August floods are fewer than early summer floods because the importance of large storm frontal rainfall decreases, soil moisture is usually relatively low, and vegetation cover is at a maximum as agricultural crops mature. Although the shift to cool season atmospheric regimes in September is associated with increased precipitation and flooding, the magnitudes and frequencies of floods are relatively small because, as in July and August, soil moisture is usually relatively low and vegetation cover is at a maximum. The latter two factors also explain the low frequency of October and November floods, although reduced precipitation totals then also account for part of the reduction in magnitudes and frequencies of floods.

Statistical analyses show that magnitudes of annual maximum floods correlate best with magnitudes of winter snow depth and early summer rainfall. Nonsignificant statistical relationships between annual maximum floods and precipitation during the preceding summer and fall indicate that most annual maximum floods do not have a long memory of antecedent moisture conditions. However, the very large annual maximum floods that were more than 0.5 standard deviations above the mean were favored when above average winter precipitation was preceded by above average precipitation during the preceding summer. As might be expected, years with flood magnitudes falling more than 0.5 standard deviations below the mean tend to be relatively warm and dry, especially in summer. There is a tendency for above average winter temperatures and below average spring and summer temperatures to favor larger annual maximum floods, but relationships are weak.

The magnitudes of floods in the Upper Mississippi Valley reflect their season occurrence. The largest floods in watersheds that drain less than several hundred square kilometers tend to occur in summer following high-intensity and/or long-duration rainfalls, even though the most frequent annual maximum flood is related to snowmelt. Large differences in magnitudes of floods of a given recurrence interval occur between snowmelt and summer rainfall floods in these tributary watersheds, suggesting that flood analyses should be partitioned to prevent mixing of unlike seasonal sample populations. Major temporal changes in the magnitude and frequency of floods may occur due to changes in seasonal precipitation, in spite of insignificant change in total annual precipitation. On very large watersheds differences in seasonal floods are smaller as most floods tend to be associated with a similar generating process. For example, most floods, including the largest floods, on the mainstem Mississippi River and its major tributaries tend to occur in spring during snowmelt runoff.

Short-term changes in flood seasonality suggest that the mean and variance of flood series often do not remain stationary as the time scale lengthens. Lack of stationarity in the mean and variance of flood series is of interest because most techniques for predicting magnitudes and frequencies of floods assume that the mean and variance of the flood series does not change significantly over time. A century long record of floods on the mainstem Mississippi River shows that relatively high frequencies of large floods occurred in the late 1800s before about 1895 and since about 1950, and these large floods were associated with mean annual temperatures that were slightly below the long-term mean and annual precipitation that was above the long-term mean. Whereas climatic change was determined to be primarily responsible for four statistically significant different flood episodes in the more than century long flood history on the Mississippi River, most records of flood observations are too short to reliably evaluate the role of climatic change to influence magnitudes and frequencies of floods.

Long-term Holocene proxy records of high-frequency floods, reconstructed from dimensions of relict stream channels, also indicate nonstationarity of the mean and variance of flood series. Climatic changes during the Holocene are thought to be responsible because the episodic behavior of the floods broadly matches variations in climate indicated from fossil pollen in the same region. The magnitudes of floods with an expected recurrence of 1–2 yr experienced relatively large and persistent (nonrandom) departures from contemporary long-term average flood magnitudes. These floods ranged from about 10–15% larger to 20–30% smaller than equivalent frequency modern floods. Relatively large floods were dominant between about 6000–4500 and 3000–2000 yr B.P. and briefly after about 1200 yr B.P. Relatively small floods were dominant between about 8000–6500, 4500–3000, and 2000–1200 yr B.P. These reconstructions probably represent the Holocene history of changes in winter snowfall since floods of 1–2 yr recurrence frequency are dominated by snowmelt runoff.

Long-term Holocene proxy records of overbank floods with return frequencies ranging from once in a few years to once in more than 100 years also indicate nonstationarity of the mean and variance. The water depths of these floods, which are estimated from competent flow depths required to transport the coarsest sediments found in overbank flood deposits, have a history of variation that closely resembles the flood history described above for floods with 1–2 yr recurrence frequency. The available stratigraphic history of overbank floods from about 6000 yr B.P. indicates that large depths of overbank floods occurred immediately prior to 4500 yr B.P., between 3000 and 1800 yr B.P., and between about 1000 and 500 yr B.P. Small depths of overbank floods were prevalent between about 4500 and 3000 yr B.P., 1800 and 1000 yr B.P., and a trend toward smaller depths of overbank floods has occurred since about 500 yr B.P. Because the overbank floods are mostly a result of summer flooding during occurrences of intense and/or long-duration rainfall, the similarity between the pattern of long-term variation in high- and low-frequency floods is surprising. Allowing for short-term differences in beginning and ending dates of episodes, it is apparent that both winter snowfall and early summer rainfall were responding similarly to the broad climatic changes suggested in the regional record of fossil pollen. Since contemporary flood records often show that patterns of variation in snowmelt and rainfall floods are out of phase on decadal time scales, the time scale of resolution for the Holocene records is apparently more broadly based. A similar relationship has been identified in long-term Northern Hemisphere average temperature variations. Jones and others (1982) reported all seasons showed similar trends of departure from average since the 1880s when the time scale was greater than 20 yr, but for shorter time scales they observed noticeable differences in the pattern of temperature variations between seasons.

In conclusion, the magnitudes and frequencies of floods in the Upper Mississippi Valley show nonrandom behavior during historical and Holocene (postglacial) time. Correlation of flood characteristics with records of historical climate and with fossil pollen indicates that climatic change is the principal cause of the nonrandom behavior. Changes in vegetation, responding to climatic change, probably contributed modestly to flood responses, but the strong association of flood magnitudes with the amount of winter snowfall and the amount of early summer rainfall indicates that annual maximum and most large floods tend to respond best to the direct effects of climatic activity. This assertion is supported by the lack of statistical significance between annual maximum flood magnitudes and antecedent environmental conditions. Although many probably believe that humid climate regions, such as the Upper Mississippi Valley, are places where geomorphic and hydrologic activity are insensitive to perceived subtle climatic change, the results of this evaluation show that relatively large climatically driven adjustments in magnitudes and frequencies of floods have occurred during Holocene and historical time. It is important to note that the climatic changes described here represent a magnitude that approximates climatic changes estimated as possible in response to increasing atmospheric carbon dioxide between now and 2100 (Kerr, 1983). Although climatic and watershed environmental conditions in future years cannot replicate past Holocene environments, the Holocene and historical hydrologic responses to climatic variations may provide useful guides for speculation about hydrologic responses to possible future climatic changes. It seems reasonable to question the assumption of stationarity of the mean and variance, which forms the basis for use of many equations for hydrologic forcasting. The assumption commonly is unrealistic in the Upper Mississippi Valley when the time scale expands beyond more than a century, and probably is nearly always unrealistic there when time scales involve more than a few thousand years.

ACKNOWLEDGMENTS

This research was supported by the National Science Foundation, Grant EAR-8306171. Cartographic assistance was provided by the University of Wisconsin College of Letters and Science Support Fund administered by the Geography Department's Cartography Laboratory. I thank Richard Dunning and Frank Magilligan who assisted with field investigations of Holocene paleofloods. I also thank P. C. Patton and D. F. Ritter for very helpful comments on a first draft of the manuscript.

REFERENCES

Baker, V. R. (1984). Flood sedimentation in bedrock fluvial systems. *Mem.—Can. Soc. Pet. Geol.* **10**, 87–98.

Baker, V. R., and Ritter, D. F. (1975). Competence of rivers to transport coarse bedload material. *Geol. Soc. Am. Bull.* **86**, 975–978.

Baker, V. R., Kochel, R. C., Patton, P. C., and Pickup, G. (1983). Palaeohydrologic analysis of Holocene flood slack-water sediments. *Spec. Publ. Int. Assoc. Sedimentol.* **6**, 229–239.

Bartlein, P. J., and Webb, T. (1982). Holocene climatic changes estimated from pollen data from the northern Midwest. *In* "Quaternary History of the Driftless Area" (J. C. Knox, L. Clayton, and D. M. Mickelson, eds.), Wis. Geol. Surv. Guide Book No. 5, pp. 83–87, Wisconsin Geological Survey.

Bartlein, P. J., Webb, T., and Fleri, E. (1984). Holocene climatic change in the northern Midwest: Pollen-derived estimates. *Quat. Res.* **22**, 361–374.

Belt, C. B. (1975). The 1973 flood and man's constriction of the Mississippi River. *Science* **189**, 681–684.

Bernabo, J. C., and Webb, T. (1977). Changing patterns in the Holocene pollen record of northeastern North America: A mapped summary. *Quat. Res.* **8**, 64–96.

Borchert, J. R. (1950). The climate of the central North American grassland. *Ann. Assoc. Am. Geogr.* **40**, 1–39.

Bras, R. L. (1980). The benefits of uncertainty: Probability in short and long term hydrologic forecasting. *In* "Improved Hydrologic Forecasting—Why and How," Proc. Eng. Found. Conf., 1979, pp. 141–159. Am. Soc. Civ. Eng., New York.

Bryson, R. A. (1966). Air masses, streamlines, and the boreal forest. *Geogr. Bull.* **8**, 228–269.

Costa, J. E. (1978). Holocene stratigraphy in flood frequency analysis. *Water Resour. Res.* **14**, 626–632.

Costa, J. E. (1983). Paleohydraulic reconstruction of flash-flood peaks from boulder deposits in the Colorado Front Range. *Geol. Soc. Am. Bull.* **94**, 986–1004.

Davis, A. M. (1977). The prairie-deciduous ecotone in the Upper Middle West. *Ann. Assoc. Am. Geogr.* **67**, 204–213.

Fahnestock, R. K. (1963). Morphology and hydrology of a glacial stream—White River, Mount Rainier, Washington. *Geol. Surv. Prof. Pap. (U.S.)* **422-A**, A1–A70.

Jones, P. D., Wigley, T. M. L., and Kelly, P. M. (1982). Variations in surface air temperatures. Part I. Northern Hemisphere, 1881–1980. *Mon. Weather Rev.* **110**, 59–70.

Kellerhals, R. (1967). Stable channels with gravel-paved beds. *J. Waterways, Harbors Coastal Eng. Div., Am. Soc. Civ. Eng.* **93**, 63–84.

Kerr, R. A. (1983). Carbon dioxide and a changing climate. *Science* **222**, 491.

King, J. E. (1981). Late quaternary vegetational history of Illinois. *Ecol. Monogr.* **51**, 43–62.

Knox, J. C. (1980). Geomorphic evidence of frequent and extreme floods. *In* "Improved Hydrologic Forecasting—Why and How,"Proc. Eng. Found. Conf., 1979, pp. 220–238. Am. Soc. Civ. Eng., New York.

Knox, J. C. (1983). Responses of river systems to Holocene climates. *In* "Late-Quaternary Environments of the United States" (H. E. Wright, Jr., ed.), vol. 2, pp. 26–41. Univ. of Minnesota Press, Minneapolis.

Knox, J. C. (1984). Fluvial responses to small scale climate changes. *In* "Developments and Applications of Geomorphology" (J. E. Costa and P. J. Fleisher, eds.), pp. 318–342. Springer-Verlag, Berlin and New York.

Knox, J. C. (1985). Responses of floods to Holocene climatic change in the Upper Mississippi Valley. *Quat. Res.* **23**, 287–300.

Knox, J. C., Bartlein, P. J., Hirschboeck, K. K., and Muckenhirn, R. J. (1975). The response of floods and sediment yields to climatic variation and land use in the Upper Mississippi Valley. *Univ. Wis. Inst. Environ. Stud. Rep.* **52**, 1–76.

Kochel, R. C., and Baker, V. R. (1982). Paleoflood hydrology. *Science* **215**, 353–361.

Lane, E. W., and Carlson, E. J. (1953). Some factors affecting the stability of canals constructed in coarse granular materials. *Proc. Int. Assoc. Hydraul. Res. Congr., 5th*, pp. 37–48.

Maher, L. J. (1982). The palynology of Devils Lake, Sauk County, Wisconsin. *In* "Quaternary History of the Driftless Area" (J. C. Knox, L. Clayton, and D. M. Mickelson, eds.), Wis. Geol. Surv. Field Trip Guide Book No. 5, pp. 119–135.

Patton, P. C., Baker, V. R., and Kochel, R. C. (1979). Slack-water deposits: a geomorphic technique for the interpretation of fluvial paleohydrology. *In* "Adjustments of the Fluvial System" (D. D. Rhodes and G. P. Williams, eds.), pp. 225–252. Kendall-Hunt Publ. Co., Dubuque, Iowa.

Scott, K. M., and Gravlee, G. C., Jr. (1968). Flood surge on the Rubicon River, California—hydrology, hydraulics and boulder transport. *Geol. Surv. Prof. Pap. (U.S.)* **422-M**, M1–M40.

Sorenson, C. J. (1977). Reconstructed Holocene bioclimates. *Ann. Assoc. Am. Geogr.* **67**, 214–222.

Sorenson, C. J., Knox, J. C., Larsen, J. A., and Bryson, R. A. (1971). Paleosols and the forest border in Keewatin, N. W. T. *Quat. Res.* **1**, 468–473.

Thornbury, W. D. (1965). "Regional Geomorphology of the United States." Wiley, New York.

Upper Mississippi River Comprehensive Basin Study Coordinating Committee (1970). "Upper Mississippi River Comprehensive Basin Study, Appendix C: Climatology and Meteorology," Vol. 3, pp. C1–C56. U.S. Army Corps of Engineers, St. Paul, Minnesota.

U.S. Water Resources Council (1981). "Guidelines for Determining Flood Flow Frequency," Bull. No. 17B. Hydrology Committee, Washington, D.C.

Williams, G. P. (1978). Bank-full discharge of rivers. *Water Resour. Res.* **14**, 1141–1154.

Williams, G. P. (1983a). Paleohydrological methods and some examples from Swedish fluvial environments. I. Cobble and boulder deposits. *Geogr. Ann.* **65**, 227–243.

Williams, G. P. (1983b). Improper use of regression equations in earth sciences. *Geology* **11**, 195–197.

Williams, G. P. (1984). Paleohydrological methods and some examples from Swedish fluvial environments. II. River meanders. *Geogr. Ann.* **66**, 89–102.

Winkler, M. G. (1985). Late-glacial and Holocene environmental history of south-central Wisconsin: A study of upland and wetland ecosystems. Ph.D. Thesis, University of Wisconsin, Madison.

Wisler, C. O., and Brater, E. F. (1959). "Hydrology," 2nd ed. Wiley, New York.

Wolman, M. G., and Eiler, J. P. (1958). Reconnaissance study of erosion and deposition produced by the flood of August 1955 in Connecticut. *Trans., Am. Geophys. Union* **39**, 1–14.

Wolman, M. G., and Leopold, L. B. (1957). River flood plains: Some observations on their formation. *Geol. Surv. Prof. Pap. (U.S.)* **282-C**, 87–109.

Wright, H. E., Jr. (1976). The dynamic nature of Holocene vegetation: A problem in paleoclimatology, biogeography, and stratigraphic nomenclature. *Quat. Res.* **6**, 581–596.

Wright, H. E., Jr., Winter, T. C., and Patten, H. L. (1963). Two pollen diagrams from southeastern Minnesota: Problems in the late and postglacial vegetational history. *Geol. Soc. Am. Bull.* **74**, 1371–1396.

18

LARGE FLOODS AS GEOMORPHIC EVENTS IN THE HUMID TROPICS

AVIJIT GUPTA
Department of Geography, National University of Singapore, Singapore

INTRODUCTION

In a study of the effectiveness of climate in watershed geomorphology, Wolman and Gerson came to the conclusion that the frequency of high-magnitude formative events is apparently high for humid tropical regions. This attributes considerable importance to the role played by high-magnitude episodic events in tropical geomorphology, especially as denudation caused by a major storm in montane wet tropics may approach the mean annual erosional rate (Wolman and Gerson, 1978). One wonders whether such episodic events are also reflected in stream channel morphology and behavior. If the frequency of such events is comparatively high for the humid tropics, as suspected by Wolman and Gerson (1978), the recovery time between two such events would be brief, and the flood-related forms may turn out to be characteristic channel features (Brunsden and Thornes, 1979) instead of being postflood transients. The objective of this chapter is to examine this possibility.

High-magnitude flood response is prompted by certain physiographic factors: hillslope morphology, soil and rock types, drainage density, and the size of the watershed (Baker, 1977). When suitable basin physiography is combined with flood size and flood periodicity, flood effects in the channels are not only heightened but also show a measure of permanency (Gupta, 1983). A pronounced disparity between the competence and efficiency of large floods and postflood flows would tend to perpetuate the flood effects. It has been suggested that a ratio of flood peak to mean discharge or a similar measure should be indicative of the effect on channel morphology by large floods (Schumm, 1977).

In the tropics there are stream systems where flood forms may persist over time as a result of the combination of both the climatic and physiographic factors already mentioned (Gupta, 1983). This hypothesis is in contrast to the traditional belief in geomorphology that streams in the tropics are unimportant as denudational agents (Thomas, 1974). On the other hand the role of episodic mass movements has been frequently recognized (Simonett, 1967; Mousinho de Meis and da Silva, 1968; So, 1971; Jones, 1973). Starkel's account of the 1968 storm-related events of the Darjeeling Himalayas is an excellent appraisal of the importance of high-magnitude floods in the humid tropics (Starkel, 1972a). Recently, more work has focused on the geomorphic role of large floods in the tropics (Gole and Chitale, 1966; Coleman, 1969; Starkel, 1972b; Temple and Sundborg, 1972; Gupta, 1975; Brunsden et al., 1981; Goswami, 1982; Bristow, 1987). Nevertheless, it is correct to state that the case studies from the tropics are fascinating but few. This chapter reviews the available literature in order to determine (1) whether one should expect at least semipermanent erosional and depositional flood features in some streams and (2) if so, what is the nature of the forms and sediment.

In spite of the recurrence of large floods in at least part of the humid tropics, long-term flood records are difficult to come by. It is therefore hard to define a "large flood" adequately, either by the recurrence interval or by some kind of standardized measure such as unit discharge ($m^3/s/km^2$). For this discussion a large flood is perceived as a high flow for which the stream channel is a manifestly inadequate transportation system and whose passage involves at least the lower part of the valley flat. In some places its stage is marked by terracelike features (Gupta, 1975; Brunsden et al., 1981), and in eastern Jamaica it has been possible to associate such a terrace with a 10-yr recurrence interval (Gupta, 1975). Such a recurrence interval is perhaps an acceptable definition for a large flood as a starting point in the investigation.

The tropics is defined by meteorologists as the zone between the subtropical anticyclones near 30° north and south latitudes (Reynolds, 1985). The humid tropics is taken as that part of the tropical world where, but for human interference, forests would prevail. Such forests may vary from a rain forest to a monsoon forest or a savanna forest (Richards, 1957), and we should expect a lower limit of about 1000 mm of rain annually. Usually the rainfall in the case studies discussed is much heavier, and as expected some of the rain falls with an intensity and magnitude high enough to cause large floods. This chapter does not deal with flooding of tropical high mountain streams from summer snowmelts, nor does it concentrate on the effect of a single flood with an extremely high recurrence interval like 100 yr. It concentrates on the effect of the periodic occurrences of large floods in river valleys resulting from episodic meteorological events. Such floods are large in magnitude and occur on a regular frequency of several years.

CLIMATOLOGIC BACKGROUND OF LARGE FLOODS

Rainfall in the tropics is associated with several types of synoptic phenomena: (1) tropical cyclones, (2) the monsoon system of winds, (3) the easterly waves, and (4) the Inter-Tropical Convergence Zone (ITCZ). A high annual total may arise out of a combination of any of these phenomena, but episodic high-magnitude rainfall events are usually from the first three.

The most impressive high-magnitude events in the tropics are the tropical cyclones. Such storms tend to occur in certain geographical areas, and Table 1 lists the annual frequency of such storms according to their geographic distribution. Tropical cyclones are important as a source of precipitation north of 10°N due to the varying nature of the Coriolis force and related duration of the cyclonic vorticity and as hurricanes can form only over water whose mean temperature is above 26–27°C. This usually restricts hurricane formation to summer and to certain parts of the tropical oceans. Very few comparable disturbances develop beyond 30° of latitude.

Hurricanes have caused some of the world's heaviest rainfall (Table 2), especially when storms have stalled or orographic barriers have been confronted. Decaying storms are more likely to cause flooding than intense ones in rapid motion. Tropical cyclones also tend to concentrate over specific geographic areas (Fig. 1). For example, over a period of 70 yr (1884–1953), 22 storms on average occurred annually in an area off the coasts of South China, the Philippines, and Vietnam bounded by latitudes 5°–30° N and longitudes 105°–150° E. In coastal areas of Vietnam, typhoons may cause 400 mm of rain to fall in 24 hr (Nieuwolt, 1981), and in other parts of the world even higher figures have been recorded (Table 3).

Detailed regional studies, where available, indicate the expected nature of such high rainfall. In an area of over 280 km^2 in eastern Jamaica, Lirios (1969) estimated the 5-yr 24-hr rainfall to be between 250 and 350 mm. For a 25-yr recurrence interval his estimates rise to figures between 350 and 600 mm (Lirios, 1969). Vickers (1967) has estimated nearly 450 mm for a maximum 24-hr rainfall for a small watershed in the southeastern parts of the island. Other calculations have shown very similar figures (Evans, 1972; Gupta, 1983). If such figures are typical, then episodic rainfall of extremely high magnitude can occur in the Caribbean and adjoining areas from time to time.

During the southwestern monsoon period (June to September), the eastern coast of India is affected by a series of depressions. Widespread rain occurs in the southwestern quadrant of such depressions (Rao, 1981), the rain being

TABLE 1 Estimated Annual Frequency of Tropical Cyclones

Geographical Area	Frequency According to Simpson and Riehl (1981)	Reynolds (1985)
Northwest Pacific	22	26
Northeast Pacific	15	13
Australian seas	5	10
Northwest Atlantic	8	9
Southern Indian Ocean	5	8
Northern Indian Ocean	8	6
South Pacific	5	6

Note: The general pattern of frequency distribution is comparable, although the estimated figures vary.

TABLE 2 Some Very Heavy Rainfall in the Humid Tropics

Date	Location	Amount (mm)	Cause	Source
Mar. 11–19, 1952	Cilaos, Reunion	4130		Landsberg, in Flores and Balagot (1969)
June 9–16, 1876	Cherrapunji, India	3388	Monsoon depression	Jennings (1950)
June 24–30, 1931	Cherrapunji, India	3213	Monsoon depression	Jennings (1950)
Jan. 22–25, 1960	Bowden Pen, Jamaica	2789	Frontal	Vickers (1967)
Nov. 4–11, 1909	Cinchona, Jamaica	2287	Hurricane	Jamaica Weather Reports (1909)
July 14–20, 1911	Baguio, Philippines	2210	Tropical storm	Jennings (1950)
July 18–20, 1913	Funkiko, Taiwan	2071	Tropical storm	Jennings (1950)
Oct. 3–8, 1963	Tacajo, Cuba	2025	Hurricane Flora	
Oct. 5–7, 1963	Silver Hill, Jamaica	1524	Hurricane Flora	Vickers (1967)

periodically concentrated by precipitation pulses of higher intensity. There have been 565 monsoon depressions over a period of 80 yr (1890–1970), which gives an average of 7 per year. According to another estimate (Sikka, 1977), 8–10 storms occur in an average season with a lifetime between 2–5 days. Monsoonal storms up to 9 days long have occasionally been recorded. Such storms are usually large, 1000–1300 km across, with a 7- to 9-km-deep cyclonic circulation with an anticyclonic outflow above (Sikka, 1977). Those episodic high rainfall events arrive in the middle of a wet period, the southwest monsoon, when the soil is saturated and the rivers already high.

Some of the monsoon depressions grow into more powerful tropical cyclones. Such cyclones, however, are not entirely absent at other times of the year. Over the Indian subcontinent, in the 70 yr between 1891 and 1960, 396 such storms took place with an average of 6 a year. About 70% of these storms, like the less powerful monsoon depressions, occur between June and November, a time of saturated ground and high streamflow. Together with the monsoon depressions already mentioned, these are responsible for flooding of streams at intervals.

Easterly waves are perturbations in the easterly trade winds that in places may cause copious rainfall, especially if convergence takes place in the vicinity of favorable topography (Snow, 1976). Easterly waves have given rise to heavy rainfall in the Caribbean (Vickers, 1967) and in the eastern Philippines along orographic barriers (Flores and Balagot, 1969). A number of easterly waves may lead to formation of hurricanes (Alaka, 1976). Stationary thundershowers may also produce copious rainfall. The May 1960 flooding in the Virgin Islands was due to stationary thundershowers that produced 390 mm of rain in 3 days (Portig, 1976). Other features like orographic barriers in

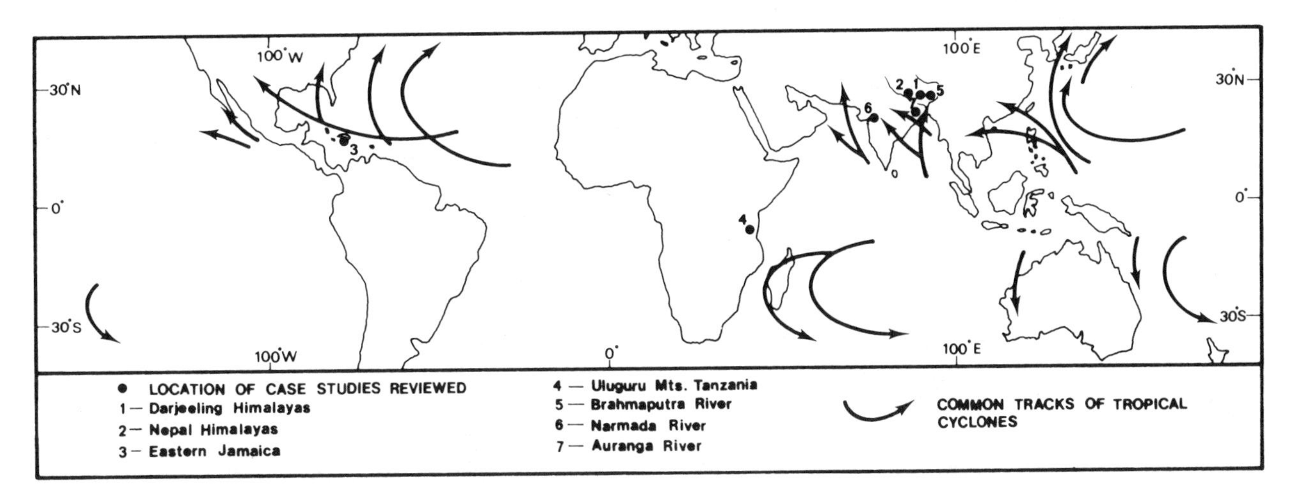

FIGURE 1. Map showing the location of case studies reviewed in this chapter and generalized tracks of tropical cyclones.

TABLE 3 Some Very Heavy 1-Day Rainfall in the Humid Tropics

Date	Location	Amount (mm)	Cause	Source
Mar. 15–16, 1952	Cilaos, Reunion	1870		Landsberg, footnote in Flores and Balagot (1969)
Sept. 10–11, 1963	Pai Shih, Taiwan	1248	Typhoon	Landsberg, footnote in Flores and Balagot (1969)
July 14–15, 1911	Baguio, Philippines	1168	Typhoon	Flores and Balagot (1969)
Jan. 22, 1960	Bowden Pen, Jamaica	977	Frontal	Vickers (1967)
Jan. 23, 1960	Bowden Pen, Jamaica	1109	Frontal	Vickers (1967)
June 14, 1876	Cherrapunji, India	1036	Monsoon depression	Das (1968)
Aug. 31, 1911	Funkiko, Taiwan	1034		Jennings (1950)
Nov. 6, 1909	Silver Hill, Jamaica	775	Incipient hurricane	Vickers (1967)
May 25, 1889	Cinchona, Jamaica	718	Mixed	Vickers (1967)
	Rochambeau, Fr. Guyana	596		Snow (1976)
	Nkhotakota, Malawi	570		Torrance (1972)
	Bombay, India	548		Rao (1981)
	Jaffna, Sri Lanka	520		Rao (1981)
	Zomba, Malawi	509		Torrance (1972)
	Diego-Suarez, Madagascar	508		Griffiths and Ranaivoson (1972)
	Santo Domingo, Dom. Republic	508		Portig (1976)
	Darjeeling, India	493	Monsoon depression	Rao (1981)
	Dehradun, India	490		Dhar et al. (1984)
	Bhuj, India	468		Rao (1981)
	Akyab, Burma	465		Nieuwolt (1981)
	Tana Tave, Madagascar	442		Griffiths and Ranaivoson (1972)
April 10, 1955	Rungwe, Tanzania	425		Temple and Rapp (1972)
Oct. 1966	Vango, Kenya	404		Griffiths (1972)

1. There is some discrepancy in the data: Some are 24-hr precipitation figures, whereas others are maximum daily amounts.
2. One also has to consider some variation in the quality of the data.
3. Only *some* examples above 400 mm have been included, but the table probably suffices to illustrate that in certain parts of the world high-magnitude rainfall is to be expected.
4. It may be inferred that in most cases the rainfall continued for more than one day.
5. Rainfall figures rounded off to the nearest millimeter.

the path of southward-moving cold fronts may also cause high-magnitude rainfall and consequential flooding, especially in the Caribbean area.

AREAS WITH HIGH FLOOD POTENTIAL IN THE TROPICS

The geographic locations experiencing high-magnitude rainfall with a limited recurrence interval are therefore mostly between the 10° and 30° latitudes. Within these areas, again, certain regions are more likely to be affected, particularly where orographic barriers run across the frequent paths of tropical storms as in the Philippines or in eastern Jamaica. One may also recall Hurricane Hazel over Puerto Rico in October 1954 or the March 1959 hurricanes in Madagascar (Simpson and Riehl, 1981).

It is near impossible to determine areas of high flood potential on the basis of stream gauging records because such records for large floods in the humid tropics are extremely limited both in distribution and duration. Even where stream gauging records have existed for a period, there is often a gap regarding larger floods. For example, the Tranquility gauging station over the Buff Bay River in northeastern Jamaica failed during the October 1963 flood, which was caused by rain from Hurricane Flora. Several stations in the tropics do have long-term stream records like the Brahmaputra at Pandu (drainage area 424,309 km^2), where it carries a mean annual flood discharge of 51,156 m^3/s or 0.12 $m^3/s/km^2$ (Goswami, 1982). But these gauging records are usually over large rivers, and records for smaller streams are extremely rare.

The recurrence of high-magnitude rainfall is better documented. Assuming that the occurrence of high-precipitation storms and large floods are related, one may expect periodic occurrences of such floods in the geographic areas described earlier. It is difficult to specifically map these zones, but rivers with at least semipermanent flood morphology probably occur in parts of the following areas, provided the necessary geologic and physiographic conditions as described by Baker (1977) are present:

1. River valleys of East Asia, especially where transverse orographic barriers exist as in Taiwan and the Philippines.
2. Parts of Southeast Asia, especially Vietnam, Sumatra and Java (orographic barriers against monsoon rainfall), and Burma.
3. The more humid parts of the Indian subcontinent.
4. Madagascar and neighboring parts of coastal East Africa (orographic barriers against tropical storm tracks).
5. North and northeast Australia.
6. Islands and coastal areas of the Caribbean, the Gulf of Mexico, and tropical or semitropical North Atlantic.

This is an extremely tentative list, and it is not suggested that all the rivers in these regions display flood characteristics. Even in the type area of eastern Jamaica, such characteristics are absent from a few rivers. But it is possible (1) that given the suitable geologic and physiographic environment, flood effects (Gupta, 1983; Table 4) will be present in at least some rivers and (2) that there might be a regional correlation between the size and periodicity of the large floods and the persistence of flood morphology. For example, in the two areas with suitable environment known to the author, eastern Jamaica and eastern India, such morphology can be recognized in a large number of streams.

Figure 2 shows some flood discharges of Indian rivers plotted against the contributory drainage area, to prepare a diagram akin to a flood envelope curve. The case studies from which the data have been collected are referred to in Table 4. Even such an arbitrary collection of flood data from the available literature produces an envelope curve with values similar to that for the United States as shown by Moss and others (1978). There is obviously a need for further investigations with better data than compiled for this chapter. There is also a gap in the middle of the range of drainage area sizes. Even such crude data indicate that Indian rivers are flooded to a considerable extent, as one would expect from the climatology of the area. Figure 3 is a flood probability analysis of an Indian river for which long-term records are available (Ahuja and Majumdar, 1959). The 100-yr flood is about 2.5 times that of the 2.33-yr flood from annual maximum series.

FLOOD GEOMORPHOLOGY IN THE TROPICS

In spite of periodic recurrence of large floods in parts of the humid tropics, specific accounts of flood geomorphology are uncommon. There are passing references to the size of a large flood in the engineering or sedimentological literature, but the geomorphological work of a large flood is seldom adequately described. Such studies also tend to cover terrain of high relief rather than lower plains.

TABLE 4 Flood Discharge of Some Indian Rivers

River	Station	Date	Area (km^2)	Discharge (m^3/s)	Unit Discharge (m^3/s/ km^2)	Source
1. Brahmaputra	Pandu	1962	424,309	72,748 (highest on record)	0.17	Goswami (1982)
2. Mahanadi			132,090	46,000	0.35	Starkel (1972b)
3. Narmada			93,180	65,100	0.7	Starkel (1972b)
4. Kosi		1968	59,540	25,840	0.43	Starkel (1972b)
5. Sabarmati			54,160	10,874	0.2	Starkel (1972b)
6. Damodar			22,000	18,410	0.84	Starkel (1972b)
7. Baitarani	Akhuapada	1927	11,360	9,207 (85-yr flood)		Ahuja and Majumdar (1959)
8. Tista	Brigde	1968	7,200	27,500	3.8	Starkel (1972b)
9. Joba Khas			161	1,601	9.93	Starkel (1972b)
10. Gish			161	630	3.92	Starkel (1972b)
11. Little Rangit	Pulbazar	1968	75	5,625	75	Starkel (1972b)
12. Lish		1952	49.2	255	5.18	Starkel (1972b)
13. Pachim	Ringtong	1968	8.4	120	14.3	Starkel (1972b)
14. Posam	Poobong	1968	5	780	156	Starkel (1972b)

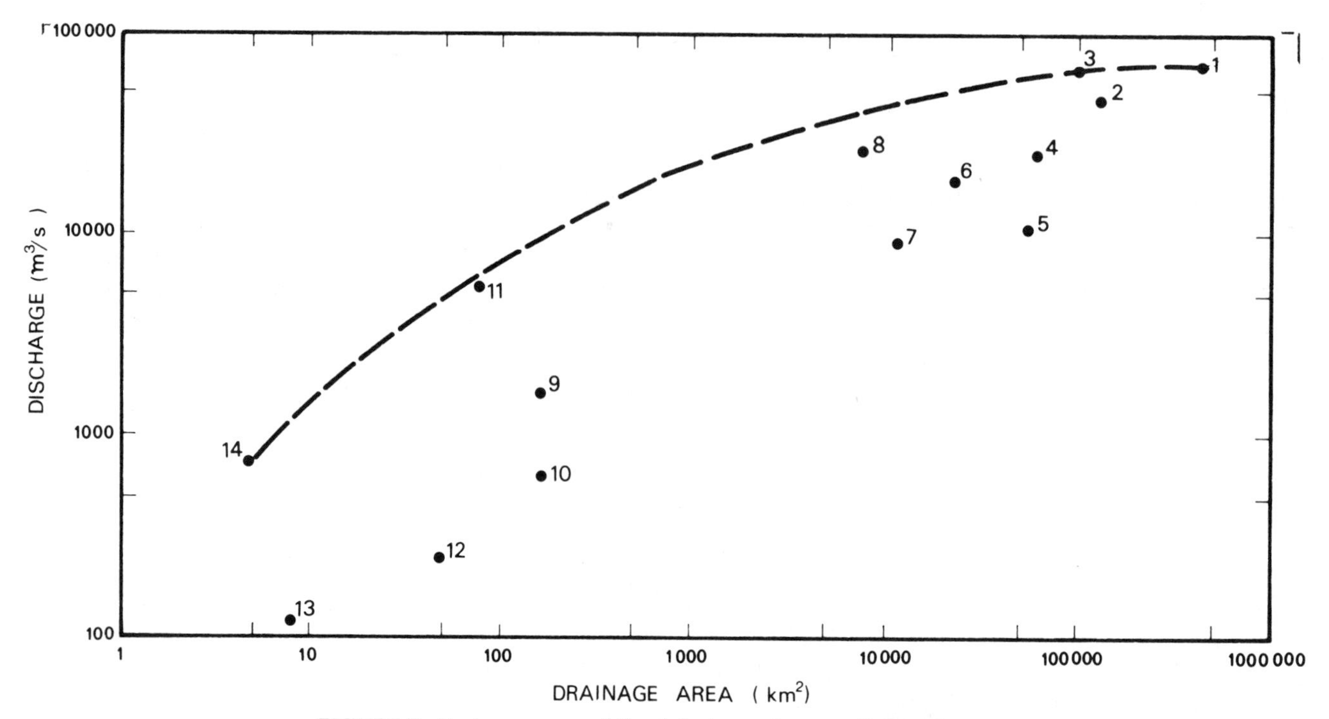

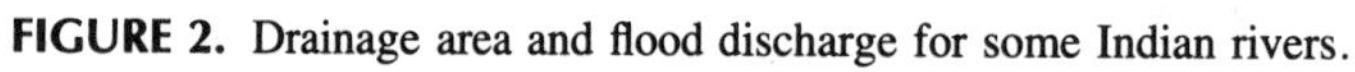

FIGURE 2. Drainage area and flood discharge for some Indian rivers.

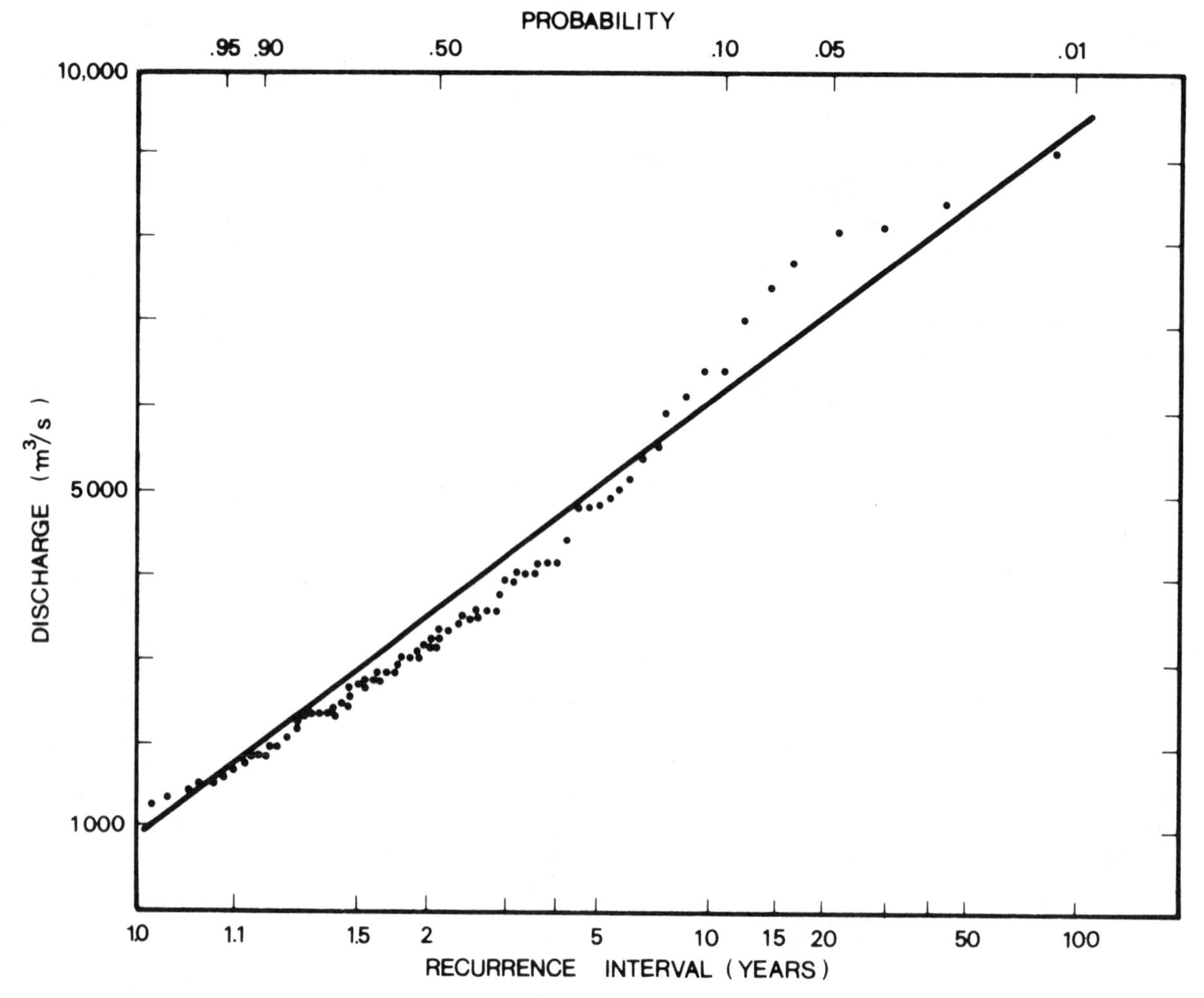

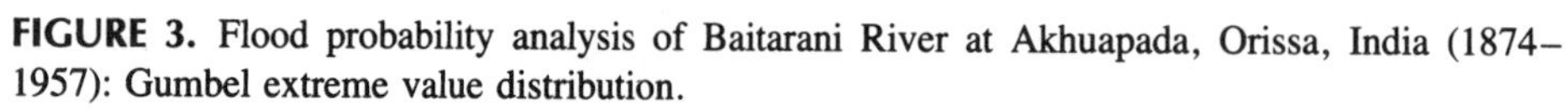

FIGURE 3. Flood probability analysis of Baitarani River at Akhuapada, Orissa, India (1874–1957): Gumbel extreme value distribution.

Good descriptive accounts of flood-related forms and processes are, however, available from India, Jamaica, and Tanzania. In this section summary accounts of such description will be presented with a view toward determining the nature and permanency of flood-related forms and sediment.

Indian Subcontinent Examples: Rivers in High-Relief Setting

Starkel (1972a) and Brunsden and others (1981) have described rivers in the high tropical mountains of India and Nepal that are periodically affected by large floods. There is a surprising similarity between their descriptions. Starkel described the effects of one incidence of high-magnitude rainfall from a Bay of Bengal tropical cyclone that deposited 700–1100 mm of water between October 2 and 5, 1968, in the Darjeeling Himalayas. Here the Lower Himalayas are a series of ridges at crest altitudes of 2200–3000 m, separated by 1000–1500-m-deep valleys. The major rivers, the Tista and the Great Rangit, have channels at only 100- to 300-m elevation. The slope is generally steep, convex to rectilinear on weathered rocks, with weak structural lineations. The annual rainfall of Darjeeling is 2758 mm, over 90% of which arrives between May and October (Rao, 1981), a period that includes days of exceptionally heavy rain. As a result the slopes prominently display mass movement scars, and the supply of sediment to the swollen rivers at the bottom of the valleys is high (Fig. 4). A wide channel, usually in rock, occupies the valley bottom, along with rock debris, 5–10 m-high terraces, and tributary cones. Boulders between 5 and 10 m are common. High-magnitude rainfall events like that of 1968 are superimposed on the general climatic pattern. Events similar to those of 1968 occurred in 1899, 1934, and 1950. The heaviest 3-day rainfall for Darjeeling at 2127 m is 965 mm, and Kurseong at about 1500 m has even higher figures. The 24-hr record there is 640 mm in 1968, while in 1950 it only took 36 hr to collect over 950 mm of rain.

The 1968 storm occurred toward the end of the wet monsoon when both soil and rivers were near their carrying capacities. Besides rock and debris fall, debris slides and avalanches, mudflows and the displacement of liquified soils by piping took place. The stream channels demonstrated extreme competence and capacity. Starkel has referred to small channels carrying boulders 3–5 m in diameter in general and up to 12 m in favorable locations. In fact, unit discharges apparently reached as high as 14.3 $m^3/s/km^2$ and 156 $m^3/s/km^2$ from watersheds of 8.4 and 5 km^2, respectively.

Headwater channels were deepened up to 2–3 m in rock, and the tributaries also widened their banks. Rivers like the Little Rangit with a watershed of 75 km^2 and a channel 25–100 m wide, increased the width in places by another 80–200 m. The tributaries supplied a massive amount of detritus to the larger rivers, which surprisingly were erosive enough to remove current valley mouth cones and even to dissect or rework older fans at such locations. An 8- to 10-m-high terrace, made of material ranging from boulders to sand, was created in valleys and in places was "heaped up with fan and tongues of masses from slopes" (Starkel, 1972a, p. 133). After the flood, streams with steep gradients often ended up with a valley flat filled with 2 to 3 m rounded chipped blocks of imbricated gneiss from which the finer fraction has been removed.

FIGURE 4. Landforms of the Darjeeling Himalayas: steep slopes, landslide scars, and deep valleys.

Starkel is of the opinion that high-magnitude events like this occur once in 20–25 yr in the Darjeeling Himalayas. Considerable mass movement on slopes and channel modification in the valleys also take place, albeit on a smaller scale, during the southwestern monsoon when 100–200 mm of rain may fall in 24 hr. However, the slopes recover and vegetation rapidly colonizes denuded patches. Lower down during such periods valley mouth alluvial fans are recreated, and depositional features are formed in the channels over and around the superflood boulders. Channel recovery is carried out by large but common flood discharges during the rainy season. An estimate of the nature of the work involved can be made because, according to Starkel, in ordinary monsoon year about 600–700 m^3/km^2 is denuded from these mountains, whereas this figure rises to 100,000 m^3/km^2 for a year like 1968 (Starkel, 1972a), an increase of over 150 times.

Very similar channel forms occur in the low Himalayas of eastern Nepal (Brunsden et al., 1981) where the physiography is comparable but annual rainfall is less (1000–2000 mm). Even then the 24-hr precipitation has been estimated to be as high as 114, 252, 332, and 450 mm for 2, 10, 25, and 100 yr, respectively. Figure 5, generalized from Brunsden and others (1981), indicates the general geomorphology of the river valleys in this area, where the valley floors consist of "a sediment filled bed (track) containing everything from silt to very coarse boulders (>2m), which acts as a continuous conveyor belt transportation system" (Brunsden et al., 1981, p. 51). The two major streams described were the Tamur and its tributary, the Leoti Khola. The unit discharge for tributary channels draining watersheds of tens of square kilometers may reach about 10 $m^3/s/km^2$ for a 10-yr flood, and twice that amount for a 100-yr one.

The recovery from such flooding is carried out by medium and high flows of the southwestern monsoon as in the Darjeeling Himalayas. Even an hourly rain of about 25 mm moves water and sediment in the channels effectively to ameliorate the effect of a large flood. Interestingly enough,

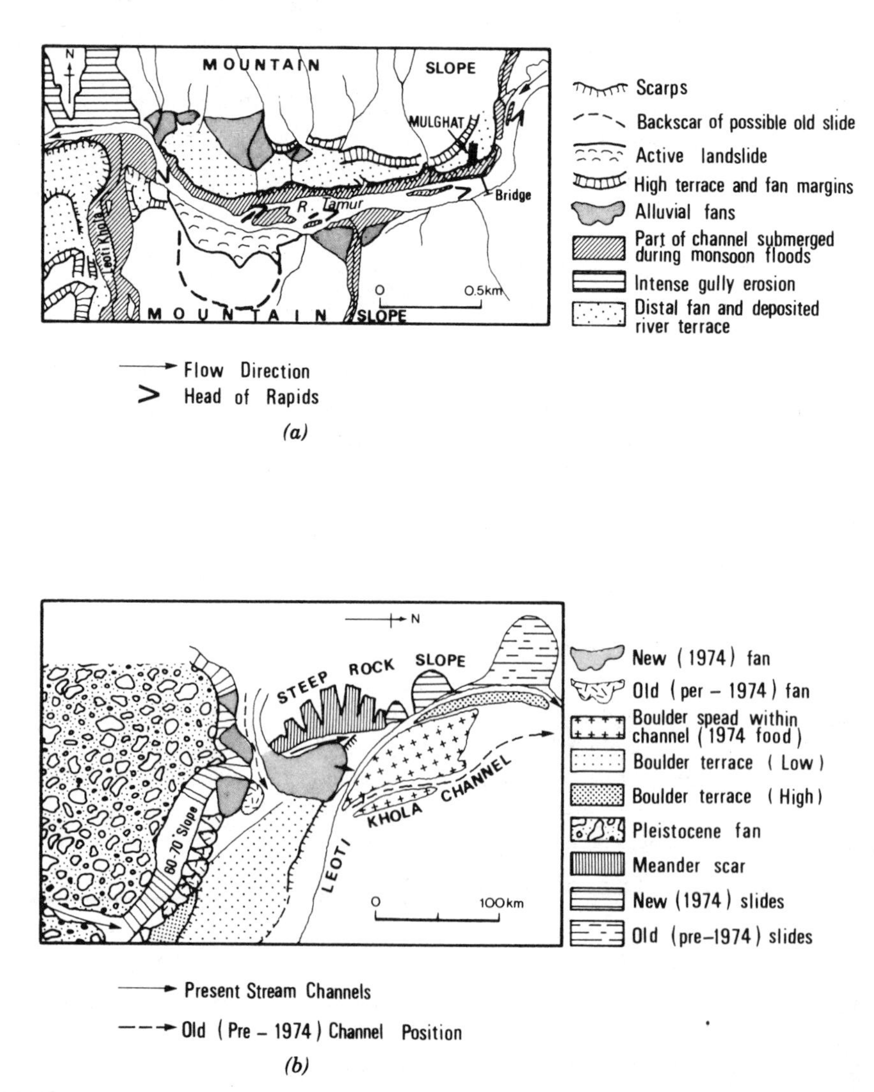

FIGURE 5. Flood geomorphology of river valleys from the Nepal Himalayas (generalized from Brunsden et al., 1981): (*a*) Physiography of the Tamur Valley near Mulghat; (*b*) left bank of the Leoti Khola showing changes due to the 1974 flood.

the bankfull discharges for the major stream Tamur arrive once every 20 yr on average, which is astonishingly different from the usual expectation of 1–2 yr. Is it possible that the flows in between two high-magnitude events are capable of resorting the sediment but not of reducing the channel section? The channel functions not only as a transporting conduit of water and sediment but also as a storehouse of sediment. Such sediment storage in bars, islands, terraces, and unvegetated fans is very effectively moved in near-bankful conditions every 20 yr or so and partly during almost every wet monsoon. Sediment stored in higher fans and vegetated flood terraces is affected at infrequent intervals only. The effects of the 1974 flood with a recurrence interval of 21 yr on the master stream Tamur and 10 on the tributary Leoti Khola were well documented as the flood happened between two field visits.

During the July 1974 flood in the Leoti Khola the velocities were estimated to be up to 3–4 m/s, enough to move 2 m boulders. The channel was widened and terraces and fans undercut up to 10 m. Elongated bars were built in the channel and features described as "boulder sheets" or "terrace plains" next to it with sediment supplied from bank erosion, landslides, and tributaries. The sediment transport was considerable, and a kinematic debris wave advanced across the channel bed, creating steps and rough bar forms. Finer material was removed and the stream courses forced to shift as a result of fan extensions at tributary mouths. The flood effects and the magnitude frequency pattern were similar to the 1968 situation in the Darjeeling Himalayas.

Indian Examples: Rivers from Plains

Very little geomorphic information exists on flood effects of tropical rivers from plains or from a rolling landscape. Three detailed descriptions of the Brahmaputra River contain some information, but mostly on monsoon flooding at low recurrence intervals (Coleman, 1969; Goswami, 1982; Bristow, 1987). This is not surprising as the Brahmaputra, one of the world's most heavily laden rivers, carries an annual average suspended load of 344.5 million tonnes at Pandu (drainage area 424,309 km^2), over 95% of which is transported between May and October. During flood peaks daily suspended-load transport may rise to 26.5 million tonnes (Goswami, 1982). Such figures, one has to remember, do not include the bedload. The work of the river during monsoon floods have been described by Coleman (1969) in terms of bank failure and channel migration, flood sedimentation, and formation of bed forms, natural levees, and crevasse splays. It is unclear whether there are forms and sediment in the valley that are related to floods of larger recurrence interval or whether recovery from such floods is carried out by the sediment-laden waters of each wet monsoon, especially if the debris load is increased at the time from the periodic earthquake-generated landslips in the Himalayan part of the basin (Goswami, 1982). The Brahmaputra tends to widen its banks in a dramatic fashion. Goswami (1982) has referred to increases between 2 and 3 km in 10 yr (1971–1981). One wonders whether such increases are achieved by steady increments or by dramatic shifts in large floods.

Bedi and Vaidyanadhan (1982) describe the Narmada River on the Gujarat coast of India. There are prominent mid-channel islands with thick sand and silt deposits 8 m above the winter low water level, flood scars, and sand splays. There is also an unexplained low terrace (T_2 of Bedi and Vaidyanathan, 1982) and two-level point bars.

Certain flood features were found in the river Auranga, a seventh-order stream at the northeastern corner of the Deccan Peninsula of India. The Auranga has a seasonal discharge related to the monsoon, and over 80% of the annual rainfall arrives between June and September. During this period the watershed also receives episodic high rainfall as shown by the daily records for local rain gauges. The pattern is in accordance with the previous discussion on monsoon rainfall. The Auranga also transports a high amount of sediment. All the river channels in the basin are large reservoirs of coarse sediment, mainly coarse sand and fine pebbles, derived from Archean granite-gneiss, Gondwana-age sedimentary rocks, and a paleo-valley-fill alluvium. During the times of high flow a large amount of sediment reaches the Auranga at multiple points along its course and travels down the channel.

The channel has a three-tier physiography (Fig. 6) during the low season. There are (1) a braided sector located near the thalweg and presumably destroyed during the southwestern monsoon, (2) point bars at higher levels at channel bends that are only partially submerged during monsoon high flows, and (3) a higher bar (0.75–1.5 m above the point bar and 2–3 m above the dry season thalweg) found at some of the channel bends on the convex side. This bar (Fig. 7) is clearly separated from the point bar by a riser, contains coarse material up to small boulders (the inset point bar below this consists of sand with fine pebbles), and displays local armoring. Flood chutes, 0.3–0.8 m deep, dissect its surface, and coarse clasts are splayed in a linear fashion with or without mud drapes along the sides of the chutes. Chute bars are locally present at the downstream ends of such flood chutes. The location of this high bar is slightly asymmetrical with reference to the axis of the river bend. The height, general coarseness, common structureless sedimentation pattern, increased competence, and surface chutes indicate a flood origin for the bar. The flood bar is built during high-magnitude floods when the point bars are submerged and at least partially destroyed, and the river takes a straighter course with a high-water stage. The deposition of material coarser than normally carried occurs at a high level next to the bank, but at

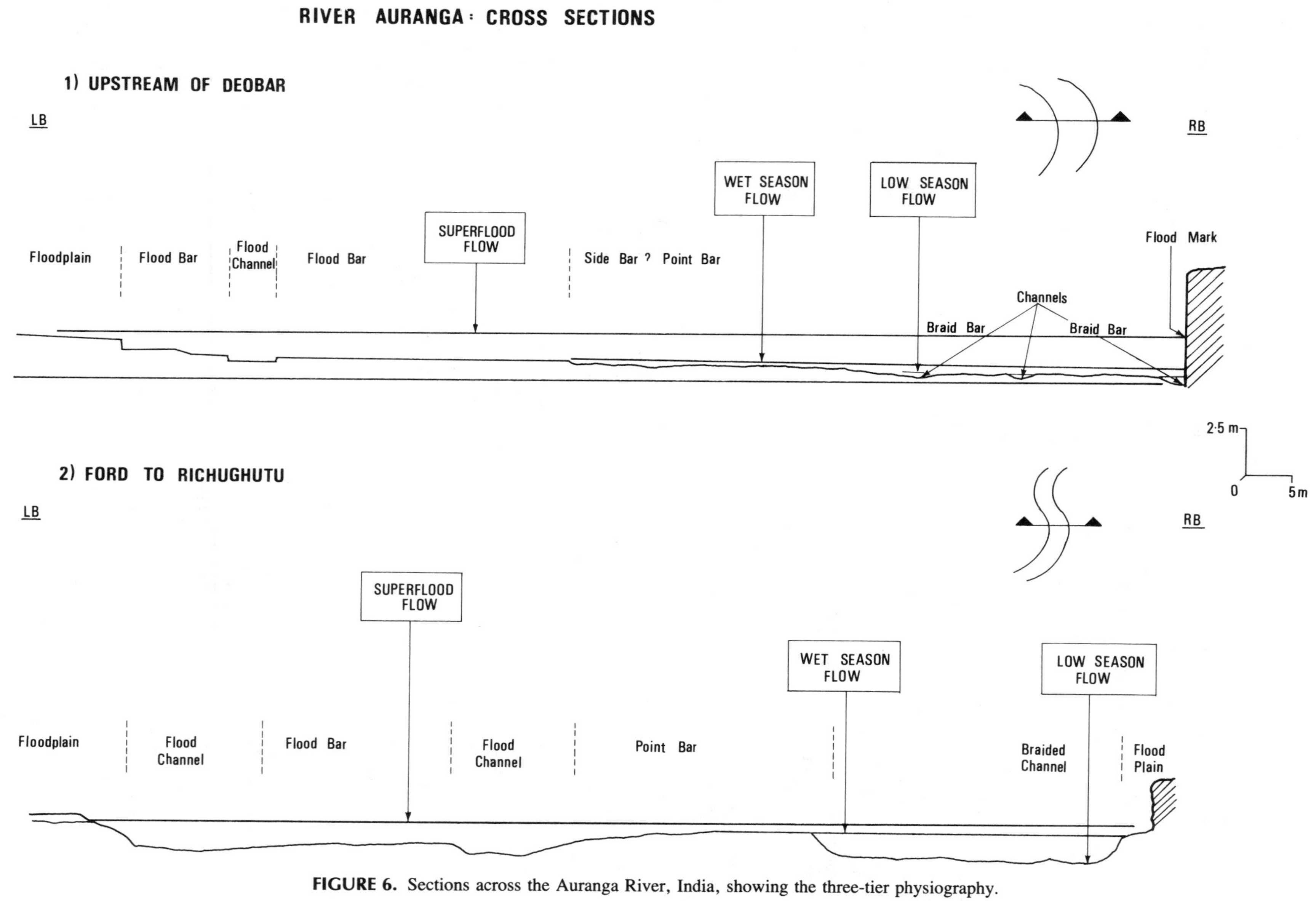

FIGURE 6. Sections across the Auranga River, India, showing the three-tier physiography.

FIGURE 7. Geomorphic features in the Auranga channel: (*a*) the three-tier physiography: (1) the low-flow braided channel, (2) the point bar, (3) the flood bar; (*b*) the flood bar with a prominent step.

locations determined by the overall river course geometry. Similar bars have been observed in Australia (V. R. Baker, personal communication). The Auranga River therefore displays both flood and "common" flow-related forms. Such a characteristic is seen in most of the rivers of the region. A detailed geomorphologic study of the Auranga River is being prepared.

Flood Geomorphology from Eastern Jamaica

In the Blue Mountains of eastern Jamaica, with storm rainfall comparable to the Darjeeling Himalayas, rivers in flood move very large volumes of coarse material. Wood (1977) studied the Hope River, east of Kingston. At the wide shallow study section the partially braided channel was approximately 32 m wide, had a slope of 0.013 and coarse bed material, and drained a catchment area of 52 km^2. The Hope at this section transported 8125 tonnes of suspended sediment in just over 30 hr in a flood with a recurrence interval of 4–5 yr. The amount of suspended sediment removed in smaller flood was insignificant by comparison. The 1-yr flood in comparison moved only 347 tonnes of suspended material. The peak velocity and discharge for the 4- to 5-yr flood were 2.5 m/s and 20 m^3/s (0.38 $m^3/s/km^2$), respectively. The coarse bedload, if it could have been measured in such a stream, would surely have shown the importance of the larger flood as in the case of suspended sediment. It is interesting to note that the October 1963 hurricane flood in the Blue Mountains reached 2.9 $m^3/s/km^2$ from a watershed of 214.9 km^2 in the only gauge that survived at Fellowship on the Rio Grande. As an extreme example the 1909 hurricane flood on the Yallahs River at Mahogany Vale (63.5 km^2) was estimated to have been 37 $m^3/s/km^2$ at peak.

The flood-related forms and sediment have been described for two other rivers, the Yallahs and the Buff Bay (Gupta, 1975), from the Blue Mountains of eastern Jamaica: an area of frequent high-magnitude rainfall as described earlier (Vickers, 1967; Lirios, 1969). The valleys show a local relief of 450–600 m, with steep slopes between 20° and 30° coming down to sharply demarcated valley floors. Mass movements and high velocity slope wash are common and frequent, their intensity rising with the destruction of natural vegetation. In October 1963, after Hurricane Flora, 40,000 m^3 of greywackes and shale slid into the Yallahs River at Mahogany Vale. Such landslips and high-magnitude floods usually happen together at many places along the channel and are responsible for the extreme coarseness of the alluvium and certain valley bottom forms.

The coarse alluvium is chiefly pebbles and cobbles with sand and silt in sheltered places. Unusually large boulders, between 1 and 2 m, occur scattered along the stream channels and on the floodplain. Some of the boulders are up to several meters across (Gupta, 1983). The valley forms include the channel, floodplain, and a set of terraces that rise 1.75–3 m above the floodplain. Big flood channels and up to 2.5 m-high mid-channel bars occur on top of such terraces. The terraces are made of subangular to subrounded pebbles, cobbles, boulders, and sand (Fig. 8). The stratigraphy varies between alternating layers of coarse sand and pebbles and about one meter of sand overlying several meters of pebbles. On top there is a pebble armor. The only sedimentary structures present are imbrications in pebbles and indistinct laminations in silt at sheltered places. The terraces are covered by thorny, scrubby trees about 3–4.5 m high, in contrast with the bare floodplain on one side and tall trees and dense undergrowth on the other (Fig. 9).

It has been postulated (Gupta, 1975) that large floods, which occur with a recurrence interval of about 10 yr, destroy the terraces and the vegetation on top. Rebuilt and modified terraces emerge during the waning flood stages, leaving the stream with a wide channel at the base of the terrace. The wide channel at this stage is clear of all but very coarse cobbles and boulders. Smaller floods and near-bankfull flows rebuild the floodplain, reduce the channel

FIGURE 8. The nature of flood sediment, terrace edge, the Yallahs Valley. Note young trees on top.

to a more expected size, and deposit bars within the channel perimeter. The relatively smaller sized material—sand, pebbles, and cobbles—is transported in all floods, though with much larger efficiency in very large ones that might even sweep long reaches of the channel free of such sediment. If the October 1963 flood is taken as typical, the valley bottom modification continues for several days.

The Tanzanian Example: The 1970 flood in the Uluguru Mountains

The combination of flood effects in stream channels with mass movements on slopes was observed in the western Uluguru Mountains of Tanzania, after about 100 mm of rain in a day in February 1970 from a tropical cyclone developed east of Madagascar (Temple and Rapp, 1972). The flood discharge was calculated as 230.5 m^3/s or 2.28 $m^3/s/km^2$ for a catchment of 101 km^2. This was an extremely flashy flood, but the sediment supply to the stream was considerable from mass movement, water action on the slopes, and erosion in the channel. The tributary channels were severely eroded, even minor ones moving boulders up to 2 tonnes, and the main stream was partially choked with coarse sediment. However, the authors concentrated on describing the landslides, and the flood response of the channels was not described in the same detail; but a general picture of active and at least temporary flood modification of the channels does emerge. The denudation rate calculated was about 14 mm from the slopes for the area of maximum damage. Considerable modifications of stream channels must have taken place, especially as the authors equate this figure to about 20–25 yr of normal denudation from the area. The storm of 1970, however, was not an isolated event, and furthermore partial recovery of the slide scars was assumed to take place in 2–5 yr. One may perhaps extend such a notion of periodic massive alteration and recovery to the stream channels also.

FIGURE 9. Flood terrace in the Yallahs Valley, Jamaica, under young trees beyond bare floodplain.

DISCUSSION

Stevens and others (1975) have shown, with data from Venezuelan streams, that channel size and pattern depend on various factors, including size of the peak flood. Therefore the form of the channel may depend partly on flood history and not entirely on currently occurring events. In spite of the present shortcomings of the available information, one should expect periodic high-magnitude flooding in certain parts of the humid tropics. Furthermore, when such flood history is combined with suitable physiography, resulting in the supply of plentiful coarse sediment to the flooded channels, the streams are typified by certain channel forms and sedimentation pattern. Flood effects on the channel and on the floodplain usually include (1) widening of channel, (2) erosion of bars and the formation of chutes, (3) scouring of floodplain, (4) increase in competence, (5) deposition of coarse gravel in channel, (6) building of transverse gravel waves and gravel bars, (7) deposition on floodplain, often of coarse gravel, (8) formation of a terracelike feature, (9) formation of levees in coarse material, and (10) destruction of vegetation (Gupta, 1983). In extreme cases, such as Brunsden and others (1981) have described, a kinematic wave of coarse material moves down the conveyor belt of the channel, building rough steps and bars to mark its passage.

Although the number of case studies is limited, it seems likely that the nature and behavior of a group of rivers in the humid tropics are dependent partly on the passage of large floods at periodic intervals down their channels. This hypothesis should be tested with additional studies of flood-prone rivers in the tropics. There are some very good reasons why this is necessary. Our knowledge of rivers is derived to a great extent from case studies in either the humid eastern or the drier western part of the United States. In the humid temperate areas the effect of even very large floods with recurrence interval of more than 100 yr is often removed in months or at best years (Wolman and Gerson, 1978). Case studies by Wolman and Eiler (1958), Hack and Goodlett (1960), Costa (1974), and Gupta and Fox (1974) on different rivers are in support of this conclusion. In semi-arid areas the channel recovery is slower, but the frequency of high-magnitude floods is low. The permanency of flood effects depends on the ability of major flood events to transport sediment that exceeds the competence of smaller flows. Such catastrophic stream channel response requires high relief, torrential rainfall and high resistance of the channel (Baker, 1977; Wolman and Gerson, 1978). The permanency is heightened by flood recurrence at short intervals.

In certain parts of the humid tropics as listed earlier, flood effects due to rainfall from recurring high-magnitude storms persist in varying degree. The prominance of the persisting forms will be higher in areas of high relief and coarser valley sediment. Though the number of case studies is limited, it seems that the effects of such floods are more prominent in rivers like Tamur and the Yallahs rather than the Auranga. It should be stressed, however, that in the case of the Auranga, the flood forms and sediment are clearly recognizable. It is evident that such stream channels cannot entirely be explained by the standard concept of dominant discharge. If we consider the case of the Auranga, where the flood response though clear is not as spectacular as in the rivers of the Himalayas or the Blue Mountains, the river system seems to be controlled by both bankful or comparable flow and very large flood discharges. For basins with greater relief and coarser sediment, the importance of catastrophic flood response rises even further.

Given the limited number of case studies on stream channel response to large tropical floods, it is possible at this stage only to formulate the hypothesis that high-magnitude floods are extremely important as channel- and valley-forming events in certain parts of the humid tropics. Perhaps in the future, with a greater number of reliable reports on rivers of such areas, it would be possible to test this hypothesis and even to examine the variation of the importance of the role of high-magnitude floods vis-à-vis the bankfull discharge across the humid tropics. There are at least two instances where the flood-prone tropical rivers differ from the standard model. First, the valleys persistently show the effect of both bankfull flow and large flood discharges on alluvial forms and sediment. Second, a large part of the work of the stream is apparently carried out in the high-magnitude floods, even when computed over a large period of time. The physiographic manifestations of such behaviors are one or more of the following:

1. The lower part of the valley flat is a storage place for abundant amounts of alluvium of sand size or coarser.
2. Geologic and topographic conditions permitting, the flood channel cross section is unusually wide with a high width–depth ratio. The flood channel section is recognizable by the absence of vegetation, or a vegetation of a particular type, and the presence of bars and terraces in coarse material.
3. A terrace, sharply differentiated from the floodplain but low enough to be at least partially submerged from time to time, is present.
4. Large bars, flood channels, and chute bars are present in the channel or on top of the floodplain or the terrace. On the surface the material of the bars is comparatively coarser than the rest of the alluvium.
5. A substantial part of the upper portion of the valley fill displays sedimentary structures associated with high-velocity flow, even up to upper flow regime conditions.
6. The nature of the vegetation in the lower parts of the valley flat indicates periodic flood destruction.
7. The forms related to more frequent and lower magnitude flows are confined to the lower parts of the channel, are smaller in size, and are made of relatively finer material, at least in terms of measurement of competence.

This is a list of diagnostic features found in many rivers of the humid tropics indicative of the occurrence of large floods at the interval of several years. A large number of case studies on rivers of this type are required at present to determine whether such features are at least regional in character and whether the list is truly diagnostic of the occurrence of high-magnitude floods on a periodic basis. Given the present state of our knowledge regarding the rivers of the humid tropics, it will be both fascinating and proper to examine the flood-prone rivers in greater detail.

ACKNOWLEDGMENT

I would like to thank V. R. Baker and D. Brunsden for encouraging me to generalize on the effect of large floods on tropical rivers. The field work in the Auranga basin was carried out while on sabbatical leave from the National University of Singapore. I am indebted to A. Dutt and M. Das for introducing me to the Auranga River. The chapter was typed by Lim Kim Leng and the cartographic work done by Lee Li Kheng. An earlier draft was critically read by A. Fraser Gupta, J. Pakiam, and G. S. P. Thomas. I have also benefitted from the careful reviews of Gerald Nanson and Peter Patton.

REFERENCES

Ahuja, P. R., and Majumdar, K. C. (1959). A sampling approach to the estimation of floods. *Indian J. Power River Valley Dev.* **9**, 23–49.

Alaka, M. A. (1976). Climatology of Atlantic tropical storms and hurricanes. *In* "Climates of Central and South America: World Survey of Climatology" (W. Schwerdtfeger, ed.), Vol. 12, pp. 479–509. Elsevier, Amsterdam.

Baker, V. R. (1977). Stream-channel response to floods, with examples from central Texas. *Geol. Soc. Am. Bull.* **88**, 1057–1071.

Bedi, N., and Vaidyanadhan, R. (1982). Effect of neotectonics on the morphology of the Narmada river in Gujarat: Western India. *Z. Geomorphol.* [N.S.] **26**, 87–102.

Bristow, C. S. (1987). Brahmaputra River: Channel migration and deposition. *In* "Recent Developments in Fluvial Sedimentology" (F. C. Ethridge, R. M. Flores and M. D. Harvey, eds.), SEPM Special Publication 39, 63–74.

Brunsden, D., and Thornes, J. B. (1979). Landscape sensitivity and change. *Trans., Inst. Br. Geogr., New Ser.* **4**, 463–484.

Brunsden, D., Jones, D. K. C., Martin, R. P., and Doornkamp, J. C. (1981). The geomorphological character of part of the Low Himalaya of Eastern Nepal. *Z. Geomorphol.* [N.S.] **37**, 25–72.

Coleman, J. M. (1969). Brahmaputra River: Channel processes and sedimentation. *Sediment. Geol.* **8**, 129–239.

Costa, J. E. (1974). Response and recovery of a piedmont watershed from tropical storm Agnes, June 1972. *Water Resour. Res.* **10**, 106–112.

Das, P. K. (1968). "The Monsoons." National Book Trust, New Delhi, India.

Dhar, O. N., Kulkarni, A. K., and Sangam, R. B. (1984). Some aspects of winter and monsoon rainfall distribution over the Garhwal-Kumaun Himalayas—a brief appraisal. *Himalayan Res. Dev.* **2**, 10–19.

Evans, C. J. (1972). Estimates of maximum 24-hour rainfall amounts for return periods from 5 to 100 years. *Sci. Res. Counc. Jamaica J.* **3**, 25–45.

Flores, J. F., and Balagot, V. F. (1969). Climate of the Philippines. *In* "Climates of Northern and Eastern Asia: World Survey of Climatology" (H. Arakawa, ed.), Vol. 8, pp. 159–213. Elsevier, Amsterdam.

Gole, C. V., and Chitale, S. V. (1966). Inland delta building activity of the Kosi River. *J. Hydraul. Div., Am. Soc. Civ. Eng.* **92**, 111–126.

Goswami, D. C. (1982). Brahmaputra River, Assam (India): Suspended sediment transport, valley aggradation and basin denudation. Ph.D. Dissertation, Johns Hopkins University, Baltimore, Maryland.

Griffiths, J. F. (1972). Eastern Africa. *In* "Climates of Africa: World Survey of Climatology" (J. F. Griffiths, ed.), Vol. 10, pp. 313–347. Elsevier, Amsterdam.

Griffiths, J. F., and Ranaivoson, R. (1972). Madagascar. *In* "Climates of Africa: World Survey of Climatology" (J. F. Griffiths, ed.), Vol. 10, pp. 461–499. Elsevier, Amsterdam.

Gupta, A. (1975). Stream characteristics in eastern Jamaica, an environment of seasonal flow and large floods. *Am. J. Sci.* **275**, 825–847.

Gupta, A. (1983). High-magnitude floods and stream channel response. *Spec. Publ. Int. Assoc. Sedimentol.* **6**, 219–227.

Gupta, A., and Fox, H. (1974). Effects of high-magnitude floods on channel form: A case study in Maryland Piedmont. *Water Resour. Res.* **10**, 499–509.

Hack, J. T., and Goodlett, J. C. (1960). Geomorphology and forest ecology of a mountain region in the Central Appalachians. *Geol. Surv. Prof. Pap. (U.S.)* **347**, 1–66.

Jamaica Weather Reports (1909). No. 372. Kingston.

Jennings, A. H. (1950). World's greatest observed point rainfalls. *Mon. Weather Rev.* **78**, 4–5.

Jones, F. O. (1973). Landslides of Rio de Janeiro and the Serra des Araras escarpment, Brazil. *Geol. Surv. Prof. Pap. (U.S.)* **697**, 1–42.

Lirios, J. F. (1969). "Rainfall-Intensity-Duration-Frequency Maps for Kingston and St. Andrew, Jamaica," mimeogr. Caribbean Meteorological Institute, Jamaica.

Moss, J. H., Baker, V. R., Doehring, D. O., Patton, P. C., and Wolman, M. G. (1978). "Floods and People: A Geological Perspective," Rep. Comm. Geol. Public Policy. Geol. Soc. Am., Boulder, Colorado.

Mousinho de Meis, R., and da Silva, R. (1968). Mouvements de masse récents a Rio de Janeiro: Une étude de géomorphologie dynamique. *Rev. Geomorphol. Dyn.*, **18**, 145–151.

Nieuwolt, S. (1981). The climates of continental Southeast Asia. *In* "Climates of Southern and Western Asia: World Survey of Climatology" (K. Takahashi and H. Arakawa, eds.), Vol. 9, pp. 1–66. Elsevier, Amsterdam.

Portig, W. H. (1976). The climate of Central America. *In* "Climates of Central and South America: World Survey of Climatology" (W. Schwerdtfeger, ed.), Vol. 12, pp. 405–478. Elsevier, Amsterdam.

Rao, Y. P. (1981). The climate of the Indian subcontinent. *In* "Climates of Southern and Western Asia: World Survey of Climatology" (K. Takahashi and H. Arakawa, eds.), Vol. 9, pp. 67–182. Elsevier, Amsterdam.

Reynolds, R. (1985). Tropical meteorology. *Prog. Phys. Geogr.* **9**, 157–186.

Richards, P. W. (1957). "The Tropical Rain Forest." Cambridge Univ. Press, London and New York.

Schumm, S. A. (1977). "The Fluvial System." Wiley, New York.

Sikka, D. R. (1977). Some aspects of the life history, structure and movement of monsoon depressions. *Pure Appl. Geophys.* **115**, 1501–1529.

Simonett, D. S. (1967). Landslide distribution and earthquakes in the Bewani and Torricelli Mountains, New Guinea. *In* "Landform Studies from Australia and New Guinea" (J. N. Jennings, and J. A. Mabbutt, eds.), pp. 64–84. Australian National University Press, Canberra.

Simpson, R. H., and Riehl, H. (1981). "The Hurricane and its Impact." Blackwell, Oxford.

Snow, J. W. (1976). The climate of northern South America. *In* "Climates of Central and South America: World Survey of

Climatology" (W. Schwerdtfeger, ed.), Vol. 12, pp. 295–403. Elsevier, Amsterdam.

So, C. L. (1971). Mass movements associated with the rainstorms of June 1966 in Hong Kong. *Trans., Inst. Br. Geogr.* **53**, 55–65.

Starkel, L. (1972a). The role of catastrophic rainfall in the shaping of the relief of the Lower Himalaya (Darjeeling Hills). *Geogr. Pol.* **21**, 103–147.

Starkel, L. (1972b). The modelling of monsoon areas of India as related to catastrophic rainfall. *Geogr. Pol.* **23**, 151– 173.

Stevens, M. A., Simons, D. B., and Richardson, E. V. (1975). Nonequilibrium river form. *J. Hydraul. Div. Am. Soc. Civ. Eng.* **101**, 557–566.

Temple, P. W., and Rapp, A. (1972). Landslides in the Mgeta area, Western Uluguru Mountains, Tanzania. *Geogr. Ann.* **54A**, 157–193.

Temple, P. W., and Sundborg, A. (1972). The Rufiji River, Tanzania, hydrology and sediment transport. *Geogr. Ann.* **54A**, 345–368.

Thomas, M. F. (1974). "Tropical Geomorphology." Wiley, New York.

Torrance, J. D. (1972). Malawai, Rhodesia and Zambia. *In* "Climates of Africa: World Survey of Climatology" (J. W. Griffiths, ed.), Vol. 10, pp. 409–460. Elsevier, Amsterdam.

Vickers, D. O. (1967). Very heavy and intense rainfalls in Jamaica. *Univ. West Indies Conf. Climatol. Relat. Fields, Proc.*, pp. 57– 63.

Wolman, M. G., and Eiler, J. P. (1958). Reconnaissance study of erosion and deposition produced by the flood of August 1955 in Connecticut. *Trans. Am. Geophys. Union* **39**, 1–14.

Wolman, M. G., and Gerson, R. (1978). Relative scales of time and effectiveness of climate in watershed geomorphology. *Earth Surf. Processes* **3**, 189–208.

Wood, P. A. (1977). Sediment transport in the Hope River, Jamaica: A tropical drainage basin characterized by seasonal flow. *IAHS-AISH Publ.* **122**, 149–156.

PART IV
PALEOFLOODS

INTRODUCTION

Following decades of war and revolution, the People's Republic of China was faced with the opportunity to modernize. In 1950 there were 300 hydrologic stations in China, and most of the record lengths were too short to provide useful design information (Luo Chengzheng, 1985). In order to generate information to design numerous major dam projects, China embarked on a national survey of historical flood marks (Chen Chia-Chi et al., 1975; Hua Shi-Qian, 1985). In part this realization derived from hazardous underdesign of several reservoir spillways based on flood-frequency analysis of inadequately short systematic records (Teng Wei-fen and Gu Chuan-zhi, 1985). The philosophy of the historic flood survey was to collect as much documentary information as possible to test the representativeness of the systematic gauge records and to reduce the potential error of extrapolation in frequency curves to extraordinary flood magnitudes.

Examples of Chinese historical flood documentation include studies of the Chang Jiang (Yangtze River) where the major floods of 1153, 1227, 1520, 1560, 1788, 1796, 1860, and 1870 are well documented by inscriptions and flood marks (Luo Chengzheng, 1985). The importance of such data in China is emphasized by its use in defining the spillway design discharge for the Gezhouba Dam, a $2.4 billion project now nearing completion, and the Three Gorges Dam, a proposed $7.4 billion project (Shih Winshing, 1985). Failure of these dams would put a greater total population at risk than that of the entire United States.

The Huang He (Yellow River) has an even longer historical record, extending back to a recorded flood stage in 223 A.D. The Chinese have incorporated geological studies of flood sedimentation into design studies on this river in order to independently verify historical records (Shih Fucheng et al., 1985). The Chinese have rigorous procedures to evaluate the uncertainty of indirect discharge estimation for ancient flood evidence and to incorporate dated ancient floods into the frequency analysis. They find that the largest events in the last several hundred to 2000 years exert a tremendous influence on frequency analysis (Shih Winshing, 1985).

The floods previously described are historic floods. They took place before the time of continuous modern hydrologic measurement, but their occurrence was observed, recorded, or otherwise communicated by human action for subsequent hydrologic analysis. Paleofloods are past flow events that need not have been humanly observed. Rather, their existence is indicated by various persistent phenomena that they induce on the landscape or its vegetative cover. Paleoflood hydrology attempts to reconstruct ancient floods utilizing various procedures of sedimentology, stratigraphy, geomorphology, and geobotany, combined with principles of hydraulics and hydrology. As reviewed by Costa (1986) the development of paleoflood hydrology in the United States through 1970 has come predominantly from geomorphologists and geologists rather than hydrologists.

PALEOFLOOD HYDROLOGIC RESEARCH

In geomorphic studies of paleofloods it is very important to distinguish effects of (1) low magnitude, high-frequency floods, and (2) high magnitude, low-frequency floods. Type 1 effects have been extensively studied in alluvial rivers, that is rivers with banks and beds of sediment that can be moved over a relatively broad range of flow conditions. Chapter 19 by Garnett P. Williams entitled "Paleofluvial Estimates from Dimensions of Former Channels and Meanders" reviews a major body of empirical data concerning type 1 floods and their relationships to alluvial river morphology. Williams details numerous empirical relationships that may be used in estimating paleostreamflow from the interrelated morphological parameters of paleochannels.

Type 2 effects are of interest for the reconstruction of rare, large floods. One line of research on this topic employs sediment transport relationships, such as described by Komar in Chapter 6, to estimate paleoflood parameters from studies of coarse-grained flood deposits. Particle sizes are generally related to shear stress, velocity, or stream power. Examples of this approach include work by Baker (1974), Baker and Ritter (1975), Church (1978), Bradley and Mears (1980), Costa (1983), and Williams (1983). Williams (1984) summarizes most of the relevant equations. As discussed by

Church (1978), this approach to paleoflood analysis can be subject to large errors.

The twofold division of paleoflood studies can also be considered from an ecogeomorphic perspective. Low magnitude, high-frequency flow events (type 1) can be studied through the use of tree ring widths as proxy records of streamflow (Stockton and Boggess, 1980, 1983; Smith and Stockton, 1981). High magnitude, low-frequency flow events (type 2) leave their direct imprint on bottomland vegetation communities. Chapter 20 by Cliff R. Hupp, "Plant Ecological Aspects of Flood Geomorphology and Paleoflood History," provides a comprehensive review of type 2 flood effects on plants. Some settings, such as Passage Creek in northwestern Virginia, provide very detailed ecogeomorphic reconstructions of floods. In general, tree ring chronologies allow paleoflood record lengths of 400 to 500 years or less, as limited by the ages of extant floodplain or channel trees.

The most accurate and detailed reconstructions of large paleofloods over long time periods are achieved in studies of certain stable-boundary fluvial reaches characterized by slackwater deposits and paleostage indicators (SWD-PSI). Chapter 21 by R. C. Kochel and V. R. Baker, "Paleoflood Analysis Using Slackwater Deposits," summarizes recent experience with SWD-PSI studies. The chapter discusses paleoflood slackwater deposition at tributary mouths, an especially common field situation. Emphasis is placed on deposit interpretation for preservation, completeness, sedimentology, channel stability, and paleoflood ages. The processes of slackwater sedimentation are discussed in Chapter 8.

The most important use of SWD-PSI studies is in paleoflood frequency analysis. Chapter 22 by R. C. Kochel entitled "Extending Stream Records with Slackwater Paleoflood Hydrology: Example from West Texas" illustrates this application for the Pecos and Devils rivers. The systematic flood record for the Pecos illustrates the classic dilemma posed by one or two unusually large outliers in an annual flood series. In this case, the recurrence interval for the great flood of 1954 was specified by incorporating SWD-PSI paleoflood data into the flood frequency analysis.

SWD-PSI paleoflood hydrology has benefited from two areas of recent technological advancement. The first area is geochronology, which has yielded more precise age determinations for a variety of datable materials in flood deposits. Kochel and Baker in Chapter 21 discuss this topic, but it is certain that additional advances can be expected in the near future. For example, thermoluminescence dating (Wintle and Huntley, 1982) might be applied to windblown quartz-rich sediments that are sometimes interbedded with slackwater deposits in arid environments. Radiocarbon dating by tandem accelerator mass spectrometry (Taylor and others, 1984) has immense potential for dating paleofloods.

A second important area of technological advancement has come in the use of hydraulic flow modeling to calculate paleodischarges. Chapter 23 by Jim E. O'Connor and Robert H. Webb, "Hydraulic Modeling for Paleoflood Analysis," provides an up-to-date review of step-backwater discharge calculations, as used in paleoflood hydrology. New computer flow models make these calculations easy to perform as a part of SWD-PSI paleoflood investigations. Because hydraulic calculations can be performed independently of the high-water indicator surveys, this approach allows a better means of specifying possible error in the analysis. The field problems of SWD-PSI geomorphology and interpretation can be separated from analytical problems in specifying the hydraulics of a study reach.

Chapter 24 by Webb, O'Connor, and Baker describes an in-depth paleoflood hydrologic study of the Escalante River in south-central Utah. Typical of many streams in the arid or semi-arid western United States, Escalante River floods have heretofore been documented by a fragmentary, short-term systematic record. In contrast, Chapter 24 provides historical data, tree ring data, and SWD-PSI data for as many as 20 paleofloods. Hydraulic flow modeling procedures, described in Chapter 23, are used to calculate paleoflood discharges. Of particular interest are the silt lines that are interpreted as precise paleostage indicators for ancient floods. Channel stability for bedrock reaches of the Escalante River is demonstrated with historic photographs.

Of considerable importance for the Escalante River is the apparent clustering of flood events. Large floods occurred approximately 1000 years ago, about 500 years ago, and in historic times. The largest flows in the last 2000 years were comparable to the largest floods observed on similar-sized drainage basins in the region. The paleoflood and historical flood data were incorporated into a flood frequency analysis according to the standard procedures adopted by U.S. governmental agencies (U.S. Water Resources Council, 1982).

Chapter 24 underscores the importance of documenting all potential error sources in paleoflood hydrologic investigation. As such studies move from the scientific to the engineering arena, there will be increased emphasis on the accuracy of risk evaluation based on paleoflood data. Hosking and Wallis (1986) demonstrated that a single, inaccurate estimate of a large paleoflood may be of little use in improving an estimate of flood frequency. In contrast, Stedinger and Cohn (1986) find that, with proper structuring of the data, the mere knowledge of flood exceedence or nonexceedence of a censoring level may result in considerable improvement of flood frequency estimates. As shown in this section, SWD-PSI paleoflood studies can yield multiple paleoflood estimates that are quite accurate both in age and magnitude. Stedinger and Baker (1987) consider these qualities most important in achieving useful flood frequency information from paleoflood studies.

Following United States convention, the frequency analyses employed by Kochel (Chapter 22) and by Webb et al. (Chapter 24) utilized procedures outlined by the U.S. Water Resources Council (1982). The SWD-PSI data are treated as "historical floods" and combined with gauging station records using a log-Pearson Type III analysis. Recent work by Stedinger and Cohn (1986) shows that the U.S. Water Resources Council (1982) procedure is relatively inefficient at extracting useful information from historical and paleoflood data. Stedinger and Cohn (1986) advocate the use of maximum likelihood analysis for paleoflood data, and work on this new approach was in progress at the time of this writing.

APPLIED PALEOFLOOD HYDROLOGY

There is an accelerating trend in the recognition and use of paleoflood data by U.S. agencies. On April 4, 1986, Colorado State House Bill No. 1186 was signed into law, requiring, when appropriate, the use of geologic and vegetative studies to establish probable future hazardous surface-water flows in relation to reservoir design and construction. This act followed from the use of paleoflood data in evaluating flood hazards and dam safety in the Colorado Front (Costa, 1978; Jarrett and Costa, 1982; Jarrett, 1987).

Paleoflood hydrology has recently been evaluated by several high level advisory committees in the United States. A National Research Council report "Safety of Dams" advocates the analysis of physical evidence of large paleofloods to provide objective evidence of the likelihood and frequency of larger floods than can be documented by gauged flow records (National Research Council, 1985, p. 235). The report notes that stratigraphic and geomorphic evidence of extraordinary paleofloods have the potential of illustrating what size floods can occur. Such paleoflood data should be used to demonstrate that calculated PMF values, used in dam safety design, are credible and are neither unreasonably large nor small (National Research Council, 1985, p. 235). Because high-quality paleoflood data cannot be obtained at all potential dam sites, paleoflood hydrologic techniques cannot always be used to construct a safety evaluation flood. Nevertheless, where such data can be obtained, they demonstrate what magnitude floods are and are not possible.

The United States has generally adopted hydrometeorological procedures for dam spillway design. U.S. dams must be designed to withstand the discharge of a probable maximum flood (PMF). A PMF is determined by using a rainfall–runoff model for a particular drainage basin receiving the probable maximum precipitation (PMP). This latter concept is a key determinant in risk analysis for dams (National Research Council, 1985). The Work Group on PMF Risk Assessment (1986) was concerned with the feasibility of assigning a probability to the probable maximum flood (PMF). Although the committee concluded that no method was adequate for that task, it observed that some approaches in paleoflood hydrology might yield interesting data. The report states, "This method allows for an evaluation of major flood events over the past hundreds to thousands of years and may provide guidance on extending the frequency curve beyond the range of the gage data."

As a part of the process of selecting a high-level radioactive waste repository, the U.S. Nuclear Waste Policy Act of 1982 requires the submission of site characterization plans (SCP) for candidate sites. An SCP is also required by the U.S. Nuclear Regulatory Commission for licensing the disposal of high-level wastes in geologic repositories. Hydrology is a paramount concern on the SCP. According to the Department of Energy (1985, p. 20) the SCP will describe the following:

> . . . the flood history of the candidate area and site. . . . The data used will be based of measurements from gaging stations and on inferences from the geologic record. The probable maximum flood and its relation to the planned facilities will be estimated, and the potential for future flooding of the site will be discussed. . . . Geologic evidence of Pleistocene and Holocene flooding used to assess future flood potential will also be described.

Paleoflood hydrology is part of a broader scientific endeavor that utilizes stratigraphic geology in natural hazards evaluation (Baker, 1982). Studies of the Quaternary stratigraphic record may be used to reconstruct magnitudes and times of earthquakes, volcanic eruptions, mass movements, coastal hurricanes, and other cataclysmic natural phenomena. Many of the same research techniques and scientific questions apply to all these phenomena. Critical needs for advancing this broader research endeavor are the same as those applying to paleoflood hydrology:

1. Accurate transfer functions are needed that relate the appropriate geological evidence of the paleohazard to the quantitative magnitude of the formative process.
2. The spatial applicability of the various hazard reconstruction procedures needs to be defined.
3. Regimes of the hazard in time and space, as might be induced by climate, tectonism, or physiography, need to be identified.
4. Better understanding is needed of the cataclysmic geological processes responsible for the hazards.
5. The appropriate probability concepts need to be related to the hazards of interest in risk analysis.

As pointed out by Wolman (1982), public policy has not yet responded to make effective use of the whole broad class of geologic paleohazard evaluation. The standard benefit–cost procedures used in design considerations are closely linked to frequency analysis based on sampling observed or directly measured processes. That such processes may not be representative of the rare, great cataclysms represented in longer geologic records is not a concern of such analysis. The irony is that, in the public debate over responses to potential hazards, the documented occurrence of an ancient (but real) cataclysmic process is likely to have more impact than is discussion of various hypothetical frequency distributions.

REFERENCES

Baker, V. R. (1974). Paleohydraulic interpretation of Quaternary alluvium near Golden, Colorado. *Quat. Res.* **4**, 94–112.

Baker, V. R. (1982). Stratigraphic geology and natural hazards evaluation. *Geol. Soc. Am. Abstr. Programs,* **14**, 438.

Baker, V. R., and Ritter, D. F. (1975). Competence of rivers to transport coarse bedload material. *Geol. Soc. Am. Bull.* **86**, 975–978.

Bradley, W. C., and Mears, A. I. (1980). Calculations of flows needed to transport coarse fraction of Boulder Creek alluvium at Boulder, Colorado. *Geol. Soc. Am. Bull.* **91** (Part II), p. 1057–1090.

Chen Chia-Chi, Yeh Yung-Yi, and Tan Wei-Yan (1974). The important role of historical flood data in the estimation of spillway design floods. *Scientia Sinica* **18** (5), 669–680.

Church, M. (1978). Palaeohydrological reconstructions from a Holocene valley fill. *In* "Fluvial sedimentology." (A. D. Miall, ed.), pp. 743–772 Canadian Society of Petroleum Geologists Memoir 5, Calgary, Alberta.

Costa, J. E. (1978). Holocene stratigraphy in flood-frequency analysis. *Water Resour. Res.* **14**, 626–632.

Costa, J. E. (1983). Paleohydraulic reconstruction of flash-flood peaks from boulder deposits in the Colorado Front Range. *Geol. Soc. Am. Bull.* **94**, 986–1004.

Costa, J. E. (1986). A history of paleoflood hydrology in the United States, 1800–1970. *EOS* **67**, 425–430.

Hosking, J. R. M., and Wallis, J. R. (1986). Paleoflood hydrology and flood frequency analysis. *Water Resour. Res.* **22**, 543–550.

Hua Shi-Qian (1985). A general survey of flood frequency analysis in China. *Proc. U.S.-China Bilateral Symp. Anal. Extraordinary Flood Events,* Nanjing, China.

Jarrett, R. D. (1987). Flood hydrology of foothill and mountain streams in Colorado. Ph.D. dissertation, Colorado State Univ., Fort Collins, Colorado, (unpublished).

Jarrett, R. D., and Costa, J. E. (1982). Multidisciplinary approach to the flood hydrology of foothill streams in Colorado. *In* "Report on International Symposium on Hydrometeorology." pp. 560–565. Am. Water Res. Assoc., Denver, Colorado.

Luo Chengzheng (1985). A survey of historical flood and its regionalization in China. Proceedings of the U.S.-China Bilateral Symposium on the Analysis of Extraordinary Flood Events, Nanjing, China.

National Research Council (1985). "Safety of Dams: Flood and earthquake criteria." National Academy Press, Washington, D.C.

Shih Fucheng, Yi Yuanjun, and Han Manhua (1985). Investigation and verification of extraordinarily large floods of the Yellow River. *Proc. U.S.-China Bilateral Symp. Anal. Extraordinary Flood Events*, Nanjing, China.

Shih Winshing (1985). Application of historic flood in the design of Three Gorges Project. *Proc. U.S.-China Bilateral Symp. Anal. Extraordinary Flood Events*, Nanjing China.

Smith, L. P., and Stockton, C. W. (1981). Reconstructed stream flow for the Salt and Verde Rivers from tree-ring data. *Water Resour. Bull.* **17**, 939–947.

Stedinger, J. N. R., and Baker, V. R. (1987). Surface water hydrology: Historical and paleoflood information. *Rev. Geophy.* **25**, 119–124.

Stedinger, J. R., and Cohn, T. A. (1986). The value of historical and paleoflood information in flood frequency analysis. *Water Resour. Res.* **22**, 785–793.

Stockton, C. W., and Boggess, W. R. (1980). Augmentation of hydrologic records using tree rings. *In* "Improved hydrologic forecasting: Why and How." pp. 239–265, American Society of Civil Engineers, New York.

Stockton, C. W., and Boggess, W. R. (1983). Tree-ring data: Valuable tool for reconstructing annual and seasonal streamflow and determining long-term trends. *In* "Improving estimates from flood studies." pp. 10–17 Trans. Res. Rec. No. 922, Transportation Research Board, National Academy of Sciences, Washington, D.C.

Taylor, R. E., Donahue, D. J., Zabel, T. H., Damon, P. E., and Jull, A. J. T. (1984). Radiocarbon dating by particle accelerators: an archaeological perspective. *In* "Archaeological chemistry—III." p. 333–356 (J. B. Lambert, ed.) American Chemical Society Advances in Chemistry Series No. 205.

Teng Wei-fen and Gu Chuan-zhi (1985). China experience on estimation of design floods. *Proc. U.S.-China Bilateral Symp. Anal. Extraordinary Flood Events*, Nanjing, China.

U.S. Department of Energy, Office of Civilian Radioactive Waste Management (1985). "Annotated Outline for Site Characterization Plans." U.S. Department of Energy, Washington, D.C.

U.S. Water Resources Council (1982). Guidelines for determining flood flow frequency. Bull. 17B (Revised) U.S. Water Resources Council.

Williams, G. P. (1983). Paleohydrological methods and some examples from Swedish fluvial environments, I—cobble and boulder deposits. *Geograf. Ann.* **65A**, 227–243.

Williams, G. P. (1984). Paleohydrological equations for rivers. *In* "Developments and Applications of Geomorphology." pp. 343–367, (J. E. Costa and P. J. Fleischer, eds.),Springer-Verlag, Berlin.

Wintle, A. G., and Huntley, D. J. (1982). Thermoluminescence dating of sediments. *Quat. Sci. Rev.* **1**, 31–53.

Wolman, M. G. (1982). Probability of natural hazards and public policy implications. *Geol. Soc. Am. Abstr. Programs* **14**, (7), 649.

Work Group on PMF Risk Assessment (1986). Feasibility of assigning a probability to the Probable Maximum Flood. Rep. of Hydrology Subcommittee, U.S. Interagency Advisory Committee on Water Data, Washington, D.C.

19

PALEOFLUVIAL ESTIMATES FROM DIMENSIONS OF FORMER CHANNELS AND MEANDERS

GARNETT P. WILLIAMS
U.S. Geological Survey, Denver, Colorado

INTRODUCTION

Paleohydrology is the study of prehistoric waters. Fluvial paleohydrology, now a popular research topic, is that branch of paleohydrology that deals with the characteristics of former rivers and their channels. The interpretation of past fluvial environments is important for understanding the past, for evaluating past and present trends, and possibly for predicting the future. Paleohydrology can provide useful information for choosing sites to store hazardous wastes (Foley et al., 1984) and for exploring economically valuable deposits in ancient fluvial sediments. Equations developed for paleohydrology often can be used for hydrologic estimates at modern ungauged sites. The rapidly growing interest and activity in fluvial paleohydrology stem largely from the pioneering work of G. A. Dury and S. A. Schumm in the mid-1960s.

Paleochannels most commonly appear in the form of (1) exposed cross sections, (2) abandoned channels on the earth's surface, and (3) (rarely) exhumed channels. Based on such exposures, paleofluvial estimates (estimates of the former channel's streamflow and channel characteristics) can be made from (1) paleochannel bed sediments (particle sizes, dune heights, etc.) combined with sediment transport concepts, (2) paleochannel cross-section features (e.g., bankfull width), (3) paleochannel planform properties (sinuosity, meander wavelength, etc.), and (4) paleodrainage basin features (stream length, basin area, etc.). This chapter reviews the second and third categories—channel cross section and planform methods (see also Ethridge and Schumm, 1978; Church, 1981; Gardner, 1983; Gardiner, 1983; Maizels, 1983; Williams, 1984a; Foley et al., 1984).

A paleofluvial variable in some instances can be measured directly; in other instances it must be estimated from established morphologic or hydraulic relations; and in still other instances, its value cannot be reconstructed at all (Maizels, 1983). Established relations that are of practical use in nearly all cases are empirical; theoretical and semitheoretical relations generally have not yet been sufficiently tested.

PALEOCHANNEL FEATURES

Channel Dimensions

Many geomorphological and engineering equations or methods are unsuitable for paleofluvial applications because such equations or methods require water discharge (and in some instances sediment discharge and other unknowns) as independent or given variables. Examples are Chang's (1979) iterative method and standard hydraulic geometry equations. Not uncommonly, researchers inadvisedly have tried to circumvent this problem by first estimating or guessing values for the required input variables and then inserting those estimates into an empirical equation to determine a channel dimension. The preferred alternative is to use values that have been measured rather than estimated, for example, other channel dimensions or channel planform characteristics. This alternative approach will be emphasized here. All variables in this chapter are expressed in metric units; logs are to base 10.

Bankfull Width. Channel (bankfull) width, W_b, in many paleofluvial studies is measured directly. Where a direct measurement cannot be made, choices of ways to estimate width are very limited, especially if the ancient channel was not meandering.

The only available method that uses other cross-sectional dimensions of the channel to estimate W_b applies only to meandering channels with sinuosity (Si) ≥ 1.70. Sinuosity is the ratio of channel length to length of meander belt axis. Leeder (1973) presented an empirical equation for estimating W_b from channel maximum depth, D_{max} (elevation difference between lowest banktop and thalweg at a cross section). This relation is

$$W_b = 6.8D_{max}^{1.54} \quad (1)$$

with a standard error of estimate SE of 0.35 log unit. The applicable range is $1 \leq D_{max} \leq 37$ m. To reiterate, three requirements should be fulfilled to apply this equation: (1) The paleochannel was meandering; (2) the paleochannel's sinuosity was at least 1.70; and (3) D_{max} can be measured. There are several problems in measuring D_{max}, as discussed in the section on Maximum Depth.

Where paths of former channels are discernible on aerial photographs or are otherwise exposed in plan view, several equations requiring measured meander characteristics are available. Building on the research of earlier authors, Inglis (1947) and Leopold and Wolman (1960) were among the first to give mathematical relations between W_b and meander geometry. Width, however, was the independent variable, rendering the equations of uncertain value for estimating width due to the required rearrangement. Williams (1986) used Leopold and Wolman's (1960) data plus considerable additional data to derive empirical equations for estimating W_b from any one of meander wavelength L_m, meander arc distance L_a, meander belt width B, or loop radius of curvature, R_c [Table 1, Equations (2)–(5)]. Meander geometry for this purpose was measured using the definitions indicated by Leopold and Wolman (1960, Fig. 1). The data represent various physiographic environments in the United States, India and Pakistan, Canada, Sweden, Australia, and elsewhere. The equations apply only to channels having Si $\geq$ 1.20.

A rather popular technique for estimating channel width requires an exposed cross section of a point bar. This method is the so-called two-thirds rule and rests on the assumption that a point bar at the bend apex extends about two-thirds of the distance across the channel. Apparently attributable to Allen (1966, p. 166), this relation does not seem to be based on any large, representative set of field measurements. Allen (1970) later replaced his two-thirds coefficient with a value in the range of 0.70–0.95. Cotter (1971) arrived at a value of 0.60–0.80. A variety of flow, channel, and sediment characteristics seems to influence the dimensions of a point bar (Bridge, 1978, Table 1;

TABLE 1 Equations for Estimating Channel Dimensions on Sinuous Channels

Equation Number	Equation[a]	Number of Data Points	Applicable Range	Standard Error (Log Units)	r^2
(2)	$W_b = 0.17L_m^{0.89}$	191	$8 \leq L_m \leq 23{,}200$ m	0.194	0.93
(3)	$W_b = 0.23L_a^{0.89}$	102	$5 \leq L_a \leq 13{,}300$ m	0.189	0.94
(4)	$W_b = 0.27B^{0.89}$	153	$3 \leq B \leq 13{,}700$ m	0.211	0.92
(5)	$W_b = 0.71R_c^{0.89}$	79	$2.6 \leq R_c \leq 3{,}600$ m	0.170	0.94
(6)	$A_b = 0.0054L_m^{1.53}$	66	$10 \leq L_m \leq 23{,}200$ m	0.300	0.92
(7)	$A_b = 0.0085L_a^{1.53}$	41	$6 \leq L_a \leq 13{,}300$ m	0.382	0.91
(8)	$A_b = 0.012B^{1.53}$	63	$5 \leq B \leq 11{,}600$ m	0.287	0.93
(9)	$A_b = 0.067R_c^{1.53}$	28	$2 \leq R_c \leq 3{,}600$ m	0.379	0.93
(10)	$D_b = 0.12W_b^{0.69}$	67	$1.5 \leq W_b \leq 4{,}000$ m	0.287	0.66
(11)	$D_b = 0.09W_b^{0.59}Si^{1.46}$	66	$1.5 \leq W_b \leq 4{,}000$ m $1.20 \leq SI \leq 2.60$	0.239	0.73
(12)	$D_b = 0.027L_m^{0.66}$	66	$10 \leq L_m \leq 23{,}200$ m	0.244	0.73
(13)	$D_b = 0.036L_a^{0.66}$	41	$7 \leq L_a \leq 13{,}300$ m	0.238	0.81
(14)	$D_b = 0.037B^{0.66}$	63	$5 \leq B \leq 11{,}600$ m	0.210	0.81
(15)	$D_b = 0.085R_c^{0.66}$	28	$2.6 \leq R_c \leq 3{,}600$ m	0.285	0.81

SOURCE: Williams, 1986.

[a] Reduced major axis. W_b = channel (bankfull) width; A_b = bankfull cross-sectional area; D_b = bankfull mean depth; L_m = meander wavelength; L_a = meander arc distance; B = meander belt width; R_c = loop radius of curvature; Si = channel sinuosity; r = correlation coefficient.

Bridge and Diemer, 1983). For example, other factors being constant, point-bar width varies considerably with channel distance around the bend, attaining a maximum in the vicinity of the apex and a minimum in the vicinity of the crossover. Moreover, the location of an exposed cross section relative to the bend apex may be very difficult to determine. Hence the two-thirds relation must be considered an approximate rule of thumb, at best. On the other hand, in some circumstances the investigator may not have anything else available.

Bankfull Cross-Sectional Area. Estimates of bankfull cross-sectional area, A_b, as a measure of channel size have not received much attention in paleofluvial studies. It can be measured directly wherever a complete cross section of a paleochannel is exposed.

If the planform of a paleomeander is exposed, A_b can be estimated from any of the same meander characteristics as can channel width [Table 1, Equations (6)—(9)]. Otherwise, there is a notable lack of reliable methods for estimating A_b.

Bankfull Mean Depth. Where A_b and W_b can be measured, bankfull mean depth, D_b, can be computed by its definition as A_b/W_b. Otherwise, methods using measured data as input consist only of empirical relations developed for undivided, sinuous rivers for which the planform is exposed. Again, therefore, even planform methods are lacking for rivers of other patterns, such as braided or straight.

For meandering channels of Si $\geq$ 1.20, mean depth can be estimated from bankfull width or, with less standard error, from width and channel sinuosity [Table 1, Equations (10)–(11)]. The former relation has a standard error of 0.287 log unit, the latter 0.239 log unit. Though valid conceptually, both relations may have an element of spuriousness in that A_b/W_b ($=D_b$) is related to W_b.

As with channel width and cross-sectional area, mean depth on meandering channels (Si $\geq$ 1.20) can be estimated from any of the four meander characteristics, namely, L_m, L_a, B, and R_c [Table 1, Equations (12)–(15)]. Standard errors for these relations range from 0.210 to 0.285 log unit. These equations do not distinguish between the mean depth of a meander bend and that of a straight reach.

Maximum Depth. Schumm (1968, 1972), Leeder (1973) and other authors have dealt with maximum channel depth, D_{max} (thalweg depth). Where a paleochannel cross section is only partly exposed but does seem to reveal the banktop and the low point of the streambed, D_{max} would be directly measurable (assuming no tectonism and no significant compaction with time); mean depth, on the other hand, would not be.

A common paleofluvial situation where D_{max} supposedly can be measured is from the exposed cross section of part of a point bar. Here, D_{max} is taken as the vertical distance from the base of the sigmoidal cross-stratified deposit to its top (Ethridge and Schumm, 1978). This method carries the same risks as the previously described two-thirds rule for channel width, plus other risks. For instance, the height of lateral-accretion surfaces varies with distance along the channel, ranging from close to zero at the crossover to a maximum at the bend apex; assessment of the relative position of an exposed cross section around a bend can be very difficult. Equating coarse-member (sandstone) thickness to D_{max}, as is done by some authors, involves questionable assumptions about the point-bar depositional process and about possible postdepositional compaction. (The latter factor also could influence the accuracy of other measured cross-section dimensions.)

Leeder's (1973) equation involving D_{max} and W_b for channels having Si > 1.70 has D_{max} as the given variable; however, the equation can be rearranged to solve for D_{max} because his fitted line is the reduced major axis. This produces

$$D_{max} = 0.29 W_b^{0.65} \quad (16)$$

The applicable range is about $8 \leq W_b \leq 1600$ m.

Some authors (e.g., Cotter, 1971; Morton and Donaldson, 1978; Ethridge and Schumm, 1978), having a value of D_{max} for a meander bend, have used or advocated an adjustment factor to estimate D_{max} for a neighboring straight reach. Use of a universal adjustment factor probably is an oversimplified approach; the difference between the two depth values seems to depend on channel sinuosity (or bend radius of curvature), bed sediment sizes, and other factors. Recent work based on models, such as the model of Engelund as modified by Bridge (1984), shows more promise.

Channel Width–Depth Ratio. Width–depth ratio is an approximate and popular indicator of channel shape and, for convenience, will be discussed here rather than elsewhere. The best way to obtain this ratio for paleochannels is to measure width and depth directly. Alternatively, at least two predictive equations are available. Both of these are based on Schumm's (1960, 1963, 1968) work and, therefore, use D_{max} rather than D_b.

The first equation stems from Schumm's (1963) equation giving channel sinuosity as a function of channel width–depth ratio, that is, with the latter as the independent variable. Although the data for this equation were not published, Williams (1984a) took other Schumm data and derived the expression

$$W_b/D_{max} = 74 \text{ Si}^{-2.94} \quad (17)$$

with SE = 0.227 log unit. The applicable range is $1.05 \leq Si \leq 2.5$. The data were derived from 33 river sites in the Great Plains of the west-central United States plus 10 sites on the Murrumbidgee River of Australia (Schumm, 1968, pp. 13, 45). Thus the equation can be used only for a meandering channel that is sufficiently exposed in plan view to allow a measurement of sinuosity. Until wider applicability is established, the equation probably should not be used on areas other than those for which it was derived (see, e.g., Pickup and Warner, 1984).

The second equation requires Schumm's M (percent silt-clay in channel perimeter). Schumm (1960, 1968) gave two empirical equations with W_b/D_{max} as a function of M. Both equations are for virtually the same environment, but the coefficients are different. Williams (1984a) combined the two data sets and performed a least-squares regression to arrive at

$$W_b/D_{max} = 148M^{-0.87} \qquad (18)$$

for $0.5 \leq M \leq 89\%$; SE = 0.213 log unit. The data were derived from 91 sites on Great Plains streams plus 10 sites on the Murrumbidgee River. The potential of M in paleofluvial studies has yet to be fully explored; three concerns are that M (1) can be very difficult to determine on paleochannels, (2) may be spuriously defined to some extent in that channel width and depth are included on both sides of the defining equation (Melton, 1961; Miller and Onesti, 1979), and (3) may not apply to environments other than that for which it was deduced (Hickin, 1983; Pickup and Warner, 1984).

Channel Slope

Channel slope, S, may be of interest only from a physiographic viewpoint or it may be needed for use in a hydraulic equation. In the latter case the energy gradient rather than channel slope should be used, but the two commonly are assumed to be approximately equal. This assumption may be questionable, but there doesn't seem to be much choice in most paleofluvial studies. A reliable value for channel slope (let alone energy gradient) is hard to get in many paleofluvial studies. In many instances a reach of the paleochannel is not exposed; even where a reach is exposed, tectonism may have affected the original slope.

Where paleochannel slope cannot be measured directly and slope is to be estimated on the basis of measured channel features, only one equation seems to be available. This is Schumm's (1972) Great Plains–Murrumbidgee River relation requiring W_b and D_{max}. In simplified form (Williams, 1984a), this is

$$S = 0.0020W_b^{-0.06}D_{max}^{-0.91} \qquad (19)$$

with slope S in meters per meter. SE = 0.175 log unit; the applicable range is $8 \leq W_b \leq 244$ m and $0.7 \leq D_{max} \leq 8$ m. The statistics for this equation indicate that the predictive contribution of W_b is insignificant compared to that of D_{max}. That is, for these particular data, an equation probably could be derived for estimating slope on the basis of D_{max} alone. Such equations, of course, are not meant to imply a dependent–independent relationship; rather, they are designed to permit the estimation of the value of one variable, given a value of the predictor variable.

Where equations requiring measured values are not available, there may be no choice but to resort to methods that use estimated flow variables and other information as input. Chang's (1979) iterative method is an example. Alternatively, a mean velocity formula such as that of Chezy might be rearranged to solve for S. This would yield

$$S = (V/CD^{0.5})^{0.5} \qquad (20)$$

where V = mean flow velocity, C = a resistance coefficient, and D = mean flow depth. In the same category is Schumm's (1968) Great Plains relation, as reformulated by Williams (1984a):

$$S = 0.0036M^{-0.38}Q_{max}^{-0.32} \qquad (21)$$

where Q_{max} = "mean annual flood" (discussed later).

Meander Geometry

Some features within an exposed paleochannel cross section might suggest a meandering pattern of the former stream. Much further work on modern streams is needed to establish such links more conclusively. Examples of possibly indicative features (from Jackson, 1978) are substantial mud content in the coarse member of a lateral-accretion deposit, a thick fine member of such a deposit, and asymmetric channel fills with much mud; other examples (from Bridge, 1985) are low proportion of channel fills relative to lateral-accretion deposits, and small grain size of channel fill relative to that of lateral-accretion deposits. Where paleochannel cross sections show such evidence or where a fragment of a meander bend is discernible from an aerial view, empirical relations can be used to reconstruct probable values of the meander wavelength, arc distance, belt width, and bend radius of curvature.

Estimation of Meander Geometry from Channel Size. Just as channel dimensions can be estimated from meander geometry (Table 1), the converse also is true. Inglis (1947), Leopold and Wolman (1960), and Dury (1976), for example, produced simple power laws that give meander wavelength

TABLE 2 Equations for Estimating Meander Geometry from Channel Size

Equation Number	Equation[a]	Number of Data Points	Applicable Range	Standard Error (Log Units)	r^2
(22)	$L_m = 7.5W_b^{1.12}$	191	$1.5 \leqslant W_b \leqslant 4{,}000$ m	0.219	0.93
(23)	$L_a = 5.1W_b^{1.12}$	102	$1.5 \leqslant W_b \leqslant 2{,}000$ m	0.220	0.94
(24)	$B = 4.3W_b^{1.12}$	153	$1.5 \leqslant W_b \leqslant 4{,}000$ m	0.241	0.92
(25)	$R_c = 1.5W_b^{1.12}$	79	$1.5 \leqslant W_b \leqslant 2{,}000$ m	0.182	0.94
(26)	$L_m = 30A_b^{0.65}$	66	$0.04 \leqslant A_b \leqslant 20{,}900$ m^2	0.202	0.92
(27)	$L_a = 22A_b^{0.65}$	41	$0.04 \leqslant A_b \leqslant 20{,}900$ m^2	0.246	0.91
(28)	$B = 18A_b^{0.65}$	63	$0.04 \leqslant A_b \leqslant 20{,}900$ m^2	0.194	0.93
(29)	$R_c = 5.8A_b^{0.65}$	28	$0.04 \leqslant A_b \leqslant 20{,}900$ m^2	0.234	0.93
(30)	$L_m = 240D_b^{1.52}$	66	$0.03 \leqslant D_b \leqslant 18$ m	0.391	0.73
(31)	$L_a = 160D_b^{1.52}$	41	$0.03 \leqslant D_b \leqslant 17.6$ m	0.354	0.81
(32)	$B = 148D_b^{1.52}$	63	$0.03 \leqslant D_b \leqslant 18$ m	0.339	0.81
(33)	$R_c = 42D_b^{1.52}$	28	$0.03 \leqslant D_b \leqslant 17.6$ m	0.399	0.81

SOURCE: Williams, 1986.

[a] Reduced major axis. W_b = channel (bankfull) width; A_b = bankfull cross-sectional area; D_b = bankfull mean depth; L_m = meander wavelength; L_a = meander arc distance; B = meander belt width; R_c = loop radius of curvature; r = correlation coefficient.

as a function of channel width. The expanded data set used by Williams (1986) provided the basis for updated versions of this and other types of relations (Table 2). Any one of W_b, A_b, or D_b can be used to estimate L_m, L_a, B, and R_c, for channels of Si $\geqslant 1.20$. In paleofluvial studies there is often a considerable risk in using equations of this sort because the available field information might be insufficient to indicate whether the former stream was meandering (Bridge, 1985).

Schumm (1972), again using 33 Great Plains sites and 3 Murrumbidgee River sites, developed an equation for meander wavelength as a function of channel width and width–depth ratio. The simplified metric equivalent of this equation, determined by least-squares regression of the data (Williams, 1984a), is

$$L_m = 15.0W_b^{1.17}D_{max}^{-0.51} \qquad (34)$$

SE = 0.223 log unit; the equation applies to $8 \leqslant W_b \leqslant 244$ m and $0.7 \leqslant D_{max} \leqslant 8$ m.

Channel sinuosity also can be estimated for Schumm's same two geographic areas (Great Plains and Murrumbidgee River) using channel dimensions. Schumm (1963) gave the relation

$$\text{Si} = 3.5(W_b/D_{max})^{-0.27} \qquad (35)$$

for his Great Plains data, for $2.5 \leqslant (W_b/D_{max}) \leqslant 84$ (SE = 0.061 log unit). Nearly the same equation obtains if data in Schumm's 1968 paper (including the 10 Murrumbidgee River sites) are used (Williams, 1984a). Some authors unjustifiably have applied Schumm's equations to geographic areas other than those represented by his data.

Estimation of Meander Geometry from Other Meander Features. Since the beginning of the twentieth century there has been a gradually increasing interest in the interrelationships among meander geometry features. For example, a large radius of curvature is associated with a large wavelength, and vice versa. From known empirical relationships, the paleohydrologist can take one given meander characteristic (L_m, L_a, B, or R_c) and estimate average values of the other three.

An example of such known empirical relationships is the Leopold and Wolman (1960) equation between meander wavelength and radius of curvature. Hey (1976) algebraically manipulated this and similar equations, combined the results with meander data from two rivers in England, and presented relations among meander features in which arc angle is a critical variable. Williams (1986) on the other hand analyzed a relatively large sample of rivers and could not confirm any basis for including arc angle in the relations. Until further evidence on arc angle appears, the safest approach at present seems to be to use the simpler relations (Table 3.)

A variety of empirical equations (Williams, 1984a) is available for estimating meander wavelength and, to a lesser extent, other meander features from various water discharge statistics. For paleofluvial studies these equations probably will be much less accurate than methods described above and will not be discussed here.

TABLE 3 Equations for Estimating Meander Geometry from Other Meander Features

Equation Number	Equation[a]	Number of Data Points	Applicable Range	Standard Error (Log Units)	r^2
(36)	$L_m = 1.25L_a$	102	$5.5 \leq L_a \leq 13{,}300$ m	0.118	0.98
(37)	$L_m = 1.63B$	155	$3.7 \leq B \leq 13{,}700$ m	0.116	0.98
(38)	$L_m = 4.53R_c$	78	$2.6 \leq R_c \leq 3{,}600$ m	0.083	0.99
(39)	$L_a = 0.80L_m$	102	$8 \leq L_m \leq 16{,}500$ m	0.119	0.98
(40)	$L_a = 1.29B$	102	$3.7 \leq B \leq 10{,}000$ m	0.115	0.98
(41)	$L_a = 3.77R_c$	78	$2.6 \leq R_c \leq 3{,}600$ m	0.131	0.97
(42)	$B = 0.61L_m$	155	$8 \leq L_m \leq 23{,}200$ m	0.117	0.98
(43)	$B = 0.78L_a$	102	$5.5 \leq L_a \leq 13{,}300$ m	0.117	0.98
(44)	$B = 2.88R_c$	78	$2.6 \leq R_c \leq 3{,}600$ m	0.152	0.96
(45)	$R_c = 0.22L_m$	78	$10 \leq L_m \leq 16{,}500$ m	0.082	0.99
(46)	$R_c = 0.26L_a$	78	$6.8 \leq L_a \leq 13{,}300$ m	0.128	0.97
(47)	$R_c = 0.35B$	78	$5 \leq B \leq 10{,}000$ m	0.149	0.96

SOURCE: Williams, 1986.

[a] Reduced major axis. W_b = channel (bankfull) width; A_b = bankfull cross-sectional area; D_b = bankfull mean depth; L_m = meander wavelength; L_a = meander arc distance; B = meander belt width; R_c = loop radius of curvature; r = correlation coefficient.

PALEOSTREAMFLOW

Mean Flow Velocity

Estimates of mean flow velocity, V, pertain to a specified hydraulic radius, R, or mean flow depth, D. One of the simplest available equations is that of Lacey (1934):

$$V = 11D^{0.67}S^{0.33} \tag{48}$$

in which V is in meters per second; D, in meters, here replaces the R that Lacey used; and S is water surface slope (dimensionless). This empirical equation was based on 188 observations on canals and rivers; flows were high enough to move the bed material, and all channels had bed sediments coarser than fine silt. Bray (1979) tested the equation on measured, high, in-bank discharges (corresponding to the 2-yr discharge) on 67 Canadian gravel and cobble rivers (median bed material diameters from 19 to 145 mm); he found that the accuracy—a standard deviation of about 30%—was as good as that for various other velocity equations. The equation has the outstanding advantage of not requiring a resistance coefficient.

Probably the most popular flow formula is

$$V = R^{0.67}S^{0.50}/n \tag{49}$$

where n is a resistance coefficient. This formula usually is called the Manning formula but more accurately deserves to be called the Gauckler–Manning formula (Williams, 1970). (It never was proposed in the above form by Manning nor by anyone else; instead, it evolved by a gradual historical process in which various engineers took curious liberties with Manning's 1891 paper. Briefly, Manning was the fifth of at least 10 people—the first of whom was P. G. Gauckler in 1867—to propose a formula giving V as a function of $R^{0.67}S^{0.50}$; the formula Manning gave did not contain n; and Manning really presented as his main formula a more complicated equation involving barometric pressure, while suggesting his form of Equation (49) as a distant second choice). The range of applicability claimed by Manning was velocities of 0.05 to 6.4 m/s, depths of 0.01–31 m, slopes of almost flat (less than one-half inch per mile!) to 0.23, and roughness coefficients that translate to n values of 0.03–0.09. However, Manning did not actually present basic data, and the actual applicable range is unknown, even though the equation has been applied liberally (and with acceptable results) to a broad range of conditions. The equation pertains to uniform flow and to channels in which the cross-sectional geometry is constant along the reach.

The main difficulty with the Gauckler–Manning equation is the need of an n value. The standard method of determining n (Chow, 1959; Benson and Dalrymple, 1967) begins with the selection of a base value for n, according to the bed material particle sizes. If the channel has a sand bed, the base n is chosen to reflect not only the sandy nature of the bed but also the bed forms for the flow of interest. Typical base values of n for sand bed channels range from about 0.01 to 0.035 and are given in Albertson et al. (1960, p. 342) and Benson and Dalrymple (1967, p. 22); base values for boulder channels can be as high as 0.07. Further increments of n, representing the other roughness-related factors of channel cross-section irregularities, depth of flow, vegetation, and channel alignment, are added to the base n to get the total or final n value. These additional four increments can double or triple the base n and therefore are very important; extreme n values as high as 0.4 in fact have been reported. Evaluating some of these roughness features may not be possible in some paleofluvial studies, in which case estimates of n can be off by a significant amount.

Trained engineers using the standard method can select n values with an accuracy of plus or minus 15% (Barnes, 1967, p. 3); for inexperienced investigators the likely error cannot be specified. For a given R and S the error in n fosters an identical error in V [Equation (49)].

Barnes (1967) provided a collection of photographs of coarse-bedded natural channels for which n values were computed using the Gauckler–Manning equation with on-site (mostly postflood) measurements of V, R, and S. Such photographs can be used both to estimate n and as a supplementary aid or check on n values determined by the standard method.

Various empirical and theoretical equations have been proposed over the years for computing roughness coefficients directly. These equations involve different combinations of V, D (or R), S, and bed material grain size. They do not directly consider the contributions of bed forms, channel irregularities, vegetation, and channel alignment, but they may do so implicitly to some unknown extent in that a few data sets may have covered some unknown range of these factors. Summaries of many of these equations are given in Chow (1959), Task Force (1963), and Limerinos (1970).

A close counterpart to the Gauckler–Manning equation is the Darcy–Weisbach equation. When arranged to solve for V, this is

$$V = (8gRS/f)^{1/2} \tag{50}$$

where g is acceleration due to gravity and f is friction factor. Hand et al. (1969) and many other authors have applied this equation in paleofluvial studies.

In the rare circumstance that a reach of channel and some high-water marks from a flood are preserved, instantaneous peak flow velocity might be estimated from standard surface water indirect methods, such as slope area, slope conveyance, step-backwater, and related techniques (Barnes and Davidian, 1978) (see O'Connor and others, this volume). These methods are sophisticated applications of the Gauckler–Manning formula; two key requirements, therefore, are values for S and n.

Rotnicki (1983) examined 1352 measurements for the Odra River basin in Poland and proposed that the Gauckler–Manning equation be modified to

$$V = (0.791/n)R^{2/3}S^{1/2} + 0.141 \tag{51}$$

The standard error was 12%. He stated that paleofluvial application should be limited to lowland rivers with sandy or gravelly beds in moderate climates, mean velocities of 0.1–1.3 m/s, slopes of 0.00001–0.01, and n values of 0.01–0.06. All S values for his data were computed from topographic maps—a method of questionable accuracy for this purpose.

Water Discharge

Instantaneous Discharges. An instantaneous discharge is a discharge occurring at a particular instant of time. Examples are the maximum discharge during a flood; the discharge associated with any given water level within a channel, including the bankfull level; and the discharge associated with the movement of a fossil bedform or of a sediment particle.

Instantaneous paleodischarges at an exposed channel cross section possibly might be estimated from the Gauckler–Manning equation. Although the original equation is for mean velocity, both sides of the equation commonly are multiplied by cross-sectional flow area, A, to get water discharge, Q. This results in

$$Q = AR^{2/3}S^{1/2}/n \tag{52}$$

Rotnicki (1983), using the 1352 observations already mentioned, presented a modified version of this equation, namely,

$$Q = (0.921/n)AR^{2/3}S^{1/2} + 2.362 \tag{53}$$

Standard error varies from 7 to 26%, depending on Q. Range of application and other considerations are the same as for his mean velocity formula [Equation (51)].

At-a-station hydraulic geometry relations (Leopold and Maddock, 1953) give water surface width and mean flow depth as power functions of instantaneous discharge. These relations are not particularly suitable (nor were they ever intended) for estimating paleodischarges. In addition to the variable to be estimated being in the independent position, a large and mostly unpredictable natural variability in the relations from one site to another (Lowham, 1976; Park, 1977) has not yet been resolved. This variability is large enough that average relations for a large geographic region (used by some investigators) have been criticized as meaningless (e.g., Knighton, 1975; Richards, 1977; Rhodes, 1977).

Bankfull discharge, Q_b, can be estimated if values can be determined for S and A_b. Williams (1978), using 233 river cross sections from a variety of environments, developed the empirical relation

$$Q_b = 4.0A_b^{1.21}S^{0.28} \tag{54}$$

The range of applicability is $0.5 \leq Q_b \leq 28{,}320$ m^3/s,

$0.7 \leqslant A_b \leqslant 8510$ m^2, and $0.000041 \leqslant S \leqslant 0.081$; SE = 0.174 log unit of Q_b.

For braided channels Cheetham (1980) analyzed Leopold and Wolman's (1957) data and arrived at

$$Q_b = 0.000585 S^{-2.01} \quad (55)$$

for $0.000066 \leqslant S \leqslant 0.0073$. The standard error is 0.396 log unit—rather large, probably because other key variables are excluded.

Long-Term Statistical Discharges. A statistical discharge is a discharge determined from many measured instantaneous discharges for a stream cross section. Examples are a "dominant" discharge, an average discharge, and the discharge having a 2-yr recurrence frequency. In fluvial geomorphology, characteristic channel properties (bankfull width, bankfull mean depth, etc.) of the site are also determined; a plot then is made for a group of stream sites, showing the relation between a channel feature and the statistical discharge. A diagram of this sort will here be called a "multisite" plot. It is the same as the standard "regime-" or "downstream hydraulic geometry" plot, or the "regional" plot of Lowham (1982), except in terminology. ["Multisite" is used here because the data in many cases are not confined to a single river, canal, or region (Ferguson, 1981); data from various tributaries commonly are included on the same graph, and in some instances rivers and canals from all over the world have been represented on one plot.]

Not only must all plotted data be for the same selected statistical discharge, but the same definition of the associated channel property must be used throughout. The most common cross-section property is bankfull width. Unfortunately, "bankfull" is defined differently by different investigators (Williams, 1978). A few investigators (e.g., Osterkamp and Hedman, 1982) prefer an "active-channel" concept; the banktop elevation of their active channel, as shown in Hedman and Osterkamp (1982), seems to me to correspond to that of the floodplain of Leopold et al. (1964) or to that of the active floodplain of Williams (1978). However, there are possible differences, nevertheless, in how various investigators define "bankfull" and "channel width." Wahl (1984) gives further discussion.

Multisite relations do not necessarily imply a dependent–independent or cause-and-effect relation between the variables involved (Riggs, 1978). For paleofluvial purposes (estimation of paleodischarges), multisite relations must be formulated with discharge as the dependent variable (Williams, 1983a,b). This requirement renders most of the discharge equations in the literature unsuitable for paleofluvial estimations. Similarly unsuitable are equations derived by improper algebraic manipulation of best-fit equations (Williams and Troutman, 1987).

Average Discharge. The statistical discharge estimated most commonly in paleofluvial studies, probably because of availability of data on which to formulate modern relationships, is the average discharge. (Perhaps inaccurately, this commonly is called "mean annual discharge" by many geomorphologists.) It is the steady discharge occurring continuously for an average day of flow. The U.S. Geological Survey, responsible for collecting much of the streamflow data in the United States, computes average discharge of a perennial stream in three steps, each involving a mean discharge. First, the mean flow rate during each day of a given year is determined from the streamflow records. Second, the mean discharge during each year is calculated by adding the mean flow rate for each day of that year and dividing by 365. Third, these average rates computed for each year are added for the number of years of record and divided by that number of years.

In spite of the popularity of average discharge, few authors have offered equations suitable for paleofluvial estimates. To a considerable extent, this is because most equations have been derived using discharge as an independent or given variable, rather than as the variable to be estimated. The simplest equation, based on 252 sites of various environments from the west-central United States, is that of Osterkamp and Hedman (1982):

$$\overline{Q} = 0.027 W_b^{1.71} \quad (56)$$

where $\overline{Q}$ = average discharge in cubic meters per second. Here SE = 0.31 log unit; the applicable range is $0.8 \leqslant W_b \leqslant 430$ m.

Osterkamp and Hedman (1982), like many other investigators, found that the sizes of bed and bank material are relevant. However, within the range of bed material sizes (silt to cobbles) in their streams, the effect of this factor varied in importance and could not be included as a second independent variable in a multiple-regression equation. The solution, which provided improved estimates of Q, was to establish categories on the basis of bed and bank material, with a separate equation for each category. Where such analyses of boundary sediment are obtainable, Osterkamp and Hedman's (1982) equations (Table 4) are likely to provide closer estimates of $\overline{Q}$ than Equation (56).

Schumm (1972, p. 102) presented an equation that, when simplified and recomputed by Williams (1984a), is

$$\overline{Q} = 0.029 W_b^{1.28} D_{max}^{1.10} \quad (64)$$

This equation, with standard error of 0.36 log units, is based on 33 sites in the Great Plains of the United States plus 3 sites on the Murrumbidgee River of Australia.

In regard to tropical humid regions, Pickup and Warner (1984) studied two rivers in New Guinea and derived the following:

TABLE 4 Equations for Estimating Average Discharge $\overline{Q}$ for Various Categories of Bed and Bank Material

Category	Channel Sediment Characteristics[a] SC_{bd} (%)	SC_{bk} (%)	d_{50} (mm)	Equation Number	Expression for $\overline{Q}$	Number of Data Points	Standard Error (Log Units)
High silt-clay bed	61–100	—	<2.0	(57)	$0.031W_b^{2.12}$	15	0.15
Medium silt-clay bed	31–60	—	<2.0	(58)	$0.033W_b^{1.76}$	17	0.23
Low silt-clay bed	11–30	—	<2.0	(59)	$0.031W_b^{1.73}$	30	0.33
Sand bed, silt banks	1–10	70–100	<2.0	(60)	$0.027W_b^{1.69}$	33	0.23
Sand bed, sand banks	1–10	1–69	<2.0	(61)	$0.029W_b^{1.62}$	96	0.30
Gravel bed	—	—	2–64	(62)	$0.023W_b^{1.81}$	42	0.22
Cobble bed	—	—	>64	(63)	$0.024W_b^{1.84}$	19	0.11

SOURCE: Osterkamp and Hedman, 1982.

[a] SC_{bd} = silt-clay content of bed material; SC_{bk} = silt-clay content of bank material; d_{50} = diameter size of particles for which 50% of the distribution is finer; W_b = active-channel width.

$$\overline{Q} = 0.013W_b^{2.13} \quad (r^2 = 0.71) \quad (65)$$

$$\overline{Q} = 54.9D_b^{1.58} \quad (r^2 = 0.45) \quad (66)$$

$$\overline{Q} = 0.21A_b^{1.17} \quad (r^2 = 0.90) \quad (67)$$

The approximate applicable range of widths is $185 \leq W_b \leq 340$ m; the ranges for depth and flow area were not given. Also, the method of defining bankfull depth, D_b, is not clear in this instance.

Carlston (1965) produced an equation relating meander wavelength L_m to $\overline{Q}$ for 31 rivers, mostly from the central United States. Least-squares regression of those data with $\overline{Q}$ as the dependent variable yields

$$\overline{Q} = 0.000017L_m^{2.15} \quad (68)$$

(Williams, 1984a). The standard error is 0.11 log unit, and the applicable range is $145 \leq L_m \leq 15{,}500$ m.

Mean Annual Peak Flow. One of the discharges occasionally estimated in paleofluvial studies is the mean annual peak flow or "mean annual flood." This is not necessarily defined in the same way by everyone. All definitions are based on the same fundamental item of information, namely, the single highest instantaneous discharge for each year of record at the streamflow gauging station of interest. The simplest and most understandable definition is the arithmetic mean of these yearly maximum instantaneous discharges (Dunne and Leopold, 1978, p. 313). Alternatively, the geometric mean (the antilog of the mean of the logs of the annual peak flows) has been used (e.g., Lowham, 1982). However, when arranged according to magnitude of discharge, some sets of peak discharge data have a rather skewed distribution, even when logs are used. In such instances an average value is less meaningful; recurrence interval definitions become preferable because these definitions reflect the skewness to a greater extent.

A gauging station's annual peak flow data, which are arranged according to magnitude of discharge, may follow any of a variety of theoretical or unique distributions (Riggs, 1968; Reich, 1976). One common distribution is the Gumbel Type I extreme value distribution. If a series of values follows this distribution, then the arithmetic mean is the discharge corresponding to the 2.33-yr recurrence interval. Most hydrologists therefore define mean annual flood as the discharge corresponding to the 2.33-yr recurrence interval on the Gumbel Type I plot (regardless of how closely the data adhere to that distribution). More recently (U.S. Interagency Advisory Committee, 1981), the log-Pearson Type III distribution has begun to gain favor. On this distribution the arithmetic mean is the discharge having a 2.0-yr recurrence interval. In many instances this value is not significantly different from the 2.33-yr discharge determined from a Gumbel Type I plot.

The annual peak flows at a station could range from well within the streambanks to complete inundations of the entire valley. (*Mean annual flood* therefore is a misleading term, because "flood" implies overbank flow.) Also, peak flows are known to vary with climate and physiography. Thus, in using an empirical multisite equation for the paleofluvial estimation of average yearly peak discharge, at least three considerations should be kept in mind: (1) the method used to compute mean annual peak flow for the multisite equation; (2) the possibility of a mixture of flow stages (within banks and over the banks) represented in the annual peak flow data on which the equation is based; and (3) the climate and physiography of the region where the data for the equation were collected. Even by paleofluvial standards, estimation of mean annual peak discharge probably has a substantial error, especially when dealing with a physiographic region different from that on which the estimating equation is based.

Two multisite equations, both of them for rather restricted environments and both of them based on channel size, have been derived to estimate mean annual peak flow. One is Schumm's (1972) Great Plains–Murrumbidgee River equation:

$$Q_{2.33} = 2.66 W_b^{0.90} D_{max}^{0.68} \qquad (69)$$

As the subscript indicates, Schumm used the 2.33-yr recurrence interval discharge (Gumbel Type I plot) for mean annual peak flow. The standard error is 0.22 log unit.

The second such equation, using the same statistical definition of discharge as that used by Schumm, was determined by Mosley (1979) for 63 sites on the South Island of New Zealand:

$$Q_{2.33} = 3.74 A_b^{1.015} (D_{max}/R)^{-0.515} \qquad (70)$$

with SE = 0.182 log unit. The range of independent variables is $1.4 \leq A_b \leq 1630$ m^2 and $1.2 \leq (D_{max}/R) \leq 4$.

Williams (1984b) derived empirical equations for estimating mean annual peak flows and average discharges from mean radius of curvature for 19 meandering rivers in Sweden. The standard errors, however, are rather large.

Other Statistical Discharges. Carlston (1965) measured meander wavelengths for 28 gauged reaches, mostly in the central United States, and related these wavelengths to $Q_{1.5}$, the discharge having a 1.5-yr return period (considered by him to approximate bankfull discharge). A least-squares regression with $Q_{1.5}$ as the dependent variable (Williams, 1984a) yielded

$$Q_{1.5} = 0.011 L_m^{1.54} \qquad (71)$$

with SE = 0.171 log unit. The applicable range of wavelengths is from 145 to 15,500 m.

Knox (1985) analyzed data for 15 streamflow gauging stations in Wisconsin and derived the following empirical multisite relation between the discharge having a 1.58-yr recurrence interval ($Q_{1.58}$) and the cross-sectional flow area ($A_{1.58}$) associated with that discharge:

$$Q_{1.58} = 2.42 A_{1.58}^{0.72} \qquad (72)$$

The standard error is 0.157 log unit, and the applicable range of cross-sectional flow areas is about from 4.5 to 165 m^2. To apply this equation to relict channels, Knox assumed that the flow area associated with $Q_{1.58}$ = the bankfull flow area = the flow area corresponding to the top of fining upward (lateral accretion) sediments in buried point bars.

Osterkamp and Hedman (1982), in the analysis already mentioned involving 252 rivers in the west-central United States, developed predictive equations for discharges having a 2-, 5-, 10-, 25-, 50-, and 100-yr recurrence interval. Categorizing the data on the basis of bed-and-bank sediment brought about a smaller standard error for some groups, but not for others, compared to the general equation for the entire data set. The diagnostic variable again is active-channel width, and minor improvement in prediction occurs if channel gradient also is available. For Q_2 and Q_5 the general equations based on width alone are

$$Q_2 = 1.9 W_b^{1.22} \qquad (73)$$

(SE = 0.41 log unit) and

$$Q_5 = 5.8 W_b^{1.10} \qquad (74)$$

(SE = 0.42 log unit). Applicable widths range from 0.8 to 430 m.

Other Flow Variables

Aside from water discharge and mean velocity, other flow variables commonly estimated in paleofluvial studies are the instantaneous values of mean flow depth, bed shear stress, stream power, and the critical mean velocity needed to move a given sediment particle. These variables generally are estimated with sedimentological techniques rather than with channel dimension or channel planform methods. A possible exception in future research could be an estimation of stream power because only Q and S are involved.

ERRORS IN PALEOFLUVIAL ESTIMATES

Paleofluvial estimates commonly are not very accurate for several reasons. One reason for error is the inaccuracy or deficiency of the available equations. The empirical equations are based on limited data—limited in regard to quantity, quality, and geographic coverage. Some of the relations may be affected by possibly erroneous assumptions, for instance in regard to channel stability and the presence and influence of bedrock outcrops.

A second reason for error is environmental misapplication of available equations, that is, use of an equation for conditions in which the applicability is unknown or inappropriate. In many cases the researcher is aware of this problem but has no other equations. Some examples involve different channel patterns, different climatic environments, and perennial versus intermittent streams.

A third general source of error is unreliable paleofluvial input data. For instance, the size and shape of an exposed paleochannel cross section may not adequately represent the general reach of the paleochannel. The cross section may be at an atypical site along the reach or it may have changed, via compaction or tectonism, since the time it conveyed water. Even if the cross section is typical, accurate measurements may be very difficult to get; estimated dimensions based on rules of thumb may need to be used instead.

Improper algebraic manipulation of empirical equations for paleofluvial purposes is a fourth common source of error (Williams, 1983a,b). (Some manipulations are valid, but many others are not, as discussed by Williams and Troutman, 1987.) Part of this problem is the compounding of errors whereby a value estimated by one equation is used as input to make an estimate from another equation, as pointed out by Ethridge and Schumm (1978).

A fifth reason for error is a paleohydrologist's misunderstanding of (or lack of agreement on) the definition of certain variables in an equation. Channel bed width and bankfull width often are interchanged. Mean depth has mistakenly been used in place of maximum (thalweg) depth, and vice versa. Various definitions of "bankfull channel width" are used (Williams, 1978).

CONCLUSIONS

Emphasis in this review has been on methods in which the independent variables used to make paleofluvial estimates have been directly measured. Channel cross-sectional dimensions in many studies need not be estimated, as they commonly can be measured from an exposure. Where they cannot be measured directly, some empirical equations are available for estimating the cross-sectional size of meandering channels using meander geometry. No methods that involve only measured planform features seem to be available for estimating cross-sectional dimensions of streams of other patterns (braided and straight).

Paleochannel slope, where it cannot be measured directly, is one of the hardest variables to estimate. A great deal of research is needed in this area. The only equation that does not require previously estimated water discharge and other estimated variables seems to be Schumm's (1972) Great Plains–Murrumbidgee River relation, requiring W_b and D_{max}.

Probable values of average meander wavelength, arc distance, belt width, and radius of curvature can be estimated from any one of these planform features or from the channel's average cross-sectional area, top width, or mean depth. However, meander characteristics can vary considerably along a reach, and values for a particular site can differ significantly from the estimated average.

Mean flow velocity at any selected water stage can be approximated with the Lacey, Gauckler–Manning, Darcy–Weisbach, or Chezy equations, as long as S and (except for Lacey's equation) the resistance coefficient can be determined with sufficient accuracy. The same method can be used for instantaneous discharges. For bankfull discharge an alternative empirical equation requiring A_b and S is also available.

Long-term statistical discharges that have been of interest to paleohydrologists include average discharge, mean annual peak flow, and several others. Channel dimension methods of estimating these discharges have largely been limited to the west-central United States. The few relations based on meander features also apply only to rather local regions, until demonstrated otherwise. Planform methods for channels other than meandering have received virtually no research attention at all.

Many established empirical equations of the regime and hydraulic geometry type, relating a channel feature to water discharge, are of uncertain use in paleofluvial studies. If the channel feature is to be estimated, then a value for discharge must be obtained (estimated) first. This leads to making an estimate based on an estimate—obviously not a preferred approach. On the other hand, if discharge is the variable to be estimated, the necessary rearrangement of the equation adds an unknown and potentially serious error into the estimate. For equations based on data with very little scatter, the error could be negligible, but this situation is not too common.

Aside from the obvious need of more good equations, two paleofluvial problems in need of further research are how to determine the pattern of nonmeandering former streams and whether a paleostream was perennial or not. These are parts of the more general problem of determining paleoenvironment. Much work remains to be done on methods of paleofluvial reconstruction.

ACKNOWLEDGMENTS

I thank John S. Bridge, Waite R. Osterkamp, and James C. Knox for many helpful and constructive comments on the manuscript.

NOMENCLATURE

f	Darcy–Weisbach friction factor
g	Acceleration due to gravity
n	Resistance coefficient in Gauckler–Manning equation (Manning's n)
r	Correlation coefficient
A	Cross-sectional flow area
A_b	Channel (bankfull) cross-sectional area
$A_{1.58}$	Cross-sectional flow area associated with the discharge having a 1.58-yr recurrence frequency
B	Meander belt width
C	Resistance coefficient in Chezy equation
D	Mean flow depth A/W

D_b	Channel (bankfull) mean depth A_b/W_b
D_{max}	Channel maximum depth (elevation difference between lowest banktop and thalweg)
L_a	Meander arc distance
L_m	Meander wavelength
M	Schumm's (1960) percent silt-clay in channel perimeter
Q	Instantaneous water discharge
$\overline{Q}$	Average discharge
Q_b	Instantaneous bankfull discharge
Q_{max}	Mean annual peak discharge
$Q_{1.5 \cdots 100}$	Discharge that, on the average, recurs every $1.5 \cdots 100$ years
R	Hydraulic radius (= cross-sectional flow area divided by wetted perimeter)
R_c	Radius of curvature of a meander loop
S	Channel slope
SE	Standard error of the estimate
Si	Channel sinuosity
V	Mean flow velocity
W	Water surface width
W_b	Channel (bankfull) top width

REFERENCES

Albertson, M. L., Barton, J. R., and Simons, D. B. (1960). "Fluid Mechanics for Engineers." Prentice-Hall, Englewood Cliffs, New Jersey.

Allen, J. R. L. (1966). On bed forms and paleocurrents. *Sedimentology* **6** (3), 153–190.

Allen, J. R. L. (1970). A quantitative model of grain size and sedimentary structures in lateral deposits. *Geol. J.* **7** (Part 1), 129–146.

Barnes, H. H., Jr. (1967). Roughness characteristics of natural channels. *Geol. Surv. Water-Supply Pap. (U.S.)* **1849**, 1–213.

Barnes, H. H., Jr., and Davidian, J. (1978). Indirect methods. *In* "Hydrometry: Principles and Practice" (R. W. Herschy, ed.), pp. 149–204. Wiley, New York.

Benson, M. A., and Dalrymple, T. (1967). General field and office procedures for indirect discharge measurements. (*Tech. Water-Resour. Invest.* (*U.S. Geol. Surv.*) Book 3, Ch. A1, pp. 1–30.

Bray, D. I. (1979). Estimating average velocity in gravel-bed rivers. *J. Hydraul. Div. Am. Soc. Civ. Eng.* **105** (HY9), 1103–1122.

Bridge, J. S. (1978). Palaeohydraulic interpretation using mathematical models of contemporary flow and sedimentation in meandering channels. *Mem.—Can. Soc. Pet. Geol.* **5**, 723–742.

Bridge, J. S. (1984). Flow and sedimentary processes in river bends: Comparison of field observations and theory. *In* "River Meandering" (C. M. Elliott, ed.), pp. 857–872. Am. Soc. Civ. Eng., New York.

Bridge, J. S. (1985). Paleochannel patterns inferred from alluvial deposits: A critical evaluation. *J. Sediment. Petrol.* **55** (4), 579–589.

Bridge, J. S., and Diemer, J. A. (1983). Quantitative interpretation of an evolving ancient river system. *Sedimentology* **30** (5), 599–623.

Carlston, C. A. (1965). The relation of free meander geometry to stream discharge and its geomorphic implications. *Am. J. Sci.* **263**, 864–885.

Chang, H. H. (1979). Minimum stream power and river channel patterns. *J. Hydrol.* **41**, 303–327.

Cheetham, G. H. (1980). Late Quaternary palaeohydrology—The Kennet Valley case-study. *In* "The Shaping of Southern England" (D. K. C. Jones, ed.), Inst. Br. Geogr., Spec. Publ. No. 11, pp. 203–223. Academic Press, New York.

Chow, V. T. (1959). "Open-channel Hydraulics." McGraw-Hill, New York.

Church, M. (1981). Reconstruction of the hydrological and climatic conditions of past fluvial environments. *Tech. Bull.—Br. Geomorphol. Res. Group* **28**, 50–79.

Cotter, E. (1971). Paleoflow characteristics of a Late Cretaceous river in Utah from analysis of sedimentary structures in the Ferron Sandstone. *J. Sediment. Petrol.* **41** (1), 129–138.

Dunne, T., and Leopold, L. B. (1978). "Water in Environmental Planning." Freeman, San Francisco, California.

Dury, G. H. (1976). Discharge prediction, present and former, from channel dimensions. *J. Hydrol.* **30** (11), 219–245.

Ethridge, F. G., and Schumm, S. A. (1978). Reconstructing paleochannel morphologic and flow characteristics—methodology, limitations, and assessment. *Mem.—Can. Soc. Pet. Geol.* **5**, 703–721.

Ferguson, R. I. (1981). Channel form and channel changes. *In* "British Rivers" (J. Lewin, ed.), pp. 90–125. Allen & Unwin, London.

Foley, M. G., Doesburg, J. M., and Zimmerman, D. A. (1984). Paleohydrologic techniques with environmental applications for siting hazardous waste facilities. *Mem.—Can. Soc. Pet. Geol.* **10**, 99–108.

Gardiner, V. (1983). Drainage networks and paleohydrology. *In* "Background to Palaeohydrology" (K. J. Gregory, ed.), pp. 257–277. Wiley, New York.

Gardner, T. W. (1983). Paleohydrology and paleomorphology of a Carboniferous, meandering, fluvial sandstone. *J. Sediment. Petrol.* **53** (3), 991–1005.

Gauckler, P. G. (1867). Etudes théoriques et pratiques sur l'écoulement et le mouvement des eaux. *C. R. Hebd. Seances Acad. Sci.* **64**, 818–822.

Hand, B. M., Wessel, J. M., and Hayes, M. O. (1969). Antidunes in the Mount Toby Conglomerate (Triassic), Massachusetts. *J. Sediment. Petrol.* **39** (4), 1310–1316.

Hedman, E. R., and Osterkamp, W. R. (1982). Streamflow characteristics related to channel geometry of streams in western United States. *Geol. Surv. Water-Supply Pap. (U.S.)* **2193**, 1–17.

Hey, R. D. (1976). Geometry of river meanders. *Nature (London)* **262**, 482–484.

Hickin, E. J. (1983). River channel changes: Retrospect and prospect. *Spec. Publ. Int. Assoc. Sedimentol.* **6**, 61–83.

Inglis, C. C. (1947). Meanders and their bearing on river training. *Inst. Civ. Eng. Marit. Waterways Eng. Div. Sess. 1946–1947*, pp. 3–54.

Jackson, R. G., II (1978). Preliminary evaluation of lithofacies models for meandering alluvial streams. *Mem.—Can. Soc. Pet. Geol.* **5**, 543–576.

Knighton, A. D. (1975). Variations in at-a-station hydraulic geometry. *Am. J. Sci.* **275**, 186–218.

Knox, J. C. (1985). Responses of floods to Holocene climatic change in the Upper Mississippi Valley, USA. *Quat. Res.* **23**, 287–300.

Lacey, G. (1934). Uniform flow in alluvial rivers and canals. Minutes of *Proc.—Inst. Civ. Eng.* **237**, Part 1, 421–453.

Leeder, M. R. (1973). Fluviatile fining-upwards cycles and the magnitude of paleochannels. *Geol. Mag.* **110** (3), 265–276.

Leopold, L. B., and Maddock, T., Jr. (1953). The hydraulic geometry of stream channels and some physiographic implications. *Geol. Surv. Prof. Pap. (U.S.)* **252**, 1–56.

Leopold, L. B., and Wolman, M. G. (1957). River channel patterns—braided, meandering and straight. *Geol. Surv. Prof. Pap. (U.S.)* **282-B**, 39–85.

Leopold, L. B., and Wolman, M. G. (1960). River meanders. *Geol. Soc. Am. Bull.* **71** (6), 769–794.

Leopold, L. B., Wolman, M. G., and Miller, J. P. (1964). "Fluvial Processes in Geomorphology." Freeman, San Francisco, California.

Limerinos, J. T. (1970). Determination of the Manning coefficient from measured bed roughness in natural channels. *Geol. Surv. Water-Supply Pap. (U.S.)* **1898-B**, 1–47.

Lowham, H. W. (1976). Techniques for estimating flow characteristics of Wyoming streams. *Geol. Surv. Water-Resour. Invest. (U.S.)* **76-112**, 1–83.

Lowham, H. W. (1982). Streamflows and channels of the Green River basin, Wyoming. *Geol. Surv. Water-Resour. Invest. (U.S.)* **81-71**, 1–73.

Maizels, J. K. (1983). Proglacial channel systems: Change and thresholds for change over long, intermediate and short time-scales. *Spec. Publ.—Int. Assoc. Sedimentol.* **6**, 251–266.

Manning, R. (1891). On the flow of water in open channels and pipes. *Trans. Inst. Civ. Eng. Irel. 54th/55th Sess.* **20**, 161–207.

Melton, M. A. (1961). Discussion of the effect of sediment type on the shape and stratification of some modern fluvial deposits. *Am. J. Sci.* **259**, 231-233, and reply, pp. 234–239.

Miller, T. K., and Onesti, L. J. (1979). The relationship between channel shape and sediment characteristics in the channel perimeter. *Geol. Soc. Am. Bull.* **90** (Part 1), 301–304.

Morton, R. A., and Donaldson, A. C. (1978). The Guadalupe River and delta of Texas—a modern analogue for some ancient fluvial-deltaic systems. *Mem.—Can. Soc. Pet. Geol.* **5**, 773–787.

Mosley, M. P. (1979). Prediction of hydrologic variables from channel morphology, South Island rivers. *J. Hydrol. (N.Z.)* **18** (2), 109–120.

Osterkamp, W. R., and Hedman, E. R. (1982). Perennial-streamflow characteristics related to channel geometry and sediment in Missouri River basin. *Geol. Surv. Prof. Pap. (U.S.)* **1242**, 1–37.

Park, C. C. (1977). World-wide variations in hydraulic geometry exponents of stream channels—an analysis and some observations. *J. Hydrol.* **33**, 133–146.

Pickup, G., and Warner, R. F. (1984). Geomorphology of tropical rivers. II. Channel adjustment to sediment load and discharge in the Fly and Lower Purari, Papua New Guinea. *Catena* (Cremlingen, Destedt, Ger.), *Suppl.* **5**, 19–41.

Reich, B. M. (1976). Magnitude and frequency of floods. Cleveland, Ohio. *CRC Crit. Rev. Environ. Control* **6**, 297–348.

Rhodes, D. D. (1977). The b-f-m diagram—graphical representation and interpretation of at-a-station hydraulic geometry. *Am. J. Sci.* **277**, 73–96.

Richards, K. S. (1977). Channel and flow geometry—a geomorphological perspective. *Prog. Phys. Geogr.* **1** (1), 65–102.

Riggs, H. C. (1968). Frequency curves. *Tech. Water-Resour. Invest. (U.S. Geol. Surv.)*, Book 4, Ch. A2, pp. 1–15.

Riggs, H. C. (1978). Streamflow characteristics from channel size. *J. Hydraul. Div. Am. Soc. Civ. Eng.* **104** (HY1), 87–96.

Rotnicki, K. (1983). Modelling past discharges of meandering rivers. *In* "Background to Palaeohydrology" (K. J. Gregory, ed.), pp. 321–354. Wiley, New York.

Schumm, S. A. (1960). The shape of alluvial channels in relation to sediment type. *Geol. Surv. Prof. Pap. (U.S.)* **352-B**, 17–30.

Schumm, S. A. (1963). Sinuosity of alluvial rivers on the Great Plains. *Geol. Soc. Am. Bull.* **74**, 1089–1100.

Schumm, S. A. (1968). River adjustment to altered hydrologic regimen—Murrumbidgee River and paleochannels, Australia. *Geol. Surv. Prof. Pap. (U.S.)* **598**, 1–65.

Schumm, S. A. (1972). Fluvial paleochannels. *Spec. Publ.—Soc. Econ. Paleontol. Mineral.* **16**, 98–107.

Task Force on Friction Factors in Open Channels (1963). Friction factors in open channels. *J. Hydraul. Div. Am. Soc. Civ. Eng.* **89** (HY2), 97–143.

U.S. Interagency Advisory Committee on Water Data (1981). "Guidelines for Determining Flood Flow Frequency," Bull. No. 17B. Hydrology Subcommittee, Washington, D.C.

Wahl, K. L. (1984). Evolution of the use of channel cross-section properties for estimating streamflow characteristics. *Geol. Surv. Water-Supply Pap. (U.S.)* **2262**, 53–66.

Williams, G. P. (1970). Manning formula—a misnomer? *J. Hydraul. Div. Am. Soc. Civ. Eng.* **96** (HY1), 193–200.

Williams, G. P. (1978). Bankfull discharge of rivers. *Water Resour. Res.* **14** (6), 1141–1154.

Williams, G. P. (1983a). Improper use of regression equations in earth sciences. *Geology* **11**, 195-197.

Williams, G. P. (1983b). *Geology* **12**, 125–127.

Williams, G. P. (1984a). Paleohydrologic equations for rivers. *In* "Developments and Applications of Geomorphology" (J. E. Costa and P. J. Fleisher, eds.), pp. 343–367. Springer-Verlag, Berlin and New York.

Williams, G. P. (1984b). Paleohydrological methods and some examples from Swedish fluvial environments. II. River meanders. *Geogr. Ann.* **66A**, 89–102.

Williams, G. P. (1986). River meanders and channel size. *J. Hydrol.* **88**, 147–164.

Williams, G. P., and Troutman, B. M. (1987). Algebraic manipulation of equations of best-fit straight lines. *In* "Use and Abuse of Statistical Methods in the Earth Sciences" (W. B. Size, ed.), pp. 129–141. Oxford Univ. Press, London and New York.

20

PLANT ECOLOGICAL ASPECTS OF FLOOD GEOMORPHOLOGY AND PALEOFLOOD HISTORY

CLIFF R. HUPP
U. S. Geological Survey, Nashville, Tennessee

INTRODUCTION

The magnitude, frequency, and duration of flooding can limit or influence most aspects of vegetation life history. Furthermore, infrequent floods leave long-term evidence of their passage via dendrogeomorphic features on bottomlands (Sigafoos, 1964). The role of water in determining plant distribution patterns is incompletely understood; however, the importance of periodic flooding on bottomland vegetation patterns is becoming increasingly appreciated (White, 1979; McIntosh, 1980; Hupp and Osterkamp, 1985).

The material presented in this chapter largely pertains to the vegetation and fluvial features that develop along relatively high-energy streams with no tidal influence. Slow, meandering, coastal plain streams with broad valley bottoms (brown- and black-water rivers), as described by Wharton et al. (1982), may operate under different hydrogeomorphic conditions and are not treated here.

Floods have two long-term effects on bottomland woody vegetation: first, periodic floods of varying magnitude affect vegetation patterns, including creating "new" areas such as point bars for vegetation establishment such that suites of bottomland species with varying tolerances to flooding develop as bands parallel to the stream channel. Second, infrequent floods damage (usually not destroy) the bottomland plants such that their growth form reveals the effects of past floods either as outwardly evident stem deformations or as anomalous growth patterns in their serial tree ring sequence (typically both). This chapter is divided into two sections covering these two rather distinct subjects separately. Both vegetation patterns (discrete species distributions) and vegetation damage allow for the interpretation of the hydrogeomorphic conditions on a bottomland site. Thus, bottomland vegetation can be useful in reconstructing flood history and in flood prediction, particularly where hydrologic records are short or lacking.

BOTTOMLAND VEGETATION PATTERNS

Bottomlands, those areas subjected to periodic floods, develop characteristic fluvial landforms, each with associated distinct hydrogeomorphic processes (Wolman and Leopold, 1957; Leopold et al., 1964). Flow duration, flood frequency, flood intensity, depositional environments, and variation within each parameter on different geomorphic levels within a bottomland are aspects of the hydrogeomorphic processes which help shape fluvial landforms. Although many factors may affect species distribution patterns across a bottomland, the particular landform with its attendant hydrogeomorphic characteristics appear to be the most influential. (Sigafoos, 1961; Hupp and Osterkamp, 1985). In this chapter the term *bottomland* refers to all fluvially generated landforms and the vegetation they support. These landforms occur as terraces high in the valley section and, in order of descending elevation, proceed through floodplains, various riparian features, channel bars down to the channel bed (Fig. 1). Floodplains are flooded, on the average once every 2 yr (Wolman and Leopold, 1957); more frequent flooding occurs on the riparian surfaces below the floodplain

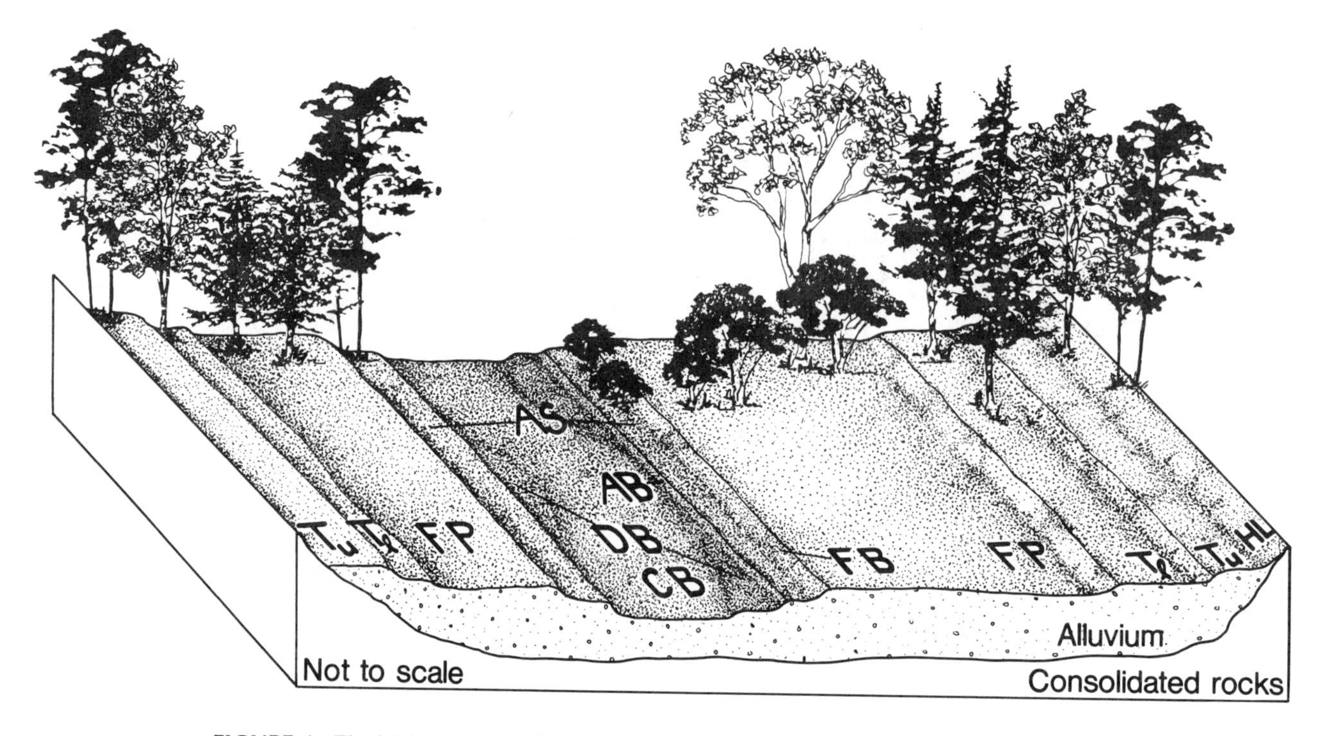

FIGURE 1. Fluvial landforms showing channel bed (CB), depositional bar (DB), active-channel bank (AB), active-channel shelf (AS) (channel shelf), floodplain bank (FB), floodplain (FP), low terrace (T_l), upper terrace (T_u), hillslope (HL). From Osterkamp and Hupp (1984) and Hupp and Osterkamp (1985).

elevation, less frequent flooding is associated with terraces. These hydrologic distinctions among the fluvial landforms and their significance relative to bottomland vegetation patterns will be discussed more fully in following sections. Studies have related the geometries of alluvial channels to the characteristics of water and sediment discharge (e.g., Lane et al., 1982; Osterkamp and Hedman, 1982). Aspects of the relations between vegetation and hydrogeomorphology have been summarized in Bedinger (1979), Wharton et al. (1982), and Hupp and Osterkamp (1985). The species included in any patterns of bottomland vegetation distribution are, of course, determined in part by regional floristics and climatic variation. However, through the work of many over much of North America, it is reasonable to state that consistent characteristic bottomland vegetation patterns develop in response to prevailing hydrogeomorphic conditions. It is stressed that clear definitions of fluvial landforms are necessary for interpreting bottomland vegetation patterns. These definitions must be couched in terms of flow characteristics such as flow duration and flood frequency. The hydrologic definitions provide the only independent parameters consistent on all perennial streams. A problem with much past plant ecological work is that many researchers have considered all fluvial features to be the floodplain. Another problem is the use of the word "riparian," which often is used in a cavalier fashion. Riparian refers to banks and as such should be reserved for bank or channel-shelf features and vegetation.

Separating the factors that influence bottomland vegetation patterns is difficult because most are distinctly interdependent. The following discussion attempts to separate the physical factors along lines drawn by previous literature and this author's own bias.

Fluvial Landforms

The floodplain is perhaps the most well-known fluvial landform (Fig. 2). Although a range of definitions have been applied to the term (Williams, 1978), the floodplain is generally accepted to be flooded once every 1–3 yr (once every 2 yr on the average); the water elevation just necessary to reach the floodplain is termed *bankfull stage* (Wolman, 1955). Terraces (T) may occur for a number of reasons (Howard et al., 1968); many are former floodplains where the stream has degraded the lower parts of the valley bottom sufficiently to preclude frequent terrace inundation (Fig. 2). All terraces are flooded less frequently than the floodplain, some are so high in elevation that inundation is extremely improbable. Terms for geomorphic features lower than the floodplain have received less attention. Fluvial landforms, particularly the floodplain and lower features, form in response to the hydraulics of fluvial systems, re-

FIGURE 2. View of channel features on (*a*) Passage Creek, Virginia, and (*b*) Potomac River near Chain Bridge, Washington, D.C. Note the broad shelf area on the Potomac River (*a*) and taller floodplain (FP) forest behind channel shelf (AS) vegetation (*a*) and (*b*).

gardless of whether streams are ephemeral, intermittent, or perennial (Hedman and Osterkamp, 1982).

Vegetation modifies or helps delineate width, height, and stability of fluvial landforms (Fig. 1). Characteristic species assemblages associated with different landforms can be used to identify and accentuate breaks in slope separating these alluvial features (Osterkamp and Hupp, 1984). Vegetation has been particularly useful for delineating riparian landforms below the floodplain. The active-channel shelf (AS) (Osterkamp and Hupp, 1984) is a horizontal to gently sloping surface (Fig. 1) that extends the usually short distance between the break in the relatively steep bank slope and the lower limit of persistent woody vegetation that marks the channel bed or depositional bar (Fig. 1).

Bottomland forests offer an exceptional opportunity to assess the effects of periodic flood disturbance on vegetation, owing to the availability of sometimes long and detailed records for flood frequency and magnitude. White (1979) and Pickett (1980) have suggested that disturbance characteristics play a major role in the development of many vegetation patterns. Periodic flooding may control certain riparian-vegetation types that persist as "nonequilibrium" systems (Pickett, 1980) without the loss of species compositional integrity through succession. It may be misleading to refer to bottomland forests as nonequililbrated because they may be well adjusted to periodic destructive flooding. This problem is, however, more semantic than conceptual. Previous studies in eastern United States have suggested that characteristic variations in bottomland forest composition are maintained by periodic flooding (Hosner and Minckler, 1963) and are not plant communities in various states of recovery since the last destructive flood (Sigafoos, 1961; Yanosky, 1982a; Hupp, 1983a). A detailed flood history along Passage Creek by Hupp (1982), using tree ring analyses, suggests that succession after flooding has little to do with vegetation patterns on the fluvial landforms. Apparently the plant distributions are at least in part controlled by inundation frequency and susceptibility of plants to damage by the destructive flooding. As examples, two shrubs on the active-channel shelf (AS, Fig. 1), alder (*Alnus serrulata*) and red willow (*Cornus amomum*) are relatively resistant to destruction by flooding owing to small, highly resilient stems and the ability to sprout rapidly from flood-damaged stumps (Hupp, 1983a). Conversely, flowering dogwood (*Cornus florida*) and some species of oak (*Quercus*) and hickory (*Carya*), which commonly grow on terraces but not on lower features (Fig. 1), may be intolerant of repeated flood damage or inundation. Flood plain (FP, Fig. 1) species, such as bitternut (*Carya cordiformis*) and black walnut (*Juglans nigra*) are probably less tolerant of destructive flooding than species of the active-channel shelf (AS, Fig. 1) but more tolerant of periodic inundation than are terrace (T_l T_u, Fig. 1) species.

That many species are indicative of bottomlands has long been recognized, however, fewer studies have related intra-bottomland species patterns to distinct hydrogeomorphic conditions. In northern Virginia, Osterkamp and Hupp (1984) and Hupp and Osterkamp (1985) have documented regionally consistent and discrete relations between bottomland vegetation patterns and several fluvial landforms (Fig. 2 and Table 1). Depositional bars (DB, Fig. 1), inundated about 40% of the time (Hedman et al., 1972), rarely have persistent woody vegetation; however, they may be densely covered in water willow (*Justicia mariana*). The active-channel shelf (AS, Fig. 1), inundated between 10 and 25% of the time (Osterkamp and Hupp, 1984), supports a characteristic low shrub thicket (riparian shrub forest; Hupp, 1984). Typical species of the active-channel shelf (AS, Fig. 1) include alder (*Alnus serrulata*), winterberry (*Ilex verticilata*), red willow (*Cornus amomum*), black willow (*Salix nigra*), and ninebark (*Physocarpus opulifolius*). The floodplain, with a 1- to 3-yr flood re-

TABLE 1 Hydrogeomorphic Relations Between Bottomland Vegetation Type and Fluvial Landforms

Fluvial Landform	Vegetation Type	Percent of Time Inundated	Flood Frequency
Depositional bar	Herbaceous species	about 40%	—
Channel shelf	Riparian shrubs	10–25%	—
Floodplain	Floodplain forest	—	1–3 yr
Terrace	Terrace assemblage	—	3 yr

SOURCE: After Hupp and Osterkamp, 1985.

currence interval, has an often diverse flora but supports one or more of the following indicator species: black walnut (*Juglans nigra*), American elm (*Ulmus americana*), bitternut (*Carya cordiformis*), silver maple (*Acer saccarhinum*), and hackberry (*Celtis occidentalis*). Terraces (T, Fig. 1) flooded less frequently than once every 3 yr have at least some individuals of upland oaks and hickories. The landform indicator species relation is summarized in Table 2. Species such as sycamore (*Platanus occidentalis*), boxelder (*Acer negundo*), ironwood (*Carpinus caroliniana*), and green ash (*Fraxinus pennsylvanica*) are good indicators of floodplains and channel shelves but do not completely discriminate between the two landforms.

The fluvial landforms, per se, do not affect vegetation patterns; it is the fluvial-geomorphic processes (episodes of inundation, erosion, and deposition) that shape the fluvial landforms. Landforms then support different assemblages of vegetation that must come to terms with these processes. Therefore, it is felt that a process-oriented explanation of bottomland species distribution is needed to understand

TABLE 2 Typical Vegetation Types on Northern Virginia Fluvial Landforms

Depositional Bar	Terrace—Terrace Assemblage
Herbaceous vegetation, commonly water willow, woody species largely absent; occasional willow, sycamore, or cottonwood seedlings	*Amelanchier arborea* (shadbush)
Channel Shelf—Riparian Shrubs	*Carya tomentosa*[a] (mockernut)
Alnus serrulata[a] (alder)	*Fraxinus americana* (white ash)
Cornus amomum[a] (red willow)	*Pinus virginiana*[a] (Va. pine)
Cephalanthus occidentalis (buttonbush)	*Sassafras albidum* (sassafras)
Ilex verticilata (black alder)	*Quercus prinus* (chestnut oak)
Physocarpus opulifolius (ninebark)	*Carya glabra*[b] (pignut)
Viburnum dentatum (arrow wood)	*Cercis canadensis*[b] (redbud)
Vitis riparia (riverbank grape)	*Cornus florida*[b] (dogwood)
Acer negundo[b] (box elder)	*Kalmia latifolia*[b] (Mt. laurel)
Populus deltoides[b] (cottonwood)	*Quercus alba*[b] (white oak)
Salix nigra[b] (black willow)	*Quercus rubra*[b] (red oak)
Ulmus rubra[b] (slippery elm)	*Quercus velutina*[b] (black oak)
Floodplain—Floodplain Forest	
Carya cordiformis[a] (bitternut)	
Celtis occidentalis (hackberry)	
Juglans nigra[a] (black walnut)	
Staphylea trifolia (bladdernut)	
Ulmus americana[a] (Am. elm)	
Betula nigra[b] (river birch)	
Carpinus caroliniana[b] (ironwood)	
Fraxinus pennsylvanica[b] (green ash)	
Lindera benzoin[b] (spicebush)	
Platanus occidentalis[b] (sycamore)	

[a] Species with widespread distribution.

[b] Common on indicated landform, however, frequently important on other fluvial features.

more fully the field of bottomland forest ecology and to allow practical applications (flood frequency estimation) of indicator sets of species.

Frequency and Duration of Inundation

The vegetation types (herbaceous depositional bar species, riparian shrubs, floodplain forest, and terrace assemblage) represent a gradient largely controlled by variation in frequency and duration of inundation from the channel bed to the terrace (Fig. 1, Table 1). Species composition and growth form display marked variation along this gradient much like those described by Sigafoos (1961), where he described parallel bands of sediment type and associated vegetation related to flood frequency. Furthermore, the effects of destructive flooding (battering by flood debris, bedload, or ice) on vegetation declines toward the terrace levels. This decrease in flooding effects may be due to higher elevation (less frequent flooding) and intervening vegetation (Hupp, 1983a). Proceeding from the terraces, typical species are capable of surviving increasing levels of destructive flooding and duration of inundation. Yanosky (1982a,b) has also reported that a gradient of flood intensity (stage, velocity, and turbulence) results in vegetation patterning. The rapid sprouting of shrub stems from flood-damaged trunks and root stocks on channel shelves and other riparian features indicates a resiliency for recovery from periodic destructive flooding. Floodplain species are known to withstand a high duration of periodic inundation relative to upland species that may occur on terraces (Hall and Smith, 1955; Hosner, 1960; Broadfoot and Williston, 1973; Yanosky, 1983).

Along parts of the Potomac River near Washington, D.C., Sigafoos (1961) found vegetation patterns and growth habit to be related to the return period of peak discharges. Osterkamp and Hupp (1984) determined that the elevations delimited by Sigafoos (1961) conformed to discharge frequencies associated with active-channel shelf (AS), floodplain (FP), and terrace (T) levels (Fig. 3). In British Columbia, Teversham and Slaymaker (1976) found the distributional patterns of several species to be related to elevation above the channel and duration of inundation along stable reaches. They also integrated flood frequency with dominant sediment size of substrate and developed a model where vegetation could be used to estimate flood frequency (Fig. 4). Hosner and Minckler (1963), Chambliss and Nixon (1975), Nixon et al. (1977), and Swanson and Lienkaemper (1982) found similar vegetation patterns related to elevation above the channel bed. Baker (1976) also described different assemblages of tree species associated with different flood levels in Texas. Wistendahl (1958) presents a detailed account of bottomland vegetation along the Raritan River in New Jersey, where he found that slight differences in topography and flooding have a marked in-

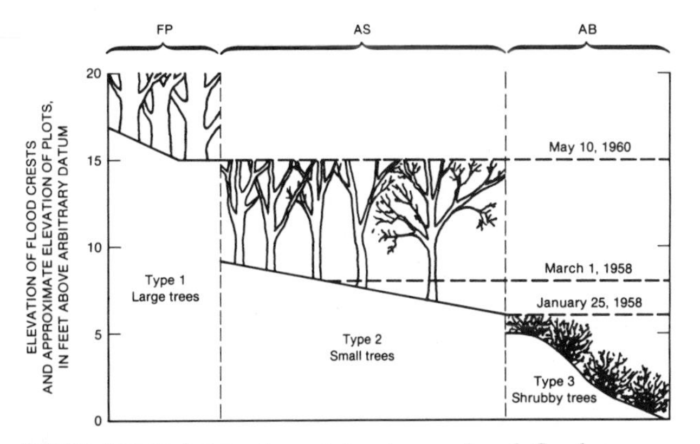

FIGURE 3. Relation of vegetation type and peak flood stage on the Potomac River bottomland near Washington, D.C. Breaks in slope, indicated by dashed vertical lines, conform to flood plain (FP), and channel shelf areas (AS) and active-channel bank (AB). Photograph in Figure 2*b* is a plan view of this cross section. (Modified from Sigafoos, 1961.)

fluence on vegetation patterns (Fig. 5). Bell and del Moral (1977) describe a relation between species distribution and flood frequency in Illinois.

Flooding represents a major limiting factor in vegetation establishment from the floodplain down to the channel bed. A trend for increasingly smaller statured species occurs

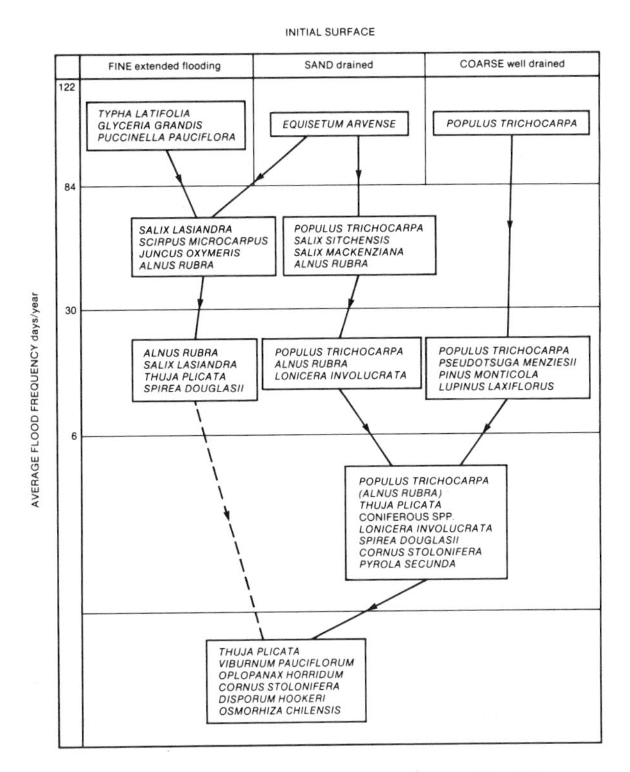

FIGURE 4. Model of relations among vegetation, flood frequency, and initial surface sediment. (Modified from Teversham and Slaymaker, 1976.)

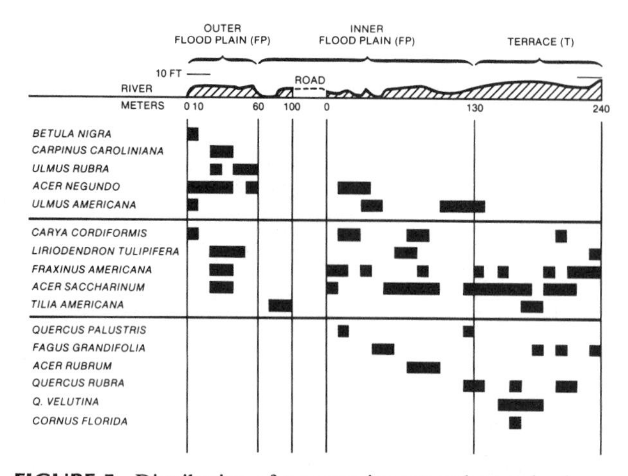

FIGURE 5. Distribution of tree species across bottomland geomorphic features on the Raritan River, New Jersey. (Modified from "The Flood Plain of the Raritan River, New Jersey" by W. A. Wistendahl, *Ecological Monographs*, 1958, **28**, 129–153. Copyright © 1958 by Ecological Society of America. Reprinted by permission.)

from the floodplain bank to the channel bed (Fig. 1); ultimately herbaceous species dominate at the channel edge (Hupp, 1983a). Die back of herbaceous plants in the fall, leaving only the perennating rootstocks, may protect these plants from high winter and early spring stages. The resprouting of stems on the channel shelf indicates a "resiliency" expressed as rapid recovery from periodic destructive flooding.

The dynamics of the seedling stage of vegetation is probably the least understood and yet one of the most important determinants of later vegetation patterns (Harper, 1977). Some hydrogeomorphic condition or group of conditions that reflect fluvial processes must exist before bottomland species can occupy an area. For example, sycamore (*Platanus occidentalis*) presence seems to be related to a minimum or threshold volume of alluvium that has been fluvially reworked into a mineral soil seedbed that the species needs for germination (Sigafoos, 1976). The seedbed for any species also has to exceed some minimum size to ensure survival, given the vagaries of seed dispersal, germination, and flood damage on the valley floor (Zimmermann and Thom, 1982).

In Maryland, species such as black willow (*Salix nigra*), cottonwood (*Populus deltoides*), box elder (*Acer negundo*), and silver maple (*Acer saccharinum*) characterize low-gradient and high-order streams (third- or fourth-order), with fine-grained alluvium (Zimmermann, 1980). But on erodible bedrock or saprolite these species may occupy first- or second-order streams (Zimmermann, 1980). Schneider and Goodlett (1962) found similar trends in the Yellow River basin of Georgia associated with the presence of permanent aquifers. They observed a distinct break in certain bottomland species distributions that coincided with the disappearance of the aquifer during severe droughts.

Relatively high water availability in upland humid regions tends to make species distributional patterns subtle at the scale of landforms (Zimmermann, 1980). The work of Zimmermann (1969) in Arizona has shown that clear relations can be documented in semi-arid regions. He found very strong correlations between several bottomland species, baseflow characteristics, and storm runoff, which could be used to predict hydrologic conditions without intensive hydrologic monitoring (Fig. 6). Similarly, Shreve (1915) and Kassas and Girgis (1964) found that the number and types of bottomland species coincide with the amount of runoff from infrequent heavy and sustained storms in arid regions.

Age of Bottomland Surface

Several authors have suggested that the age of a fluvial surface may be an important determinant in bottomland vegetation patterns. Although most of these authors acknowledge a relation with elevation (thus flood frequency), a plant successional influence is strongly implied. Notable among these studies are Hefley (1937), Ware and Penfound (1949), Fonda (1974), Hickin and Nanson (1975), Johnson et al. (1976), and Nanson and Beach (1977).

The term *succession* has been overworked in ecological literature. To some it implies the classical directional change in community structure through time, independent of environmental conditions. Others use succession to mean a gradual development of vegetation in response to external environmental change; these authors imply some sort of equilibrium between vegetation and environment at each stage of development. The latter is probably the best interpretation and the one implied in the present study.

Terraces in the Hoh River valley, Washington, were investigated by Fonda (1974). These terraces vary in age, dominant sediment, and vegetation. Fonda (1974) suggests that soil age and moisture may be the leading determinants of vegetation patterns and that successional development over various lengths of time has led to distinct plant "communities." The distributional patterns of six tree species over four "terraces" is shown in Figure 7.

Scroll topography adjacent to the Beatton River, British Columbia, has been investigated by Nanson and Beach (1977). They showed that increasing age of ridges and swales away from the channel could be related to differences in vegetation patterns and rates of sedimentation (Fig. 8). As in Fonda (1974), Nanson and Beach (1977) imply that successional development following deposition is an important factor in bottomland species patterns, at least along migrating channels.

The placement of bottomland species distribution into some successional framework has occurred in a number

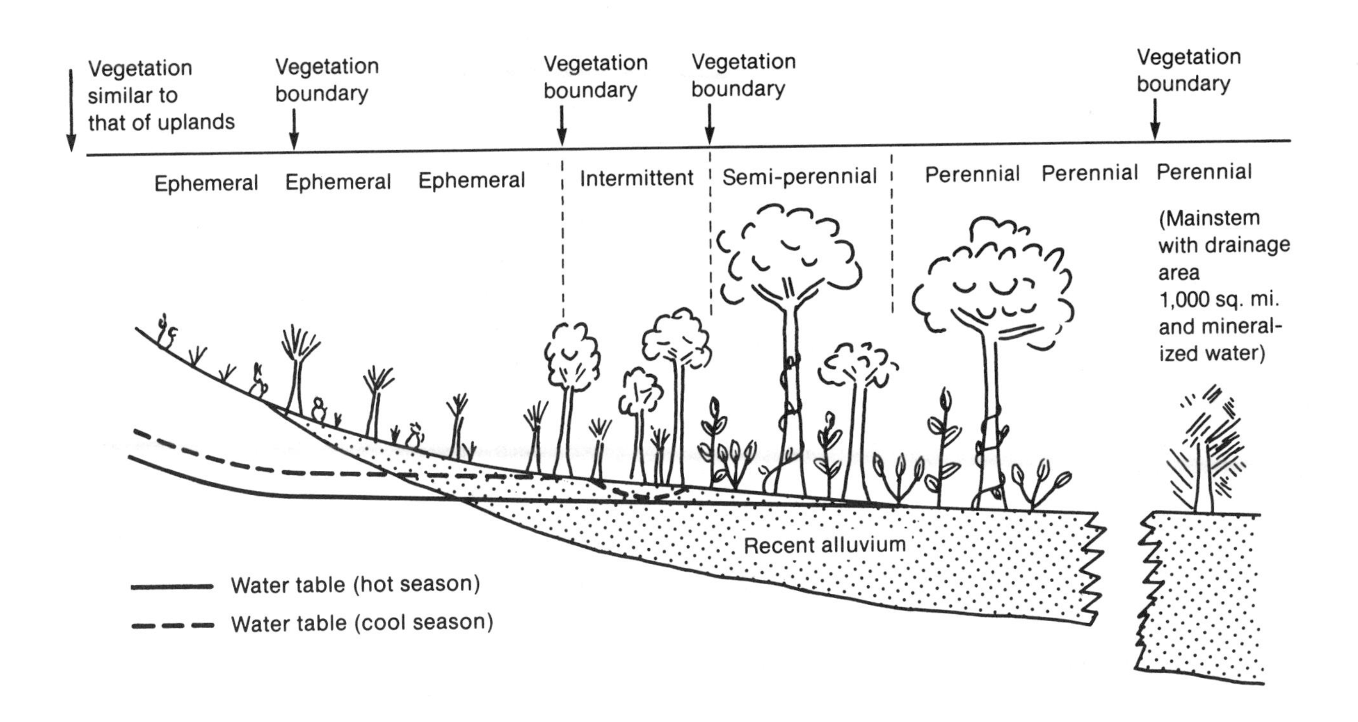

FIGURE 6. Cross section of valley floor vegetation and seasonal hydrologic conditions in southern Arizona. (Modified from Zimmerman, 1980.)

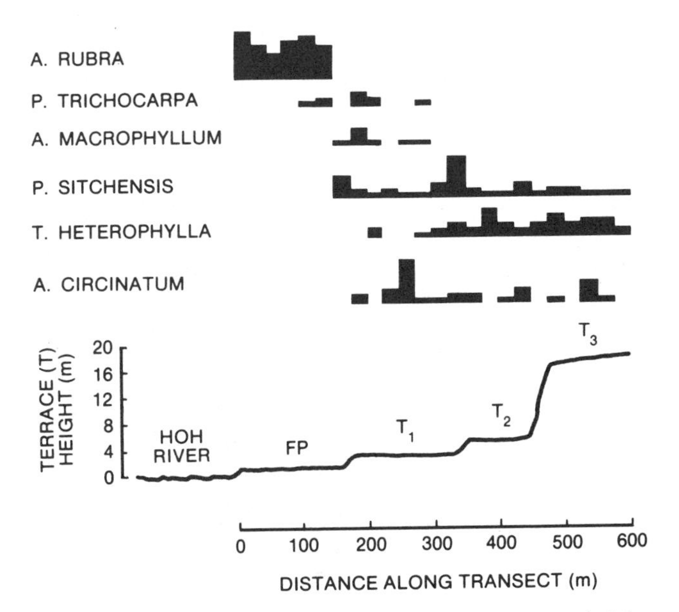

FIGURE 7. Vegetation distribution in relation to terrace height profile and distance from Hoh River, Washington. (Modified from "Forest Succession in Relation to River Terrace Development in Olympic National Park, Washington" by R. W. Fonda, *Ecology*, 1974, **55**, 927–942. Copyright © 1974 by Ecological Society of America. Reprinted by permission.)

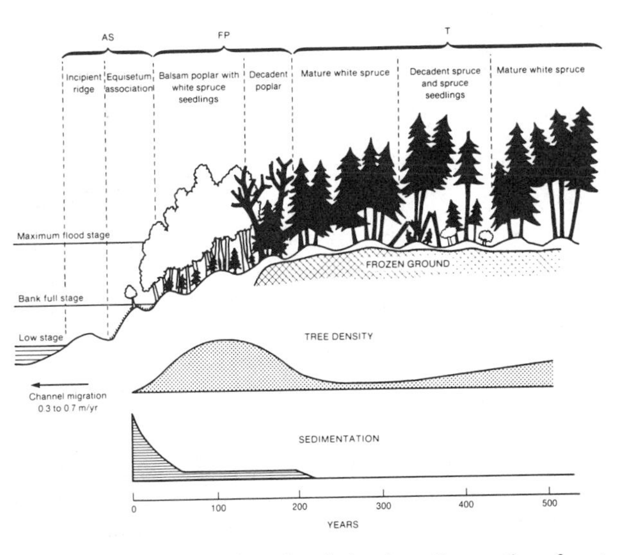

FIGURE 8. Representation of variation in sedimentation, forest cover, understory species richness, tree density, dense moss cover, frozen ground, and flood stages along the Beatton River, British Columbia. Direction of active-channel migration is indicated. (Modified from Nanson and Beach, 1977. Copyright © 1977 by Blackwell Scientific Publications Limited.)

of reports; these include Shaver and Denison (1928), Weaver (1960), Lindsey et al. (1961), Hosner and Minckler (1963), Robertson et al. (1978), and Frye and Quinn (1979). A minor role for succession, in the classical sense, in bottomland species patterns has been suggested by Sigafoos (1961, 1964), Goodlett (1969), Drury and Nisbet (1973), Yanosky (1982a), Hupp (1982, 1983a), and Hupp and Osterkamp (1985). Natural flood disturbance is a common occurrence on bottomlands and may, in the long run, maintain vegetation patterns (Pickett, 1980; White, 1979). Bottomland vegetation, adapted to periodic floods, may be regulated by variable flooding tolerances. A reason for the lack of consensus on the role of succession may stem from the fact that studies with reasonably good evidence of succession were conducted on streams with actively migrating channels (Nanson and Beach, 1977) or are affected by prolonged periods of aggradation or degradation (Fonda, 1974). Such conditions periodically create "new" surfaces for vegetation establishment. Conversely more stable streams, with a smaller sediment load, may have apparently similar vegetation patterns unrelated to successional stages.

Stream Gradient

Few plant ecologists have addressed the effects of variation in stream gradient on bottomland vegetation patterns (Hupp, 1982). Stream gradient, the slope or gradient of the channel along a given reach, influences many aspects of fluvial geomorphic and hydrologic processes (Hack, 1957; Lane, 1957; Kilpatrick and Barnes, 1964; Leopold et al., 1964; Osterkamp, 1978; Bull, 1979; Emmett, 1982).

Cowles (1901) described bottomland vegetation variation in Illinois that coincided with variation in stream gradient. Cowles tied vegetation patterns to physiographic forms and processes, noting a trend in vegetation composition from xerophytic (upland) to mesophytic (bottomland), which coincided with stream development from initial ravines (steep gradient) to meandering reaches (gentle gradient). The destructive action of currents and depositional processes were cited as probable causes of variation in the vegetation patterns (Cowles, 1901). Similar trends were found along Passage Creek in northern Virginia (Hupp, 1982), where a lithologic discontinuity created an anomalously high stream gradient and a corresponding shift to upland vegetation.

Stream gradient has a significant influence on the factors that determine discharge along a reach (channel width and depth and water velocity). Given a relatively constant discharge along a reach, variation in gradient changes the proportion each factor contributes to discharge. Changes in width, depth, or velocity are related to variation in channel geometry and flood intensity. Increased stream gradient usually increases the stream power with concomitant increases in channel width, decreases in channel depth, and increases in channel roughness (Leopold et al., 1964). These changes usually create greater local flow velocities and turbulence through the reach of increased gradient particularly during flood stage (Bull, 1979; Hupp, 1982). High flow velocity, turbulence, and transported debris are conducive to greater amounts of flood damage to the adjacent bottomland than along reaches with more gentle gradients (Sigafoos, 1964; Hupp, 1982; Yanosky, 1982a).

Along Passage Creek in northern Virginia, Hupp (1982) investigated the effects of increased stream gradient on fluvial landforms and vegetation patterns. Passage Creek, after meandering through a reach of erodible shale, contacts a very resistant sandstone and then flows through a steep gorge with steep gradients, coarse bed material, and altered channel geometry. The relative proportion of channel shelf area increases in the gorge along with its flood-adapted vegetation. The floodplain is nearly absent in the gorge and vegetation immediately above bankfull level is decidedly more upland in composition (Fig. 9, Table 3). Damage to trees from destructive floods is considerably more evident in the gorge reaches than in the reaches above the Passage Creek gorge.

In addition to high flow velocity and turbulence in gorges, this increase in flood intensity may also partly result from a greater amount of vegetated surface area below bankfull (Fig. 9) than along more gentle reaches (Hupp, 1982). Thus, the effects of floods can be expected to be more damaging in areas of steep gradients than those of gentle gradients. A summary of geomorphic and botanical changes associated with increased gradients is presented in Table 4.

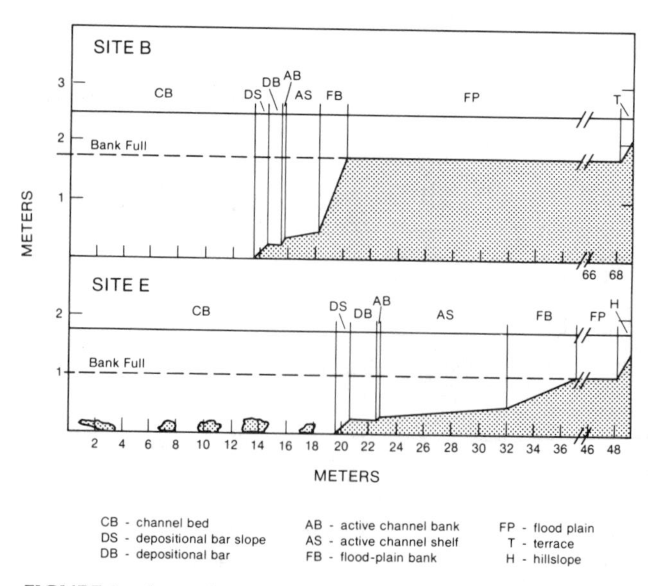

FIGURE 9. Generalized cross section of Passage Creek, Virginia, along a high-gradient reach (site B) and a low-gradient reach (site E). Note greater area of channel shelf along high-gradient reach (area below bankfull discharge) than on low-gradient reach. Landforms are indicated (from Hupp, 1982).

TABLE 3 Importance Values for Floodplain Tree Species along Low-Gradient and High-Gradient Reaches of Passage Creek, Virginia[a]

Species	Importance Value
SITE B (LOW GRADIENT, GI = 119, H′ = 2.67)	
Liriodendron tulipifera (tulip tree)	62
Platanus occidentalis (sycamore)	36
Ulmus americana (Am. elm)	29
Acer rubrum (red maple)	25
Carya cordiformis (bitternut)	22
Fraxinus pennsylvanica (green ash)	13
Betula lenta (sweet birch)	5
SITE E (HIGH GRADIENT REACH, GI = 769, *H′*= 3.47)	
Quercus rubra (red oak)	40
Nyssa sylvatica (black gum)	28
Acer rubrum (red maple)	23
Pinus strobus (white pine)	21
Tsuga canadensis (E. hemlock)	18
Platanus occidentalis (sycamore)	17
Fraxinus pennsylvanica (green ash)	14
Liriodendron tulipifera (tulip tree)	11
Betula lenta (sweet birch)	6
Juniperus virginiana (E. red cedar)	5
Tilia heterophylla (basswood)	5
Robinia pseudoacacia (black locust)	3
Ulmus americana (Am. elm)	3
Populus deltoides (cottonwood)	3
Carya cordiformis (bitternut)	3

SOURCE: After Hupp, 1982.

[a] GI is gradient index. Importance value is the sum of relative values of basal area and stem density. H′ is the Shannon–Wienner diversity index.

There was an increase in species diversity associated with the increase in gradient on Passage Creek (Tables 3 and 4). R. S. Sigafoos (personal communication, 1980, unpublished data), after studying much of the Potomac River floodplain in the piedmont, found that the Chain Bridge area near Washington, D.C. (Fig. 2), had the greatest number of species. This is an area of poorly developed banks with a high frequency of flooding and is somewhat analogous to the Passage Creek gorge in that it is underlain by resistant rock and has a high stream gradient. Hack and Goodlett (1960) found more species in flood-damaged areas of the floodplain than nondamaged areas in the Shenandoah Mountains of Virginia. They attributed this higher diversity to periodic inundation by seed-carrying floodwater. A greater frequency of seed-carrying floods may, in part, be responsible for the higher diversity in the area of the gorge along Passage Creek. However, recent research in disturbance-prone areas has indicated that periodic disturbance may set up conditions conducive to greater diversity by creating open areas and maintaining small patches of the system in various states of vegetational recovery (Fox, 1977; Forcier, 1975; Huston, 1979).

Channel shelves (AS), a feature more pronounced in high-gradient reaches, were shown to occur farther upstream than floodplains (FP) (Hupp, 1986a), as would be expected with increasingly greater gradients in the headward direction. Osterkamp and Hupp (1984) suggest that different sets of hydrogeomorphic processes form the floodplain and the channel shelf, and that different suites of vegetation (floodplain forest vs. channel shelf shrubs) characterize each. Along any given reach, processes that favor the development of either the floodplain or the channel shelf will dominate, such that the formation of one feature develops at least partially at the expense of the other. Stream gradient appears to be a critical factor determining which set of processes dominate (Hupp, 1982, 1986a).

Hydrogeomorphic processes responsible for the development of floodplains are characterized by overbank flow and deposition of relatively fine-grained sediment. Thus most floodplain development occurs during relatively infrequent episodes, on the average, once every 1–3 yr. Gradients necessary for deposition of fine-grained sediment are generally low, allowing for sufficiently slow flow velocities. Hydrogeomorphic processes responsible for the

TABLE 4 Summary of Geomorphic and Botanical Changes Corresponding to a Lithologic Change to More Resistant Bedrock on Passage Creek, Virginia

Parameter	Change
Gradient index	Increase
Stream channel width	Increase
Stream channel depth	Decrease
Median size of bed material	Increase
Channel pattern	Meandering to anabranched
Flood damage to vegetation	Increase
Species diversity (*H′*)	Increase
Basal area	Increase
Vegetation zonation, parallel to stream	Weak to strong better developed
Forest type	Floodplain to riparian shrub

SOURCE: Hupp, 1982.

development of channel shelves (AS) are not well understood. Steep gradients and the deposition of sand are associated with well-developed channel shelves. Flow durations below 10–20% and high discharges tend to shape channel shelves, however, resilient woody plants tend to ameliorate erosive streamflow.

Kilpatrick and Barnes (1964) suggested that channel gradient had a great influence on the character, elevation, and extent of "benches" along southern piedmont streams. They found that deposition on floodplains was strongly related to channel gradient. Furthermore, Kilpatrick and Barnes (1964) found that steep-gradient streams contained the mean annual flood within the floodplain banks, which would enhance channel shelf development. Whereas, mild-gradient streams usually had their floodplains inundated by the mean annual flood (Kilpatrick and Barnes, 1964). This suggests a regime shift away from floodplain processes on steep-gradient streams as proposed by Hupp (1982) and Osterkamp and Hupp (1984). This regime shift can be detected through brief botanical reconnaisance; bottomlands with extensive channel shelf vegetation can be expected to suffer greater damage from flooding than bottomlands with little channel shelf vegetation.

Depositional Environment

Rates and types of sediment deposition have been shown to be important environmental factors affecting vegetation patterns. Also included here is sediment scour, which has an obvious negative effect on vegetation establishment and growth. Most bottomland substrates available for seedling establishment are alluvial, thus hydraulic sorting of size clasts and rate of bed, bar, or bank accretion may be important geomorphic processes that influence bottomland vegetation. Many riparian species are restricted to a narrow range of sediment types that allow for successful seed germination. Furthermore, relatively high, natural or otherwise, sedimentation rates limit the number of species that can occupy certain bottomland areas. Most woody plants must maintain their actively absorbing root zones in the upper 15 cm of soil (Coile, 1937; Bilan, 1960; Kramer and Kozlowski, 1979). Thus, only species capable of rapid root growth along newly buried stems may occupy bottomland areas periodically affected by frequent or large amounts of deposition or active-channel migration. Flooding regime is intimately associated with the depositional environment along most streams with adequate sediment supply.

Shaver and Denison (1928) describe a development of vegetation associated with channel bar growth in central Tennessee. Bars, composed of clasts from sand size to cobbles, when high enough to be above most water stages, were suitable for woody vegetation establishment; the only tree species to establish on these "bars" were black willow (*Salix nigra*), sycamore (*Platanus occidentalis*), silver maple (*Acer saccharinum*), and green ash (*Fraxinus pennsylvanica*). They also found that banks subjected to active accretion by sand rarely had woody species other than silver maple, hackberry (*Celtis occidentalis*), box elder (*Acer negundo*), or sycamore. Sycamore has been shown to require a mineral soil seedbed (usually sand) for germination (Sigafoos, 1976). Natural layering, the flood training of supple stems with subsequent burial and sprouting (Sigafoos, 1964; Everitt, 1968), occurs in sycamores growing on active bars along streams affected by high sedimentation rates associated with strip mining (Bryan and Hupp, 1984). Therefore, sycamore has both germination characteristics and the ability to survive high sedimentation rates and associated flooding, which make it an important species in active depositional environments.

Cottonwood (*Populus*) is another genus whose ecological requirements are similar to sycamore, although the genus is more important along western and midwestern streams. Everitt (1968) showed that the plains cottonwood (*Populus sargentii*) forms extensive, single-species stands along the point bars of the Little Missouri River in North Dakota. Flood training and natural layering on river bars formed episodically, permitted this species to grow where most other species were environmentally excluded (Everitt, 1968). Black cottonwood (*Populus trichocarpa*), in the Pacific Northwest, becomes established where bare and moist mineral (sand) soil is exposed (Smith, 1955). Such conditions occur naturally along actively migrating channels with substantial point bar deposition.

Saltcedar (*Tamarix chinensis*), an exotic species, occurs in riparian environments throughout much of the American southwest. Like sycamore and cottonwood, saltcedar is capable of natural layering associated with sediment deposition rates (Everitt, 1980). However, this species appears to germinate on gently sloping river banks where slowly receding floods produce slackwater deposits largely of silts and clays (Everitt, 1980). Quiet-water deposits are also associated with a common eastern riparian species, river birch (*Betula nigra*). Wolfe and Pittillo (1977) in North Carolina found that river birch stands thrived in protected riparian areas where the alluvium was largely composed of high portions of clay, silt, and organic matter, whereas areas with high proportions of sand usually supported little or no river birch. Similarly, Hupp and Simon (1986) found river birch to be a common "pioneer" riparian species along previously modified aggrading reaches of west Tennessee streams where the dominant accreted sediment was silt; on aggrading banks composed primarily of sand, box elder, silver maple, and alder (*Alnus serrulata*) were principal establishing species with river birch largely absent.

Actively degrading or scoured banks (cut banks), typically on outside bends (concave) usually have no establishing woody vegetation (Hupp and Simon, 1986). Fluvial or mass-wasting processes remove the substrate faster than seedlings can become established. Rapid accretion on point bars or banks limit species patterns such that only a few are adapted to the high sedimentation rates. These same species are usually excluded from floodplains, with less sedimentation, through competition with shade tolerant species (Everitt, 1968, 1980; Hupp and Simon, 1986). The end result is a distinct vegetation pattern produced by distinct variation in depositional environment and specific plant adaptations.

PALEOFLOOD HISTORY RECONSTRUCTION

Perhaps the most important extrinsic factor in bottomland systems is flooding—from prolonged inundation characteristic of large, low-gradient basins to short-period (flashy), destructive floods more characteristic of relatively small, montane basins. Thus, the knowledge of flooding characteristics and magnitude/frequency information is of great utility to students of bottomland systems. Flood prediction depends on knowledge of magnitude and frequency of past floods for any given stream. This information is obtained by analysis of records compiled at gauging stations and historical records of high stages. Ages of trees on bottomland landforms and of scars and sprouts from flood-damaged stems (Sigafoos, 1964; Harrison and Reid, 1967; Helley and LaMarche, 1973; Hupp and Sigafoos, 1982; Hupp, 1983b), and differences in properties of wood anatomy related to flooding (Yanosky, 1982a) yield dates of past floods. Such dates may be used to extend flooding history where short or no records exist. Long flood records provided by botanical evidence of floods permit more accurate estimations than from short records; thus, procedures that extend flood records are potentially useful. Damage to trees during flooding is often severe, but the proportion of trees killed may be low. High water velocities, turbulence, and large quantities of transported debris are characteristics of floods that are most damaging to bottomland vegetation (Sigafoos, 1964). However, even relatively gentle inundation can cause datable anomalies in intra-ring structure (Yanosky, 1982a, 1983).

The mechanism of tree growth is well documented in botanical literature. Sigafoos (1964), Phipps (1967), and Yanosky (1982b) have described tree growth in relation to surface water hydrology. The annual increment of tree growth has been the basis of many studies using tree rings in documentation of the magnitude and frequency for important hydrologic and geomorphic events (Sigafoos and Hendricks, 1961; Sigafoos, 1964; Harrison and Reid, 1967; LaMarche, 1968; Alestalo, 1971; Helley and LaMarche, 1973; Shroder, 1978; Yanosky 1982a,b; Hupp, 1983a, 1984).

Trees grow vertically only from the tips of branches. Radial growth (growth in girth) is accomplished by a vascular cambium beneath the bark that forms a cone of new wood each year. This cone is called a tree ring when viewed in cross section. A count of tree rings from an increment core taken from the base of the stem yields the age of the stem so long as each year is represented by a single ring. The standard dendrochronological technique of cross dating (Stokes and Smiley, 1968; Cleaveland, 1980) is used to ensure that dating errors are not introduced by the possible occurrence of missing or multiple rings.

Types of Botanical Evidence of Floods

Four basic types of botanical evidence of geomorphic events, floods in this case, can be identified (Hupp, 1983b, 1986b): (1) corrasion scars (2) adventitious sprouts (3) tree age, and (4) ring anomalies (Fig. 10). Detailed descriptions of each follows.

Corrasion Scars. Scars are the most conspicuous evidence of past flooding on riparian trees and shrubs. Currently, the most reliable and accurate method of tree-ring-determined

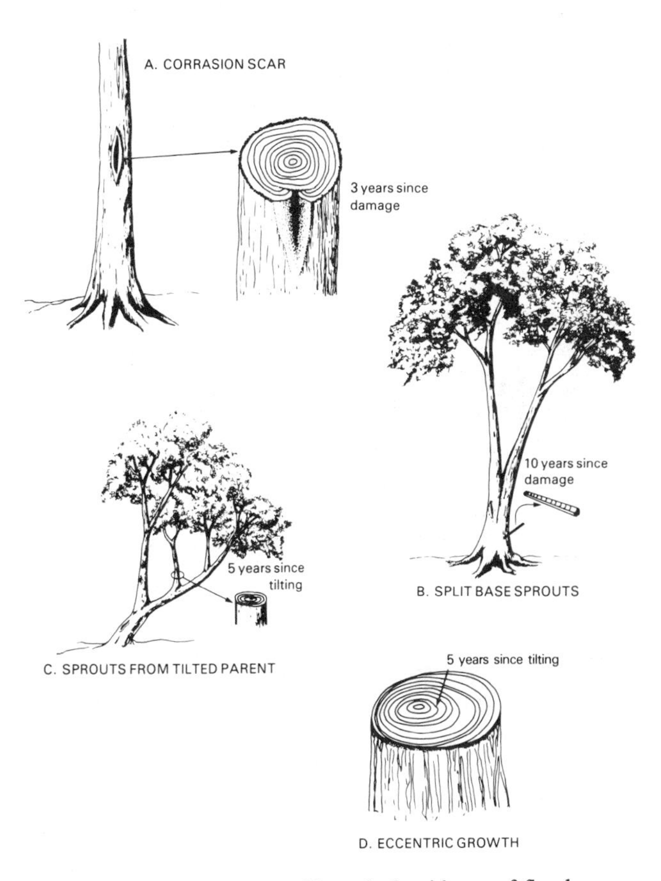

FIGURE 10. Types of botanical evidence of floods.

dates of flooding is the analysis of increment cores or cross sections through scars. Analysis of these samples yields the exact year of a flood. Owing to the differences in wood produced during the early and late parts of the growth season, it is often possible to determine the season when the scar was formed. Corrasion destroys the cambium (wood-producing tissue) in the area of impact, thus growth is stopped in the damaged tissue, and the event is recorded as undamaged tissue grows in annual increments around and over the scar (Fig. 10*a*). The scar, in time, is covered by this growth (callus). Evidence of the scar may remain for several years owing to the presence of this callus tissue. After many years, outward evidence of the scar may disappear, and the usefulness of scars as a dating technique diminishes, unless the tree is cut to reveal a scar deep in the wood. Maximum scar heights on damaged trees also allow for the estimation of minimum peak stage (Fig. 11).

FIGURE 11. Flood-scarred sycamore along Passage Creek, Virginia. Exposed scars on tree provide a record of four different floods, back to 1942. Height of scars provide a minimum stage for floods.

Adventitious Sprouts. Sprouts from broken or inclined stems are easily determined in the field and can date a flood by coring at the base of the sprout. This type of evidence, after a few years, has the appearance of a split base (Fig. 10*b*) or vertically growing sprouts from a tilted main stem that has been trained in a downstream direction (Fig. 10*c*). Flood training of riparian trees and shrubs is a highly visible feature along streams subjected to periodic high-velocity floods. Use of the term *sprout* implies growth form or habit and not youth, as some sprouts may be hundreds of years old. The age of these sprouts is usually within one year of the flood that caused the inclination or decapitation of the main stem. The center ring occurs in the first year of growth and must be included in the increment core. The sprout may begin growth either the same year of flooding or during the growth season immediately after the flood. Some trees may bear evidence of several floods. The accuracy obtained by coring sprouts is exceeded only by corrasion scar analysis, which often has limited use in dating older events.

Tree Age. Flood-deposited sediments or flood-scoured areas provide sites where vegetation can become established. For example, flood-deposited longitudinal bars or point bars (Everitt, 1968; Bryan and Hupp, 1984) may be rapidly colonized by certain woody species. The age of trees and shrubs growing on these "new" surfaces indicate a minimum time since initial deposition or scour (Fig. 10*c* and 10*d*). When all of the oldest trees on a deposit are within a few years of each other in age, it is reasonable to assume that the age of the oldest is close to the age of the deposit. Along actively migrating channels, tree age can be used to estimate channel shifts through time. Of course, some flood deposits are much older than the trees growing on them. The population age structure can indicate whether the deposit is young enough to be dated dendrochronologically (Sigafoos and Hendricks, 1961; Hupp, 1984). This technique has been used by Helley and LaMarche (1973) to extend the historic flood record for certain streams in northern California.

Ring Anomalies. Floods may affect trees without leaving outward evidence as described above. This occurs as abrupt changes in ring width or as various alterations in intraring tissue (Yanosky, 1983). Trees must be cored and have the cores analysed, usually microscopically, to detect and analyze anomalous ring patterns.

Eccentric ring patterns occur when a tree is tilted from vertical. Slight inclinations can induce eccentric ring patterns without causing the formation of adventitious sprouts. Abrupt tilting of a tree by a flood results in subsequent rings that are wide on one side of the trunk while on the opposite side the same rings are relatively narrow (Fig. 10*d*). When this pattern occurs after concentric ring production, the date of the onset of the eccentric growth is usually within one year of the event. This line of evidence is particularly useful in areas where conifers are the dominant species because even severe tilting will not induce adventitious sprouts in conifers (Hupp, 1984).

Suppression and release tree-ring sequences appear as abrupt shifts in ring width. This type of sequence can be caused by a number of events such as fire, disease, or windthrow. In general, suppression is the result of competition between the specimen tree and its neighboring trees. Should these competitors be removed or damaged, the specimen tree is said to be released, and a period of greater growth (wide rings) follows. Trees growing adjacent to the destructive path of a flood that removed or severely damaged neighboring trees may be released after the flood. Conversely, trees partially buried by flood-deposited sediments may have a period of suppressed growth while new roots are growing into the new soil surface. These types of evidence for a particular date must be used with a measure of caution, and substantial repetition of the date in other specimens must be found. Furthermore, trees unaffected by flooding must be cored for controls, to rule out climatic fluctuations.

Intra-ring abnormalities have been shown to be correlated with floods of magnitudes below that necessary for substantial overt evidence (Yanosky, 1983). These abnormalities occur as a zone of enlarged vessels between the annual ring boundaries (Fig. 12). Thus, the position of these "flood rings," relative to earlywood or latewood, will also allow for more precise dates of occurrence; sometimes within a few weeks (Yanosky, 1983). The height of the tree and position on the bottomland can also provide an estimate of the minimum peak stage of the flood.

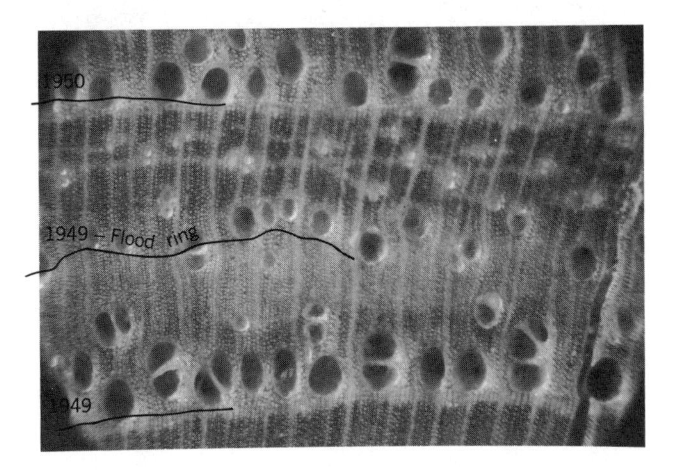

FIGURE 12. Cross section of green ash. Annual rings are indicated at the earlywood pore zone. Intra-annual flood ring (1949) is shown with arrow (from Yanosky, 1983; photograph by T. M. Yanosky).

Sediment Deposition and Erosion

Substantial amounts of sediment can be deposited or eroded in alluvial areas during infrequent floods. Rates and amounts of sediment deposition, scour, and associated channel shifting can be inferred using woody vegetation. Sigafoos (1964) documented changes in sediment levels on the Potomac River floodplain and along small tributaries through a dendrochronological study of trees affected by sediment deposition or scour. The dates of flooding from tree rings and deposition have been correlated with dates of flooding determined from Potomac River gauging stations (Sigafoos, 1964).

Sediment carried by floods may surround the bases of upright bottomland trees or partly cover trees tilted by flooding. Portions of trees that remain buried for at least one growing season will undergo changes in wood structure that, through laboratory analyses, allow for accurate determinations of date of deposition (Sigafoos, 1964). Similarly, tree stems and roots previously covered by sediment can be analyzed to determine the date of sediment scour.

The presence of adventitious roots above the initial root collar and natural layering allows for the determination of depth of sedimentation subsequent to germination (Fig. 13). Flood training and subsequent burial through lateral accretion of cottonwoods allowed Everitt (1968) to develop a history of channel migration for the Little Missouri River in North Dakota. Everitt (1968) dated valley-floor surfaces (Fig. 14) and estimated the rate of lateral accretion along a meandering reach for a 200-yr period through analysis of buried stems and adventitious roots.

Analysis of woody plants affected by sediment deposition or erosion is particularly useful in estimating sediment load in basins disturbed by strip mining. Increased sediment load, channel widening, and channel bar development, in response to strip mining, have been documented in an east Tennessee basin by Bryan and Hupp (1984). The vertical growth of a channel bar over a 3-yr period is shown in Figure 15; a sycamore sapling demonstrates the effects and depth of sediment deposition on bar vegetation (Fig. 16). Threshold discharges for bar development were determined through tree ring analysis and gauging station records. Results of Bryan and Hupp (1984) and Osterkamp et al. (1984) suggest that as much as 20% of total sediment load in heavily mined basins may be tractive, transported by peak discharges above thresholds for active bar development.

Knowledge of the site preference of alder (*Alnus serrulata*) to the stream edge (Hupp and Osterkamp, 1985) allowed for the determination of the amount and rate of channel migration along Smoky Creek, Tennessee (Bryan and Hupp, 1984). In places of active bar development, usually just downstream of the confluence with a tributary, some bars have grown laterally as much as 20 m in 25 yr,

FIGURE 13. Green ash on Potomac River floodplain. Tree was tilted by one flood partly buried by another (note root formation on inclined stem) and exposed by a third (from Sigafoos, 1964; photograph by R. S. Sigafoos).

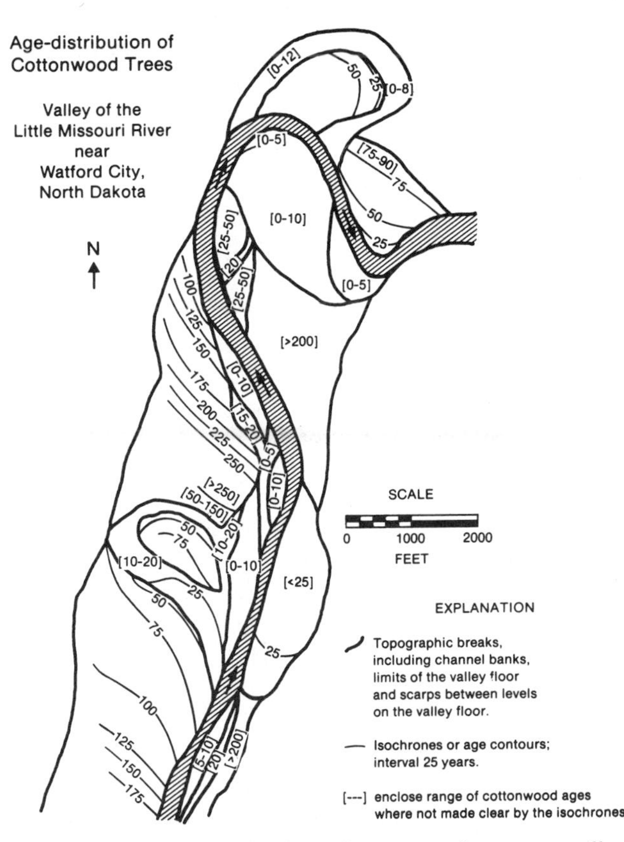

FIGURE 14. Age distribution of cottonwood trees on valley floor. (Modified from Everitt, 1968.)

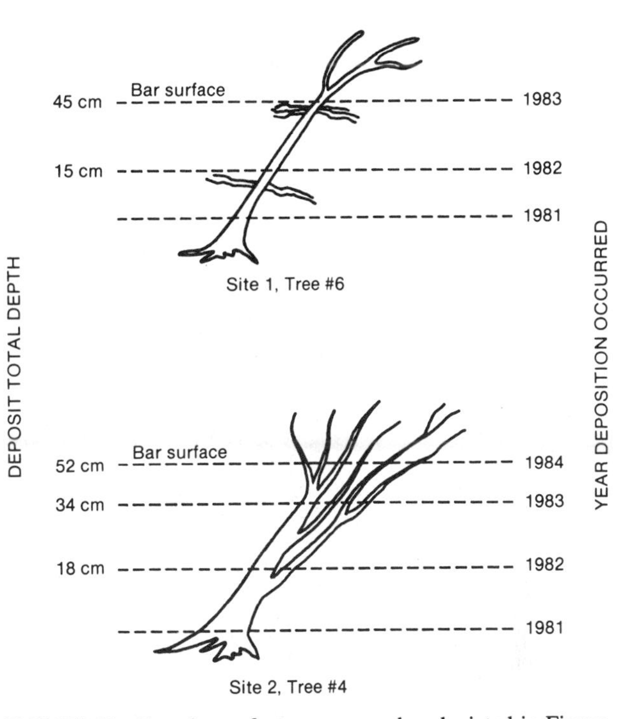

FIGURE 16. Drawings of sycamores on bar depicted in Figure 15. Zones of adventitious roots and stem sprouting indicate episodic deposition events; year and sediment depth are shown (from Bryan and Hupp, 1984).

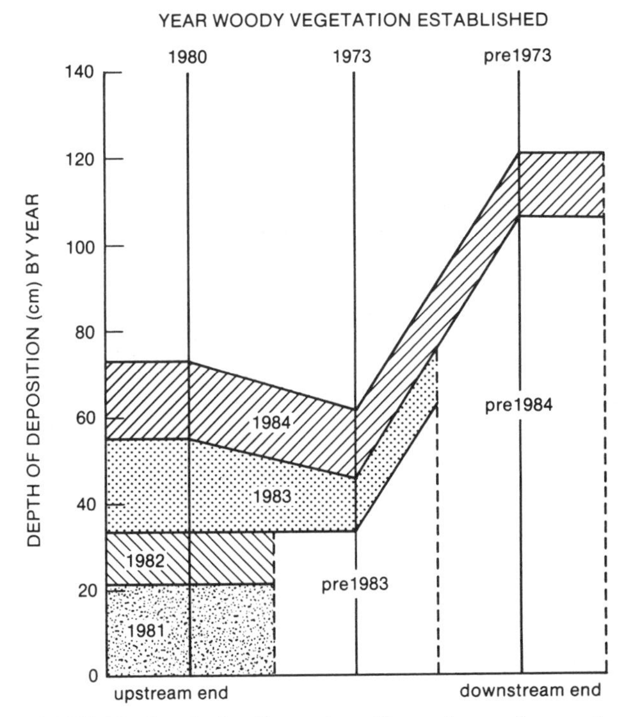

FIGURE 15. Depth of sediment deposition and year of vegetation establishment along a bar in Smoky Creek, east Tennessee. Bar is 125 m from upstream to downstream ends (from Bryan and Hupp, 1984).

based on minimum ages of alders (Fig. 17). A concomitant erosion of the opposite bank has normally occurred.

Paleofloods

Paleofloods are floods that occurred before the time of continuous hydrological records or direct measurements on a given stream and can have occurred from less than a decade to millions of years ago (Costa, 1984). Sigafoos (1964) showed that floodplain trees yield evidence of past flooding from tree ages, trunk scars, and differences in properties of wood related to flooding. Dates of flooding determined from trees along the Potomac River correlated closely with those documented by gauge and historical records (Sigafoos, 1964). Similarly, Hupp (1983b) showed dates of high peak stages correlated closely with dates of flooding determined through dendrochronologic evidence on Passage Creek, Virginia. A flood frequency graph (Fig. 18) was produced by Harrison and Reid (1967) through analysis of tree scars (Fig. 11) in North Dakota. They analyzed the age and heights of tree scars to determine the magnitude and frequency relation where streamflow records were short or lacking. Helley and LaMarche (1973) de-

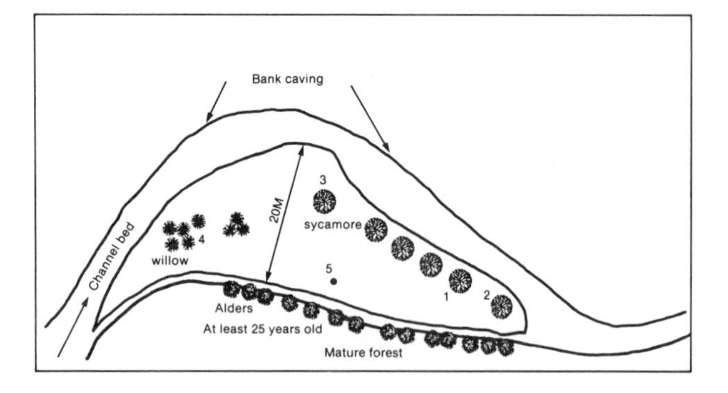

FIGURE 17. Plan view of bar and vegetation shown in Figures 15 and 16. Bar growth is upstream and toward top of figure. Location of sampled sycamores and line of alder is indicated (from Bryan and Hupp, 1984).

veloped a 400-yr record of large floods for parts of northern California through the analysis of tree ages and radiocarbon dates of flood-entrained wood.

Statistical methods for estimating flood frequency where gauge records are short or lacking consist of constructions, often from surrogate records and frequently provide no new information. Some of the methods include multiple-regression analyses relating flood peak discharges to basin and climatic characteristics and rainfall–runoff models utilizing precipitation, infiltration, and evaporation data. Bottomland trees, with flood-related deformations in their stems or wood provide direct, on-site, evidence of past floods. Analysis of botanical evidence of floods will not replace other methods for estimation of flood frequency, but when used with various hydrologic methods it can provide support and physical evidence of past floods. An example of this interdisciplinary approach to flood frequency study follows.

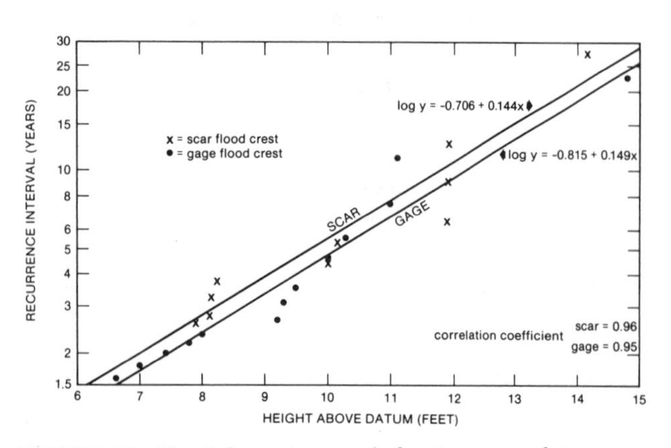

FIGURE 18. Flood frequency graph from gauge and tree scar data for the Turtle River, North Dakota. (Modified from Harrison and Reid, 1967.)

Passage Creek Flood History Reconstruction

Site Description. Passage Creek drains the northern part of Massanutten Mountain in northwestern Virginia. The basin has an area of 230 km^2 above the confluence of Passage Creek with North Fork Shenandoah River. Most of the basin is contained in a synclinal valley, Fort Valley, which is structurally controlled by ridges of Silurian Massanutten Sandstone. The stream flows through a gorge near the downstream end of the drainage basin. The sites selected for study of flood deformations in trees and their ages consist of the gorge, adjacent parts of the bottomland above the gorge, and an alluvial fan below the gorge. These sites were selected because most of the forest in this area has not been disturbed recently by lumbering or agriculture and because evidence of floods in the vegetation is common in the gorge and adjacent reaches. Most of the study area is in George Washington National Forest.

The floodplain in the study area is dissected locally with flood channels, aside from the main channel, that carry water for flow events exceeding the 10–15% flow duration (Hupp, 1982). These higher flows produce a braided stream pattern; the water surface for extreme flows is continuous across the valley bottom.

Streamflow in Passage Creek has been recorded continuously since 1932 at a site 2.1 km above its mouth, near Buckton, Virginia, and 3.5 km below the study area. According to surface water records of the U.S. Geological Survey, the area of Passage Creek basin above the gauge is 227 km^2. The average discharge for 45 yr of record is 1.88 m^3/s. Maximum discharge for the period of record was 588 m^3/s, recorded on October 15, 1942. A flood associated with Hurricane Agnes (June 1972) left much evidence on the riparian vegetation within the gorge. Maximum discharge during this flood was 360 m^3/s, and flows of this magnitude have a recurrence interval of about 12.5 yr.

Tree Ring Flood Dates and Calibration. Over 200 trees and shrubs bearing evidence of flood damage were cored or sectioned to determine dates of flooding. These dates were then chronologically tabulated (Table 5). A flood date was incorporated into a plot of possible dates when indicated by three or more samples from different trees or shrubs. This plot was then compared to the annual flood series of the gauging station (since 1932) to determine coincidence of flood years. The period of record was used as a check, or calibration period, to determine the plausibility of earlier flood years indicated by tree-ring analysis.

Many of the flood dates indicated by tree and scar ages were prior to the Passage Creek record. Supporting evidence of earlier floods was sought in the stage records of nearby North Fork Shenandoah River, which has a longer period of record. North Fork Shenandoah River near Strasburg, Virginia, has been gauged continuously since 1926. Shen-

TABLE 5 Flood Dates from Botanical Evidence and Type of Botanical Evidence on Passage Creek, Virginia

	Botanical Evidence Type		
Date	Corrasion Scar	Inclined Stem or Stump Sprout	Landform Age
1972	X	X	X
1955	X	X	
1942	X	X	X
1936	X	X	X
1924		X	
1910	X	X	
1902		X	
1897		X	X
1889		X	X
1878		X	
1870		X	X
1866		X	
1858		X	
1852		X	X
1836			X
1825			X
1720			X

andoah River at Riverton, Virginia, has annual peak stages extending intermittently back to 1870.

All dendrochronologically determined flood dates, back to 1870, correspond with high peak flows listed in the gauge or stage records from Passage Creek (back to 1936) and from the Shenandoah River gauges (back to 1870). Thus, a correlation exists between the botanical evidence and gauging station records, as has been shown elsewhere (Sigafoos, 1964). However, at present, most botanical evidence cannot allow for precise inferences as to the relative magnitude of a dated event.

Regression Analysis. Peak stages from the Shenandoah River gauge at Riverton, Virginia, were regressed against those of Passage Creek, from 1933 to 1980, to determine if peak flows of the Shenandoah River, in the area of its confluence with Passage Creek, are reasonable estimators of Passage Creek peak stages. Stage data were chosen for use in the regression analysis because the longest record, from the Riverton gauge, has only stage data. The rating curves at the gauges have not changed substantially during their periods of record. All three gauge sites are under bedrock control. Thus, it is assumed that discharge at the gauging stations has remained relatively time stable for their periods of record. Regression analysis results were used to project an estimated series of flood stages for Passage Creek. The dates of floods indicated by the regression were checked to confirm whether they actually occurred on Passage Creek through botanical evidence.

A linear relation was defined between Passage Creek stage data and stage data from each of the Shenandoah River stations. Shenandoah River at Riverton stage data accounted for 82.8% ($R^2 = 0.828$) of the variance of the Passage Creek record. Results of this regression analysis were used to estimate stage for flood dates indicated by botanical evidence between 1870 and 1932. These flood dates coincided with flooding on the Shenandoah River. The combination of Passage Creek flood dates from botanical evidence with peak flows determined from the gauge record and the Passage Creek–Shenandoah River regression analysis is illustrated in Figure 19. All years of flooding indicated by botanical evidence coincided with high peak flows estimated from the regression. Thus, botanical evidence allows for verification of flood dates predicted through regression analyses. The second- through fifth-highest stages of this historic record occurred during the gauging station period of record. The highest stage of the historic record is slightly higher than the 1942 peak. Analysis of Figure 19 shows that the minimum stage necessary to affect botanical evidence of a flood along Passage Creek is about 3.4 m. This stage corresponds to a discharge of 155 m^3/s, determined from the 1981 stage-discharge rating for Passage Creek near Buckton, Virginia.

Dates of flood damage to vegetation prior to 1870 stand without supporting gauge records. The correlation of flood dates determined from gauge records with dates determined by tree ring analysis suggests that flood dates prior to 1870 are valid. It is inferred that these floods had a minimum stage of 3.4 m (155 m^3/s), based on a minimum necessary to affect damage to vegetation (Fig. 19).

Flood Frequency. Peak stages of Passage Creek estimated from the regression analysis, which coincided with botanical evidence of high stage, were transformed to discharge and processed through standard flood frequency analysis (log-

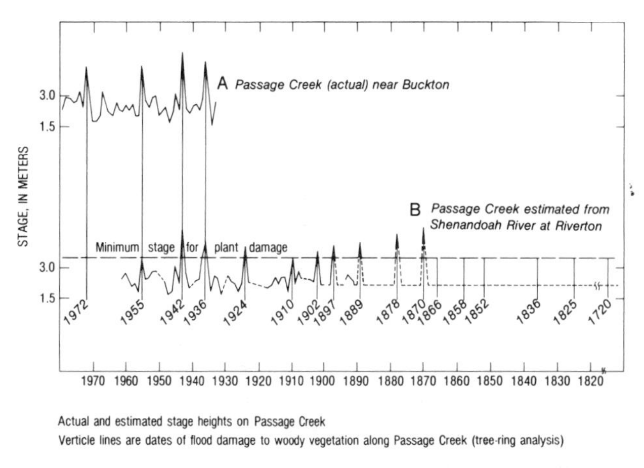

FIGURE 19. Reconstructed flood frequency on Passage Creek. Actual and estimated stage heights are shown. Dated vertical lines are based on botanical evidence of floods. Minimum stage to affect vegetation damage on floodplain is indicated. Floods prior to 1870 are not supported by gauge records.

Pearson Type III, U.S. Water Resources Council, 1981) as historic peaks. Only peaks that were assumed with confidence to be highest for the given year were used. The historically adjusted flood frequency curve was then plotted beside the flood frequency curve determined from gauging station records alone, in order to observe any variation as a result of adding botanical-historic data.

Most streamflow records are substantially shorter than the Passage Creek record of 53 yr. This record was broken into six 7-yr periods, and each period was processed through standard flood frequency analysis. Each of the six periods of record was then processed a second time with five botanically estimated peaks to determine the effect of these peaks on flood frequency curves from short records.

Flood-frequency analysis of peak flow discharges is illustrated in Figure 20. Five historic peaks, determined through tree-ring analysis, are included in one of the curves (Fig. 20), and the flood-frequency curve without the addition of historical data is shown in the other curve (Fig. 20). The similarity of the two curves indicate a close match in flood frequency predicted between the systematic record and the record developed with botanical evidence of floods, which suggests that the reliability of flood information from tree rings is good and can be used where the systematic records from gauging stations are short. The Passage Creek record, when broken into six 7-yr periods exhibits, as expected, a wide divergence of estimated flood frequency between the periods (Fig. 21*a*). This is particularly true for the important infrequent flows, where the estimated

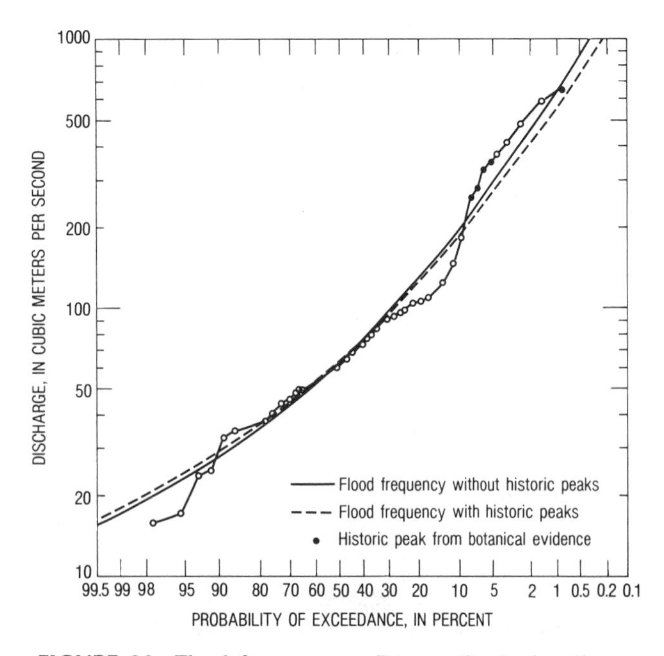

FIGURE 20. Flood frequency on Passage Creek, log-Pierson Type III model. Dashed line includes botanically derived dates of historic floods.

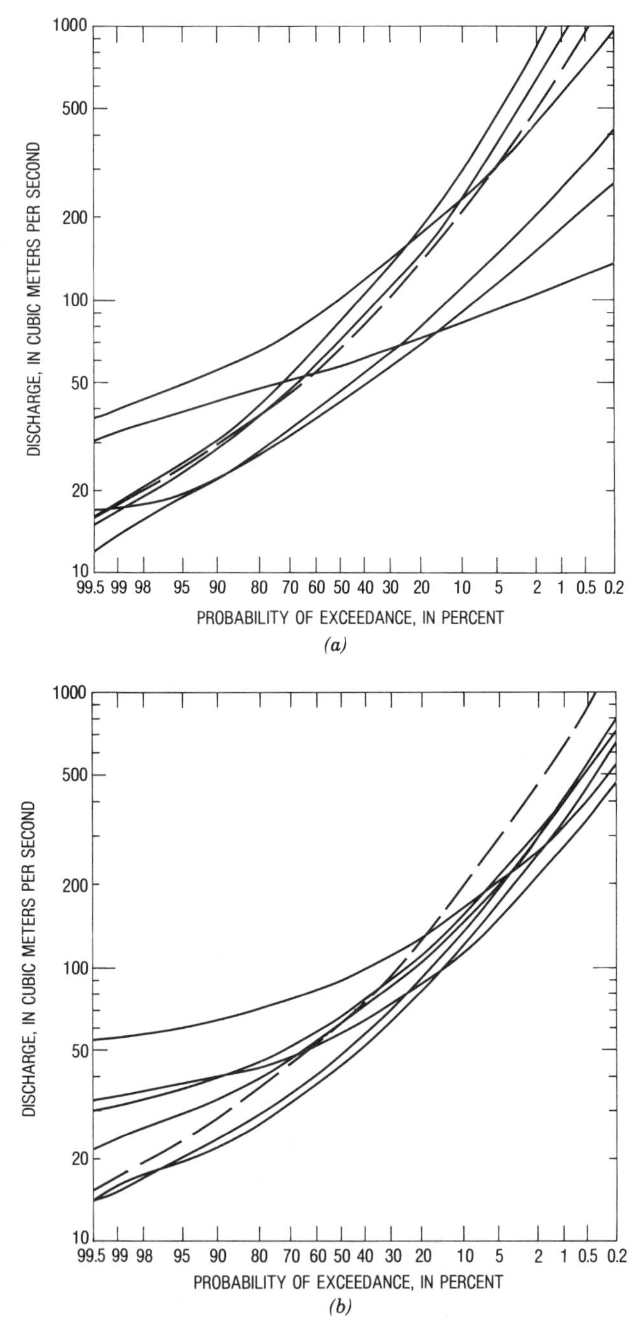

FIGURE 21. Flood frequency based on (*a*) six 7-yr periods without botanically derived historic peaks and (*b*) with the adition of botanically derived historic peaks.

500-yr flood ranged in discharge from 134 to 2951 m^3/s. The addition of the historical-botanical data to these six periods substantially narrowed the divergence of the flood frequency curves at the low-frequency end (Fig. 21*b*) and shows a closer fit to the actual flood frequency-curve developed over the entire period of record. Estimated discharges of the 500-yr flood ranged from 473 to 802 m^3/s for the six periods when historical-botanical data were added. The mean difference (absolute value) between the actual flood-frequency curve (systematic record) and that from the broken record was reduced with the addition of historical-botanical information for both the 10- and 100-yr floods; mean difference at the 10-yr flood was 82.2 m^3/s (SD = 35.9) prior to addition of historical-botanical information and 63.3 m^3/s (SD = 20.4) after the addition of the historical-botanical information; mean difference at the 100-yr flood was 343.3 m^3/s (SD = 168.0) prior to the addition of the historical-botanical information and 243.3 m^3/s (SD = 51.3) after the addition of the historical-information (Fig. 21*b*). The addition of historical-botanical flood information also reduced the discharge for a given return interval in the entire record analysis and in each of the broken, short-period analyses (Figs. 20 and 21). The botanical evidence of floods along Passage Creek, when combined with the regression results from the Shenandoah River, provides a 63-yr extension of the timing and magnitude of major floods. These data provide the information necessary for incorporation into a standard flood-frequency analysis thereby yielding a flood record from the year 1870 (Fig. 19). Between 1720 and 1870, botanical evidence indicated six additional dates. Thus, botanical evidence alone allows for a reconstruction of flood dates over a period of 265 yr (Fig. 19).

Analysis of flood-damaged vegetation supplements other methods of flood frequency estimation. The mechanism of plant growth provides for accurate dating of past floods through tree ring analysis. This technique permits the extension of flooding history and provides physical evidence of past floods without resorting to solely indirect techniques. Current methods in dendrochronologic analysis of flood-damaged vegetation are adequate to produce a historic record of flooding dates (Table 5), but are generally not well suited for estimating the magnitude of a given flood. Because both frequency and magnitude data are necessary to produce a flood frequency graph, botanical evidence must be combined with some method that can estimate magnitude.

The utility of this type of flood record extension is limited to high-magnitude, infrequent flood events, such as hurricane-generated floods. Infrequent floods, however, are difficult to predict with short gauge records. Frequent flow events can usually be predicted from relatively short gauge records. Most gauged streams do not have periods of record longer than 50 yr. However, large streams, such as the Shenandoah River, commonly have intermittent stage or historical records of large floods that extend back substantially further than 50 yr. The addition of historical-botanical data to existing gauge records is of clear value, substantially improving flood frequency prediction, where gauge records are short. These techniques may be applied to numerous streams that lack or have short flood records, as long as woody vegetation occurs along the stream.

SUMMARY

Floods and the landforms generated by floods affect distinct species distribution patterns and leave lasting evidence of their passage in the wood of trees. Analyses of bottomland vegetation yield useful information concerning flow duration and the magnitude and frequency of large floods. Woody vegetation, when present, provides relatively detailed information on the location and rates of sediment deposition or scour, rates of channel migration, and relative bank stability. Botanical evidence of floods can be an important tool in most studies of flood geomorphology and paleoflood history. Reliable prediction of infrequent hydrologic events is critical where high flood stages may be destructive to human interests. Process-oriented interdisciplinary studies that integrate plant ecology with hydrogeomorphic study have and should continue to allow for an understanding of bottomland systems unattainable in single-discipline approaches. It is hoped that this chapter has illustrated the value of combining plant ecology and tree ring study with the study of flood geomorphology, to provide a methodology and approach to interpretation that may fuel further research. The field is wide open.

ACKNOWLEDGMENTS

The author wishes to express his sincere appreciation to John T. Hack and Robert S. Sigafoos who have provided guidance, inspiration, and friendship over many years. Special thanks are due John E. Costa and Waite R. Osterkamp without whose support this work would not have been accomplished. For critical review the author acknowledges Benjamin L. Everitt (especially), Bruce Hayden, A. G. Scott, Frederick J. Swanson, and the section editor R. Craig Kochel. Special thanks to Loretta Severin who helped put this work into print, several times.

REFERENCES

Alestalo, J. (1971). Dendrochronological interpretation of geomorphic process. *Fennia* **105**, 1–140.

Baker, V. R. (1976). Hydrogeomorphic methods for the regional evaluation of flood hazards. *Environ. Geol.* **1**, 261–281.

Bedinger, M. S. (1979). Forests and flooding with special reference to the White River and Ouachita River basins, Arkansas. *Water Resour. Invest. (U.S. Geol. Surv.)* **79-68**, 58.

Bell, D. T., and del Moral, R. (1977). Vegetation gradients in the streamside forest of Hickory Creek, Wil County, Illinois. *Bull. Torrey Bot. Club* **104**, 127–135.

Bilan, M. V. (1960). "Root Development of Loblolly Pine Seedlings in Modified Environments," Bull. No. 4. Stephen F. Austin State College, Dept. of Forestry, Nacogdoches, Texas.

Broadfoot, W. M., and Williston, H. L. (1973). Flooding effects on Southern forests. *J. For.* **71**, 584–587.

Bryan, B. A., and Hupp, C. R. (1984). Dendrogeomorphic evidence of channel morphology changes in an east Tennessee coal area stream. *EOS, Trans., Am. Geophys. Union* **65** (45), 891.

Bull, W. B. (1979). Threshold of critical power in streams. *Geol. Soc. Am. Bull.* **90**, 453–464.

Chambliss, L. F., and Nixon, E. D. (1975). Woody vegetation-soil relations in a bottomland forest of east Texas. *Tex. J. Sci.* **26**, 407–416.

Cleaveland, M. K. (1980). Dating tree rings in Eastern United States. *Va. Polytech. Inst. State Univ.* **FWS-2-80**, 110–124.

Coile, T. S. (1937). Distribution of tree roots in North Carolina Piedmont soils. *J. For.* **35**, 247–257.

Costa, J. E. (1984). Fluvial paleoflood hydrology. *Trans., Am. Geophys. Union* **65**, 890.

Cowles, H. C. (1901). The physiographic ecology of Chicago and vicinity; a study of the origin, development, and classification of plant societies. *Bot. Gaz. (Chicago)* **31** (2), 73–182.

Drury, W. H., and Nisbet, I. C. T. (1973). Succession. *J. Arnold Arbor., Harv. Univ.* **54**, 331–368.

Emmett, W. W. (1982). Field data describing the movement and storage of sediment in the East Fork River, Wyoming. Part III. River hydraulics and sediment transport 1980. *Geol. Surv. Open-File Rep. (U.S.)* **82-359**, 1–289.

Everitt, B. L. (1968). Use of cottonwood in an investigation of the recent history of a floodplain. *Am. J. Sci.* **266**, 417–439.

Everitt, B. L. (1980). Ecology of saltcedar—a plea for research. *Environ. Geol.* **3**, 77–84.

Fonda, R. W. (1974). Forest succession in relation to river terrace development in Olympic National Park, Washington. *Ecology* **55**, 927–942.

Forcier, R. B. (1975). Reproductive strategies and the co-occurrence of climax true species. *Science* **189**, 808–810.

Fox, J. F. (1977). Alternation and coexistence of tree species. *Am. Nat.* **111**, 69–89.

Frye, R. J., and Quinn, J. A. (1979). Forest development in relation to topography and soils in a flood-plain of the Raritan River, New Jersey. *Bull. Torrey Bot. Club* **106**, 334–345.

Goodlett, J. C. (1969). Vegetation and the equilibrium concept of landscape. *In* "Essays in Plant Geography and Ecology" (K. N. Greenidge, ed.), pp. 33–44. Nova Scotia Museum, Halifax.

Hack, J. T. (1957). Studies of longitudinal stream profiles in Virginia and Maryland. *Geol. Surv. Prof. Pap. (U.S.)* **294-B**, 45–97.

Hack, J. T., and Goodlett, J. C. (1960). Geomorphology and forest ecology of a mountain region in the central Appalachians. *Geol. Surv. Prof. Pap. (U.S.)* **347**, 1–66.

Hall, T. F., and Smith, G. E. (1955). Effect of flooding on woody plants, West Sandy Dewatering Project, Kentucky Reservoir. *J. For.* **53**, 281–285.

Harper, J. L. (1977). "Population Biology of Plants." Academic Press, London.

Harrison, S. S., and Reid, J. R. (1967). A flood-frequency graph based on tree-scar data. *Proc. N.D. Acad. Sci.* **21**, 23–33.

Hedman, E. R., and Osterkamp, W. R. (1982). Streamflow characteristics related to channel geometry of streams in Western United States. *Geol. Surv. Prof. Pap. (U.S.)* **2193**, 1–17.

Hedman, E. R., Moore, D. O., and Livingston, R. K. (1972). Selected streamflow characteristics as related to channel geometry of perennial streams in Colorado. *Geol. Surv. Open-File Rep., (U.S.) 1–14.*

Hefley, H. M. (1937). Ecological studies on the Canadian River floodplain in Cleveland County, Oklahoma. *Ecol. Monogr.* **7**, 346–402.

Helley, E. J., and LaMarche, V. C., Jr. (1973). Historic flood information for northern California streams from geological and botanical evidence. *Geol. Surv. Prof. Pap. (U.S.)* **485-E**, 1–16.

Hickin, E. J., and Nanson, G. C. (1975). The character of channel migration on the Beatton River, Northeast British Columbia Canada. *Geol. Soc. Am. Bull.* **86**, 487–494.

Hosner, J. F. (1960). Relative tolerance to complete inundation of fourteen bottomland tree species. *For. Sci.* **6** (3), 246–251.

Hosner, J. F., and Minckler, L. S. (1963). Bottomland hardwood forests of southern Illinois. *Ecology* **44**, 29–41.

Howard, A. D., Fairbridge, R. W., and Quinn, J. H. (1968). Terraces, fluvial introduction. *In* "The Encyclopedia of Geomorphology" (R. Fairbridge, ed.), pp. 1117–1123. Reinhold, New York.

Hupp, C. R. (1982). Stream-grade variation and riparian-forest ecology along Passage Creek, Virginia. *Bull. Torrey Bot. Club* **109**, 488–499.

Hupp, C. R. (1983a). Vegetation patterns in the Passage Creek gorge, Virginia. *Castanea* **48**, 62–72.

Hupp, C. R. (1983b). Geo-botanical evidence of Late Quaternary mass wasting in block field areas of Virginia. *Earth Surf. Processes Landforms* **8**, 439–450.

Hupp, C. R. (1984). Dendrogeomorphic evidence of debris flow frequency and magnitude at Mount Shasta, California. *Environ. Geol. Water Sci.* **6** (2), 121–128.

Hupp, C. R. (1986a). The headward extent of fluvial landforms and associated vegetation on Massanutten Mountain, Virginia. *Earth Surf. Processes Landforms* 11, 545–555.

Hupp, C. R. (1986b). Botanical evidence of floods and paleoflood frequency. Paper presented at the International Symposium on Flood Frequency and Risk Analysis, Baton Rouge, Louisiana, May, 1986 (in press).

Hupp C. R., and Osterkamp, W. R. (1985). Bottomland vegetation distribution along Passage Creek, Virginia, in relation to fluvial landforms. *Ecology* **66**, No. 3.

Hupp, C. R., and Sigafoos, R. S. (1982). Plant growth and block-field movement in Virginia. *U.S. For. Serv., Gen. Tech. Rep.* **PNW-141**, 78–85.

Hupp, C. R., and Simon, A. (1986). Riparian vegetation as indicator of bank-slope development, stability and sediment storage on modified West Tennessee streams. *Proc. Fed. Inter-Agency Sediment Conf., 4th, 1986*, Vol. 2, pp. 5-83–5-92.

Huston, M. (1979). A general hypothesis of species diversity. *Am. Nat.* **113**, 81–101.

Johnson, W. C., Burgess, R. L., and Keammerer, W. R. (1976). Forest overstory vegetation and environment on the Missouri River Floodplain in North Dakota. *Ecological Monographs* **46**, 59–84.

Kassas, M., and Girgis, W. A. (1964). Habitat and plant communities in the Egyptian Desert. V. The Limestone plateau. *J. Ecol.* **52**, 107–119.

Kilpatrick, F. A., and Barnes, H. H., Jr. (1964). Channel geometry of Piedmont streams as related to frequency of floods. *Geol. Surv. Prof. Pap. (U.S.)* **422-E**, 1–10.

Kramer, P. J., and Kozlowski, T. T. (1979). "Physiology of Woody Plants." Academic Press, New York.

LaMarche, V. C., Jr. (1968). Rates of slope degradation as determined from botanical evidence, White Mountains, California. *Geol. Surv. Prof. Pap. (U.S.)* **352-I**, 1341–1377.

Lane, E. W. (1957). A study of the shape of channels formed by natural streams flowing in erodible materials. *U.S. Army Eng. Div., Mo. River, M.R.D. Sediment Ser.* **9**, 1–106.

Lane, L. J., Chang, H. H., Graf, W. L., Grissinger, E. H., Guy, H. P., Osterkamp, W. R., Parker, G., and Trimble, S. W. (1982). Relations between morphology of small streams and sediment yield: Task Committee Report. *J. Hydraul. Div., Am. Soc. Civ. Eng.* **108**, 1328–1365.

Leopold, L. B., Wolman, M. G., and Miller, J. P. (1964). "Fluvial Processes in Geomorphology." Freeman, San Francisco, California.

Lindsey, A. A., Petty, R. O., Sterling, D. K., and Van Asdall, W. (1961). Vegetation and environment along the Wabash and Tippecanoe Rivers. *Ecol. Monogr.* **31**, 105–156.

McIntosh, R. P. (1980). The relationship between succession and the recovery process in ecosystems. *In* "The Recovery Process in Damaged Ecosystems" (J. Cairnes, ed.), pp. 11–62. Ann Arbor Sci. Publ., Ann Arbor, Michigan.

Nanson, G. C., and Beach, H. F. (1977). Forest succession and sedimentation on a meandering river floodplain, northeast British Columbia, Canada. *J. Biogeogr.* **4**, 229–251.

Nixon, E. S., Willet, R. L., and Cox, P. W. (1977). Woody vegetation in a virgin forest in an eastern Texas river bottomland. *Castanea* **42**, 227–236.

Osterkamp, W. R. (1978). Gradient, discharge, and particle size relations of alluvial channels in Kansas, with observations on braiding. *Am. J. Sci.* **278**, 1253–1268.

Osterkamp, W. R., and Hedman, E. R. (1982). Perennial-streamflow characteristics related to channel geometry and sediment in Missouri River basin. *Geol. Surv. Prof. Pap. (U.S.)* **1242**, 1–56.

Osterkamp, W. R., and Hupp, C. R. (1984). Geomorphic and vegetative characteristics along three northern Virginia streams. *Geol. Soc. Am. Bull.* **95**, 1093–1101.

Osterkamp, W. R., Carey, W. P., Hupp, C. R., and Bryan, B. A. (1984). Movement of tractive sediment from disturbed lands. *J. Hydraul. Div., Am. Soc. Civ. Eng.* **110**, 59–63.

Phipps, R. L. (1967). Annual growth of suppressed chestnut oak and red maple, a basis for hydrologic inference. *Geol. Surv. Prof. Pap. (U.S.)* **485-C**, 1–27.

Pickett, S. T. A. (1980). Non-equilibrium coexistance of plants. *Bull. Torrey Bot. Club* **107**, 238–248.

Robertson, P. A., Weaver, G. T., and Cavanaugh, J. A. (1978). Vegetation and tree species patterns at the north terminus of the southern flood-plain forest. *Ecol. Monogr.* **48**, 249–267.

Schneider, W. J., and Goodlett, J. C. (1962). Portrayal of drainage and vegetation on topographic maps. *U.S. Geol. Surv. Water Resour. Div. Rep. SW*, pp. 1–63.

Shaver, J. M., and Denison, M. (1928). Plant succession along Mill Creek. *Tenn. Acad. Sci.* **3** (4), 5–13.

Shreve, F. (1915). The vegetation of a desert mountain range as conditioned by climatic factors. *Carnegie Inst. Washington Publ.* **217**, 1–112.

Shroder, J. F., Jr. (1978). Dendrogeomorphological analysis of mass movement of Table Cliffs Plateau, Utah. *Quat. Res.* **9**, 168–185.

Sigafoos, R. S. (1961). Vegetation in relation to flood frequency near Washington, D.C. *Geol. Surv. Prof. Pap. (U.S.)* **424-C**, 248–249.

Sigafoos, R. S. (1964). Botanical evidence of floods and floodplain deposition. *Geol. Surv. Prof. Pap. (U.S.)* **485-A**, 1–35.

Sigafoos, R. S. (1976). Relations among surficial material, light intensity, and sycamore seed germination along the Potomac River near Washington, D.C. *J. Res. U.S. Geol. Surv.* **4**, 733–736.

Sigafoos, R. S., and Hendricks, E. L. (1961). Botanical evidence of the modern history of Nisqually Glacier, Washington. *Geol. Surv. Prof. Pap. (U.S.)* **387-A**, 1–20.

Smith, J. H. (1955). Some factors indicative of site quality for black cottonwood (*Populus trichocarpa*, Torr. Gray). *J. For.* **55** (8) 578–580.

Stokes, M. A., and Smiley, T. L. (1968). "An Introduction to Tree-ring Dating." Univ. of Chicago Press, Chicago, Illinois.

Swanson, F. J., and Lienkaemper, G. W. (1982). Interaction among fluvial processes, forest vegetation, and aquatic ecosystems, South Fork Hoh River, Olympic Natl. Park. *In* "Ecological Research in National Parks of the Pacific Northwest" (E. E. Starkey, J. F. Franklin, and J. W. Matthews, eds.), pp. 30–34.

Teversham, J. M., and Slaymaker, J. (1976). Vegetation composition in relation to flood frequency in Lillooet River Valley, British Columbia. *Catena (Cremlingen-Destedt, Ger.)* **3**, 191–201.

U.S. Water Resources Council (1981). "Guidelines for Determining Flood Flow Frequency." Bull. No. 17B. Washington, D.C.

Ware, G. H., and Penfound, W. T. (1949). The vegetation of the lower levels of the floodplain of the South Canadian River in central Oklahoma. *Ecology* **30**, 478–484.

Weaver, J. E. (1960). Floodplain vegetation of the central Missouri Valley and contacts of woodland with prairie. *Ecol. Monogr.* **30**, 37–64.

Wharton, C. H., Kitchens, W. M., Pendleton, E. C., and Sipe, T. W. (1982). The ecology of bottomland hardwood swamps of the Southeast: A community profile. *U.S. Fish Wildl. Serv. Off. Biol. Serv. Program [Tech. Rep.] FWS/OBS* **FWS/OBS/81-37**, 1–133.

White, P. S. (1979). Pattern, process, and natural disturbance in vegetation. *Bot. Rev.* **45**, 229–299.

Williams, G. P. (1978). Bank-full discharge of rivers. *Water Resour. Res.* **14** (6), 1141–1154.

Wistendahl, W. A. (1958). The flood plain of the Raritan River, New Jersey. *Ecol. Monogr.* **28**, 129–153.

Wolfe, C. B., Jr., and Pittillo, J. D. (1977). Some ecological factors influencing the distribution of *Betula nigra* L. in western North Carolina. *Castanea* **42**, 18–30.

Wolman, M. G. (1955). The natural channel of Brandywine Creek, Pennsylvania. *Geol. Surv. Prof. Pap. (U.S.)* **271**,

Wolman, M. G., and Leopold, L. B. (1957). River floodplains; some observations on their formation. *Geol. Surv. Prof. Pap. (U.S.)* **282-C**, 87–109.

Yanosky, T. M. (1982a). Effects of flooding upon woody vegetation along parts of the Potomac River floodplain. *Geol. Surv. Prof. Pap. (U.S.)* **1206**, 1–21.

Yanosky, T. M. (1982b). Hydrologic inferences from ring widths of flood-damaged trees, Potomac River, Maryland. *Environ. Geol. Water Sci.* **4**, 43–52.

Yanosky, T. M. (1983). Evidence of floods on the Potomac River from anatomical abnormalities in the wood of flood-plain trees. *Geol. Surv. Prof. Pap. (U.S.)* **1296**, 1–42.

Zimmermann, R. C. (1969). Plant ecology of an arid basin, Tres Alamos-Redington area, southeastern Arizona. *Geol. Surv. Prof. Pap. (U.S.)* **485-D**, 1–51.

Zimmermann, R. C. (1980). "Physiographic Plant Geography: The Study of Plant Distributions as Related to Geohydrology and Dynamic Geomorphology." Private printing, Burlington, Vermont.

Zimmermann, R. C., and Thom, B. G. (1982). Physiographic plant geography. *Prog. Phys. Geogr.* **6**, 45–59.

21

PALEOFLOOD ANALYSIS USING SLACKWATER DEPOSITS

R. CRAIG KOCHEL
Department of Geology, Southern Illinois University, Carbondale, Illinois

and

VICTOR R. BAKER
Department of Geosciences, University of Arizona, Tucson, Arizona

INTRODUCTION

Paleohydrologists have adopted two general approaches in the reconstruction of ancient flow histories of rivers. One approach focuses upon the estimation of mean hydrologic conditions such as mean annual flow, bankfull flow, and other flows with relatively high exceedance frequency. Mean discharge paleohydrologic studies are largely based on comparing physical characteristics of paleochannels, that is, channel geometry, channel morphology, and sedimentology with modern analogs, aimed at providing estimates of mean annual discharge and paleoclimatologic conditions (Ethridge and Schumm, 1978; Williams, this volume).

The other major approach in paleohydrologic research focuses on the characterization of discrete flood events, typically the magnitude and frequency of exceptionally large floods. Such studies utilize a variety of techniques for estimating paleoflood peak discharge: (1) indices of flow strength (velocity, shear stress, stream power) derived from studies of flood-transported gravel and boulders (Costa, 1983; Williams, 1983; Komar, this volume); (2) erosion of tributary debris fans (Costa, 1978); (3) floodplain vegetation (Hupp, this volume); and (4) slackwater sediments (Kochel and Baker, 1982; Kochel et al., 1982; Baker et al. 1983; Patton et al. 1979; Webb et al., this volume; Ely and Baker, 1985; Partridge and Baker, 1987). The slackwater sediment approach is becoming widely used and, for the appropriate settings, may provide the most abundant and accurate source of data for determining the recent geologic history of multiple large flood events in a river basin. With the appropriate field studies, slackwater sediments can often be analyzed in stratigraphic suites that contain continuous paleoflood records over thousands of years.

Slackwater deposits are relatively fine-grained (usually fine sand and coarse silt) flood sediments deposited in floodplain areas that are sheltered from high-velocity flood flows. Bretz (1929) first described slackwater deposits formed in the mouths of tributaries that were backflooded by the catastrophic Pleistocene Missoula floods in the Channeled Scabland of western Washington (Baker and Bunker, 1985). Missoula flood slackwater sediments range in size from coarse gravel to coarse silt and were deposited in backflooded tributaries extending tens of miles upstream from their main channel sources (Bretz, 1929; Baker, 1973; Patton et al., 1979). Although Bretz did not use them in a hydrological sense, later investigators analyzed Missoula flood slackwater sediments to estimate the number and chronology of Pleistocene floods (Baker, 1978; Bunker, 1982; Waitt, 1980, 1984, 1985; Baker and Bunker, 1985).

Slackwater sediments have been recognized by many investigators in a wide variety of physiographic and climatic settings outside the Channeled Scabland (Fig. 1). Holocene slackwater deposits have been described most commonly at the junctions of major rivers and minor tributaries in bedrock canyon settings. Studies of tributary mouth and bedrock cave slackwater sediment sites have thus far proven the most successful in paleoflood hydrologic studies.

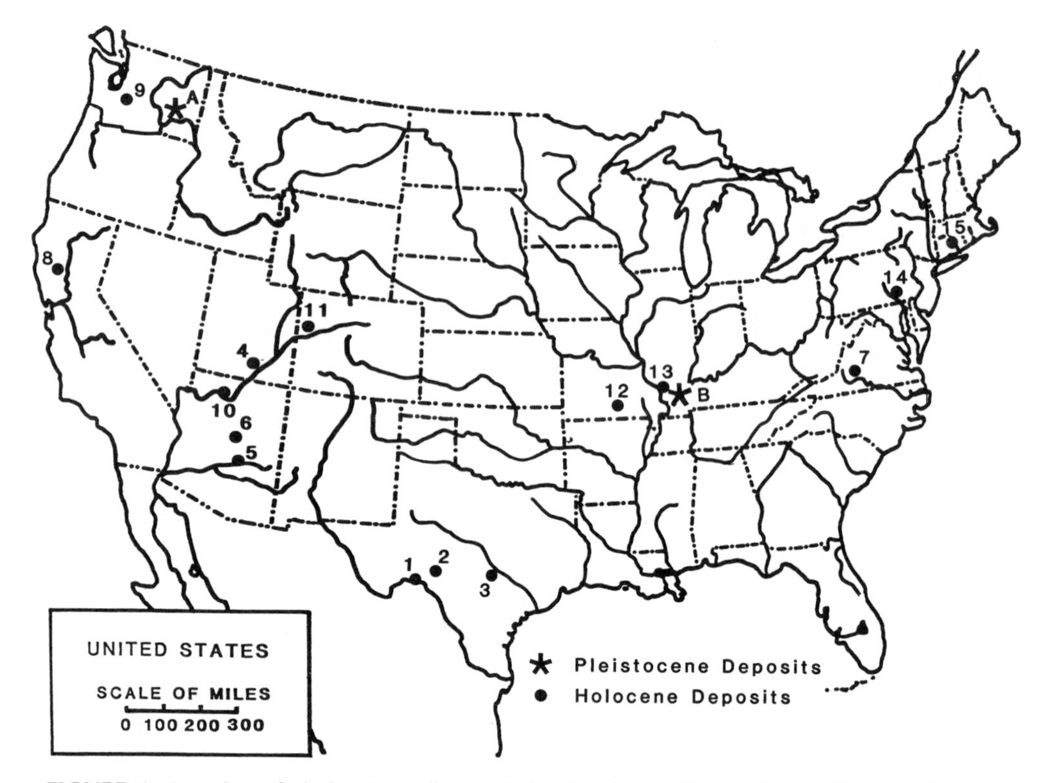

FIGURE 1. Location of slackwater sediment study sites in the United States. Sites are from published studies, investigations in progress, or based on our field reconnaissance. Most of the sites are in the mouths of tributaries to the following rivers: (1) Rio Grande (Kochel, 1980); (2) Pecos and Devils Rivers (Kochel, 1980; Kochel et al., 1982); (3) Pedernales River (Patton et al., 1979); (4) Escalante River (Webb et al., this volume); (5) Salt River (Partridge and Baker, 1987); (6) Verde River (Ely and Baker, 1985); (7) Rockfish River (Kochel and Johnson, 1984); (8) Coffee Creek (Stewart and LaMarche, 1967); (9) Skagit River (Stewart and Bodhaine, 1961); (10) Colorado River (Howard and Dolan, 1981); (11) Colorado River (this study); (12) Black River (this study); (13) Mississippi River (this study); (14) Susquehanna River (Moss and Kochel, 1978); (15) Housatonic River (Patton and Handsman, 1984); and Pleistocene deposits: (A) Channeled Scabland (Baker, 1973) and (B) Mississippi and Ohio rivers (Willman and Frye, 1980).

This chapter summarizes our recent experience in paleoflood hydrology. We describe tributary mouth sedimentation sites in west Texas to illustrate slackwater paleoflood hydrology. Kochel (this volume) discusses the application of the west Texas slackwater paleoflood data to environmental management of rivers.

PHYSICAL CHARACTERISTICS OF SLACKWATER DEPOSITS

Slackwater sediments accumulate in floodplain and valley regions where flood flow velocities are minimal during the time that inundation occurs. Slackwater sediments are rapidly deposited from suspension in sediment-laden waters at localities where the flow becomes separated from the main thread of flood flow. Four geomorphic situations (Fig. 2) provide favorable conditions for the deposition and continued accumulation of slackwater sediment sequences: (1) tributary mouths; (2) shallow caves along the bedrock channel walls; (3) downstream from major bedrock and/or talus obstructions and in areas of dramatic channel widening; and (4) as overbank accumulations on high terraces.

Slackwater Flooding of Tributary Mouths

Slackwater sedimentation occurs in virtually all river systems but is most pronounced in narrow, deep, entrenched bedrock canyons where small increases in flood discharge are accompanied by large increases in flood stage (Baker and Kochel, this volume). Washload and suspended sediment conveyed by intense flood flows are deposited high along canyon walls, in bedrock caves, and far upstream in tributary

FIGURE 2. Sites favorable for the accumulation and preservation of stratigraphic sequences of slackwater sediments. (*a*) Pedernales River, Texas, showing backflooding into tributary mouths during tropical storm Amelia, August 1978. (*b*) Tributary mouth slackwater deposit in Cedar Canyon along the Pecos River. Quartz sand from numerous Pecos floods has accumulated at this site, now outlined by the darker tone resulting from vegetation by mesquite. (*c*) Arrow shows a reverse slack eddy at a channel expansion at Pedernales Falls, Texas, during the 1978 Pedernales River flood. (*d*) Arrow points to the new layer of slackwater sediment at site (*c*), photographed 14 days later. (*e*) Bedrock caves (Fate Bell Rockshelter) along the walls of Seminole Canyon (Kochel, 1982), where slackwater sands have accumulated. (*f*) Slackwater sediments in the lee of bedrock spurs and talus in the Escalante River basin, Utah.

mouths. The highest such deposits provide a record of high-magnitude, low-frequency floods.

Tributary mouth slackwater sedimentation is most pronounced when there are significant temporal differences in peak flow stage between the tributary and the mainstream. This hydrologic situation is common for tributaries with small catchments entering mainstreams with large drainage areas (Kochel, 1980). The temporal separation of tributary and mainstream floods are often generated by intense, localized rainfall over a small portion of the drainage while little or no flooding occurs in the tributary basin. A classic example of this was the 1954 flood on the lower Pecos River that produced the thickest and most extensive slackwater sediment unit in lower Pecos tributary mouths during the last 10,000 yr. Contributing drainage area for this flood was less than 10% of the basin (Kochel, this volume).

Tributary Backflooding. Backflooding of tributary mouths occurs during most major floods on mainstream canyon rivers. Powerful surges of sediment-laden water move up the tributaries, depositing slackwater sediments rapidly as backflood velocity decreases. These deposits can normally be traced up-tributary as distally thinning wedges of sand and silt draped over the tributary channel and floodplain surfaces up to the maximum extent of backflooding (Fig. 3).

If the rainfall responsible for the mainstream flood is regional in extent, tributary flooding may also occur. In such cases the tributary flood usually precedes mainstream backflooding, and a sedimentary couplet results from the two flood events (Fig. 4). In tributary mouths along the Susquehanna River, Hurricane Agnes flood slackwater couplets were composed of a basal tributary-derived gravel overlain by silt and clay backflooded by the Susquehanna (Moss and Kochel, 1978). Proof that the mud was from the Susquehanna came from intercalated coal, which does not outcrop in the tributary basins. The basal portion of each couplet is tributary sediment. Tributary sediment is typically coarse grained because of the steeper tributary gradient, and it is of local provenance due to the smaller drainage area of the tributary. Overlying the coarse sediments are the fine-grained slackwater deposits from mainstream backflooding. Slackwater sediment mineralogies show distal provenance and display paleocurrent indicators indicating up-tributary flow. Pecos River slackwater sediments (Figs. 2*b*, and 3) are composed of fine-grained quartz sand derived from Permian sandstone outcrops hundreds of kilometers upstream from the depositional sites. Pecos River tributary sediments are composed of coarse limestone gravel and silt because their basins are underlain entirely by Cretaceous limestones of the Edwards Plateau. In settings where rivers cut across the axes of geologic structures, distinctly different mineralogies of mainstream and tributary sediment normally result. Hence, such cross-axial rivers provide optimal situations for clearly distinguishing mainstream and tributary floods.

The duration of backflooding varies greatly among rivers, depending on mainstream hydrograph characteristics and tributary gradients. Along the lower Pecos River, slackwater inundation periods for major floods last a few hours to a day or two. At the other extreme, Mississippi River backflooding of southern Illinois tributaries typically continues for weeks or months during major floods. Mississippi River floods during 1984–1985 inundated tributary mouths for periods of between 3 and 6 weeks, depositing slackwater sediments kilometers upstream in tributaries and their tributaries. Figure 5 shows a slackwater site in southern Illinois after a flood in November 1984.

Preservation of Continuous Slackwater Sequences. A new slackwater sedimentation unit is deposited with each successive flood whose stage equals or exceeds the preexisting level of slackwater sediment accumulation. Caution must be used in interpreting slackwater sediment stratigraphy at sites where couplets are common because significant erosion of older sediments may have occurred during the deposition of tributary sediments. Figure 6 shows the slackwater stratigraphies observed in two neighboring tributary mouth sites along the lower Pecos River. The Davis Canyon stratigraphy is interpreted to contain six slackwater floods. The most recent flood (Fig. 6, unit 6) is the 1954 event. The next older flood occurred at least 700 yr B.P. In neighboring Rowland Canyon, nine floods are preserved, all of which occurred in the last 40 yr. The isohyetal map from the 1954 flood indicates that the local area received large rainfalls during that storm. The resulting tributary flood in Rowland Canyon may have completely removed the pre-existing slackwater sediments. Sedimentation from the 1954 flood and subsequent events account for the observed slackwater sequence. The disparity of paleoflood interpretations that could result from studies based on only one or two sites along a river would be enormous because of this problem of tributary flushing. Such problems are minimized by using correlations between multiple sites and by careful efforts in locating slackwater stratigraphies in locations protected from erosive tributary flooding. This example illustrates the necessity of correlations using radiocarbon dates in addition to physical stratigraphic methods.

Successful paleoflood reconstruction over long periods of time depends on the presence of continuous stratigraphic sequences of slackwater sediments. Long stratigraphic sequences occur where abundant slackwater sedimentation occurs during major floods, and where the sedimentation site is shielded from erosive portions of subsequent tributary and mainstream flows. Important factors affecting the potential for the formation and preservation of slackwater deposits include: (1) the tributary junction angle, (2) the

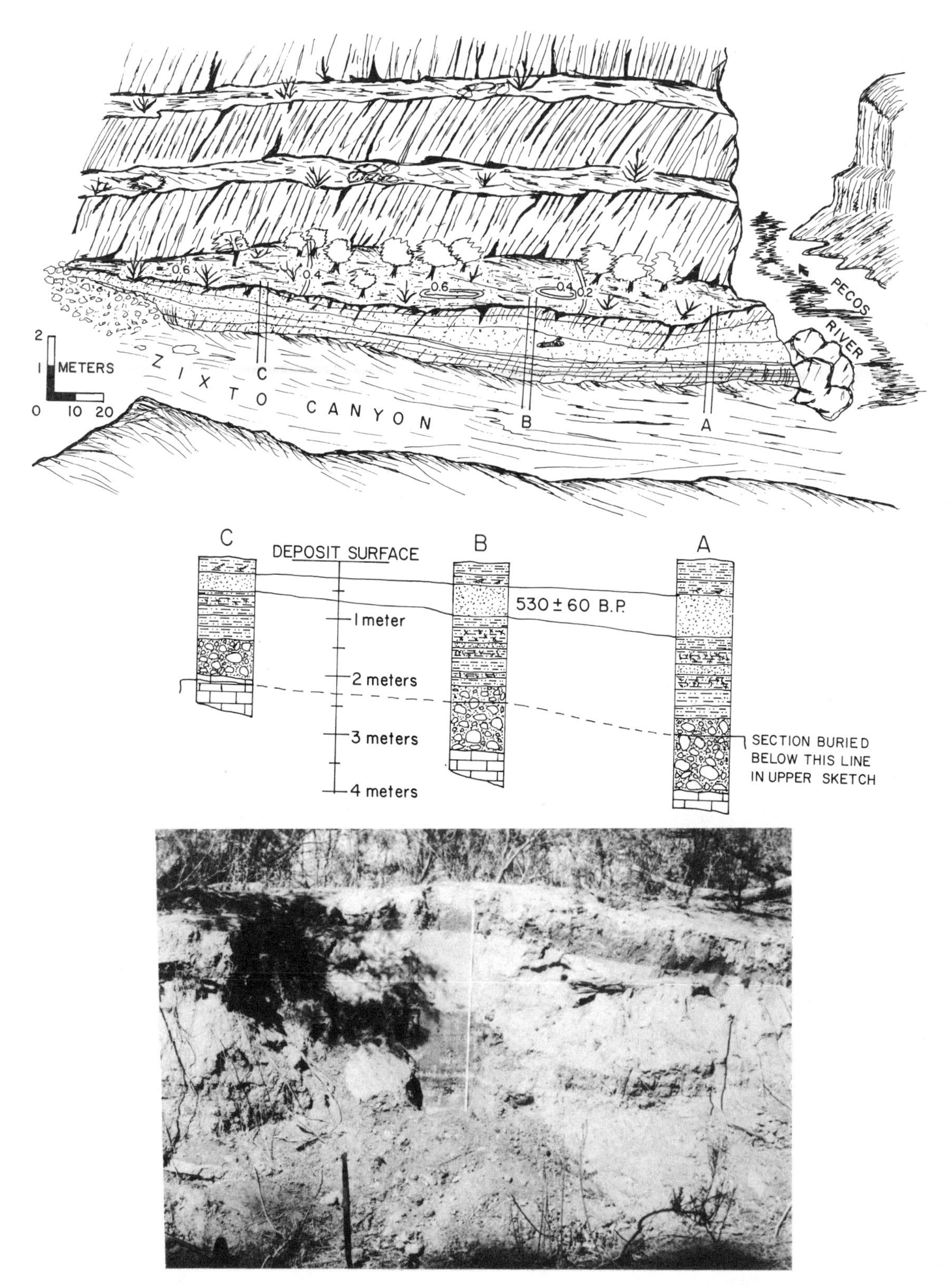

FIGURE 3. Slackwater sediments in the mouth of Zixto Canyon along the Pecos River. Limestone tributary gravel occurs in the lower 1–1.5 m of the section. Six major floods deposited quartz sand by backflooding from the Pecos in units that fine and thin up Zixto Canyon. The photograph shows four of these units exposed in a trench cut near section B. The log in the third layer from the top has been dated at 530 ± 60 yr B.P. Overlying this unit are the sediments from recent floods in 1954 and 1974. (From Kochel and Baker, 1982. Copyright 1982 by the American Association for the Advancement of Science.)

FIGURE 4. Tributary and mainstream slackwater sedimentary couplet formed during Hurricane Agnes flooding on the Susquehanna River, Pennsylvania, in June 1972. The arrow shows the contact between coarse schist gravel from the Tucquan Creek tributary, which was deposited first. After the Tucquan gravel was deposited, the Susquehanna River backflooded into this site and deposited 20 cm of coal-rich slackwater silt on top.

FIGURE 5. Tributary mouth slackwater sedimentation site in Little Grand Canyon along the Mississippi River in southern Illinois. This canyon is typical of the small, entrenched canyons along the Mississippi bluffs between Chester and Cairo, Illinois. (*a*) Near the mouth of the canyon, trees contain slackwater corrasion scars on their downstream sides caused by impact of floating debris during backflooding by the Mississippi. View is looking upstream. Note also the high-water line on the trees (arrows) from a flood in November 1984, a few weeks before the photograph was taken.

FIGURE 5. *(Continued).* (*b*) Numerous slackwater sediment couplets of tributary and Mississippi sediment preserved in a large bedrock cave at this site.

mainstream gradient, (3) the tributary gradient, (4) tributary drainage basin morphometry, and (5) site location with respect to tributary channel morphology.

The first requirement necessary to produce thick accumulations of slackwater sediments is that the mainstream be able to backflood efficiently into tributary mouths. Field studies in southwest Texas (Kochel, 1980) and our recent flume experiments indicate that maximum backflood discharges occur when tributary junction angles are between 55° and 125° to the mainstream (Baker and Kochel, this volume; Kochel and Ritter, 1987). Mainstream floodwaters tend to bypass tributaries with particularly acute junction angles (less than 45°). On the other hand, where tributaries join the mainstream at angles greater than 130°, mainstream flood flows can blast up tributary channels at high velocity. Slackwater sediments are deposited during these blasts from the mainstream, but the slackwater flows are usually so powerful that any pre-existing slackwater sediments are eroded. In these high-angle settings, continuous stratigraphies are rare, but any preserved deposits are useful for documenting the most recent slackwater flood (Fig. 7*a*). Figure 7*b* shows Rio Grande slackwater sediments deposited in a tributary mouth by the October 1978 flood. The 1978 flood stage was unable to inundate surfaces of pre-existing slackwater stratigraphies; hence, the 1978 sediments were rapidly eroded during 1979 by low-magnitude tributary flows, leaving no permanent record of the 1978 flood.

Our flume experiments show that mainstream gradient is also an important factor affecting the volume of backflooding in tributary mouths (Baker and Kochel, this volume; Kochel and Ritter, 1987). Backflooding discharges appear to decrease as mainstream gradient increases. Steep mainstream gradients cause floodwater to bypass tributary mouths because of excessively high velocity. Surveys of slackwater sediment thickness for individual floods and cumulative flood thickness in southwest Texas support this observation (Kochel and Ritter, 1987).

Tributary basin and channel characteristics are also important factors affecting the accumulation of continuous slackwater sediment sequences. Tributary basins that have morphometries conducive to relatively flashy hydrograph characteristics are prone to floods that destroy accumulations of slackwater sediments in their mouths. High ruggedness number, high first-order channel frequency, and steep tributary gradient appear to be the least favorable tributary basin characteristics for the preservation of thick slackwater sequences (Kochel, 1980). These characteristics are similar to those that Patton and Baker (1976) found related best to streams having flashy flood hydrographs.

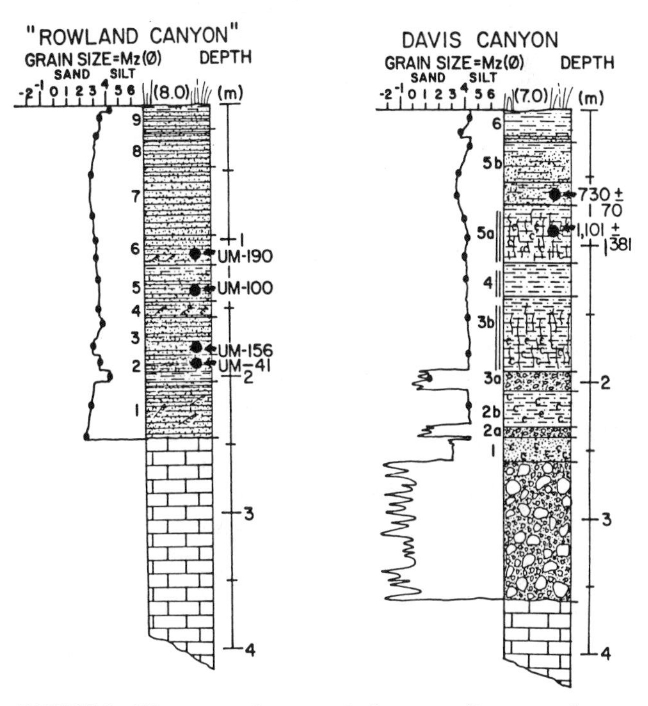

FIGURE 6. Discrepancy between slackwater sediment stratigraphies in neighboring tributary canyons along the Pecos River near Pandale, Texas. Erosion by a post-1945 flood (probably 1954) flushed pre-existing slackwater sediments from Rowland Canyon because the accumulation site was not as well protected from tributary flood flows as the site in Davis Canyon.

FIGURE 7. Discontinuous slackwater sediments. (*a*) Photograph of the mouth of Harkell Canyon along the lower Pecos River. Because of the high angle juncture with the Pecos River (about 150°) the Pecos River blasts into Harkell Canyon (Pecos is shown by the arrow) during flood and scours pre-existing slackwater sediments. Scattered remnants of the 1974 sand are visible but are being rapidly removed by tributary and mainstream floods. (*b*) Inset terrace drape of slackwater sand deposited in Jabelina Canyon along the Rio Grande near Sanderson, Texas, by a flood in October 1978. The flood was estimated as the 10-yr event, but peak stage was insufficient to inundate pre-existing stratigraphic accumulations of slackwater sediment in tributary mouths. The 1978 sands were rapidly eroded in 1979 by subsequent small floods on the tributaries and will not be recorded in the slackwater record.

Finally, the location of the slackwater sedimentation site with respect to tributary channel geometry also affects preservation potential. The most favorable sites in the Texas rivers occur along the insides of tributary meander bends a few tens of meters from the mainstream and just downstream from major bedrock or talus obstructions to the tributary channel flow. Tributary flood velocities are generally minimal in these areas.

Sedimentology

Slackwater deposits in the Channeled Scabland are rhythmically bedded (Baker, 1973; Patton et al., 1979; Waitt, 1980). Typical Scabland rhythmites grade upward from gravel to fine sand. These were probably deposited by density current-like processes in the large flood-generated pondings that were created in Scabland basins upstream from structurally controlled contracted openings (Baker, 1978). Due to space limitations, we will not be discussing the Scabland slackwater sediments. In contrast, most Holocene slackwater sediments, like those along the Pecos River, are devoid of rhythmites.

Sedimentary Structures. Pecos River slackwater sedimentation units range in thickness from a few centimeters to 1.5 m. Slackwater sediments typically fine upward from coarse sand to silt, but grading is usually poorly developed. Many slackwater sedimentation units are structureless, indicating rapid deposition from suspension. Horizontal lamination is the dominant sedimentary structure where primary structures are visible (Fig. 8). Sand laminae in Pecos River sediments range from 0.5 to 3 cm in thickness and are separated by millimeter-scale silt partings. Our flume observations showed that backflooding during major floods is pulsatory. Backflood waves moved up tributary mouths with irregular periodicity during our experiments. The silt partings may be marking the separations between these up-tributary pulses moving through the backwater. Horizontal laminations are considered to form as the result of the migration of bedforms such as small ripples during rapid sedimentation.

Cross-bedding also occurs at scales from ripple drift features a few centimeters high to foresets with amplitudes up to 50 cm (Fig. 8*b*). Cross-beds occur most frequently near the base of thick sandy slackwater units. Most foresets indicate paleoflow directions up the tributary canyons away from the Pecos River. However, bimodal cross-bedding is sometimes observed. Down-canyon cross-beds are generally smaller in scale than up-canyon ones and are most common in the upper part of slackwater sediment units. Down-canyon cross-beds probably form during the waning stages of backflooding as water drains back into the mainstream channel. Bimodal sets occasionally occur as herringbone cross-beds similar to those observed in tidal environments. The down-canyon component of herringbone cross-beds

in Pecos River sands are very small compared to the up-canyon structures and may have been formed by periodic oscillations like those observed in flume experiments (Baker and Kochel, this volume).

Discriminating Flood Events. Discrimination of individual flood events in slackwater sequences is based on radiocarbon dating, the recognition of abrupt vertical grain size changes, presence of paleosols, changes in sediment induration, reversals of mean grain size trends, color changes, buried mudcracks, colluvial horizons, and in some cases interbedded coarse tributary alluvium (couplets). Most of the slackwater sedimentation units observed in southwest Texas are capped by silt drapes or fine-grained organic detritus that was probably concentrated in the upper few meters of floodwaters (Fig. 8).

Stratigraphic relationships can be clearly defined in slackwater deposits in Texas, Utah, Arizona, and other semi-arid regions, but such relationships may be less distinct in humid regions for several reasons. Foremost is the increased bioturbation by roots that occurs in humid regions because of higher vegetation density. In addition, the accelerated rate of pedogenesis in humid regions tends to homogenize slackwater sediments more rapidly than in arid regions. The net result is that greater care must be taken when interpreting humid slackwater sediments. Some floods may go undetected if strata become homogenized.

Problems of slackwater sediment analysis in humid regions can be minimized by carefully selecting study sites that have suffered minimal disturbance by vegetation and by using greater numbers of sites for correlation purposes. Reconnaissance studies of slackwater sediment stratigraphies in southern Missouri and southern Illinois indicate that sites can be found in humid regions where stratigraphic disturbance is minimal. Figure 5*b* shows numerous Mississippi River slackwater and tributary sediment couplets observed at a slackwater site in a bedrock cave near the mouth of Little Grand Canyon, Illinois. Vegetation disturbance was minimal at this site because of its location in the cave.

Grain Size and Thickness. The thickness and grain size of individual slackwater sedimentation units and the cumulative pile of slackwater sediments at protected sites vary significantly along a given river like the Pecos. Sediment

FIGURE 8. Sedimentology of slackwater sediments in Pecos River tributary mouth sites. (*a*) Thick-bedded, structureless sand like this shown in Still Canyon typifies most of the sites. Individual flood units are separated by the darker silt drapes. (*b*) Horizontally bedded units are also common like these in Ladder Canyon. Note the large cross-beds in the 1954 flood layer near the center indicating up-tributary flow. The fine-grained organic deposits at the top of the 1974 unit (top) are common at the tops of units as floating debris settles out. These organics provide the best radiocarbon dates for their associated floods.

thickness and grain size between units at a site are expected to vary with the hydraulic characteristics of the flow and the geometry of the accumulation site and tributary canyon. If these relationships can be quantified, it would greatly increase the paleoflood information measurable from the study of slackwater deposits. The following discussions will be restricted to protected tributary mouth sites where interruption of slackwater sedimentation is unaffected by subsequent flows.

Assuming a sediment-charged mainstem flood, the controls on the resulting slackwater sediment thickness are similar to the factors affecting the cumulative sediment thickness from multiple floods at a site. These controls include: (1) the peak stage and duration of the flood, (2) the tributary junction angle, (3) cross-section of the mainstream and tributary channels, (4) mainstream and tributary gradient, (5) base elevation of the slackwater site above the mainstream channel, and (6) the synchronism between tributary and mainstream flooding. Our flume studies confirmed that the thickness of slackwater sediments at a site varies directly with duration and peak flood stage. However, there remain significant between-site variations in thickness due to the other factors.

Figure 9 is a generalized summary of the effects of other factors affecting variations in between-site slackwater sediment thickness based on field studies along the Pecos River (Kochel, 1980; Kochel et al., 1982) and flume experiments discussed in more detail by Baker and Kochel (this volume) and by Kochel and Ritter (1987). Cross-sectional area of the mainstream and tributary canyons affect the local peak stage and hence the elevation of slackwater sedimentation in tributary mouths. Cross-sectional area also affects the velocity of tributary backflooding. In general, slackwater sediments become increasingly thick as canyon widths decrease and slackwater stages increase (Fig. 9*a*). As junction angles approach 90°, the thickness of slackwater sediments increases (Fig. 9*b*), showing the optimization of backflooding and minimization of mainstream erosion of slackwater sediments. Figure 9*d* shows that slackwater sediment thickness is greatest when mainstream gradient is lowest because bypassing of tributary mouths occurs when mainstream velocity becomes too high in these steep reaches.

Figure 9*c* suggests that during a given flood, slackwater sediment thickness will be greater at sites whose preflood sediment surface is closest to the elevation of the mainstream

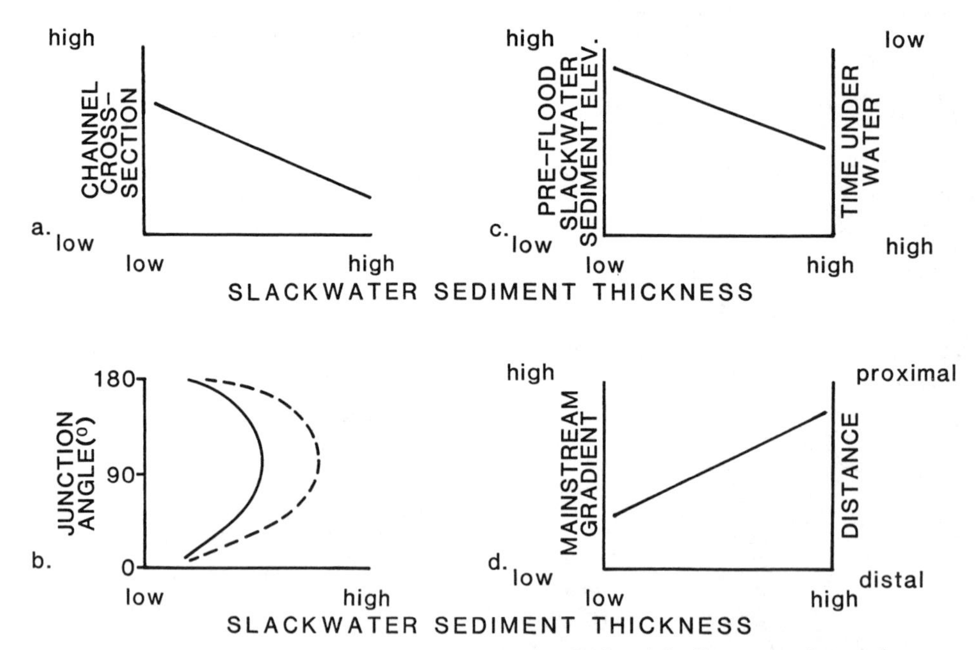

FIGURE 9. Generalized relationships observed in the field and in flume experiments between slackwater sediment thickness and controlling variables. (*a*) With decreasing mainstream and tributary cross section, relative flood stage increases and backflood velocity increases, resulting in thicker slackwater sediments. (*b*) Slackwater sediments are thickest where tributary mainstream junction angles approach 90°. Low angles limit backflooding because of poor mainstream access to the site, while high-angle geometries allow the mainstream to erode previous accumulations of sediment in the tributary mouths. (*c*) Single flood and cumulative slackwater sediment thickness tends to be thicker at sites whose surface of accumulation is closest to the mainstream channel. (*d*) Steep mainstream gradient results in increased mainstream velocity and negligible tributary backflooding occurs.

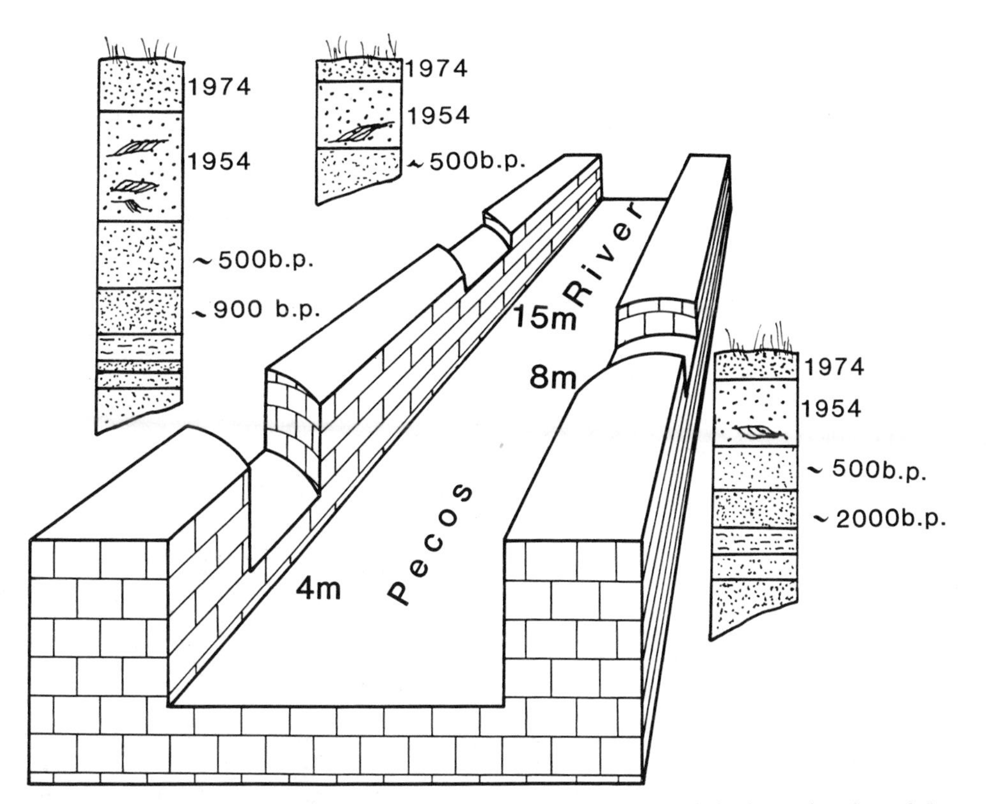

FIGURE 10. Generalized relationships between the Pecos River and the base elevation of the tributary mouth slackwater sequences. During a given flood (and similar channel geometries) thicker units will result in sites whose depositional surface is closest to that of the Pecos channel. These variations indicate that when using thickness as an index of relative flood magnitude, studies should be restricted to at-a-site investigations. However, correlations between neighboring sites are extremely useful in narrowing estimates of paleoflood stage as discussed in the text. For example, the flood of 900 B.P. appears in two sites, but not in the highest site. Hence, its stage range can be better documented.

channel. Comparisons of slackwater sediment thickness in 18 sites along the lower Pecos River suggest that this may be the most important of the four factors represented in Figure 9. Figure 10 shows how at-a-site thickness varies directly with the elevation of the site above the Pecos channel for four major paleofloods. Sediments are thickest at sites nearest the base level of the main channel.

At a site, slackwater sediments thin away from the mainstream. Individual units pinch out up the tributary, thereby marking the approximate peak stage associated with each flood. This lateral variance in thickness probably accounts for much of the scatter that occurs in attempts to view relationships between slackwater sediment thickness and controlling factors. For example, it is very difficult to ensure that between-site comparisons are made at similar locations in each tributary canyon, with respect to distance from the mainstream.

In summary, slackwater sediment thickness can be used as an approximate index of the relative magnitude of paleofloods at a site, but it should not be used in this way between sites. Figure 11 shows these relationships for several slackwater sites along the lower Pecos. Good relationships like these, however, are not common because of the variety of interacting controls on sediment thickness.

Controls affecting the grain size of slackwater sediments appear to be even more complex than those affecting sediment thickness. Mean slackwater sediment size is partially inherited from the lithology of the mainstream upstream from the depositional site. In addition, grain size is controlled by the fluvial regime of the main river. At opposite ends of the spectrum of mean slackwater sediment size are the Pleistocene Scabland deposits, which include coarse gravel, and the Mississippi River deposits in southern Illinois, which are dominantly very fine sand and coarse silt. The size of Scabland deposits were limited by the competence of backwater flows. Tractive processes transported gravels up to tens of meters in diameter in the main Scabland channels (Baker, 1973). In contrast, the size of Mississippi

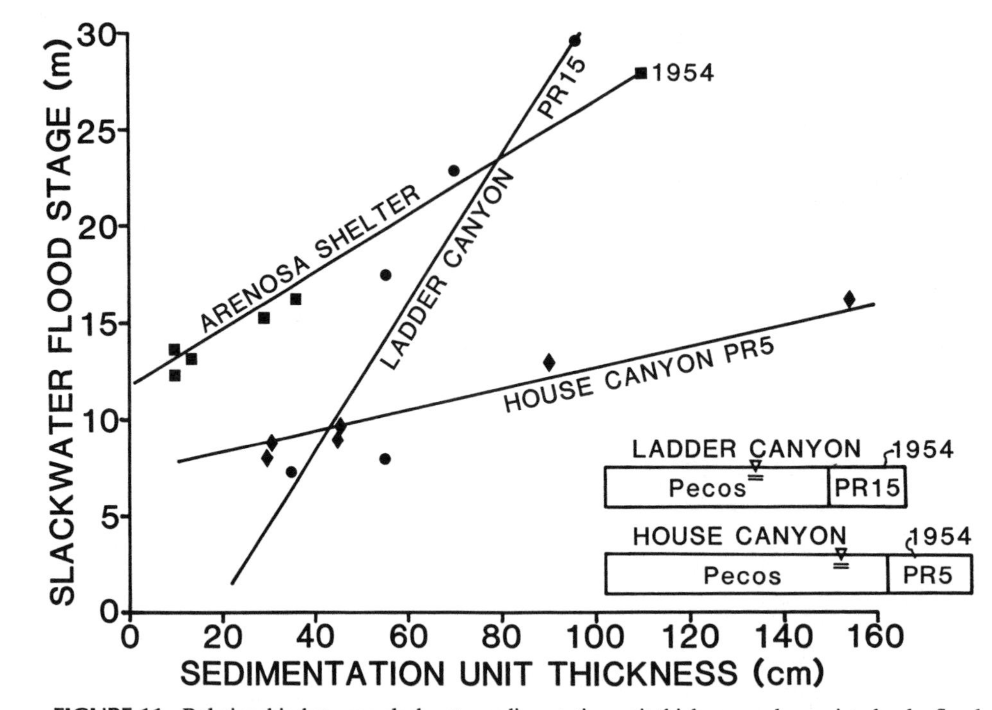

FIGURE 11. Relationship between slackwater sedimentation unit thickness and associated paleoflood stage for three sites along the lower Pecos River. Sketches in the upper right show relative cross sections of the Pecos and tributary channel for the two tributary mouth sites (Arenosa is a cave site). Note that the slope of these lines appears to be controlled by the geometry of the flood cross sections.

River slackwater sediments is limited less by competence than by the maximum size of sediments available in the mainstream. In southern Illinois, Mississippi River point bar sediments are well-sorted medium sand.

Between-flood variance in grain size is usually minor. For example, in the lower Pecos River sites, slackwater sediments range from coarse silt to coarse sand. Most of the between-flood variation is probably accounted for by one or more of the following: (1) distance from the mainstream; (2) competence of the backflow, which is regulated partly by canyon geometry as discussed above; (3) character of the flood hydrograph; and (4) the preflood elevation of slackwater sediments. Figure 12 summarizes the expected relationships between mean slackwater sediment grain size and these variables.

Despite these expectations, limitations on the availability of certain grain sizes for suspended transport may limit the range of sizes transported to slackwater sedimentation sites. For the Pecos River study we found a trend of increasing grain size with paleoflood discharge (Fig. 13), but the range of variance is only from very fine sand (3.8 ϕ) to fine sand (2.9 ϕ). This difference is measurable but requires sieving samples at 1/4 ϕ intervals. Similar trends have been observed at the Arenosa Shelter (cave site) on the Pecos River (Patton and Dibble, 1982). The 1954 flood that filled Arenosa Shelter is the thickest and coarsest of the slackwater units preserved. In addition, the 1954 sediment was deposited on top of the slackwater sequence, suggesting that it may have been the largest flood recorded at that site, where the paleoflood record extends for 10,000 yr (see Kochel, this volume).

Within a flood unit slackwater sediments tend to fine away from the mainstream. The lateral variation is most pronounced in coarse slackwater sediments as in the Channeled Scabland (Bunker, 1982) but is observed in most cases (Fig. 14). These trends were also observed in our flume experiments. Because slackwater sediments are dominantly deposited from suspension, large lateral variations in mean grain size and sorting should not be expected.

The scale of sedimentary structures, that is, up-tributary cross-beds, are useful in determining relative paleoflood magnitude at a site. The 1954 slackwater unit along the Pecos River always contains the largest scale cross-beds. This was also interpreted as the largest paleoflood based on sediment thickness (Kochel et al., 1982). Kochel and Ritter (1987) also observed a direct correlation between bedform size and peak flood discharge in flume experiments.

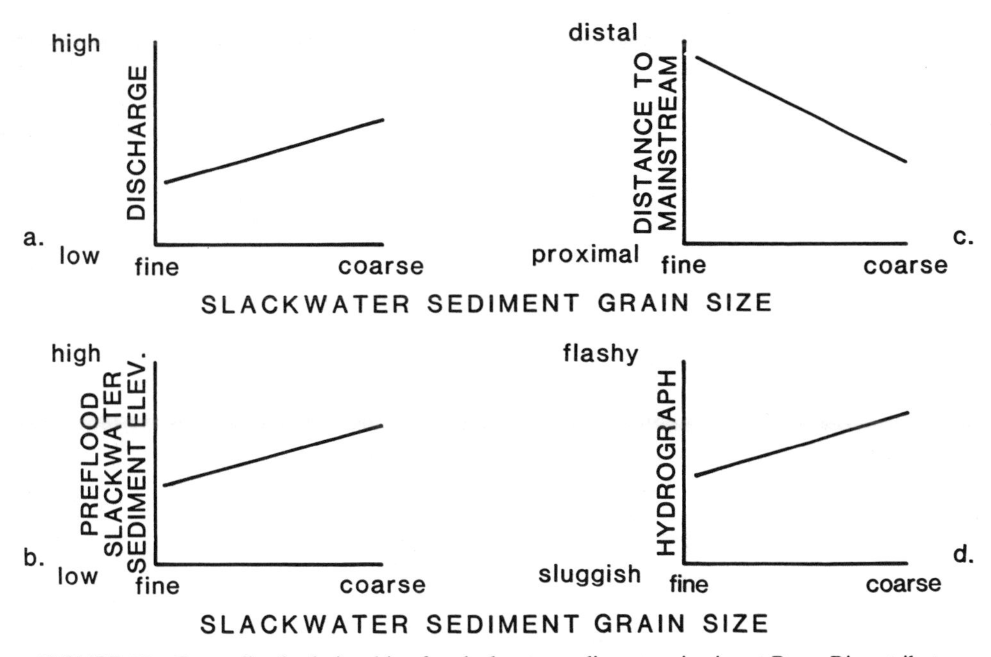

FIGURE 12. Generalized relationships for slackwater sediment grain size at Pecos River tributary mouth sites. These relationships are not as well defined as those for thickness shown in Figure 9. (*a*) Larger discharges may produce increased backflooding velocities, resulting in coarser slackwater sediments provided the suspended load is not limited by supply. (*b*) Higher stage (and discharge) flows required to overtop high slackwater sites may be transporting coarser sediments in suspension in these flows. (*c*) Sediment size varies depending on where a unit is sampled with distance from the mainstream. Distal fining is common. (*d*) Preliminary flume experiments with variable hydrograph characteristics resulted in coarser sediments during flashier floods.

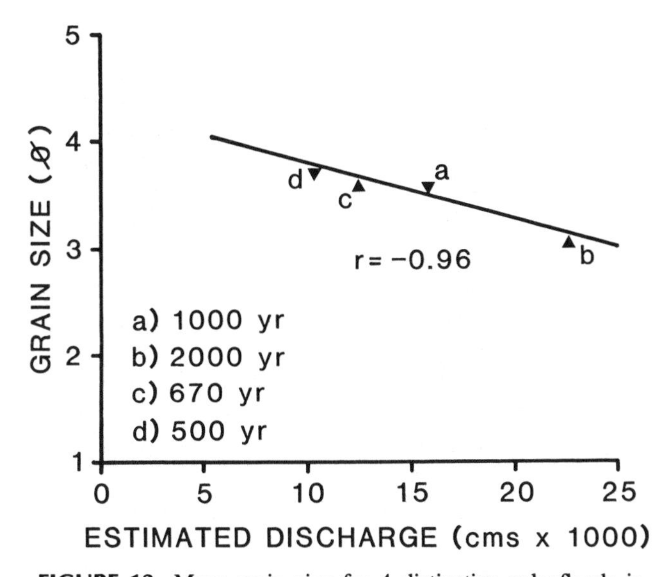

FIGURE 13. Mean grain size for 4 distinctive paleofloods is plotted using an average from 15 Pecos River slackwater sites between Pandale and Langtry. Although there is not much range in grain size, there does appear to be a definable trend indicating coarsening sediment with increasing paleoflood discharge.

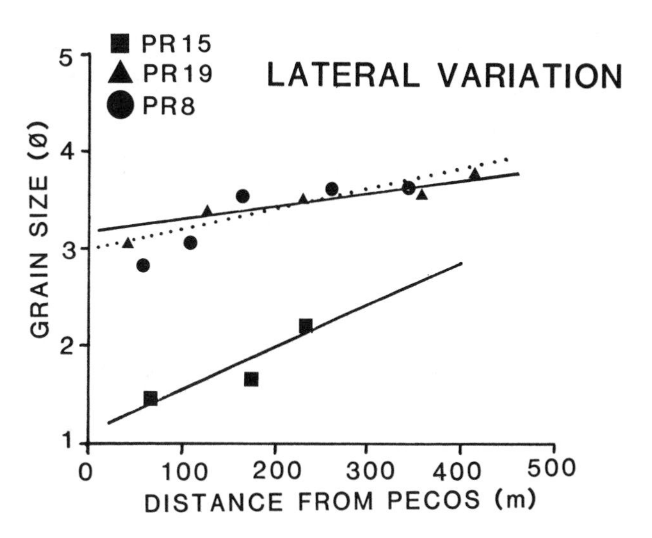

FIGURE 14. Fining of tributary mouth slackwater sand away from the Pecos River at three sites. The samples were taken from the 1954 flood unit at all sites.

PALEOFLOOD STAGE AND SLACKWATER DEPOSITS

Paleoflood reconstruction requires a method of estimating paleoflood stage hundreds or thousands of years after a flood has occurred. Slackwater deposits can be used to approximate paleoflood peak stage from which estimates of paleoflood discharge are possible. In addition, paleoflood histories can be established using slackwater techniques because slackwater deposits can be studied in continuous stratigraphic sequences to provide data on major floods experienced by the river over thousands of years.

The use of any form of high-water indicator to estimate paleoflood discharge requires the acceptance of several qualifying assumptions. First, the slackwater deposits must be associated with the modern flood regime of the river. If evidence is gathered that suggests that climatic conditions have varied significantly over the duration of time represented by the deposits, then the record should be treated in parcels separated by the climatic variations. Significant climatic variations may cause adjustments in flood frequency that would be best treated separately from the modern conditions. In cases where historical large floods are known to have inundated slackwater sites, like those along the Pecos River, then this assumption is appropriate.

A second important assumption is that the channel cross sections of the mainstream and tributary have remained relatively stable, that is, there has been insignificant scour or fill in the channel during flood events. This assumption is probably valid for bedrock rivers like the Pecos (Fig. 15) and some Australian rivers (Baker et al., 1983, 1985). However, severe difficulties occur for alluvial river reaches. Great care must be taken to locate the most stable sites in alluvial channels where bedrock control can be documented. Field observations by Baker (1977), Tinkler (1971), and Shepherd (1979) indicate that bedrock reaches in central Texas are minimally affected by erosion during floods, and that most of the erosion is localized around meander bends. In addition, there is little alluvium available in these bedrock channel reaches for significant filling to occur. These statements are supported by observations made by local residents in Texas and by comparisons of aerial photographs taken before and after the large 1974 flood on the Pecos River (Kochel, 1980). Slight errors that may result from erosion of bedrock reaches are minimized by the great stages attained by these rare flood flows in the narrow canyons. For example, the 1954 flood stage on the lower Pecos River exceeded 27 m at many locations. Therefore, errors of even a few meters due to subsequent erosion of the channel would only result in overestimation of the paleoflood stage by less than 5%.

FIGURE 15. Typical view of the stable bedrock canyon of the lower Pecos River. The depth of the canyon here near Shumla is about 30–35 m. Note the lack of alluvium in the channel.

The third assumption is that there has been negligible channel aggradation or degradation over the time period represented by the slackwater sediments. Studies of the geomorphology (i.e., terraces) of the river are required to provide estimates of the rate of regional downcutting. Along the lower Pecos and Rio Grande, Evans (1962) observed Pleistocene terraces located between 3 and 4 m above the present channel. In addition, the base of most of the tributary mouth slackwater sediment sequences in Texas and Australia contain coarse tributary alluvium averaging 0.5–1.5 m thick (see Figs. 3 and 6). Based on these observations, the probable maximum Pecos River downcutting occurring during the Holocene was about 3 m. With the tremendous flood stages, above 27 m, the error introduced in stage (and discharge) estimations for paleofloods would be insignificant.

The final, but most important, assumption necessary for using slackwater sediments to estimate paleoflood stage is that their elevation records the maximum peak flood stage. In general an individual slackwater deposit elevation represents a minimal peak stage marker, resulting in a conservative estimate of paleoflood discharge. Scattered slackwater sediments can usually be found along channel walls and floodplains up to the exact limit of the peak stage. However, unless these sediments are located on continuous stratigraphic sequences, they are only of use in reconstructing the stage of one particular event. Due to the rapid erosion of loose sediments, these deposits may only provide information from the most recent flood (see Fig. 7*b*). For example, on the Pecos River, sand from the 1954 flood can still be seen at many locations in joints and alcoves along channel walls up to 27 m above the Pecos River. Complete sequences of slackwater sediments usually occur at levels below the peak flood stage.

Improved accuracy of determining peak stage of stratigraphic accumulations of slackwater sediments is possible

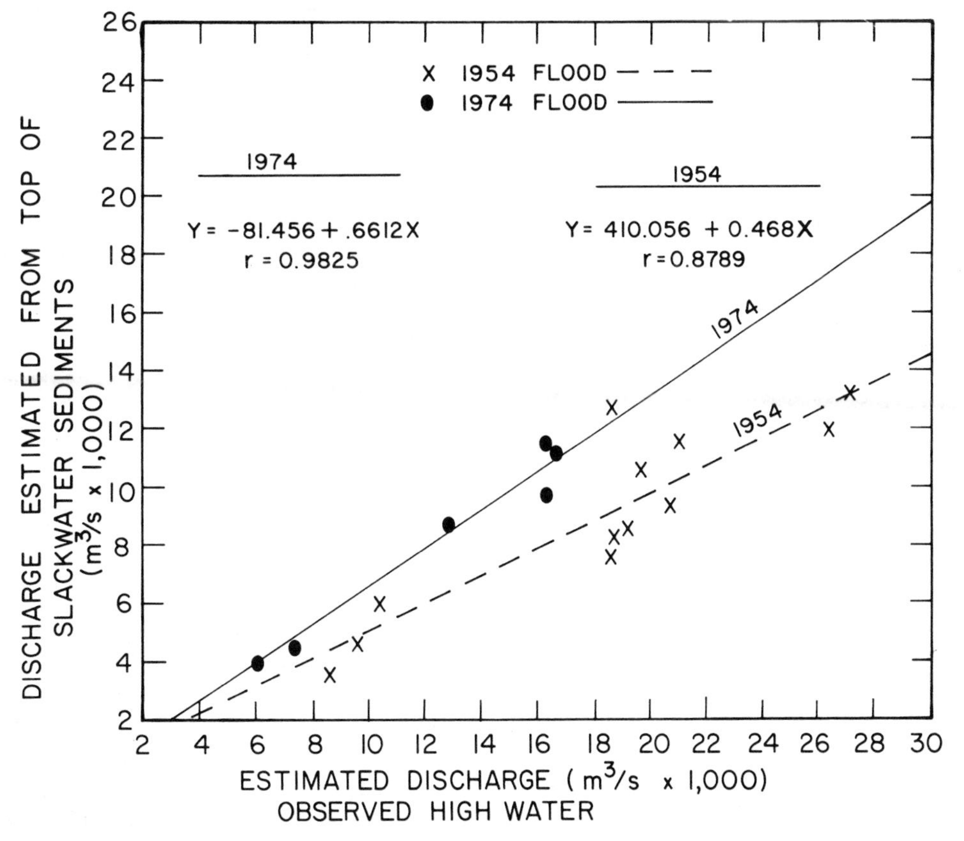

FIGURE 16. Relationship of the stage of historical floods in 1974 and 1954 to the elevation of slackwater sediments in tributary mouth sites along the lower Pecos River. When historical observations are available, adjustments can be made locally to reduce errors in underestimating paleoflood stage. (From Kochel et al., 1982. Copyright by the American Geophysical Union.)

using several techniques. First, if there have been any historical large floods along the river where high-water marks can be documented by debris lines or by local residents, corrections can be made that are locally applicable for interpreting the differences between water stage and maximum slackwater sediment elevation (Fig. 16).

Better estimates of paleoflood stage from slackwater sediments can be obtained by tracing flood units up the tributaries to determine their elevation where they pinch out. The maximum elevations of these pinch-outs more closely correlate with peak flood stage in studies along the Pecos River (Kochel, 1980). Kochel and Ritter (1987) found that this assumption was supported in preliminary studies of slackwater sedimentation in flume experiments. Finally, the use of correlations between multiple sites with different base elevations above the mainstream can greatly reduce the uncertainty of peak stage estimates (see Fig. 10). The accuracy of peak stage estimates improves directly with the number of sites used in correlation efforts.

Finally, it may be possible to correlate slackwater deposits to preserved high-water mark indicators, such as silt lines and erosional trim lines in valley-side soils. Such correlations have been demonstrated in the canyons of south-central Utah (Webb et al., this volume; O'Connor et al., 1986) and central Arizona (Ely and Baker, 1985; Partridge and Baker, 1987).

PALEOFLOOD DISCHARGE ESTIMATES

Once the peak stages of various paleofloods have been designated from studies of slackwater deposits, silt lines, and flood scars, the investigator must transform these data to paleodischarge estimates. This can be accomplished by several hydraulic procedures. However, each requires an accurate survey of the study reach to precisely characterize the channel geometry (Williams and Costa, this volume). Cross sections should be surveyed perpendicular to paleoflow directions, and they should be chosen to characterize all expansions, constrictions, and changes in slope for the study reach. The peak slackwater stage must be extrapolated back to the mainstream and used for flow depth at surveyed cross sections of the mainstream near the tributary mouth.

However, estimates must be made of energy slope and roughness.

In most slackwater settings it is rare to have enough precise control to accurately map the flood water surface slope. Accurate measurements are sometimes possible for recent paleofloods by using nonstratigraphic occurrences of slackwater sediments and other flood indicators. For the slope area procedure (Dalrymple and Benson, 1967) cross sections must be located at sites of high-water indicators in order to approximate the paleoflood water surface slope. In such cases it is important that a sufficiently long reach be surveyed to minimize the average deviations of energy slope from channel bed slope. More extensive discussions of paleoflood slope area procedures are provided by Baker (1973) and Kochel and others (1982). Slope area calculations based on slackwater sediments from the 1954 and 1974 Pecos River floods were within 5% of the estimates made by the U.S. Boundary and Water Commission gauging station on the lower Pecos (Kochel, 1980). Although estimates of channel roughness may vary between investigators, error in roughness estimation becomes less important in the deep flood flows experienced in narrow, deep bedrock rivers. Roughness estimates for the bedrock channels in west Texas vary between 0.035 and 0.040.

Application of the slope area method is not appropriate where slackwater sites are separated by complex sequences of deep pools and channel constrictions like those that occur in the Katherine Gorge in Australia (Baker et al., 1985). In these settings flow calculations require extensive profile analysis to include the effects of backwater and flow transitions. Such analysis can be performed with the aid of computer flow models (Feldman, 1981; Hydrologic Engineering Center, 1982). These models generate a variety of profiles for known discharges in surveyed channels. When compared to paleoflood stage indicators, the modeling can be used to evaluate paleoflood discharges. O'Connor and Webb (this volume) discuss the application of flow modeling in paleoflood analysis. Recent applications of the procedure include Ely and Baker (1985), O'Connor and others (1986), Webb and others (this volume), and Partridge and Baker (1987).

PALEOFLOOD AGES

Once the relative sequence and discharges of paleofloods are determined along a river reach, the absolute flood chronology can be established using radiocarbon dating. A conventional radiocarbon date (Stuiver and Polach, 1977) is derived from laboratory determination of the remaining present-day ^{14}C activity in an appropriate organic material (Faure, 1977). This activity is then compared to an atmospheric ^{14}C level that is assumed to have been constant in the past. Samples collected for radiocarbon analysis require thorough pretreatment, including both the physical separation of rootlets and inorganic materials and appropriate treatment with reagents. Relatively minor amounts of contaminants may yield major discrepancies in the ultimate dates (Polach and Golson, 1966).

Table 1 lists typical materials subjected to radiocarbon analyses in paleoflood studies. Radiocarbon ages based on wood samples may not always correspond exactly with the precise timing of the flood that deposited the sediments containing the wood. The discrepancy between the age of the wood and the flood date is amplified in semi-arid climates because wood can remain undecomposed on the surface for many years after the plant dies. Therefore, when the wood is picked up by floodwaters for transportation to a slackwater depositional site, it might have been dead for 10 to perhaps 200 yr. The durability of large pieces of wood and charcoal also can lead to problems because these materials can survive reworking of older flood deposits and become incorporated into a new deposit (Fig. 17).

If considerable time passes between the successive inundation of slackwater sediment surfaces, pedogenesis will occur. Radiocarbon techniques can be used to obtain mean residence time dates for the soil-forming interval between successive floods (Geyh et al., 1971). When soils are used, three fractions should be dated: (1) total soil humus, (2) soluble humic acids, and (3) insoluble residue. When all three fractions yield similar ages, contamination is considered minor, and the date can be used to provide a minimal estimate for the time elapsed between the deposition of the sediment containing the paleosol and the overlying sediment (Kochel et al., 1982).

Whenever possible, the most suitable material for dating paleofloods is the layer of fine-grained organic detritus that often occurs in the upper few centimeters of the sedimentation unit (see Fig. 8). These fine-grained organic materials tend to decompose rapidly if exposed at the surface. Their rapid burial in slackwater sediments will usually result in a sample whose radiocarbon age is nearly synchronous with the flood event (Kochel, 1980).

A major advance in radiocarbon dating has recently been achieved through the use of the tandem accelerator mass spectrometer (TAMS). This instrument does not measure ^{14}C by decay counting as in conventional practice. Rather, carbon ions are accelerated to energies of several million volts. At these energies the interfering molecular ions can be discriminated on the basis of their atomic weights. Thus, the device directly determines ^{14}C, ^{13}C, and ^{12}C.

The University of Arizona operates a TAMS that has achieved high analytical accuracy (Donahue et al., 1983). The system requires transformation of carbon samples to targets that will be sputtered by a cesium ion source for injection into the TAMS (Jull et al., 1983). Targets typically use 1–2 mg of elemental carbon, but as little as 300 μm

TABLE 1 Dating Slackwater Deposits

Type of Material	Dating Technique	Problems, Notes[a]
Prehistoric, artifacts	Archeological studies	Mixing and cultural time transgression (up to 100's of years)
Trees, logs	Dendrochronology	Can cross-reference with radiocarbon; shorter time span (minimal, often none)
Logs, wood	Radiocarbon dating	May be older than flood or reworked; decomposes slowly in arid areas (10's to 100's of years)
Charcoal	Radiocarbon dating	Same as logs and wood, sometimes worse (10's to 100's of years)
Organics in paleosols	Radiocarbon dating	Minimal age, mean residence time (100's of years)
Fine-grained organics	Radiocarbon dating	May be contaminated by modern rootlets; best date for flood (0 to 10's of years)

[a] Parentheses indicate expected errors possible in dating the actual occurrence of the flood.

of carbon may be used. Thus, a revolutionary advantage of TAMS dating is the very small required sample size (Taylor et al., 1984). Individual seeds, blebs of charcoal, and other small organic particles can now be dated with high accuracy.

Another new procedure is the use of the "post bomb" radiocarbon chronology. Anomalously high ^{14}C activity generated by extensive nuclear tests up to August 5, 1963 (Test Ban Treaty) results in so-called ultramodern dates for radiocarbon samples younger than 1950. The artificially high ^{14}C activity provides a method of calibrating very precise ultramodern radiocarbon analyses. Reference to the appropriate ^{14}C concentration curve (Fig. 18) may make it possible to "date" samples younger than 1950 to the precise year. Paleoflood studies employing this technique may be useful in remote areas where conventional hydrologic data are lacking. Baker and others (1985) discuss applications in the Northern Territory, Australia.

It is important to emphasize that radiocarbon analytical laboratories report sample ages in *radiocarbon years* before the present not in *calendar years*. For purposes of radiocarbon analyses, the "present" is considered calendar year 1950. Moreover, secular changes in $^{12}C/^{14}C$ reservoir characteristics have occurred, resulting in significant deviations of radiocarbon years from calendar years. Calibration curves exist for the last 7500 yr (Damon et al., 1974), but these must be used with caution because they retain the uncertainty characteristics of conventional radiocarbon dates. Dates taken from a calibration curve should be expressed as *near-calendar years*.

In general, paleoflood studies concern ages that do not pose major problems in relating calendar years to radiocarbon years. The critical date is the one defining the total length of the flood record. Since most paleoflood records are at least 1000 yr in length, radiocarbon counting errors are small in comparison. An exception, however, is the period 1950 to 350 radiocarbon years before present. Changes in the global carbon budget from the industrial revolution have resulted in complexities for this period (Stuiver, 1982). In general it is best to treat dates from this period as

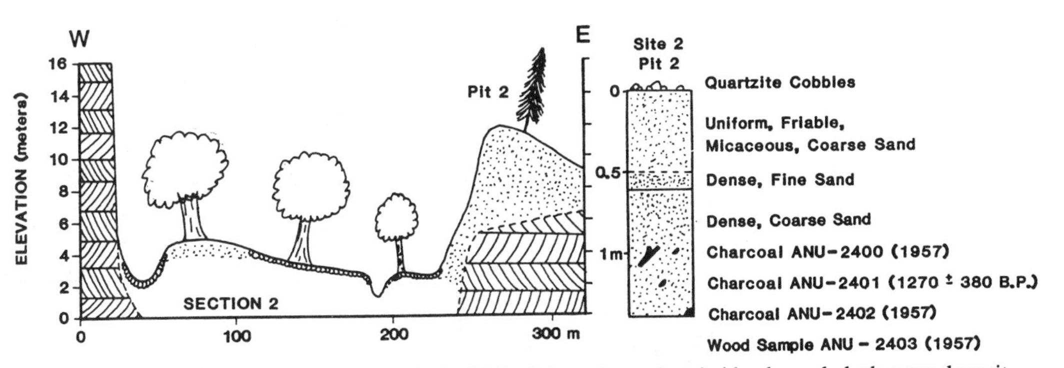

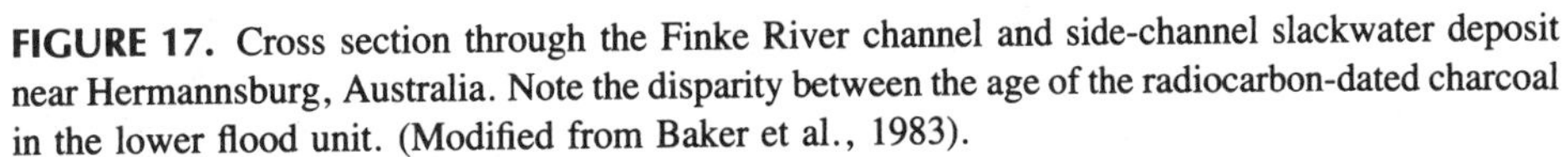

FIGURE 17. Cross section through the Finke River channel and side-channel slackwater deposit near Hermannsburg, Australia. Note the disparity between the age of the radiocarbon-dated charcoal in the lower flood unit. (Modified from Baker et al., 1983).

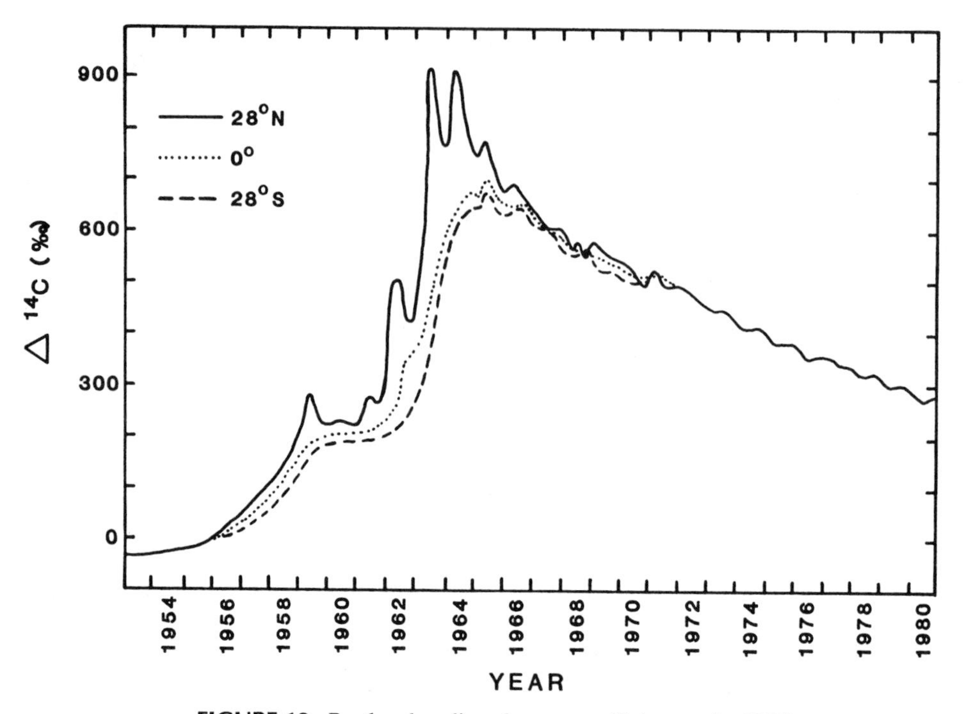

FIGURE 18. Postbomb radiocarbon curve (Baker et al., 1985).

suspect, unless confirmed by other geochronologic techniques (Webb et al., this volume).

Correlations of 18 slackwater sites along the Pecos River of west Texas resulted in a 10,000-yr record of paleoflooding. More details of the slackwater paleoflood record in west Texas are provided by Kochel (1980; this volume) and Kochel and others (1982). Radiocarbon dates are required in frequency analyses to provide time intervals for the application of recurrence interval computations. However, every flood unit need not be dated if stratigraphic correlations can be made between slackwater sites.

CONCLUSION

Slackwater sedimentation is the dominant form of fine-grained flood sedimentation in narrow, deep bedrock fluvial systems (Baker and Kochel, this volume). Tributary mouth slackwater deposits appear to be the most common. Mainstream floods can be distinguished from tributary floods using observations of the mineralogy and sedimentology of the deposits. The maximum elevation of a slackwater sediment unit records a conservative level of paleoflood stage that can be used to estimate discharge with the slope area method or step-backwater flow modeling procedures, depending on the complexity of channel morphology. Because slackwater sediments can usually be found in stratigraphic sequences, they can be used to collect a rather continuous long-term record of paleoflooding for a river, particularly when multiple sites are available for correlation studies.

Standard statistical hydrological techniques of estimating the frequency of rare, large-magnitude floods become increasingly unreliable as the recurrence interval of the flood exceeds the length of historical gauging records. Paleoflood slackwater data can extend flow records by thousands of years, making estimates of the frequency of large floods much more reliable (Kochel, this volume).

Continued flume studies and quantitative field observations of slackwater sediments are needed to refine the relationships between the following: (1) peak stage and slackwater sediment stage, (2) peak discharge and slackwater sediment thickness, and (3) peak discharge and slackwater sediment grain size. Promising trends have resulted from our preliminary flume experiments, some of which led to better understanding of the field observations.

Although problems exist with applying slackwater paleoflood techniques to humid alluvial rivers, the technique will probably be successful when used cautiously. Many alluvial rivers in humid areas have bedrock reaches along their course that can be used to obtain the conditions necessary for minimizing temporal instability of channel cross sections. We are now studying slackwater sites in Missouri, Illinois, and Virginia where these conditions are satisfied. In many humid regions slackwater sediments occur in tributary mouth locations as well as in bedrock caves. We have observed slackwater sites in reconnaissance and/or detailed studies throughout the United States (see Kochel, this volume). Most of these sites contain abundant stable bedrock reaches and channel wall caves. The sites with the greatest promise for producing long paleoflood records will be those with minimal vegetation and pedogenesis and sites with cross-axial drainage patterns where mainstream and tributary mineralogies can be easily distinguished.

ACKNOWLEDGMENTS

Peter C. Patton worked closely with us on the development of the slackwater method. David S. Dibble, formerly of the Texas Archaeological Survey, lent freely of his time in discussing the stratigraphies and correlations we made between Arenosa and Devils Mouth sites. Salvatore Valastro, Jr., performed and/or provided instruction in the radiocarbon analyses and provided much help and guidance in these matters. Field research was supported by the National Science Foundation, Surficial Processes Program, Grants EAR 77-23025, EAR-8119981, EAR-8229982, EAR-8100391, and EAR-8300183. A. Long, J. E. O'Connor, G. Pickup, H. A. Polach, and R. H. Webb shared many ideas with us in numerous discussions as we developed the conclusions reported herein. We thank the people of west Texas for their kindness and access to their land. Finally, we thank the following for able field assistance: P. C. Patton, B. L. Kochel, and L. L. Kochel.

REFERENCES

Baker, V. R. (1973). Paleohydrology and sedimentology of Lake Missoula flooding in eastern Washington. *Spec. Pap.—Geol. Soc. Am.* **144**, 1–79.

Baker, V. R. (1977). Stream channel response to floods with examples from central Texas. *Geol. Soc. Am. Bull.* **88**, 1057–1070.

Baker, V. R. (1978). Large-scale erosional and depositional features of the Channeled Scabland. *In* "The Channeled Scabland" (V. R. Baker and D. Nummedal, eds.), pp. 81–115. Natl. Aeronaut. Space Admin., Washington, D.C.

Baker, V. R., and Bunker, R. C. (1985). Cataclysmic late Pleistocene flooding from glacial Lake Missoula: A review. *Quat. Sci. Rev.* **4**, 1–41.

Baker, V. R., Kochel, R. C., Patton, P. C., and Pickup, G. (1983). Palaeohydrologic analysis of Holocene flood slackwater sediments. *Spec. Publ. Int. Assoc. Sedimentol.* **6**, 229–239.

Baker, V. R., Pickup, G., and Polach, H. A. (1985). Radiocarbon dating of flood events, Katherine Gorge, Northern Territory, Australia. *Geology* **13**, 344–347.

Bretz, J H. (1929). Valley deposits immediately east of the Channeled Scablands of Washington. *J. Geol.* **37**, 393–427.

Bunker, R. C. (1982). Evidence of multiple late Wisconsin floods from glacial Lake Missoula in Badger Coulee, Washington. *Quat. Res.* **18**, 17–31.

Costa, J. E. (1978). Colorado Big Thompson flood: Geologic evidence of a rare hydrologic event. *Geology* **6**, 617–620.

Costa, J. E. (1983). Paleohydraulic reconstruction of flash-flood peaks from boulder deposits in the Colorado Front Range. *Geol. Soc. Am. Bull.* **94**, 986–1004.

Dalrymple, T., and Benson, M. A. (1967). Measurement of peak discharge by the slope-area method. *Tech. Water Resour. Res. Div. (U.S. Geol. Surv.)*, Book 3, Ch. A-2.

Damon, P. E., Ferguson, C. W., Long, A., and Wallick, E. I. (1974). Dendrochronologic calibration of the radiocarbon time scale. *Am. Antiq.* **39**, 350–365.

Donahue, D. J., Jull, A. J. T., Zabel, T. H., and Damon, P. E. (1983). The use of accelerators for archaeological dating. *Nucl. Instrum. Methods Phys. Res.* **218**, 425–429.

Ely, L. L., and Baker, V. R. (1985). Reconstructing paleoflood hydrology with slackwater deposits: Verde River, Arizona. *Phys. Geogr.* **5**, 103–126.

Ethridge, F. G., and Schumm, S. A. (1978). Reconstructing paleochannel morphologic and flow characteristics: Methodology, limitations, and assessment. *Mem.—Can. Soc. Pet. Geol.* **5**, 703–721.

Evans, G. L. (1962). Notes on terraces of the Rio Grande, Falcon-Zapata area, Texas. *Bull. Tex. Archaeol. Soc.* **32**, 33–45.

Faure, G. (1977). "Principles of Isotope Geology." Wiley, New York.

Feldman, A. D. (1981). HEC models for water resources system simulation: Theory and experience. *Adv. Hydrosci.* **12**, 297–423.

Geyh, M. A., Benzler, J. H., and Roeschmann, G. (1971). Problems of dating Pleistocene and Holocene soils by radiometric methods. *In* "Paleopedology: Origin, Nature, and Dating of Paleosols" (D. H. Yaalon, ed.), pp. 63–75. Israel Univ. Press, Jerusalem.

Howard, A. D., and Dolan, R. (1981). Geomorphology of the Colorado River in the Grand Canyon. *J. Geol.* **89**, 269–298.

Hydrologic Engineering Center (1982). "HEC-2 Water Surface Profiles: Program User's Manual," U.S. Army Corps of Engineers, Davis, California.

Jull, A. J. T., Donahue, D. J., and Zabel, T. H. (1983). Target preparation for radiocarbon dating by tandem accelerator mass spectrometry. *Nucl. Instrum. Methods Phys. Res.* **218**, 509–514.

Kochel, R. C. (1980). Interpretation of flood paleohydrology using slackwater deposits, lower Pecos and Devils Rivers, southwest Texas. Ph.D. Dissertation, University of Texas, Austin.

Kochel, R. C. (1982). Quaternary geomorphology of Seminole Canyon State Historical Park, Val Verde County, Texas. *In* "Seminole Canyon: The Art and the Archaeology" (S. A. Turpin, ed.), Austin, Tex. Archaeol. Surv. Res. Rep. No. 83, pp. 227–276.

Kochel, R. C., and Baker, V. R. (1982). Paleoflood hydrology. *Science* **215**, 353–361.

Kochel, R. C., and Johnson, R. A. (1984). Geomorphology and sedimentology of humid-temperate alluvial fans, central Virginia. *Mem.—Can. Soc. Pet. Geol.* **10**, 109–122.

Kochel, R. C., and Ritter, D. F. (1987). Implications of flume experiments on the interpretation of slackwater paleoflood sediments. *In* "Regional Flood Frequency Analysis" (V. P. Singh, ed.), pp. 371–390. Reidel, Boston.

Kochel, R. C., Baker, V. R., and Patton, P. C. (1982). Paleohydrology of southwestern Texas. *Water Resour. Res.* **18**, 1165–1183.

Moss, J. H., and Kochel, R. C. (1978). Unexpected geomorphic effects of the Hurricane Agnes storm and flood, Conestoga drainage basin, southeastern Pennsylvania. *J. Geol.* **86**, 1–11.

O'Connor, J. E., and Webb, R. H., and Baker, V. R. (1986). Paleohydrology of pool and riffle pattern development, Boulder Creek, Utah. *Geol. Soc. Am. Bull.* **97**, 410–420.

Partridge, J., and Baker, V. R. (1987). Paleoflood hydrology of the Salt River, Arizona. *Earth Surf. Processes Landforms* **12**, 109–125.

Patton, P. C., and Baker, V. R. (1976). Morphometry and flood potential of drainage basins subject to diverse hydrogeomorphic controls. *Water Resour. Res.* **12**, 941–952.

Patton, P. C., and Dibble, D. S. (1982). Archeologic and geomorphic evidence for the paleohydrologic record of the Pecos River in west Texas. *Am. J. Sci.* **282**, 97–121.

Patton, P. C., and Handsman, R. G. (1984). Paleoflood record for Housatonic River basin, western Connecticut. *EOS, Trans. Am. Geophys. Union* **65** (45), 891.

Patton, P. C., Baker, V. R., and Kochel, R. C. (1979). Slackwater deposits: A geomorphic technique for the interpretation of fluvial paleohydrology. *In* "Adjustments of the Fluvial System" (D. D. Rhodes and G. P. Williams, eds.), pp. 225–252. Kendall-Hunt Publ. Co., Dubuque, Iowa.

Polach, H. A., and Golson, J. (1966). "Collection of Specimens for Radiocarbon Dating and Interpretation of Results, Aust. Inst. Aboriginal Stud. Manual No. 2, pp. 1–42. Australian National University, Canberra.

Shepherd, R. G. (1979). River channel and sediment responses to lithology and stream capture, Sandy Creek drainage, central Texas. *In* "Adjustments of the Fluvial System" (D. D. Rhodes and G. P. Williams, eds.), pp. 255–275. Kendall-Hunt Publ. Co., Dubuque, Iowa.

Stewart, J. E., and Bodhaine, G. L. (1961). Floods in the Skagit River basin, Washington. *Geol. Surv. Water-Supply Pap. (U.S.)* **1527**, 1–66.

Stewart, J. H., and LaMarche, V. C. (1967). Erosion and deposition in the flood of December 1964 on Coffee Creek, Trinity County, California. *Geol. Surv. Prof. Pap. (U.S.)* **422-K**, 1–22.

Stuiver, M. (1982). A high-precision calibration of the A.D. radiocarbon time scale. *Radiocarbon* **24**, 1–26.

Stuiver, M., and Polach, H. A. (1977). Discussion—reporting of ^{14}C data. *Radiocarbon* **19**, 355–363.

Taylor, R. E., Donahue, D. J., Zabel, T. H., Damon, P. E., and Jull, A. J. T. (1984). Radiocarbon dating by particle accelerators: An archaeological perspective. *Adv. Chem. Ser.* **205**, 333–356.

Tinkler, K. J. (1971). Active valley meanders in south-central Texas and their wider implications. *Geol. Soc. Am. Bull.* **82**, 1783–1800.

Waitt, R. B., (1980). About forty last-glacial Lake Missoula jökulhlaups through southern Washington. *J. Geol.* **88**, 653–679.

Waitt, R. B., Jr. (1984). Periodic jokulhlaups from Pleistocene glacial Lake Missoula—new evidence from varved sediment in northern Idaho and Washington. *Quat. Res.* **22**, 46–58.

Waitt, R. B., Jr. (1985). Case for periodic, colossal jökulhlaups from Pleistocene glacial Lake Missoula. *Geol. Soc. Am. Bull.* **96**, 1271–1286.

Williams, G. P. (1983). Paleohydrological methods and some examples from Swedish fluvial environments. I. Cobble and boulder deposits. *Geogr. Ann.* **65A**, 227–243.

Willman, H. B., and Frye, J. C. (1980). The glacial boundary in southern Illinois. *Circ.—Ill. State Geol. Surv.* **511**, 1–22.

22

EXTENDING STREAM RECORDS WITH SLACKWATER PALEOFLOOD HYDROLOGY: EXAMPLES FROM WEST TEXAS

R. CRAIG KOCHEL
Department of Geology, Southern Illinois University, Carbondale, Illinois

INTRODUCTION

Conventional statistical approaches for estimating the recurrence intervals of floods become less reliable with increasing flood return interval. Application of the most widely used techniques such as the Weibull method, the Gumbel extreme value theory (Gumbel, 1957), and the log-Pearson Type III method (U.S. Water Resources Council, 1981) result in exceedingly large discrepancies for increasing recurrence intervals estimated for floods with less than 2% exceedance probability (greater than the 50-yr flood). These problems are especially apparent for semi-arid rivers with highly skewed annual flood distributions and, in particular, for rivers with large outliers in their flood records.

Significant engineering design problems result from the variety of interpretations that can be made from rivers with large flood outliers (U.S. Water Resources Council, 1981). For example, if an exceedingly large flood occurs during the span of systematic measurements, its heavy weighting in frequency analyses could result in tremendous overdesign of flood control structures because its true recurrence interval may be much greater than implied in this analysis.

Figure 1 schematically demonstrates how significant the accuracy of estimation of flood recurrence interval can be to the costs incurred over the expected life span of hydrologic structures such as bridges, dams, culverts, or overflow channels. The initial investment costs are largely controlled by estimates of how much discharge is expected during the design flood such as the 100-yr flood. Construction costs can vary enormously depending on what flood frequency method is used, especially for rivers that are characterized by highly skewed annual peak flood distributions (see Kochel and Baker, 1982). Likewise, the replacement costs also depend on accurate determinations of flood frequency. Underestimation of flood discharge and frequency will result in frequent damage and repair costs. Normally, the replacement cost curve would be expected to decrease with increasing flood recurrence interval because frequency of floods decreases. However, if a critical flood discharge occurs, complete destruction of the structure may result, hence, this curve may increase rapidly at some return frequency.

An example of this problem exists in central Virginia where Hurricane Camille dropped up to 75 cm of rain in about 8 hr in August 1969 (Williams and Guy, 1973). According to standard engineering design criteria, local farmers now wishing to construct small stream impoundments are expected to design structures able to contain the 1969 flood discharge because that flow occurred with the systematic record. Therefore, a central Virginia farmer desiring a small pond might be required to build a dam with a 200-m-high spillway, flooding his entire property. Available flood frequency techniques have no reliable way of estimating the true geologic recurrence interval of the 1969 flood. Sedimentologic studies of debris fans activated in 1969 show that similar floods have occurred in the Holocene, but with an average recurrence on the order of 3000–4000 yr at a site (Kochel and Johnson, 1984). Clearly, the small farm impoundments need not be designed to contain such a flood.

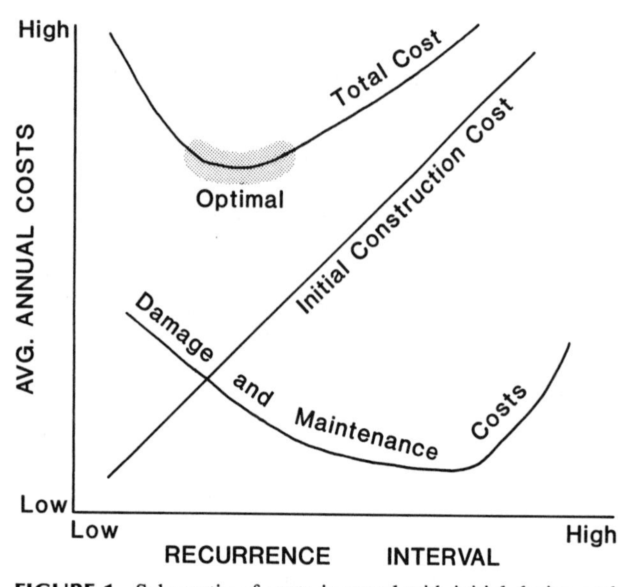

FIGURE 1. Schematic of costs incurred with initial design and maintenance of hydrologic structures. See text for discussion. (Modified from Costa and Baker, 1981).

Ideally, the design of hydrologic structures should target within the optimum zone (Fig. 1) where minimal initial construction costs and maintenance costs would occur. To achieve these design criteria, accurate estimates of flood frequency and magnitude are required. The differences in total costs may be millions of dollars for major reservoir projects and could also affect many lives.

Slackwater paleoflood techniques (Kochel and Baker, this volume; Kochel et al., 1982) can extend flood frequency records of large floods over thousands of years, thus providing the required length of record to assess the true return periods of extreme events. Paleoflood techniques have not been used to treat much debated questions concerning the theoretical distribution of floods. Paleoflood hydrology has mainly been used to extend the length of historical records of large floods using geomorphologic and stratigraphic techniques. This chapter will discuss an example of the application of paleohydrologic techniques to the interpretation of flood frequency in southwest Texas.

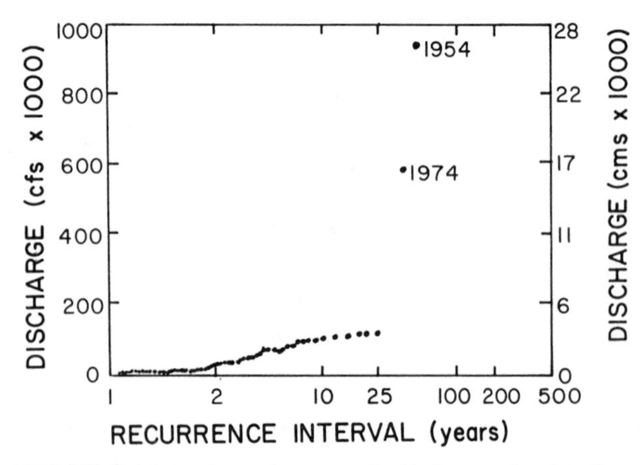

FIGURE 2. Annual maximum peak discharges for the Pecos River near Langtry, Texas, between 1900 and 1977. The two outliers are from the floods of 1954 and 1974. The remainder of the data plots along a well-defined curve. Engineering design decisions will vary dramatically based on how these outliers are treated. (Data were obtained from International Boundary and Water Commission Bulletins, 1930–1977).

HISTORICAL FLOODING IN SOUTHWEST TEXAS

An excellent example of the application of this method is the flood record of the Pecos and Devils rivers of southwest Texas (Fig. 2). The 1954 flood on the Pecos River exceeded

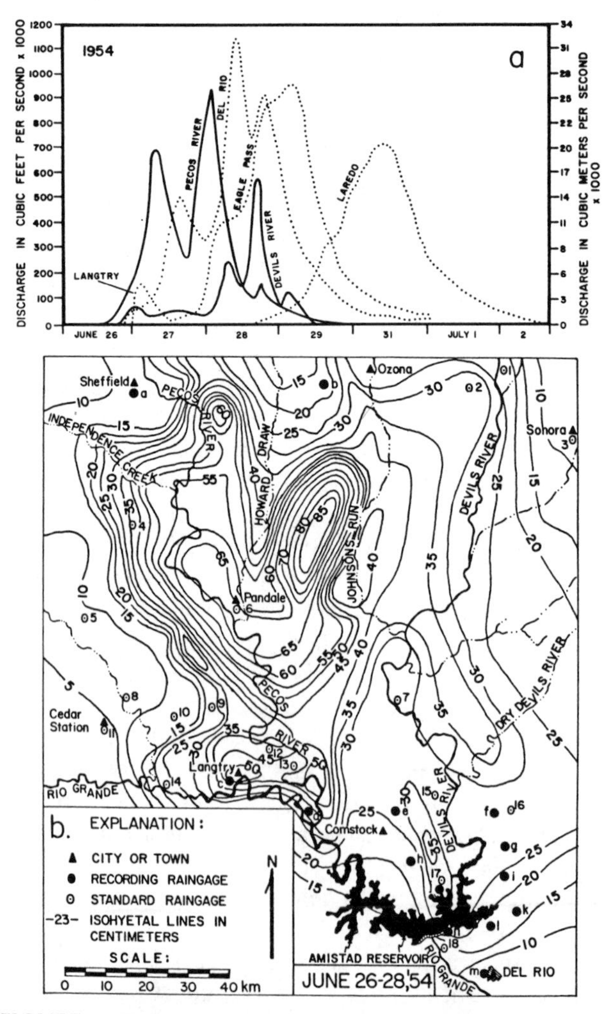

FIGURE 3. Flood of June 26–28, 1954, on the Pecos and Devils rivers in southwest Texas. The isoheytal map (*b*) shows the rainfall from Hurricane Alice that was restricted to the area between Sheffield and Del Rio, Texas. Resulting flood hydrographs on the Rio Grande, Pecos, and Devils rivers are shown in (*a*).

all previously recorded floods by almost an order of magnitude. The immensity of this event resulted in considerable concern as to whether statistical procedures based on historical data could accurately predict catastrophic floods. However, slackwater paleoflood analyses showed that the recurrence interval of the 1954 flood is at least 2000 yr (Kochel et al., 1982; Patton and Dibble, 1982).

Southwest Texas has one of the most catastrophic rainfall–runoff regimes in the United States (Patton and Baker, 1976). Gauging records have been kept on the lower Pecos River near Langtry since 1900 (International Boundary and Water Commission, 1930–1977). The resulting flood frequency curve for the Pecos River is shown in Figure 2. Note the outliers from the 1954 and 1974 floods. None of the statistical methods of treating flood frequency data would have come close to predicting the 1954 flood even from the long 54-yr record that existed in 1953. Most gauging stations have even shorter records from which to predict flood frequency.

Most of the major floods in southwest and south-central Texas result from tropical storms originating in the Gulf of Mexico. Catastrophic rainfalls are triggered in response to interaction of the tropical moisture with extratropical cyclones and/or orographic effects produced by the Balcones Escarpment (Caracena and Fritsch, 1983). The 1954 flood was produced when Hurricane Alice moved up the Rio Grande valley and stalled over the divides of the lower Pecos and Devils rivers (Fig. 3). A discharge of 27,400 m^3/s resulted at the Langtry gauge from over 100 cm of rainfall in less than 48 hr. The contributing area (Fig. 4) for the flood was only the lower 9500 km^2 of the 91,000 km^2 basin (Patton et al., 1979). The gauging station at Sheffield, Texas, 170 km upstream, recorded negligible flow increase during the 1954 event, while lower Pecos flood stages exceeded normal levels by 25–30 m. Similar conditions occurred in 1974 when flood stages exceeded 15 m downstream from Sheffield.

The dilemma faced by users of conventional flood frequency techniques is apparent. These problems became real for engineers constructing the Amistad Dam (Fig. 5) in the 1960s on the Rio Grande downstream from its confluence with the Pecos and Devils rivers. Extensions of the Pecos River flood data by conventional methods result in estimates of the frequency of the 1954 flood ranging

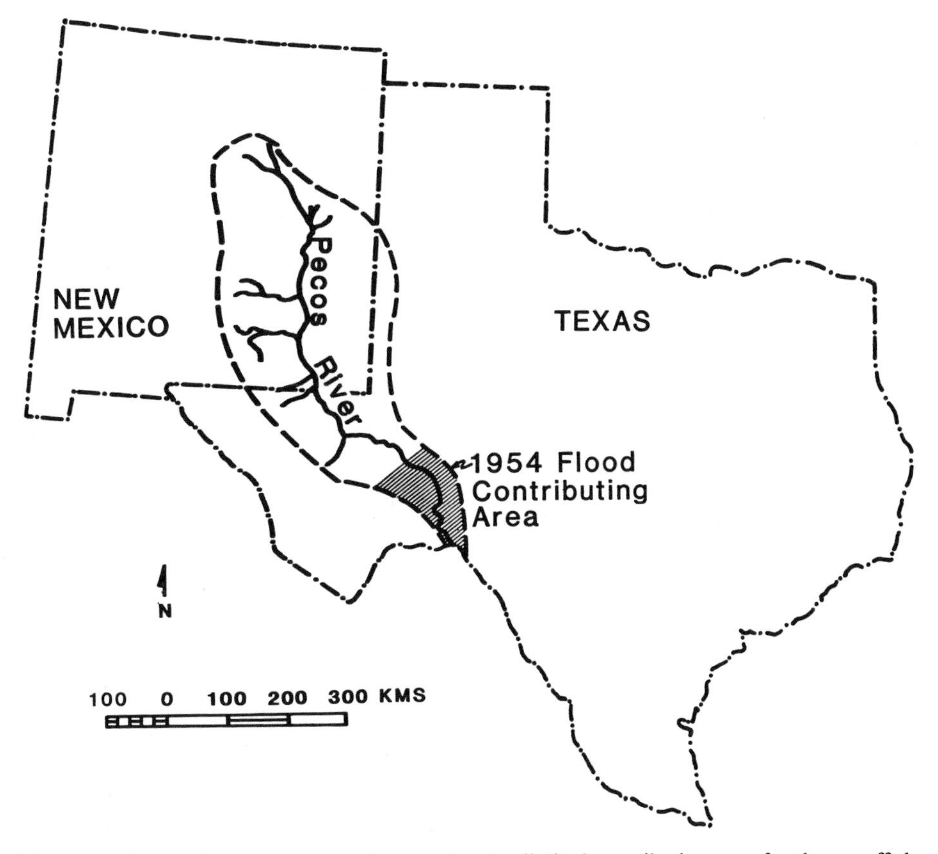

FIGURE 4. Pecos River drainage basin showing the limited contributing area for the runoff that produced the 1954 flood on the lower Pecos. The gauge at Sheffield is located at the upstream edge of the stippled region. The gauge near Langtry, which recorded over 27,000 m^3/s, is about 10 km upstream from the Rio Grande.

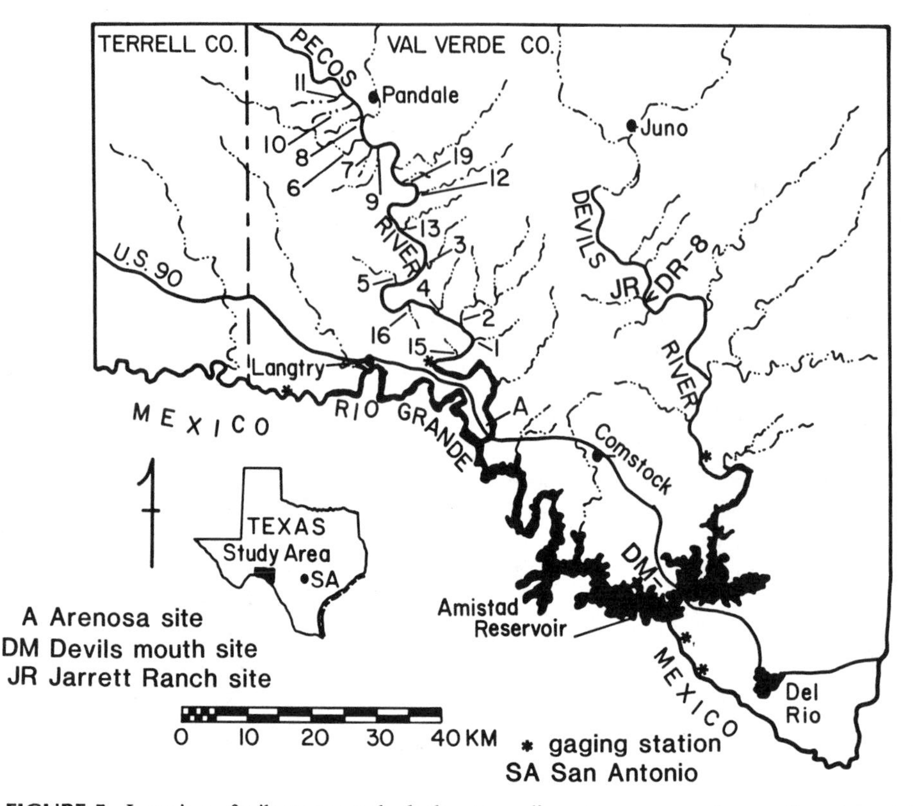

FIGURE 5. Location of tributary mouth slackwater sediment sequences where radiocarbon dates have been obtained. (From Kochel et al., 1982. Copyright by the American Geophysical Union.) Table 4 lists the radiocarbon dates used in paleoflood studies along the Pecos and Devils rivers (Kochel, 1980; Kochel et al., 1982). Table 5 shows summarized slope-area data at selected sites along the lower Pecos.

from 86 to 20,000,000 yr depending on how the outliers are treated. Application of other techniques results in a similar variety of estimates for the return interval of the outliers (see Fig. 8).

APPLICATION OF SLACKWATER PALEOFLOOD DATA

Kochel (1980) studied tributary mouth slackwater sediments in 18 sites along the lower Pecos River between Pandale and Langtry. Figure 5 shows a map of stratigraphic slackwater sediment sequences along the Pecos and Devils rivers where radiocarbon dates have been obtained (Kochel and Baker, 1982). The addition of slackwater sediment data from archeological sites near the mouths of these rivers (Arenosa and Devils mouth sites) resulted in a 10,000-yr record of paleoflooding for these rivers (Patton and Dibble, 1982; Kochel et al., 1982; Sorrow, 1968).

Pecos River Paleoflood History

Correlations of tributary mouth slackwater sites resulted in a 2000-yr extension of the lower Pecos River flood record using techniques described by Kochel and Baker (this volume). Fine-grained slackwater flood sediments are deposited along channel walls and in backflooded tributaries by major rivers during large floods (Kochel and Baker, 1982).

Tributary Mouth Slackwater Sites. Slackwater sites in the mouths of numerous Pecos River tributaries contain records of paleoflooding by the Pecos. Pecos River sands are easily distinguished from tributary sediments by their distinctive mineralogy, grain size, and paleocurrent indicators (see Kochel and Baker, this volume). The elevation of slackwater sediments can be used as an indicator of peak paleoflood stage, from which indirect estimates of paleoflood discharge can be made (see Kochel and Baker, this volume). Kochel and Ritter (1987) showed that correlation of flood stage and slackwater sediment elevation is probably valid in a series of flume experiments in cement channels used to simulate bedrock. Radiocarbon-dated sequences were used to determine the frequency of the four major floods that occurred along the lower Pecos River during the last 2000 yr (Table 1). In this analysis the 1954 flood recurrence interval is estimated to be at least 2000 yr because it resulted in the thickest and coarsest sedi-

TABLE 1 Lower Pecos Paleofloods

Paleoflood	$Q_{max}(m^3/s)^a$	Mean Grain Size (ϕ)	Recurrence Interval (yr)
A.D. 1974	16,000	3.34	1000 ± 70
A.D. 1954	23,000	2.93	2000 ± 100[b]
450–550 B.P.	11,000	3.40	670 ± 50
800–1000 B.P.	10,000	3.46	500 ± 40

[a] Estimates at House Canyon (PR5).
[b] Thickest, coarsest in 10,000-yr sequence.

mentation unit during this period (Kochel and Baker, this volume). The 1974 flood is estimated to be the 1000-yr flood. Figure 6 shows a summary correlation of several tributary mouth slackwater sites containing a paleoflood record up to 2000 yr long.

Addition of Archeological Data from Arenosa Shelter. In the 1960s the Texas Archeological Survey (then the Texas Archeological Salvage Project) conducted extensive excavations of alluvial sites along the lower Pecos and Devils rivers prior to their inundation by the Amistad Reservoir in 1969. Arenosa Shelter is located near the mouth of the Pecos River, downstream from the U.S. 90 bridge (Fig. 5). Excavations at Arenosa revealed an extensive sequence of interbedded cultural units and slackwater flood sediments from the Rio Grande and Pecos River (Dibble, 1967). Patton (1977) and Patton and Dibble (1982) used this well-dated stratigraphy (Fig. 7, left) to reconstruct a 10,000-yr record of paleoflooding along the lower Pecos River.

Table 2 shows the paleoflood frequency analysis that results from combining the tributary mouth slackwater sediment record with the Arenosa data. Radiocarbon dates were used to provide intervals (n values as historical floods) for use in the following Weibull plotting formula:

$$\mathrm{RI} = \frac{n + 1}{m}$$

where RI = recurrence interval in years
n = number of years of record
m = magnitude rank of annual maximum discharges

The intervals were established from breaks in the 10,000-yr record selected according to probable variations in Holocene climate in west Texas (see Table 3), as indicated by palynological data (Bryant, 1969; Bryant and Larson, 1968; Bryant and Shafer, 1977).

Interpretations of Texas Holocene climatic variations can be inferred from palynological data and from paleoflood studies done by Baker and Penteado-Orellana (1977) along the Colorado River in central Texas (Table 3). Patton and Dibble (1982) recognized that the climatic variations affected the flood character of the Pecos River. From Figure 7 and Table 3 one can see that during relatively humid late Pleistocene conditions, the Pecos River experienced frequent floods. These floods at Arenosa were recorded as thin units of fine-grained sand. Arid conditions between 9000 and 3000 B.P. resulted in far less frequent Pecos River floods, but these floods left thicker, coarse-grained slackwater sediments in Arenosa Shelter. Between 3000 and 2000 yr B.P., relatively humid climatic conditions returned to west Texas (Table 3). Arenosa slackwater sediments during this period show more frequent flooding, but flood units are thin and fine-grained (Patton and Dibble, 1982). The only flood recorded at Arenosa by slackwater sediments during the last 2000 years of arid conditions was the 1954 flood. The 1954 flood is the thickest and coarsest of all floods in the section. Kochel and Baker (this volume) demonstrated the direct correlation between slackwater sediment thickness and grain size with paleoflood discharge at a site. Therefore, the Arenosa data suggest that during arid climates, like today, floods along the Pecos were infrequent, but of large magnitude. During more humid climates, Pecos River floods were more frequent, but of lower magnitude.

Figure 8 illustrates the tremendous range of flood frequency curves that result from different analytical methods applied to the Pecos River flood data. Data for Figure 8 are shown in Tables 4 and 5. Curves 8 and 9 (Fig. 8) resulted from incorporating slackwater paleoflood data into the flood frequency analysis. Curve 8 was calculated from Table 2 using the log-Pearson Type III method plus the slackwater flood record, but ignoring the systematic gauging record. Curve 9 was calculated by combining the gauging record with the slackwater data and using the log-Pearson Type III method. This approach used the slackwater data from Tables 2 and 4 as historical outliers according to procedures outlined by U.S. Water Resources Council (1981). Curve 9 probably represents the best blend of historical and paleoflood data and indicates that the 1954 flood has a recurrence interval of approximately 10,000 yr.

Devils River Paleoflood History

Large floods along the Devils River (Fig. 5) are generally produced by tropical storms producing localized catastrophic rainfall like those affecting the Pecos River and Rio Grande. However, the tributary mouth slackwater record for the Devils River is much more complex than that of the Pecos River. Part of this complexity occurs because the entire Devils River basin is underlain by the Edwards Limestone. Hence, there are insignificant differences between the mineralogies of mainstream and tributary sediments in the slackwater sites. However, grain size can usually be used to distinguish Devils River mainstream floods from tributary floods so that a paleoflood stratigraphy can be developed.

Virtually all of the five slackwater sites studied along the Devils River contain evidence of significant tributary

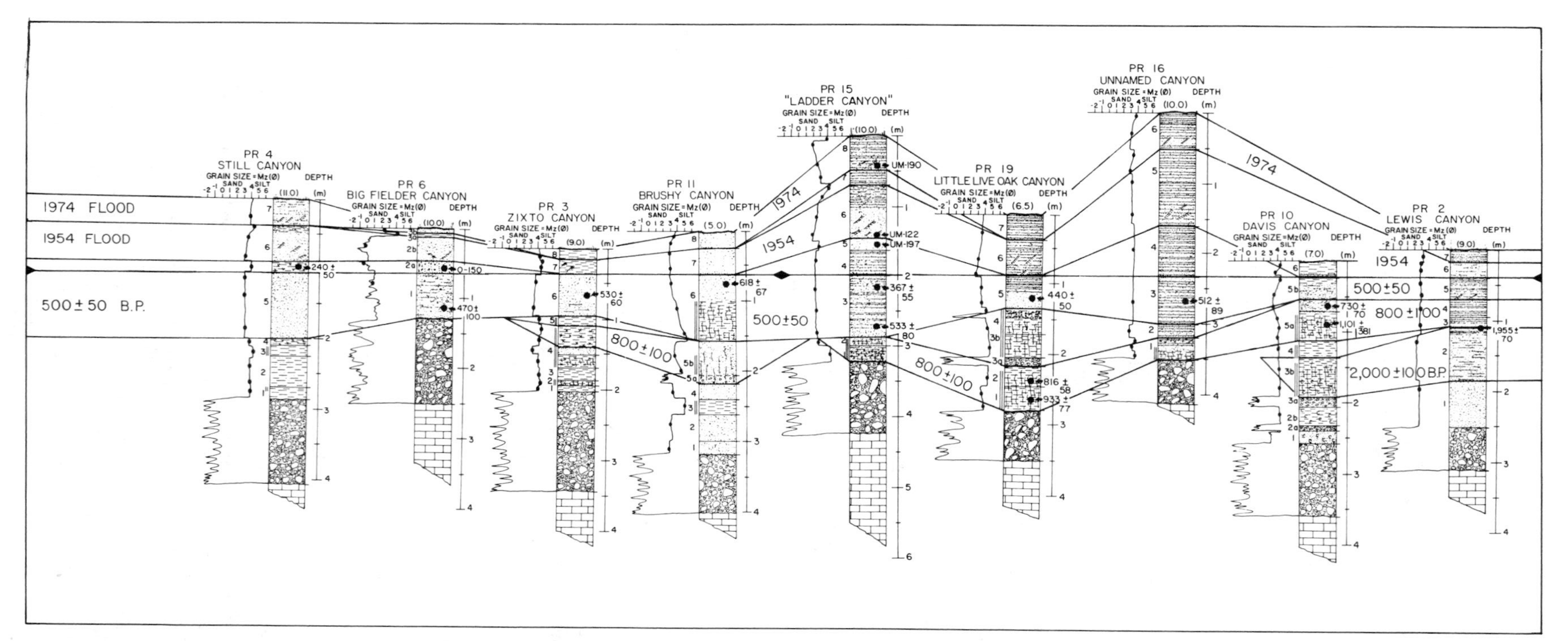

FIGURE 6. Summary of radiocarbon-dated stratigraphies of tributary mouth slackwater sites along the lower Pecos River. PR number refers to locations in Figure 5. Correlations are first based on radiocarbon dates and secondarily upon paleosols, colluvial horizons, grain size trends, and sediment induration as discussed by Kochel and Baker (this volume). These sections are hung along the flood of approximately 500 yr B.P., which is found at all sites. (From Kochel and Baker, 1982. Copyright 1982 by the American Association for the Advancement of Science.)

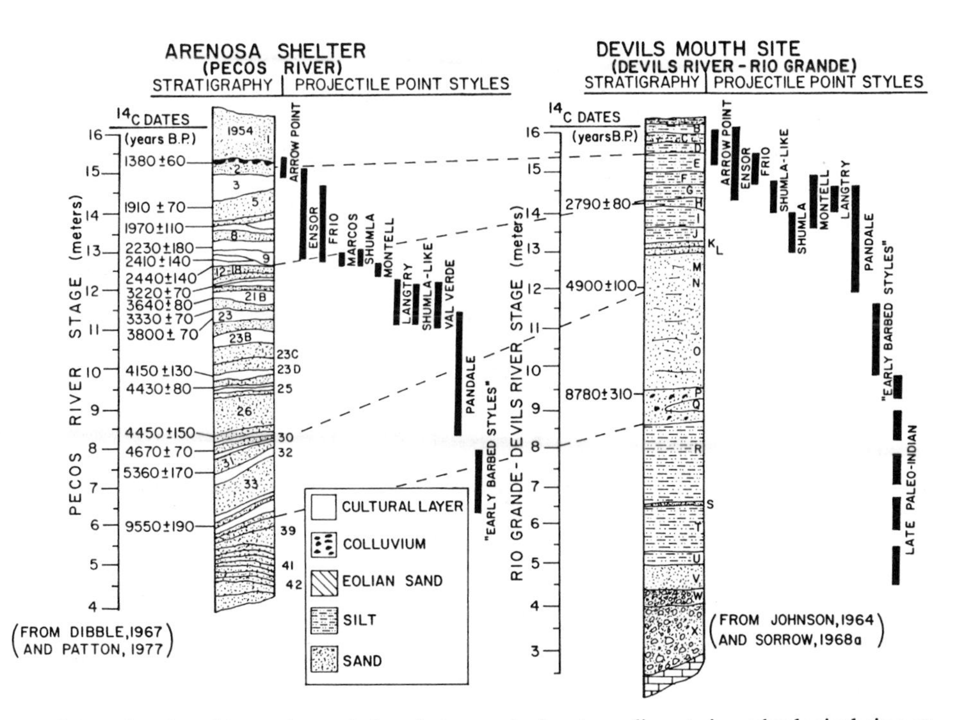

FIGURE 7. Stratigraphies and correlations between slackwater sediments in archeological sites at Arenosa Shelter on the Pecos River and Devils Mouth sites on the Devils River. Arenosa stratigraphy was modified from Dibble (1967) and Patton (1977). Devils Mouth site stratigraphy was modified from Johnson (1964) Sorrow (1968) and from photographs and discussion with D. S. Dibble (personal communication, 1980). Correlations based on projectile points were established through discussions with D.S. Dibble. (From Kochel et al., 1982. Copyright by the American Geophysical Union.)

TABLE 2 Recurrence Interval Calculations for Arenosa Shelter

Stratigraphic Layers	Number of Floods	Time Interval (yr B.P.)	River Stage (m)	Estimated Discharge (m^3/s)	Recurrence[a] Interval (yr)
5	3	1970 ± 70 to present	14.8	11,320	630–660–680
4			15.4	12,450	950–985–1020
1			24.4	27,735	1900–1970–2040
27, 26A, 26, 24	4	4450 ± 150 to 4150 ± 150	10.1	5,095	0–75–150
23C, 23A	4	4150 ± 150 to 3350 ± 85	11.0	6,095	140–200–250
22B, 21A	2	3320 ± 70 to 3600 ± 70	11.4	6,790	120–190–260
19–11	5	3220 ± 70 to 2230 ± 80	12.6	8,490	170–200–230
8	1	2440 ± 140 to 1970 ± 1110	14.1	10,755	220–470–720

[a] Recurrence intervals in the upper three strata (5, 4, 1) calculated by the formula: $n + 1/m$, where n = years, m = magnitude ranking. (Modified from Patton and Dibble, 1982.)

TABLE 3 Holocene Climate and Fluvial Character in Texas

Time (yr B.P.)	Pollen and Fauna Southwest Texas[a]	Fluvial Character, Colorado River, Central Texas[b]	Lower Pecos River Flood Character	Climate
0	Mesquite, cactus grass, pine	Low sinuosity channel, coarse grained	Large and infrequent, floods, coarse, thick layers	Semi-arid, warm, dry
2,000	Pine, grass, bison[c]	High-sinuosity channel, fine-grained sediments	Small and frequent, floods; fine, thin, layers	More humid
3,000	Mesquite, cactus pine, herbivores	Low sinuosity, coarse grained	Large, infrequent, floods	Warm, dry
7,000	Arboreal pollen bison[c]	Intermediate	Intermediate	More humid
9,000	Grass, pine	Low-sinuosity, coarse-grained	Large, infrequent floods	Warm, dry
11,000	Grass, pine, spruce, large herbivores	?	Small, frequent floods	Cool, humid

[a] Bryant (1969).
[b] Baker and Penteado-Orellana (1977).
[c] Dibble and Lorraine (1968).

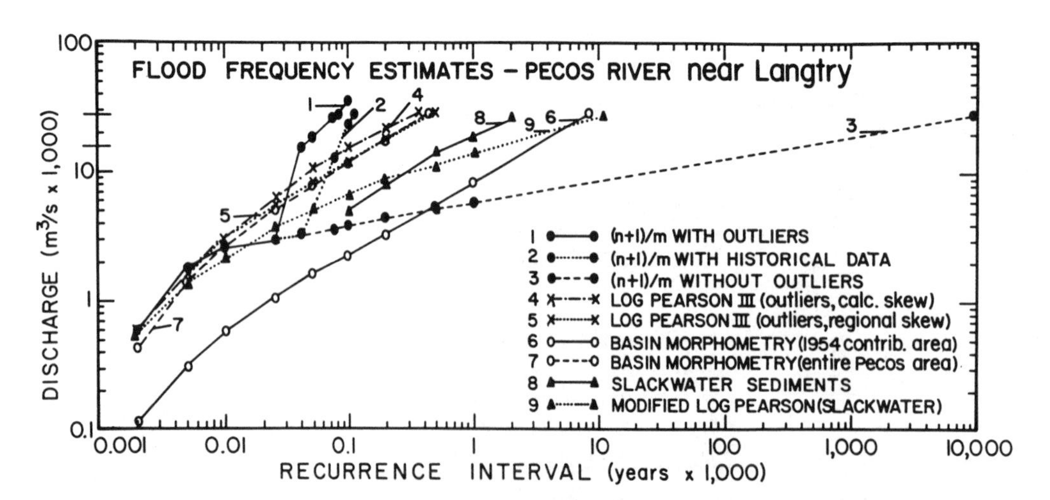

FIGURE 8. Range of flood frequency curves for the Pecos River near Langtry. See text for discussion. (From Kochel et al., 1982. Copyright by the American Geophysical Union.)

TABLE 4 Radiocarbon Dates

U.Tx Lab No.	Sample No.	Canyon	Stratum[a]	Sample[b] Nature	^{14}C (‰)	^{13}C (‰)	Uncorrected[c] ^{14}C Date (yr B.P.)	Corrected[d] ^{14}C Date (yr B.P.)
LOWER PECOS RIVER								
3192	PR 3/13	Zixto	6	a	−60.5 ± 6.1		500 ± 50	530 ± 60
3195	PR 2/12	Lewis	2	b	−213.4 ± 3.9		1,940 ± 60	1,955 ± 70
3196	PR 4/15	Still	5–6	b	−29.3 ± 5.0		240 ± 50	240 ± 50
3482	PR 6/7	Big Fielder	2a	b	0 ± 20.9	−26.28	0 ± 150	0 ± 150
3483	PR 6/8	Big Fielder	1	b	−57.1 ± 9.0	−26.25	470 ± 100	479 ± 126
3484	PR 7/23	"Rowland"	2	b	+43.3 ± 10.7	−25.72	UM	UM
3485	PR 7/24	"Rowland"	3	b	+153.5 ± 13.5	−23.48	UM	UM
3486	PR 7/27	"Rowland"	5	b	+101.6 ± 6.8	−25.88	UM	UM
3487	PR 7/29	"Rowland"	6	b	+191.1 ± 8.1	−25.40	UM	UM
3488	PR 8/11	Cash	4b	b	+181.5 ± 5.4	−26.56	UM	UM
3489	PR 8/14	Cash	4b	b	+152.8 ± 6.5	−26.28	UM	UM
3490	PR 10/20	Davis	5a	a	−125.3 ± 27.7	−22.59	1,080 ± 380	1,101 ± 381
3491	PR 10/21	Davis	5b	b	−86.4 ± 5.9	−23.14	730 ± 70	730 ± 70
3492	PR 11/20	Brushy	6	a	−73.3 ± 5.3	−26.31	610 ± 60	618 ± 67
3679	PR 15/20	"Ladder"	8	b	+192.8 ± 10.0	−26.14	UM	UM
3680	PR 15/21	"Ladder"	6	b	+124.4 ± 10.5	−25.85	UM	UM
3681	PR 15/22	"Ladder"	5	b	+198.6 ± 8.1	−25.5	UM	UM
3682	PR 15/23	"Ladder"	3	b	−37.1 ± 4.9	−22.50	300 ± 40	367 ± 55
3683	PR 15/24	"Ladder"	3	a	−58.2 ± 6.0	−23.62	480 ± 70	533 ± 80
3715a	PR 19/20	L. Live Oak	5	c	−53.0 ± 5.5	−24.80	450 ± 50	440 ± 50
3715b	PR 19/20	L. Live Oak	5	e	−40.2 ± 9.9	−25.32	330 ± 90	366 ± 98
3717	PR 19/21	L. Live Oak	2	c	−89.3 ± 10.9	−22.87	790 ± 60	816 ± 58
3718	PR 19/22	L. Live Oak	1	c	−101.5 ± 14.3	−22.05	910 ± 70	933 ± 77
DEVILS RIVER								
3193	DR 5c/1	Thompson	1	f	+265.6 ± 5.2		UM	UM
3194	DR 5c/2	Thompson	2	a	−8.0 ± 5.1		60 ± 50	60 ± 50
3891	DR 8/1	Jarrett	4	c	−303.4 ± 4.1	−20.50	2,910 ± 70	3,240 ± 80
3892A	DR 8/2	Jarrett	2	g	−387.3 ± 16.2	−23.10	3,940 ± 70	4,505 ± 130
3892B	DR 8/2	Jarrett	2	d			5,610 ± 60	5,610 ± 60
3893	DR 8/3	Jarrett	5	c	−141.6 ± 4.7	−21.60	1,230 ± 60	1,265 ± 80

[a] Refers to strata on described sections.
[b] Nature of samples: (a) = wood; (b) = fine-grained wood, seeds, etc.; (c) = buried soil horizon; (d) = shell material (mostly gastropods); (e) = humic acid fraction of soil; (f) = leaf mat; (g) = charcoal.
[c] UM = ultramodern sample (post-A-bomb era).
[d] Corrected for half-life 5730 yr, ^{13}C, and dendrochronology.

TABLE 5 Comparison of Flood Frequency Estimates

Flood Frequency Methods	Discharges and Recurrence Intervals (RI) Q_2	Q_5	Q_{10}	Q_{25}	Q_{50}	Q_{100}	Q_{200}	Q_{500}	Q_{1000}	1954 Flood RI (yr)
$(n + 1)/m$, with outliers	560	1,817	2,515	3,075	3,250	35,000				10,000,000
Without outliers	560	1,817	2,515	3,075	3,250	5,480				80
With historical data	560	1,817	2,515	3,075	3,250	27,000				150
Schroeder and Massey (1977) (U.S.G.S.)	450[a]	1,400	3,800	5,200	7,900	12,000				450
Morphometric regressions	115[b]	325	570	1,100	1,600	2,300				10,000,000+

(Table continues on p. 386)

TABLE 5 *(Continued)*

Flood Frequency Methods	Discharges and Recurrence Intervals (RI)									1954 Flood RI (yr)
	Q_2	Q_5	Q_{10}	Q_{25}	Q_{50}	Q_{100}	Q_{200}	Q_{500}	Q_{1000}	
Log-Pearson Type III										
With regional skew and outliers	600	1,730	3,065	5,745	8,7224	12,811				400
Calculated skew and outliers	575	1,702	3,142	6,289	10,092	15,715				300
Slackwater Deposits										
Deposits only						5,000	8,400	16,000	18,000	2,000
Modified log-Pearson with slackwater data[c]	586	1,458	2,309	3,733	5,965	6,647	8,588	11,442		11,000

[a] Calculated with entire Pecos River drainage.
[b] Calculated with 1954 flood area only. Log-Pearson data obtained with FREQFLO program of the Center for Research in Water Resources, University of Texas at Austin (written by Beard and Ford).
[c] Modified by using slackwater flood data as historical outliers.

flood erosion in their stratigraphies. Most of the slackwater sequences contain couplets like those described by Kochel and Baker (this volume). Sequences like these (Fig. 9) must be viewed with greater caution than the Pecos examples because tributary erosion may have removed significant portions of the Holocene slackwater section. Devils River tributary basins are likely to produce flashier flows than average Pecos River tributaries. The greater drainage efficiency of Devils River basins may account for the increased problems with tributary erosion (Kochel, 1980).

Jarrett Ranch Tributary Mouth Site. A slackwater deposit at the mouth of a small tributary on the Jarrett Ranch (Figs. 9 and 10) contains the best documented tributary

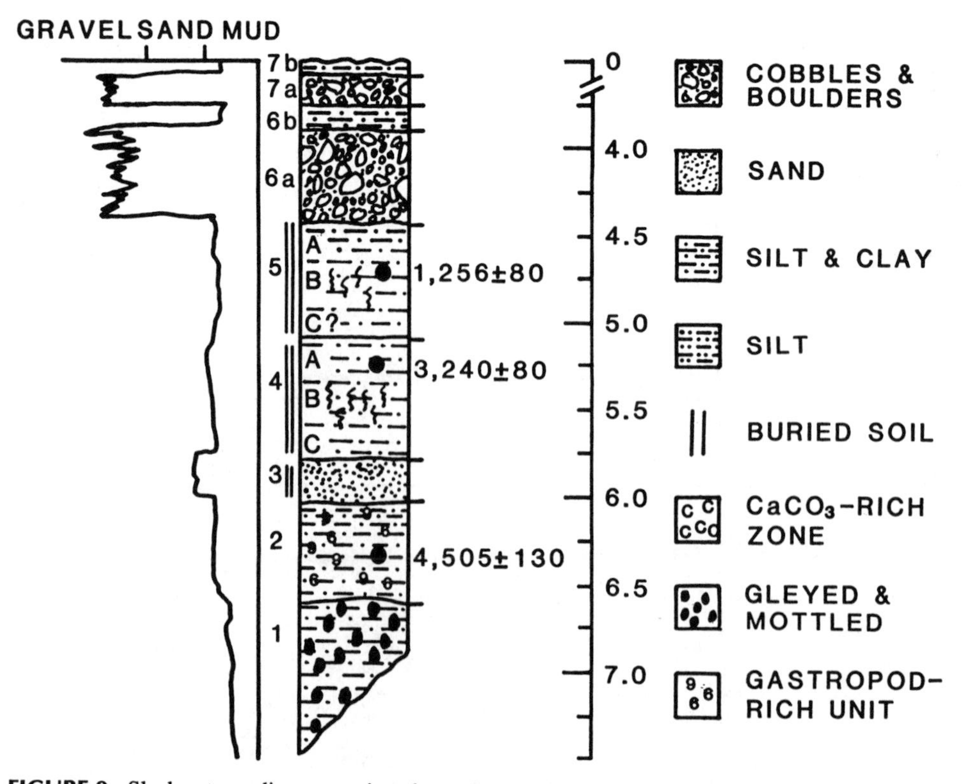

FIGURE 9. Slackwater sediment stratigraphy at the mouth of a small tributary to the Devils River on Jarrett Ranch (Fig. 5, site DR-8).

FIGURE 10. Jarrett Ranch slackwater site. Water at the base is low-flow backwater from the Devils River located to the right. The height of this section is 8 m from the water to the base of the trees. Note the coarse gravel in the upper few meters and also near the center of the section. The gravel is deposited when the large bar along the west side of the Devils River migrates over the site. Episodic erosion during bar emplacement may have removed significant portions of the slackwater section.

mouth site along the Devils River. The basal unit in the Jarrett Ranch sequence is a gleyed and mottled silty clay underlying a gastropod-rich silt dated at 4500 yr B.P. The lower two units probably represent temporary ponds created in the tributary mouth after a major flood, particularly if a main-channel bar had encroached over the mouth to facilitate local ponding. Large gravel bars are significant features of the Devils River bedrock channel morphology along this reach today (Fig. 10). Figure 11 and Table 6 show the Devils River cross section and paleoflood frequency estimations made for the Jarrett Ranch site.

Devils Mouth Site. Archeological excavations of an extensive terrace slackwater site near the mouth of the Devils River were conducted in the late 1950s and 1960s (Johnson, 1964; Sorrow, 1968). Figure 7 shows correlations made between the Devils Mouth site and the Arenosa site on the Pecos River. The paucity of radiocarbon dates from this site restricts efforts to refine paleoflood frequency estimates to the same degree as was done for the Pecos River. However, rudimentary correlations can be made based on the few radiocarbon dates and by using projectile point styles in the Devils Mouth site sediments (D.S. Dibble, personal communication, 1980). Almost twice as many floods were recorded at the Arenosa site during the interval between 5000 and 2400 yr B.P. than at Devils Mouth site. A much better correlation exists between flooding in the Devils and Pecos rivers between 9000 and 5000 yr B.P. If the abundance of Pecos River floods between 2000 and 3000 yr B.P. resulted from more humid conditions, then it is possible that floods at the Devils Mouth site during that interval were not large enough to inundate the terrace site. Greater discharge would have been required to inundate this site based on where the elevation of its surface was during that time (Kochel, 1980). The better correlation during more arid periods of the Holocene suggests that only the very large floods were able to inundate the Devils Mouth site.

The upper three floods at the Devils Mouth site (Fig. 7, units B, D, and E) probably correspond to the upper three floods at the Jarrett Ranch site (Fig. 9, units 5, 6a, and 7ab). Historical and paleoflood data suggest that these

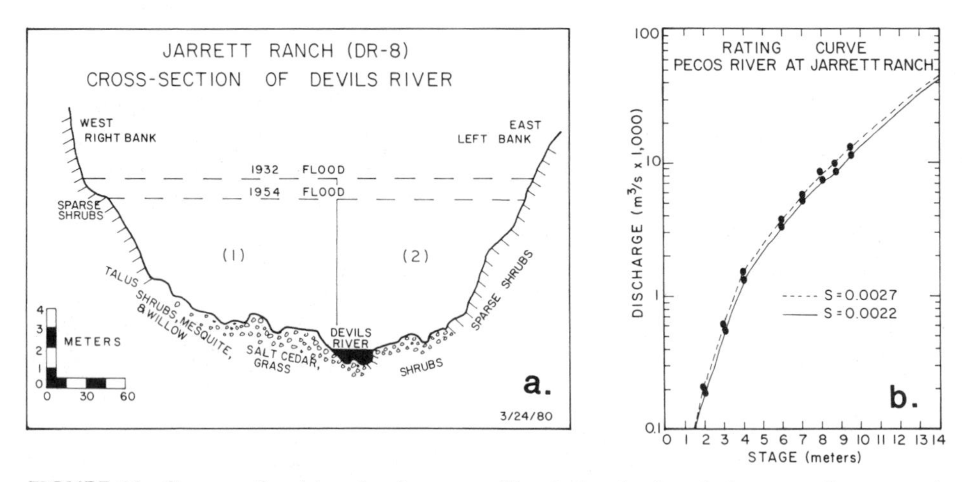

FIGURE 11. Cross section (*a*) and rating curve (*b*) calculated using slackwater sediments at the Jarrett Ranch site. Debris lines and observations of local residents fixed the precise location of the high water for the floods of 1932 and 1954. Similar cross sections for the Pecos River sites discussed earlier can be found in Kochel and others (1982).

TABLE 6 Recurrence Intervals and Discharges, Devils River at Jarrett Ranch

Flood Rank	Time Interval (yr)	Recurrence Interval (yr)	Stage (m)	Discharge (m^3/s)
	1,265 ± 80 to present			
1		1,185–1,265–1,345	11.0	20,000[a]
2		590–630–670	9.5	16,700
3		390–420–450	8.5	16,300
4		300–320–340	8.0	13,300
	4,505 ± 130 to 1,265 ± 80 B.P.			
1		3,030–3,240–3,450	11.0	20,000[a]
2		1,515–1,620–1,725	9.0	18,000[a]

[a] Rough estimate of discharge based on comparison of deposit thickness with 1932 and 1954 flood deposits.

three floods occurred in 1954, 1932, and 1265 yr B.P., in order from top to bottom.

CONCLUSION

Slackwater sediments can usually be found along most rivers that have significant bedrock reaches (Kochel and Baker, this volume). Because they often occur in stratigraphic sequences, they can be used to extend historical records of flooding along rivers for thousands of years. Long records are necessary when attempting to assess the return intervals of rare, large floods. In addition, this technique can be used to establish paleoflood histories of rivers lacking gauging data or those that have very short gauging records. These studies are most easily done in semi-arid rivers where bedrock channels are common and where conventional techniques have difficulties with outliers in these flashy rivers. These kinds of data would be very beneficial to hydrological projects being planned in the rapidly expanding regions of the southwestern United States where large problems are common due to our poor understanding of the dynamics of flooding in semi-arid rivers (Committee on Natural Disasters, National Research Council, 1982, 1984).

Figure 12 shows the approximate ranges of reliable flood frequency estimates using various historical and paleoflood techniques. Slackwater paleoflood techniques can provide estimates of the discharge and frequency of large floods over greater temporal ranges than other methods. In addition, the slackwater technique can be used to estimate discharges of large floods and obtain continuous records over hundreds to thousands of years.

Although slackwater investigations have thus far been most successful in semi-arid regions, slackwater sedimentation is common in humid regions as well. Kochel and Baker (this volume) discuss sites throughout the Midwest and eastern United States where studies are currently in

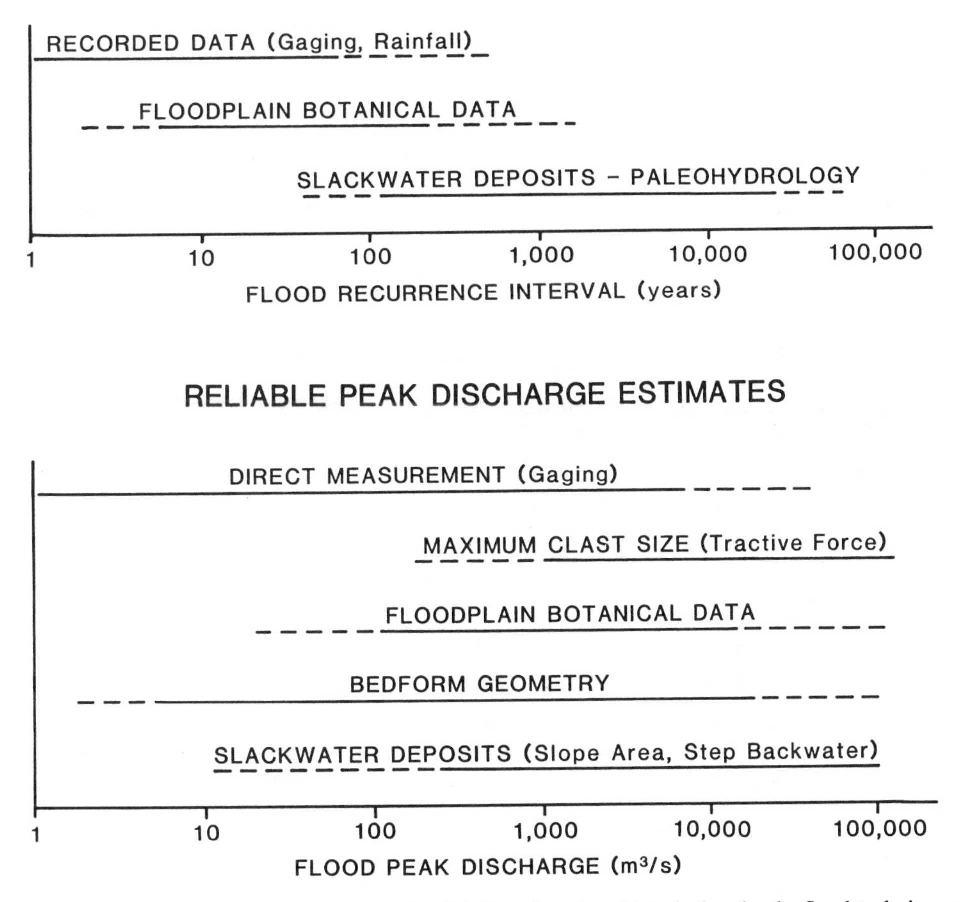

FIGURE 12. Range of applicability and reliability of various historical and paleoflood techniques for estimating paleoflood discharge and frequency.

progress. The selection of suitable sites is admittedly more difficult in humid regions, but with careful investigation, the method should be able to significantly extend flood records of many of those rivers as well. Figure 13 shows areas throughout the United States where geology and channel conditions have proven favorable for the occurrence and preservation of slackwater sediments. Most of the sites in humid areas contain adequate stable bedrock reaches suitable for the application of slackwater techniques for extending records of rare, high-magnitude floods.

Paleohydrologic estimates of flood discharge and frequency contain some error. Our studies in west Texas (Kochel et al., 1982) and elsewhere indicate that these errors are generally less than 20%. Flood control projects should therefore consider incorporating paleohydrologic data into their design. Although paleohydrologic estimates contain some error, they provide excellent order-of-magnitude estimates of the frequency and magnitude of expected river discharges. This kind of analysis, which expands the record of flood history over thousands of years, is required to obtain good estimates of recurrence intervals of floods whose return period significantly exceeds the period of gauging observations. In addition, paleohydrologic investigations such as the slackwater technique can be done rapidly and inexpensively compared to the costs of constructing major flood control structures.

ACKNOWLEDGMENTS

I thank P. C. Patton and V. R. Baker who worked closely with me on developing the techniques used in the slackwater method. D. S. Dibble provided aid in interpreting the archeological data. S. Valastro did the radiocarbon dates and aided with their interpretation. Research was supported by NSF Grant EAR 77-23025, awarded to V. R. Baker. Thanks to the ranchers of west Texas for access to the sites and to P. C. Patton, B. L. Kochel, and L. L. Kochel for field assistance. DeAnn Kirk and Andy Mason drafted many of the figures. Betty Atwood did the word processing.

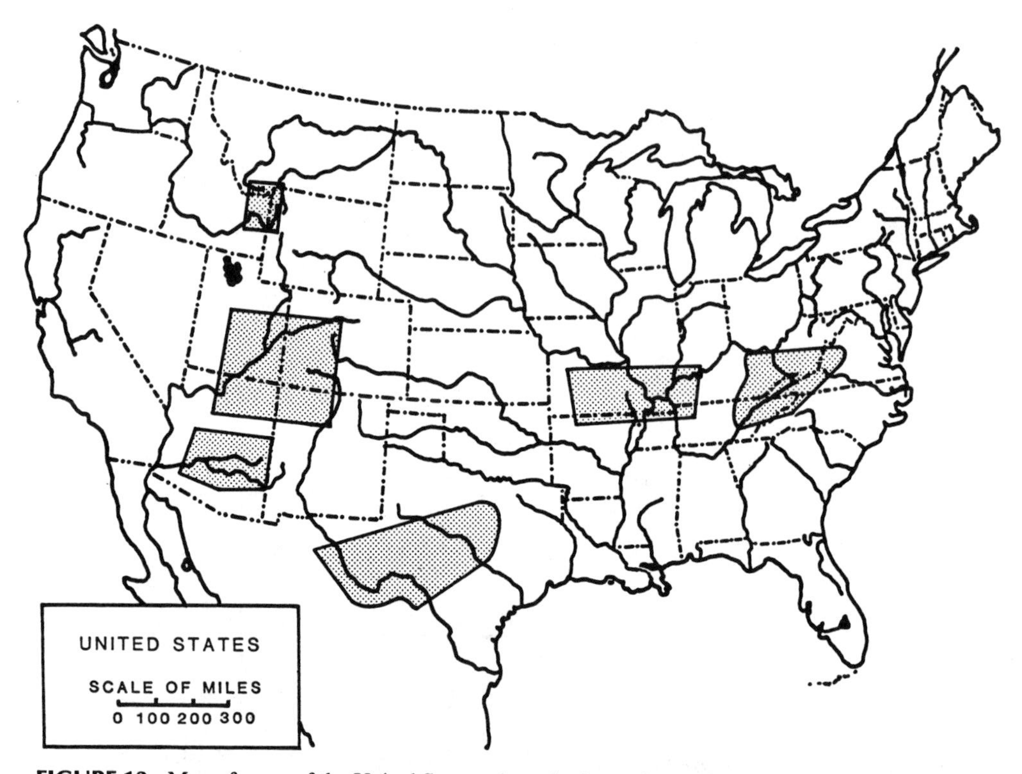

FIGURE 13. Map of areas of the United States where the formation and preservation of slackwater sediments is favorable. Reconnaissance studies and detailed studies have located sites where conditions are appropriate for the application of slackwater paleoflood techniques in all these areas.

REFERENCES

Baker, V. R., and Penteado-Orellana, M. M. (1977). Adjustment to Quaternary climatic change by the Colorado River in central Texas. *J. Geol.* **85**, 3935–422.

Baker, V. R., Kochel, R. C., and Patton, P. C. (1979). Long-term flood frequency analysis using geological data. *IAHS-AISH Publ.* **128**, 3–9.

Baker, V. R., Kochel, R. C., Patton, P. C., and Pickup, G. (1983). Palaeohydrologic analysis of Holocene flood slack-water sediments. *Spec. Publ. Int. Assoc. Sedimentol.* **6**, 229–239.

Bryant, V. M. (1969). Late full-glacial and post-glacial pollen analysis of Texas Sediments. Ph.D. Dissertation, University of Texas, Austin.

Bryant, V. M., and Larson, D. A. (1968). Pollen analysis of the Devils Mouth Site, Val Verde County, Texas. *In* "The Devils Mouth Site, the Third Season, 1967" (W. Sorrow, ed.), Austin, Tex. Archeol. Salvage Proj., No. 14, pp. 57–70.

Bryant, V. M., and Shafer, H. J. (1977). The late Quaternary paleoenvironment of Texas: A model for the archeologists. *Bull. Tex. Archeol. Soc.* **47**, 1–25.

Caracena, F., and Fritsch, J. M. (1983). Focusing mechanisms in the Texas Hill Country flash floods of 1978. *Mon. Weather Rev.* **111**, 2319–2332.

Committee on Natural Disasters, National Research Council (1982). "The Austin, Texas, Flood of May 24–25, 1981." Natl. Acad. Press, Washington, D.C.

Committee on Natural Disasters, National Research Council (1984). "The Tucson, Arizona, Flood of October 1983." Natl. Acad. Press, Washington, D.C.

Costa, J. E., and Baker, V. R. (1981). "Surficial Geology: Building with the Earth." Wiley, New York.

Dibble, D. S. (1967). "Excavations at Arenosa Shelter, 1965–66;" Preliminary report to National Park Service. Texas Archeological Survey, Austin.

Dibble, D. S., and Lorraine, D. (1968). Bonfire shelter: A stratified bison kill site, Val Verde County, Texas. *Tex. Mus. Misc. Pap.* No. 1.

Gumbel, E. J. (1957). Statistical theory of floods and droughts. *Inst. Water Eng. J.* **12**, 157–184.

International Boundary and Water Commission (1930–1977). "Flow of the Rio Grande and Related Data," Water Bulls., Nos. 1–47.

Johnson, L. (1964). "The Devils Mouth Site: A Stratified Campsite at Amistad Reservoir," Val Verde County, Texas, Archeol. Ser., No. 6. Dept. of Anthropology, Austin, Texas.

Kochel, R. C. (1980). Interpretation of flood paleohydrology using slackwater deposits, lower Pecos and Devils Rivers, southwest Texas. Ph.D. Dissertation, University of Texas, Austin.

Kochel, R. C., and Baker, V. R. (1982). Paleoflood hydrology. *Science* **215**, 353–361.

Kochel, R. C., and Johnson, R. A. (1984). Geomorphology and sedimentology of humid-temperate alluvial fans, central Virginia. *Mem.—Can. Soc. Pet. Geol.* **10**, 109–122.

Kochel, R. C., and Ritter, D. F. (1987). Implications of flume experiments on the interpretation of slackwater paleoflood sediments. *In* "Regional Flood Frequency Analysis" (V. P. Singh, ed.), pp. 371–390. Reidel, Boston.

Kochel, R. C., Baker, V. R., and Patton, P. C. (1982). Paleohydrology of southwestern Texas. *Water Resour. Res.* **18**, 1165–1183.

Patton, P. C. (1977). Geomorphic criteria for estimating the magnitude and frequency of flooding in central Texas. Ph.D. Dissertation, University of Texas, Austin.

Patton, P. C., and Baker, V. R. (1976). Morphometry and floods in small drainage basins subject to diverse hydrogeomorphic controls. *Water Resour. Res.* **12**, 941–952.

Patton, P. C., and Dibble, D. S. (1982). Archeologic and geomorphic evidence for the paleohydrologic record of the Pecos River in west Texas. *Am. J. Sci.* **282**, 97–121.

Patton, P. C., Baker, V. R., and Kochel, R. C. (1979). Slackwater deposits: A geomorphic technique for the interpretation of fluvial paleohydrology. *In* "Adjustments of the Fluvial System" (D. D. Rhodes and G. P. Williams, eds.), pp. 225–252. Kendall-Hunt Publ. Co., Dubuque, Iowa.

Shroeder, E. E., and Massey, B. C. (1977). Technique for estimating the magnitude and frequency of floods in Texas. *Water Resour. Invest. (U.S. Geol. Surv.)* **77-110**, 1–22.

Sorrow, W. M., ed. (1968). "The Devils Mouth Site: The Third Season, 1967," Pap. Austin, Tex. Archeol. Salvage Proj., No. 14.

U.S. Water Resources Council (1981). "Guidelines for Determining Flood-flow Frequency," Bull. No. 17B. U.S. Water Resour. Counc., Washington, D.C.

Williams, G. P., and Guy, H. P. (1973). Erosional and depositional aspects of Hurricane Camille in Virginia, 1969. *Geol. Surv. Prof. Pap. (U.S.)* **804**.

23

HYDRAULIC MODELING FOR PALEOFLOOD ANALYSIS

JIM E. O'CONNOR

ROBERT H. WEBB

Department of Geosciences, University of Arizona, Tucson, Arizona

INTRODUCTION

Studies of prehistoric floods (paleofloods) have proven to be effective in determining flood frequency and magnitude relationships for certain types of fluvial systems. A technique of paleoflood analysis that has shown considerable utility is the analysis of geologic evidence for paleoflood stages (Baker et al., 1979, 1983; Patton et al., 1979; Kochel et al., 1982; O'Connor et al., 1986). Slackwater sediments, silt and debris lines, and scour lines all serve as excellent paleostage indicators. If such features can be dated and related to flood discharges, they provide a powerful tool in reconstructing the timing and magnitude of large flows in the past (Kochel and Baker, this volume).

A key element in paleoflood analysis is the transformation of stage information into accurate discharge estimates. The aim of this section is to illustrate how computer routines for step-backwater water surface elevation calculations, designed primarily for floodplain hazard analysis, can be utilized to increase the accuracy of paleoflood discharge determinations. Hydraulic flow modeling can also provide insights into large-discharge flow hydraulics, information that can enhance an understanding of flood geomorphology (O'Connor et al., 1986).

Previous Methods

The earliest scientific paleohydrologic discharge estimates were made by Bretz (1925), who used the Chezy equation to determine outburst flood magnitudes from Pleistocene-aged glacial Lake Missoula in the Channeled Scabland of eastern Washington. The Chezy equation and its commonly used variation, the Manning equation, were developed specifically for uniform flow, a criterion rarely met in natural stream systems. Bretz (1925) recognized the inadequacy of his estimates, but at that time more sophisticated techniques were not available. Subsequent workers in the scablands (Baker, 1973) and elsewhere (Baker et al., 1979; Patton et al., 1979; Kochel et al., 1982) did have access to hydraulic descriptions of nonuniform flow that could be utilized for improved paleoflood discharge estimates. Primary among these is the slope-area method of peak discharge estimation described by Dalrymple and Benson (1967). This procedure can be applied to reaches of steady, gradually varied flow where there are multiple sites of high-water indicators. The method represents an improvement over the direct use of the Chezy and Manning equations in that it does, within limits, account for flow energy losses associated with variations in channel geometry and roughness along a stream reach. However, the slope-area procedure requires that cross sections used in the analysis be restricted to sites of high-water indicators.

Although sound in principle and still widely used, the slope-area method seems to consistently yield discharge estimates that are high relative to direct discharge measurements (Jarrett, 1984), or more variable than discharges calculated using more sophisticated hydraulic-modeling techniques (Webb, 1985). In paleohydrologic applications discharge overestimation is probably a result of bias in the locations of preserved high-water indicators. Commonly used paleostage criteria, such as slackwater sediments, debris accumulations, and silt lines, are generally deposited and preserved at sites of flow expansion or separation. These features are much less common in reaches of constricted flow. Because cross sections used in the slope-area method must correspond to sites of paleostage evidence, a bias is generated toward larger cross sections and greater calculated conveyances. Moreover, flow energy losses associated with any intervening constrictions and/or expansions will not be accounted for, further enhancing the overestimation of discharges.

The development of open channel flow step-backwater methods for use with high-speed computers has provided the opportunity to increase the accuracy of paleohydrologic studies through more precise modeling of large-discharge flow conditions. There are several variations of these computer models available; all are similar in theoretical basis but vary slightly in computational methods and accessory features. Widely available models in the United States include the U.S. Army Corps of Engineers' HEC-2 Water Surface Profiles Computer Program (Hydrologic Engineering Center, 1982), the U.S. Geological Survey's Step-Backwater Model E431 (Shearman, 1976), and the Soil Conservation Service's WSP2 Computer Program (Soil Conservation Service, 1976). If properly used, these hydraulic flow models produce accurate energy-balanced water surface profiles for known discharges in surveyed channels. Geologic evidence of paleoflood stages is compared to model-generated profiles in order to obtain relatively precise estimates of paleoflood discharges (Baker, 1984; Ely and Baker, 1985; O'Connor et al., 1984, 1986; Partridge and Baker, 1987; Webb, 1985; Webb and Baker, 1984).

HYDRAULIC MODELING

Theory

Hydraulic step-backwater routines assume flow conditions that are steady with time and gradually varied in space. To predict water surface profiles associated with gradually varied flows, a necessary assumption is that "the head loss at a section [of gradually varied flow] is the same for a uniform flow having the velocity and hydraulic radius of the section" (Chow, 1959, p. 217). This permits the use of uniform-flow formulas to evaluate the energy slope at each cross section; that is, the uniform-flow friction coefficients are assumed to be applicable to varied flow conditions (Feldman, 1981). Under these conditions the one-dimensional energy equation is appropriate for solution of the flow profiles (Chow, 1959, p. 267) (see Fig. 1 for defining illustration):

$$Z_1 + Y_1 + \frac{\alpha_1 v_1^2}{2g} = Z_2 + Y_2 + \frac{\alpha_2 v_2^2}{2g} + h_e \quad (1)$$

Equation (1) expresses conservation of mechanical energy for gradually varied flow within an incremental reach of

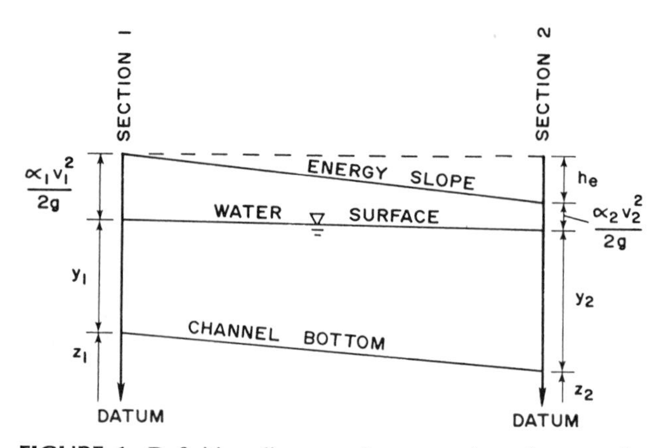

FIGURE 1. Definition diagram of conservation of energy for gradually varied flow for small channel slopes: z is the elevation of the channel above an arbitrary datum, y is the flow depth, v is the mean flow velocity, g is gravitational acceleration, α is the velocity head coefficient accounting for nonuniform velocity distribution in a subdivided channel, and h_e is the head loss between cross sections. Note that in natural stream systems, the water surface slope may deviate locally from both the channel slope and the energy slope depending on the local channel geometry.

channel. The sum of a flow's potential and kinetic energy must equal that of a downstream cross section less any energy (head) losses between sections. Head loss (h_e) is subdivided into frictional losses created by flow boundary roughness elements, eddy losses associated with turbulence, and flow separation generated by channel constrictions and expansions. Friction losses are evaluated by a variation of the Manning equation:

$$S_f = \frac{n^2 v^2}{R^{4/3}} \quad (2)$$

where S_f is the local friction slope, n is the Manning's friction coefficient (estimated by the analyst), v is flow velocity, and R is the hydraulic radius. Eddy losses are usually calculated as a function of the change in velocity head between sections:

$$\text{Eddy loss} = k \left| \frac{\alpha_1 v_1^2}{2g} - \frac{\alpha_2 v_2^2}{2g} \right| \quad (3)$$

where k is taken to equal 0.0–0.1 for gradually narrowing reaches ($\alpha_1 v_1^2/2g < \alpha_2 v_2^2/2g$) and 0.2–0.5 for expanding reaches ($\alpha_1 v_1^2/2g < \alpha_2 v_2^2/2g$) (Chow, 1959, p. 267; Dalrymple and Benson, 1967; Hydrologic Engineering Center, 1982). Note that some step-backwater routines do not have the capability to evaluate eddy losses directly; therefore, this type of energy loss must also be accounted for in selection of local Manning's n coefficients.

The use of the Manning equation to evaluate local friction slopes allows a solution to Equation (1) given initial stage, discharge conditions, and a known channel geometry. The computational approach is the "standard step" method described by Chow (1959, pp. 274–280). This is an iterative procedure where successive attempts are made at determining an energy-balanced water surface elevation at a section of unknown flow depth given the cross-sectional geometry and knowledge of the flow conditions at an adjacent cross section. The final estimated water surface elevation at the unknown section must have an associated energy that equals the total energy of the cross section of known flow conditions, less any calculated head losses between them. The newly predicted water surface elevation is then taken as known for the next incremental step along the reach. In this manner a water surface profile can be calculated for an indefinite length of channel provided stage and discharge conditions are given at an initial cross section and the assumptions of gradually varied flow are satisfied. Chow (1959) and Feldman (1981) provide more detailed explanations of the solution procedure.

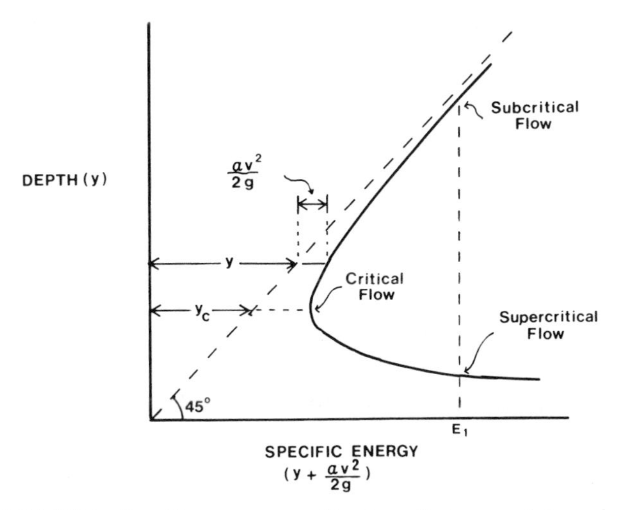

FIGURE 2. Specific energy curve. For given discharge and channel geometry conditions, the specific energy ($y + \alpha v^2/2g$) of a flow is related to flow depth. In most situations (e.g., E_1) there are two alternate depths for a flow of a certain specific energy (subcritical and supercritical flow conditions). For flow conditions of minimum specific energy, the flow is considered critical, and there is one possible depth of flow ($y_2 = \alpha v^2/g$). Modified from Chow (1959).

Application to Natural Channels

The assumptions and limitations of open channel flow hydraulic-modeling techniques impose constraints on their use for natural channels if accurate results are desired. The basic assumptions of steady, gradually varied flow imply that for short channel increments: "(1) The flow is steady, that is, the hydraulic characteristics of the flow remain constant for the time interval under consideration; and (2) that streamlines are parallel; that is, hydrostatic distribution of pressure prevails over the channel section" (Chow, 1959, p. 217). In application these assumptions require the following: (1) The discharge of concern is (was) of sufficient duration to have simultaneously affected the entire modeled reach. (Step-backwater techniques are not appropriate for modeling flood waves that are short with respect to the length of the study reach.) (2) Channel cross sections are separated into short enough increments so that the flow characteristics do not vary significantly between sections. For exceedingly complex channel geometries, the assumption of one-dimensional gradually varied flow may not be valid regardless of the lengths of the modeled channel increments.

In addition, presently available step-backwater routines cannot be used to reliably reconstruct flows in channels with deformable boundaries. The channel geometry at the time of the flow of concern must be known or approximated. In alluvial streams where there may be substantial scour, deposition, and bank migration during large discharges, hydraulic step-backwater routines probably do not yield accurate results. Best results are achieved for flows in channels with rigid boundaries.

In addition to initial stage and discharge conditions, flow regime must also be specified. For given discharge and flow energy conditions, there are generally two combinations of flow depth and velocity that result in equal total energy at a section (Fig. 2). Supercritical flows are those dominated by inertial forces ($v > \sqrt{gy}$), while subcritical flows are those primarily influenced by gravitational forces ($\sqrt{gy} > v$). The distinction is pragmatically important for hydraulic-modeling purposes because the type of flow regime experienced by the stream dictates the direction in which the water surface profile is computed. Numeric stability in the solution procedure requires that subcritical flows are calculated in the upstream direction and supercritical flows should be computed in the downstream direction. As Chow (1959, p. 265) warns, "step computations carried in the wrong direction tend inevitably to diverge from the correct profile."

The flow type experienced at a channel section depends on local cross-sectional geometry and channel slope. For large-magnitude paleofloods studied in bedrock-controlled rivers in the southwestern United States, flow is primarily subcritical (Ely and Baker, 1985; Partridge and Baker, 1987; O'Connor et al., 1986). Critical or supercritical conditions may locally occur at channel constrictions, at sites of local channel steepening, or immediately downstream, of locales of hydraulic damming. Best results are usually obtained by modeling flows as subcritical. If resulting flow depths suggest that some portions of the study reach (two or more adjacent cross sections) experience supercritical flow, separate profiles can be computed in the downstream direction for those sections. The generated profiles can be connected to create a single, reach-long water surface profile (Fig. 3). At present, there is no rational method to account

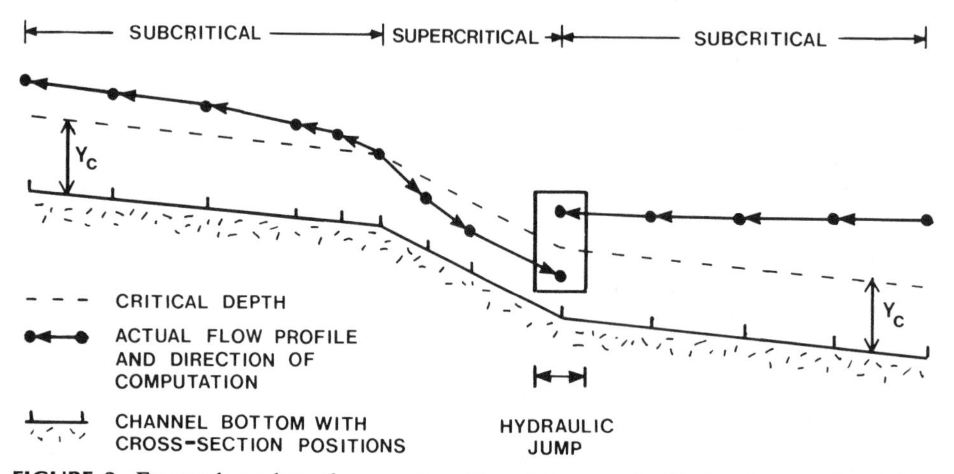

FIGURE 3. For reaches where there are transitions between subcritical and supercritical flow, separate profiles can be computed (in the direction indicated) for each subreach of similar flow regime. These independently generated profiles can be connected to yield a reach-long water surface profile. Energy losses between subreaches of different flow regimes cannot be adequately modeled; especially the intense energy losses associated with hydraulic jumps from supercritical to subcritical flow depths. Modified from Hydrologic Engineering Center (1982).

for energy losses associated with hydraulic jumps between reaches of supercritical and subcritical flow in natural channels.

The most important requirement for accurate hydraulic modeling is precise characterization of channel geometry. A major advantage of step-backwater hydraulic-modeling techniques is that cross sections can be chosen solely on the basis of channel configuration and are not restricted to sites of high-water indicators. This is important because eddy losses associated with channel expansions and constrictions are a major source of energy loss between sections, especially in rivers with nondeformable boundaries. Cross-section locations should be chosen carefully; they should characterize all expansions, constrictions, and changes of slope within the study reach (Fig. 4). Cross sections should be surveyed perpendicular to the presumed direction of flow; hence, cross-section orientations may vary with the discharge being modeled and may not necessarily relate to the low-water channel. For some models, spacing between cross sections need not be equal; long reaches of more uniform conditions require fewer cross sections than do reaches of complex flow conditions. For smaller streams an on-site survey will produce the best results. For larger river systems high-resolution topographic maps can provide adequate information and allow for considerable savings in time and effort without reducing accuracy (Dawdy and Motayed, 1979).

The regions of effective flow (flow in the downstream direction) should be distinguished from portions of the channel that do not actively participate in conveying the discharge. Regions of ineffective flow include areas behind bedrock spurs into the channel and in tributary mouths where there is flow separation and development of large-scale eddies. These portions of the channel only provide for storage of floodwaters and therefore should be excluded from step-backwater analyses. Determining the exact extent of effective flow is a field judgment and can be a source of significant error. Extensive slackwater deposits, debris accumulations, and protected groves of vegetation may provide clues to potential areas of ineffective flow.

Although discharge, initial flow depth, and energy loss coefficients are all variables that must be specified to produce a water surface profile, discharge is the primary profile-controlling variable. This is especially the case in narrow, bedrock-confined streams where flow depths are large in relation to flow width. Profiles initiated at equal discharges, but different stages, will invariably converge after a few computed steps along the channel (Fig. 5). Profiles generated with equal discharges, but utilizing various reasonable estimates of Manning's n and eddy loss coefficients, will differ slightly in water surface slope. Nevertheless, such effects are generally small in comparison to the importance of discharge on the computed water surface profile (Fig. 6). In addition, flow stages are independent of energy loss coefficients at sections experiencing critical flow.

PALEOFLOOD ANALYSIS

Appropriately used, step-backwater methods produce water surface profiles that are principally a function of the assumed discharge and channel geometry. Water surface profiles

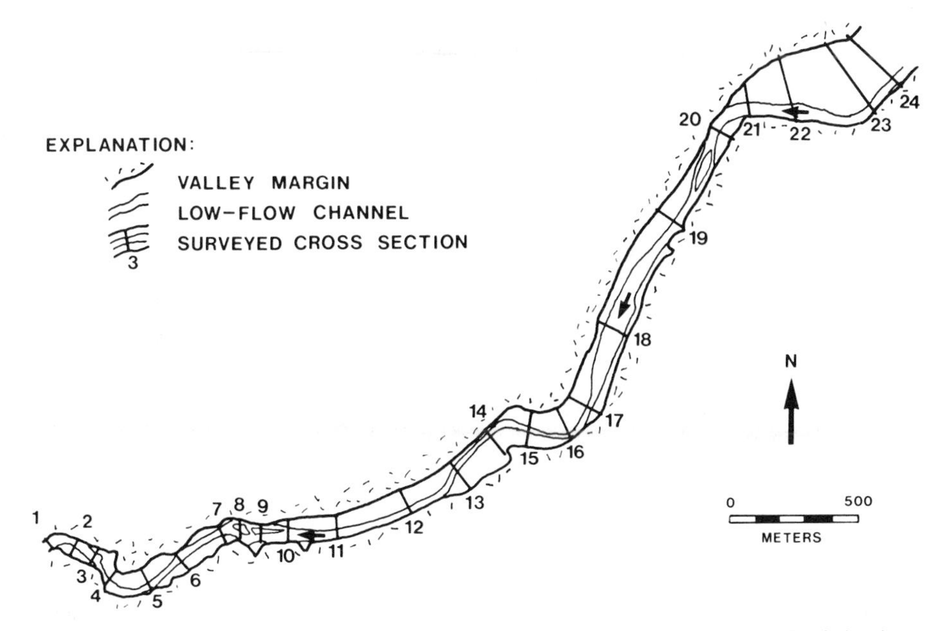

FIGURE 4. Cross-section locations for a reach selected for paleoflood analysis on the Salt River of central Arizona. Cross sections were surveyed by level and rod. The map was prepared from the survey information in conjunction with aerial photographs and topographic maps. Note that the cross-section positions are not uniformly spaced; reaches of greater hydraulic complexity contain more closely spaced cross sections than reaches of simpler geometry. Modified from Partridge and Baker (1987).

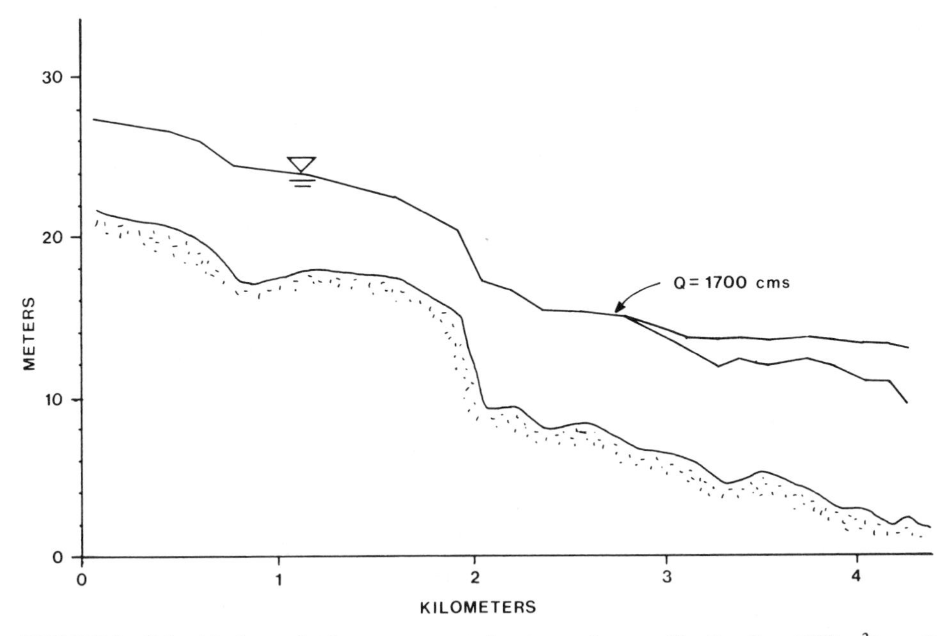

FIGURE 5. Subcritical step-backwater-generated water surface profiles for $Q = 1700$ m^3/s on the Salt River. The profiles were computed with identical data except for initially specified stage elevations that varied by approximately 2 m. Beginning stage conditions appear to only affect the generated water surface profiles for the first few modeled cross sections. Water surface profiles for the upstream portion of the study reach are independent of starting stage conditions. Modified from Partridge and Baker (1987).

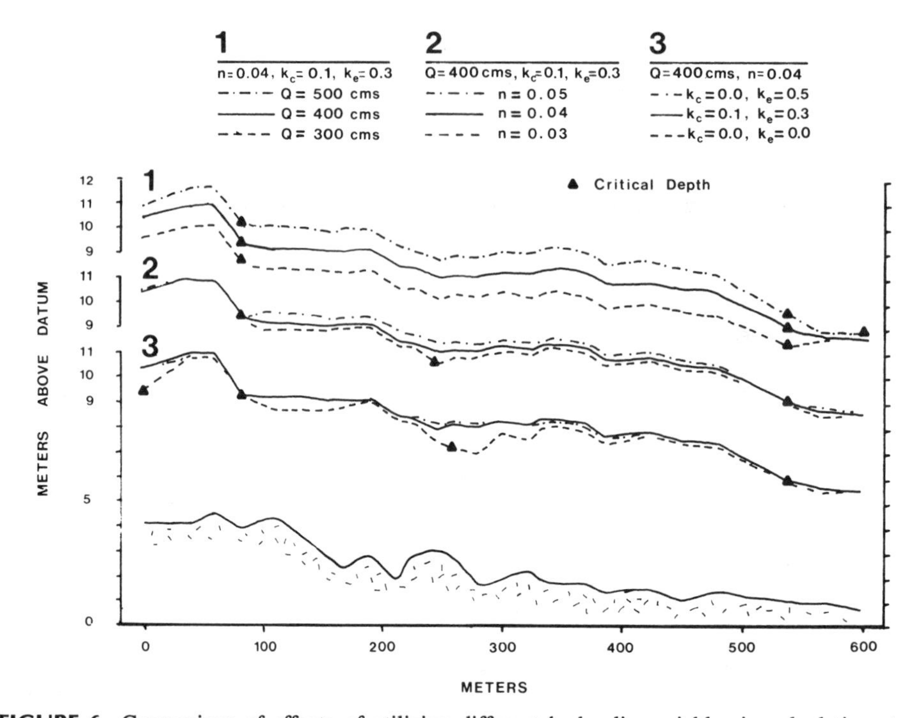

FIGURE 6. Comparison of effects of utilizing different hydraulic variables in calculating step-backwater water surface profiles for Boulder Creek, southcentral Utah. The choice of either the U.S. Army Corps of Engineers' (Hydrologic Engineering Center, 1982) or the U.S. Geological Survey's (Shearman, 1976) recommended values for the eddy-loss coefficients (0.1, 0.0 and 0.3, 0.5 for contractions and expansions, respectively) has little effect on the final profile form. However, assuming no energy losses associated with changing channel geometries results in a calculated profile that, in places, deviates significantly from the other two, indicating the potential importance of this type of energy loss in certain channel environments. Varying Manning's n coefficients and discharge by similar proportions indicates that for Boulder Creek, discharge is the primary profile-controlling variable. The variations in water surface elevations caused by even rather large uncertainties in Manning's n are generally less than 25% of the differences produced by varying discharge by the same degree. All the profiles were initiated at the same stage, so strict comparisons should be confined to approximately the upstream three-fourths of the reach, where the flow profiles are essentially independent of the specified stage (for the Q = 500 m^3/s profiles, critical depth was assumed for the initial stage).

defined by geologic evidence of flow stage can be compared to independently generated step-backwater profiles associated with various modeled discharges to yield a "best-match" paleoflood discharge estimate. The accuracy of estimated paleoflood discharges is reflected in the degree of agreement between the generated water surface profile and the profile defined by geologic criteria. This in turn will depend on factors such as the nature of the reach selected for paleoflood analysis, the quality and quantity of the available field information, and the method of analysis.

Study Reach Selection

The most important criteria in selecting suitable reaches for paleoflood analysis are: (1) the existence of multiple paleostage indicators preserved along the reach and (2) a channel environment that is conducive to accurate step-backwater hydraulic-modeling techniques. Although paleoflood discharge estimates can be made on the basis of a single paleostage indicator, confidence in the discharge determinations is enhanced when a step-backwater-generated profile is matched by several correlative high-water indicators. Systematic over- or underestimation of roughness and eddy loss coefficients can result in computed water surface profiles that possess gradients that are either too steep or too shallow, respectively, when compared to real flow profiles. This uncertainty can only be reduced when a paleoflood water surface profile is defined independently by multiple sites of stage indicators along the study reach. In addition, certain types of paleostage evidence, such as slackwater deposits and debris accumulations, may represent minimum high-water indicators as they may have been emplaced beneath an unknown depth of water. Paleoflood discharge estimates based on solitary evidence of this nature may be low. With greater information about the actual water surface profile, a more precise match can be attained with a generated water surface profile and its associated discharge.

Study reaches should be selected so that hydraulic-modeling techniques will yield the most accurate possible results. Channel boundaries should be stable, both horizontally and vertically. Flow conditions should not be exceedingly complex. Hydraulically simple reaches can be more accurately modeled than can more complex reaches for which the assumptions of gradually varied flow may be untenable. The number of possible transitions between flow regimes (subcritical to supercritical and vice versa) and their associated unaccountable energy losses should be kept to a minimum. The modeled reach should be long enough so that the change in water surface elevation along the reach is great with respect to uncertainties in the elevations of the high-water indicators (Dalrymple and Benson, 1967). At a minimum the length of the study reach should be several times greater than the valley width to ensure that the preserved geologic evidence defines a distinct water surface profile. The modeled reach should also be long enough to include channel geometries immediately above or below the study site that might affect flow conditions at the sites of paleostage indicators. Additionally, this will minimize the effects of an incorrect estimate of flow stage at the initial cross section. These guidelines should be considered somewhat flexible in that only in rare circumstances will all the above criteria be fully satisfied. The nature and scale of a particular study and the degree of accuracy desired will dictate the study reaches that may be considered suitable.

Field Measurements

In addition to the reach geometry, information required from the field includes the surveyed positions and elevations of the paleostage indicators as well as estimates of Manning's n values. The most precise matching of step-backwater-generated water surface profiles to profiles defined by paleostage indicators can be achieved when key paleostage indicators have immediately adjacent channel cross sections. Otherwise, generated water surface profiles must be matched by interpolation to sites of paleostage evidence between cross sections. Geologic evidence of multiple paleofloods must be resolved before individual paleoflood water surface profiles can be defined. The usual methods of delineating separate flood deposits at a site and correlating flood evidence between sites include stratigraphic analysis, flood deposit morphology, and a variety of absolute and relative dating techniques (Baker et al., 1983).

Although the choices of Manning's n coefficients often do not substantially affect step-backwater-generated profiles in narrow channels, the coefficients should be evaluated in a reasonable and consistent manner. Perhaps the easiest method of assessing channel roughness is comparison of on-site observations or photographs with illustrations in Barnes (1967). More objective and rigorous approaches to evaluating bed roughness are described by Limerinos (1970), Thompson and Campbell (1979), and Bathurst (1982). Roughness coefficients should be estimated with respect to the flow conditions most likely experienced during the discharge of interest. Vegetation effects may vary with discharge and possibly with time, as will the relative roughness imposed by bed materials.

Paleoflood Discharge Estimation

Once various correlated paleoflood high-water indicators are identified within a surveyed stream reach, an attempt can be made to determine the discharge that produced the geologically defined paleoflood water surface profile. Various water surface profiles are calculated for the reach until a discharge is obtained that produces an energy-balanced water surface profile that approximates the profile defined by the high-water indicators (Fig. 7). Step-backwater-generated profiles that parallel, but are above or below, the paleostage evidence generally result from discharge estimates that are high or low, respectively. Successively improved discharge estimates should eventually produce a water surface profile similar to that of the actual discharge.

Although discharge is the most important variable in determining the form of a generated water surface profile, adjustment of some of the other input hydraulic parameters can often yield profiles that more closely match the paleoflood water surface profile and furnish more precise information about paleoflood flow conditions. Generated profiles that are too steep relative to the geologic evidence generally result from systematic overestimation of energy loss coefficients (Manning's n and/or eddy loss coefficients). Likewise, generated profiles that are flat in comparison with the stage evidence probably reflect underestimated values of energy loss coefficients. Computed profiles that match the paleostage evidence fairly precisely along most of the length of the study reach, except for the first few modeled cross sections, are often the product of inaccurate estimates of initial stage conditions. They may also occur when channel geometries outside the modeled reach affect flow conditions within the study area.

Commonly, a generated profile will be attained that precisely matches the paleostage evidence over the length of the study reach except at a few isolated locations. These singular discrepancies can result from local flow-modeling errors, unrepresentative stage indicators, or erroneous correlations between flood evidence. Incorrect judgments about

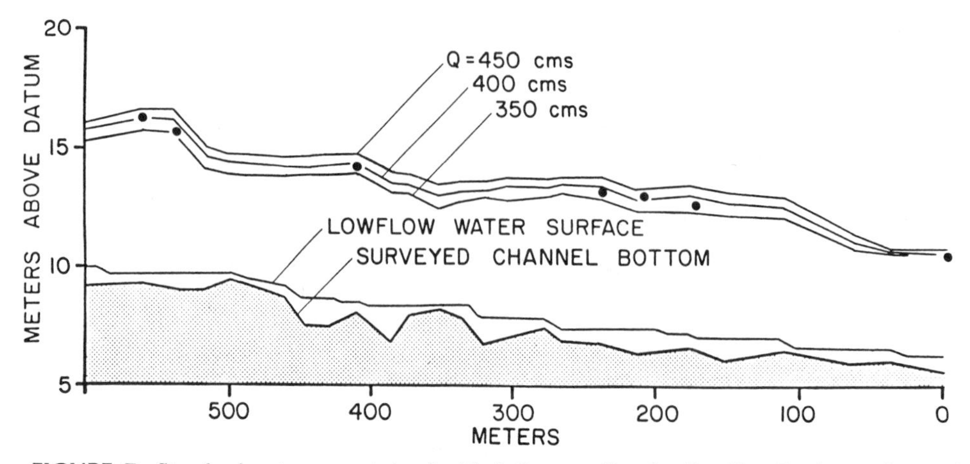

FIGURE 7. Step-backwater-generated subcritical flow profiles for Boulder Creek, southcentral Utah. A 400-m^3/s discharge appears to best match the correlated paleoflood indicators (slackwater sediments and silt lines). The 350- and 450-m^3/s generated profiles bracket the paleoflood water surface profile defined by the geologic evidence, suggesting a degree of precision on the order of $\pm$10–15% for the estimated discharge. The 350- and 400-m^3/s profiles initiated at identical stages for the beginning (downstream) cross section. Critical depth was assumed for the 450-m^3/s water surface profile. Modified from O'Connor and others (1986).

effective flow area within a particular cross section or series of cross sections will result in local deviations from the correct flow profile. Abrupt changes in flow regime cannot be modeled adequately and will result in computed profiles dissimilar to the actual water surface profile. In addition, because step-backwater routines model flows one-dimensionally, an *average* (with respect to cross-section width) water surface elevation is calculated for each cross section. At the outsides and insides of channel bends the actual water surface elevation will be higher and lower, respectively, due to the three-dimensional nature of real flows and associated super-elevation effects. Flood evidence emplaced at these sites may deviate above or below a one-dimensionally computed flow profile by some fraction of the flow's velocity head, depending on its location relative to the channel geometry.

As noted previously, some paleostage indicators may not always accurately represent actual water surface elevations. Scour lines and silt lines probably do accurately indicate high-water levels, but other commonly used paleostage indicators, such as slackwater sediments and debris accumulations, only represent minimum paleoflood stages. Depending on the depth of water overlying these features during emplacement, such evidence may not adequately represent high-water levels along some portions of a reach. Postemplacement erosion can also render flood deposits unrepresentative of actual water surface elevations. Many of these potential problems can be identified and minimized by careful analysis and selection of the paleostage evidence employed.

The precision of a "best matching" discharge can be assessed by determining the range of discharges necessary to "bracket" the paleostage indicators (Fig. 7). For reaches that are hydraulically simple and contain representative and accurately measured paleostage indicators, the precision, determined in this manner, of the estimated discharges should be on the order of $\pm$10%. The precision of paleoflood discharge estimates will be less for reaches that are of greater hydraulic complexity and/or contain less representative stage indicators. Due to slightly different computational methods, generated water surface profiles are slightly dependent on the step-backwater routine utilized. These differences are minor; water surface profiles generated from these commonly used step-backwater routines with identical data varied by a maximum of 4% of the calculated flow depths (Dawdy and Motayed, 1979). Uncertainties of this magnitude can be considered minimal with respect to other sources of potential error in paleoflood analysis.

The accuracy of flood discharge estimates can only be evaluated if there exist independent direct-discharge measurements. This can be achieved when reaches selected for paleoflood analysis contain geologic evidence of large historic flows that have associated nearby gauge records (assuming, of course, that the gauged discharges are accurate). Discharge estimates determined in the manner described here can be compared to the recorded discharges in order to assess the accuracy of the paleoflood analysis methods at a particular location. For large recorded flows (recorded flows greater than 50% of the maximum flow evidenced in the geologic record) on the Salt and Verde

TABLE 1 Comparison of Gauged Discharges with Those Determined by Hydraulic Modeling Techniques

Location	Year	Gauged Discharge[a] (m^3/s)	Hydraulically Modeled Bracketing Discharges (m^3/s)	"Best Matching" Modeled Discharge[b] (m^3/s)
Salt River[c]	1983	1730	1350–2050 (2300)[d]	1700
Salt River[c]	1980	2800	2300–2900 (3700)[d]	2500
Salt River[c]	1952	3140	2500–3250 (4500)[d]	2900
Verde River[d]	1980	2690	—	2700
Verde River[d]	1951	2310	—	1950

[a] U.S.G.S. gauges (#94985, #95085) located 5 and 8 km downstream, respectively, from the Salt and Verde rivers study reaches.
[b] Modeled discharge that produced a generated water surface profile that most closely matched the geologic evidence for flood stage over the lengths of the study reaches.
[c] Data from Partridge and Baker (1987).
[d] Anomalously high bracketing discharges (values in parentheses) occur at a site of complex flow conditions. More reasonable limits were determined by not considering this site in the analysis.
[e] Data from Ely and Baker (1985).

rivers of central Arizona, discharge estimates determined from geologic evidence were generally in close agreement to nearby gauged records of peak discharges (Table 1) (Ely and Baker, 1985; Partridge and Baker, 1987).

CONCLUSION

Hydraulic step-backwater models are powerful tools for reconstructing paleoflood flow conditions. These models are an improvement over previously used paleoflood discharge determination methods in that they more accurately account for flow energy losses experienced by discharges in natural channels. Water surface profiles generated by step-backwater routines are primarily dependent on the modeled discharge, the channel geometry, and, to a minor extent, the flow resistance elements. Computed water surface profiles can be matched to paleoflood water surface profiles defined independently by geologic evidence to determine paleoflood discharges and flow conditions. Best results are achieved for hydraulically simple reaches in stable channel systems that contain several representative paleoflood high-water indicators. Accurate paleoflood discharge estimates should prove to be a valuable aid to planners and designers concerned about potential peak discharges at a site, as well as to geomorphologists concerned about quantifying fluvial processes associated with extreme-magnitude flow events.

REFERENCES

Baker, V. R. (1973). Paleohydrology and sedimentology of Lake Missoula flooding in eastern Washington. *Spec. Pap.—Geol. Soc. Am.* **144**, 1–79.

Baker, V. R. (1984). Recent paleoflood hydrology studies in arid and semi-arid environments. *EOS, Trans. Am. Geophys. Union* **65**, 893 (abstr.).

Baker, V. R., Kochel, R. C., and Patton, P. C. (1979). Long-term flood frequency analysis using geological data. *IAHS-AISH Publ.* **128**, 3–9.

Baker, V. R., Kochel, R. C., Patton, P. C., and Pickup, G. (1983). Paleohydrologic analysis of Holocene flood slack-water sediments. *Spec. Publ. Int. Assoc. Sedimentol.* **6**, 229–239.

Barnes, H. H. (1967). Roughness characteristics of natural channels. *Geol. Surv. Water-Supply Pap. (U.S.)* **1849**, 1–213.

Bathurst, J. C. (1982). Flow resistance in boulder-bed streams. *In* "Gravel-bed Rivers" (R. D. Hey, J. C. Bathurst, and C. R. Thorne, eds.), pp. 443–465. Wiley, New York.

Bretz, J. H. (1925). The Spokane flood beyond the channeled scablands. *J. Geol.* **33**, 97–115, 236–259.

Chow, V. T. (1959). "Open-channel Hydraulics." McGraw-Hill, New York.

Dalrymple, T., and Benson, M. A. (1967). Measurement of peak discharge by the slope-area method. *Tech. Water Resour. Res. Div. (U.S. Geol. Surv.)* Book 3, Ch. A-2, pp. 1–12.

Dawdy, O. R., and Motayed, A. K. (1979). Uncertainties in determination of flood profiles. *In* "Inputs for Risk Analysis in Water Resources," Proc. Int. Symp. Risk Reliab. Water Resour., 1978, pp. 193–208. Water Resource Publications, Fort Collins, Colorado.

Ely, L. L., and Baker, V. R. (1985). Reconstructing paleoflood hydrology with slackwater deposits: Verde River, Arizona. *Phys. Geog.* **6**, 103–126.

Feldman, A. D. (1981). HEC models for water resources system simulation: Theory and experience. *Adv. Hydrosci.* **12**, 297–423.

Hydrologic Engineering Center (1982). "HEC-2 Water Surface Profile: Program User's Manual." U.S. Army Corps of Engineers, Davis, California.

Jarrett, R. D. (1984). Evaluation of methods of estimating paleofloods on high-gradient streams. *EOS, Trans. Am. Geophys. Union* **65**, 893 (abstr.).

Kochel, R. C., Baker, V. R., and Patton, P. C. (1982). Paleohydrology of southwestern Texas. *Water Resour. Res.* **18**, 1165–1183.

Limerinos, J. T. (1970). Determination of the Manning coefficient from measured bed roughness in natural channels. *Geol. Surv. Water-Supply Pap. (U.S.)* **1898-B**, 1–47.

O'Connor, J. E., Webb, R. H., and Baker, V. R. (1984). The relationship of pool and riffle pattern development to large magnitude flow hydraulics within a canyonland stream system. *Geol. Soc. Am., Abstr. Programs* **9**, 612.

O'Connor, J. E., Webb, R. H., and Baker, V. R. (1986). Paleohydrology of pool-and-riffle pattern development: Boulder Creek, Utah. *Geol. Soc. Am. Bull* **97**, 410–420.

Partridge, J. B., and Baker, V. R. (1987). Paleoflood hydrology of the Salt River, central Arizona. *Earth Surface Processes Landforms*, **12**, 109–125.

Patton, P. C., Baker, V. R., and Kochel, R. C. (1979). Slackwater deposits: A geomorphic technique for the interpretation of fluvial paleohydrology. *In* "Adjustments of the Fluvial System" (D. D. Rhodes and G. P. Williams, eds.), pp. 225–253. Kendall/Hunt Publ. Co., Dubuque, Iowa.

Shearman, J. O. (1976). Computer applications for step-backwater and floodway analysis. *Geol. Surv. Open-File Rep. (U.S.)* **76-499**.

Soil Conservation Service (1976). "WSP2 Computer Program;" Tech. Release No. 61. Soil Conserv. Serv., Washington, D.C.

Thompson, S. M., and Campbell, P. L. (1979). Hydraulics of a large channel paved with boulders. *J. Hydraul. Res.* **17**, 341–359.

Webb, R. H. (1985). Late Holocene flood history of the Escalante River, south-central Utah. Ph.D. Dissertation, University of Arizona, Tucson.

Webb, R. H., and Baker, V. R. (1984). Flooding and alluvial processes on the Escalante River, Utah. *EOS, Trans. Am. Geophys. Union* **65**, 893 (abstr).

24

PALEOHYDROLOGIC RECONSTRUCTION OF FLOOD FREQUENCY ON THE ESCALANTE RIVER, SOUTH-CENTRAL UTAH

ROBERT H. WEBB
U.S. Geological Survey, Tucson, Arizona

JIM E. O'CONNOR

VICTOR R. BAKER
Department of Geosciences, University of Arizona, Tucson, Arizona

INTRODUCTION

One of the most important problems in the hydrology of the southwestern United States, and in other arid and semi-arid regions of the world, is the estimation of flood frequency at ungauged sites or sites with short gauging records. Frequent damage from large floods that greatly exceed the 100-yr flood (e.g., Saarinen et al., 1984) underscores the need for improved flood frequency analysis in this region of sporadic but occasionally intense rainfall. Recent development of paleohydrologic methods for calculation of prehistoric flood discharges from "slackwater deposits" (Kochel and Baker, 1982) provides a method for supplementing systematic gauging records in the evaluation of flood frequency.

The Escalante River in south-central Utah (Fig. 1) provides an excellent setting for the use of paleohydrologic methods to reconstruct flood frequency. This river, with a fragmentary 27-yr gauging record, flows through 135 km of bedrock canyons that control channel width and depth at high stages. Numerous rock shelters and alcoves along the river provide sites conducive to deposition of flood sediments and protection from weathering. The semi-arid climate of the drainage basin is conducive to flash-flood generation (Butler and Marsell, 1972), allowing rapid formation of slackwater deposits from high sediment concentration discharges. These deposits are used to augment historic and gauging-record accounts of large floods on the Escalante River.

GEOLOGIC AND ENVIRONMENTAL SETTING

The Escalante River is a major southeast-flowing tributary of the Colorado River in south-central Utah (Fig. 1). The headwaters of the basin are developed predominantly on Triassic to Upper Cretaceous sedimentary rocks (Hackman and Wyant, 1973), and downstream of Escalante, the river flows into a deep canyon superimposed on a monocline of Triassic(?) and Jurassic Navajo Sandstone. The drainage basin has a warm, semi-arid climate characterized by high variability in annual precipitation. Precipitation at Escalante (Fig. 1) averages 295 mm per year for an 82-yr record. Strong orographic gradients in temperature and precipitation occur: Lower elevation sites experience an estimated 12°C mean annual temperature and 250 mm or less mean annual precipitation, and, at high elevation, the Aquarius Plateau (Fig. 1) experiences an estimated 2°C mean annual temperature and 575 mm mean annual precipitation. The vegetation ranges from desert shrub assemblages below 1700 m through pinyon and juniper woodland between 1800 and 2300 m to spruce-fir assemblage above 2700 m. The riverine

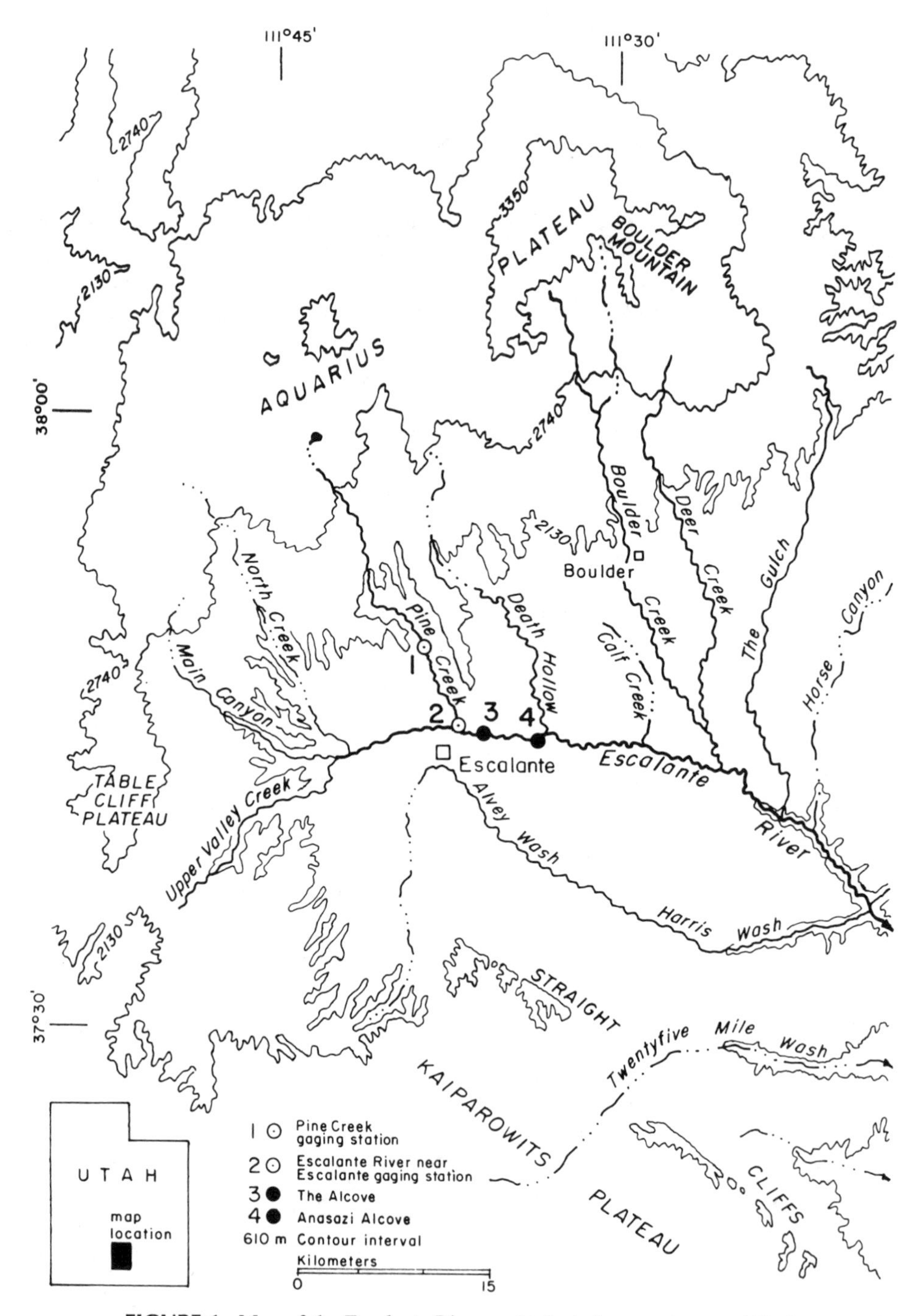

FIGURE 1. Map of the Escalante River and tributaries, south-central Utah.

environment of the Escalante River supports dense stands of native cottonwood and willow and nonnative salt cedar and Russian olive (Irvine and West, 1979).

Stream Gauging Records

The stream gauging record for the Escalante River is short and discontinuous. The U.S. Geological Survey maintains only two gauging stations in the drainage basin on the Escalante River near Escalante (09337500) and at Pine Creek (09337000) (Fig. 1). The Escalante River continuous gauging station, with a drainage area of 810 km^2, has a mean annual discharge of 0.44 m^3/s with a maximum recorded peak discharge of 98 m^3/s for the 27-yr record (1943–1955; 1972–1985). A log-Pearson Type III flood frequency analysis (Water Resources Council, 1981) indicates a 100-yr flood for this station of 197 m^3/s (Thomas and Lindskov, 1983). The 100-yr flood for the Pine Creek gauging record, with a record of 32 yr and a drainage basin area of 200 km^2, is 49 m^3/s (Thomas and Lindskov, 1983).

Records of several discontinued continuous and crest-stage gauge stations are available for the basin. Continuous or crest-stage gauge stations were present from 1951 to 1955 on Birch Creek, North Creek, Boulder Creek, Deer Creek, and the Escalante River at mouth. The continuous station on the Escalante River at mouth (drainage area of 4950 km^2) recorded a 410-m^3/s flood on August 23, 1953, the largest flood recorded in the basin. Crest-stage gauge stations were present from 1959 to 1974 on Deer Creek, Upper Valley Creek, Birch Creek, and Harris Wash. This network of gauge stations, combined with the other two continuous gauge stations, provides a fragmentary record of flood events in the drainage basin after 1943.

Historic Floods on the Escalante River

Historical information, newspaper accounts, and growth suppressions in trees indicate the dates for several large historic floods. Prior to 1909 the Escalante River was narrow and flowed through an unincised floodplain (Webb, 1985). The first large flood in the history of Escalante (settled in 1875) occurred on August 29–31, 1909. This flood, of unknown discharge, destroyed the first gauging station (U.S. Geological Survey, 1910) and began the erosion of an arroyo upstream of Escalante (Webb, 1985). Growth suppressions in ponderosa pine adjacent to Pine Creek (Fig. 1), indicative of channel-changing flood events, were infrequent between 1700 and 1909 (Laing and Stockton, 1976), which was the year of the most pervasive growth suppressions in the 270-yr record (Fig. 2). Growth suppressions in cottonwood trees on Twentyfive Mile Wash (Fig. 1) also occurred in 1909 (S. Clark, oral communication, 1985). A total of 91 mm of rainfall fell at Escalante during the 4-day storm, which caused the most widespread regional flooding recorded in the history of southern Utah (Webb, 1985).

The last large flood on the Escalante River occurred in 1932 and is well documented in newspaper accounts and the rainfall record. The summer of 1932 is the wettest of the 80-yr rainfall record for Escalante (1902–1984) with a total rainfall of 248 mm for July and August. A large flood on August 27 swept a bridge 100 m downstream, destroyed the power station at Escalante, and destroyed a house on the floodplain (*Garfield County News*, September 2, 1932). No floods of this magnitude have been reported since 1932.

ANALYSIS OF FLOOD DEPOSITS

Methods

Two sites were studied between June 1983 and July 1985 to quantify flood discharges downstream of the Escalante River near Escalante gauging station (Fig. 1). High-water marks and slackwater deposits were found at sites we informally named the Alcove and Anasazi Alcove (Fig. 1). At the Alcove, 3.2 km downstream of the Escalante River gauge station, flood sediments were deposited under low-velocity conditions in a large cave (alcove) upstream of a contraction. At Anasazi Alcove, 7.8 km downstream of the gauge station, flood sediments were deposited under a 100-m-long section of overhanging rock upstream of a contraction. Silt lines have been preserved at both sites on the overhanging canyon walls, and both sites are on the outside of river bends. No major tributaries enter the Escalante River between the gauge station and the flood deposit sites (Fig. 1); therefore, any flood data derived from these sites should directly augment the upstream gauging record.

Cross sections at both sites were surveyed at flood deposits and high-water marks, and intermediate cross sections were surveyed to include channel irregularities that create non-uniform flow conditions. Flood marks adjacent to expected high-velocity flow, such as silt lines or flood deposits in small rock cavities or overhangs, were considered as representative water-surface elevations. At the Alcove the elevation of one deposit was higher than nearby flood marks considered representative of water surface elevations, and lack of sedimentary structures in the deposit suggested low water velocities. Therefore, the elevation of this deposit

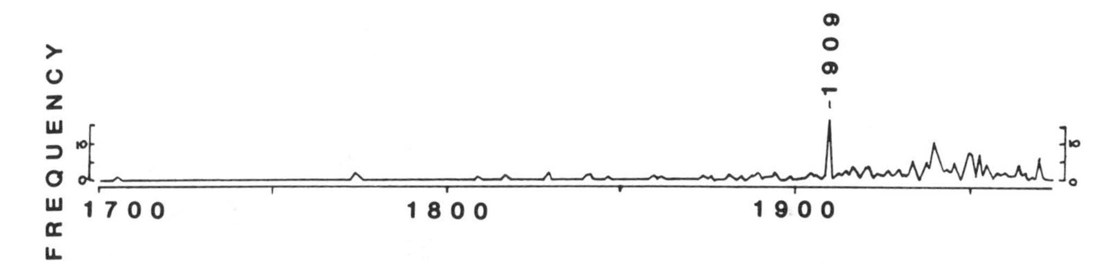

FIGURE 2. Frequency of tree ring growth suppressions indicative of the occurrence of channel-changing floods on Pine Creek, Escalante River basin (after Laing and Stockton, 1976).

was considered as less than or equal to the energy-grade line elevation.

Discharges were determined by matching elevations of high-water marks with a water-surface profile calculated using step-backwater equations incorporated in the U.S. Army Corps of Engineers computer program HEC-2 (Feldman, 1981). A Manning *n* value of 0.035 for both sites was estimated by comparison with published photographs of reaches with known Manning *n* values (Barnes, 1967). The expansion and contraction coefficients used in HEC-2 are 0.3 and 0.1 (Hydrologic Engineering Center, 1982), as opposed to 0.5 and 0.0 used in U.S. Geological Survey models (Davidian, 1984). Because the use of the different coefficients could caused minor discrepancies in the water surface profile calculations (Motayed and Dawdy, 1979), this factor was evaluated in the study.

Radiocarbon (^{14}C) dating was used as an absolute dating tool for flood deposits older than 1650 A.D.; ^{14}C activity is not a unique function of time for the period of approximately 1650–1950 A.D. (0–300 yr B.P.) (Stuiver, 1982). Therefore, all laboratory-reported radiocarbon dates younger than 300 yr B.P. used in this chapter are considered "modern." The nature of material used for ^{14}C dating presents another difficulty. Transported organic material, particularly charcoal, may be considerably older than the flood and therefore is not associated temporally with the event (Blong and Gillespie, 1978). Hence, small organic fragments, principally leaves and twigs probably dating from the growing season preceding the flood event, were the preferred material for dating in this study.

Radiocarbon samples collected from the flood deposits were rated for association with the time of the flood event and are denoted parenthetically in the text on a scale from 1 (best) to 5 (worst) by the following criteria:

1. Flood-transported organic material, consisting primarily of leaves and twigs, which would have a short residence time in the environment prior to entrainment in the flood.
2. Charcoal from in situ burned horizons or hearths where the wood burned was of local origin.
3. Flood-transported wood, which may be significantly older than the flood event.
4. Flood-transported charcoal, which may have been redeposited from older sediments.
5. Organic material not in direct association with the flood deposit but that provides an age constraint for the event.

An example radiocarbon date would be 2080 ± 70 (2) yr B.P., meaning a date of 2080 yr before 1950, a measurement standard deviation of 70 yr, and a rating of (2) (in situ charcoal). A list of the radiocarbon dates used in this study appears in Table 1.

TABLE 1 Radiocarbon Dates from the Escalante River, South-Central Utah

Depth (cm)	Lab Number[a]	Radiocarbon Date (yr B.P.)	δ^{13}C	Material Dated[b]	Rating[c]	Notes
		The Alcove, km 38.3				
LEFT INSET (LEFT SIDE OF FIG. 3)						
64	A-3988	Modern	−22.8	O	(1)	
96	A-3793	150 ± 80	−24.4	O	(1)	
135	A-4003	210 ± 90	−23.9	O	(1)	
MAIN DEPOSIT (RIGHT SIDE OF FIG. 3)						
20	A-4001	980 ± 80	−23.1	O	(1)	
64	A-4002	1150 ± 110	−23.4	O	(1)	
135	A-4054	1210 ± 100	−24.4	O	(1)	
160	A-3987	1680 ± 60	−23.8	W	(3)	Age reversal
190	A-3468	1480 ± 100	−25.5	C	(2)	
220	A-3794	2080 ± 70	−22.9	C	(2)	
DOWNSTREAM TERRACE (FIG. 4)						
60	A-4019	430 ± 100	−24.3	O	(1)	
90	A-4004	550 ± 90	−24.2	O	(1)	

Depth (cm)	Lab Number[a]	Radiocarbon Date (yr B.P.)	δ^{13}C	Material Dated[b]	Rating[c]	Notes
		Anasazi Alcove, km 42.9				
DOWNSTREAM SECTION (RIGHT SIDE OF FIG. 5)						
61	A-4056	350 ± 110	−23.4	O	(1)	Cattle dung
92	TX-5107	260 ± 90	—	O	(1)	
144	TX-5105	1400 ± 90	—	O,C	(2)	
180	TX-5106	2590 ± 860	—	O	(1)	
187	TX-5104	310 ± 70	—	O,C	(4)	Inset deposit
UPSTREAM SECTION (LEFT SIDE OF FIG. 5)						
65	TX-5102	1030 ± 70	—	CH	(2)	
130	A-4057	1500 ± 110	−25.5	CH	(2)	
220	TX-5103	830 ± 60	—	CH	(2)	Age reversal
CENTER SECTION (MIDDLE OF FIG. 5)						
50–70	A-4058	1100 ± 120	−9.3	O	(5)	Aboriginal *Zea mays*

[a] (A) University of Arizona; (TX) University of Texas at Austin.

[b] (C) charcoal; (CH) charcoal from local hearth; (W) transported wood; (O) transported leaves and twigs.

[c] Ratings were (1) transported leaves and twigs directly entrained in the deposit; (2) charcoal from local hearth or burned horizon; (3) transported wood, probably significantly older than the deposit to be dated; (4) transported and possibly redeposited charcoal; (5) organic material not in direct association with a deposit but that provides an age constraint.

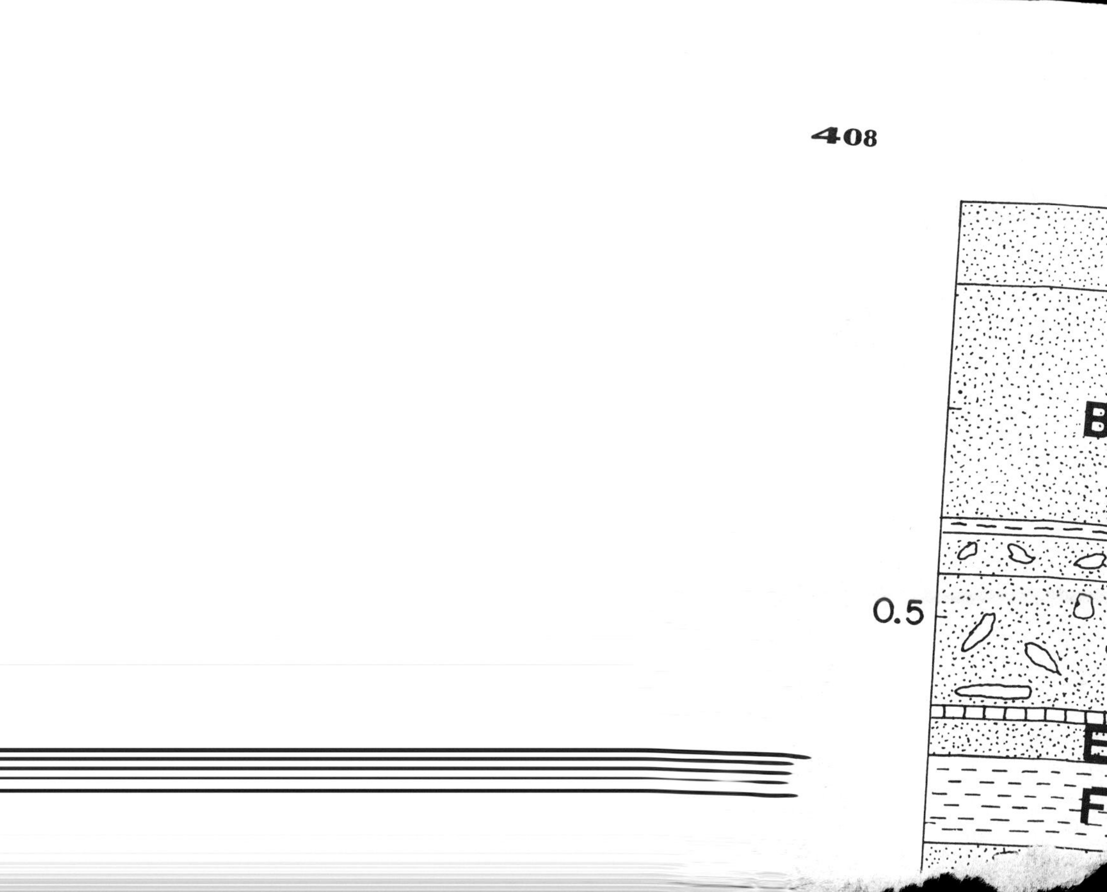
0.5

1932 AD
1909 AD
980±80
1150±110
1210±100
1680±60
1480±100
2080±70

, south-central
s indicate silty
ble 1 for more

two flood layers (E and F
...dated a

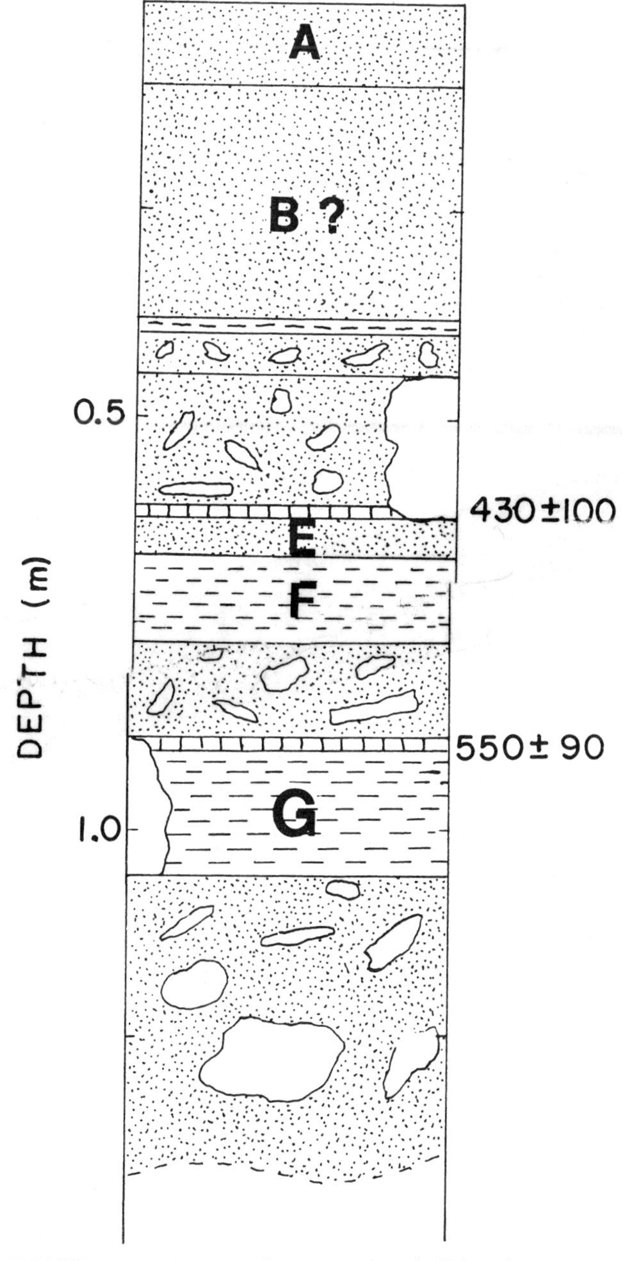

FIGURE 4. Stratigraphy of a terrace deposit 200 m downstream from the Alcove, Escalante River, south-central Utah. Stippling indicates fine sand; dashes indicate silty fine sand to very fine sand; and irregular blocks indicate colluvium. Radiocarbon dates are in ^{14}C yr B.P.; see text and Table 1 for more information.

top of the 1932 flood layer. These layers may represent either discrete flood events or multiple peaks of the same flood. A silt line for the 1932 flood is preserved beneath the silt line for the ~1000 yr B.P. flood (layer H).

The stratigraphy downstream at Anasazi Alcove is complicated but correlative with the stratigraphy at the Alcove. The flood deposit is separated into three distinct sections by waterfall scour holes that are filled with recent sediments (Fig. 5). Four "modern" flood layers uncomformably overlie older layers in the center and downstream sections (Fig. 5). These flood deposits, 10 YR 6/3 colored fine sand to sand with a basal date of 260 ± 110 yr B.P. (1), are correlated with flood layers A–D at the Alcove (Table 2). Cattle dung in direct association with a radiocarbon date of 350 ± 110 (1) yr B.P. at the top of layer C (Fig. 5) indicates that at least two of these floods are historic. The age reversal (a 350 ± 110 yr B.P. date overlying a 260 ± 110 yr B.P. date) and direct association of a 350 yr B.P. date with cattle dung (post-1875) illustrates the problem with radiocarbon dating young flood deposits.

Three flood layers of fine sand correlative with layers H–J at the Alcove were deposited in the upstream section between 900 and 1100 yr B.P. (Table 2). The lower two layers (I and J) contain abundant flaked rock, hearths, and bone indicative of aboriginal occupation surfaces. A radiocarbon date of 1030 ± 70 (2) yr B.P. indicates that layers H and I were deposited between 1000 and 1100 yr B.P. (Fig. 5). Layer H contains pottery sherds, rock flakes, and a grinding stone (metate); these artifacts are attributable to Kayenta Anasazi Indians who abandoned the region before 750–900 yr B.P. (Dee Hardy, Boulder-Anasazi State Park, written communication, 1985). Flood layer H was probably deposited before 900–1000 yr B.P., and its top is the highest high-water mark at Anasazi Alcove. Evidence for flood layers E, F, and G (400–600 yr B.P.) was not found at Anasazi Alcove.

The deposits older than 1100 yr B.P. are complicated at this site because of erosional contacts, aboriginal disturbances (particularly in the upstream section; Fig. 5), and variability in the thickness and grain size of the deposits. The lower parts of all three sections are bracketed by radiocarbon dates of 1100 ± 150 (5) (on corn cobs in an aboriginal cyst) and 2590 ± 860 (1) yr B.P. The deposits dated between 2600 and 1100 yr B.P. are generally thinner and finer grained than the younger deposits. Spurious radiocarbon dates at the bottom of the upstream and downstream sections (830 ± 60 and 310 ± 70 yr B.P., respectively; Fig. 5) are attributable to aboriginal disturbances and younger sediments inset against the older deposit.

Paleoflood History

The stratigraphy at the Alcove and Anasazi Alcove suggests a history of flooding on the Escalante River (Table 2). Relatively small floods, in a channel with a potentially higher bed elevation, occurred between about 2000 and 1300 yr B.P. Lack of correlation between the two sites

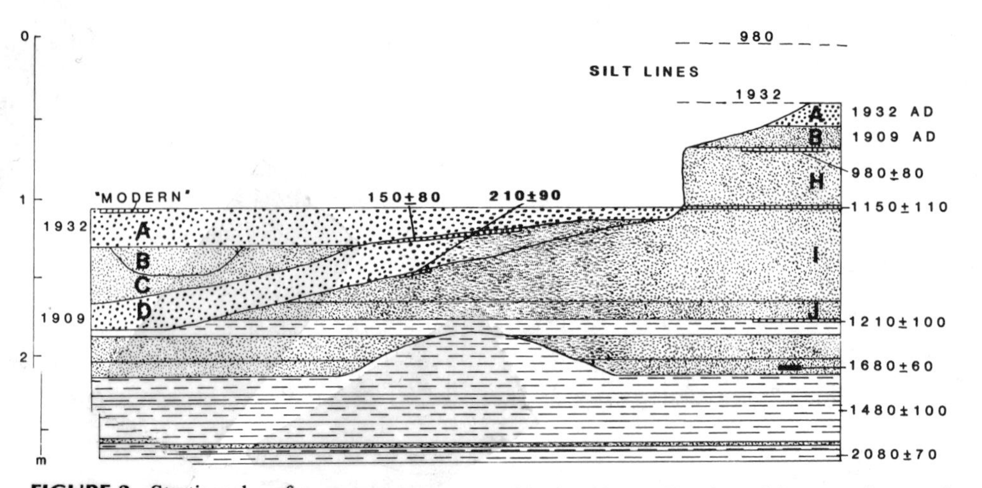

FIGURE 3. Stratigraphy of flood deposits preserved in the Alcove, Escalante River, south-central Utah. Coarse stippling indicates sand; fine stippling indicates fine sand; and dashes indicate silty fine sand to very fine sand. Radiocarbon dates are in ^{14}C yr B.P.; see text and Table 1 for more information.

Stratigraphy of Flood Deposits

Flood deposits at the Alcove and Anasazi Alcove provide evidence for flood events during the last 2100 yr (Figs. 3 and 4; Table 2). Large floods during the last 1200 yr have occurred in three discrete time periods: four historic floods, three floods between 400 and 600 yr B.P., and three floods between 1000 and 1200 yr B.P. The erosional inset of younger flood deposits into older deposits (Figs. 3 and 5) suggest that a preservation threshold may occur wherein evidence for floods higher than a certain stage are preserved in characteristic slackwater deposits.

The two highest flood layers and the four inset layers shown on the left side of Figure 3 were deposited during recent floods. The two highest layers (A and D) on the right side of Figure 3 are correlated on the basis of lithology, grain size, and color (Munsell color of 10 YR 6/3) with the highest (A) and lowest (B) layers of the inset (Table 2). Three radiocarbon dates—210 ± 90 (1), 150 ± 80 (1), and "modern" (near 1950) (1)—indicate a "modern" age for these floods (Fig. 3) but do not provide an absolute age control because resolution is not possible for radiocarbon dates younger than 300 yr B.P. Because floods were infrequent on Pine Creek between 1700 and 1909 (Fig. 2), and the 1909 flood was the first large flood after settlement in 1875, these four flood layers are presumably post-1909 with the lowermost (D) representing the 1909 flood. Flood layer A has graffiti dated "1936" carved into its top, and probably represents the 1932 flood.

The stratigraphy of a terrace deposit 200 m downstream provides the next oldest flood evidence preserved at the Alcove. Three flood deposits are preserved in the upper 40 cm (Fig. 4); the uppermost is correlated with layer A and the lower two are also presumably "modern." Colluvium separates these deposits from two flood layers (E and F) at 78 cm, the uppermost (E) radiometrically dated at 430 ± 100 (1) yr B.P. A flood deposit (G) dated at 550 ± 90 (1) yr B.P. is sandwiched between two colluvial deposits at 90–105 cm depth (Fig. 4). The height of the 400-to 600-yr-old flood deposits indicates that the stages for these floods were lower than for the "modern" floods, assuming a similar channel morphology for the respective events.

Three large floods occurred between 1000 and 1200 yr B.P. Two layers (H and I; Fig. 3) have radiocarbon dates of 980 ± 80 (1) and 1150 ± 110 (1) yr B.P., and a third overlies a layer with a 1210 ± 100 (1) yr B.P. date. These layers are of fine sand with a color of 7.5 YR 6/4 (Table 2). In a three-dimensional view of the deposit (not shown), the top of layer H has a connection with a high silt line; this silt line is shown 35 cm above the deposit in Figure 3. The silt line coats old rockfalls and crosses an Anasazi pictograph on the back wall of the Alcove.

The relationships among the elevations of the flood layers are indicative of the relative sizes of the floods, assuming similar channel morphologies for each event. A silt line associated with layer H (~1000 yr B.P.) is the highest high-water mark at the Alcove. The 1932 flood layer (A) overlies the other four flood layers and is 14 cm higher than the 1909 flood layer (D). The silt line coincident with layer A (Fig. 3) abruptly drops 35 cm several meters downstream, suggesting that this part of the deposit may represent the energy-grade elevation or surge waves. Layers B and C sandwiched between the 1909 and 1932 flood layers have an inset relationship wherein sediments from the younger flood were deposited in pockets eroded from the older flood layer (Fig. 3); the tops of both layers are at approximately the same elevation, 32 cm beneath the

TABLE 2 Summary of Paleoflood Evidence at the Alcove and Anasazi Alcove, Escalante River, South-Central Utah

Time Period	Number of Flood Layers		Characteristics of Flood Layers		
	The Alcove	Anasazi Alcove	Flood Layer[a]	Munsell Color[b]	Grain Size[c]
"Modern" (0–300 yr B.P.)	4	4	A	10 YR 7–6/4	ms
			B	10 YR 6/4–3	fs
			C	10 YR 6/4–3	fs
			D	10 YR 6/4–3	f–ms
400–600 yr B.P.	3	0	E	10 YR 6/3	fs
			F	10 YR 6/3	vfs
			G	10 YR 6/3	si/vfs
1000–1200 yr B.P.	3	3	H	7.5 YR 6/4	fs
			I	7.5 YR 6/4	fs
			J	7.5 YR 6/4	fs
Older than 1200 yr B.P.	10	>10	—	7.5–10 YR 6–5/4	Mostly si/fs to si/vfs

[a] See Figures 3, 4, and 5 and text.
[b] Munsell colors are given as value hue/chroma.
[c] (ms) medium sand, (fs) fine sand, (vfs) very fine sand, (si) indicates significant (>20%) silt content.

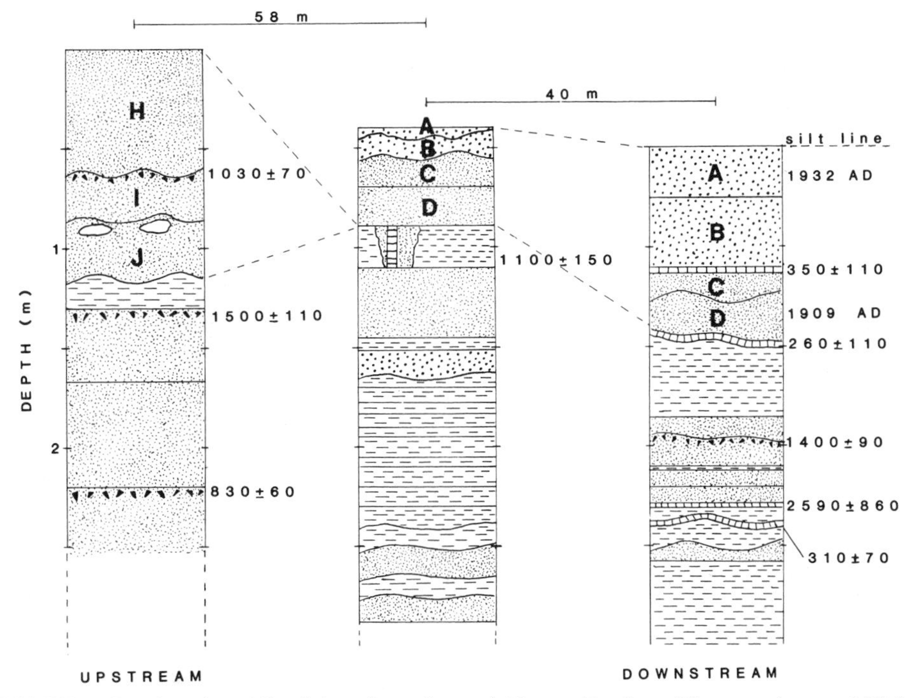

FIGURE 5. Stratigraphy of flood deposits at Anasazi Alcove, Escalante River, south-central Utah. Coarse stippling indicates sand; fine stippling indicates fine sand; dashes indicate silty fine to very fine sand; rocks indicate aboriginal hearths; and wavy lines indicate erosional contacts. Radiocarbon dates are in ^{14}C yr B.P.; see text and Table 1 for more information.

precludes further interpretation of these layers. The magnitude of large floods increased between 1200 and 1000 yr B.P., with three flood layers preserved; the highest stage flood at both sites occurred during this period. Between 900 and 600 yr B.P. no flood layers were preserved in the stratigraphy. Evidence for three lower stage floods between 600 and 400 yr B.P. is preserved at the Alcove but not at Anasazi Alcove. Following another hiatus, four major floods have occurred in modern times. The last large flood occurred in 1932, and the first flood presumably occurred in 1909 on the basis of dendrochronological evidence (Fig. 2) and lack of upstream arroyos prior to 1909 (Webb, 1985).

The dates of the other two floods recorded in the modern flood deposits are not as readily determined as the 1909 and 1932 flood layers. We assume that these deposits represent two distinct floods and not multiple peaks of the same event. Years for other possible floods are 1914, 1916, 1921, and 1927. Rainfall for July 1914 totaled 134 mm, the second highest July rainfall total (behind 1932) at Escalante. On October 6, 1916, 86 mm of rainfall fell after a wet summer. Rainfall totaling more than 25 mm fell during a thunderstorm on August 31, 1921, causing closure of roads into the basin (Woolley, 1946). Rainfall totaling 81 mm fell during a storm on September 13, 1927, after a wet summer. October 1916 and September 1927 are the most likely dates for the floods because of the large, 24-hr rainfall totals following wet summers. No other verification for flooding during these years could be obtained.

Stability of Cross Sections

Calculation of the discharges associated with the deposition of the flood sediments requires the assumption that the surveyed channel cross sections represent the cross sections during the flood events. The presence of old trees on the floodplain, desert varnish on canyon walls, and photogrammetric analysis of a historical photograph indicate the validity of this assumption. Three cottonwood trees growing within 1.5 m of the low-water surface at the Alcove had ring counts of 183, 174, and 187, indicating all were in excess of 170 yr old (S. Clark, University of Arizona, oral communication, 1985). An additional 30–40 cottonwood trees of similar age occur between the gauging station and Anasazi Alcove. The presence of these trees indicates that only minimal shifts in the low-water channel occurred during the last 170 yr. No changes in the intensity of desert varnish were observed on exposed canyon walls down to within 1 m of the low-water surface, suggesting a long period of stability in the bedrock walls.

Historical photographs provide additional evidence of channel stability. In 1872 Jack Hillers, photographer for the Powell Survey (Fowler, 1972), photographed a downstream stereo view of the Escalante River below Pine Creek and 100 m downstream of the Escalante River near Escalante gauge station (left photograph shown in Fig. 6*a*). The channel in the 1872 Hillers photograph is masked by a dense growth of willows, young cottonwood trees, and ponderosa pine trees, but an alluvial terrace and riparian vegetation mark its location. A matching stereo view (left photograph shown in Fig. 6*b*) taken in 1984 indicates that the channel location has not changed, although minor changes have occurred in the bedrock walls and plant distributions during the intervening 112 yr. Loose talus and dead shrubs common to both photographs indicate that the high-water channel has not changed. Both the 1872 and 1984 channels are constrained by bedrock walls, and a cottonwood tree still grows on the floodplain. The alluvial terrace visible in the center of Figure 6*a* was still present (although obscured by a large cottonwood tree in Fig. 6*b*) in 1984.

The two sets of stereo photographs were photogrammetrically analyzed to assess channel changes. The coordinates of prominent rock points (Fig. 6) were used to calculate distances and elevations on the channel floodplain (right foreground of Fig. 6*a* and 6*b*). The edge of riparian vegetation in Figure 6*a* defines the left terrace edge in 1872, while basalt boulders define the left terrace edge in 1984. Using photogrammetric formulas (Malde, 1973), the minimum increase in distance between the bedrock wall on the right bank and the left terrace ranges from 7.9 to 9.4 m, and the maximum increase in depth below the terrace ranges from 1.2 to 2.0 m. The depth increase is probably less than this amount because the 1872 channel, not visible in Figure 6*a*, had an unknown but significant depth, as the 1872 terrace (Fig. 6*a*) indicates. The 1909 flood reportedly widened the channel an unknown amount and scoured the bed 1 m at the gauge station 100 m upstream, although no survey data are available (U.S. Geological Survey, 1910).

Additional channel scour during peak discharges should not have occurred at either the Alcove or Anasazi Alcove. Bedrock is exposed in the channel at Anasazi Alcove and several sites between Anasazi Alcove and the Alcove. Basalt boulders with an average diameter of 75 cm locally armor parts of the channel sides and bed in this reach. Therefore, the magnitude of channel change causing uncertainty in discharge calculations is an approximate 8-m widening and 1-m lowering of the low-water channel, probably as a result of the 1909 flood.

Discharge Calculations

The silt line and flood deposits were used to calculate the discharges for the 1932 flood at the Alcove. The terrace deposit top (Fig. 4) provided a downstream water surface elevation, and two small flood deposits perched in cavities provided upstream water surface elevations. The elevation

FIGURE 6. Photographs of the Escalante River 100 m downstream of the gauging station. (*a*) Left photograph of a stereo view taken in June 1872 by Jack Hillers (photograph 670; U.S. Geological Survey Photo Library, Denver, CO). "1" indicates the terrace deposit. "2" indicates a new alcove forming on the canyon walls. White dots indicate the location of the channel. "C" indicates a control point common with the 1984 photograph. Arrows indicate where width and depth were estimated by comparison with (b). (*b*) Left photograph of a stereo view taken in October 1984 by Robert Webb (R. M. Turner stake 1207). "2" indicates a new alcove on the canyon wall. "c" indicates a control point common with the 1872 photograph. Additional control points not common width the 1872 photograph but used to rectify the 1984 photograph were located at shrubs and rocks in the foreground.

of the silt line in the Alcove was assumed to represent the elevation of the energy grade line at its contact with the bedrock wall (right side of Fig. 3), and the 1932 deposit (A) 40 m upstream was assumed to represent a water surface elevation. Ten cross sections were surveyed in this reach and spaced to account for channel expansions and contractions (Table 3).

Step-backwater calculations using a discharge of 600 m^3/s produced a water surface profile that best matched the stage evidence for the 1932 flood (Fig. 7*a*). Water surface profiles were calculated from three different initial conditions at the downstream cross section: (1) the initial water surface elevation was set equal to critical depth, (2) the initial friction slope was set equal to the slope of the two downstream high-water marks, and (3) the initial water surface elevation was set equal to the downstream high-water mark. The three water surface profiles converge 160 m upstream and match five high-water marks and the energy grade line elevation (Fig. 7*a*).

We varied the coefficients in the step-backwater model HEC-2 to check the potential range in discharges that could be responsible for the 1932 flood evidence. The Manning n value was increased from 0.035 to 0.045, the largest roughness coefficient we considered applicable to this channel; a discharge of 500 m^3/s produced a water surface profile similar with Figure 7*a*. Variation in the expansion

TABLE 3 Hydraulic Parameters Associated with Discharge Calculations for the 1932 Flood at the Alcove, Escalante River, Utah

Site Characteristics

Description: Deposits on both sides of a 120° river bend 3.2 km downstream of the Escalante River gauging station.
Drainage area: approximately 820 km^2
Length of reach: 477 m
Average channel topwidth: 70 m
Average channel slope: 0.012
Number of cross sections: 10
Number of high-water marks: 7
Number of energy-grade elevations: 1
Manning n value: 0.035
Discharge: 600 m^3/s

Reach Properties

Section	Area (m^2)	Mean Depth (m)	Mean Velocity (m/s)	Length between Sections (m)	Froude Number
Downstream					
1	218	6.36	2.75	—	0.35
2	191	6.45	3.15	39	0.40
3	162	5.50	3.71	59	0.50
4	131	5.20	4.58	66	0.64
5	145	5.20	4.15	28	0.58
6	223	5.07	2.69	35	0.38
7	125	4.37	4.81	36	0.74
8	110	4.55	5.46	56	0.82
9	185	5.31	3.25	54	0.45
10	189	4.74	3.17	49	0.47
Upstream					

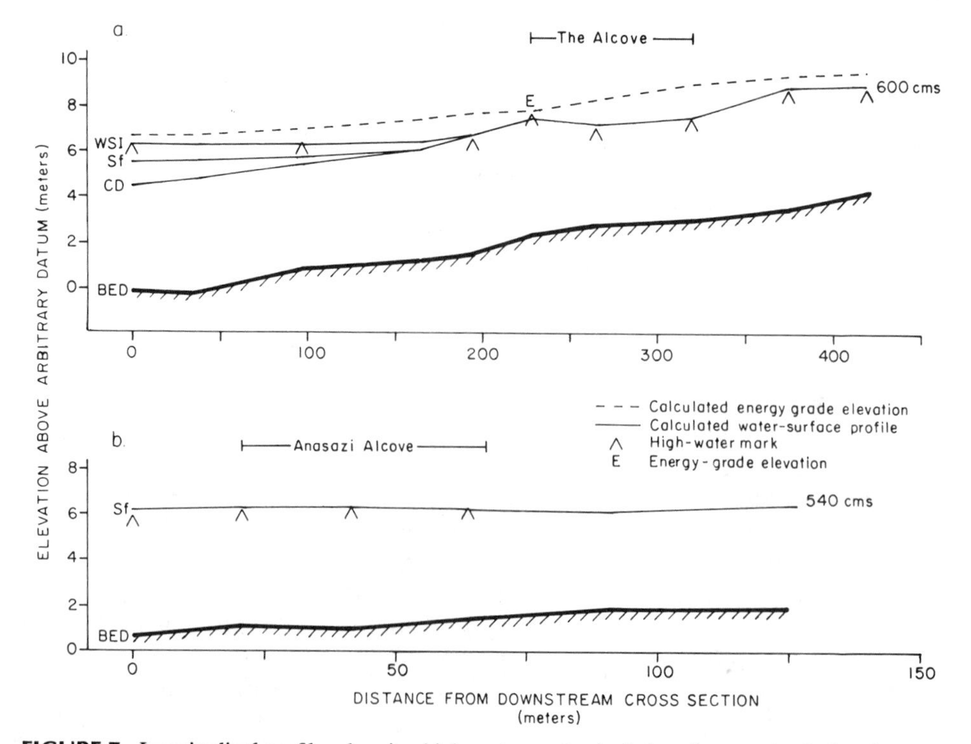

FIGURE 7. Longitudinal profiles showing high-water marks, bed elevations, and calculated water-surface profiles for the 1932 flood at (*a*) the Alcove and (*b*) Anasazi Alcove.

and contraction coefficients to 0.5 and 0.0, respectively, increased the discharge to 620 m^3/s. Although not an exhaustive evaluation of accuracy, these results show the potential uncertainty in the estimate of discharge for the 1932 flood.

Uncertainty resulting from potential channel change during the floods was a greater source of error. Based on the analysis of the historic photograph (Fig. 6*a*), the bed elevation over an 8-m width of low-water channel was raised 1 m in all cross sections. A Manning *n* value of 0.045 was used to simulate the expected roughness prior to 1909. A discharge of 500 m^3/s produced a similar water surface profile with Figure 7*a*. This channel configuration would be expected only during the 1909 flood, as the changes occurred during that event.

A stage-rating curve based on a Manning *n* value of 0.035 was developed to estimate discharges required to deposit several of the other flood layers at the Alcove, assuming a similar channel morphology. The discharge for the 980 yr B.P. silt line was an estimated 720 m^3/s. The discharges for the 1927(?), 1916(?), and 1909 floods were estimated at 550, 550, and 570 m^3/s, respectively. Alternatively, if the 1909 peak discharge occurred before the channel change, the 1909 discharge could be as low as 500 m^3/s. The uncertainty in these discharges is large because only one high-water mark is available for each layer.

At Anasazi Alcove, step-backwater calculations for the 1932 flood were not conclusive. This site had six cross sections and four high-water marks in a 60-m reach (Table 4); no high-water marks were found downstream. The top of the deposit has a slope of 0.002 while the channel slope is 0.009, indicating that a backwater surface is forming behind a downstream constriction. The step-backwater method, using a discharge of 540 m^3/s and the deposit slope as a friction slope, produced a water surface profile approximating three of four high-water marks for the 1932 flood (Fig. 7*b*). However, use of either critical depth or the downstream high-water mark as the starting water surface elevation did not produce a converging water surface profile for this short reach. Hence, the discharge of 540 m^3/s is not considered accurate, and discharges for other flood layers were not estimated.

FLOOD FREQUENCY ON THE ESCALANTE RIVER

Regional Analysis of Flood Magnitudes

The paleohydrologic analyses of flood deposits revealed that large pregauging record floods have occurred in distinct periods over the last 2100 yr. These floods had discharges five to seven times larger than the largest flood recorded

TABLE 4 Hydraulic Parameters Associated with Discharge Calculations for the 1932 Flood at Anasazi Alcove, Escalante River, Utah

Site Characteristics

Description: Deposits on north side of 30° river band, 7.8 km downstream of the Escalante River gauging station.
Drainage area: approximately 825 km^2
Length of reach: 128 m
Average channel topwidth: 48 m
Average channel slope: 0.008
Number of cross sections: 6
Number of high-water marks: 4
Manning *n* value: 0.035
Discharge: 540 m^3/s (not accepted)

Reach Properties

Section	Area (m^2)	Mean Depth (m)	Mean Velocity (m/s)	Length between Sections (m)	Froude Number
Downstream					
1	172	5.63	3.13	21	0.42
2	174	5.18	3.09	21	0.43
3	154	5.32	3.50	22	0.48
4	145	4.79	3.73	22	0.54
5	119	4.23	4.52	28	0.70
6	140	4.68	3.85	35	0.57
Upstream					

at the Escalante River gauge. To check for a regional precedent for the floods, an envelope curve was developed for the Colorado Plateau. Envelope curves are developed from plotting an upper-bounding envelope on the largest floods experienced in a region. Because they are developed from data collected from a short (less than 100 yr) observation period, envelope curves represent the maximum flood of experience rather than a physical limitation on flood magnitudes, and a finite probability remains for the occurrence of a larger flood. Envelope curves have been developed for specific regions of the conterminous United States (Crippen and Bue, 1977) and specifically for Arizona (Malvick, 1980).

We developed an envelope curve from 2093 station-years of floods measured at 143 stations on the Colorado Plateau, plotted by drainage basin area (Fig. 8). The five largest peak discharges of record for each gauging record longer than 5 yr, and all discharges for all other sites, were obtained from U.S. Geological Survey Water–Supply Papers, Butler and Marsell (1972), and Crippen and Bue (1977). The region studied was bounded by the Little Colorado River on the south, the Arizona–New Mexico and Utah–Colorado borders on the east, the Uinta and Wasatch Ranges on the north and northwest, and the Nevada border and drainage divide of the Colorado River tributaries on the west and southwest. A list of the gauging stations used appears in Webb (1985).

The plotted data suggest a slightly different curve than Crippen and Bue's (1977) curve for region 14, which was areally larger than but included the region studied here. Crippen and Bue used six floods to define their curve, and the largest flood (1420 m^3/s on Clear Creek, Arizona, on April 4, 1929) was from a drainage south of the area defined here. Two floods on less than 10 km^2 drainages (Dry Canyon and The Gap, Butler and Marsell, 1972) plotted above Crippen and Bue's curve (Fig. 8). Therefore, another envelope curve was plotted to better represent the data. Floods near the new curve (all from southern Utah) are:

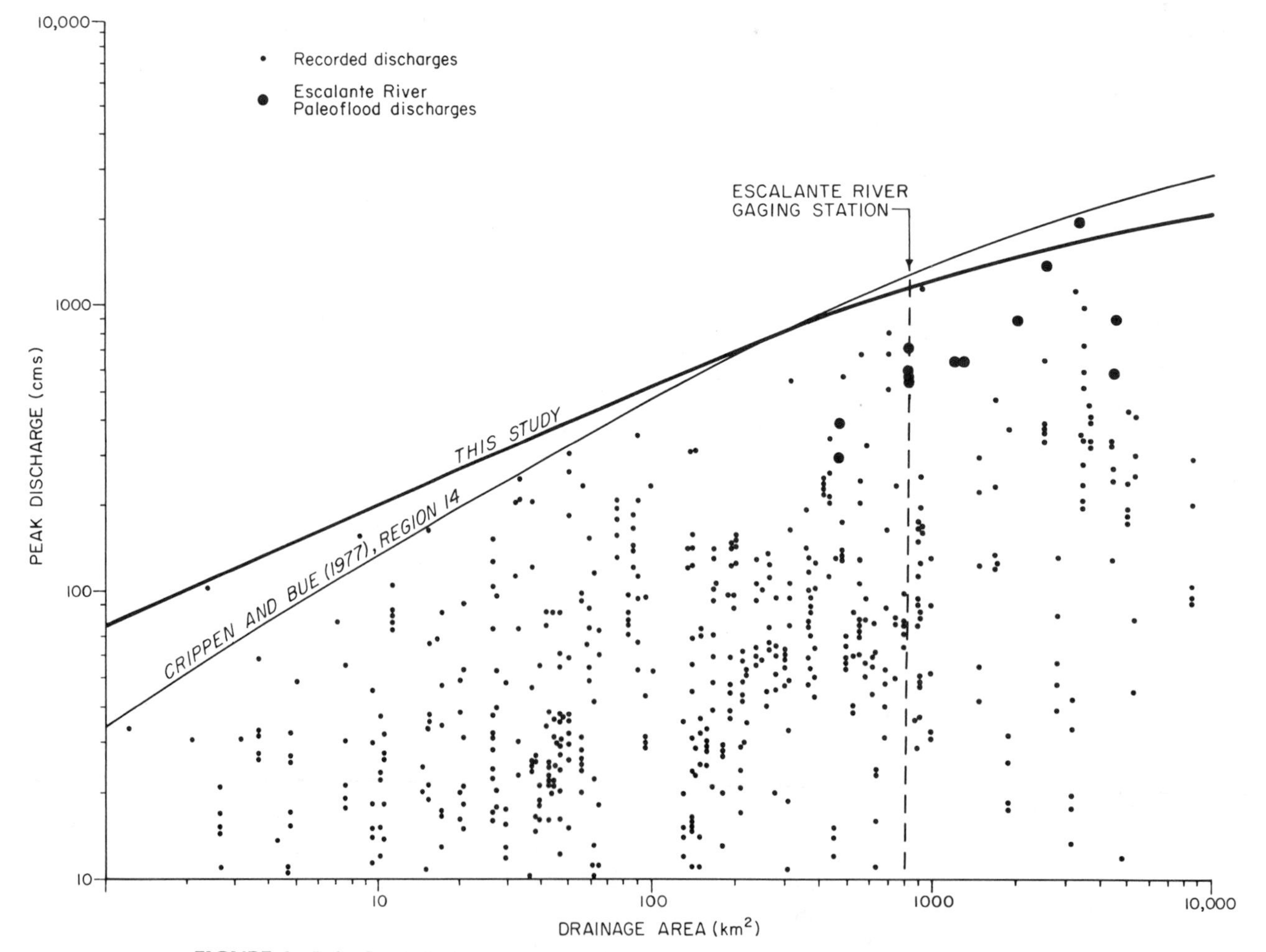

FIGURE 8. Paleoflood discharges compared with envelope curves for the largest floods on the Colorado Plateau (from Webb, 1985). Two paleoflood discharges were taken from O'Connor and others (1986) and 7 from Webb (1985).

Site	Drainage Area (km^2)	Discharge (m^3/s)	Date
Dry Canyon near Cedar City	2.3	104	Aug. 17, 1965
The Gap near St. George	8.2	160	Aug. 12, 1964
Hog Canyon near Kanab	48.0	307	Aug. 12, 1964
Cottonwood Creek near Bluff	884.0	1190	Aug. 1, 1968

The accuracy of these discharges was not checked. The envelope curve can be expressed approximately as

$$Q_e = 60 * A^{(0.58 - 0.02 \ln A)} \quad (1)$$

where Q_e is the envelope curve flood in cubic meters per second and A is the drainage basin area in kilometers squared. The envelope curve flood for the drainage area of the Escalante River gauging station is about 1200 m^3/s (Fig. 8).

The paleoflood discharges estimated from the step-backwater method, including two from O'Connor and others (1986) for Boulder Creek (Fig. 1) and seven additional paleoflood discharges from the Escalante River (Webb, 1985), were found to plot very close to the envelope curve (Fig. 8). The fact that all paleofloods plot below Crippen and Bue's (1977) curve for region 14 and all but one below the new curve indicates that these floods are of a reasonable magnitude for the Escalante River.

Flood Frequency on the Escalante River

The close proximity of the Alcove to the gauging station (Fig. 1) allows incorporation of the historic flood information in a flood-frequency analysis for the Escalante River. The Escalante River gauging record and historic flood discharges are given in Table 5. A generalized skew coefficient of −0.2 with associated mean square error of 0.81 deter-

TABLE 5 Discharges Used in Log-Pearson Type III Analysis of the Escalante River near Escalante Gauging Record, with Historic Discharges Estimated from Flood Deposits in the Alcove and Anasazi Alcove

Systematic Record
09337500
Escalante River near Escalante, Utah
Drainage area: 810 km^2
Annual Flood Peak Discharges

Year	Discharge (m^3/s)	Year	Discharge (m^3/s)	Year	Discharge (m^3/s)
1943	50	1952	35	1977	22
1944	56	1953	98	1978	1
1945	24	1954	15	1979	7
1946	15	1955	84	1980	40
1947	31	1972	4	1981	78
1948	9	1973	72	1982	6
1949	79	1974	5	1983	5
1950	5	1975	6	1984	28
1951	44	1976	31	1985	38

Mean logarithm	1.29
Standard deviation	0.51
Computed skew	−0.63
Generalized skew	−0.23
Adopted Skew	−0.53

Historic Flood Discharges

Year	Discharge (m^3/s)
1932	600
1927 (?)	550
1916 (?)	550
1909	500–570

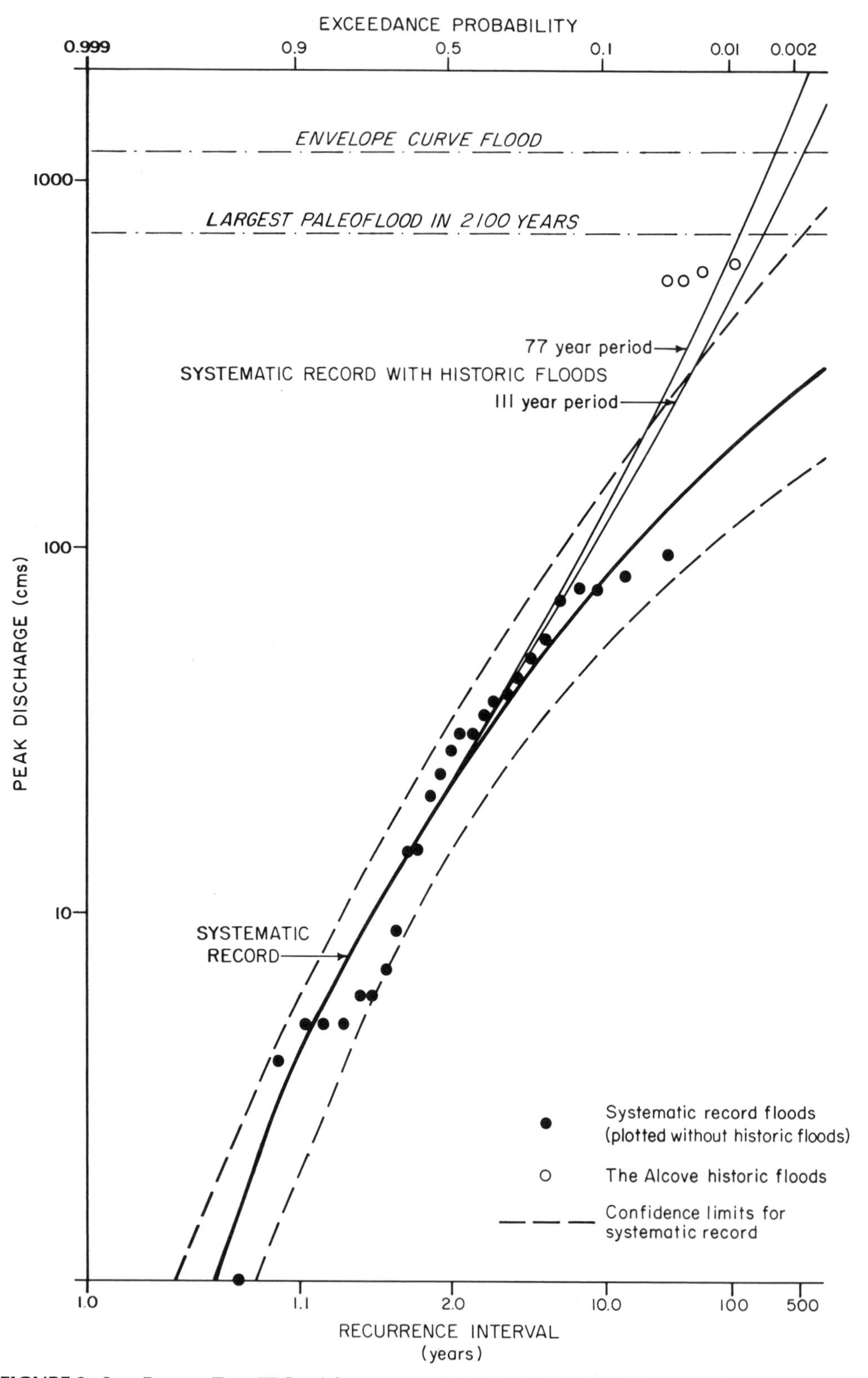

FIGURE 9. Log-Pearson Type III flood-frequency analysis for the Escalante River gauging record. The systematic record was modified with historic flood discharges from the Alcove representing two time periods; see text. The systematic record and historic floods used appear in Table 5 and are presented by Weibull plotting positions.

mined from a regional flood-frequency analysis (Webb, 1985) was used to weight the station skew coefficient. Using the log-Pearson Type III distribution, the 100-yr flood was calculated from the 27-yr record as 193 m^3/s (Fig. 9), consistent with Thomas and Lindskov's (1983) 100-yr floods of 197 m^3/s calculated from a 22-yr record and about 200 m^3/s calculated from their regional regression equations.

The four historic discharges determined from flood deposits at the Alcove were combined with the gauge record as "historic floods" (Water Resources Council, 1981). Flood dates of 1909, 1916(?), 1927(?), and 1932 were used as the basis for a historical record length of 77 yr. The estimated 100-yr flood increased to 630 m^3/s, and the station skew coefficient changed from -0.53 to $+0.34$, with addition of these four floods (Fig. 9). Because the four floods are known to be the largest since settlement in 1875, a record length of 111 yr was also used. The resulting 100-yr flood is 480 m^3/s (Fig. 9). Use of 500 m^3/s for the 1909 flood resulted in a 10 m^3/s lower 100-yr flood estimate.

CONCLUSIONS

Flood deposits along the Escalante River, at sites 3.2 and 7.8 km downstream from a gauging station with no intervening tributaries, were used to reconstruct a history of large floods for the last 2100 yr. Large floods occurred between 1100 and 980 yr B.P., between 600 and 400 yr B.P., and in historic times. Paleoflood discharges were determined for a 1932 flood by matching high-water marks with a water-surface profile calculated using the step-backwater method; other discharges were determined from a stage-rating curve. The largest paleoflood in 2100 yr was estimated to be 720 m^3/s at the Alcove. The discharges were comparable to the largest floods from similar-sized drainage basins on the Colorado Plateau, and plotted near a regional flood envelope curve. Flood frequency information for the Escalante River gauge (27 yr of record) was augmented using the discharges for the four large historic floods. The computed 100-yr flood increased from 193 to 480–630 m^3/s with the addition of the historic floods and a variable historic record length.

Incorporation of paleoflood (or historic flood) information in flood-frequency analysis involves a synthesis of dates of events and historic flood discharges, each of which involves some error or uncertainty in determination, with a systematic record. Consequently, possible sources of error and uncertainty and their influence on the results of flood-frequency analysis are of great importance. The date of events or more importantly the period of record encompassing the events has a marked influence on flood-frequency analysis (see Fig. 9). Dating each individual event, a sometimes formidable task, is not as important as defining a record length for the events. Because this involves dating intervals of no floods as well as flood events, considerable uncertainty in the flood-frequency analysis is likely to result from this problem.

Dates for paleoflood events and, more importantly, the record length, involve uncertainty regardless of the method used. Radiocarbon dates, the most widely used dating method, are inconclusive between about 1650 and 1950 A.D. (see Stuiver, 1982), and older dates are given with a minimum standard deviation of about ± 40 yr. Analyses of growth suppressions in riparian trees (Laing and Stockton, 1976) provide frequency information for floods capable of damaging (but not removing) trees from the floodplain. Dating of floods from newspaper accounts is somewhat subjective because only large, property-damaging floods are recorded in isolated areas such as the Escalante River basin.

Paleoflood discharge calculations are subject to numerous errors and uncertainties. Site selection is based on the availability of paleoflood evidence, which usually is preserved in sites with poor hydraulic characteristics. In the cases of the Alcove and Anasazi Alcove, both sites are upstream of constrictions with highly nonuniform flow in the channel. Estimation of channel roughness and morphology for events 1000 yr in the past introduces a nonquantifiable error in the calculations; however, use of historical photographs (Fig. 6) and choice of channels with bedrock control of bed elevations alleviates some of the uncertainty.

Paleoflood research would benefit from future research on dating techniques and an error analysis of paleoflood discharge calculations. Dating problems may be alleviated or minimized using maximum-likelihood techniques (e.g., Cohn, 1984) but definition of the period of record remains a problem. Some assessment of measurement error, possibly similar to that of Benson and Dalrymple (1967), is needed for paleoflood discharges. Both sources of error must be understood before paleoflood data can be properly incorporated in flood frequency analyses.

ACKNOWLEDGMENTS

We thank C. Angell, K. Katzer, L. Marshall, J. Rohrbaugh, J. Shenk, S. Smith, and T. Yocum for their assistance in the field work. G. Sherrill and T. Gillette of the National Park Service provided logistical assistance and advice. E. Alvey of Escalante, Utah, gave needed historical and geographical information. S. Clark of the University of Arizona examined the tree ring cores from cottonwood trees. W. Kirby, A. Spieker, R. Roeske, and B. Reich critically reviewed the manuscript. This work was supported partially under NSF grants EAR-81-19981 and EAR-83-00183 and a States Water Resources Research Program Grant to V. R. Baker.

REFERENCES

Barnes, H. H., Jr. (1967). Roughness characteristics of natural channels. *Geol. Surv. Water-Supply Pap. (U.S.)* **1849**, 1–213.

Benson, M. A., and Dalrymple, T. (1967). General field and office procedures for indirect discharge measurements. *Tech. Water-Resour. Invest. (U.S. Geol. Surv.)*, Book 3, Ch. A1, pp. 1–30.

Blong, R. J., and Gillespie, R. (1978). Fluvially transported charcoal gives erroneous ^{14}C ages for recent deposits. *Nature (London)* **271**, 739–741.

Butler, E., and Marsell, R. E. (1972). Cloudburst floods in Utah, 1939-69. *U.S. Geol. Surv., Utah Div. Nat. Resour. Coop., Invest. Rep.* **11**, 1–103.

Cohn T. A. (1984). The incorporation of historical information in flood frequency analysis. M.S. Thesis, Cornell University, Ithaca, New York.

Crippen, J. R., and Bue, C. D. (1977). Maximum floodflows in the conterminous United States. *Geol. Surv. Water-Supply Pap. (U.S.)* **1887**, 1–52.

Davidian, J. (1984). Computation of water-surface profiles in open channels. *Tech. Water-Resour. Invest. (U.S. Geol. Surv.)*, Book 3, Ch. A14, pp. 1–48.

Feldman, A. D. (1981). HEC models for water resources system simulation, theory and experience. *Adv. Hydrosci.* **12**, 297–423.

Fowler, D. D. (1972). "Photographed all the best scenery," Jack Hillers's diary of the Powell Expeditions, 1871-1875. Univ. of Utah Press, Salt Lake City.

Hackman, R. J., and Wyant, D. G. (1973). Geology, structure, and uranium deposits of the Escalante quadrangle, Utah and Arizona. *U.S. Geol. Surv. Misc. Invest. Ser.*, Map I-744, Scale 1:250,000, 2 sheets.

Hydrologic Engineering Center (1982). "HEC-2, Water Surface Profiles: Program User's Manual." U.S. Army Corps of Engineers, Davis, California.

Irvine, J. R., and West, N. E. (1979). Riparian tree species distribution and succession along the lower Escalante River, Utah. *Southwest. Nat.* **24**, 331–346.

Kochel, R. C., and Baker, V. R. (1982). Paleoflood hydrology. *Science* **215**, 353–361.

Laing, D., and Stockton, C. W. (1976). "Riparian Dendrochronology: A Method for Determining Flood Histories of Ungaged Watersheds," Final Report to the Office of Water Resources Technology, Proj. A-058-ARIZ (unpublished).

Malde, H. E. (1973). Geologic bench marks by terrestrial photography. *J. Res. U.S. Geol. Surv.* **1**, 193–206.

Malvick, A. J. (1980). "A Magnitude-Frequency-Area Relation for Floods in Arizona, Gen. Rep. No. 2. University of Arizona, Engineering Experiment Station, Tucson.

Motayed, A., and Dawdy, D. R. (1979). Uncertainties in step-backwater analysis. *J. Hydraul. Div., Am. Soc. Civ. Eng.* **105**, 617–622.

O'Connor, J. E., Baker, V. R., and Webb, R. H. (1986). Paleohydrology of pool and riffle pattern development, Boulder Creek, Utah. *Geol. Soc. Am. Bull.* **97**, 410–420.

Saarinen, T. F., Baker, V. R., Durrenberger, R., and Maddock, T., Jr. (1984). "The Tucson, Arizona, Flood of October 1983." Nat. Acad. Press, Washington, D.C.

Stuiver, M. (1982). A high-precision calibration of the AD radiocarbon time scale. *Radiocarbon* **24**, 1–26.

Thomas, B. E., and Lindskov, K. L. (1983). Methods for estimating peak discharge and flood boundaries of streams in Utah. *Geol. Surv. Water-Resour. Invest. Rep. (U.S.)* **83-4129**, 1–77.

U.S. Geological Survey (1910). Surface water supply. Part IX. Colorado River basin. *Geol. Surv. Water-Supply Pap. (U.S.)* **269**, 183–184.

Water Resources Council (1981). "Guidelines for Determining Flood Flow Frequency." Bull. No. 17B. U.S. Water Resour. Counc., Washington, D.C.

Webb, R. H. (1985). Late Holocene flooding on the Escalante River, south-central Utah. Ph.D Thesis, University of Arizona, Tucson.

PART V

ENVIRONMENTAL MANAGEMENT

Sichuan Province in the central southwest of China has long been the most productive agricultural area of the country. A population of more than 100 million people lives in a fertile plateau extending north from the Chang Jiang (Yangtze River), centered on the capital Chengdu. The wealth of this province derives from enlightened management and pragmatic engineering during the early third century B.C. Prior to that time the Min River ravaged the Plain of Chengdu with irregular floods of its braided channel. The river was tamed by excavating two great canals and distributing water by the Dujiangyan irrigation scheme. The works were conceived by governor Li Bing of the state of Qin, and they were completed by his son Li Erlang. To this day a temple beside the river honors these two engineers.

In ancient China the control of floods was the noblest of all tasks. In the twentieth century B.C. the task for managing the Yellow River (Huang He) Valley fell to Yu, son of K'wan. K'wan had built dikes to hold in the river's water. When the river silted its bed and overflowed the dikes, disaster occurred. King Shun had K'wan executed.

Pang (1985) presents an account quoted from Chinese legend telling of the heroics of Yu:

> It was said that Yu retreated up to the Neng Mountain to contemplate his mission. While saddened by his father's fate and weighted down by the herculean task before him, it suddenly dawned on him that the solution to the problem was not to build ever higher dikes but to dredge the rivers and channel their flow. Yu traveled all over the land to survey its geomorphology. Having understood the topography he then led the people in (1) dredging the rivers and clearing the obstacles to their flow, (2) increasing the number of outlets by building channels, (3) draining the tributaries by guiding them into larger ones, lowlands or reservoirs. Although some dikes were still constructed the emphasis this time was to facilitate drainage. Yu was as diligent as he was ingenious. It was said that he toiled alongside the people, and did not go home once during the 13 years he spent controlling the floods, although having walked past his house three times.

As reward for his deeds King Shun first made Yu a high official and subsequently acceded him to the throne. Yu was to found the Xia Dynasty, the first hereditary dynasty of China.

Today neither the rewards nor the punishments for flood hydrologists match those of ancient China. Nevertheless, the problems still require the creativity shown by King Yu. The thesis of this book is that such creativity requires both an engineering and a scientific approach.

The ultimate goal for a *scientific* flood hydrologist is to understand flood-hydrologic processes. Any applications of such understanding are certainly desirable, though of secondary importance. The ultimate goal of the engineering flood hydrologist, on the other hand, is to predict future design conditions; processes that lead to such conditions are of secondary interest (Linsley, 1986). The irony of this distinction is that the results of the secondary aspects of the scientific study may be more critical in some applications than would be results from studies primarily directed at those applications. Of course, the scientific studies are not intended for optimizing design. Rather, they merely offer the potential for identifying important aspects of the system that may have been overlooked in the engineering design process.

Chapter 25 by Thomas Dunne illustrates some of the ways in which fluvial geomorphology can play a role in flood control planning. As in other aspects of flood study, geomorphology provides the critical perspective on fluvial systems over temporal and spatial scales that extend well beyond those considered in local engineering design problems. Channel changes and sediment problems may be ignored by engineering studies, but these are central to any geomorphic analysis. Until mathematical modeling achieves the ability to predict accurately such phenomena from a basis of scientific understanding flood control planners will ignore geomorphological analysis at their peril.

Nowhere is the peril of inadequate flood control planning more evident than in the case of a dam failure. In Chapter 26, John E. Costa discusses the failure of both natural and man-made dams. Because dam failures are capable of releasing much more water and sediment more suddenly than most natural rainfall–runoff floods, their effects on fluvial systems can be most profound. Moreover, if distinctive morphologic and sedimentologic features can be recognized for past dam failure floods, such data may be useful in assessing potential future catastrophes. Flood hazards in areas prone to landslide damming of rivers might easily be inadequately estimated by a flood frequency analysis of rainfall–runoff streamflow data. Again, the lack of an adequate geomorphic study could lead to disaster.

Why are not more geomorphological studies done in relation to flood hazard evaluation? Klemeš (1987) provides part of the answer:

> The basic thing that the engineer needs in connection with flood related planning and design is protection—not so much from floods but from accusations that his flood measures were (1) underdesigned (in cases where flood damage was not completely avoided), or (2) overdesigned (in cases where it was).

Thus, given the inadequacies of predictive models, engineers are faced with an impossible dilemma in flood control planning. To achieve their needed protection from accusations they must concentrate on the establishment of standards or regulations based on "generally accepted principles" and collective judgment. Guidebooks or handbooks are used to outline the standard procedures to be used in a given situation. If Nature fails to read these books and disaster strikes, then the design engineer is protected by the regulatory constraints under which he or she had to work.

Louis Agassiz, a nineteenth century glacial geomorphologist, once observed, "Study nature, not books." Such studies can produce problems in the regulatory environments that are artificially created for flood hazard areas. Studies of the natural system may identify anomalies that were unforeseen at the time the regulations were established. For such reasons, it is clear that geomorphic studies must be used in parallel with engineering studies. Inputs from both areas of expertise should be combined at a higher level of management.

The need for managers to integrate engineering and geomorphic studies is illustrated in Chapter 27, the final chapter of this volume. Peter L. Kresan discusses the Tucson, Arizona, flood of October 1983, and its implications for floodplain management. This example reinforces the general discussion of floods on arid-region rivers by William L. Graf (Chapter 14). The Tucson flood raised numerous questions for local planners (Baker, 1984). Along the Santa Cruz River, the flood was the largest to occur in 70 years of continuous observation. Considerable controversy surrounds the specification of its recurrence interval, an issue made significant by the use of the 100-year flood level for regulatory purposes. Paradoxically, however, the flows on the Tucson reaches of the Santa Cruz did not go overbank as preflood surveys had predicted. Instead, the 1983 flood encountered a narrow, deep arroyo, and spectacular scour of channel banks accommodated the flood discharge. Houses collapsed into the river as bank recession undercut their foundations. Bridges were rendered useless as the river banks receded into and beyond their abutments.

The flood experience of Tucson, Arizona, shows that geomorphology should be as integral a part of flood management as is hydrology, engineering, and jurisprudence. The Santa Cruz River, like many streams in the southwestern United States (Graf, 1985), experienced a complex metamorphosis over the past century. Its heritage of arroyo formation, vegetation removal, and headcut recession created a nonstationarity in the flow series that was available for analysis to define the regulatory flood in Tucson. Moreover, the bank protective works required by regulation to accommodate Tucson's future floods actually exacerbate erosion during present floods by enhancing the erosion of adjacent unprotected banks (Baker, 1984).

In its cycles of disaster, followed by population growth, flood control, and scientific study, Tucson is beginning the pattern displayed by a much larger southwestern city, Los Angeles (Cooke, 1984). As the natural environment is transformed to an engineered one, we clearly do not avoid the need for better understanding of fluvial processes. Indeed the transformation itself creates new challenges for understanding the interaction of floods and landscape.

REFERENCES

Baker, V. R. (1984). Questions raised by the Tucson Flood of 1983. Proc. Amer. Water Res. Assoc., Arizona Sec. **14**, 211–219.

Cooke, R. U. (1984). "Geomorphological Hazards in Los Angeles." George Allen and Unwin, London.

Graf, W. L. (1985). The Colorado River: Instability and Basin Management." Association of American Geographers, Washington, D.C.

Klemeš, V. (1987). Hydrological and engineering relevance of flood frequency analysis. *In* "Hydrologic Frequency Modeling." (V. P. Singh, ed.), Reidel, Boston 1–18.

Linsley, R. K. (1986). Flood estimates: How good are they? *Water Resour. Res.* **22**, 159S–164S.

Pang, K. D. (1985). Extraordinary floods in early Chinese history and their absolute dates. *Proc. U.S.-China Bilateral Symp. Anal. Extraordinary Flood Events*, Nanjing, China.

25

GEOMORPHOLOGIC CONTRIBUTIONS TO FLOOD CONTROL PLANNING

THOMAS DUNNE

Department of Geological Sciences, University of Washington, Seattle, Washington

FLOOD CONTROL STRATEGIES

The enormity of damage caused by floods and the cost of mitigating their effects have stimulated engineers, geomorphologists, and planners to develop a wide range of flood control strategies appropriate to different physical and cultural settings. The methods most commonly employed for the reduction of flood damage are:

1. Flood warning and emergency action
2. Impoundments
3. Channel alteration and stabilization
4. Diversion and storage of floodwaters above and below ground
5. Land management for soil and water conservation
6. Control of land use on floodplains

Geomorphology can contribute to the choice and design of these strategies. The possibilities range from the correlation of flood potential with simple descriptive indices developed in the early days of quantitative geomorphology to more sophisticated analyses of flow and sediment transport processes. Most applications of geomorphology arise because of the interaction among flooding, sedimentation, and channel behavior. This chapter reviews some of the uses of geomorphology in flood control planning and suggests that inclusion of geomorphologic studies leads to more successful and stable schemes.

Flood Warning and Emergency Action

The primary responsibility for flood prediction, forecasting, and response planning rests with meteorologists, hydrologists, and planners. However, there are geomorphologic aspects of the flood prediction problem (Reich, 1971; Orsborn, 1976). Physiographic regions have different flood potentials because of variations in elevation and associated climate, drainage density, channel gradients, and width of valley floor. Therefore, in the prediction of flood discharges it is useful to stratify large drainage basins into homogeneous subdivisions, and to design hydrometeorologic networks, and to monitor and route floods from each region, combining them downstream. These regional differences are already taken into account in some hydrologic procedures such as the use of regional flood frequency curves (Dalrymple, 1960; Wiard, 1962) and synthetic unit hydrographs (Snyder, 1938). However, with widespread availability of satellite imagery for mapping the large-scale geomorphic features of even remote areas, and recent developments in automatic monitoring and telemetry, there is potential for improving flood prediction and warning through systems designed partly on a geomorphic base.

At a slightly greater level of complexity, Kirkby (1976) and Valdes et al. (1979) have demonstrated how the structure of a drainage network influences flood hydrographs and can be used for their prediction. Much research remains to be done by geomorphologists and hydrologists to follow these suggestions.

On a smaller scale, within flood-prone areas there is often considerable uncertainty about the consequences of a predicted discharge. Subtle changes in channel characteristics can alter the timing and magnitudes of flood waves. Campbell and others (1972) have made some illustrative computations of the effect of channel straightening on flood hydrographs. Burkham (1976) documented how a series of large storms at the head of Safford Valley, Arizona, during 1914–1927 scoured a wide, deep, and straight channel along which flood waves passed quickly with relatively little transformation. During 1930–1964, floods were generated mainly in the lower part of the drainage basin and were smaller and more turbid than during the preceding period. As a consequence, floodplain accretion accelerated, salt cedar encroached, and the channel became smaller and more sinuous. A larger proportion of the flood discharges traveled slowly over the densely vegetated floodplain, and flood waves were strongly attenuated. Reversal of this trend could increase flood hazard downstream once more.

Changes of channel position may lead to the breaching of banks, causing floodwaters to spill in an unexpected direction. Such changes can be anticipated for the purposes of predicting inundated areas and planning evacuation routes by detailed examination of floodplain topography in the field and on aerial photographs. Examples of such investigations are described by Popov and Gavrin (1970) and by Velikanova and Yarnykh (1970), who present maps of their results. The hazardous area can be updated routinely in this way, and allowance for the dynamic nature of the flood-prone areas of large alluvial valleys enhances the value of flood maps. Remote sensing, such as the use of LANDSAT imagery of large floodplains, is particularly suitable for this task.

A special type of flood hazard results from the potential for catastrophe breaching of natural and artificial dams. Costa (1985) has marshaled geomorphologic and hydrologic evidence of such floods as an aid in their prediction. Dunne and Fairchild (1984a,b) also used geomorphologic field evidence to analyze the potential flood hazard due to breaching of lakes impounded by landslide debris after the 1980 eruption of the Mount St. Helens volcano.

Impoundments

The role and limitations of storage reservoirs in flood control has been discussed by many authors (e.g., Leopold and Maddock, 1954; Dunne and Leopold, 1978). The following problems exist: (1) dam siting and safety, especially in geologically young, seismically active mountains; (2) the storage volume required to attenuate very large, season-long floods sometimes occurring in monsoon lands; (3) the flooding of valuable agricultural land; (4) the rapid reduction of reservoir effectiveness by sedimentation in some regions; and (5) effects on the river channel and valley floor upstream and downstream of the reservoir. However, in many flood control problems, large reservoirs must be considered, and geomorphologic studies contribute to the prediction of sedimentation rates, design floods, and the effects of dam construction on the channel.

Sedimentation Rates. The usual analysis of probable sedimentation rates in reservoirs is inadequate, especially in tropical and subtropical countries and in mountains. The rate of infilling is usually assessed from a short flow record and a small number of suspended-sediment samples collected during a few years between project proposal and construction. Sediment samples collected earlier than this are frequently obtained by nonstandard procedures and are not interpretable. The two sets of records are usually much too short to sample the variability of sediment transport, and the important high flows, which carry most of the sediment, are often underrepresented because of their rarity and difficulties of sampling. Yet after brief review of the region, such data are used for computing the design life of reservoirs and frequently lead to overly optimistic economic analysis.

Even at stations at which a large number of samples have been collected, there may be considerable uncertainty in the definition of the sediment rating curve because of the large scatter of points. For example, in Figure 1*a*, it is difficult to judge whether the rating curve should be extrapolated linearly or whether the slope of the curve should decline with increasing discharge. Even greater uncertainties develop when sediment rating curves shift dramatically through time (Fig. 1*b*).

On the Tana River in central Kenya, three studies of sediment yield at a proposed dam site had concluded that the economic life of the associated reservoir would exceed 100 yr. Dunne and Ongweny (1976) pointed out that this yield was far less than the sum of yields from the tributaries to the Tana. Yet there was no field evidence for extensive deposition along the main stem. They estimated the sediment influx to the channel to be several times greater than anticipated. The suspended-sediment rating curve (based on 53 samples collected over 13 yr and fitted by regression) underestimated sediment transport when extrapolated to high discharges. Ongweny (1978) then instituted a program of suspended-sediment sampling in the Tana and its tributary basins, and he demonstrated that the reservoir would be filled within a period of 30–35 yr under average weather conditions and within 20–25 yr if a run of wet years were to occur. The economic life would be even shorter. Ongweny used the method of Borland and Miller (1958) to calculate that almost all of the dead storage volume of the reservoir and 16–25% of the live storage volume should be filled 8 yr after impoundment. His work, which has been confirmed by reservoir surveys, illustrated the value of improving sediment yield computations based on short records at a single location with field observations and geomorphologic interpretation of the basin sediment budget. His field observations and plot experiments also isolated some of the major sediment sources upon which conservation could be focused to prolong the life of the reservoir, as well as to obtain other benefits.

Much more attention needs to be paid to the prediction of reservoir sedimentation rates, especially in tropical and subtropical mountain regions where erosion rates are gen-

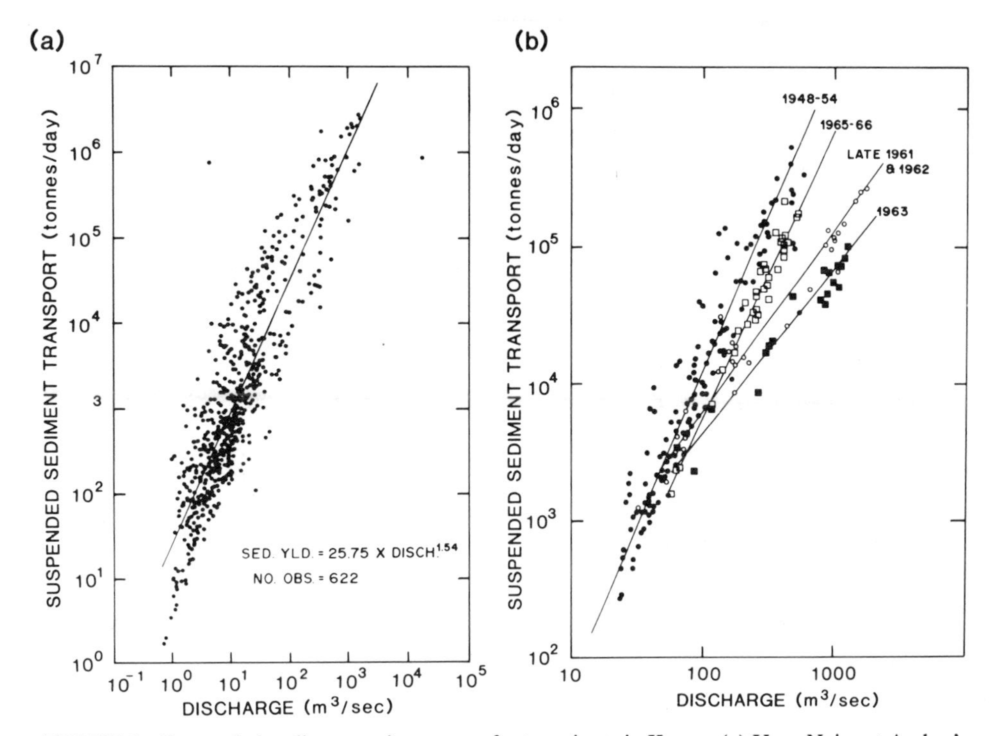

FIGURE 1. Suspended-sediment rating curves for two rivers in Kenya: (*a*) Uaso Nyiro at Archer's Post, indicating the large amount of scatter around a rating curve based on samples collected during more than a decade and the uncertainty involved in linear extrapolation of a best-fit regression line. (*b*) Tana River at Garissa, showing radical shifts due to variations of weather, ground cover, and animal stocking rates. Data from Kenya Ministry of Water Development, Nairobi.

erally high. Because sediment sampling records can give misleading results, more detailed and innovative analysis is required of the stochastic aspects of flow and sediment transport, using techniques such as those described by Fiering and Jackson (1971), to extract the maximum amount of information from limited records and to document the uncertainty in estimates of mean sediment transport.

However, it is not likely that large numbers of monitoring stations will be maintained for many years on all of the rivers with a potential for impoundment. This is especially true for measurements of bedload, which is usually deposited near the inlet of a flood control reservoir where it reduces the live storage volume. Because the empirical record of sediment transport is usually sparse, and the popular computational methods (see Vanoni, 1975) are notoriously unreliable without some independent evidence, it is important to develop some geomorphologic methods for assessing annual sediment fluxes. Lustig (1965), for example, used landscape morphometry as a basis for transferring sediment yields between drainage basins to estimate the life of a proposed reservoir. In some rivers with favorable geomorphologic circumstances, all of the gravel bedload may be deposited in a particular reach, so that the bedload yield can be estimated by repeated survey of the channel cross section (Griffiths, 1979). In other cases it may be possible to quantify the entire sediment influx to a river using the sediment budget techniques described by Dietrich and others (1981) and Reid and others (1981).

A small-scale example is provided by an estimate of the probable rate of filling of a small reservoir on the South Fork Snoqualmie River, Washington. The river drains a 212-km^2 forested basin in the Cascade Mountains. Average sediment influx to stream channels from mass wasting was estimated by measuring on aerial photographs the density and area of landslide scars supplying sediment to channels over a period of 17 yr. During field work, a volume–area relationship was developed on a sample of these scars, and the grain size composition of the sediment forming the margin of the scars was measured. Thus, it was possible to compute the average annual contribution of each grain size from landslides. Sediment yields and grain sizes eroded by water from roads were estimated by extrapolation of measurements made along roads in the Olympic Mountains of western Washington (Reid et al. 1981). The estimated influx was then partitioned into bedload and suspended load on the basis of the virtual absence of sand in point bars along the South Fork Snoqualmie.

The resulting yields of bedload and suspended load could be checked against independent evidence. Nelson (1971) estimated annual suspended-sediment yield on the

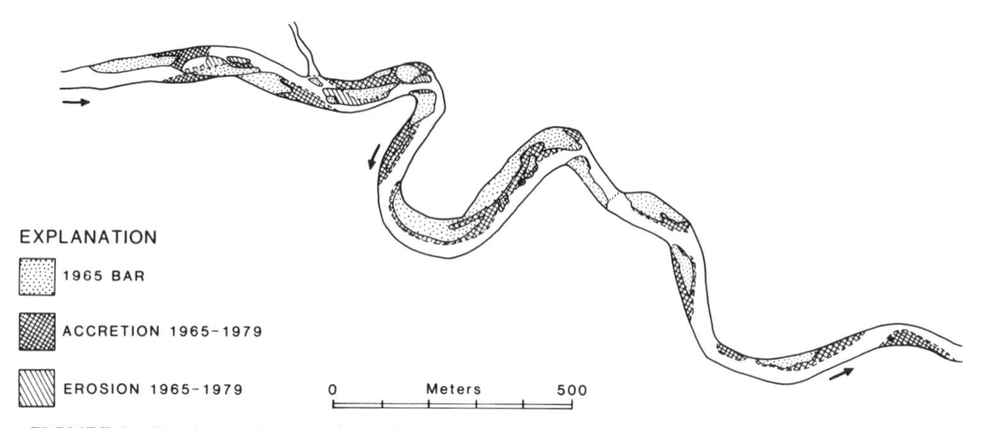

FIGURE 2. Erosion and deposition of gravel bars along a portion of a 12-km-long zone of gravel deposition in the South Fork Snoqualmie River, Washington, 1965–1979. The bars were mapped from aerial photographs at similar low flows. The outer banks of the channel were stabilized artificially throughout the period. The mapped areas of growth were converted to volumes of deposition with the aid of surveyed cross sections.

basis of 2 yr of sampling and flow measurements. Measurements of bed material texture along the South Fork showed that all of the gravel bedload accumulates on bars in a low-gradient reach, and the rate of accumulation indicates the annual basin yield of gravel. The rate of accretion and erosion of channel bars along the lower 12 km of the river was measured on aerial photographs taken 14 yr apart and at similar low flows (Fig. 2). Change in bar area was converted to the weight of deposition by surveying channel cross sections and using a typical bulk density of poorly sorted gravels. The various estimates of the average annual sediment load are compared in Table 1. They are close enough to encourage the further use of geomorphologic checks and extrapolations of sediment yield measurements.

Another application of geomorphologic methods in sediment yield prediction was necessary after the 1980 eruption of Mt. St. Helens into the Toutle River valley, and the same methods could be used after other devastations that increase sediment yields. Immediately after the eruption a vast amount of erodible sediment was emplaced in the upper Toutle valley and was eroded and transported to the lower Toutle and Cowlitz valleys, where its deposition choked the channel, reducing bankfull capacity and causing rapid channel shifting. During the first posteruption winter, the only flood control response possible was diking and dredging, but in later years there has been discussion of the value of impounding all sediment within the upper Toutle valley by means of a high earthfill dam. A short-term estimate of the probable rates of erosion in the upper Toutle valley and of deposition along the Cowlitz channel was made 6 months after the eruption on the basis of a sediment budget and suspended-sediment samples collected by the U.S. Geological Survey. The approach involved the following activities: aerial photogrammetry to map and measure various sediment sources and sinks within the basin; monitoring of erosion rates in the field; measurement of grain sizes of sediment eroded and deposited at various places along the channel; and computations of sediment

TABLE 1 Comparison of the Average Annual Sediment Load of the 212 km^2 South Fork Snoqualmie from Methods Described in the Text

Method	Suspended Load (tonnes/yr)	Bedload (tonnes/yr)
Sediment budget[a]	42,060	2,630
Suspended sediment[b] sampling by Nelson (1971) and his estimate that bedload = 5% of total load	59,700	2,980
Measured rate of bed material accumulation[c]	—	2,380

[a] Mass wasting supply averaged over 17 yr.
[b] Suspended-sediment sampling over 2 yr.
[c] Bed material accumulation averaged over 12 yr.

transport and deposition. The methods and judgments used were described and critiqued by Dunne and Fairchild (1984a,b), and the updated field results were described by Collins and others (1983), Collins and Dunne (1986), and Lehre and others (1983). The sediment budget of the Toutle valley could be updated and projected into the future for the purpose of assessing the necessary size of the proposed reservoir or of the cost of continued dredging in the Cowlitz River.

When an estimate of sediment transport has been made for a proposed reservoir site, it is useful to examine the geomorphologic context of that sediment flux rate. Where are the major sources of sediment? Are they strongly affected by land use or can they be so affected in the future? What is the pattern of channel sedimentation upstream from the site for which the estimate was made? For reasons that are not well understood, some sediment temporarily accumulates in large waves along some river reaches and then is scoured and transported downstream over a period of years, presumably after some threshold of channel gradient or width is attained by the accumulating wedge of sediment. Griffiths (1979) has documented the migration of large waves of bedload in the Waimakariri River, New Zealand. If measurements of bedload transport were made for only a few years, beginning immediately upstream or downstream of such a wave, one could obtain values of transport that were not representative of the long-term mean flux rate. To discern whether such wedges of sediment are confounding long-term estimates requires examination of channel migration and depositional forms through long reaches of valley above the measurement site. The observations can be made on sequences of aerial photographs covering a time interval before and during the sampling period. Trimble (1977, 1983) has made detailed stratigraphic and sedimentologic studies of the effect of sediment storage in valleys, fans, and footslopes on the downstream sediment yield of a drainage basin undergoing land-use changes. Failure to take such storage changes into account can lead to large underestimates or overestimates of sediment yields and the effectiveness and longevity of flood control reservoirs.

Prediction of Design Floods. Estimation of the magnitude and frequency of large floods that might damage a flood control dam is usually based on statistical analyses of relatively short streamflow records. The problems resulting from this technique include the difficulty of extrapolating flood frequency curves based on small, frequent floods; the difficulty of agreeing upon which theoretical probability distribution best fits the recorded data as a basis for extrapolation; the effect of runs of years of high and low flood potential associated with weather fluctuations; and the relatively large number of ungauged rivers. Where the instrumental record is short or absent, it is usual to design spillways to accommodate the maximum probable flood, which must be calculated on the basis of the maximum probable rainfall, a judgment about rainfall–runoff relations in the basin, and some means of computing the temporal distribution of storm runoff from the basin. Binnie and Mansell-Moulin (1966) describe an example for a large dam in Pakistan. The method is fraught with many uncertainties and begs for empirical confirmation.

The simplest applications of geomorphology to flood prediction involve regional statistical relationships between channel geometry, especially the active-channel width, and flood discharges of various frequencies (e.g., Osterkamp and Hedman, 1982). Such correlations allow the transfer of flood predictions from gauged basins to ungauged sites. However, they are subject to large errors when rare floods are estimated from short records.

Geomorphologic studies have provided ways of extending the record of floods far beyond the length of instrumental observations. The earliest studies of this type involved mapping the distribution of fine-textured, bedded flood deposits to define the highest levels of floods before gauging stations were established. Jahns (1947) examined the distribution of overbank sediments deposited by the 1936 and 1938 floods along the Connecticut River valley and found no evidence of higher floods during the period of European settlement and probably for several hundred years prior to invasion. From such evidence, outliers in a flood record can be assigned a more realistic recurrence interval than is possible from the duration of the instrumental record alone. Later studies along steeper, gravel-bedded rivers recognized local bar deposits, gravel levees, and flood channels, which could be dated by radiocarbon measurements on buried debris and tree stumps or by dendrochronology (Helley and LaMarche, 1973). This kind of work has been combined with strictly botanical methods of extending flood records (Sigafoos, 1964; Yanosky, 1983).

The most extensive and longest geomorphologic flood records are provided by slackwater sediments, which accumulate in embayments, at tributary junctions, and in other portions of a valley floor where velocity declines and suspended sediment settles from overbank flows (Kochel and Baker, 1982). Each successive flood with a stage higher than earlier floods adds a layer to the sequence of sediments. The age of the sediment and associated flood can be obtained from the radiocarbon content of buried wood, fine organic detritus, or the humus of paleosols. These dates are used for correlation between sites along a river valley, and for indicating the duration of the entire sedimentary record. The water level indicated by a sediment layer is used to compute a flood discharge by the slope-area method. These discharges are then ranked and their recurrence intervals can be computed in the normal way. Kochel and Baker (1982) used this method to construct a 2000-yr-long record of floods.

Another contribution of geomorphology to spillway and reservoir design is the recognition that in some physiographic situations the most damaging flood may not originate as runoff from rainfall or snowmelt but from a glacial outburst (jokulhlaup) or the catastrophic breaching of a natural lake impounded by landslide or other debris or the rapid melting of a snowpack by a pyroclastic flow or the triggering of a large debris flow. Such events are most common in glaciated, tectonically active mountains or around active volcanoes. They leave sediments and other evidence, which although not unequivocal, can usually be recognized and interpreted genetically. Glacial outburst floods may leave particularly coarse sediments and large bars in outwash sediments (Birkeland, 1968; Church, 1972). Birkeland used such sedimentologic evidence together with field evidence of flow depth and slope to compute peak discharges. Clague and Mathews (1973) have compiled field evidence of the magnitude of jokulhlaups, and Clarke and Mathews (1981) have calibrated a predictive model of these flood hydrographs against field data.

The design of flood control dams around active volcanoes requires that some estimate be made of potential peak discharge, volume, and other properties of mudflows. Such flows or more dilute muddy floods could be generated around volcanoes by a number of mechanisms, including rapid snowmelt as a consequence of a pyroclastic flow, pyroclastic surge, or stream explosion; liquefaction of saturated, unconsolidated sediments as a result of shaking during an eruption; and a rapid breach of debris dams impounding lakes. The deposits of such mudflows can be mapped to give a general idea of the presence of such hazards (e.g., Mullineaux and Crandell, 1962), but the deposits give no indication of peak discharge, and their elevation above the present valley floor may be misleading if the channel bed has been lowered since deposition. The number of these deposits occurring within a datable period gives a rough indication of the average frequency of large mudflows. However, because it is difficult to relate individual deposits to their generating mechanisms and because the configuration of the cone and of surrounding lakes may have changed significantly during that period, it may not be obvious how the stratigraphic record can be extrapolated into the future without some interpretation.

Dunne and Fairchild (1984a,b) and Fairchild (1985) considered the various mechanisms that might generate mudflows or muddy floods around the Mount St. Helens volcano. They first considered the probable volume, peak discharge, and location of flows that might occur due to pyroclastic flows eroding snowpacks and to the catastrophic breaching of lake barriers. These initial hydrographs were then routed downvalley (see Fig. 3*b*,and 3*c*), using a Muskingum flood-routing model that had been calibrated against mudflow hydrographs resulting from the eruption of May 18, 1980 (Fig. 3*a*). The hydrographs were constructed from geomorphic evidence at six locations on the South Fork Toutle River and three on the North Fork Toutle River (Wigmosta, 1983; Fairchild, 1985). These geomorphologic studies also yielded density and rheologic parameters for the 1980 mudflows (Wigmosta, 1983) and for mudflows of other grain size distribution and water content (Fairchild, 1985). Such information is useful for designing engineering structures to withstand impact and shear forces from mudflows.

Channel Changes above and below Impoundments. Rivers entering a reservoir deposit all or most of their load. The coarser fractions of the load are deposited near the mouth of the stream, either in the live storage zone of the reservoir or in the backwater zone of the stream channel. Below the dam, clear water may scour the channel bed and banks. The best studied case of this development occurred after the closure of Hoover Dam on the Colorado River. The channel was degraded for more than 500 km downstream, rendering useless many intake structures for municipal supply and irrigation, and undermining bridge foundations. At Yuma, Arizona, 560 km downstream of the dam, the river bed was lowered 2.8 m. In gravel bed streams of western Washington, the trapping of gravel behind hydroelectric dams and the scouring of grain sizes finer than cobbles from sediments downstream have led to a coarsening of the channel bed so that the gravel bars are now too coarse for salmon spawning. In some gravel bed streams in the same region, the opposite situation exists. Before impoundment, large floods carried away coarse sediment entering the river from steep tributaries. The lowered peak discharges cannot remove this sediment, which accumulates as alluvial fans, raising the bed of the stream and in some places causing the river to develop a braided, unstable channel.

Williams and Wolman (1984) summarized the literature on case studies of the downstream effects of dams on alluvial rivers and compiled the results into a statistical summary of the changes observed in bed elevation, width, bed material texture, sediment concentration, and load as functions of downstream distance and time since dam closure. The results, though mainly from the subhumid western United States, exhibit enough scatter to emphasize that useful predictions require consideration of the sediment sources and transporting flows in particular rivers. An estimate of sediment influx and texture can be made by the methods referred to above, and sediment can be routed downstream to predict sediment flux, scour, and fill by methods proposed by Komura and Simons (1967) and the U.S. Army Corps of Engineers (1977). However, there remains an important difficulty in the prediction of channel width by these one-dimensional computations of sediment transport. Two-dimensional methods of sediment routing are now being developed, but they incorporate little of the

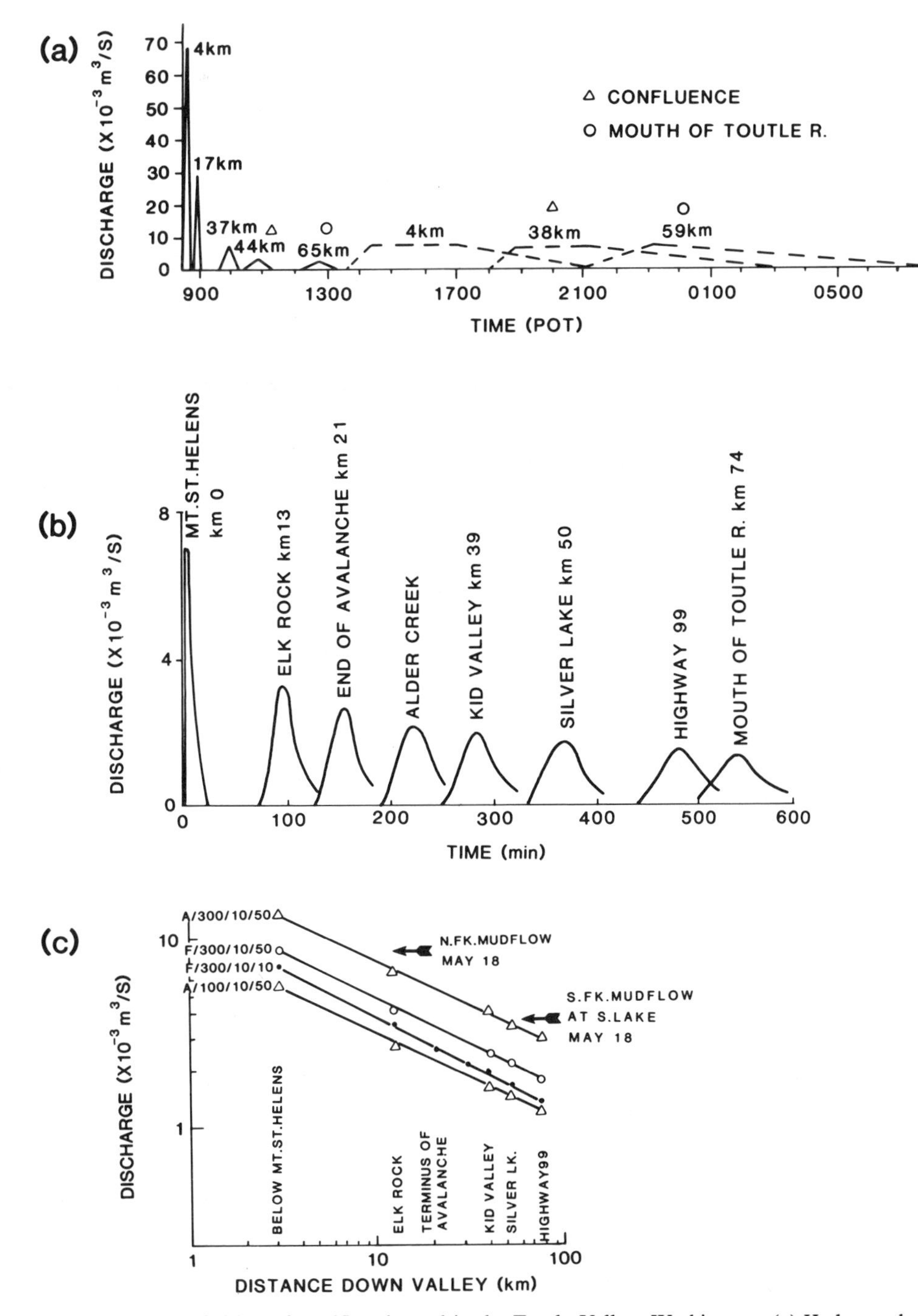

FIGURE 3. Definition of mudflow hazard in the Toutle Valley, Washington. (*a*) Hydrographs of 1980 mudflows in the South Fork (solid lines) and North Fork (dashed lines) Toutle valley reconstructed on the basis of field evidence. Source: Fairchild (1985). (*b*) Computed hydrographs of a mudflow generated by a pyroclastic flow onto a snowpack on the slopes of Mount St. Helens volcano (km 0) and routed along the North Fork Toutle valley. Source: Dunne and Fairchild (1984a,b). (*c*) Downstream variation of peak discharge for mudflows originating as pyroclastic flows on Mount St. Helens. The sequence of symbols refer to various assumptions made to generate the mudflows, as follows: month of average snowpack used; width of pyroclastic flow track (m); speed of mudflow on the volcano slopes (m/s); depth of snowmelt under the depositional fan of the pyroclastic flow (cm of water). Also shown are field estimates of peak mudflow discharges during the 1980 eruption. These estimates are based on geomorphic evidence. Source: Dunne and Fairchild (1984a,b).

complexity governing adjustments in channel width to changes of flow and sediment transport. Leopold and Bull (1979) discussed the simultaneous adjustment of width, slope, and hydraulic roughness that results from changes of base level and deposition above a dam, and they summarized the scanty field data on this process. However, much remains to be understood about these aspects of river channel response.

In the absence of theoretical techniques prediction of the geomorphologic consequences of dam construction and flood reduction relies on empirical relationships and qualitative arguments, which are nevertheless useful predictors. For example, Lane (1955) proposed that an alluvial river attains an equilibrium that can be described by the following proportionality between flood magnitude (Q), bed material flux (Q_b), bed material grain size (D), and channel slope (S):

$$Q \cdot S \sim Q_b \cdot D$$

Despite the imprecise definition of the terms, this relationship is remarkably useful in the initial, qualitative stages of predicting the downstream effects of flood control reservoirs. For example, if flood flows are reduced moderately and sediment load drastically, Lane's relationship suggests that the river would respond by reducing its slope and increasing the texture of its bed material by selective transport and armoring. The relationship does not predict the magnitude of these changes or of associated effects that might be expected, but it suggests critical variables to be measured in field studies and analyzed by more formal modeling. Schumm (1969, 1977, pp. 133–137) extended Lane's approach on the basis of empirical equations developed by himself and others relating channel form, flow, and sediment properties. This more extensive analysis, heavily weighted by data from sand-bedded streams in subhumid environments, allows qualitative consideration of the effects of dam construction on channel width, depth, gradient, sinuosity, meander wavelength, and bed material texture. The relations need to be extended into regions with gravel-bedded rivers.

More quantitative prediction of the effects of dam construction on channels upstream and downstream awaits a more rigorous, quantitative theory of the mutual adjustment among the hydraulic and sedimentologic parameters that control channel form in cross section and in plan.

Channel Alteration and Stabilization

River channels are frequently deepened, widened, straightened, or diked to increase their capacity. In other cases the main aim in flood control is to prevent deposition within the channel and to confine the channel so that it will not migrate laterally and shift the flood-prone zone into occupied parts of the valley floor. Such alterations require the design of stable river channels.

Hydraulic engineers have developed a set of empirical equations, relevant mainly to sediment-transporting canals, which describe the width, depth, and slope of straight, stable channels as functions of discharge and the grain size of bed and bank materials. These "regime" equations are summarized by Blench (1969) for rivers with fine bed material and by Kellerhals (1967) for gravel bed channels. Geomorphologists have extended this approach to meandering and braided rivers through definition of the hydraulic geometry: a set of empirical equations representing the variation of width, depth, mean velocity, water surface slope, hydraulic roughness, and sediment concentration as functions of discharge, as it varies temporally at a single cross section and spatially downstream at a single discharge frequency (Leopold and Maddock, 1953). Schumm (1977) has extended such relationships to meander wavelength, amplitude, and sinuosity and has incorporated the effect of bed and bank materials. It is also possible to use regional relationships between each of these hydraulic and morphologic characteristics and the drainage area, which is a surrogate for both flood discharge and sediment supply (Fig. 4). Channel alterations for flood control are more likely to remain stable if the designed channels conform approximately to such empirical relationships.

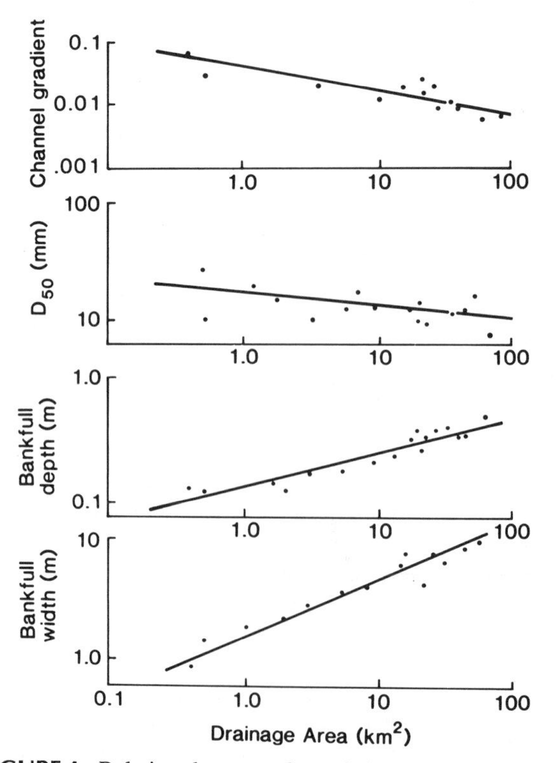

FIGURE 4. Relations between channel characteristics and drainage area for streams in forested drainage basins of the Puget Lowland, western Washington. Source: Madej (1982).

Radical alteration of river geometry will usually require that the channel be rendered immobile. Otherwise, it will regain its original form by erosion and deposition to adjust its cross-sectional area and sinuosity (Keller, 1976). Severe erosion and deposition usually occur if the discharge–slope relationship is upset in channels with easily erodible beds, either by channel straightening or by the focusing of flow from several braided channels into a single channel. Daniels (1960) and Ruhe (1971) documented how the straightening and steepening of Willow River in Iowa provoked deep entrenchment (to a maximum of 14 m), widening of the channel, and lowering of the base levels of tributary drainage networks (Fig. 5). The downstream end of the entrenched channel was later filled by the accelerated sediment flux. Such changes are irreversible in the short term and extremely expensive to stabilize. Yet the application of current geomorphologic knowledge can indicate how to avoid and to minimize such problems (Schumm et al., 1984).

The planform aspects of rivers are the most difficult characteristics to predict. One is often limited to a strictly empirical approach of examining the position of a channel on sequences of aerial photographs and developing some locally relevant insights into its behavior. Some meander forms are more stable than others and can be used as models for designing channel widths, meander wavelengths, gradients, and sinuosities to which a channel should be coaxed by means of dredging, diking, groynes, or bank afforestation. Winkley and others (1984) outline some geomorphologic principles that have proved useful in stabilizing rivers in the horizontal plane.

Geomorphologic interpretation of alluvial features and historical evidence can also be used to assess the likelihood of successful diking and channel stabilization. Figure 6 illustrates the positions of the channel along a reach of the Green River, Washington, at three times during this century. The channel positions were obtained from an early topographic map and two sets of aerial photographs. The record of shifting could be extended through dendrochronology on older point bar deposits as illustrated by Everitt (1968) and by Hickin and Nanson (1975). Maps of this kind provide several important types of information for river engineering.

For example, Figure 6 indicates that the river moves across its floodplain by gradual lateral shifting: mid-channel bar formation and avulsion are rare. Second, it is possible to measure rates of lateral migration. When such rates are plotted on a graph (Fig. 7*a*) that also indicates the downstream pattern of boundary shear stress, it is clear that lateral migration is most intense along reaches that experience a rapid downstream decrease in boundary shear stress. The ability of the river to carry coarse sediment declines with decreasing boundary shear stress. The result is a general downstream decrease in bed material size between kilometers 67 and 40 (Fig. 7*b*), accumulation of sediment on point bars, and the rapid shifting of the channel (Fig. 7*a*) in response to their growth. Diking along this reach would be very costly and unlikely to succeed.

Between kilometers 40 and 19, boundary shear stress remains constant or increases, and there is little or no deposition of the sand and silt entering the reach. In this zone shifting is very slow and dikes have been stable. In the lower reach, where shear stress again declines, there is deposition and slow shifting that causes some damage to dikes, but the stresses on the channel boundaries are much lower than those above kilometer 40.

Diversion and Storage of Floodwater

Where other means of control are impossible and where conditions are suitable, a portion of a flood may be diverted into a low-lying area of floodplain, by opening gates in a natural levee or artificial dike. The strategy may involve simply holding the water in a surface depression until the flood peak has passed, and then opening a gate further downstream to allow the water to flow back into the channel. In other cases the valley alluvium may be so permeable that a significant proportion of the diverted floodwater will infiltrate over succeeding weeks and be stored below ground, recharging the groundwater for use during the low-flow season. The groundwater table may even be lowered deliberately by pumping during the dry season as part of a coordinated program of water supply and flood control. Both surface and subsurface storage of floodwater require conditions that are not ubiquitous on large floodplains and consequently require detailed geomorphologic and hydrologic analysis.

Large, unoccupied depressions, such as former river channels, drained ox-bow depressions, and backwater swamps behind natural levees offer the best possibilities for temporary storage, but to effect a significant reduction in flood flow the depression must be very large. For example, a diversion of 500 m^3/s for one day would fill a 10-km^2 depression to a depth of 4.3 m. Depressions of this depth are not common on large floodplains. If depressions are not occupied by cultivated lands and villages, they tend to be thickly forested, which makes an assessment of the storage volume and consequences of diversion extremely difficult. In such conditions the assessment of surface detention opportunities should involve a detailed geomorphologic analysis of subtle alluvial landforms. This might include aerial photograph interpretation with conventional photography, radar or LANDSAT imagery; photogrammetry, field mapping and topographic survey of critical areas; and perhaps dye or float tracing and aerial photography during floods, when the potential storage sites are inundated and it is possible to trace flow patterns through

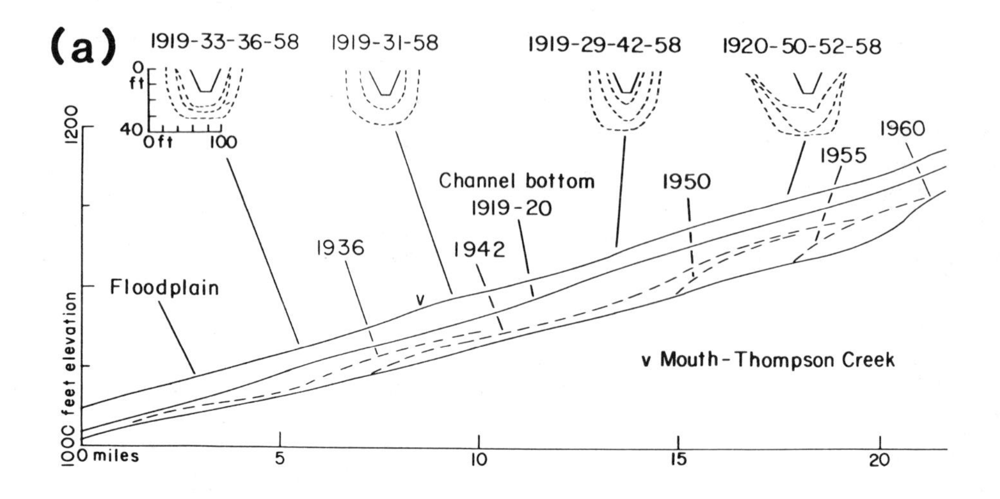

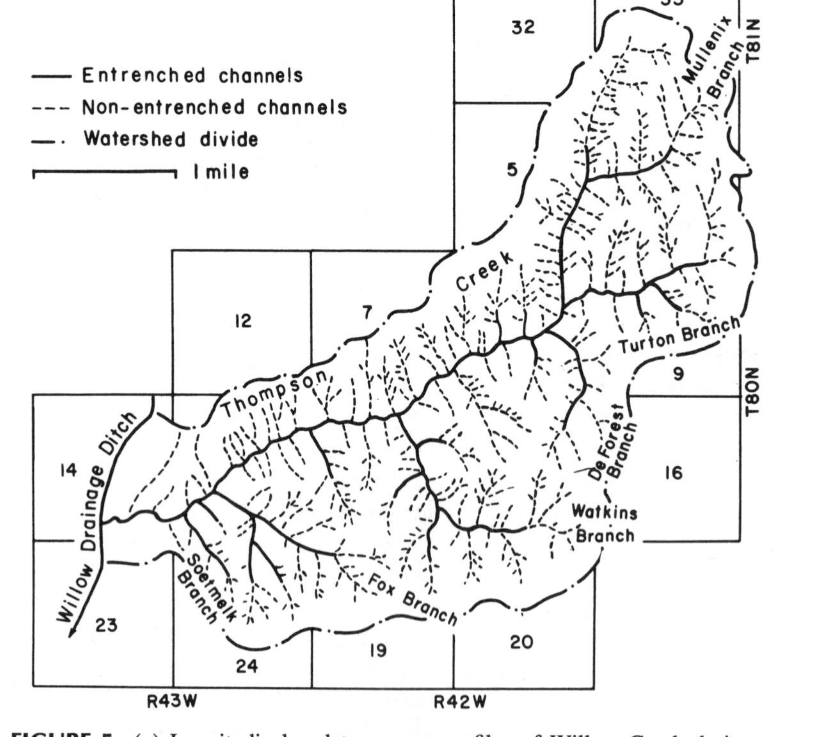

FIGURE 5. (*a*) Longitudinal and transverse profiles of Willow Creek drainage at various times after channelization, which occurred between 1906 and 1920. Source: Daniels (1960). (*b*) Extension of entrenched channels through the drainage basin of a tributary of Willow Creek. Source: Ruhe (1971).

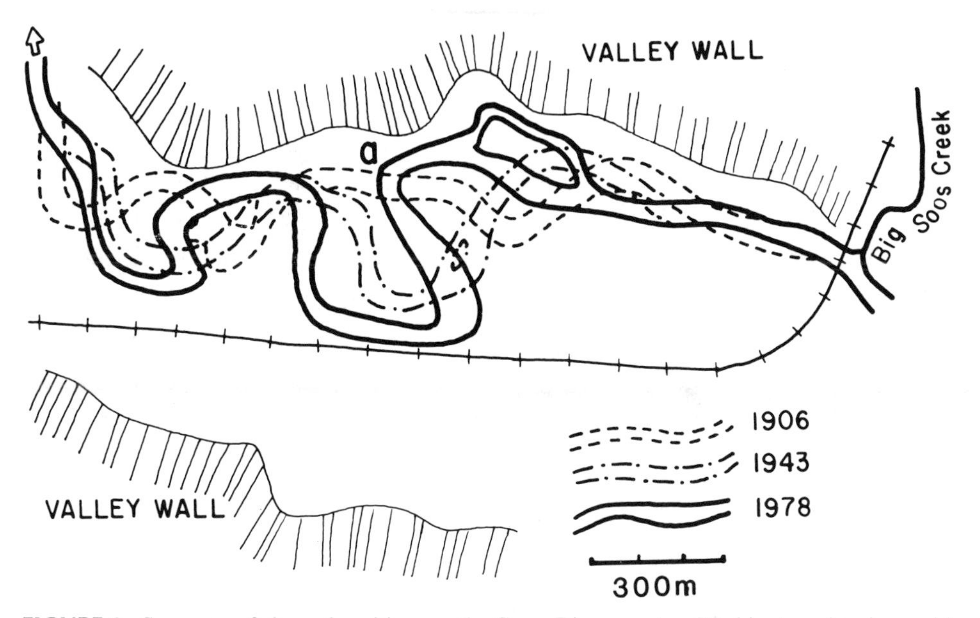

FIGURE 6. Sequence of channel positions on the Green River, western Washington, showing rapid lateral shifting and downstream migration of meanders. The maps were constructed from two sets of aerial photographs and a topographic survey.

the complex microtopography. Popov and Gavrin (1970) and Moore and North (1974) present examples.

The conditions for successful groundwater recharge are even more stringent because not only is a large surface depression required but it must be floored with thick, highly permeable alluvium to allow infiltration within a useful period of time. Fine sediment settling out of floodwater may reduce the infiltration capacity of the detention basin and must be excavated periodically. Much lowland river alluvium is not very permeable because of its fine texture, but there are some opportunities in sandy or bouldery alluvium. Such zones often occur in reaches of declining shear stress at the transition from hilly terrain to a lowland (e.g., between kilometers 40 and 67 in Fig. 7*b*), where thick sequences of sand or gravel are deposited by braided streams or by rapidly shifting, meandering rivers that have extensive point bar deposits. Other potential recharge zones exist in sandy or gravelly river channel, levee, and splay deposits, which have a coarser texture, better sorting and higher permeability than the surrounding overbank and backswamp deposits, or in alluvium deposited under earlier hydrologic regimes. The sedimentology and stratigraphy of favorable situations for groundwater recharge are illustrated by the case studies summarized by Reineck and Singh (1980, pp. 257–314) and Miall (1978).

Estimating the potential for groundwater recharge would require a mapping and drilling program and the cooperation of a hydrogeologist, a sedimentologist, and a geomorphologist to interpret subsurface alluvial forms and to quantify the hydrologic properties of the sediment. Maps of hydraulic conductivity (Fig. 8) would be needed, and could be superimposed on maps of topography and land use to define useful recharge areas. It is also necessary to take account of groundwater conditions during the flood season. Water tables rise early in the wet season in permeable deposits that receive small natural inflows. It may be necessary to draw down the water table quickly by pumping from a proposed subsurface reservoir, and it may be uneconomical to install the necessary pumping capacity, unless there is a local means of using such water at high cost during the wet season.

Land Management for Soil and Water Conservation

Soil and water conservation does not significantly reduce the size of flood peaks generated by long rainy periods on large drainage basins. It is more effective during short, intense rainstorms on small basins. Hoyt and Langbein (1955) and Leopold and Maddock (1954) discussed this issue in some detail, and their conclusions have been verified by more recent empirical and theoretical studies. However, in some cases where deposition of sediment in river channels causes aggradation and channel shifting, a reduction of soil erosion through watershed management may facilitate river engineering for flood control. Before an expensive watershed management program is relied upon for flood control benefits, several geomorphologic issues should be clarified.

The first issue involves the nature and distribution of sediment sources and the extent to which they are affected by land use. For this purpose a sediment budget defining

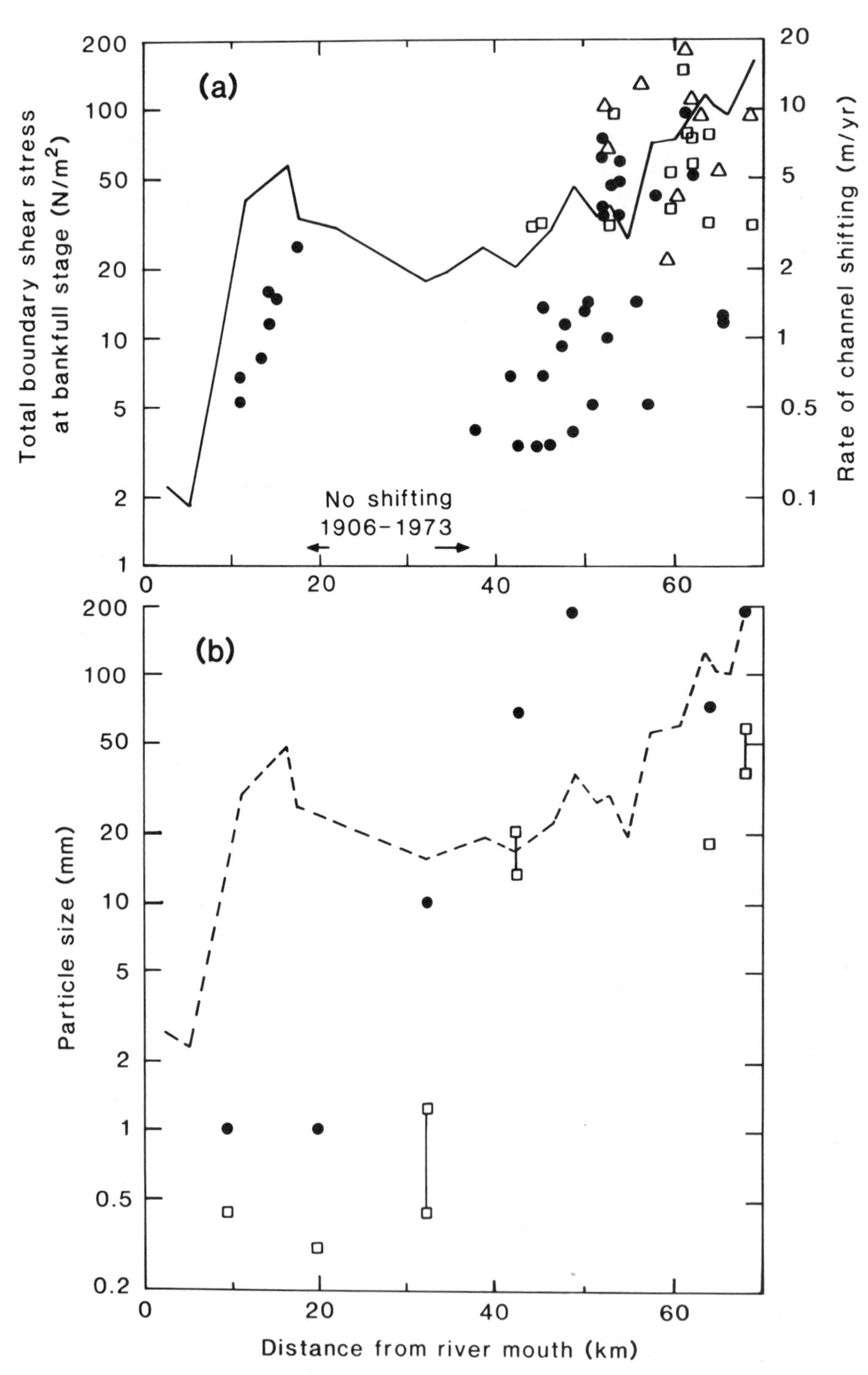

FIGURE 7. (*a*) Downstream variation of maximum total boundary shear stress (solid line) at bankfull discharge in the Green River, computed from thalweg flow depth and gradient, and the rate of channel shifting during three time periods: 1960–1943 (solid circles), 1943–1973 (triangles), and 1968–1978 (squares). (*b*) Downstream variation of computed stream competence (dashed line), and the sizes of the largest (solid circles) and mean (squares) particle sizes found on the bed in each measured reach. Connected squares indicate two sample means from the same reach.

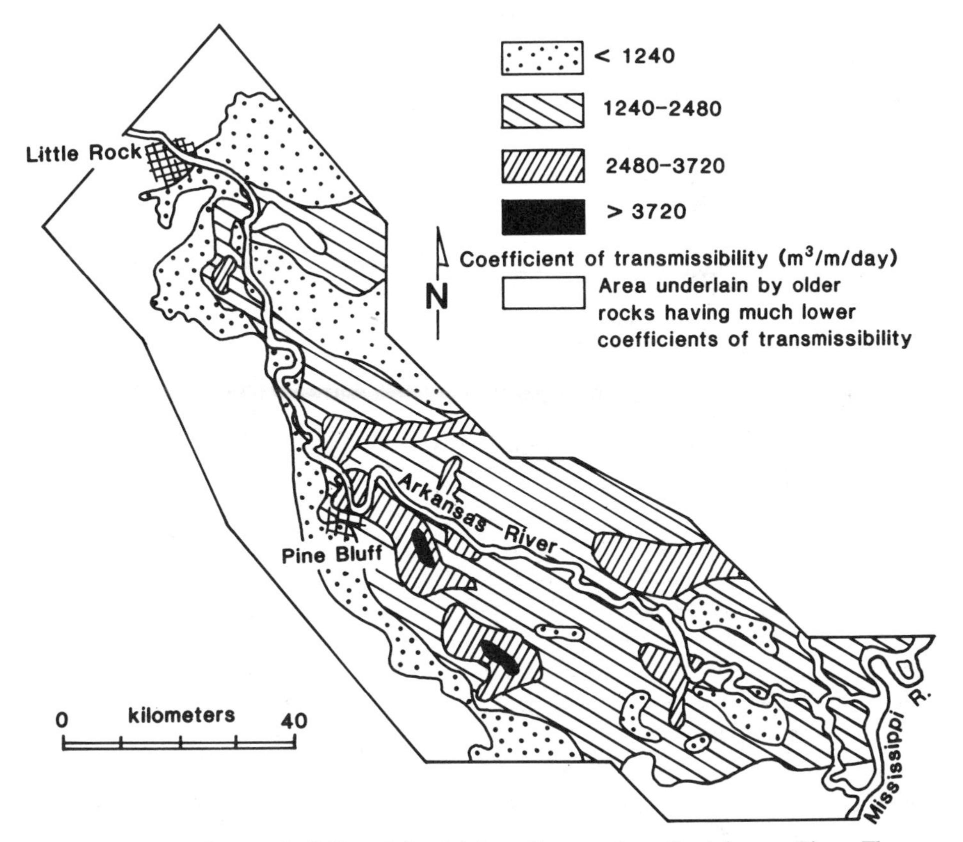

FIGURE 8. Map of transmissibility of floodplain sediments along the Arkansas River. The map was based on measurements in the down-valley direction, but similar measurements of permeability and depth could be made in the vertical direction to estimate the capacity of the sediments for infiltration and storage of floodwaters. Source: Bedinger and Emmett (1963).

the rates of sediment production by individual processes on disturbed and undisturbed areas may be useful, as discussed above. Large natural inputs of sediment from glaciers and landslides, for example, cannot be reduced by land management. If individual sediment sources cannot be isolated with the resources available, then it is usually possible to follow the approach used by Ongweny (1978) and Dunne (1979) of measuring sediment yields from subcatchments with various dominant land uses and to interpret the results in the light of field observation and plot experiments on erosion processes.

A second issue concerns the size distribution of sediment entering the channel. Some accelerated erosion processes, such as sheetwash, selectively remove finer soil particles, which may become washload when they enter major stream channels and travel to estuarine and deltaic zones. Thus, a relatively large reduction in the supply of fine sediment may not reduce channel instability if the latter is associated with the deposition of sand and gravel from landsliding or the undermining of glacial outwash deposits. Conclusions about this aspect of the problem should be based on geomorphologic field work, involving the mapping of sediment textures along the main channel, sampling of the grain size distributions of major sediment sources, and lithological or other studies of the provenance of sediment.

Finally, it is necessary to clarify whether there is a relatively simple relationship between apparent sediment sources and the sediment deposited in major river channels. This problem is not always as obvious as it seems. For many years high sedimentation rates in the rivers of the Canterbury Plains, New Zealand, were blamed on sheep grazing and disturbance of the natural forest vegetation by "noxious," introduced wild animals. With the accumulation of measurements of sediment transport along rivers (Griffiths, 1979) and of sediment mobilization on the disturbed hillsides, it is becoming apparent that the high sedimentation rates and channel instability in the plains result from undermining of glacial terraces and similar coarse deposits, and is mainly a natural phenomenon not accelerated by land use. Conversely, Haggett (1961) and Trimble (1977) have demonstrated that long after reafforestation or soil conservation practices have diminished erosion on agricultural land, large amounts of sediment stored on footslopes, fans, and floodplains may continue to wash downstream. Geomorphologic mapping and measurement of sediment sources is again required to clarify the situation.

Control of Land Use on Floodplains

The general aim of land-use control on floodplains is to limit occupance of flood-prone areas by activities that suffer heavily from flooding. Control may be accomplished by forbidding certain activities or by specifying standards of construction for buildings. Further details are given by Dunne and Leopold (1978, Ch. 11).

Planning of land-use control depends on methods for delineating the area that would be inundated by floods of chosen frequencies. Various hydrologic techniques, such as field or aerial mapping during or immediately after a large flood and a combination of flood routing and topographic surveying (Wiitala et al., 1961), are useful for small areas of valuable land but are too slow and expensive for large floodplains (Dingman and Platt, 1977). However, the delineation of flood-prone areas can be facilitated and accelerated through use of geomorphologic techniques.

Wolman (1971) has summarized methods of mapping geomorphic and pedologic features that can be correlated to areas flooded with a specific frequency. For example, Jahns (1947) and Costa (1974a) mapped alluvial deposits associated with particular large floods, and Costa (1974b) outlined various types of evidence for recognizing prehistoric floods. Along many valley floors it is then possible to recognize topographic features such as terraces or the currently forming floodplain, which are flooded by events of known magnitude and frequency, and to map them quickly and cheaply. Witwer (1966) and Cain and Beatty (1968) have shown that characteristic soil types are also associated with such surfaces and can also be used to outline areas subject to flooding with a definable frequency.

These methods are particularly useful where the flood-prone zone is relatively stable over time. However, on many large floodplains, the flood hazard changes as the channel shifts across the valley floor. Through subtle alterations of valley floor hydraulics, channel migration may alter the total area flooded by a particular discharge. More commonly, migration, enlargement, or filling of channels causes only a redistribution in the depth and velocity of floodwaters and the susceptibility of land to scouring or undermining. Each of these changes affects the susceptibility of structures or crops to inundation, the viability of roads and bridges across the floodplain, and the possibility of evacuating people in the event of a flood warning. Thus, prediction of the location and behavior of the channel is an important requirement for flood damage abatement in some valleys. For example, Coleman's (1969) study of the Brahmaputra floodplain revealed channels that now receive only small discharges during major floods but that will probably become the site of future diversions of the main channel.

Figure 9 shows a reach of the Yakima River, Washington, along which there was a need to predict possible channel locations over the succeeding 25 yr as part of a land-use plan for the valley floor. The intention was to minimize flood damage by locating facilities where they would not be inundated after channel migration. Sequences of aerial photographs spanning 35 yr were superimposed, and maps of channel changes, similar to Figure 6, indicated that channel migration occurs by two mechanisms. The first

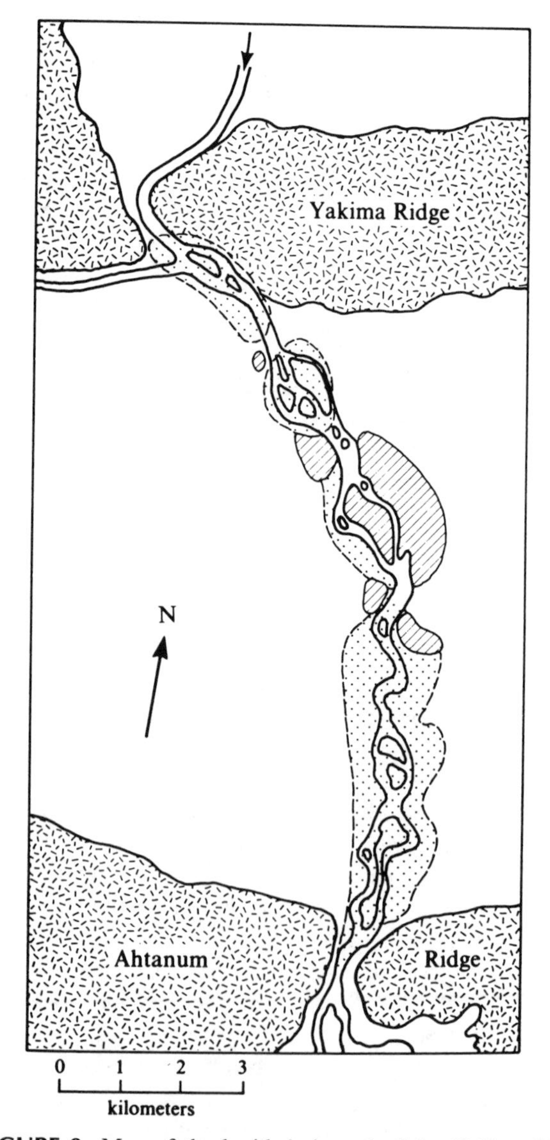

FIGURE 9. Map of the braided channel of the Yakima River near Yakima in eastern Washington, showing relatively stable zones of the floodplain (lined areas) not threatened by channel migration in the next 25 yr and threatened areas (dotted). The map was used as a basis for planning the siting of facilities.

consists of rather uniform lateral shifting of river bends at an average rate of 10 m/yr. The other mechanism of migration involves sudden avulsion by which a portion of the flood flow breaches the bankline and follows a new course, sometimes in a radically different direction from the original channel, and sometimes taking advantage of a former main channel or a smaller braided channel downstream. Succeeding floods may exploit the breach until the new channel is enlarged and conveys all or most of the flood. There is often a vigorous readjustment of channel morphology immediately above and below the diversion as the river alters its width and sinuosity to accommodate the new discharge. Severe bank erosion results. Such avulsions increase flood hazard because they can suddenly divert dangerous, fast, deep flows into zones that were formerly above water or were inundated only by shallow, slow drainage. Structures can be undermined and scoured away, and people and livestock can be trapped and drowned. The danger is enhanced because of the suddenness of breaching and the difficulty of its prediction.

Avulsion is a common, natural process along large, fast, braided rivers with weak bank materials, but along the Yakima the process has been accentuated by gravel mining in the valley floor. Deep pits are excavated, and after abandonment they are isolated from the river only by gravel dykes. Figure 10 indicates a location at which the river invaded a gravel pit, flowed out of its downstream end into an old channel, and re-entered the main channel via a second pit 1800 m downstream. The river was suddenly diverted 600 m to the west and is now undercutting the embankment of a major highway built with gravel from the pits. Subsequent readjustments of the new channel have isolated buildings situated on terraces above flood level. Similar channel diversions can result from alterations of the floodplain topography through irrigation and other engineering works unless a detailed geomorphologic analysis of the microtopography, hydraulics, and history of the floodplain is undertaken to define zones that should be avoided or strengthened. Although it is only possible to make short-term predictions of channel migration on a braided river of this kind, a longer term "worst-case" analysis is also possible. A map of zones threatened by channel migration over the succeeding 25 yr (Fig. 9) was drawn by examining the channel banks and valley floor on sequences of aerial photographs and in the field for likely avulsion sites. Some were obvious, and the path of the river after diversion could easily be foreseen; a few were more equivocal but were included in the interest of conservative design. After the probable diversions were mapped it was conservatively assumed that they would all occur in the next few years, and recent directions and rates of gradual shifting from the diverted courses were projected over the next 25 yr to define the hazard. The prediction was insensitive to the assumed timing of breaches because avulsion usually causes much larger shifts than persistent migration on this river.

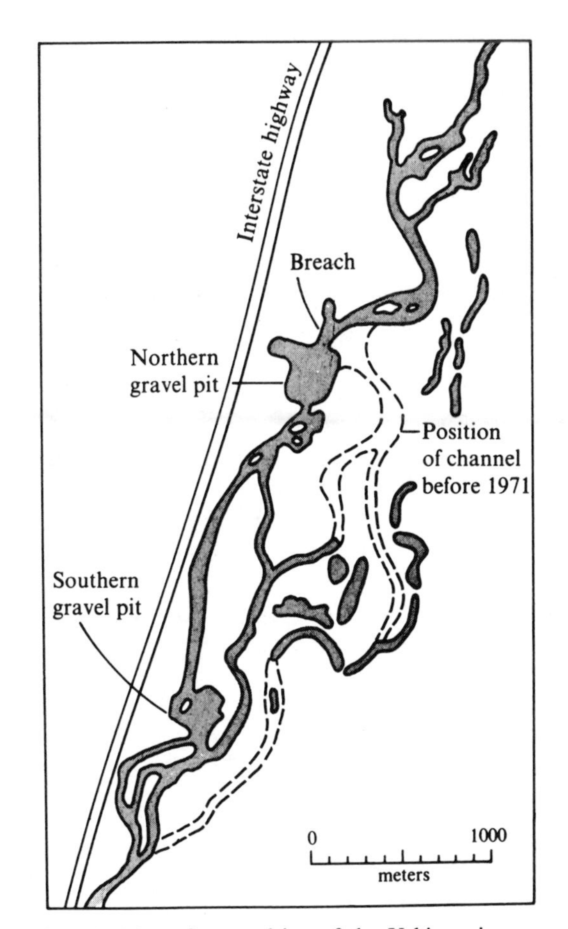

FIGURE 10. Map of an avulsion of the Yakima river caused when the channel breached a dyke surrounding the northern gravel pit, and floodwaters drained out of the southern end of the pit and into another pit downstream before re-entering the main river course. The shaded zone is now the main river channel, while the former course (dashed lines) is occupied only during infrequent, large floods and is being filled with fine sediment from local sources.

CONCLUSION

Fluvial geomorphology can play a more extensive role in flood control planning than is commonly realized. The most useful applications are those that recognize the relationships between the control of floodwater and the control of sediment and of channel form and activity. Geomorphologic studies are also valuable for extending information on flood magnitudes, sedimentation, and channel stability over long periods of time and between river basins. Since mathematical modeling of these phenomena is still relatively crude and unreliable, geomorphology is the only means of providing the required historical background for flood and sediment control.

REFERENCES

Bedinger, M. S., and Emmett, L. F. (1963). Mapping transmissibility of alluvium in the Lower Arkansas River, Arkansas. *Geol. Surv. Prof. Pap. (U.S.)* **475-C**, C-188–C190.

Binnie, G. M., and Mansell-Moulin, M. (1966). The estimated probable maximum storm and flood on the Jhelum River, a tributary of the Indus. *In* "River Flood Hydrology," pp. 189–210. Inst. Civ. Eng., London.

Birkeland, P. W. (1968). Mean velocities and boulder transport during Tahoe-age floods of the Truckee River, California-Nevada. *Geol. Soc. Am. Bull.* **79**, 137–142.

Blench, T. (1969). "Mobile Bed Fluviology." Univ. of Alberta Press, Edmonton.

Borland, W. M., and Miller, C. R. (1958). Distribution of sediment in large reservoirs. *J. Hydraul. Div., Am. Soc. Civ. Eng.* **84**, 1587-1–1587-18.

Burkham, D. E. (1976). Effects of changes in an alluvial channel on the timing, magnitude, and transformation of flood waves, southeastern Arizona. *Geol. Surv. Prof. Pap. (U.S.)* **655-K**, 1–25.

Cain, J. M., and Beatty, M. T. (1968). The use of soil maps in the delineation of floodplains. *Water Resour. Res.* **4** 173–182.

Campbell, K. L., Kumar, S., and Johnson, H. P. (1972). Stream straightening effects on flood-runoff characteristics. *Trans. ASAE* **15**, 94–98.

Church, M. A. (1972). Baffin Island sandurs: A study of Arctic fluvial processes. *Can. Geol. Surv. Bull.* **216**.

Clague, J. J., and Mathews, W. H. (1973). The magnitude of jokulhlaups. *J. Glaciol.* **12**, 501–504.

Clarke, G. K. C., and Mathews, W. H. (1981). Estimates of the magnitude of glacier outburst floods from Lake Donjek, Yukon Territory, Canada. *Can. J. Earth Sci.* **18**, 1452–1463.

Coleman, J. M. (1969). Brahmaputra River: Channel processes and sedimentation. *Sediment. Geol.* **3**, 129–139.

Collins, B. D., and Dunne, T. (1986). Erosion of tephra from the 1980 eruptions of Mount St. Helens. *Geol. Soc. Am. Bull.* (in press).

Collins, B. D., Dunne, T., and Lehre, A. K. (1983). Erosion of tephra-covered hillslopes north of Mount St. Helens, Washington, May 1980–May 1981. *Z. Geomorphol., Suppl.* **46**, 143–163.

Costa, J. E. (1974a). Response and recovery of a piedmont watershed from Tropical Storm Agnes, June 1972. *Water Resour. Res.* **10**, 106–112.

Costa, J. E. (1974b). Stratigraphic, morphologic, and pedologic evidence of large floods in humid environments. *Geology* **2**, 301–303.

Costa, J. E. (1985). Floods from dam failures. *Geol. Surv. Open-File Rep. (U.S.)* **85-560**, 1–54.

Dalrymple, T. (1960). Flood-frequency analysis. *Geol. Surv. Water-Supply Pap. (U.S.)* **1543-A**, 1–80.

Daniels, R. B. (1960). Entrenchment of the Willow Creek drainage ditch, Harrison County, Iowa. *Am. J. Sci.* **258**, 161–176.

Dietrich, W. E., Dunne, T., Humphrey, N. F., and Reid, L. M. (1981). Construction of sediment budgets for drainage basins. *U.S. For. Serv. Gen. Tech. Rep.* **PNW-141**, 5–23.

Dingman, S. L., and Platt, R. H. (1977). Floodplain zoning: Implications of hydrologic and legal uncertainty. *Water Resour. Res.* **13**, 519–523.

Dunne, T. (1979). Sediment yield and land use in tropical catchments. *J. Hydrol.* **42**, 281–300.

Dunne, T., and Fairchild, L. H. (1984a). Estimation of flood and sedimentation hazards around Mount St. Helens. I. *Shin-Sabo (J. Eros. Control Eng. Soc. Jpn.)* **36**, 12–22.

Dunne, T., and Fairchild, L. H. (1984b). Estimation of flood and sedimentation hazards around Mount St. Helens. II. *Shin-Sabo* (*J. Eros. Control Eng. Soc. Jpn.*) **37**, 13–22.

Dunne, T., and Fairchild, L. H. (1984b). Estimation of flood and sedimentation hazards around Mount St. Helens. II. *Shin-Sabo* (*J. Eros. Control Eng. Soc. Jpn.*) **37**, 13–22.

Dunne, T., and Leopold, L. B. (1978). "Water in Environmental Planning," p. 808. Freeman, San Francisco, California.

Dunne, T., and Ongweny, G. S. (1976). A new estimate of sediment yields in the Upper Tana catchment. *Kenya Geogr.* **2**, 109–126.

Fairchild, L. H. (1985). Lahars at Mount St. Helens, Washington. Ph.D. Thesis, University of Washington, Seattle.

Fiering, M. B., and Jackson, B. B. (1971). "Synthetic Streamflows," Water Resour. Monogr. No. 1. Am. Geophys. Union, Washington, D.C.

Griffiths, G. A. (1979). Recent sedimentation history of the Waimakariri River, New Zealand. *J. Hydrol. (N. Z.)* **18**, 6–28.

Haggett, P. (1961). Land use and sediment yield in an old plantation tract of the Serra do Mar, Brazil. *Geogr. J.* **127**, 50–62.

Helley, E. J., and LaMarche, V. C. (1973). Historic flood information for Northern California streams from geological and botanical evidence. *Geol. Surv. Prof. Pap. (U.S.)* **485-E**, 1–16.

Hickin, E. J., and Nanson, G. C. (1975). The character of channel migration on the Beatton River, northeast British Columbia, Canada. *Geol. Soc. Am. Bull.* **86**, 487–494.

Hoyt, W. G., and Langbein, W. B. (1955). "Floods." Princeton Univ. Press, Princeton, New Jersey.

Jahns, R. H. (1947). Geologic features of the Connecticut Valley, Massachusetts, as related to recent floods. *Geol. Surv. Water-Supply Pap. (U.S.)* **996**, 1–158.

Keller, E. A. (1976). Channelization: Environmental, geomorphic, and engineering aspects. *In* "Geomorphology and Engineering" (D.R. Coates, ed.), pp. 115–140. Allen & Unwin, London.

Kellerhals, R. (1967). Stable channels with gravel-paved beds. *J. Hydraul. Div., Am. Soc. Civ. Eng.* **93**, 63-84.

Kirkby, M. J. (1976). Tests of the random network model and its application to basin hydrology. *Earth Surf. Processes* **1**, 197–212.

Kochel, R. C., and Baker, V. R. (1982). Paleoflood hydrology. *Science* **215**, 353–361.

Komura, S., and Simons, D. B. (1967). River bed degradation below dams. *J. Hydraul. Div., Am. Soc. Civ. Eng.* **93**, 1–14.

Lane, E. W. (1955). The importance of fluvial morphology in hydraulic engineering. *Proc. Am. Soc. Civ. Eng.* **81**, 745-1–745-17.

Lehre, A. K., Collins, B. D., and Dunne, T. (1983). Post-eruption sediment budget for the North Fork Toutle River drainage, June 1980–June 1981. *Z. Geomorphol., Suppl.* **46**, 143–163.

Leopold, L. B., and Bull, W. B. (1979). Baselevel, aggradation, and grade. *Proc. Am. Philos. Soc.* **123**, 168–202.

Leopold, L. B., and Maddock, T. M. (1953). The hydraulic geometry of stream channels and some physiographic implications. *Geol. Surv. Prof. Pap. (U.S.)* **252**, 1–56.

Leopold, L. B., and Maddock, T. M. (1954). "The Flood-Control Controversy," p. 278. Ronald Press, New York.

Lustig, L. K. (1965). Sediment yield of the Castaic watershed, Western Los Angeles County, California—a quantitative geomorphic approach. *Geol. Surv. Prof. Pap. (U.S.)* **422-F**, 1–23.

Madej, M. A. (1982). Sediment transport and channel changes in an aggrading stream in the Puget Lowland, Washington. *U.S. For. Serv. Gen. Tech. Rep.* **PNW-141**, 97–108.

Miall, A. D., ed. (1978). "Fluvial Sedimentology." Can. Soc. Pet. Geol., Calgary.

Moore, G. K., and North, G. W. (1974). Flood inundation in the southeastern United States from aircraft and satellite imagery. *Water Resour. Bull.* **10**, 1082–1096.

Mullineaux, D. R., and Crandell, D. R. (1962). Recent lahars from Mount St. Helens, Washington. *Geol. Soc. Am. Bull.* **73**, 855–870.

Nelson, L. M. (1971). Sediment transport by streams in the Snohomish River basin, Washington, October 1967–July 1969. *Geol. Surv. Open-File Rep. (U.S.)* pp. 1–44.

Ongweny, G. S. (1978). Erosion and sediment transport in the Upper Tana catchment. Ph.D. Thesis, University of Nairobi, Nairobi.

Orsborn, J. F. (1976). Drainage basin characteristics applied to hydraulic design and water-resources management. *In* "Geomorphology and Engineering" (D.R. Coates, ed.), pp. 141–171. Allen & Unwin, London.

Osterkamp, W. R., and Hedman, E. G. (1982). Perennial-streamflow characteristics related to channel geometry and sediment in Missouri River basin. *Geol. Surv. Prof. Pap. (U.S.)* **1242**.

Popov, I. V., and Gavrin, Y. S. (1970). Use of aerial photography in evaluating the flooding and emptying of river flood plains and the development of flood-plain currents. *Sov. Hydrol.* **5**, 413–425.

Reich, B. M. (1971). Land surface form in flood hydrology. *In* (D.R. Coates, ed.), "Environmental Geomorphology" Publ. Geomorphol., pp. 49-70. State University of New York, Binghamton.

Reid, L. M., Dunne, T., and Cederholm, C. J. (1981). Application of sediment budget studies to the evaluation of logging road impact. *J. Hydrol. (N. Z.)* **20**, p. 49–60.

Reineck, H. E., and Singh, L. B. (1980). "Depositional Sedimentary Environments." Springer-Verlag, Berlin and New York.

Ruhe, R. V. (1971). Stream regimen and man's manipulation. *In* "Environmental Geomorphology" (D.R. Coates, ed.), Publ. Geomorphol., pp. 9–23. State University of New York, Binghamton.

Schumm, S. A. (1969). River metamorphosis. *J. Hydraul. Div., Am. Soc. Civ. Eng.* **95**, 255–273.

Schumm, S. A. (1977). "The Fluvial System." Wiley, New York.

Schumm, S. A., Harvey, M. D., and Watson, D. C. (1984). "Incised Channels." Water Resources Publications, Littleton, Colorado.

Sigafoos, R. S. (1964). Botanical evidence of floods and flood-plain deposition. *Geol. Surv. Prof. Pap. (U.S.)* **485-A**, 1–35.

Snyder, F. F. (1938). Synthetic unit hydrographs. *EOS, Trans. Am. Geophys. Union* **19**, 447–454.

Trimble, S. W. (1977). The fallacy of stream equilibrium in contemporary denudation studies. *Am. J. Sci.* **277**, 876–887.

Trimble, S. W. (1983). A sediment budget for Coon Creek basin in the driftless area, Wisconsin, 1853-1977. *Am. J. Sci.* **283**, 454–474.

U.S. Army Corps of Engineers (1977). "HEC-6, Scour and Deposition in Rivers and Reservoirs." Hydrologic Engineering Center, Davis, California.

Valdes, J. B., Fiallo, Y., and Rodriguez-Iturbe, I. (1979). A rainfall-runoff analysis of the geomorphologic IUH. *Water Resour. Res.* **15**, 1421–1434.

Vanoni, V. A., ed. (1975). "Sedimentation Engineering." Am. Soc. Civ. Eng., New York.

Velikanova, Z. M., and Yarnykh, N. A. (1970). Field investigations of the hydraulics of a flood plain during a high flood. *Sov. Hydrol.* **5**, 426–440.

Wiard, L. (1962). Floods in New Mexico: their magnitude and frequency. *Geol. Surv. Circ. (U.S.)* **464**, 1–13.

Wigmosta, M. S. (1983). Rheology and flow dynamics of the Toutle River debris flows from Mount St. Helens. M.S. Thesis, University of Washington, Seattle.

Wiitala, S. W., Jetter, K. R., and Somerville, A. J. (1961). Hydraulic and hydrologic aspects of floodplain planning. *Geol. Surv. Water-Supply Pap. (U.S.)* **1526**, 1–69.

Williams, G. P., and Wolman, M. G. (1984). Downstream effects of dams on alluvial rivers. *Geol. Surv. Prof. Pap. (U.S.)* **1286**, 1–63.

Winkley, B. R., Schumm, S. A., Mahmood, K., Lamb, M. S., and Linder, W. M. (1984). New developments in the protection of irrigation, drainage, and flood-control structures on rivers. *Proc. Symp. Int. Comm. Irrig. Drain.*, pp. 69-111.

Witwer, D. B. (1966). Soils and their role in planning a suburban county. *In* "Soil Surveys and Land Use Planning" (L. Bartelli et al., eds.), pp. 15-30. Soil Sci. Soc. Am., Madison, Wisconsin.

Wolman, M. G. (1971). Evaluating alternative techniques for floodplain mapping. *Water Resour. Res.* **7**, 1383-1392.

Yanosky, T. M. (1983). Evidence of floods on the Potomac River from anatomical abnormalities in the wood of flood-plain trees. *Geol. Surv. Prof. Pap. (U.S.)* **1296**, 1–42.

26

FLOODS FROM DAM FAILURES

JOHN E. COSTA
U.S. Geological Survey, Cascades Volcano Observatory, Vancouver, Washington

INTRODUCTION

The purpose of this chapter is to summarize, in a general way, information about the hydrology and geomorphology of floods produced from the failure of natural and constructed dams. Floods resulting from dam failures usually are much larger than those originating from snowmelt or rainfall. The sudden release of large quantities of stored water from a breached dam can cause great destruction to property and potential loss of life downstream. The volumes and magnitudes of some floods from dam failures may be unprecedented in the cultural and natural history of the affected valley. Tremendous quantities of sediment and debris eroded, transported, and deposited downstream can cause significant modifications to channels and valley morphology. Floods from dam failures have produced shear stresses and unit stream powers comparable to the largest rainfall–runoff or snowmelt floods ever measured in the United States (J. E. Costa, U.S. Geological Survey, unpublished data, 1985).

Dams generally can be categorized as constructed dams and natural dams. Within each category there are numerous types.

CONSTRUCTED DAMS AND CAUSES OF FAILURES

The construction of dams to create reservoirs for water supply was one of the earliest engineering undertakings. Remains of one of the oldest constructed dams in the world, believed to date from ca. 2900 B.C., still survive in the Wadi El-Garawi about 32 km south of Cairo, Egypt. This rubble-masonry structure was 107 m long, 11 m high, and had a reservoir capacity of 570,000 m^3 (Jansen, 1980). The Romans built many stone dams throughout their empire, the earliest near Toledo, Spain, sometime after 193 B.C. In the United States the first dams were constructed to impound water to run gristmills and sawmills. One of the earliest dams was erected in 1623 to run the first sawmill in America on the Piscataqua River at South Windham, Maine. On the western coast of the United States, Jesuits constructed one of the earliest dams on the San Diego River in 1770 (Jansen, 1980).

In the United States the National Dam Inspection Program (Public Law 92-367) compiled data on about 68,000 dams that were either more than 7.6 m high or impounded at least 61,650 m^3 at maximum water storage elevation. Earth- and rockfill dams by far constitute the largest percentage of dam types (93%). Their characteristics consist of an impermeable barrier made from compaction of fine-grained materials, combined with a mass of earth and rock material to impound water. An earthfill dam is an embankment dam in which more than 50% of the total volume is formed of compacted fine-grained material obtained from a borrow area. A rockfill dam is an embankment dam in which more than 50% of the total volume consists of compacted or dumped pervious natural or crushed rock. Foundation requirements are less stringent than for other types of dams.

A gravity dam, constructed of concrete or masonry or both, relies on its weight for stability. Gravity dams represented 5% of the investigated dams. An arch dam (0.4% of the sample) is a concrete or masonry dam that is curved in plan so as to transmit the major proportion of the water stress to the bedrock abutments. Buttress dams (0.3% of the sample) consist of a watertight upstream face supported at intervals on the downstream side by a series of buttresses.

Of the 8639 dams inspected by the U.S. Army Corps of Engineers by September 1981, one-third were declared unsafe. The primary defect of 82% of the dams was inadequate spillway capacity. The two largest groups of owners of unsafe dams are one or more individuals (27.4%) and state, city, or county governments (26.4%) (Morrison, 1982). By May 1982 no corrective measures had been initiated at 64% of the unsafe dams, principally because of the owners' lack of resources (Committee on the Safety of Existing Dams, 1983).

Ever since the earliest dams were built, there have been dam failures. The history and description of the failure of some famous constructed dams throughout the world can be found in Jansen (1980). There have been approximately 2000 failures of constructed dams throughout the world since the twelfth century (Jansen, 1980), and many thousands more failures of natural dams. During the last 100 yr, there have been about 200 significant failures of constructed dams, in which more than 11,100 people died; 6800 lives were lost in three failures alone: Vaiont, Italy, 1963, (2600); South Fork (Johnstown) Pennsylvania, 1889 (2200); and Machhu II, India, 1974 (2000+) (Jansen, 1980). Financial losses associated with dam failures are probably inestimable.

Several investigations have attempted to summarize the causes of major dam failures throughout the world. In 1961 the Spanish publication *Revista de Obras Públicas* presented the results of an investigation of 1620 major dams (as reported in Gruner, 1963). In the 145 yr between 1799 and 1944, 308 dams suffered serious accidents or failures. Of these, 57% were fill or embankment dams, 23% were gravity dams, 3% were arch dams, and the remaining 17% were other types.

Reported causes of failure were foundation failure (51%), including uneven settlement and earthquakes; inadequate spillways (23%); piping and seepage (7%), including high pore pressures and embankment slips; and various other causes (19%), including improper construction, acts of war, defective materials, and incorrect operation.

In a study of more than 300 dam failures throughout the world, Biswas and Chatterjee (1971) reported that about 35% were a result of floods exceeding the spillway capacity, and 25% resulted from foundation problems such as seepage, piping, excess pore pressure, inadequate cut-off, fault movement, settlement, or rock slides. The remaining 40% of the failures were caused by various problems, including improper design or construction, inferior materials, misuse, wave action, and acts of war.

The International Commission on Large Dams (1973) compiled a survey of dams more than 15 m in height that failed between 1900 and 1973. The three main causes of failure were overtopping (inadequate spillway), foundation defects, and piping (Fig. 1). The major cause of failure of concrete dams was foundation failure (53%), and the major cause of failure of fill or embankment dams was piping and seepage (38%). Piping and seepage failures did not occur with any concrete dam. For all dams built, failures by overtopping due to an inadequate spillway (34%), foundation defects (30%), and piping and seepage (28%) have about the same rate of occurrence (Fig. 1). These data clearly indicate that the greatest risks of a dam failing

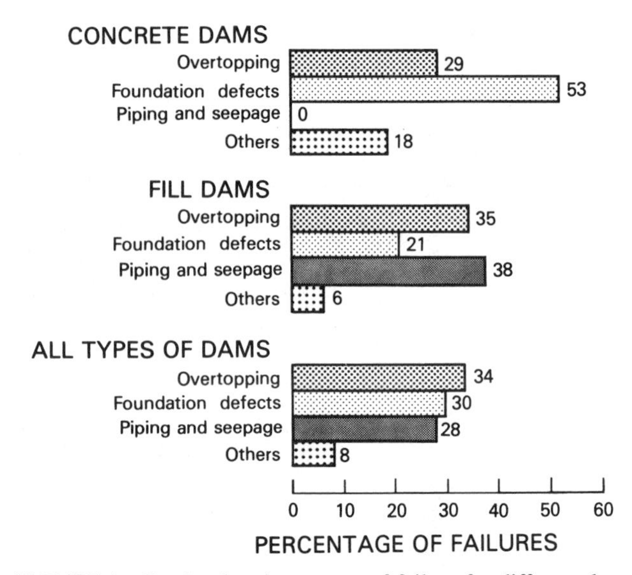

FIGURE 1. Graphs showing causes of failure for different dam types more than 15 m high (International Commission on Large Dams, 1973).

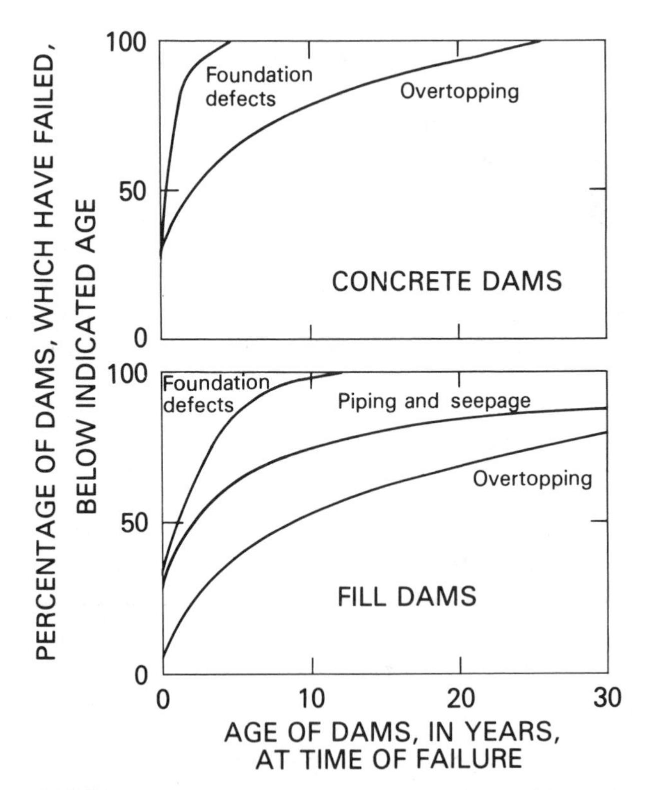

FIGURE 2. Graphs showing age of dams at time of failure, for different dam types more than 15 m high (International Commission on Large Dams, 1973).

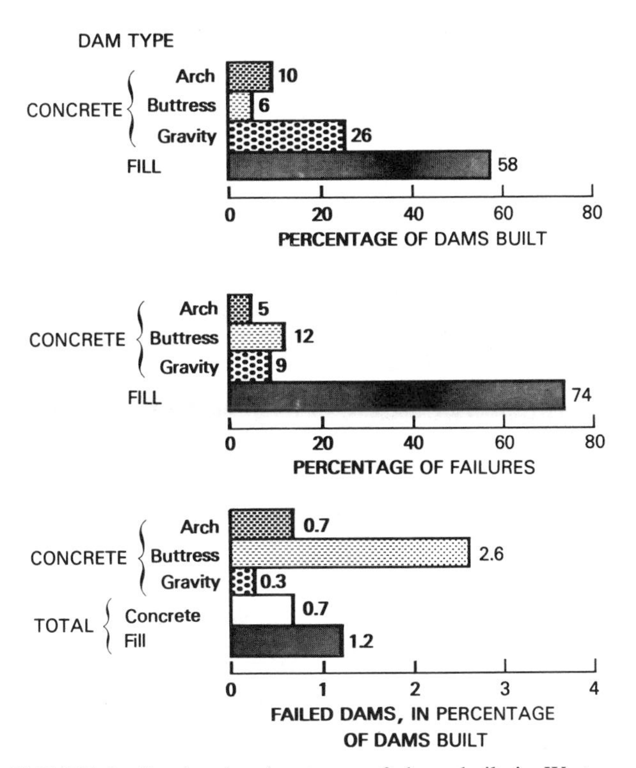

FIGURE 3. Graphs showing types of dams built in Western Europe and the United States, and their failure percentage, between 1900 and 1969 (International Commission on Large Dams, 1973).

originate from ignorance of the magnitude and frequency of extreme floods and uncertainties of the geologic setting.

The incidence of the causes of dam failures as a function of the dam's age at the time of failure is shown in Figure 2. Foundation failures occur early in a dam's history, whereas other causes take relatively longer to develop. A very large percentage of all dam failures occurs during initial filling because this is the time when design or construction flaws or latent site defects will appear.

The percentage of different kinds of dams built in western Europe and the United States between 1900 and 1969 is shown in Figure 3. Fill or embankment dams (50%) were the most numerous types built; gravity dams were second (26%). The percentage of failures that occurred for the four different types of dams is summarized in Figure 3. Nearly 75% of the dams that failed were fill dams. However, data in Figure 3 also show the failed dams as a percentage of the dams built and indicates that gravity dams were the safest, followed by arch and fill dams. Buttress dams have the poorest safety record; however, they were the least used.

The risk of dam failures in the United States is approximately 3×10^{-4} to 7×10^{-4} per dam-year (Baecher et al., 1980). Worldwide dam failure rates are estimated to be 2×10^{-4} to 4×10^{-4} per dam-year (Baecher et al., 1980). The seemingly higher dam failure rate in the United States probably is an artifact of better and more comprehensive data on failures than exists in most of the rest of the world. When major dam failures are considered, the United States failure rate is 0.8×10^{-4} per dam-year, and the world failure rate is 2×10^{-4} per dam-year (Mark and Stuart-Alexander, 1977). There are many uncertainties in these kinds of estimates, and the only consistent conclusion available is that an order-of-magnitude estimate of risk of dam failure anywhere in the world is approximately 10^{-4} per dam-year. This failure rate has been shown to be a significant factor in the cost–benefit analysis of large dams where there is potential for large loss of life (Mark and Stuart-Alexander, 1977; Rose, 1978).

In the United States between 1963 and 1983, the average annual loss of life from dam failures was about 14 deaths (Table 1). This figure compares with 200 deaths per year from flooding, 25 deaths from mass movements, 12 deaths from earthquakes, and 6 deaths from tsunamis (Costa and Baker, 1981, p. 462). Interestingly, the failure of constructed dams less than 15 m in height caused about 90% of all dam failure fatalities during this period. On the basis of a sample of 20 dams that failed in the United States during this century, Wayne Graham (U.S. Bureau of Reclamation, written communication, 1984) found that the average number of fatalities per dam failure was 19 times greater when there was inadequate or no warning than when there was adequate warning. The effects of successful warning for some recent dam failures and the Big Thompson River, Colorado, flash flood of 1976 are given in Table 2. The data show that in the absence of early detection and warning, a significant percentage of the people exposed to the flood hazard can lose their lives. For example, the failure of the relatively low Laurel Run Dam in Pennsylvania (12.8 m high) at 4:00 a.m. claimed the lives of about 1 out of every 4 people potentially exposed to floodwaters because there was no warning. The failure of the 84-m high Teton Dam, Idaho, at 11:57 a.m. was preceded by a warning, and only about 1 out of every 3000 people potentially exposed to the floodwaters died.

Obviously, there is great benefit to be derived from early warning of the failure of a dam. Some relatively inexpensive actions that communities downstream from hazardous dams could undertake include development of emergency action plans, establishment of an early-warning and notification system, preparation of evacuation plans, stockpiling repair materials, locating local repair forces, training operation personnel, and increasing inspection frequency. However, risk can never be completely eliminated. In 1969, a 76-yr-old earth dam near Wheatland, Wyoming, failed without warning, flooding more than 40 km^2 of cropland. The failure occurred in dry weather less than 10 hr after the dam had been inspected and found safe (Anonymous, 1969).

TABLE 1 Loss of Life and Property Damage from Notable U.S. Dam Failures, 1963–1983

Name and Location of Dam	Date of Failure	Number of Lives Lost	Damages
Mohegan Park, CT	Mar. 1963	6	$3 million (1963 dollars)
Little Deer Creek, UT	June 1963	1	Many summer cabins damaged.
Baldwin Hills, CA	Dec. 1963	5	41 houses destroyed, 986 houses damaged, 100 apartment buildings damaged.
Swift, MT	June 1964	19	Unknown.
Lower Two Medicine, MT	June 1964	9	Unknown.
Lee Lake, MA	Mar. 1968	2	6 houses destroyed, 20 houses damaged, 1 manufacturing plant damaged or destroyed.
Buffalo Creek, WV	Feb. 1972	125	546 houses destroyed, 538 houses damaged.
Lake "O" Hills, AR	Apr. 1972	1	Unknown.
Canyon Lake, SD	June 1972	33[a]	Unable to separate damage due to failure from damage caused by natural flooding.
Bear Wallow, NC	Feb. 1976	4	1 house destroyed.
Teton, ID	June 1976	11	771 houses destroyed, 3002 houses damaged, 246 businesses damaged or destroyed.
Laurel Run, PA	July 1977	39	6 houses destroyed, 19 houses damaged.
Sandy Run and 5 others, PA	July 1977	5	Unknown
Kelly Barnes, GA	Nov. 1979	39	9 houses, 18 house trailers, and 2 college buildings destroyed; 6 houses, 5 college buildings damaged.
Swimming Pool, NY	1979	4	Unknown.
About 20 dams in CT	June 1982	0	Unknown.
Lawn Lake, CO	July 1982	3	18 bridges destroyed, 117 businesses damaged, 108 houses damaged, campgrounds, fisheries, powerplant damaged.
DMAD, UT	June 1983	1	Unknown.

SOURCE: Graham, 1983.

[a]Lives that would not have been lost if dam had not failed.

TABLE 2 Comparison of Warning Success for Selected Dam Failures and Flash Floods

Event	Early Detection and Warning	Potential Loss of Life	Actual Loss of Life	Fatality Rate (%)
Big Thompson, CO (flash flood)	No	2,500	139	5.6
Laurel Run Dam, PA	No	150	39	25.0
Kelly Barnes Dam, GA	No	200	39	20.0
Buffalo Creek, WV	Some	4,000	125	3.1
Teton Dam, ID	Yes	35,000	11	<0.1
Southern Connecticut, June 1982 (20 dams failed)	Yes	Unknown	0	0
Lawn Lake, CO	Yes	4,000	3	<0.1
DMAD, UT	Yes	500	1	0.2

SOURCE: Graham, 1983.

DAM BREAK MODELS

Many types of dam break models exist, ranging from simple computations based on historical dam failure data that can be performed manually to complex models that require computer analyses. The purpose of each model is to predict the characteristics (such as peak discharge or stage, volume, and flood wave travel time) of a dam failure flood.

The simplest estimation of the peak discharge and attenuation downstream from a dam failure involves empirical data from historic dam failures. Much of the available data on peak discharges from failures of constructed dams is summarized in Table 3. The simplest and earliest relations to be developed involve characteristics of the dam and reservoir. Kirkpatrick (1977) plotted data on height of dam (arithmetic) versus peak discharge (log) for 21 actual and hypothetical dam failures and drew what appears to be an average curve through the data points. His equation is:

$$Q_{max} = 2.297(H + 1)^{2.5} \quad (1)$$

where Q_{max} is peak discharge in cubic feet per second, and H is height of dam in feet. This relation later was revised by the U.S. Soil Conservation Service (1981) using data from 13 actual dam failures, and plotted as a power function that appears to be an enveloping curve, although three data points are above the curve:

$$Q_{max} = 65H^{1.85} \quad (2)$$

Dam height versus peak discharge for 31 failures of constructed dams between 1.8 and 84 m high is plotted in Figure 4 and listed in Table 3. An envelope curve for flood peaks from all the constructed dams has the equation

$$Q_{max} = 48H^{1.63} \quad (3)$$

where Q_{max} is peak discharge in cubic meters per second, and H is dam height in meters. This envelope curve is not plotted in Figure 4.

Accuracy of many peak discharge estimates from dam failures is questionable, and errors of one order of magnitude may exist. If a rapid, conservative assessment of the potential peak discharge from a failed dam is desired, then the envelope curve [Equation (3)] could be used. If the purpose is to compare flood peaks from failures of different types of dams or to reconstruct past flood peaks from old, failed dams for paleohydrological or sedimentological investigations, then a regression equation is more appropriate. A regression equation using dam height as the independent variable has been developed from the data in Table 3, and plotted in Figure 4:

$$Q_{max} = 10.5H^{1.87} \qquad r^2 = 0.80 \quad (4)$$

and the standard error (SE) is 82%.

A plot of reservoir volume at time of failure versus peak discharge is shown in Figure 5. Sensitivity studies have indicated that reservoir volume as well as dam height are critical factors in the magnitude of dam failure hydrographs (Hagen, 1982; Petrascheck and Sydler, 1984). An envelope curve encompassing all the data points except two (Malpasset Dam, France, point 15, and Cascade Lake Dam, Colorado, point 30) has the following form:

$$Q_{max} = 2950V^{0.57} \quad (5)$$

where V is reservoir volume at time of failure, in $m^3 \times 10^6$. This envelope curve is not shown in Figure 5.

The Malpasset Dam was a high concrete arch structure and Cascade Lake Dam was a concrete gravity dam. These kinds of dams are more susceptible to rapid failure than most of the other types. The sequence of the toppling failure of Cascade Lake Dam, Colorado, is shown in Figure 6. The inflow flood peak leading to the failure was 45% of the outflow peak following the failure.

Including the Malpasset and Cascade Lake Dam failures, the envelope curve is

$$Q_{max} = 4000V^{0.57} \quad (6)$$

This envelope curve is not shown in Figure 5.

Hagen (1982) and the Committee on the Safety of Existing Dams (1983) developed a criterion for estimating peak discharge based upon the product of dam height (H) and reservoir volume (V). This product ($H \times V$) is the dam factor and is a crude index of the energy expenditure at the dam when it fails. Hagen's equation, based on seven data points excluding the Malpasset failure, is

$$Q_{max} = 370(HV)^{0.5} \quad (7)$$

and including the Malpasset failure, is

$$Q_{max} = 530(HV)^{0.5} \quad (8)$$

where V is reservoir volume in acre-feet, and H is dam height in feet.

Using 29 data points covering a much broader range of reservoir heights and volumes from Table 3, the envelope curve (not plotted) is

$$Q_{max} = 1150(HV)^{0.44} \quad (9)$$

and the regression equation is

$$Q_{max} = 325(HV)^{0.42} \qquad r^2 = 0.75 \qquad SE = 95\% \quad (10)$$

where V is reservoir volume in $m^3 \times 10^6$, and H is dam height in meters. The regression curve is plotted in Figure 7.

MacDonald and Langridge-Monopolis (1984) produced a similar relationship to Hagen's (1982), except they relate the product of outflow volume of water during the failure and the difference in elevation of the peak reservoir water surface and breach base with peak outflow discharge. The relation is very similar to that developed by Hagen (1982).

Simple theoretical estimates of dam break hydrographs originated with Ritter (1892) who used the approximate Saint-Venant equation but assumed rectangular, horizontal channels with no frictional or turbulent resistance to the

TABLE 3 Constructed Dams That Have Failed

Index Number and Name of Dam	Date of Failure	Type of Dam	Height H (m)
1. Bradfield Dam, England	March 11, 1864	Earth	29.0
2. Mill River, MA	May 16, 1874	Earth and masonry	13.1
3. South Fork (Johnstown) PA	May 31, 1889	Earth and rockfill	22.0
4. Austin, TX	April 7, 1900	Masonry gravity	20.7
5. Hatchtown, UT	May 25, 1914	Rockfill	15.9
6. Davis Reservoir, CA	1914	Earth	11.9
7. Goose Creek, SC	1916	Earth	6.1
8. Schaeffer, CO	June 5, 1921	Earth	27.4
9. Apishapa, CO	Aug. 22, 1923	Earth	32.0
10. French Landing, MI	April 13, 1925	Earth with stone on face	8.5
11. St. Francis, CA	March 12, 1928	Concrete gravity	56.4
12. Castlewood, CO	Aug. 3, 1933	Rockfill	21.3
13. Fred Burr, MT	1948	Earth	10.4
14. Frenchman Creek, MT	April 15, 1952	Earth	12.5
15. Malpasset, France	Dec. 2, 1959	Concrete arch	61.0
16. Oros, Brazil	March 26, 1960	Earth	35.4
17. Little Deer Creek, UT	June 16, 1963	Earth	21.3
18. Baldwin Hills, CA	Dec. 23, 1963	Earth	18.0
19. Swift, MT	June 8, 1964	Rockfill	47.9
20. Lower Two Medicine, MT	June 8, 1964	Earth	11.0
21. Hell Hole, CA	Dec. 23, 1964	Rockfill	30.5
22. Buffalo Creek, WV	Feb. 26, 1972	Coal waste	14.0
23. Teton, ID	June 5, 1976	Earth	83.8
24. Laurel Run, PA	July 20, 1977	Earth	12.8
25. Otto Run, PA	July 20, 1977	Earth	5.8
26. South Fork tributary, PA	July 20, 1977	Earth	1.8
27. North Branch tributary, PA	July 20, 1977	Earth	5.5
28. Kelly Barnes, GA	Nov. 6, 1977	Earth over original rock crib	10.4
29. Lawn Lake, CO	July 15, 1982	Earth	7.9
30. Cascade Lake, CO	July 15, 1982	Concrete gravity	5.2
31. DMAD, UT	June 6, 1983	Earth	8.8

TABLE 3 (Continued)

Volume at Time of Failure V ($m^3 \times 10^6$)	Dam Factor ($H \times V$)	Peak Outflow (m^3/s)	Peak Determination	Reference
3.2	93	1,130	?	Jansen, 1980
2.5	32.8	1,645	?	Jansen, 1980; Barrows, 1948
14.2	312	5,700–8,500	?	Jansen, 1980
?	—	6,684	Slope-area measurement	Freeman and Alsop, 1941
16.8	267	3,100–7,000	Drawdown rate, 1 hr avg.	Office of the State Engineer, 1913-1914
58	690	510	?	MacDonald and Langridge-Monopolis, 1984
10.6	65	566	?	MacDonald and Langridge-Monopolis, 1984
4.4	121	4,300–4,900	Slope-area measurement 13 km downstream; twice avg. outflow during ½ hr required to drain reservoir	Follansbee and Jones, 1922
22.8	730	6,853	Drawdown rate, 15 min avg.	Anonymous, 1923
?	—	929	Drawdown rate, 1-hr period	Anonymous, 1925
47.1	2,656	12,744–19,824	Slope-conveyence	Outland, 1977
6.2	132	3,568	Drawdown rate 15-min period	Field, 1933
0.75	7.8	654	Slope-area measurement	W. Graham, personal communication, 1981
21.0	263	1,416	?	MacDonald and Langridge-Monopolis, 1984
22.0	1,342	28,320	?	Jansen, 1980
703	24,886	9,629	?	Jansen, 1980
1.2	25.6	1,331	?	Rostvedt et al., 1968
0.91	16.4	991	Slope-area ? measurement	Jessup, 1964
42.3	2,026	24,950	Slope-area measurement 27 km downstream	Boner and Stermitz, 1967
25.8	284	1,798	Slope-area measurement	Boner and Stermitz, 1967
30.6	933	8,780	Gauging station 75 km downstream	Scott and Gravlee, 1968 (Table 1)
0.5	7.0	1,416	Slope-area measurement	Davies et al., 1972
311	26,020	65,136	Slope-area measurement	Ray and Kjelström, 1978
0.38	4.9	1,048	Slope-area measurement	Brua, 1978
0.0074	0.043	60	Slope-area measurement	Brua, 1978
0.0038	0.0066	122	Slope-area measurement	Brua, 1978
0.022	0.121	29.5	Slope-area measurement	Brua, 1978
0.78	8.1	680	Slope-area measurement	Sanders and Sauer, 1979
0.84	6.6	510	Dam-break model	Jarrett and Costa, 1986
0.031	0.16	453	Dam-break model	Jarrett and Costa, 1986
19.7	173	793	Drawdown rate	Roger Walker, Utah Water Commissioner, personal communication, 1985

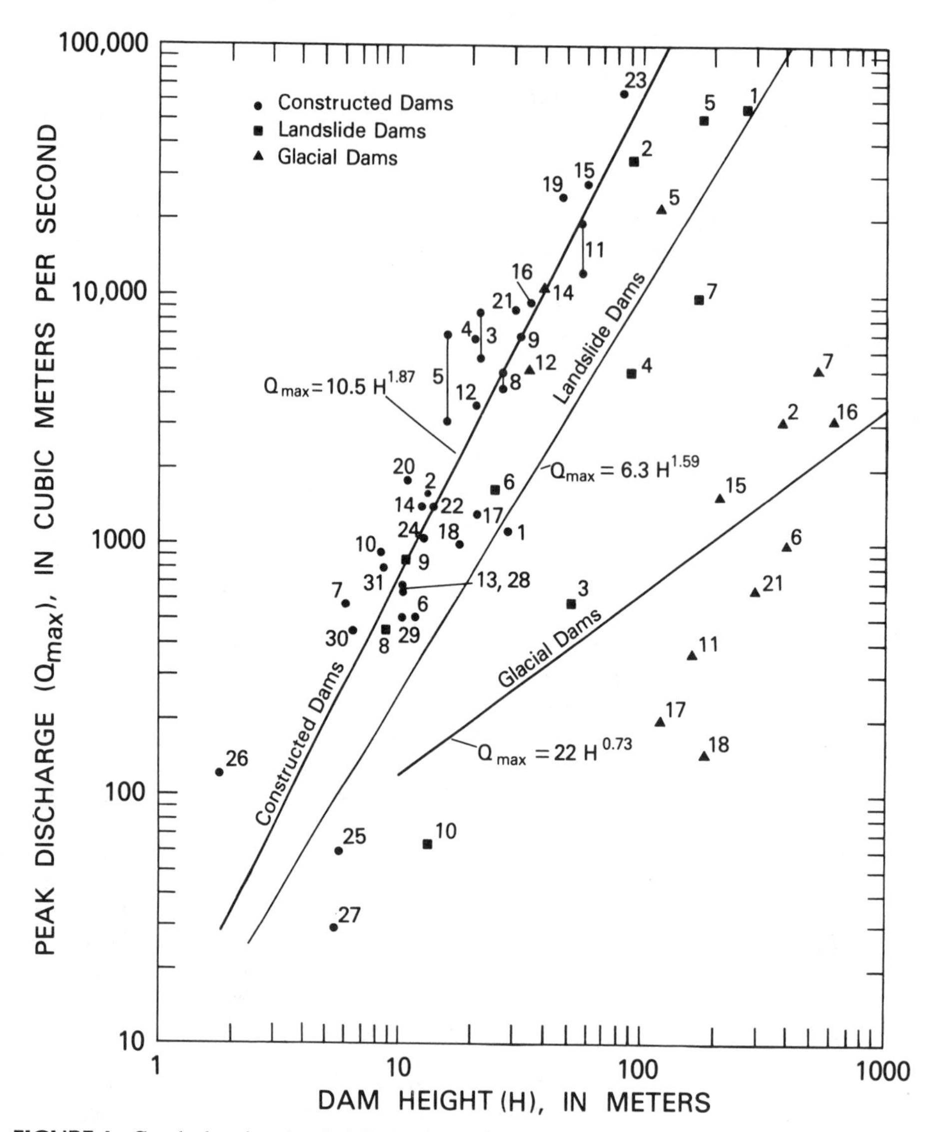

FIGURE 4. Graph showing dam height (H, in meters) versus peak discharge (Q_{max}, in cubic meters per second) for constructed, glacial, and landslide dams with numbers keyed to Tables 3, 5, and 6.

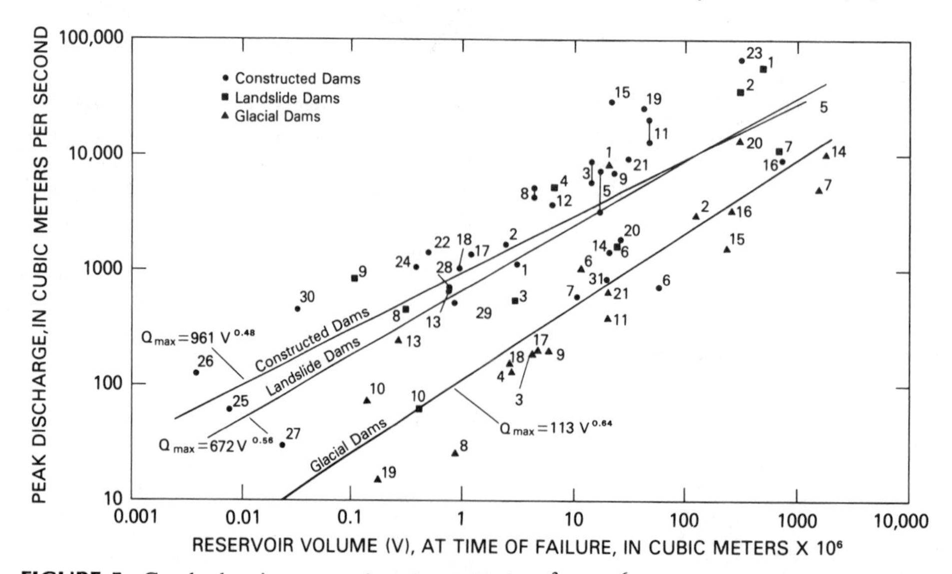

FIGURE 5. Graph showing reservoir volume (V, in $m^3 \times 10^6$) versus peak discharge (Q_{max} in cubic meters per second) for constructed, glacial, and landslide dams with numbers keyed to Tables 3, 5, and 6.

FIGURE 6. Photographs showing sequence of Cascade Lake Dam failure, Colorado, July 15, 1982 (from Jarrett and Costa, 1986).

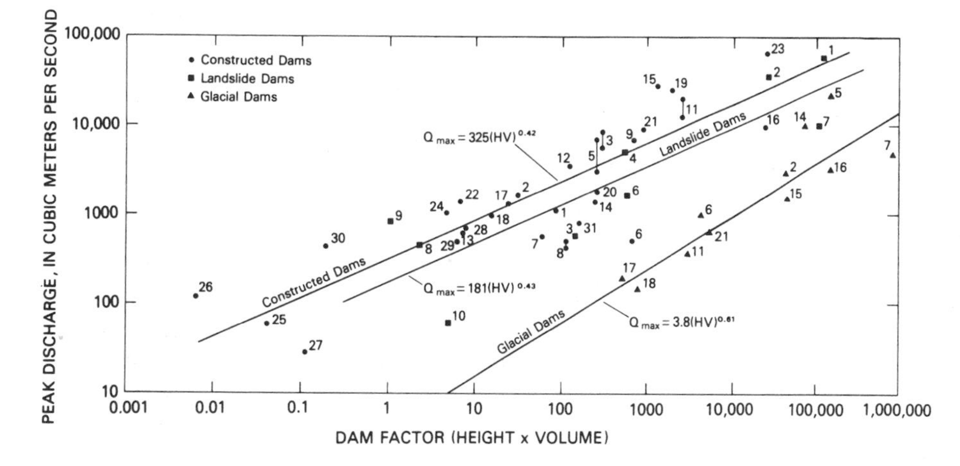

FIGURE 7. Graph showing dam factor ($H \times V$) versus peak discharge (Q_{max}) for failures of different kinds of dams, with numbers keyed to Tables 3, 5 and 6.

unsteady flow. Maximum flood discharge following a dam break is approximated by

$$q = {}^{8}/_{27}\, g^{1/2}\, Y^{3/2} \tag{11}$$

where q = breach unit width discharge
g = gravitational acceleration
Y = reservoir depth upstream of the dam before failure

Because most dam breaches are trapezoidal in shape (MacDonald and Langridge-Monopolis, 1984), the equation has been expanded by Price and others (1977) to

$$Q_{\max} = {}^{8}/_{27}\, g^{1/2}\, Y^{3/2}(0.4b + 0.6T) \tag{12}$$

where $Q_{\max}$ = maximum discharge
b = width of breach base
T = top width of breach at initial water level.

Because this equation ignores frictional and turbulent resistance, computed peak discharges tend to be larger than peak discharges determined by slope area or drawdown rate methods. Equations (11) and (12) can be rewritten as a simple energy conservation equation for instantaneous disappearance of the dam, where slope and resistance do not initially matter [$v = {}^{8}/_{27}\,(gy)^{1/2}$], where v is velocity (Michael Church, written communication, 1985).

An important control on the downstream flood hydrograph from a failed dam is breach characteristics, including size, shape, and time of formation of the breach. MacDonald and Langridge-Monopolis (1984) present a large amount of data on the size and shape of the breach from failed constructed dams. Breach shapes tend to be trapezoidal, with top width four times the dam height and bottom width two times the dam height (McMahon, 1981). Of the three breach characteristics, shape has the least influence on the flood hydrograph.

Time of breach formation can be estimated from $t \leq (1008\ \mathrm{SA})/\mathrm{W(H)}^{1/2}$ where t is time for breach formation, in seconds, SA is surface area of reservoir, in acres, W is average breach width, in feet, and H is depth of water at time of failure, in feet (Fread, reported in McMahon, 1981). Time of breach formation becomes increasingly insignificant as reservoir volume becomes very large. Size of breach, especially average breach width (W) is a very important variable (Petrascheck and Sydler, 1984). Size of breach is most difficult to estimate for concrete gravity and buttress dams, while for earthfill dams it seems to fall between one-half and three times the height of the dam (Johnson and Illes, 1976). For concrete arch dams breach width is probably the full width of the dam.

Peak discharge resulting from a dam failure can be estimated from a time-dependent relationship:

$$Q_{\mathrm{p}} = 3.1W\,[c/(t + c/\sqrt{H})]^3\ ; \qquad c = 23.4\ \mathrm{SA}/W \tag{13}$$

where Q_{p} = peak discharge
W = average breach width, in feet
c = arbitrary constant
t = time in hours of breach formation
H = height in feet of the dam
SA = reservoir surface area (acres) at the dam crest (Wetmore and Fread, 1981)

With this equation peak discharge sensitivity can be computed for various estimated values of t. For a conservative estimate, the maximum expected breach width and minimum expected breach time would be used to estimate peak discharge. Equations (11)–(13) are general forms of the broad-crested wier formula (Brater and King, 1976).

Within the last decade numerous computer programs have been developed to simulate dam-break hydrographs. Peak discharges, depths, and areas inundated downstream need to be known to minimize loss of life and property. Two popular examples are the HEC-1 program of the Corps of Engineers and the National Weather Service DAMBRK model (Fread, 1980). The National Weather Service DAMBRK model, modified by Land (1980b), uses a hydraulic routing procedure based on a nonlinear implicit finite-difference algorithm for the equations of continuity and momentum. References to other programs can be found in Land (1980a,b). The purpose of these models is to predict the behavior of floodwaters released from a dam failure. The initial outflow hydrograph from a failed dam usually is approximated by a triangle. After the dam break outflow hydrograph is determined by one of the methods described previously, the hydrograph must be routed through the downstream valley. The models usually require river cross sections, Manning's n values, and upstream and downstream boundary conditions. Model output should include prediction of flood wave travel time, peak discharges and volumes at different locations downstream, and inundation areas.

Land (1980a) makes some interesting comparisons among four dam-break flood-wave models by using data from three actual dam failures and provides suggestions for finding the most accurate, stable, and economical models to use. Dam failure models are constrained by inaccuracies in estimates of breaching characteristics such as timing, size, and shape; by estimations of roughness coefficients, volume losses, debris, and sediment effects; and by channel hydraulics inadequately described by one-dimensional flow equations. Consequently results of dam break models can have large and significant errors, and operating the more complicated models can be a difficult task (Land, 1980a).

In simulation the user specifies the timing, size, and shape of the final breach. Breach parameters have little impact on flood characteristics far downstream from the dam (Petrascheck and Sydler, 1984). Morphological characteristics of breaches in historic constructed dams are described by Johnson and Illes (1976) and MacDonald and Langridge-Monopolis (1984).

An analysis of some failed dams for which downstream hydraulic measurements were made allows an estimate of attenuation rates based on empirical data. Available downstream flow data from some failed dams are listed in Table 4. Downstream peak discharges are related to peak discharge from the dam failure and downstream distance from the dam. Attenuation rates as a percentage of upstream discharge are plotted against distance downstream in Figure 8. A conservative envelope curve that encompasses all plotted data points for constructed dams and includes steep, narrow downstream valleys is

$$Q_x = \frac{100}{10^{(0.0021x)}} \quad (14)$$

where Q_x = discharge as a percentage of the peak discharge at kilometer 0, and x = distance downstream from location of peak discharge determination, in kilometers. For broader, more open valleys a conservative empirical enveloping curve has the form

$$Q_x = \frac{100}{10^{(0.0052x)}} \quad (15)$$

TABLE 4 Attenuation Rates from Some Historic Dam-Failure Floods

Name of Dam (index number from Table 3)	Cumulative Distance Along floodplain (km)	Peak Discharge (m^3/s)	Percentage of Discharge at River Kilometer 0
Schaeffer, CO (8)	0	4,900	100
	21	4,330	88
	88	880	18
Apishapa, CO (9)	0	6,850	100
	91	1,420	21
Castlewood, CO (12)	0	3,570	100
	60	960	27
	91	425	12
Little Deer Creek, UT (17)	0	1,330	100
	17	1,100	82
	22	500	37
	80	150	11
	124	85	6.3
Swift, MT (19)	0	—	—
	35	24,950	100
	144	5,780	23
Hell Hole, CA (21)	0	—	—
	75	8,780	100
	117	7,165	82
Buffalo Creek, WV (22)	0	1,420	100
	16	370	26
	31	250	18
	39	210	15
Teton, ID (23)	0	65,140	100
	16	30,020	46
	138	2,560	3.9
	178	1,910	2.9
	284	1,515	2.3
Kelly Barnes, GA (28)	0	680	100
	5.4	405	60
	12	180	27
	16	105	15
Lawn Lake, CO (29)	0	510	100
	7.6	340	67
	10.5	205	40

SOURCE: In part from Graham, 1980.

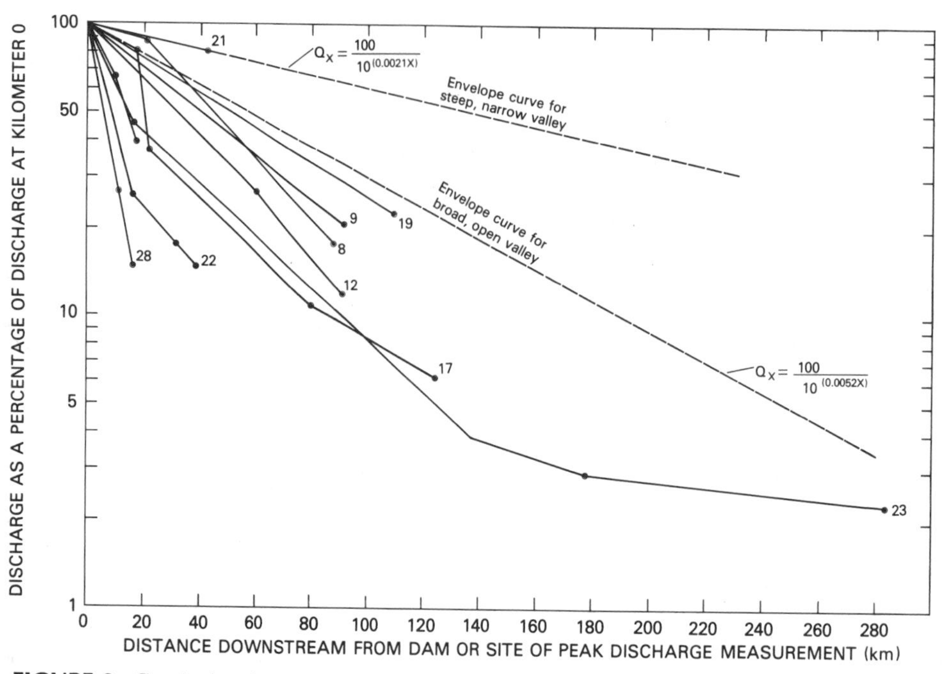

FIGURE 8. Graph showing attenuation rates of floods from selected dam failures, with numbers keyed to Table 4.

Knowledge of the valley geometry downstream should be used to modify the previous equations as necessary. Wide floodplains and high infiltration rates may lead to more rapid attenuation than the curves would indicate. Flood elevations and inundation areas can be determined from depth-discharge and depth-area curves.

NATURAL DAMS

A wide variety of types of natural dams have failed, producing large floods. Hutchinson (1957) provides a comprehensive overview of the origins of lakes and includes discussion of failures of natural dams. The most common types of natural dams that have failed producing large floods are ice dams, morainal dams, volcanic flow dams, and landslide dams.

Jökulhlaups

Jökulhlaup ("glacier burst") is an Icelandic term for a flood caused by the sudden and often catastrophic release of water impounded within or behind glacial ice (Thorarinsson, 1953). The largest flood known to have occurred on the surface of the earth, the Spokane flood in eastern Washington, originated from the sudden release of water from glacial Lake Missoula in the valley of the Clark Fork River in western Montana when an ice dam formed by a lobe of the Cordilleran Ice Sheet failed between 16,000 and 12,000 yr ago (Baker, 1973). The lake had a volume of 2×10^{12} m^3, and failure of the ice dam produced a flood with an estimated peak discharge of 21×10^6 m^3/s (Baker, 1973).

Jökulhlaups can occur in any area covered by continental or valley glaciers. They have caused large loss of life and property damage in many places throughout the world including Iceland (Thorarinsson, 1953, 1957); northern India (Hewitt, 1982); Pakistan (Nash et al., 1985); Peru (Lliboutry et al., 1977); Norway (Aitkenhead, 1960); Alaska (Post and Mayo, 1971); Switzerland, France, and Italy (Eisbacher and Clague, 1984); and Canada (Clarke, 1982; Young, 1980).

In a study of more than 50 jökulhlaups in the Alps, over 95% occurred in the months of June to September, inclusive, with maxima in June and August (Tufnell, 1984). Glacial lakes often drain periodically, which suggests that the depth of water (and consequent hydrostatic pressure) may be the primary factor controlling when a lake drains. The release of glacier-dammed lakes can occur by the formation of a drainage channel under, through, or over the ice. Several proposed mechanisms for failure of ice-dammed lakes include (1) slow plastic yielding of ice from hydrostatic pressure differences between the lake and adjacent, less dense ice; (2) raising of the ice barrier by hydrostatic flotation; (3) crack progression under combined shear stress from glacier flow and high hydrostatic pressure; (4) drainage through small, pre-existing channels at the ice–rock interface and consequent enlargement of ice tunnels by melting by heat in the lake water, and heat produced from kinetic energy of the water from the rapidly draining lake; (5) water overflowing the ice dam, generally along the margin; (6) subglacial melting by volcanic heat; and (7) weakening of the ice dam by earthquakes (Post and Mayo, 1971). Factors 1 through 4 are controlled by thickness of the ice dam, which determines the necessary pressure for flotation, and controls ice dynamical behavior such as tunnel closure rate and crevasse behavior. Lake depth, meanwhile, is limited by the elevation of the lowest bedrock divide or col.

When a lake drains, drainage tunnels freeze in the winter and runoff collects behind the ice dam in the spring and summer, during which time the ice dam may fail again. The characteristics and behavior of ice-dammed lakes can change drastically as ice advances or retreats in response to local climate variations. Lakes dammed by polar and subpolar ice in cold regions normally drain supraglacially or marginally through downmelting of the outlet channel, while ice dams in more temperate climates are more liable to sudden englacial or subglacial breaching (Blachut and Ballantyne, 1976).

Jökulhlaups can produce enormous floods (Table 5). The largest floods seem to be outbursts produced by subglacial melting of ice by volcanoes. The most intense observed floods occur from Myrdalsjokull in Iceland where peak discharges may exceed 100,000 m^3/s (Thorarinsson, 1957). Hypothetical hydrographs of two types of jökulhlaups are shown in Figure 9. Dams that burst suddenly are characterized by a steep rising limb, sharp peak, and a steep recession limb (Young, 1980). Floods produced by progressive enlargement of veins and channels by passing waters have more gradually rising limbs, sharp peaks, and steep recessional limbs (Haeberli, 1983). Some examples of actual jökulhlaup hydrographs can be found in Thorarinsson (1953) and Stone (1963). The time of year when a jökulhlaup occurs can be an important factor in determining the magnitude of the flood. In January 1969 the drainage of a glacial lake into the Kenai River, Alaska, fractured the river ice and formed large ice jams that plugged the channel, resulting in severe flooding (Post and Mayo, 1971).

The timing and potential magnitude of jökulhlaups can only be crudely estimated. If the hydrostatic flotation theory of Thorarinsson (1953) is applicable, subglacial drainage becomes possible when the hydrostatic pressure of water from a lake exceeds the ice overburden pressure in a dam. This occurs when the depth of water in the glacial lake reaches about 0.9 times the height of the ice barrier. Knowing the height of the ice barrier and monitoring depth of lake water may allow warnings to be given when critical depths are approached. Unfortunately, jökulhlaups can occur long before critical depths for flotation are reached (Mathews, 1965).

The magnitude of jökulhlaups can be estimated from the empirical relationship between lake volume and peak discharge developed by Clague and Mathews (1973):

$$Q_{max} = 75V^{0.67} \qquad r^2 = 0.96 \tag{16}$$

where Q_{max} is maximum discharge in cubic meters per second, and V is lake storage in $m^3 \times 10^6$. A more comprehensive data set that does not include the Pleistocene Lake Missoula flood is in Table 5. Using these data, the regression equation is

$$Q_{max} = 113V^{0.64} \quad r^2 = 0.80 \quad SE = 106\% \tag{17}$$

The curve is plotted in Figure 5. Because glacial dams can drain in a variety of ways, dam height does not seem to be a good indicator of peak discharge. A regression equation using data from Table 5 is

$$Q_{max} = 21.6H^{0.73} \quad r^2 = 0.08 \quad SE = 236\% \tag{18}$$

However the dam factor ($H \times V$) regression equation has a much lower standard error

$$Q_{max} = 3.8(HV)^{0.61} \quad r^2 = 0.79 \quad SE = 75\% \tag{19}$$

This curve is plotted in Figure 7.

Theoretical investigations of glacier outburst floods have been undertaken by Mathews (1973), Nye (1976), and Clarke (1982). These hydraulic-thermodynamic models are based on tunnel geometry, continuity, energy conservation, and heat transfer, and require estimation or direct measurement of several critical parameters, including roughness coefficients and lake temperatures. These models do not do equally well in all circumstances. Models of jökulhlaups are subject to many of the same difficulties and uncertainties as models of constructed dam failures. The phenomena are more complex than can easily be analyzed, and therefore models are not very reliable.

Morainal Dams

Terminal or recessional moraines in glaciated areas may be sufficiently well-preserved that they dam the stream that replaces the melting glacier. The advances and retreats of mountain glaciers in different parts of the world over the past few hundred years have created a large number of small, unstable morainal dams. These dams subsequently may fail by overtopping and erosion of the dam by large runoff events or large waves (Fig. 10). Most constructed dams are zoned to minimize leakage and riprapped to minimize surface erosion. However, morainal and landslide dams are heterogenous mixtures of a variety of particle sizes. When a morainal or landslide dam is breached, downcutting commences and erosion of a breach usually begins as headward-eroding cuts. Eventually the headward erosion reaches the impounded water, lowering the outlet and allowing more water to escape.

TABLE 5 Some Glacial Dams That Have Failed, Producing Jökulhlaups

Index Number and Name of Lake	Year of Failure	Height H (m)	Volume V ($m^3 \times 10^6$)
1. Gietro, Switzerland	1818	?	20 (outburst volume)
2. Vatnsdalur, Iceland	1898	372	120
3. Aletsch, Switzerland	1913	?	4.5 (outburst volume)
4. Albigna, Switzerland	1927	?	2.7 (outburst volume)
5. Chong Kumdan (Shyok), India	1929	120	1,350
6. Demmevatn, Norway	1937	406	11.6
7. Graenalon, Iceland	1939	535	1,500
8. Trient, Switzerland	1942	?	0.84 (outburst volume)
9. Gorner, Switzerland	1944	?	>6 (outburst volume)
10. Unter Grindelwald, Switzerland	1951	?	0.135 (outburst volume)
11. Gjanupsvatn, Iceland	1951	167	20
12. Lake George, Alaska[a]	1951	35	?
13. Ferpecle, Switzerland	1952	?	0.255 (outburst volume)
14. Lake George, Alaska[a]	1958	40	1,730
15. Tulsequah Lake, British Columbia	1958	210	229
16. Summit Lake, British Columbia	1965, 1967	620	251
17. Ekalugad Valley, Baffin Island	1967	120	4.8
18. Strupvatnet, Norway	1969	186	2.6 (outburst volume)
19. Gruben, Switzerland	1970	?	0.17 (outburst volume)
20. Chakachatna Lake, Alaska[a]	1971	?	294
21. Hazard Lake, Canada	1978	300	19.6

[a] Drains at surface along ice margin.

In 1874 a proglacial lake that had formed behind a Little Ice Age terminal moraine of the Madatschferner glacier in Austria, breached its unstable morainal dam and caused flooding in the valley below (Eisbacher and Clague, 1984, p. 131). Breaches in young morainal dams also have occurred from waves produced by large rock and ice falls into lakes and subsequent overtopping and erosion of the natural dam. This phenomenon has been described in Austria (Eisbacher and Clague, 1984, p. 131); Peru (Lliboutry et al., 1977); and Canada (Blown and Church, 1985) (Fig. 10).

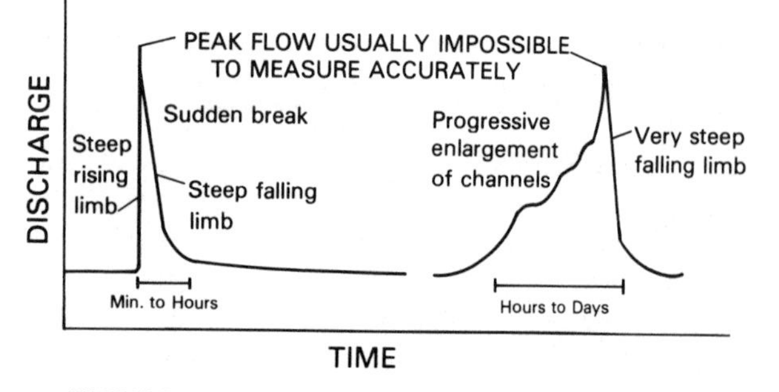

FIGURE 9. Generalized hydrographs of jökulhlaups.

Many lakes dammed by frontal moraines have no defined outlets. Outflow occurs through springs located in the moraine, and discharges vary with lake levels. In these situations, piping failures are a potential hazard. Hundreds of thousands of dollars have been spent trying to lower lake levels, and thus to minimize failures of morainal dams, in the Cordillera Blanca, Peru, following a major ice retreat in the early to mid 1900's (Lliboutry et al., 1977).

Volcanic Dams

Floods can originate from the breaching of natural dams formed by lava flows (Finch, 1937; Cotton, 1944) or pyroclastic flows (Aramaki, 1981) or from the breaching and expelling of volcanic crater lakes (Zen and Hadikusumo, 1965; Nairn et al., 1979).

On March 28 and 29, 1982, El Chichon Volcano in southern Mexico erupted. A pyroclastic flow dammed the Rio Magdalena southwest of the volcano, forming a lake

TABLE 5 *(Continued)*

Dam Factor ($H \times V$)	Peak Outflow (m^3/s)	Peak Determination	Reference
—	8,000	?	Haeberli, 1983, Table III
44,640	3,000	?	Clague and Mathews, 1973
—	195	?	Haeberli, 1983, Table II
—	128	?	Haeberli, 1983, Table II
162,000	22,650	?	Gunn, 1930; Hewitt, 1982
4,710	1,000	?	Clague and Mathews, 1973
802,500	5,000	?	Clague and Mathews, 1973
—	26	?	Haeberli, 1983, Table II
—	200	?	Haeberli, 1983, Table II
—	74.6	?	Haeberli, 1983, Table III
3,340	370	?	Clague and Mathews, 1973
—	5,200	Slope-area measurement	Stone, 1963
—	230	?	Haeberli, 1983, Table III
69,200	10,100	?	Clague and Mathews, 1973
48,090	1,556	?	Clague and Mathews, 1973
155,620	3,260	Drawdown rate	Clague and Mathews, 1973
576	200	Gauged 1 km downstream	Clague and Mathews, 1973
856	150	?	Clague and Mathews, 1973
—	15	?	Haeberli, 1983, Table II
—	13,310	Field estimate 9 km downstream	Lamke, 1972
5,880	640	Drawdown rate	Clarke, 1982

FIGURE 10A. Aerial photograph of Nostetuko Lake and Cumberland Glacier, British Columbia, Canada, in July 1977. (Photograph by J. M. Ryder, courtesy of Michael Church).

FIGURE 10B. Aerial photograph of Nostetuko Lake and Cumberland Glacier, British Columbia, Canada, in August 1983, after the morainal dam was breached. (Photograph courtesy of Michael Church).

5 km long and several million cubic meters in volume (Silva et al., 1982). On May 26, 1982, the pyroclastic dam failed, draining the lake in about 1 hr and sending a flood of very hot water downstream. Bridges, a village, and a hydroelectric plant downstream were damaged. Ten kilometers from the dam, a floodwater temperature of 82° C was measured. At a hydroelectric plant 35 km downstream, one worker was killed and three were badly burned by the hot floodwaters (52° C). The breaching of a large pyroclastic dam that was formed by the 1783 eruption of Asama Volcano, Japan, resulted in the destruction of more than 1200 houses and the loss of more than 1200 lives (Aramaki, 1981).

In 1912 a large volcanic eruption occurred in the Valley of Ten Thousand Smokes, Katmai National Park, Alaska. A cluster of phreatic craters dammed a 1.5-km-long lake atop the tuff. It is estimated that the dam failed in the summer of 1912 or 1913 (Hildreth, 1983). The flood scoured the ash-flow surface to depths of 1 to 2 m, transported 50-cm-diameter blocks of welded tuff over 20 km, and deposited 1 to 8 m of sediment in the lower parts of the valley (Hildreth, 1983).

Landslide Dams

Landslides that move into stream valleys and dam rivers are very common, yet no comprehensive investigation has ever been undertaken to investigate this worldwide phenomenon. If a landslide dam fills and overspills, rapid erosion of the landslide deposits can lead to catastrophic flood discharges and great loss of life and property. Landslide dams form in all kinds of physiographic settings, ranging from rock avalanches in steep mountainous terrains to quick-clay failures in flat river lowlands. Earthquakes appear to be a significant cause of landslides that dam valleys (Adams, 1981; Keefer, 1984). One of the earliest reported landslide dams formed in A.D. 563 in the St. Barthelemy Basin, Switzerland. The debris blocked the upper Rhone River, and the subsequent failure killed many people as floodwaters flowed into Lake Geneva (Eisbacher and Clague, 1984).

Probably the greatest landslide dam failure disaster was the Indus River landslide dam failure of 1841 (Mason, 1929). During the winter of 1840 and 1841, part of Nanga Parbat collapsed into the Indus River following an earthquake. The landslide dam formed a lake 305 m deep and 64 km long. In June 1841 the dam was breached by the Indus, and a tremendous flood resulted. At Attock, over 400 km downstream, the water rose to over 30 m deep, engulfing a Sikh army camp on the floodplain near Attock, killing 500. Hundreds of villages were washed away, and thousands of lives were lost (Mason, 1929).

Following the May 1980 eruption of Mount St. Helens, Washington, debris avalanche deposits impounded three large lakes (Coldwater Creek, South Fork Castle Creek, and Spirit Lakes) and several smaller ones (Fig. 11). Coldwater Creek Lake, had it been allowed to fill naturally, would have developed an estimated maximum volume of 123×10^6 m^3 by late 1981. If this dam had overtopped, Dunne and Fairchild (1983) estimated that a vertical erosion rate of 0.3 m/min would occur, developing a rectangular breach with width–depth ratio of two, and a maximum incision of 30 m. Clear-water peak discharge through the breach was calculated using the broad-crested weir formula to be 14,000 m^3/s within 100 min and 18,000 m^3/s after bulking up the floodwaters with easily eroded unconsolidated sediment. The U.S. Army Corps of Engineers constructed a permanent spillway for Coldwater Creek and South Fork Castle Creek lakes during the summer of 1981 that stabilized their volumes at 82.6×10^6 m^3 and 24×10^6 m^3, respectively (Schuster, 1984).

For the debris avalanche dam impounding Spirit Lake, the expected flood hydrograph from the hypothetical failure of the debris dam was simulated with the National Weather Service DAMBRK model (Fread, 1980) and appropriate revisions (Land, 1980a) for an overtopping failure (Jennings et al., 1981) and for a piping failure at lower reservoir levels (Swift and Kresch, 1983). Dunne and Fairchild (1983) employed the same procedure of estimating erosion rates, breach characteristics, and peak discharge that they used for Coldwater Creek Lake. Clear-water peak discharge values were 17,400 m^3/s, 15,000 m^3/s, and 15,600 m^3/s, respectively. It was expected that these clear-water peak flows would quickly entrain large volumes of sediment, so when the flows were routed downstream, large volumes

FIGURE 11. Photograph of Castle Lake, Mount St. Helens, Washington, dammed by a debris avalanche deposit. Debris avalanche moved from left to right in photograph. An artificial spillway was cut into the right embankment. (April 1984) (Photograph courtesy of R. L. Schuster).

of sediment were added to simulate muddy floodwaters or debris flows.

The hazard presented by the volcanic debris avalanche dam of Spirit Lake has been mitigated by constructing a permanent drainage tunnel in bedrock at a cost of $14 million (Sager et al., 1984; Schuster, 1985).

Numerous smaller lakes were formed by the unstable volcanic debris surrounding Mount St. Helens, and several breakout floods have been produced. A landslide dam that accumulated water from Castle and Maratta creeks failed on August 19, 1980. The floodwaters were reimpounded in another dam near Elk Rock. This dam was overtopped and breached on August 27, 1980, releasing 0.3×10^6 m^3 of water into the North Fork Toutle River. The estimated flood peak was 450 m^3/s, and 2.8×10^5 m^3 of material was eroded from the debris dam, depositing 2.0×10^5 m^3 of material in the channel of the North Fork Toutle River (Meier et al., 1981; Jennings et al., 1981).

Although data are scarce at the present time, it seems that landslide and volcanic dams consisting of pyroclastic debris, if they are going to fail, fail during filling more quickly than other types of landslide or volcanic dams (Michael Church, written communication, 1985). This may be related to unique density, packing, composition, or texture of pyroclastic sediments in natural dams. Obviously this is an important area for further investigation.

Landslide dams are different from constructed fill dams in several important aspects. Landslide dams are a heterogenous mixture of rock and soil whose permeability can vary greatly. Commonly landslide sediments will plug the valley and flow or slide for some distance down valley. Landslide dams are typically much wider than constructed dams. For example, the South Nation River landslide dam in Canada was 2.5. km wide (Eden et al., 1971). A transverse section of the Mayunmarca landslide dam in Peru is compared with the Oroville Dam, a large earthfill dam in California (Fig. 12). The landslide dam is higher at the abutments and almost as high as the constructed dam in the center, but over three times as wide. Landslide dams typically involve large volumes of sediment. The Madison Canyon landslide that dammed the Madison River forming Earthquake Lake in 1959, had a base width five to eight times as great as would have been used in building a rockfill dam of the same height (Knight and Bennett, 1960).

When a landslide dam is overtopped, there is commonly much more sediment and debris for water to erode before a full breach is developed. This is probably why flood peaks from failed landslide dams appear to be smaller than constructed dam failures with the same dam height and reservoir volume (Figs. 4 and 5; Table 6).

Regression equations have been developed to predict peak discharge from the landslide dam failure data in Table 6. Knowing the height of the dam in meters,

$$Q_{\max} = 6.3H^{1.59} \quad r^2 = 0.74 \quad \mathrm{SE} = 147\% \qquad (20)$$

knowing the volume (V) at time of failure, in $m^3 \times 10^6$,

$$Q_{\max} = 672V^{0.56} \quad r^2 = 0.73 \quad \mathrm{SE} = 142\% \qquad (21)$$

and finally, using the dam factor ($H \times V$),

$$Q_{\max} = 181(HV)^{0.43} \quad r^2 = 0.76 \quad \mathrm{SE} = 129\% \qquad (22)$$

These curves are plotted in Figures 4, 5, and 7.

Hydrographs resulting from landslide dam failures depend on the rate of breach erosion and reservoir characteristics. A hydrograph for the 1974 Mayunmarca landslide dam failure in Peru is shown in Figure 13.

The reasons some landslide dams fail and others do not are uncertain. Water seepage through some debris dams obviously maintains lake levels at safe depths and volumes (Adams, 1981). A large percentage of coarse boulders in a debris dam can retard or prohibit erosion if overtopping occurs and can form a natural riprapped spillway channel, with particles too coarse to move, through which the lake can drain.

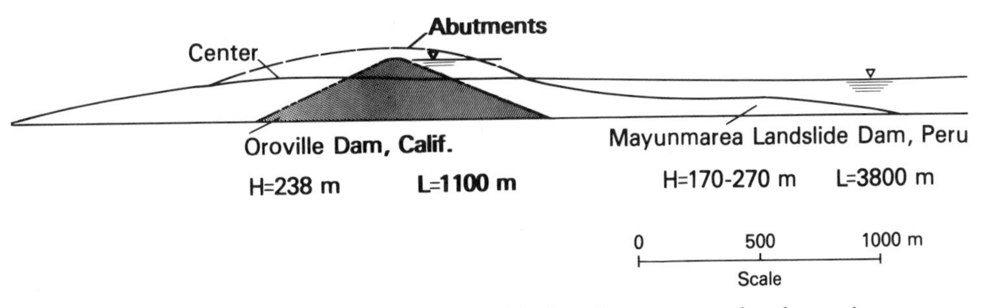

FIGURE 12. Cross section of Mayunmarca landslide dam, Peru, compared to the maximum cross section of a large earthfill dam at Oroville, California (from Lee and Duncan, 1975).

TABLE 6 Summary of Several Landslide Dams That Have Failed

Index Number and Name of Dam and River	Date of Formation	Date of Failure	Height H (m)
1. Gohna, Bireh-ganga River, India	Sept. 22, 1893	Aug. 25, 1894	274
2. Cerro Condor-Sencca, Mantaro River, Peru	Aug. 16, 1945	Oct. 28, 1945	92
3. Zepozhu, Dong River, China	?	May, 1966	51
4. Yashinkul', Isfayramsay River, USSR	?	June 18, 1966	90
5. Tanggudang, Yalong River, China	?	June, 1967	175
6. Embankment, Granite Creek tributary, Alaska	?	Aug. 10, 1971	26
7. Mayunmarca, Mantaro River, Peru	Apr. 25, 1974	June 6, 1974	170
8. Elk Rock Lake, Washington	May 18, 1980	Aug. 27, 1980	18.3
9. East Fork Hood River, Oregon	Dec. 25, 1980	Dec. 25, 1980	10.7
10. Whitehall, So. Fork American River, California	Apr. 9, 1983	Apr. 9, 1983	13.7

Swanson and others (1985) proposed that larger dams blocking smaller rivers will have a smaller probability of failure than a smaller landslide dam below a larger drainage area. In a plot of watershed area versus landslide volume for nine sites in Japan, failed landslide dams clearly cluster as a separate population of smaller landslide volumes in larger drainage areas. With more data a discriminant function between stable and failed landslide dams may be possible to ascertain, which would allow prediction of the potential for a catastrophic failure.

Today most landslide dams are viewed as serious flood threats, and extraordinary engineering feats are used to remove the dam or control the reservoir levels. In China, a large landslide dammed the Bailong River in 1981. Three hundred metric tons of dynamite were used to blast through the landslide (Li and Hu, 1982). In the Soviet Union a large landslide dammed the Zeravshan River north of Afghanistan. The ancient city of Samarkand lay directly in the flood path should the dam fail. A spillway was blasted through the landslide before the lake reached its full capacity of 490×10^6 m^3 (Anonymous, 1964). Another landslide dam formed in the previous year in Yugoslavia near the Bulgarian border. An evacuated village was already submerged in 30 m of backwater when an emergency spillway was blasted and dug into the landslide dam (Anonymous, 1963).

In the United States one of the best-known landslide dams that was modified before it overtopped or failed was

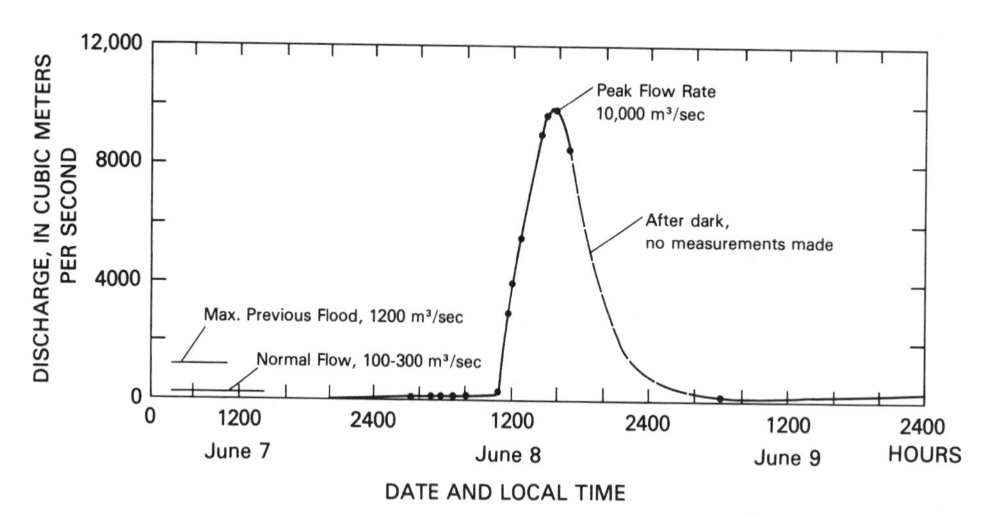

FIGURE 13. Hydrograph of discharge resulting from the 1974 failure of Mayunmarca landslide dam, Peru (from Lee and Duncan, 1975).

TABLE 6 *(Continued)*

Volume at Time of Failure V (m^3 × 10^6)	Dam Factor ($H \times V$)	Peak Outflow (m^3/s)	Peak Determination	Reference
467	128,000	56,650	Drawdown rate, 1 hr avg.	Strachey, 1894; Lubbock, 1894
301	27,700	35,400	Drawdown rate	Snow, 1964
2.7	138	560	2-hr avg.	Investigation Team of the Zepozhu Landslide-dammed River, 1967
6.6	594	5,000	Chezy formula	Glazyrin and Reyzvikh, 1968
?	—	50,000	Measured 6 km downstream	Investigation Team of the Tanggudang Landslide Dam, 1967
25.2	655	1,660	Slope-area measurement, 11 km downstream	Lamke, 1972
670	113,900	10,000	Drawdown rate, 15 min avg.	Lee and Duncan, 1975
0.31	5.67	453	Estimated	Jennings et al., 1981
0.105	1.12	850	Dam-break model; slope conveyance	Gallino and Pierson, 1985
0.4	5.5	64	Estimate, 1-hr avg.	Michael Kuehn, U.S. Forest Service, written communication

the dam created by the 1959 Madison Canyon landslide of August 17, 1959, near Yellowstone National Park in Wyoming. An emergency spillway 75 m wide and 0.8 km long, and designed to pass a discharge of 280 m^3/s, was quickly bulldozed into the landslide dam, and prevented a potentially catastrophic failure (Stermitz, 1964). More recently, in April 1983 a landslide at Thistle, Utah, dammed the Spanish Fork River and flooded the town, a major transcontinental railroad line, and two highways. The 98 × 10^6 m^3 lake that formed was eventually drained by drilling large tunnels into the right bedrock abutment (Schuster, 1985).

Landslides may also produce floods by the rapid displacement of water in existing lakes and reservoirs. One of the most infamous examples is the Vaiont, Italy, dam failure, the worst dam disaster in history (Kiersch, 1964). On October 9, 1963, over 240 × 10^6 m^3 of rock and soil 1.8 km long and 1.6 km wide slid rapidly into the reservoir, displacing the water and sending 100 m-high waves over the top of the dam. About 2600 people lost their lives in the subsequent flooding downstream. The thin arch concrete dam sustained little damage to its main shell or abutments.

A more recent example of this type of landslide flood occurred in May 1983 near Reno, Nevada (Watters, 1983). A large mass of rock on a mountain slope in the Sierra Nevada slid rapidly into a small reservoir that overflowed into a second small reservoir. The combined storage capacity of both bodies of water was estimated to have been 15,000–20,000 m^3. Four kilometers downstream the resulting flood killed one person, injured others, and damaged several houses and vehicles.

Other Types of Dam Failures

Four other unusual types of "dam failures" that resulted in large floods can be mentioned. One was the complete draining of Lake Emma, a natural glacial lake located in the San Juan Mountains above Silverton, Colorado. On June 4, 1978, water breached through a network of tunnels in an abandoned mine beneath the lake, emptying the lake (Carrara et al., 1984).

A second unusual "dam failure" occurred on September 29, 1982, in the Sierra Nevada at Lost Canyon, about 80 km northeast of Fresno, California. A major public utility company was conducting preoperational testing of a 6.7-m diameter water pipeline when the line ruptured and released a 1133 m^3/s flood down Lost Canyon. This flow was sustained for about 1 hr and was five times the estimated probable maximum flood peak for the drainage basin (Pacific Gas and Electric Company, 1983). The stream channel was scoured, steepened, and straightened, and about 1.34 × 10^6 m^3 of sediment was removed from the canyon.

The third unusual type of dam failure occurs in areas of thick peat deposits underlain by impermeable tills. During intense rainstorms, water flows at the peat-till interface and may form large, irregular-shaped mounds of water up to 4 m high under the peat. These mounds of water will eventually fail, causing what are known as "bog-bursts"

FIGURE 14. Damage from collapse of giant molasses tank at Boston Harbor, January 15, 1919. View along Atlantic Avenue. Photograph courtesy of the Boston Public Library, Print Department.

in Ireland, Scotland, and England (Colhoun et al., 1965), releasing large floods that can cause great damage.

The fourth unusual dam failure worth noting was the great Boston molasses flood of January 15, 1919. At the inner harbor a distilling company's giant tank, over 15 m tall and 86 m in diameter, was filled with 900 m^3 of molasses. Suddenly the tank burst, and 15,000 metric tons of molasses swept down nearby city streets as a thick brown wave 5 m high and moving at 16 m/s. Twenty-one people lost their lives (Hartley, 1981; Lane, 1965), (Fig. 14).

CHANNEL AND VALLEY CHANGES

The sudden release of water and sediment as a consequence of the failure of a constructed or natural dam can result in significant changes in the downstream valley and river channel. Such flows can fill valleys and overtop floodplains by 3–10 m or more. The amount of geomorphic change is controlled by the duration of the flood and the slope of the channel. If the duration of high discharge is great, profound geomorphic changes can occur. But if flood peaks are of short duration, and channel slopes small, channel changes will be minimal, as in the flat glacial lake floor of Horseshoe Park, Colorado, after the Lawn Lake Dam failure in Rocky Mountain National Park (Jarrett and Costa, 1986). Schumm (1969) presents some empirical data on stream channel response to changes in water and sediment loads. Chen and Simons (1979) attempt to apply some of these data to flume studies of simulated dam failures.

The following general characteristics have been reported following large floods from dam failures: (1) Aggradation of the valley upstream by trapped sediment. The sudden and rapid erosion of this stored sediment can have a great impact on the flood flow hydraulics following dam breaching (Chen and Simons, 1979). In Buffalo Creek, West Virginia, an estimated 153,000 m^3 of coal waste were deposited in the Buffalo Creek Valley downstream of the dam (Davies et al., 1972). (2) Additional landslides may be triggered by the rapid drawdown of ponded water when the dam is breached (Costa and Baker, 1981, pp. 266–267). (3) Down valley, large amounts of local scour and deposition can occur. In the Knik River valley in Alaska, following a jökulhlaup, 2.4 m of local scour and 1.2 m of local fill were measured (Post and Mayo, 1971). Following the Lawn Lake Dam failure in Rocky Mountain National Park, Colorado, alternating reaches of scour and deposition were observed and measured along a 6-km reach of the Roaring River, which has an average slope of 10%. Channel reaches steeper than 7–9% were extensively scoured, while less steep reaches were depositional sites. As much as 15 m of scour were measured (Jarrett and Costa, 1986). The same pattern was reported from a dam failure on a steep valley in Great Britain in 1925 (Fearnsides and Wilcockson, 1928).

Following the Teton Dam failure in 1976 in Idaho, the floodwaters eroded about one meter of weathered volcanic rocks in the channel, destroyed all trees and vegetation, and stripped topsoil along an 8-km reach of canyon below the dam (Fig. 15) (Ray and Kjelström, 1978). During the Buffalo Creek, West Virginia, dam break flood, sandstone bedrock along valley walls was eroded (Davies et al., 1972).

If a landslide dam fails, the large amount of sediment from the dam available for transport can result in widespread valley aggradation. The Madison River, Montana, aggraded to an average depth of over 9 m in 2 months following the Madison Canyon landslide and spillway construction (Hanly, 1964). Following the failure of a large landslide dam in Peru, Snow (1964) documented rapid channel aggradation for many kilometers downstream. As these flood sediments became incised by later flows, terraces formed in the valley. These terraces consisted of poorly sorted, unstratified angular fragments of landslide debris. Upstream of the dam incision formed terraces in sediments deposited in the ponded water. These terraces contained well-sorted, rounded, imbricated, and stratified stream gravels.

Sediments scoured from local surficial deposits during a dam failure can form distinctive depositional features.

FIGURE 15. Photograph of Teton River valley at Newdale, Idaho, showing area of bedrock erosion along the left bank after the Teton Dam failure in 1976. Person in center of photograph for scale.

In Rocky Mountain National Park, Colorado, the Lawn Lake Dam failure flood scoured local moraines and formed an alluvial fan of 0.17 km^2 (Jarrett and Costa, 1986). In an investigation of the potential floods accompanying the hypothetical breaching of two large debris dams formed in debris avalanche sediments from the 1980 eruption of Mount St. Helens, Dunne and Fairchild (1983) assumed that the release of a large volume of water onto loose, unconsolidated volcanic deposits would result in rapid erosion and bulking of the floodwaters. They modeled flood peaks assuming sediment concentrations of 300,000–1,180,000 mg/liter.

Augmented sediment loads in channels downstream from breached dams require steepened slopes for efficient sediment transport. This results in channels that are wide, shallow, and highly braided, such as that of the Gros Ventre River near Kelly, Wyoming following the failure of a landslide-dammed lake in 1927 (Alden, 1928).

CONCLUSION

Some of the largest floods to have occurred on Earth have resulted from the failure of dams. Both constructed dams and natural dams have failed with the resulting loss of many thousands of lives and millions of dollars in property. Estimation of peak discharges from hypothetical dam failures, and routing of those floods down-valley remains an imprecise art. The diversity of dam types, failure mechanisms, and down-valley sediments and morphology make the prediction of channel changes, scour, and deposition speculative. One point becomes clear, however: The ability to estimate the hydrology, hydraulics, and geomorphology of all types of dam failures is predicated on our knowledge of historic events and appropriate measurements and observations.

For rapid prediction purposes when loss of life or property is involved, a conservative peak discharge estimate based on envelope curves developed from historic dam failures can be made from knowledge of dam height and volume using equations (3), (5), (6), or (9). If breach dimensions can be estimated, Equations (11), (12), or (13) may be used.

For reconstructing past flood peaks from dam failures for paleohydrological or sedimentological investigations, regression equations with peak discharge as the dependent variable, probably provide a reasonable estimation if dam height and volume of stored water are known. For constructed dams Equation (4) requires knowledge of the dam height to predict peak discharge. For glacial dams and landslide dams, Equations (19) and (22) can be used to reconstruct peak discharge.

Regression equations for constructed, landslide, and glacial dams are summarized in Table 7. Dam height is

TABLE 7 Summary of Regression Equations to Predict Peak Discharge for Constructed, Landslide, and Glacial Dams

	Independent Variable[a]		
Type of Dam	Dam Height (H)	Volume (V)	Dam Factor ($H \times V$)
Constructed dams	$Q_{max} = 10.5H^{1.87}$; $r^2 = 0.80$; SE = 82%	$Q_{max} = 961V^{0.48}$; $r^2 = 0.65$; SE = 124%	$Q_{max} = 325(HV)^{0.42}$; $r^2 = 0.75$; SE = 95%
Landslide dams	$Q_{max} = 6.3H^{1.59}$; $r^2 = 0.74$; SE = 147%	$Q_{max} = 672V^{0.56}$; $r^2 = 0.73$; SE = 142%	$Q_{max} = 181(HV)^{0.43}$; $r^2 = 0.76$; SE = 129%
Glacial dams	$Q_{max} = 21.6H^{0.73}$; $r^2 = 0.08$; SE = 236%	$Q_{max} = 113V^{0.64}$; $r^2 = 0.80$; SE = 106%	$Q_{max} = 3.8(HV)^{0.61}$; $r^2 = 0.79$; SE = 75%

[a] Q_{max} in m^3/s; H in m; V in $10^6 m^3$.

the best independent variable to estimate peak discharge for constructed dams. Dam factor (height times volume) is the best independent variable for estimating flood peaks from landslide and glacial dams. For all kinds of dams, dam factor has a lower overall average standard error, and is an approximate measure of the energy expended at the dam at time of failure. Enough well-documented, intensive investigations of dam failures simply do not exist, especially for landslide dams. These are obvious areas for new and innovative hydrologic and geomorphic research, and the returns will have many practical as well as scientific merits.

ACKNOWLEDGMENTS

I would like to thank Robert L. Schuster, David C. Froehlich, Robert D. Jarrett, and Julie M. Stewart, U.S. Geological Survey, and Michael Church, University of British Columbia, for excellent constructive reviews of this chapter.

REFERENCES

Adams, J. (1981). Earthquake-dammed lakes in New Zealand. *Geology* **9**, 215–219.

Aitkenhead, N. (1960). Observations on the drainage of a glacier-dammed lake in Norway. *J. Glaciol.* **3**, 607–609.

Alden, W. C. (1928). Landslide and flood at Gros Ventre, Wyoming. *Am. Inst. Min. Metall. Eng. Trans.* **76**, 347–358.

Anonymous (1923). Failure of Apishapa earth dam in southern Colorado. II. *Eng. News-Rec.*, Sept. 13, pp. 418–424.

Anonymous (1925). Undermining causes failure of French Landing Dam. *Eng. News-Rec.*, April 30, pp. 735–736.

Anonymous (1963). Engineers fear dam won't hold. *Eng. News-Rec.*, April 11, p. 22.

Anonymous (1964). Russians blast through landslide dam. *Eng. News-Rec.*, May 7, p. 24.

Anonymous (1969). Wyoming dam fails. *Eng. News-Rec.*, July 17, p. 16.

Aramaki, S. (1981). The sequence and nature of 1783 eruption of Asama volcano. *In* "1981 IAVCEI Symposium, Arc Volcanism, Tokyo and Hakone," Abstr., pp. 11-12. Volcanology Society of Japan and International Association of Volcanology and Chemistry of the Earth's Interior, Tokyo.

Baecher, G. B., Pate, M. E., and de Neufville, R. (1980). Risk of dam failure in benefit-cost analysis. *Water Resour. Res.* **16**, 449–456.

Baker, V. R. (1973). Paleohydrology and sedimentology of Lake Missoula flooding in eastern Washington. *Spec. Pap.—Geol. Soc. Am.* **144**, 1–79.

Barrows, H. K. (1948). "Floods, Their Hydrology and Control." McGraw-Hill, New York.

Biswas, A. K., and Chatterjee, S. (1971). Dam disasters—An assessment. *Eng. J.* **54** (3), 3–8.

Blachut, S. P., and Ballantyne, C. K. (1976). "Ice-Dammed Lakes—A Critical Review of Their Nature and Behaviour," Discuss. Pap. No. 6. McMaster University, Department of Geography, Hamilton, Ontario.

Blown, I., and Church, M. (1985). Catastrophic lake drainage within the Homathko River basin, British Columbia. *Can. Geotech. J.* **22**, 551–563.

Boner, F. C., and Stermitz, F. (1967). Floods of June 1964 in northwestern Montana. *Geol. Surv. Water-Supply Pap. (U.S.)* **1840-B**, 1–242.

Brater, E. F., and King, H. W. (1976). "Handbook of Hydraulics." McGraw-Hill, New York.

Brua, S. A. (1978). Floods of July 19-20, 1977 in the Johnstown area, western Pennsylvania. *Geol. Surv. Open-File Rep. (U.S.)* **78-963**, 1–62.

Carrara, P. E., Mode, W. N., Rubin, M., and Robinson, S. W. (1984). Deglaciation and postglacial timberline in the San Juan Mountains, Colorado. *Quat. Res.* **21**, 42–55.

Chen, Y. H., and Simons, D. B. (1979). An experimental study of hydraulic and geomorphic changes in an alluvial channel induced by failure of a dam. *Water Resour. Res.* **15**, 1183–1188.

Clague, J. J., and Mathews, W. H. (1973). The magnitude of jökulhlaups. *J. Glaciol.* **12**, 501–504.

Clarke, G. K. C. (1982). Glacier outburst floods from "Hazard Lake," Yukon Territory, and the problem of flood magnitude prediction. *J. Glaciol.* **28**, 3–21.

Colhoun, E. A., Common, R., and Cruickshank, M. M. (1965). Recent bog flows and debris slides in the north of Ireland. *Sci. Proc. R. Dublin Soc., Ser. A* **2** (10), 163–174.

Committee on the Safety of Existing Dams (1983). "Safety of Existing Dams, Evaluation and Improvement." Natl. Acad. Press, Washington, D. C.

Costa, J. E., and Baker, V. R. (1981). "Surficial Geology: Building with the Earth." Wiley, New York.

Cotton, C. A. (1944). "Volcanoes as Landscape Forms." Whitcombe & Tombs. Christchurch and London.

Davies, W. E., Bailey, J. F., and Kelly, D. B. (1972). West Virginia's Buffalo Creek flood—A study of the hydrology and engineering geology. *Geol. Surv. Circ. (U.S.)* **667**, 1–32.

Dunne, T., and Fairchild, L. H. (1983). Estimation of flood and sedimentation hazards around Mount St. Helens. *Shin-Sabo (J. Eros. Control Soc. Jpn.)* **36**, 12–22.

Eden, W. J., Fletcher, E. B., and Mitchell, R. J. (1971). South Nation River landslide, 16 May 1971. *Can. Geotech. J.* **8**, 446–451.

Eisbacher, G. H., and Clague, J. J. (1984). Destructive mass movements in high mountains—Hazard and management. *Geol. Surv. Can. Pap.* **84-16**, 1–230.

Fearnsides, W. G., and Wilcockson, W. H. (1928). A topographical study of the flood-swept course of the Porth Llwyd above Dolgarrog. *Geogr. J.* **72**, 401–419.

Field, J. E. (1933). Data on Castlewood Dam failure and flood. *Eng. News-Rec.*, Sept. 7, pp. 279–280.

Finch, R. H. (1937). A tree-ring calendar for dating volcanic events, Cinder Cove, Lassen National Park, California. *Am. J. Sci.* **33**, 140–146.

Follansbee, R., and Jones, E. E. (1922). The Arkansas River flood of June 3-5, 1921. *Geol. Surv. Water-Supply Pap. (U.S.)* **487**, 1–44.

Fread, D. L. (1980). "DAMBRK—The NWS Dam-Break Flood Forecasting Model." National Weather Service, Office of Hydrology, Silver Spring, Maryland.

Freeman, G. L., and Alsop, R. B. (1941). Underpinning Austin Dam. *Eng. News-Rec.*, Jan. 30, pp. 52–57.

Gallino, G. L., and Pierson, T. C. (1985). The 1980 Polallie Creek debris flow and subsequent dam-break flood, East Fork Hood River basin, Oregon. *Geol. Surv. Water-Supply Pap. (U.S.)* **2273**, 1–22.

Glazyrin, G. Ye., and Reyzvikh, V. N. (1968). Computation of the flow hydrograph for the breach of landslide lakes. *Sov. Hydrol.* **5**, 592–596.

Graham, W. J. (1980). "Value of Inundation Maps in Dam Failure Emergencies." U.S. Bureau of Reclamation, Denver, Colorado (unpublished report).

Graham, W. J. (1983). "Dam Failure Warning Effectiveness." U.S. Bureau of Reclamation, Denver, Colorado (unpublished report).

Gruner, E. (1963). Dam disasters. *Proc. Inst. Civ. Eng.* **24**, 47–60.

Gunn, J. P. (1930). The Shyok flood, 1929. *Himalayan J.* **2**, 35–47.

Haeberli, W. (1983). Frequency and characteristics of glacier floods in the Swiss Alps. *Ann. Glaciol.* **4**, 85–90.

Hagen, V. K. (1982). Re-evaluation of design floods and dam safety. Paper presented at 14th International Commission on Large Dams Congress, Rio de Janeiro.

Hanly, T. F. (1964). Sediment studies on the Madison River after the Hebgen Lake earthquake. *Geol. Surv. Prof. Pap. (U.S.)* **435-M**, 151–158.

Hartley, J. A. (1981). Boston's great molasses flood. *Mod. Maturity* **24** (4), 16–18.

Hewitt, K. (1982). Natural dams and outburst floods of the Karakoram Himalaya. *Int. Assoc. Hydrol. Sci.* **138**, 259–269.

Hildreth, W. (1983). The compositionally zoned eruption of 1912 in the Valley of Ten Thousand Smokes, Katmai National Park, Alaska. *J. Volcanol. Geotherm. Res.* **18**, 1–56.

Hutchinson, G. E. (1957). "A Treatise on Limnology," Vol. 1. Wiley, New York.

International Commission on Large Dams (1973). "Lessons from Dam Incidents," Abridged ed. U.S. Commission on Large Dams, Boston, Massachusetts.

Investigation Team of the Tanggudang Landslide Dam (1967). "An Investigation Report on the Landslide Dam of Tanggudang on the Yalong River" (in Chinese). Chinese Science Committee, Chengdu (unpublished report).

Investigation Team of the Zepozhu Landslide-dammed River (1967). "Investigation Report of the Zepozhu Landslide-Dammed Lake, Dong River, Xichang County, Sichuan Province (in Chinese) (unpublished report).

Jansen, R. B. (1980). "Dams and Public Safety." U.S. Department of the Interior, Bureau of Reclamation, Denver, Colorado.

Jarrett, R. D., and Costa, J. E. (1986). Hydrology, geomorphology, and dam-break modeling of the July 15, 1982 Lawn Lake Dam and Cascade Lake Dam failures, Larimer County, Colorado. *Geol. Surv. Prof. Pap. (U.S.)* **1369**, 1–78.

Jennings, M. E., and Schneider, V. R., and Smith, P. E. (1981). Computer assessments of potential flood hazards from breaching of two debris dams, Toutle River and Cowlitz River systems. *Geol. Surv. Prof. Pap. (U.S.)* **1250**, 829–836.

Jessup, W. E. (1964). Baldwin Hills Dam failure. *Civ. Eng. (N.Y.)* **34**, 60–62.

Johnson, F. A., and Illes, P. (1976). A classification of dam failures. *Int. Water Power Dam Construct.*, Dec., pp. 43–45.

Keefer, D. K. (1984). Landslides caused by earthquakes. *Geol. Soc. Am. Bull.* **95**, 406–421.

Kiersch, G. A. (1964). The Vaiont River disaster. *Civ. Eng. (N.Y.)* **34**, 32–39.

Kirkpatrick, G. W. (1977). Evaluation guidelines for spillway adequacy. *In* "The Evaluation of Dam Safety," Proc. Eng. Found. Conf., pp. 395-414. Am. Soc. Civ. Eng., New York.

Knight, D. K., and Bennett, P. T. (1960). Stability of slide dam and recommendations on development of overflow spillway. *In* "Report on Flood Emergency, Madison River Slide," Appendix VI. U.S. Army Corps of Engineers, Riverdale, North Dakota.

Lamke, R. D. (1972). Floods of the summer of 1971 in south-central Alaska. *Geol. Surv. Open-File Rep. (U.S.)*, pp. 1–88.

Land, L. F. (1980a). Evaluation of selected dam-break floodwave models by using field data. *Geol. Surv. Water-Resour. Invest. (U.S.)* **80-44**, 1–54.

Land, L. F. (1980b). Mathematical simulations of the Toccoa Falls, Georgia, dam-break flood. *Water Resour. Bull.* **16**, 1041-1048.

Lane, F. W. (1965). "The Elements Rage," p. 165. Chilton Books, Philadelphia and New York.

Lee, K. L., and Duncan, J. M. (1975). "Landslide of April 25, 1974 on the Mantaro River, Peru." Natl. Acad. Sci., Washington, D.C.

Li, G.-S., and Hu, T.-F. (1982). The activity and taming of Zhouqu landslide-earthflow. *In* "Landslides and Mudflows," Rep. Alma-Ata Int. Semin., pp. 466–470. Centre of International Projects, GKNT, Moscow.

Lliboutry, L., Arnao, B. M., Pautre, A., and Schneider, B. (1977). Glaciological problems set by the control of dangerous lakes in Cordillera Blanca, Peru. I. Historical failures of morainic dams, their causes and prevention. *J. Glaciol.* **18**, 239–254.

Lubbock, G. (1894). The Gohna Lake. *Geogr. J.* **4**, 457.

MacDonald, T. C., and Langridge-Monopolis, J. (1984), Breaching characteristics of dam failures. *J. Hydraul. Eng., Am. Soc. Civ. Eng.* **110** (5), 567–586.

Mark, R. K., and Stuart-Alexander, D. E. (1977). Disasters as a necessary part of benefit-cost analyses. *Science* **197**, 1160-1162.

Mason, K. (1929). Indus floods and Shyok glaciers. *Himalayan J.* **1**, 10–29.

Mathews, W. H. (1965). Two self-dumping ice-dammed lakes in British Columbia. *Geogr. Rev.* **55**, 46-52.

Mathews, W. H. (1973). Record of two jökulhlaups. *IASH-AISH Publ.* **95**, 99–110.

McMahon, G. F. (1981). Developing dam-break flood zone ordinance. *J. Water Resour. Div. Am. Soc. Civ. Eng.*, **107** (WR2), 461–476.

Meier, M. F., Carpenter, P. J., and Janda, R. J. (1981). Hydrologic effects of Mount St. Helen's 1980 eruption. *EOS, Trans. Am. Geophys. Union* **62**, 625–626.

Morrison, R. E. (1982). United States dams-statistical data. *Assoc. Eng. Geol. Newsl.* **25**, 19–21.

Nairn, I. A., Wood, C. P., and Hewson, C. A. Y. (1979). Phreatic eruptions of Ruapehu—April 1975. *N. Z. J. Geol. Geophys.* **22**, 155–173.

Nash, D. F. T., Brunsden, D. K., Hughes, R. E., Jones, D. K. C., and Whalley, B. F. (1985). A catastrophic debris flow near Gupis, northern areas, Pakistan. *Proc. Int. Conf. Soil Mech. Found. Eng., 11th, 1985*, pp. 1–4.

Nye, J. F. (1976). Water flow in glaciers—Jökulhlaups, tunnels, and veins. *J. Glaciol.* **17**, 181–207.

Office of the State Engineer (1913-1914). "Ninth Biennial Report of State Engineer." Salt Lake City, Utah.

Outland, C. F. (1977). "Man-Made Disaster—The Story of St. Francis Dam." Arthur H. Clark Co., Glendale, California.

Pacific Gas and Electric Company (1983). "Description of the Environment Affected by the Lost Canyon Pipe Failure. PG&E Co., San Francisco, California.

Petrascheck, A. W., and Sydler, P. A. (1984). Routing of dam break floods. *Int. Water Power Dam Construct.* **36**, 29–32.

Post, A., and Mayo, L. R. (1971). Glacier dammed lakes and outburst floods in Alaska. *U.S. Geol. Surv. Hydrol. Invest. Atlas* **HA-455**, 1–10.

Price, J. T., Lowe, G. W., and Garrison, J. M. (1977). Unsteady flow modeling of dam-break waves. *In* "Dam-Break Flood Routing Model Workshop." U.S. Water Resources Council, Hydrology Committee, Bethesda, Maryland (available only from National Technical Information Service, Springfield, Virginia, PB-275-437).

Ray, H. A., and Kjelström, L. C. (1978). The flood in southeastern Idaho from the Teton Dam failure of June 5, 1976. *Geol. Surv. Open-File Rep. (U.S.)* **77-765**, 1–48.

Ritter, A. (1892). Die fortpflanzung der wasserwellen. *VDI-Z.* **36** (33), 947–954.

Rose, D. (1978). Risk of catastrophic failure of major dams. *J. Hydraul. Div. Am. Soc. Civ. Eng.* **104** (HY9), 1349–1351.

Rostvedt, J. O., et al. (1968). Summary of floods in the United States during 1963. *Geol. Surv. Water-Supply Pap. (U.S.)* **1830-B**, B84-B86.

Sager, J. W., Griffiths, J. B., and Fargo, N. J. (1984). "Spirit Lake Outlet Tunnel, Tunnel. Technol. Newsl., No. 48, pp. 1-5. National Research Council, Washington, D.C.

Sanders, C. L., and Sauer, V. B. (1979). Kelly Barnes Dam flood of November 6, 1977, near Toccoa, Georgia. *U.S. Geol. Surv. Hydrol. Invest. Atlas* **HA-613**, 2 sheets.

Schumm, S. A. (1969). River metamorphosis. *J. Hydraul. Div., Am. Soc. Civ. Eng.* **95** (HY1), 255-273.

Schuster, R. L. (1984). Effects of landslides and mudflows associated with the May 1980 eruption of Mount St. Helens, northwestern U.S.A. *J. Jpn. Landslide Soc.* **21**, 1–10.

Schuster, R. L. (1985). Landslide dams in the western United States. *Proc. Int. Conf. Field Workshop Landslides, 4th, 1985*, Tokyo, pp. 1–8.

Scott, K. M., and Gravlee, G. C., Jr. (1968). Flood surge on the Rubicon River, California—Hydrology, hydraulics, and boulder transport. *Geol. Surv. Prof. Pap. (U.S.)* **422-M**, 1–40.

Scott, K. M., and Janda, R. J. (1982). Preliminary map of lahar inundation during the Pine Creek eruptive period in the Toutle-Cowlitz River system, Mount St. Helens, Washington. *Geol. Surv. Water-Resour. Invest. Rep. (U.S.)* **82-4067**.

Silva, L., Cocheme, J. J., Canul, R., Duffield, W. A., and Tilling, R. I. (1982). "El Chichon Volcano," Bull., Vol. 7, No. 5, pp. 2–6. Scientific Event Alert Network (SEAN), Washington, D.C.

Snow, D. T. (1964). Landslide of Cerro Condor-Sencea, Department of Ayacucho, Peru. *In* "Engineering Geology Case Histories" (G. A. Kiersch, ed.), No. 5; pp. 1-6. *Geol. Soc. Am.*, Boulder, Colorado.

Soil Conservation Service (1981). "Simplified Dam-Breach Routing

Procedure, Tech. Release No. 66. U.S. Dept. of Agriculture, Washington, D.C.

Stermitz, F. (1964). Effects of the Hebgen Lake earthquake on surface water. *Geol. Surv. Prof. Pap. (U.S.)* **435-L**, 139–150.

Stone, K. H. (1963). The annual emptying of Lake George, Alaska. *Arctic* **16**, 26–40.

Strachey, R. (1894). The landslip at Gohna, in British Garwhal. *Geogr. J.* **4**, 162–170.

Swanson, F. J., Graham, R. L., and Grant, G. E. (1985). Some effects of slope movements on river channels. *In* "Proceedings of the International Symposium on Erosion, Debris Flow, and Disaster Prevention, Tsukuba, 1985." Erosion Control Engineering Society of Japan, Tsukuba. 1–6.

Swift, C. H., and Kresch, D. L. (1983). Mudflow hazards along the Toutle and Cowlitz Rivers from a hypothetical failure of Spirit Lake blockage. *Geol. Surv. Water-Resour. Invest. Rep. (U.S.)* **82-4125**, 1–10.

Thorarinsson, S. (1953). Some new aspects of the Grimsvötn problem. *J. Glaciol.* **2**, 267–275.

Thorarinsson, S. (1957). "The Jökulhlaup from the Katla Area in 1955 Compared with Other Jökulhlaups in Iceland," Misc. Pap. No. 18, pp. 21–25. Museum of Natural History, Reykjavik.

Tufuell, L. (1984). Glacier hazards: London, Longman, 97 p.

Watters, R. J. (1983). A landslide induced waterflood-debris flow. *Bull. Int. Assoc. Eng. Geol.* **28**, 177–182.

Wetmore, J. N., and Fread, D. L. (1981). The NWS simplified dam-break flood forecasting model. *Proc. Can. Hydrotech. Conf., 5th, 1981*, pp. 1–24.

Young, G. J. (1980). Monitoring glacier outburst floods. *Nord. Hydrol.* **11**, 285–300.

Zen, M. T., and Hadikusumo, D. (1965). The future danger of Mount Kelut (eastern Java-Indonesia). *Bull. Volcanol.* **28**, 275–282.

27

THE TUCSON, ARIZONA, FLOOD OF OCTOBER 1983

Implications for Land Management along Alluvial River Channels

PETER L. KRESAN
Department of Geosciences, University of Arizona, Tucson, Arizona

IMMEDIATE IMPACT OF THE FLOOD

Too Few Planned for Storm . . .

Nobody planned it that way. Nobody wanted people killed or property destroyed. But some people who wanted to make money apparently guessed wrong over the years. And a lot of people suffered over the weekend. It wasn't all fate or bad luck. Nor was it simply an act of God. It was partly the result of our head-long pursuit of growth at any cost. And this time the cost was very high.

—*Arizona Daily Star,* Oct. 5, 1983, p. 15A.

"Two amendments to the Pima County floodplain ordinance approved Tuesday by the Board of Supervisors wouldn't have passed before the recent flooding" the head of the county Department of Transportation and Flood Control District said. "When the sun was shining (the amendments) didn't stand a chance."

—*Arizona Daily Star,* Oct. 15, 1983

One immediate impact of the October 1983 flood was to jar the Tucson community and Pima County into reconsidering its land-use policy along watercourses. The debate about developing flood-prone land has focused on four issues. First, major uncertainties are inherent in estimates of the likely discharge for the regulatory (100-yr) flood. Methods applied throughout the United States for estimating recurrence intervals of floods typically assume that fluvial hydrologic and hydraulic characteristics have not changed over the period of recorded flood events. Changes in floodplain vegetation, river channel configuration, and climate may invalidate this assumption for streams in arid regions such as the southwestern United States.

Second, given an estimate for the regulatory flood discharge, flood-prone areas are determined according to national standards, which assume a stable stream channel. Yet, historic documentation and the 1983 flood experience clearly demonstrate profound morphologic changes in the alluvial channels in Tucson over the last 100 yrs. Present national standards appear to be inadequate for defining flood-prone land along alluvial channels in southwestern United States.

Third, most of the 1983 flood-related damage in the Tucson basin was due to erosion and collapse of channel banks, rather than inundation from overbank flow. Severe bank erosion occurred while the "flood" waters were contained within the channel. However, federal policy (Flood Protection Act of 1973) recognizes the hazard of erosion only in association with unusually high levels of water, as during a flash flood.

Fourth, the primary approach in Pima County to flood hazard control is based on structural "solutions," like channelization and soil cement channel bank protection. An alternative would be to use federal, state, and local funds to purchase flood-prone land.

The magnitude and timing of the October 1983 Tucson flood fortuitously focused the debate about land-use policies along the watercourses in southern Arizona (Peirce and

Kresan, 1984). Most of the land along the rivers through Tucson is privately owned and just beginning to be developed. However, development is progressing rapidly as Pima County experiences unprecedented growth. The experience of the Tucson flood of 1983 emphasizes the continuing need for study, debate, and re-evaluation of floodplain management policies.

> The flooding that swept the region three years ago left 13 people dead and millions of dollars in damages—much of it in Pima County.
>
> Today, roughly two-thirds of the 1983 damage is repaired, county officials say.
>
> But three more years of work remain before the end of a $105.8 million program to repair flood damage and safeguard against future floods.
>
> —*Arizona Daily Star*, Oct. 1, 1986, p. 9B

GEOGRAPHIC SETTING

Tucson is located in the Sonoran Desert region of the basin and range provinces, characterized by low-lying, alluvial-filled basins alternating with mountains. The elevation of Tucson is 728 m (2389 ft), where semi-arid conditions prevail. Average annual rainfall for Tucson is 28.3 cm (11.14 in.). In contrast, the surrounding mountains reach heights of 1500–3000 m (5000–10,000 ft) and precipitation on the higher peaks ranges from 76 to 100 cm (30–40 in.) per year.

TUCSON'S ALLUVIAL CHANNELS

The Santa Cruz River and its tributaries, Rillito Creek and Canada del Oro (Fig. 1) are ephemeral streams. Runoff occurs mainly as overland flow that reaches the nearest stream channel in minutes to hours from the onset of rainfall. Much of the runoff from urban and suburban Tucson is collected by Rillito Creek, which is formed by the confluence of Tanque Verde Creek and Pantano Wash. These rivers have alluvial channels, typical of the drainage systems in southwestern United States (Maddock, 1976). Channels are incised within deep fluvially deposited basin fill. Because of the low bank resistance of unconsolidated sediments, changes in shape and position of channels can be substantial. Adjustments in channel characteristics can occur in response to the frequency, magnitude, and duration of streamflow (Wolman and Brush, 1961).

Studies of the Santa Cruz River (Hays, 1984) and Rillito Creek (Pearthree, 1982) show that channel widening and floodplain destruction occur during flows with broad peaked hydrographs and relatively longer duration, typically caused by winter storms. In contrast, periods of progressive floodplain reconstruction and channel narrowing (Fig. 2) occur during low to moderate flows of shorter duration. Because streamflow is depleted rapidly by infiltration into extremely permeable streambeds, major reduction of flood peaks can occur, particularly for shorter duration flows. With prolonged flow, decreased channel bed infiltration due to bank and bed saturation results in decreased sediment concentrations and increased erosive power of the stream (Matlock, 1965; Hays, 1984). Changes in the characteristics of alluvial channels can be dramatic (Simons et al., 1986). For example, Hays (1984) documented an increase from 8.23 to 137.66% in the area of floodplain occupied by the channel from 1982–1984 along a section of the Santa Cruz near Marana, Arizona (Fig. 3).

Changes in channel location result from two different processes: (1) channel shifting (avulsion), associated with overbank flow that cuts off meanders and establishes a new active channel by scour and headward erosion of the floodplain (Figs. 4 and 5); and (2) meander migration through bank erosion (Fig. 2), which can occur during either high discharges or low discharges of long duration (Hays, 1984).

The dynamic character of alluvial channels in semi-arid regions has important implications for floodplain management policy. Changes in channel width and depth, changes in the position of the channel on the floodplain, and the slow rate of recovery to preflood conditions following flow events, all may modify the projected water surface elevations of a regulatory flow. Such changes necessitate constant revision of the 100-yr floodplain delineations, which assume fixed channel characteristics (Pearthree, 1982; Gordon, 1983; Hays, 1984; Pearthree and Baker, 1987). Thus, the 100-yr flood hazard zone, established prior to a flood by standard procedures, may be completely invalidated by a single flow event.

During October 1983 local channel migration moved the floodway, which includes the active channel and floodplain necessary for conveyance of a 100-yr flow, into zones predicted to have only the risk of overbank floods rarer than the 500-yr event (Baker, 1984) (Fig. 6). Pearthree's (1982) study of Rillito Creek concluded that 100-yr floodplain delineations will be ineffective unless historical channel changes are considered.

ENTRENCHMENT AND HISTORIC CHANGES

Widespread arroyo cutting occurred in the southwest during the late 1800s and early 1900s (Cooke and Reeves, 1976; Hastings and Turner, 1972). Entrenchment along the Tucson reach of the Santa Cruz River may have been affected by: (1) urbanization and associated channelization of river systems (Betancourt and Turner, 1987); (2) mining sand and gravel from the channel floodplain (Bull and Scott, 1974; Laursen, 1985); (3) changes in slope due to isostatic rebound associated with unloading by groundwater withdrawal

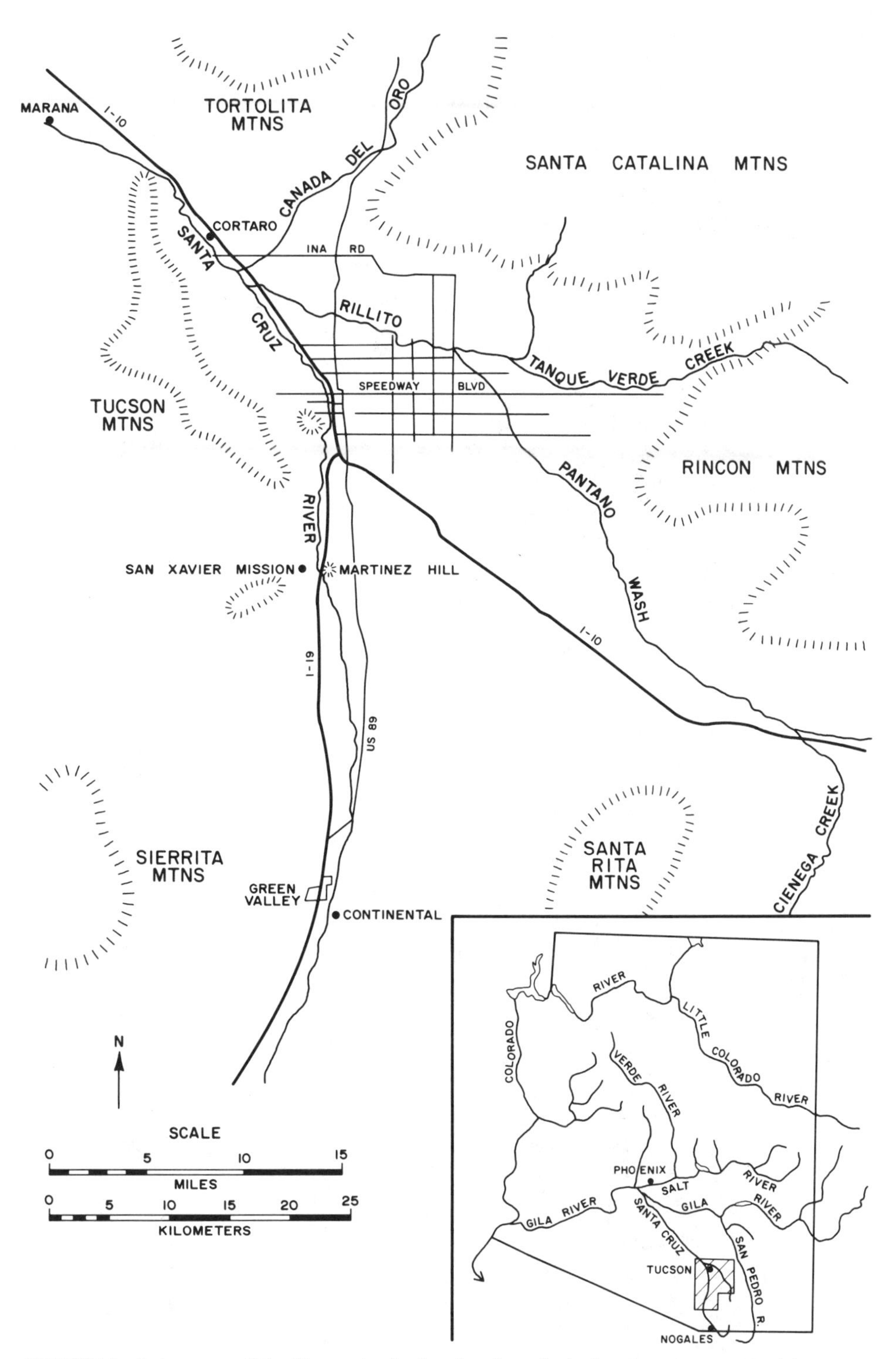

FIGURE 1. Index map of the Tucson basin showing the principal watercourses, mountain ranges, and transportation routes (Saarinen et al., 1984).

FIGURE 2. Views of the Santa Cruz River at Martinez Hill.

FIGURE 3. Aerial view to the south of the Santa Cruz River from near Marana on October 4, 1983.

FIGURE 4. Headward erosion of a section of the Santa Cruz River floodplain, scoured by overbank flow in October, 1983.

FIGURE 5. North of the confluence with Rillito Creek, overbank flow scoured a low-sinuosity channel through a vegetation-lined meander bend of the Santa Cruz River.

FIGURE 6. Kostka Ave. at the Santa Cruz River, showing bank erosion undermining streets and houses.

(Holzer, 1977, 1979); (4) changes in slope caused by ground subsidence induced by excessive withdrawal of groundwater (Winikka, 1984; S. Anderson, U.S. Geol. Surv. Water Resources Div., Tucson, personal communication, 1985); (5) destruction of stabilizing riparian vegetation in the floodplain and along the channel banks also caused by lowering of the water table (Applegate, 1981); and (6) destruction of floodplain vegetation (Reich and Davis, 1985, 1986b).

Betancourt and Turner (1987) document historic changes along the Tucson section of the Santa Cruz River, providing an especially good record of arroyo cutting. During the early 1800s the Santa Cruz was characterized by a cottonwood-lined, poorly defined channel (Fig. 7). The transition to modern conditions is shown by comparing Figures 7 and 8. Headcut erosion began along the Santa Cruz during the late 1800s (Fig. 9). The most recent and continuing phase of channel entrenchment along the Tucson section of the Santa Cruz may be attributed to channel construction associated with landfill operations, and mining of sand and gravel from the channel and floodplain for highway and urban construction (Betancourt and Turner, 1987; Laursen and Carmody, 1980; Laursen, 1985). Even though the amount of discharge carried by the Santa Cruz has most probably increased due to urbanization and channel manipulation, entrenchment has, in effect, lowered the flood crests through Tucson.

The historic record for the Santa Cruz through Tucson manifests the responsiveness of alluvial channels to the changing characteristics of the drainage basin. There is no doubt that dramatic changes in the river channel have occurred over the length of this record. Yet, the impact of historic changes in the channel and drainage basin are not typically taken into consideration as significant factors influencing estimations of the regulatory flood.

THE STORM

"This was the flood we'll remember! This flood will set the standards by which we judge the rivers' dangers."
—*Arizona Daily Star*, Oct. 17, 1983

September 1983 was the second wettest September on record. On September 31, 1983, tropical storm Octave and a low-pressure trough brought additional rainfall to southern Arizona. In Tucson, the most rainfall (6.9 cm; 2.72 in.) fell on October 1. On Sunday, October 2, a cold

FIGURE 7. Historic views from A-Mountain of the Santa Cruz River and Tucson.

FIGURE 8. Aerial view to the southwest on May 27, 1985, of the Santa Cruz River and A-Mountain/Tumamoc Hill just west of downtown Tucson.

FIGURE 9. Historic photograph of a Santa Cruz headcut as it was eroding during a minor flood in October 1889.

·ont moved across southern Arizona and triggered more recipitation. In all, 17.1 cm (6.71 in.) of rainfall, just ver half of Tucson's average annual rainfall of 28.3 cm 11.14 in.) was officially recorded from September 28 to ctober 1, 1983. Other areas in southern Arizona received s much as 36 cm (14 in.) of precipitation over this same orm period.

Smith (1986, p. 220) concludes, "eastern north Pacific opical cyclones (like Octave) have and will continue to e associated with heavy precipitation and floods in late ummer and early fall over much of the southwestern United tates. Given the rapid growth and urbanization of many ities in the southwest, these storms most likely will cause any serious floods in the future."

On October 1 river systems in southern Arizona were welling to their capacity (Fig. 10). Record flows were easured in Arizona for the San Francisco River at Clifton, ith 2550 m^3/s (90,900 cfs); the Gila River at Calva, with 250 m^3/s (15,000 cfs); the San Pedro River at Winkelman, ith 3120 m^3/s (110,000 cfs); and the Santa Cruz River : Congress Street in Tucson with 1490 m^3/s (52,700 cfs) R. H. Roeske, written communication, U.S. Geol. Surv., Vater Resour. Div., Tucson, 1986).

HYDROLOGIC CONSIDERATIONS

"The 100-year flood has come and gone, so, by all rights, Tucsonans should enjoy another century of great southwestern weather."

—postflood message sent to national media by the Metropolitan Tucson Convention and Visitor's Bureau.

The estimated flood peak of 1490 m^3/s (52,700 cfs) on e Santa Cruz at Congress Street in October 1983 exceeded y a factor of 2 any other flood recorded at that station nce 1915. It also exceeded by a factor of 1.75 the magnitude the 100-yr flood as designated in the Federal Emergency lanagement Agency (FEMA) (1982) Flood Insurance Study r Tucson. Furthermore, FEMA had estimated the reg-atory flood discharge upward from the magnitude estimated y standard procedures (Saarinen et al., 1984).

What is the 100-yr flood? Answering this question is ucial to the establishment of local floodplain manage-ent ordinances as mandated by the 1973 National Flood isaster Protection Act. Yet, as Table 1 illustrates, estimates ary significantly for the 100-yr flood discharge at this one te. Such contradictory figures lead to confusion among e formulators of land-use policy for floodplains of alluvial vers in the southwest.

Figure 11 summarizes the flow record of the Santa Cruz iver at Congress Street. Pre-1983 flood events on the anta Cruz and other southwestern rivers are discussed by Aldridge and Eychaner (1984) and by Sauer and others (1983). Various methodologies for estimating the 100-yr discharge are discussed by Reich and Renard (1981). In 1982 FEMA assigned a regulatory flow of 880 m^3/s (31,000 cfs) to the Rillito at its confluence with the Santa Cruz. Rillito Creek has a drainage basin of 2422 km^2 (935 mi^2). Yet, the Santa Cruz with a drainage basin of 5910 km^2 (2282 mi^2) above the confluence with Rillito Creek was assigned a 100-yr discharge of 850 m^3/s (30,000 cfs). Nevertheless, this 100-yr discharge estimate for the Santa Cruz agrees with the "provisional revised" 100-yr flood discharge of 880 m^3/s (31,000 cfs) determined by the U.S. Geological Survey, based on annual peak discharge records at the site, including the October 1983 flood and using standard procedures (U.S. Water Resources Council, 1981), weighted with a regional regression formula (Eychaner, 1984). Without including the October 1983 record peak flow, these standard procedures yield a 100-yr discharge of 660 m^3/s (23,200 cfs) (Eychner, 1984).

Based on a regional analysis, the U.S. Army Corps of Engineers in 1972 estimated the "intermediate regional flood" for the Santa Cruz at Tucson to be 1300 m^3/s (46,000 cfs). Before estimates assigned by FEMA were available, it was common practice to use the Corps of Engineers' "intermediate regional flood" as an approximate guideline for the 100-yr discharge (B. M. Reich, personal communication, 1985).

The Pima County Board of Supervisors passed a revised floodplain management ordinance in May 1985. For this new ordinance, the regulatory flood discharge was revised by the County Engineer to 1700 m^3/s (60,000 cfs) for the Santa Cruz above Rillito Creek. This is double the 1982 FEMA regulatory discharge. In addition, the county decided to adopt a dual standard and designate a "design flood" of 1980 m^3/s (70,000 cfs) applicable to bridges and flood control structures (Pima County, 1986a). Instead of following the standard procedures (U.S. Water Resources Council, 1981), Pima County utilized the results of a rainfall–runoff model, "calibrated" by the response of the Santa Cruz basin to the October 1983 storm (Hydrosoftware, 1984).

Brian Reich, former Assistant Tucson City Engineer for floodplain management, proposed a 100-yr discharge of at least 2700 m^3/s (96,000 cfs) for the Santa Cruz above the confluence with Rillito Creek. Reich reached this approximation using only the peak annual discharges for the last 24 yr (1960–1984) (Reich, 1985; Reich and Davis, 1985, 1986a). With the exception of the 1915 flow, the six largest flows for the Santa Cruz through Tucson occurred after 1960 (Fig. 11). This is the same period over which Pima County experienced unprecedented growth, with its population increasing from 265,700 in 1960 to 603,300 in 1984. Reich and Davis (1985, 1986b) suggested that the flow record for the last 24 yr provided a more accurate

FIGURE 10. Views of Santa Cruz River from bridge for St. Mary's Road, Tucson, Arizona, during 1983 flood (top) and immediately after flow recession (bottom).

TABLE 1 Estimates of the 100-Yr Flood Discharge for the Santa Cruz River, Tucson, above the Confluence with Rillito Creek

Estimated Discharge	Notes	Reference
1,300 m^3/s (46,000 cfs)	"Intermediate regional, flood"—regional, watershed analysis	U.S. Army Corps of Engineers, 1972, p. 16.
850 m^3/s (30,000 cfs)	Regulatory flood for, preparation of flood, insurance maps for Tucson in 1975	Reich and Davis, 1986a
1,270 m^3/s (45,000 cfs)	"Design" flood for the Rio Nuevo Project, Tucson	Simons et al., 1981
640 m^3/s (22,600 cfs)	Log-Pearson III analysis based on 61 yr, beginning in 1914	Roeske, 1978
623 m^3/s (22,000 cfs)	Log-Pearson III analysis based on annual peak flow up to 1975	Laursen and Carmody, 1980, p. 17
1,840 m^3/s (65,000 cfs)	Regional watershed analysis	Laursen and Carmody, 1980, p. 18
850 m^3/s (30,000 cfs)	Regulatory flood, adopted by the City of Tucson and accepted by FEMA	FEMA, 1982
1,420 m^3/s (50,000 cfs)	Analysis based on rainfall from October 1983 storm	Reich and Davis, 1986a
660 m^3/s (23,200 cfs)	Log-Pearson III weighted with new regional regression	Eychner, 1984
2,180 m^3/s (77,000 cfs)	Log-Pearson III/log Boughton	Boughton and Renard, 1984
1,700 m^3/s (60,000 cfs)	Rainfall–runoff model, calibrated by October 1983 storm. Adopted as the regulatory flood by Pima County in 1985	Hydrosoftware, 1984
880 m^3s (31,000 cfs)	Provisional revised estimate, determined by LP III, weighted with regional regression (Eychner, 1984), including October 1983 peak.	W. Hjalmerson, personal communication, 1985
2,830 m^3/s (100,000 cfs)	Log extreme value, based on flow record from 1960–1984	Reich and Davis, 1985
2,860 m^3/s (101,000 cfs)	Boughton and Renard equation, based on area of watershed	Reich and Davis, 1986a

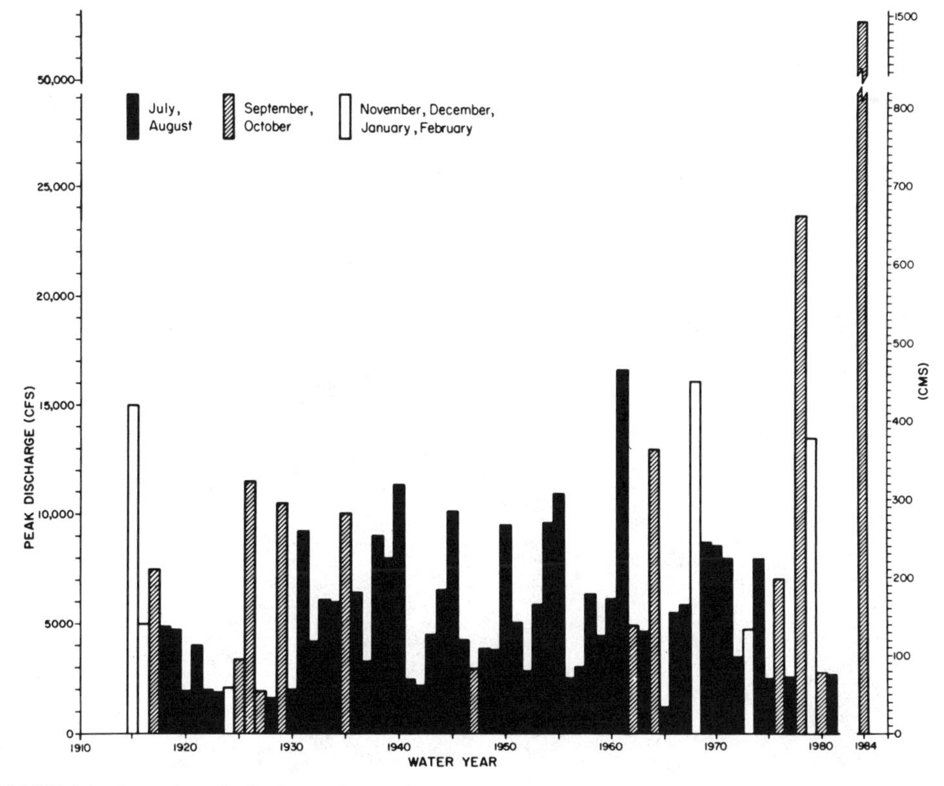

FIGURE 11. Annual peak discharge for the Santa Cruz River at Congress Street between the water years 1915 and 1984.

estimation of the flood potential along the Santa Cruz in Tucson because the impact of dramatic changes in the channel and drainage basin (Figs. 12 and 13) over this period of rapid growth is not diluted by the record from earlier years.

FEMA officially published in 1982 a regulatory flood discharge of 850 m^3/s (30,000 cfs) for the Santa Cruz at Tucson. The U.S. Geological Survey used this estimate for the preparation of floodplain maps for the City of Tucson. The record discharge, damage, and dislocation caused by the October 1983 flood shocked Pima County (1985) into re-evaluating the regulations for managing flood-prone land. The response was to revise the floodplain management ordinance and amend the regulatory discharge to 1700 m^3/s (60,000 cfs) for the Santa Cruz River at Tucson.

In summary, there are major uncertainties in estimating the likely discharge for the regulatory flood because of the nonstationarity of the flood record. Changes in the watershed, channel characteristics, floodplain and channel bank vegetation, and possibly climate (Hirschboeck, 1985) may contribute to changes in the amount and transmission of discharge through the alluvial drainage system. Such uncertainties lead to inevitable problems in delineating the floodway for a regulatory flood. Reich and Davis (1986b, p.12) suggest that "it may be better to predict future floods from shorter post-change periods; and to temper the estimates with safety factors."

FLOODING

In contrast to the entrenched reaches through Tucson, overbank flows occurred during the 1983 flood along the Santa Cruz River northwest of Tucson, downstream from the confluence with Rillito Creek and Canada del Oro (Fig. 14). Over the last century channel entrenchment and bank erosion along the Tucson reach of the Santa Cruz River provided increased sediment supply for areas downstream, such as Marana, which is currently a zone of aggradation (Betancourt and Turner, 1987). During the October 1983 flood, Saarinen and others (1984) concluded that "sediment transported from the extensive areas of erosion upstream contributed to aggradation in the Marana area. The aggradation, in turn, led to more extensive inundation."

Floodwaters reached Marana High School, located 3.2 km (2 mi) from the river channel. All river crossings at Marana were washed out and about 2000 residents were

FIGURE 12. View to the southeast of the Santa Cruz River from just south of Martinez Hill, upstream of the City of Tucson.

FIGURE 13. Historic photograph (Arnold, 1940), ca. 1940, of a mesquite–hackberry–greythorn community near San Xavier Mission.

evacuated (Fig. 15). The *Tucson Citizen* (October 4, 1983, p. 5A) reported:

> Only two national flood insurance policies were issued in the Marana area before flooding inundated the whole area, because the town officials "didn't believe it floods here," a flood insurance official said. The regional administrator for the National Flood Insurance Program met in January with town officials about advising residents to take out such policies. "They told us our flood maps were inaccurate. They didn't believe it floods there.

BANK EROSION

> ***Supervisors OK Moratorium on Floodplain Development***
>
> Boundaries of floodplains and floodprone areas have changed considerably since this month's flooding, which sent some entire floodplains crashing into raging rivers. Some areas that three weeks ago were considered safe from the worst flood are now located squarely in a floodway.
>
> —*Arizona Daily Star*, October 12, 1983

Lateral erosion of the river banks caused by far the greatest damage during the October 1983 flood (Figs. 6 and 16–18). The amount of bank erosion was probably influenced by several factors, including magnitude of discharge, duration of flow, and availability of sediment for transport (U.S. Army Corps of Engineers, 1985). The silt-clay content of the bank and the amount of riparian vegetation can also influence the erodibility of the channel banks (Pearthree, 1982; Applegate, 1981). A significant reduction in riparian vegetation has probably occurred throughout the Tucson basin because of recent lowering of the groundwater table due to overdraft (Figs. 12 and 13). Incomplete artificial bank protection was found to act locally in amplifying bank erosion along Tucson's major rivers during the October 1983 flood (Baker, 1984).

Bank erosion can be a significant hazard even for moderate to large nonflood flows along the alluvial channels of the Tucson basin. Lateral bank migration has caused land outside the designated 100- to 500-yr floodplain to collapse into the channel (Fig. 6). Studies have documented shifts in channel alignment of Rillito Creek of nearly 305 m (1000 ft) since the 1940s. "During a single 100-year flood, certain areas could have a lateral migration of as much as several thousand feet" (U.S. Army Corps of Engineers, 1985, p. 74). Because flood damage assessments generally consider inundation alone, the potential flood-related damages for rivers in arid and semi-arid regions

FIGURE 14. La Puerta del Norte Trailer Court near the Santa Cruz River at the northeast end of the Tucson Mountains showing 1983 flood sedimentation.

FIGURE 15. View to the south of Marana, Arizona, on October 9, 1983.

FIGURE 16. Flood damage to buildings at First Avenue and Rillito Creek, showing effects of bank erosion.

FIGURE 17. Flood-related damage due to bank erosion along Rillito Creek at Prince Road.

FIGURE 18. Flood damage to mobile home park near Golf Links Road and Pantano Wash. Note homes that were undermined at meander bend (top center) and carried downstream (bottom center).

may be significantly underestimated (Graf, 1984). Despite this observation, bank erosion, so commonly associated with nonflood flows along semi-arid alluvial channels, is not recognized as a significant hazard in federal floodplain management regulations.

In response to the October 1983 flood, Pima County's revised floodplain management ordinance, passed in May 1985, extended to 152 m (500 ft) the setback provision for structures on all property along unprotected channel banks for major river courses. The U.S. Army Corps of Engineers (1985, p. 76) reports that Pima County "has an internal policy of requiring river mechanics or similar studies before they will look favorably upon any new structures being built within 305 m (1000 ft) of any unprotected bank along the Rillito River." Previously, a setback distance of 30 m (100 ft) was required for commercial, industrial, and rental properties and 91 m (300 ft) for owner-occupied property. Except for a 15-m (50-ft) easement, setback provisions are waived when channel banks are stabilized by soil cement.

BANK PROTECTION: SOIL CEMENT

Soil cement is a compact mixture of soil and 10–14% cement, emplaced 3 m (8 ft) thick along channel banks. For major watercourses, county specifications require that the cemented bank extend to a minimum of one foot above the 100-yr water surface level for that locality and be anchored to below the depth of scour, often approximately 2.4–3 m (8–10 ft) below the dry channel bed. Soil cement bank protection costs approximately $1–$3 million per mile. Private developers of vacant, undeveloped property, lying adjacent to a flood repair or flood hazard mitigation project, contribute at least 50% of the cost for soil cement in Pima County project areas. If the section of river is not in a Pima County Improvement District, the developer pays the total cost (Pima County, 1984).

Although soil cement successfully protected channel banks during the October 1983 flood, severe bank erosion occurred immediately downstream form every protected bank along the deeply incised reaches of the Santa Cruz River and Rillito Creek (Figs. 19–21). The National Research Council Report on the Tucson Flood of October 1983 concluded that "piecemeal bank protection does not work. Clearly the rivers must be treated as a system. The areas along streams should either be left alone or completely protected" (Saarinen et al., 1984, p. 85). "From an overall river management perspective, piecemeal bank protection generates greater channel instability than does no protection at all" (Baker, 1984, p. 211). Complete bank protection would be enormously expensive and require the added cost of grade control structures to prevent downcutting that could undermine the protective works.

IMPLICATIONS FOR LAND MANAGEMENT ALONG ALLUVIAL RIVER CHANNELS

Floodplain management policy for Tucson's river systems was definitely influenced by the impact of the October 1983 flood. Questions were raised regarding the validity of assumptions and methods used to estimate the discharge of the regulatory flood. The regulatory flood discharges for Tucson's major streams were re-evaluated and upgraded by Pima County and the City of Tucson. In addition, because damages in Tucson due to channel bank erosion far exceeded damages caused by inundation, Pima County extended the setback provisions along unprotected channel banks and developed erosion hazard boundary maps along the Santa Cruz River (Simons et al., 1986) and Rillito Creek (Simons et al., 1984). Preliminary erosion hazard maps have been compiled based on analyses of historic changes in channel position, as documented by aerial photography, present channel patterns, locations of sand and gravel operations, landfill sites, and existing or planned bank stabilization measures (Pearthree, 1985).

Pearthree (1985) pointed out that the geomorphic complexity of alluvial stream systems in the Southwest creates difficulties for implementing federal regulations that are primarily based on flooding. In fact, dynamic and abrupt channel change, including fluctuations in channel characteristics and location, has generally constituted a greater hazard for Pima County than has overbank flow.

The director of the Pima County Department of Transportation and Flood Control District, Charles Huckelberry, stated: "In the future, we must look toward active acquisition and preservation of flood hazard areas that are in their natural state." Community ownership of the natural floodplain and channel minimizes flood hazards, enhances the potential for groundwater recharge, and preserves valuable open space and natural habitat. In fact, the county is spending $8.3 million to purchase flood-prone land. But this is a small commitment compared to the "$105 million program to repair damage and to safeguard against future floods" (*Arizona Daily Star*, October 1, 1986) and the proposed projects for Rillito Creek (U.S. Army Corps of Engineers, 1985; Pima County, 1986b). Huckelberry (1985) characterized the county's present approach to flood hazards a involving mostly "structural flood control solutions," including soil cement bank protection and channelization. One proposal envisions combining flood and erosion control with an attempt to enhance groundwater recharge by impounding storm water along a section of Rillito Creek (Pima County, 1986b).

If structural solutions are to be employed for the mitigation of flood hazards, the following considerations need to be weighed. First, piecemeal emplacement of structures to control flooding and erosion generally enhances flood and erosion hazards along unprotected sections, thereby

FIGURE 19. Sudden widening of the Santa Cruz channel, occurring at the downstream terminus of bank protection.

FIGURE 20. View downstream (north) along Santa Cruz River from St. Mary's Road Bridge, Tucson.

FIGURE 21. Damage to the downstream end of the soil cement bank protection on the east (right) bank of the Santa Cruz, just north of the St. Mary's road crossing, caused by the October 1983 flow.

creating the need for additional expenditures for flood hazard control. Second, there are inherent uncertainties in estimations of the design flood discharge. Third, structural solutions to flood hazard control encourages encroachment of urban development on the geomorphic floodplain. So, after buildings are optimistically constructed in these controlled areas, the potential for damage is greatly amplified if a flood overwhelms the control structures. Finally, over the long term, construction and maintenance costs may outweigh the costs of outright purchase of the floodplain land.

CONCLUSIONS

1. Flood experience in Tucson, Arizona, shows that the dynamic nature of alluvial channels in the semi-arid southwestern United States has important implications for floodplain management policy. The potential for dramatic changes in geomorphic and hydrologic characteristics of alluvial channels, in response to a single flow event, must be considered in establishing procedures for flood hazard evaluation.

2. Erosion of channel banks represents a significant hazard for both large floods and for lower magnitude flows of long duration. Structural works, such as soil-cemented bank protection, may aggravate bank erosion along unprotected reaches immediately downstream of the protected reaches. Piecemeal bank protection produces additional erosion, which then requires additional expenditures on more bank protection plus grade control structures. Clearly, alluvial channels in semi-arid regions must be treated as total systems.

3. Delineation of flood-prone land and the design of structures to mitigate flood hazards are predicted on estimates of the magnitude of discharge probable over a reasonable period of time. Nationally standardized procedures, based on statistical analysis of the annual peak discharges, generally assume a stationary mean. Inherent in this assumption is a lack of significant temporal changes in the hydrologic and hydraulic characteristics of the stream system. Such changes, including entrenchment and channelization, destruction of floodplain and channel bank vegetation, climatic fluctuation, and urbanization, affect the stream's response to a storm. Historical documentation of changes in land-use and channel characteristics along the Santa Cruz, coupled with the flow record of the Santa Cruz River at Tucson, substantiate a nonstationary mean.

4. Given the complex behavior of alluvial channels, the uncertainty in estimating the magnitude of the peak discharge likely to occur over a given period of time, the fact that flood peaks change and usually increase as land in the drainage basin is developed, and the costs and pitfalls of structural solutions (such as channelization and bank

protection), the wisest approach to land management along alluvial river channels is to maintain natural conditions by discouraging development in the geomorphic floodplain.

ACKNOWLEDGMENTS

I wish to thank Victor Baker, Julio Betancourt, Emmett Laursen, Ed McCullough, E. Tad Nichols, Marie Pearthree, Brian Reich, John Sumner, and Ray Turner for their sincere and enthusiastic assistance.

REFERENCES

Aldridge, B. N., and Eychaner, J. H. (1984). Floods of October, 1977 in Southern Arizona and March, 1978 in Central Arizona. *Geol. Surv. Water-Supply Pap. (U.S.)* **2223**, 1–19.

Applegate, L. H. (1981). Hydraulic effects of vegetation changes along the Santa Cruz River near Tumacacori, Arizona. MS. Thesis, pp. 4–7. University of Arizona, Tucson.

Arnold, L. W. (1940). An ecological study of the vertebrate animals of the Mesquite Forest. MS. Thesis, University of Arizona, Tucson.

Baker, V. R. (1984). Questions raised by the Tucson flood of 1983. *Ariz. Nev. Acad. Sci.* **14**, 211–219.

Betancourt, J. R., and Turner, R. M. (1987). Historical arroyo-cutting and subsequent channel changes at the Congress Street Crossing, Santa Cruz River, Tucson, Arizona. *Proc., Arid Lands Today Tomorrow, 1985* (in press).

Boughton, W. C., and Renard, K. G. (1984). Flood frequency characteristics of some Arizona watersheds. *Water Resour. Bull.* **20**(5), 761–769.

Bull, W. B., and Scott, K. M. (1974). Impact of mining gravel from urban stream beds in the southwestern United States. *Geology,* **2**, 171–174.

Cooke, R. U., and Reeves, R. W. (1976). "Arroyos and Environmental Change in the American Southwest." Oxford Univ. Press (Clarendon), London and New York.

Eychner, J. H. (1984). Estimation of magnitude and frequency of floods in Pima County. *Water-Supply Res. Invest., Open-File Report (U.S.),* pp. 1–61.

Federal Emergency Management Agency (FEMA) (1982). "Flood Insurance Study, City of Tucson, Arizona, Pima County." Federal Emergency Management Agency, Washington, D.C.

Gordon, J. P. (1983). Channel changes on the lower Canada Del Oro 1936-1980 and policies of flood plain management. MS. Thesis, University of Arizona, Tucson.

Graf, W. L. (1984). A probabilistic approach to the spatial assessment of river channel instability. *Water Resour. Res.,* **20** (7), 953–962.

Hastings, J. R., and Turner, R. M. (1972). "The Changing Mile." Univ. of Arizona Press, Tucson.

Hays, M. E. (1984). Analysis of historic channel changes as a method for evaluating flood hazard in the semi-arid southwest. MS. Thesis, University of Arizona, Tucson.

Hirschboeck, K. K. (1985). Hydroclimatology of flow events in the Gila River Basin, central and southern Arizona. Ph.D. Thesis, University of Arizona, Tucson.

Holzer, T. L. (1977). Crustal uplift caused by ground water withdrawal in the lower Santa Cruz Basin, Arizona. *Geol. Soc. Am. Abstr.,* **9**(7), 1023.

Holzer, T. L. (1979). Elastic expansion of the lithosphere caused by ground water depletion. *J. Geophys. Res.,* **84**(139), 4685–4698.

Huckelberry, C. H. (1985). Interview, Jan. 8, KUAT radio. University of Arizona, Tucson.

Hydrosoftware (1984). Hydrologic evaluation of Santa Cruz Basin, Arizona. Prepared for Department of Transportation and Flood Control District, Pima County, Tucson, Arizona, October (Rep. No.84220).

Laursen, E. M. (1985). "On Flooding in Tucson." Department of Civil Engineering, University of Arizona, Tucson (unpublished).

Laursen, E. M., and Carmody, T. (1980). "Hydrology and Hydraulic Report for Irvington Road, Santa Cruz River." Ruiz Engineering Corp. for Pima County.

Maddock, T., Jr. (1976). A primer on floodplain dynamics. *J. Soil Water Conserv.,* **31**(2), 44–47.

Matlock, W. G. (1965). The effects of silt-laden water on infiltration in alluvial channels. Ph.D. Thesis, University of Arizona, Tucson.

Pearthree, M. S. (1982). Variability of the ephemeral stream channels of the Rillito Creek system, southeastern Arizona, and implications for floodplain management. M.S. Thesis, University of Arizona, Tucson.

Pearthree, M. S. (1985). Channel changes in southern Arizona—Implications for floodplain management. *Proc., Assoc. State Floodplain Managers Conf., 9th,* Nat. Hazards Res. Appl. Inf. Cent. Spec. Publ. No. 12, pp. 301–315.

Pearthree, M. S., and Baker, V. R. (1987). "Channel Change Along the Rillito Creek System of Southeastern Arizona, 1941 through 1983—Implications for Floodplain Management." Arizona Bureau of Geology and Mineral Technology, Tucson (in press).

Peirce, H. W., and Kresan, P. L. (1984). The "floods" of October, 1983. *Fieldnotes, Ariz. Bur. Geol. Miner. Technol.,* **14**(2), 1–6.

Pima County (1984). "Flood Repair and Flood Hazard Mitigation Implementation Plan," pp. 27–29. Department of Transportation and Flood Control District, Tucson, Arizona.

Pima County (1985). "Floodplain Ordinance No.1985-FCI for Pima County, Arizona." Department of Transportation and Flood Control District, Tucson, Arizona.

Pima County (1986a). "Santa Cruz River Management Plan—Policy Manual." Department of Transportation and Flood Control District, Tucson, Arizona.

Pima County (1986b). "Rillito Creek Groundwater/Flood Storage/ Natural Riverine Preservation Project." Department of Transportation and Flood Control District, Tucson, Arizona.

Reich, B. M. (1985). Santa Cruz River 1985: How bad can

floods get? *Newsl. South. Ariz. Environ. Counc.*, Winter, pp. 2–6.

Reich, B. M., and Davis, D. R. (1985). The 1983 Santa Cruz flood: How should highway engineers respond? *Transp. Res. Rec.* **1017**, 1–8.

Reich, B. M., and Davis, D. R. (1986a). Reluctance to increase the regulatory flood on a degrading river. *Proc., West. States High Risk Flood Areas Symp., Assoc. State Floodplain Managers,* March.

Reich, B. M., and Davis, D. R. (1986b). Estimating the regulatory flood on a degrading river. Presented at the International Symposium on Flood Frequency and Risk Analysis, Baton Rouge, Louisiana, May, 1986.

Reich, B. M., and Renard, K. G. (1981). Application of advances in flood frequency analysis. *Water Resour. Bull.*, **17**(1), 67–74.

Roeske, R. H. (1978). "Methods for Estimating the Magnitude and Frequency of Floods in Arizona," Department of Transportation, Rep. No. ADOT-RS-15-121. Phoenix, Arizona.

Saarinen, T.F., Baker, V.R., Durrenberger, R., and Maddock, T., Jr. (1984). "The Tucson, Arizona, Flood of October, 1983," Report for the Committee on Natural Disasters. Natl. Res. Counc., Natl. Acad. Press, Washington, D.C.

Sauer, V. B., Thomas, W. O., Jr., Stricker, V. A., and Wilson, K. V. (1983). Flood characteristics of urban watersheds in the United States. *Geol. Surv. Water-Supply Pap. (U.S.)* **2207**, 58.

Simons, D. B., Li, R., and Associates (1981). Hydraulic and geomorphic anaylsis for Rio Nuevo development project, Santa Cruz, Tucson, Arizona. Prepared for Cella Barr and Associates, Tucson, Arizona.

Simons, D. B., Li, R., and Associates (1984). River management plan for the Rillito River and major tributaries. Prepared for Pima County Department of Transportation and Flood Control District, Tucson, Arizona.

Simons, D. B., Li, R., and Associates (1986). Santa Cruz management plan technical report. Prepared for Pima County Department of Transportation and Flood Control District, Tucson, Arizona.

Smith, W. (1986). The effects of eastern north pacific tropical cyclones on the southwestern United States. *NOAA Tech. Memo.* **NWS WR-197**, 1–229.

U.S. Army Corps of Engineers (1972). "Interim Report on Survey for Flood Control." Santa Cruz River Basin, L.A. District, U.S. Army Corps of Engineers, Arizona.

U.S. Army Corps of Engineers (1985). "Rillito River and Associated Streams," Surv. Rep. Environ. Assess., December. U.S. Army Corps of Engineers, Tucson, Arizona.

U.S. Water Resources Council (1981). "Guidelines for Determining Flood Flow Frequency," Bull. No. 17B. U.S. Water Resour. Counc., Washington, D.C.

Winikka, C. C. (1984). A view of subsidence. *Fieldnotes. Ariz. Bur. Geol. Miner. Technol.* **14**(3), 1–5.

Wolman, M. G., and Brush, L. M. (1961). Factors controlling the size and shape of stream channels in coarse noncohesive sands. *Geol. Surv. Prof. Pap. (U.S.)* **282-G**, 183–210.

INDEX